Lecture Notes in Computer Science 4974

Commenced Publication in 1973
Founding and Former Series Editors:
Gerhard Goos, Juris Hartmanis, and Jan van Leeuwen

Mario Giacobini et al. (Eds.)

Applications of Evolutionary Computing

EvoWorkshops 2008: EvoCOMNET, EvoFIN, EvoHOT, EvoIASP
EvoMUSART, EvoNUM, EvoSTOC, and EvoTransLog
Naples, Italy, March 26-28, 2008
Proceedings

Volume Editors

see next page

Cover illustration: "Ammonite II" by Dennis H. Miller (2004-2005)
www.dennismiller.neu.edu

Library of Congress Control Number: Applied for

CR Subject Classification (1998): F.1, D.1, B, C.2, J.3, I.4, J.5

LNCS Sublibrary: SL 1 – Theoretical Computer Science and General Issues

ISSN 0302-9743
ISBN-10 3-540-78760-7 Springer Berlin Heidelberg New York
ISBN-13 978-3-540-78760-0 Springer Berlin Heidelberg New York

Springer is a part of Springer Science+Business Media

springer.com

Printed in Germany

Typesetting: Camera-ready by author, data conversion by Scientific Publishing Services, Chennai, India
Printed on acid-free paper SPIN: 12245185 06/3180 5 4 3 2 1 0

Volume Editors

Mario Giacobini
Dept. of Animal Production,
Epidemiology and Ecology
University of Torino, Italy
mario.giacobini@unito.it

Anthony Brabazon
School of Business
University College Dublin, Ireland
anthony.brabazon@ucd.ie

Stefano Cagnoni
Dept. of Computer Engineering
University of Parma, Italy
cagnoni@ce.unipr.it

Gianni A. Di Caro
"Dalle Molle" Institute for
Artificial Intelligence (IDSIA)
Lugano, Switzerland
gianni@idsia.ch

Rolf Drechsler
Institute of Computer Science
University of Bremen, Germany
drechsle@informatik.uni-bremen.de

Anikó Ekárt
Knowledge Engineering Research
Group, Aston University
Birmingham, UK
ekarta@aston.ac.uk

Anna I Esparcia-Alcázar
Instituto Tecnológico de Informática
Ciudad Politécnica de la Innovación
Valencia, Spain
anna@iti.upv.es

Muddassar Farooq
National University of Computer
and Emerging Sciences
Islamabad, Pakistan
muddassar.farooq@nu.edu.pk

Andreas Fink
Fac. of Economics & Social Sciences
Helmut-Schmidt-University
Hamburg, Germany
andreas.fink@hsu-hamburg.de

Jon McCormack
Clayton School of Information
Technology
Monash University, Clayton, Australia
Jon.McCormack@infotech.monash.edu.au

Michael O'Neill
School of Computer Science and
Informatics
University College Dublin, Ireland
m.oneill@ucd.ie

Juan Romero
Facultad de Informatica
University of A Coruña, Spain
jj@udc.es

Franz Rothlauf
Dept. of Information Systems
Johannes Gutenberg University
Mainz, Germany
rothlauf@uni-mainz.de

Giovanni Squillero
Dip. di Automatica e Informatica
Politecnico di Torino, Italy
giovanni.squillero@polito.it

A. Şima Uyar
Dept. of Computer Engineering
Istanbul Technical University, Turkey
etaner@cs.itu.edu.tr

Shengxiang Yang
Dept. of Computer Science
University of Leicester, UK
s.yang@mcs.le.ac.uk

Preface

Evolutionary computation (EC) techniques are efficient, nature-inspired planning and optimization methods based on the principles of natural evolution and genetics. Due to their efficiency and simple underlying principles, these methods can be used in the context of problem solving, optimization, and machine learning. A large and continuously increasing number of researchers and professionals make use of EC techniques in various application domains. This volume presents a careful selection of relevant EC examples combined with a thorough examination of the techniques used in EC. The papers in the volume illustrate the current state of the art in the application of EC and should help and inspire researchers and professionals to develop efficient EC methods for design and problem solving.

All papers in this book were presented during EvoWorkshops 2008, which consisted of a range of workshops on application-oriented aspects of EC. Since 1998, EvoWorkshops has provided a unique opportunity for EC researchers to meet and discuss application aspects of EC and has served as an important link between EC research and its application in a variety of domains. During these ten years new workshops have arisen, some have disappeared, while others have matured to become conferences of their own, such as EuroGP in 2000, EvoCOP in 2004, and EvoBIO last year.

EvoWorkshops are part of EVO*, Europe's premier co-located event in the field of evolutionary computing. EVO* 2008 was held in Naples, Italy, on 26-28 March 2008, and included, in addition to EvoWorkshops, EuroGP, the main European event dedicated to genetic programming; EvoCOP, the main European conference on EC in combinatorial optimization; and EvoBIO, the main European conference on EC and related techniques in bioinformatics and computational biology. The proceedings of all of these events, EuroGP 2008, EvoCOP 2008 and EvoBIO 2008, are also available in the LNCS series (volumes 4971, 4972, and 4973).

The central aim of the EVO* events is to provide researchers, as well as people from industry, students, and interested newcomers, with an opportunity to present new results, discuss current developments and applications, or just become acquainted with the world of EC. Moreover, it encourages and reinforces possible synergies and interactions between members of all scientific communities that may benefit from EC techniques.

EvoWorkshops 2008 consisted of the following individual workshops:

- *EvoCOMNET*, the Fifth European Workshop on the Application of Nature-Inspired Techniques to Telecommunication Networks and Other Connected Systems,
- *EvoFIN*, the Second European Workshop on Evolutionary Computation in Finance and Economics,

- *EvoHOT*, the Fourth European Workshop on Bio-inspired Heuristics for Design Automation,
- *EvoIASP*, the Tenth European Workshop on Evolutionary Computation in Image Analysis and Signal Processing,
- *EvoMUSART*, the Sixth European Workshop on Evolutionary and Biologically Inspired Music, Sound, Art and Design,
- *EvoNUM*, the First European Workshop on Bio-inspired Algorithms for Continuous Parameter Optimization,
- *EvoSTOC*, the Fifth European Workshop on Evolutionary Algorithms in Stochastic and Dynamic Environments, and
- *EvoTRANSLOG*, the Second European Workshop on Evolutionary Computation in Transportation and Logistics.

EvoCOMNET addresses the application of EC techniques to problems in distributed and connected systems such as telecommunication and computer networks, distribution and logistic networks, interpersonal and interorganizational networks, etc. To address these challenges, this workshop promotes the study and the application of strategies inspired by the observation of biological and evolutionary processes, that usually show the highly desirable characteristics of being distributed, adaptive, scalable, and robust.

EvoFIN is the first European event specifically dedicated to the applications of EC, and related natural computing methodologies, to finance and economics. Financial environments are typically hard, being dynamic, high-dimensional, noisy and co-evolutionary. These environments serve as an interesting test bed for novel evolutionary methodologies.

EvoHOT focuses on innovative heuristics and bio-inspired techniques applied to the electronic design automation. It shows the latest developments, the reports of industrial experiences, the successful attempts to *evolve* rather than *design* new solutions, and the hybridizations of traditional methodologies.

EvoIASP, the longest-running of all EvoWorkshops, which celebrates its tenth edition this year, since 1999 constituted the first international event solely dedicated to the applications of EC to image analysis and signal processing in complex domains of high industrial and social relevance.

EvoMUSART addresses all practitioners interested in the use of EC techniques for the development of creative systems. There is a growing interest in the application of these techniques in fields such as art, music, architecture and design. The goal of this workshop is to bring together researchers that use EC in this context, providing an opportunity to promote, present and discuss the latest work in the area, fostering its further developments and collaboration among researchers.

EvoNUM aims at applications of bio-inspired algorithms, and cross-fertilization between these and more classical numerical optimization algorithms, to continuous optimization problems in engineering. It deals with engineering applications where continuous parameters or functions have to be optimized, in fields such as control, chemistry, agriculture, electricity, building and construction, energy, aerospace engineering, design optimization.

EvoSTOC addresses the application of EC in stochastic and dynamic environments. This includes optimization problems with changing, noisy, and/or approximated fitness functions and optimization problems that require robust solutions. These topics recently gained increasing attention in the EC community, and EvoSTOC was the first workshop that provided a platform to present and discuss the latest research in this field.

EvoTRANSLOG deals with all aspects of the use of evolutionary computation, local search and other nature-inspired optimization and design techniques for the transportation and logistics domain. The impact of these problems on the modern economy and society has been growing steadily over the last few decades, and the workshop aims at design and optimization techniques such as evolutionary computing approaches allowing us to use computer systems for systematic design, optimization, and improvement of systems in the transportation and logistics domain.

Along the line of adapting the list of the events to the needs and demands of the researchers working in the field of evolutionary computing, EvoINTERACTION, the European Workshop on Interactive Evolution and Humanized Computational Intelligence, did not take place in 2008, but it will be run again next year. Similarly, EvoHOT was held this year, while it was not held during EVO* 2007. A new workshop was also proposed this year, EvoTHEORY, the first European Workshop on Theoretical Aspects in Artificial Evolution. Due to the limited number of submissions, the workshop's chairs, Jonathan E. Rowe and Mario Giacobini, decided not to hold the event in 2008. However, two high-quality articles that were submitted there were presented during the EVO* days and are included in this volume.

The number of submissions to EvoWorkshops 2008 was once again very high, cumulating in 133 entries (with respect to 149 in 2006 and 160 in 2007). The following table shows relevant statistics for EvoWorkshops 2008 (both short and long papers are considered in the acceptance statistics), where the statistics for the 2007 edition are also reported, except for EvoHOT whose last edition was in 2006:

year	2008			previous edition		
	submissions	accepted	ratio	submissions	accepted	ratio
EvoCOMNET	10	6	60%	44	18	40.9%
EvoFIN	15	8	53.3%	13	8	61.5%
EvoHOT	15	9	60%	9	5	55.6%
EvoIASP	26	16	61.5%	35	21	60%
EvoMUSART	31	17	54.8%	30	15	50%
EvoNUM	14	8	57.1%	-	-	-
EvoSTOC	8	4	50%	11	5	45.5%
EvoTHEORY	3	2	66.6%	-	-	-
EvoTRANSLOG	11	5	45.4%	20	8	40%
Total	133	75	56.4%	160	79	49.4%

Full papers, covering 10 pages, and short papers, covering 6 pages, were accepted for the event. The two classes of papers were either presented orally over the three conference days, or presented and discussed during a special poster session. The low acceptance rate of 56.4% for EvoWorkshops 2008 along with the significant number of submissions, is an indicator of the high quality of the articles presented at the workshops, showing the liveliness of the scientific movement in the corresponding fields.

Many people helped make EvoWorkshops 2008 a success. We would like to thank the following institutions:

- the Instituto Tecnológico de Informática in València, Spain, for hosting the EVO* website,
- Naples City Council, Italy, for supporting the local organization and their patronage of the event,
- the Centre for Emergent Computing at Napier University in Edinburgh, Scotland, for administrative help and event coordination.

We would particularly like to acknowledge the invaluable support of Professor Guido Trombetti, Rector of the University of Naples Federico II, and Professor Giuseppe Trautteur of the Department of Physical Sciences.

Even with excellent support and a perfect location, an event like EVO* would not be feasible without authors submitting their work, members of the program committees dedicating their time and energy to review the papers, and an audience. All these people deserve our gratitude.

Finally, we are grateful to all those involved in the preparation of the event, especially Jennifer Willies for her unfaltering dedication to the coordination of the event over the years. Without her support, running an event like this, with such a large number of different organizers and different opinions, would be impossible. Further thanks to the local organizers Ivanoe De Falco, Antonio Della Cioppa, and Ernesto Tarantino for making the organization of such an event possible in a place as unique as Naples. Last but surely not least, we want to specially thank Anna I. Esparcia-Alcázar, for her hard work as publicity chair of the event, and Marc Schoenauer for his continuous help in setting up and maintaining the MyReview management software.

March 2008

Mario Giacobini, Anthony Brabazon, Stefano Cagnoni
Gianni A. Di Caro, Rolf Drechsler, Anikó Ekárt
Anna I. Esparcia-Alcázar, Muddassar Farooq, Andreas Fink
Jon McCormack, Michael O'Neill, Juan Romero
Franz Rothlauf, Giovanni Squillero, A. Şima Uyar
Shengxiang Yang

Organization

EvoWorkshops 2008 was part of EVO* 2008, Europe's premier co-located event in the field of evolutionary computing, that included also the conferences EuroGP 2008, EvoCOP 2008, and EvoBIO 2008.

Organizing Committee

EvoWorkshops Chair	Mario Giacobini, University of Torino, Italy
Local Chairs	Ivanoe De Falco, ICAR-CNR, Italy Antonio Della Cioppa, University of Salerno, Italy Ernesto Tarantino, ICAR-CNR, Italy
Publicity Chair	Anna Isabel Esparcia-Alcazar, Instituto Tecnológico de Informática, València, Spain
EvoCOMNET Co-chairs	Muddassar Farooq, National University of Computer and Emerging Sciences, Pakistan Gianni A. Di Caro, IDSIA, Switzerland
EvoFIN Co-chairs	Anthony Brabazon, University College Dublin, Ireland Michael O'Neill, University College Dublin, Ireland
EvoHOT Co-chairs	Rolf Drechsler, University of Bremen, Germany Giovanni Squillero, Politecnico di Torino, Italy
EvoIASP Chair	Stefano Cagnoni, University of Parma, Italy
EvoMUSART Co-chairs	Juan Romero, University of A Coruña, Spain, Jon McCormack, Monash University, Australia
EvoNUM Co-chairs	Anikó Ekárt, Aston University, UK Anna Isabel Esparcia-Alcazar, Instituto Tecnológico de Informática, València, Spain
EvoSTOC Co-chairs	A. Şima Uyar, Istanbul Technical University, Turkey Shengxiang Yang, University of Leicester, UK
EvoTHEORY Co-chairs	Jonathan E. Rowe, University of Birmingham, UK Mario Giacobini, University of Torino, Italy
EvoTRANSLOG Co-chairs	Andreas Fink, Helmut-Schmidt-University Hamburg, Germany Franz Rothlauf, Johannes Gutenberg University of Mainz, Germany

Program Committees

EvoCOMNET Program Committee

Mehmet E. Adyin, University of Bedfordshire, UK
Uwe Aickelin, University of Nottingham, UK
Ozgur B. Akan, Middle East Technical University, Turkey
Payman Arabshahi, Washington University, USA
Peter J. Bentley, University College London, UK
Marco Dorigo, IRIDIA, Belgium
Falko Dressler, University of Erlangen, Germany
Frederick Ducatelle, IDSIA, Switzerland
Erol Gelenbe, Imperial College London, UK
Silvia Giordano, SUPSI, Switzerland
Malcolm I. Heywood, Dalhousie University, Canada
Nur-Zincir Heywood, Dalhousie University, Canada
Jin-Kao Hao, University of Angers, France
Steve Hurley, Cardiff University, UK
Byrant Julstrom, St. Cloud State University, USA
Kenji Leibnitz, Osaka University, Japan
Vittorio Maniezzo, University of Bologna, Italy
Alcherio Martinoli, EPFL Lausanne, Switzerland
Roberto Montemanni, IDSIA, Switzerland
Franz Rothlauf, Johannes Gutenberg University of Mainz, Germany
Chien-Chung Shen, University of Delaware, USA
Kwang M. Sim, Hong Kong Baptist University, Hong Kong
Rogar Whitaker, Cardiff University, UK
Lidia Yamamtoto, University of Basel, Switzerland
Franco Zambonelli, Università degli Studi di Modena e Reggio Emilia, Italy
Jie Zhang, University of Bedfordshire, UK

EvoFIN Program Committee

Eva Alfaro-Cid, Instituto Tecnológico de Informática, València, Spain
Anthony Brabazon, University College Dublin, Ireland
Shu-Heng Chen, National Chengchi University, Taiwan
Ian Dempsey, Pipeline Trading, USA
Rafal Drezewski, AGH University of Science and Technology, Poland
David Edelman, University College Dublin, Ireland
Kai Fan, University College Dublin, Ireland
Philip Hamill, University of Ulster, Ireland
Ronald Hochreiter, University of Vienna, Austria
Youwei Li, Queen's University Belfast, Ireland
Piotr Lipinski, University of Wroclaw, Poland
Dietmar Maringer, University of Essex, UK

Michael O'Neill, University College Dublin, Ireland
Robert Schafer, AGH University of Science and Technology, Poland
Kerem Senel, Bilgi University, Turkey
Chris Stephens, Universidad Nacional Autónoma de México, Mexico
Andrea G.B. Tettamanzi, Università degli Studi di Milano, Italy

EvoHOT Program Committee

Varun Aggarwal, Aspiring Minds, India
Paolo Bernardi, Politecnico di Torino, Italy
Michelangelo Grosso, Politecnico di Torino, Italy
Doina Logofatu, University of Applied Sciences, Germany
Mihai Oltean, Babes-Bolyai University, Cluj-Napoca, Romania.
Gregor Papa, Jozef Stefan Institute, Slovenia
Wilson Javier Prez Holgun, Universidad de Valle, Colombia
Danilo Ravotto, Politecnico di Torino, Italy
Ernesto Sanchez, Politecnico di Torino, Italy
Lukas Sekanina, Brno University of Technology, Czech Republic

EvoIASP Program Committee

Lucia Ballerini, European Center for Soft Computing, Spain
Bir Bhanu, University of California at Riverside, USA
Leonardo Bocchi, University of Florence, Italy
Stefano Cagnoni, University of Parma, Italy
Ela Claridge, University of Birmingham, UK
Oscar Cordon, European Center for Soft Computing, Spain
Ivanoe De Falco, ICAR CNR, Italy
Antonio Della Cioppa, University of Naples Federico II, Italy
Laura Dipietro, Massachusetts Institute of Technology, USA
Marc Ebner, University of Wuerzburg, Germany
Špela Ivekovič, University of Dundee, UK
Mario Koeppen, Kyuhsu Institute of Technology, Japan
Evelyne Lutton, INRIA, France
Luca Mussi, University of Perugia, Italy
Gustavo Olague, CICESE, Mexico
Riccardo Poli, University of Essex, UK
Stephen Smith, University of York, UK
Giovanni Squillero, Politecnico di Torino, Italy
Kiyoshi Tanaka, Shinshu University, Japan
Ankur M. Teredesai, University of Washington Tacoma, USA
Andy Tyrrell, University of York, UK
Leonardo Vanneschi, Università degli Studi di Milano-Bicocca, Italy
Mengjie Zhang, Victoria University of Wellington, New Zealand

EvoMUSART Program Committee

Alain Lioret, Paris 8 University, France
Alan Dorin, Monash University, Australia
Alejandro Pazos, University of A Coruna, Spain
Amilcar Cardoso, University of Coimbra, Portugal
Andrew Gildfind, Google Inc., Australia
Andrew Horner, University of Science & Technology, Hong Kong
Anna Ursyn, University of Northern Colorado, USA
Antonino Santos, University of A Coruña, Spain
Artemis Sanchez Moroni, Renato Archer Research Center, Brazil
Bill Manaris, College of Charleston, USA
Brian J. Ross, Brock University, Canada
Carla Farsi, University of Colorado, USA
Christa Sommerer, Institute of Advanced Media Arts and Sciences, Japan
Christian Jacob, University of Calgary, Canada
Colin Johnson, University of Kent, UK
David Hart, Independent Artist, USA
Eduardo R. Miranda, University of Plymouth, UK
Eleonora Bilotta, University of Calabria, Italy
Gary Greenfield, University of Richmond, USA
Gary Nelson, Oberlin College, USA
Gerhard Widmer, Johannes Kepler University Linz, Austria
Hans Dehlinger, Independent Artist, Germany
James McDermott, University of Limerick, Ireland
John Collomosse, University of Bath, UK
Jon Bird, University of Sussex, UK
Jonatas Manzolli, UNICAMP, Brazil
Jorge Tavares, University of Coimbra, Portugal
Luigi Pagliarini, PEAM, Italy & University of Southern Denmark
Margaret Boden, University of Sussex, UK
Maria Verstappen, Independent Artist, Netherlands
Matthew Lewis, Ohio State University, USA
Mauro Annunziato, Plancton Art Studio, Italy
Nell Tenhaaf, York University, Canada
Nicolas Monmarché, University of Tours, France
Pablo Gervás, Universidad Complutense de Madrid, Spain
Paul Brown, University of Sussex, UK
Paulo Urbano, Universidade de Lisboa, Portugal
Penousal Machado, University of Coimbra, Portugal
Peter Bentley, University College London, UK
Philip Galanter, Independent Artist, USA
Rafael Ramirez, Pompeu Fabra University, Spain
Rodney Waschka II, North Carolina State University, USA
Ruli Manurung, University of Indonesia, Indonesia
Scott Draves, Independent Artist, USA

Simon Colton, Imperial College, UK
Somnuk Phon-Amnuaisuk, Multimedia University, Malaysia
Stefano Cagnoni, University of Parma, Italy
Stephen Todd, IBM, UK
Steve DiPaola, Simon Fraser University, Canada
Tim Blackwell, Goldsmiths College, University of London, UK
William Latham, Art Games Ltd, UK

EvoNUM Program Committee

Eva Alfaro-Cid, Instituto Tecnológico de Informática, Spain
Anne Auger, INRIA, France
Wolfgang Banzhaf, Memorial University of Newfoundland, Canada
Hans-Georg Beyer, FH Vorarlberg, Austria
Xavier Blasco, Universidad Politécnica de Valencia, Spain
Ying-ping Chen, National Chiao Tung University, Taiwan
Carlos Cotta, Universidad de Málaga, Spain
Marc Ebner, Universität Würzburg, Germany
Francisco Fernández, Universidad de Extremadura, Spain
Nikolaus Hansen, INRIA, France
Bill Langdon, University of Essex, UK
JJ Merelo, Universidad de Granada, Spain
Boris Naujoks, University of Dortmund, Germany
Una-May O'Reilly, MIT, USA
Mike Preuss, University of Dortmund, Germany
Günter Rudolph, University of Dortmund, Germany
Marc Schoenauer, INRIA, France
P. N. Suganthan, Nanyang Technological University, Singapore
Ke Tang, University of Science and Technology of China, China
Darrell Whitley, Colorado State University, USA

EvoSTOC Program Committee

Dirk Arnold, Dalhousie University, Canada
Hans-Georg Beyer, Vorarlberg University of Applied Sciences, Austria
Tim Blackwell, Goldsmiths College London, UK
Juergen Branke, University of Karlsruhe, Germany
Ernesto Costa, University of Coimbra, Portugal
Yaochu Jin, Honda Research Institute Europe, Germany
Stephan Meisel, Technical University Braunschweig, Germany
Daniel Merkle, University of Leipzig, Germany
Zbigniew Michalewicz, University of Adelaide, Australia
Martin Middendorf, University of Leipzig, Germany
Ronald Morrison, Mitretek Systems, Inc., USA

Ferrante Neri, University of Jyvaskyla, Finland
Yew-Soon Ong, Nanyang Technological University, Singapore
William Rand, Northwestern University, USA
Hendrik Richter, University of Leipzig, Germany
Christian Schmidt, University of Karlsruhe, Germany
Ken Sharman, Instituto Tecnológico de Informática, València, Spain
Anabela Simões, University of Coimbra, Portugal
Renato Tinos, Universidade de Sao Paulo, Brazil

EvoTHEORY Program Committee

Hans-Georg Beyer, FH Vorarlberg, Austria
Cecilia Di Chio, Essex University, UK
Christian Igel, Ruhr-Universität Bochum, Germany
Thomas Jansen, University of Dortmund, Germany
William Langdon, Essex University, UK
Alberto Moraglio, University of Coimbra, Portugal
Riccardo Poli, Essex University, UK
Adam Prugel-Bennett, Southampton University, UK
Guenter Rudolph, University of Dortmund, Germany
Jim Smith, University of the West of England, UK
Leonardo Vanneschi, Università degli Studi di Milano-Bicocca, Italy
Michael Vose, University of Tennessee, USA
Ingo Wegener, University of Dortmund, Germany
Darrell Whitley, Colorado State University, USA
Paul Wiegand, University of Central Florida, USA
Carsten Witt, University of Dortmund, Germany

EvoTRANSLOG Program Committee

Christian Bierwirth, University of Halle-Wittenberg, Germany
Peter A.N. Bosman, Centre for Mathematics and Computer Science, Amsterdam, The Netherlands
Karl Doerner, University of Vienna, Austria
Martin Josef Geiger, University of Hohenheim, Germany
Jens Gottlieb, SAP, Germany
Hoong Chuin Lau, Singapore Management University, Singapore
Giselher Pankratz, FernUni Hagen, Germany
Christian Prins, University of Technology of Troyes, France
Agachai Sumalee, The Hong Kong Polytechnic University, Hong Kong
Theodore Tsekeris, Center of Planning and Economic Research, Athens, Greece
Stefan Voß, University of Hamburg, Germany

Sponsoring Institutions

- Research Center in Pure and Applied Mathematics, Salerno, Italy
- Institute of High Performance Computing and Networking, National Research Council, Italy
- University of Naples Federico II, Italy
- The Centre for Emergent Computing at Napier University in Edinburgh, UK

Table of Contents

EvoCOMNET Contributions

EvoFIN Contributions

EvoHOT Contributions

EvoIASP Contributions

EvoMUSART Contributions

EvoNUM Contributions

EvoSTOC Contributions

EvoTHEORY Contributions

EvoTRANSLOG Contributions

New Research in Nature Inspired Algorithms for Mobility Management in GSM Networks

Enrique Alba[1], José García-Nieto[1], Javid Taheri[2], and Albert Zomaya[2]

[1] Dept. de Lenguajes y Ciencias de la Computación, University of Málaga, ETSI Informática, Campus de Teatinos, Málaga - 29071, Spain
{eat,jnieto}@lcc.uma.es

[2] School of Information Technologies, University of Sydney, Sydney, NSW 2006, Australia
{javidt,zomaya}@it.usyd.edu.au

Abstract. Mobile Location Management (MLM) is an important and complex telecommunication problem found in mobile cellular GSM networks. Basically, this problem consists in optimizing the number and location of paging cells to find the lowest location management cost. There is a need to develop techniques capable of operating with this complexity and used to solve a wide range of location management scenarios. Nature inspired algorithms are useful in this context since they have proved to be able to manage large combinatorial search spaces efficiently. The aim of this study is to assess the performance of two different nature inspired algorithms when tackling this problem. The first technique is a recent version of Particle Swarm Optimization based on geometric ideas. This approach is customized for the MLM problem by using the concept of Hamming spaces. The second algorithm consists of a combination of the Hopfield Neural Network coupled with a Ball Dropping technique. The location management cost of a network is embedded into the parameters of the Hopfield Neural Network. Both algorithms are evaluated and compared using a series of test instances based on realistic scenarios. The results are very encouraging for current applications, and show that the proposed techniques outperform existing methods in the literature.

Keywords: Mobile Location Management, GSM Cellular Networks, Geometric Particle Swarm Optimization, Hopfield Neural Network.

1 Introduction

Mobility Management becomes a crucial issue when designing infrastructure for wireless mobile networks. In order to route incoming calls to appropriate mobile terminals, the network must keep track of the location of each mobile terminal. Mobility management requests are often initiated either by a mobile terminal movement (crossing a cell boundary) or by deterioration of the quality of a received signal in a currently allocated channel. Due to the expected increase in the usage of wireless services in the future, the next generation of mobile networks should be able to support a huge number of users and their bandwidth requirements [1,4].

M. Giacobini et al. (Eds.): EvoWorkshops 2008, LNCS 4974, pp. 1–10, 2008.

Several strategies for Mobility Management have been used in the literature being the location area (LA) scheme one of the most popular [6,11]. An analogous strategy is the *Reporting Cells* (RC) scheme suggested in [3]. In RC, a subset of cells in the network is designated as reporting cells. Each mobile terminal performs a location update only when it enters one of these reporting cells. When a call arrives, the search is confined to the reporting cell the user last reported and the neighboring bounded nonreporting cells. It was shown in [3] that finding an optimal set of reporting cells, such that the location management cost is minimized, is an NP-complete problem. For this reason, bioinspired algorithms have been commonly used to solve this problem [7,10].

In this work, we use two nature inspired algorithms to assign the reporting cells of a network following the RC scheme. The first algorithm, called Geometric Particle Swarm Optimization (GPSO), is a generalization of the Particle Swarm Optimization for virtually any solution representation, which works according to a geometric framework. The second technique combines a Hopfield Neural Network with a Ball Dropping (HNN+BD) mechanism. Our contributions are both to perform better with respect to existing works and to introduce the GPSO algorithm for solving Telecommunications problems. In addition, these two techniques are experimentally assessed and compared from different points of view such as quality of the solutions, the robustness and design issues.

The remaining of the paper is organized as follows: Section 2 briefly explains the Mobility Management problem. The two algorithms, GPSO and HNN+BD, are described in sections 3 and 4 respectively. After that, Section 5 presents a number of experiments and results that show the applicability of the proposed approaches to this problem. Finally, conclusions are drawn in Section 6.

2 The Mobility Management Problem

Basically, the Mobility (location) Management problem consists in reducing the total cost of managing a mobile cellular network. Two factors take part when calculating the total cost: the updating cost and the paging cost. The updating cost is the portion of the total cost due to location updates performed by roaming mobile terminals in the network. The paging cost is caused by the network during a location inquiry when the network tries to locate a user[1].

According to the reporting cells scheme, there are two types of cells: reporting cells (RC) and non-reporting cells (nRC). A neighborhood is assigned to each reporting cell, which consists of all nRC that must also page the user in case of an incoming call. For both RC and nRC, a *vicinity* factor is calculated representing the maximum number of reporting neighbors for each cell that must page the user (including the cell itself) in case of an incoming call. Obviously, the vicinity factor of each RC is the number of neighbors it has (see Fig. 1).

[1] Other costs like the cost of database management to register user's locations or the cost of the wired network (backbone) that connects the base stations to each other were not considered here, since these costs are assumed to be the same for all location management strategies and hence aren't contemplated in comparisons.

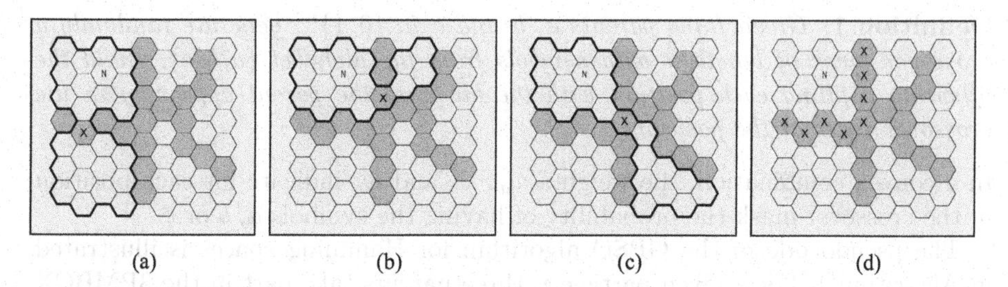

Fig. 1. Cells marked as 'N' belong to the neighborhoods of at least three RCs (grey cells). For example, the number of neighbors for cell 'X' is 25, 17, and 22 for (a), (b) and (c) respectively (25 to consider the worst case). However, if a nRC belongs to more than two neighborhoods the calculation must be done for all of them, and then, the maximum number is considered as the vicinity factor for this nRC. For example, the nRC marked as 'N' is a part of the neighborhood of all cells marked as 'X' in (d).

For nRC, the vicinity factor is calculated based on the fact that each nRC might be in the neighborhood of more than one RC, the maximum number of paging neighbors that contains such a cell is considered its vicinity factor.

Therefore, to calculate the total cost of the network location management, the general cost function is formulated as:

$$Cost = \beta \times \sum_{i \epsilon S} N_{LU}(i) + \sum_{i=0}^{N} N_P(i) \times V(i) \tag{1}$$

where, $N_{LU}(i)$ is the number of location updates for reporting cell number i, $N_P(i)$ is the number of arrived calls for cell i, $V(i)$ is the vicinity factor for cell i, S is the set of cells defined as reporting cells, and N is the total number of cells in the network. β is a constant representing the cost ratio of a location update to a paging transaction in the network (typically $\beta = 10$). This function is used either as *fitness function* by the GPSO or *energy function* by the HNN.

3 Geometric Particle Swarm Optimization

The recent Geometric Particle Swarm Optimization (GPSO) [5,2], enables us to generalize PSO to virtually any solution representation in a natural and straightforward way, extending the search to richer spaces, such as combinatorial ones. This property was demonstrated for the cases of Euclidean, Manhattan and Hamming spaces in the referenced work.

The key issue in this approach consists of using a multi-parental recombination of particles which leads to the generalization of a *mask-based crossover* operation, proving that it respects four requirements for being a *convex combination* in a certain space (see [5] for a complete explanation). This way, the mask-based crossover operation substitutes the classical *movement* in PSO, based on the *velocity* and *position update* operations, only suited for continuous spaces.

For Hamming spaces, which is the focus of this work, a *three-parent mask-based crossover* (3PMBCX) was defined in a straightforward way:

Definition 1. *Given three parents a, b and c in* $\{0,1\}^n$, *generate randomly a crossover mask of length n with symbols from the alphabet* $\{a,b,c\}$. *Build the offspring o filling each position with the bit from the parent appearing in the crossover mask at the position.*

In a convex combination, the weights w_a, w_b and w_c indicate for each position in the crossover mask the probability of having the symbols a, b or c.

The pseudocode of the GPSO algorithm for Hamming spaces is illustrated in Algorithm 1. For a given particle i, three parents take part in the 3PMBCX operator (line 13): the current position x_i, the social best position g_i and the historical best position found h_i (of this particle). The weight values w_a, w_b and w_c indicate for each element in the crossover mask the probability of having values from the parents x_i, g_i or h_i respectively. A constriction of the geometric crossover forces w_a, w_b and w_c to be non-negative and add up to one.

Algorithm 1. GPSO for Hamming spaces

```
1: S ← SwarmInitialization()
2: while not stop condition do
3:    for each particle x_i of the swarm S do
4:       evaluate(x_i)
5:       if fitness(x_i) is better than fitness(h_i) then
6:          h_i ← x_i
7:       end if
8:       if fitness(h_i) is better than fitness(g_i) then
9:          g_i ← h_i
10:      end if
11:   end for
12:   for each particle x_i of the swarm S do
13:      x_i ← 3PMBCX((x_i, w_a), (g_i, w_b), (h_i, w_c))
14:      mutate(x_i)
15:   end for
16: end while
17: Output: best solution found
```

Since the GPSO for Mobility Management was developed for Hamming space, each particle i of the swarm consists of a binary vector $x_i = (x_{i1}, x_{i2}, ..., x_{in})$ representing a reporting cell configuration, where each element x_{ij} represents a cell of the network; x_{ij} can have a value of either "0", representing a nRC, or "1", representing a RC. For example, in an 6×6 network, the particle position will have a length (n) of 36.

4 Hopfield Neural Network with Ball Dropping

In this approach, the Ball Dropping technique is used as the backbone of the algorithm that employs the HNN as its optimizer, and is inspired by the natural behavior of individual balls when they are dropped onto a non-even plate (a plate with troughs and crests). As can be expected, the balls will spontaneously move to the concave areas of the plate, and in a natural process, find the minimum of the plate. A predefined number of balls are dropped onto several random positions on the plate, which is equivalent to the random addition of a predefined number of paging cells to the current paging cell configuration of the network.

As a result, after dropping a number of balls on the plate the energy value of the network increases suddenly and the HNN optimizer tries to reduce it by moving the balls around. The following procedure summarizes the basic form of this algorithm.

Algorithm 2. Ball Dropping Mechanism

```
1: Drop a predefined number of balls onto random positions
2: repeat
3:     Shake the plate
4:     Remove unnecessary balls
5: until location of balls does not lead to any better configuration
6: Output: best solution found
```

In relation to Equation 1, the state vector of the HNN, 'X', is considered to have two different components for location updates and call arrival as follows:

$$X = [x_0 \ x_1 \ \wedge \ x_{N-1} \ x_N \ x_{N+1} \ \wedge \ x_{2N-1}]^T \tag{2}$$

where x_0 to x_{N-1} is the location updates part, x_N to x_{2N-1} is the call arrival part and 'N' is the total number of cells in the network. This HNN model is designed to represents a RC configuration network, and then, tries to modify its RCs in order to reduce the total cost gradually. To summarize this explanation, we refer the reader to [8] where other aspects like generating a initial solution generation, definition of function to modify the state vector and reduction of the number of variations are given completely.

5 Simulation Results

In this section we present the experiments conducted to evaluate and compare the proposed GPSO and HNN+BD. We firstly give some details of the test network instances used. The experiments with both algorithms are presented and analyzed afterwards. We have made 10 independent runs for each algorithm and instance. Comparisons are made from different points of view such as the performance, robustness, quality of solutions and even design issues concerning the two algorithms. Finally, comparisons with other optimizers found in the literature are encouraging since our algorithms obtain competitive solutions which even beat traditional metaheuristic techniques in the previous state of the art.

5.1 Test GSM Network Instances

In almost all of the previous research in the literature, the cell attributes of the network are generated randomly. In general, two independent attributes for each cell are considered: the number of call arrivals (NP) and the number of location updates (NLU), which are set at random according to a normal distribution. However, these numbers are highly correlated in real world scenarios. Therefore, in this work, a more robust and realistic approach is used to seed the initial solutions, and consequently, the network attributes of each cell [9]. This makes the configuration of the solutions obtained in this work to be more realistic.

Therefore, a benchmark of twelve different instances were generated here to be used for testing GPSO and HNN+BD. The numeric values shaping the test networks configurations are given in tables below[2] for future reproduction of our results.

Test-Network 4			Test-Network 5			Test-Network 6		
Cell	NLU	NP	Cell	NLU	NP	Cell	NLU	NP
0	335	97	0	373	86	0	859	659
1	944	155	1	958	155	1	1561	621
2	588	103	2	264	99	2	450	93
3	1478	500	3	571	119	3	599	98
4	897	545	4	431	132	4	535	151
5	793	495	5	451	97	5	425	138
6	646	127	6	693	153	6	1219	590
7	1159	119	7	1258	149	7	1638	137
8	1184	115	8	847	112	8	991	114
9	854	95	9	1412	173	9	646	72
10	1503	529	10	1350	163	10	587	97
11	753	140	11	711	135	11	361	94
12	744	120	12	356	81	12	559	101
13	819	103	13	951	171	13	787	110
14	542	61	14	2282	1016	14	1738	191
15	476	103	15	2276	1067	15	1433	165
16	937	117	16	1217	139	16	562	87
17	603	69	17	341	96	17	404	63
18	617	90	18	337	87	18	342	79
19	888	102	19	1210	121	19	595	97
20	452	53	20	2228	979	20	1312	164
21	581	86	21	1104	171	21	1129	92
22	773	86	22	718	99	22	884	102
23	741	125	23	362	113	23	630	138
24	693	131	24	669	119	24	306	80
25	1535	576	25	1189	158	25	593	87
26	921	128	26	1032	157	26	603	82
27	1225	73	27	620	93	27	977	136
28	1199	133	28	893	140	28	1354	122
29	710	139	29	596	112	29	1225	641
30	782	464	30	367	74	30	421	158
31	879	477	31	389	108	31	594	163
32	1553	532	32	418	120	32	689	99
33	613	68	33	220	102	33	569	115
34	1044	121	34	799	120	34	1554	631
35	400	148	35	344	117	35	733	534

Test-Network 1			Test-Network 2			Test-Network 3		
Cell	NLU	NP	Cell	NLU	NP	Cell	NLU	NP
0	452	484	0	280	353	0	488	455
1	767	377	1	762	438	1	765	290
2	360	284	2	686	599	2	271	201
3	548	518	3	617	503	3	626	475
4	591	365	4	447	403	4	550	247
5	1451	1355	5	978	560	5	1572	1479
6	816	438	6	1349	648	6	1010	377
7	574	415	7	562	431	7	635	300
8	647	366	8	608	412	8	526	240
9	989	435	9	1305	681	9	962	422
10	1105	510	10	966	508	10	1643	1545
11	736	501	11	466	408	11	642	274
12	529	470	12	664	503	12	570	485
13	423	376	13	710	530	13	249	196
14	1058	569	14	746	473	14	842	354
15	434	361	15	282	336	15	516	488

Test-Network 7			Test-Network 8			Test-Network 9		
Cell	NLU	NP	Cell	NLU	NP	Cell	NLU	NP
0	354	160	0	293	88	0	225	85
1	819	198	1	651	134	1	692	128
2	214	75	2	239	53	2	471	124
3	394	147	3	470	73	3	776	104
4	238	135	4	379	69	4	478	106
5	505	99	5	1089	435	5	1034	152
6	433	134	6	690	436	6	931	678
7	397	134	7	615	416	7	890	807
8	588	164	8	509	137	8	445	124
9	895	121	9	557	68	9	866	137
10	658	129	10	472	68	10	1068	136
11	636	121	11	481	80	11	699	112
12	462	104	12	678	100	12	737	108
13	925	134	13	860	124	13	796	120
14	1017	163	14	1229	446	14	1569	706
15	339	86	15	851	401	15	520	117
16	398	122	16	328	71	16	324	93
17	657	95	17	527	77	17	651	94
18	945	122	18	551	86	18	754	75
19	1088	161	19	708	64	19	582	83
20	826	148	20	626	109	20	552	99
21	995	130	21	640	69	21	570	98
22	687	128	22	924	108	22	809	103
23	295	114	23	507	86	23	384	92
24	324	101	24	334	74	24	330	85
25	652	153	25	1187	171	25	588	89
26	1130	142	26	868	74	26	652	117
27	2558	912	27	1324	512	27	584	89
28	1445	191	28	666	86	28	570	107
29	959	151	29	775	87	29	540	84
30	602	133	30	842	60	30	620	88
31	314	92	31	358	50	31	298	85
32	311	123	32	366	75	32	376	102
33	632	127	33	1545	149	33	659	140
34	1250	155	34	1148	92	34	604	98
35	2470	991	35	1239	420	35	577	100
36	2299	847	36	1406	469	36	522	77
37	1051	188	37	1088	104	37	558	88
38	602	140	38	1203	154	38	615	101
39	350	124	39	304	76	39	336	88
40	282	81	40	646	56	40	381	112
41	796	135	41	1215	92	41	763	129
42	1226	147	42	758	91	42	639	99
43	1076	149	43	646	103	43	565	103
44	1301	172	44	885	101	44	567	117
45	909	128	45	780	78	45	765	104
46	622	128	46	1024	169	46	641	119
47	413	105	47	307	74	47	345	96
48	367	115	48	937	477	48	566	148
49	1125	143	49	1308	544	49	1579	716
50	1053	127	50	879	110	50	852	149
51	585	126	51	682	87	51	876	104
52	701	118	52	533	62	52	789	144
53	722	109	53	527	70	53	1126	126
54	656	96	54	602	69	54	948	164
55	646	184	55	454	123	55	485	134
56	422	136	56	666	463	56	905	756
57	426	122	57	703	454	57	1000	744
58	568	142	58	1118	465	58	1100	179
59	264	138	59	353	133	59	429	83
60	480	143	60	474	67	60	902	109
61	223	92	61	258	54	61	536	114
62	734	114	62	629	131	62	706	113
63	341	153	63	273	102	63	253	102

Test-Network 10			Test-Network 11			Test-Network 12		
Cell	NLU	NP	Cell	NLU	NP	Cell	NLU	NP
0	144	83	0	461	619	0	392	562
1	304	98	1	665	584	1	551	509
2	201	66	2	534	554	2	440	466
3	266	85	3	449	89	3	441	83
4	137	100	4	172	91	4	200	49
5	206	80	5	339	84	5	430	45
6	127	79	6	201	93	6	280	90
7	393	112	7	438	89	7	347	84
8	162	46	8	186	63	8	109	30
9	187	116	9	144	64	9	98	43
10	265	82	10	542	553	10	452	502
11	552	99	11	803	515	11	723	467
12	565	83	12	884	528	12	813	440
13	467	95	13	552	75	13	721	99
14	277	114	14	388	62	14	572	60
15	444	109	15	384	68	15	643	82
16	387	95	16	417	77	16	600	92
17	752	83	17	559	95	17	547	95
18	457	76	18	403	90	18	289	77
19	271	84	19	247	60	19	205	74
20	249	80	20	233	79	20	544	441
21	468	90	21	408	90	21	842	446
22	469	74	22	550	83	22	1008	417
23	612	103	23	538	93	23	683	88
24	571	114	24	431	57	24	614	69
25	1335	678	25	604	99	25	501	85
26	802	112	26	347	65	26	702	123
27	656	87	27	404	91	27	644	95
28	731	124	28	539	75	28	469	77
29	274	86	29	290	69	29	296	64
30	367	104	30	248	103	30	617	457
31	533	125	31	540	107	31	911	412
32	429	84	32	423	76	32	989	365
33	542	83	33	526	74	33	472	69
34	1306	708	34	840	107	34	426	65
35	1308	615	35	822	152	35	306	70
36	773	120	36	404	52	36	421	76
37	468	107	37	413	68	37	482	75
38	597	81	38	501	71	38	441	67
39	374	99	39	376	113	39	276	68
40	866	780	40	608	434	40	387	74
41	1050	697	41	1120	586	41	586	82
42	523	105	42	581	90	42	591	94
43	588	113	43	449	62	43	357	67
44	687	113	44	489	70	44	321	66
45	735	132	45	489	97	45	289	47
46	634	97	46	516	96	46	318	66
47	449	99	47	592	86	47	453	58
48	595	133	48	600	67	48	454	77
49	852	699	49	703	496	49	278	81
50	852	768	50	705	573	50	294	80
51	595	97	51	693	110	51	477	83
52	507	86	52	573	99	52	514	90
53	687	101	53	525	93	53	309	48
54	728	123	54	503	86	54	265	51
55	825	154	55	503	71	55	325	73
56	628	109	56	522	78	56	348	64
57	528	91	57	642	91	57	595	102
58	1097	667	58	1076	589	58	569	80
59	894	735	59	639	490	59	383	100
60	374	82	60	380	83	60	278	66
61	523	94	61	577	100	61	455	69
62	468	73	62	466	88	62	540	81
63	891	130	63	415	94	63	438	79
64	1414	692	64	790	116	64	310	63
65	1368	669	65	841	123	65	429	82
66	653	115	66	590	81	66	473	83
67	445	88	67	437	49	67	1070	450
68	590	99	68	481	92	68	901	414
69	385	100	69	249	94	69	859	483
70	309	74	70	267	60	70	288	53
71	647	104	71	555	109	71	481	97
72	717	96	72	426	58	72	705	125
73	878	104	73	422	60	73	675	127
74	1367	653	74	640	91	74	476	47
75	602	128	75	502	75	75	629	70
76	709	100	76	535	90	76	757	90
77	603	91	77	571	95	77	1041	434
78	530	99	78	403	81	78	912	395
79	288	72	79	239	85	79	596	499
80	317	93	80	276	80	80	190	37
81	462	82	81	403	84	81	306	69
82	793	116	82	575	71	82	558	120
83	430	105	83	460	77	83	579	102
84	455	117	84	385	69	84	668	99
85	294	94	85	385	77	85	544	68
86	526	108	86	585	98	86	743	88
87	619	120	87	881	492	87	815	490
88	580	101	88	751	408	88	736	440
89	261	72	89	496	566	89	517	587
90	169	98	90	150	79	90	113	41
91	178	99	91	169	70	91	140	59
92	378	91	92	394	100	92	342	81
93	118	89	93	199	99	93	256	64
94	214	77	94	357	93	94	461	70
95	123	79	95	212	84	95	212	57
96	264	67	96	477	83	96	484	76
97	232	115	97	573	585	97	470	470
98	344	87	98	639	570	98	542	419
99	162	82	99	450	615	99	374	459

[2] Four groups of Test-Network (TN) instances: (1)TN1-2-3 with 4×4 cells; (2)TN4-5-6 with 6×6 cells; (3)TN7-8-9 with 8×8 cells; (4)TN10-11-12 with 10×10 cells. TN files are available in URL `http://oplink.lcc.uma.es/problems/mmp.html`.

5.2 Experimental Results

We have conducted different experiments with several configurations of GPSO and HNN+BD depending on the test network used. Since the two algorithms perform quite different operations, we have set the parameters (Table 1) after preliminary executions of the two algorithms (with each instance) where the computational effort in terms of time and number of evaluations was balanced.

Table 1. Parameter settings for HNN+BD and GPSO. The columns indicate: the number of dropping balls (*N.DroppBalls*) and the number of trials (*N.Trials*) for HNN+BD. For GPSO are reported: the number of particles (*N.Particles*), the crossover probability (P_{cross}), the mutation probability (P_{mut}) and the weighted values (w_a, w_b and w_c).

Test Network	HNN+BD		GPSO			
Dim.	*N.DroppBalls*	*N.Trials*	*N.Particles*	P_{cross}	P_{mut}	$w_a + w_b + w_c$
(4×4)	7	3	20	0.9	0.1	0.33+0.33+0.33
(6×6)	10	5	50			
(8×8)	15	5	100			
(10×10)	15	5	120			

After the initial experimentation, several results were obtained; they are shown in Table 2. The first column contains the number and dimension (in parenthesis) of each test network. Three values are presented for each evaluated algorithm: the best cost (out of 10 runs), the average cost (*Aver.*) of all the solutions, and the deviation (*Dev.*) percentage from the best cost.

As it can be seen from the results, the two algorithms have similar performance in almost all of the instances, although there are a few differences for the large test networks. For example, GPSO obtains better solutions in Test-Network 7 and 10, while, HNN+BD obtains a better solution in Test-Network 11. In addition, it can be noticed that the deviation percentage from the best cost is generally lower in GPSO than in HNN+BD, specially for the smaller test networks. This behavior leads us to believe that the GPSO approach is more robust than HNN+BD, but just slightly.

Table 2. Results for Test Networks obtained by HNN+BD and GPSO

Test Network	HNN+BD			GPSO		
No.(Dim.)	*Best*	*Aver.*	*Dev.*	*Best*	*Aver.*	*Dev.*
1 (4×4)	98,535	98,627	0.09%	98,535	98,535	0.00%
2 (4×4)	97,156	97,655	0.51%	97,156	97,156	0.00%
3 (4×4)	95,038	95,751	0.75%	95,038	95,038	0.00%
4 (6×6)	173,701	174,690	0.56%	173,701	174,090	0.22%
5 (6×6)	182,331	182,430	0.05%	182,331	182,331	0.00%
6 (6×6)	174,519	176,050	0.87%	174,519	175,080	0.32%
7 (8×8)	308,929	311,351	0.78%	308,401	310,062	0.53%
8 (8×8)	287,149	287,149	0.00%	287,149	287,805	0.22%
9 (8×8)	264,204	264,695	0.18%	264,204	264,475	0.10%
10 (10×10)	386,351	387,820	0.38%	385,972	387,825	0.48%
11 (10×10)	358,167	359,036	0.24%	359,191	359,928	0.20%
12 (10×10)	370,868	374,205	0.89%	370,868	373,722	0.76%

Another obvious difference between HNN+BD and GPSO lies in the behavior of each algorithm. This can be observed in Fig. 2, where we show a graphical representation of algorithm runs for the different evaluated networks. Each graph, corresponding to one of the twelve test networks, plots a representative trace of the execution of each algorithm tracking the best solution obtained versus the number of iterations. On the one hand, GPSO shows a typical behavior in evolutionary metaheuristics, that is, it tends to converge from the solutions in the initial population to an optimal reporting cell arrangement. Graphically, the GPSO operation is represented by a monotonous decreasing (minimization) curve. On the other hand, HNN+BD carries out a different searching strategy, as from the initialization, it provokes frequent shaking scenarios in the population with the purpose of gradually diversifying and intensifying the search. These "shakes" are carried out by means of the Ball Dropping technique (Section 4) when no improvement in the overall condition of the network is detected, so the frequency of this operation is variable.

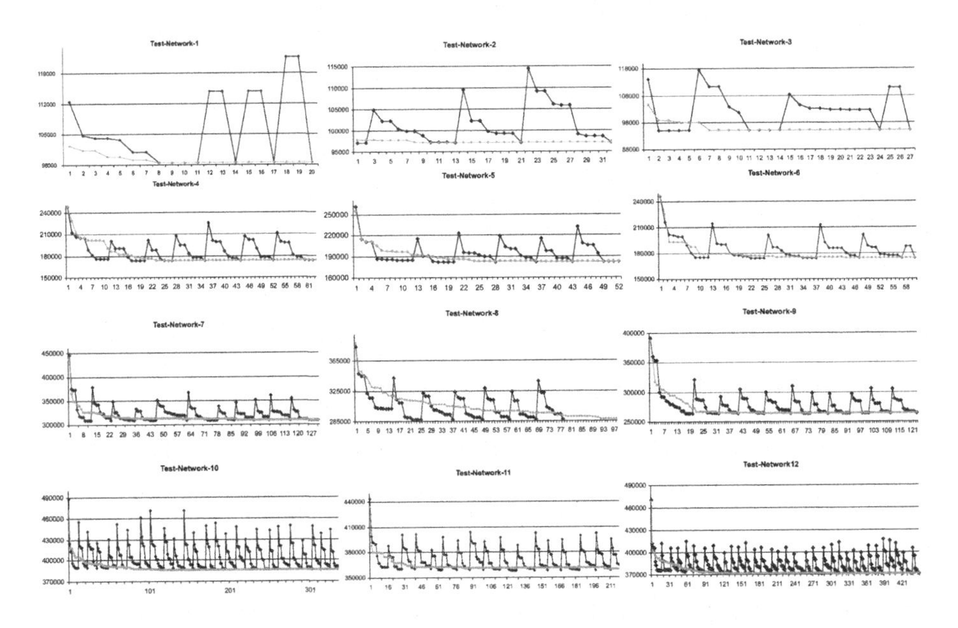

Fig. 2. Cost values level (Y axis) versus iterations (X axis) of all the test networks. Each graphic plots the energy level obtained, we track the evolution of the HNN+BD algorithm (black line with peaks and valleys), and the fitness level in the evolution of the GPSO algorithm (concave grey curve).

Evidently, as Fig. 2 shows, the number of drops in larger test networks is higher than in smaller ones, since the number of iterations required here to converge is also higher. Graphically, this behavior produces intermittent peaks and valleys in the evolution line.

From the point of view of the quality of solutions, as expected, optimal reporting cell configurations for all test networks split the network into smaller sub-networks by clustering the full area. This property can be seen in the large instances in a much clearer way than in the short ones (Fig. 3).

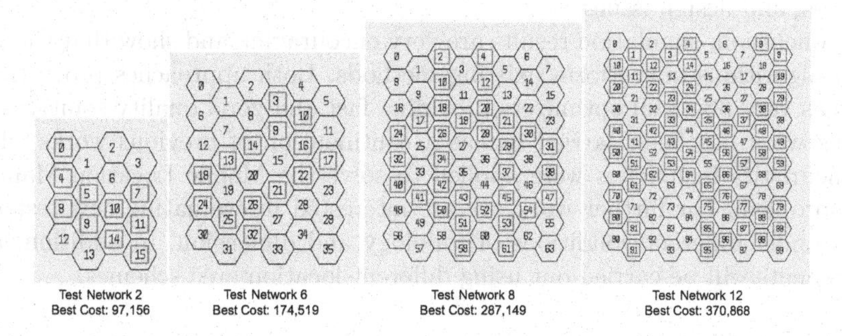

Fig. 3. Paging Cells (with squares) configurations obtained as solutions by the two algorithms (the same solutions) in Test Network 2, Test Network 6, Test Network 8 and Test Network 12. Neighborhood area clusters are easily visible in larger instances. All the legends show the Best Cost found by both algorithms.

5.3 Comparison with Other Optimizers

To the best of our knowledge a Genetic Algorithm (GA) is the only algorithm that can be compared against in this work. The modeling of the problem, the quality of the initial population, and the number of iterations are the main design issues that can affect the performance of the GA. When comparing the proposed approaches with a GA implementation given in [7], one can observe two advantages in terms of convergence and quality of solution in our two new approaches.

Despite the general good behavior of the GA, our two approaches generate a better solution when solving the Test-Network-2 (6×6 instance provided in [7]) in additional experiments. The energy value obtained by the GA is 229,556 with a total of 26 paging cells in the network, while, the cost obtained by HNN+BD in this work is 211,278 with 24 paging cells, and the GPSO obtained a cost of 214,313 with 23 paging cells. With respect to HNN+BD, a reasonable explanation for this difference could be due to the setup parameters used for the GA in [7]. However, our GPSO uses a similar setup parameters compared to the GA, providing a better solution with a smaller number of paging cells.

6 Conclusions

This paper addresses the use of two nature inspired approaches to solve the Mobile Location Management problem found in telecommunications: a new binary Particle Swarm Optimization algorithm called GPSO, and an algorithm based on a Hopfield Neural Network hybridized with the Balls Dropping Technique.

The problem is described and tackled following the Reporting Cells Scheme. In addition, the design and operation of HNN+BD and GPSO are discussed. Twelve test networks of different dimensions, generated following realistic scenarios of mobile networks, were for the first time used in this work. In addition, a comparison of the algorithms is carried out focusing on the performance, robustness, and design issues.

In conclusion, simulation results are very encouraging and show that the proposed algorithms outperform existing methods. Both approaches prove themselves as very powerful optimizers providing fast and good quality solutions.

This work has been carried out as a continuation of previous works where metaheuristics techniques were applied to solve the Mobile Location Management problem. For further work, we are interested in evaluating new test networks under different conditions of topology and dimension. In addition, new experiments will be carried out using different location area schemes.

References

1. Agrawal, D.P., Zeng, Q.-A.: Introduction to Wireless and Mobile Systems. Thomson Brooks/Cole Inc. (2003)
2. Alba, E., García-Nieto, J., Jourdan, L., Talbi, E.-G.: Gene Selection in Cancer Classification using PSO/SVM and GA/SVM Hybrid Algorithms. In: IEEE Congress on Evolutionary Computation CEC-2007, Singapore (September 2007)
3. Bar-Noy, A., Kessler, I.: Tracking mobile users in wireless communication networks. In: INFOCOM (3), pp. 1232–1239 (1993)
4. Lin, Y.-B., Chlamatac, I.: Wireless and Mobile Network Architecture. John Wiley and Sons, Chichester (2001)
5. Moraglio, A., Di Chio, C., Poli, R.: Geometric Particle Swarm Optimization. In: Ebner, M., O'Neill, M., Ekárt, A., Vanneschi, L., Esparcia-Alcázar, A.I. (eds.) EuroGP 2007. LNCS, vol. 4445, Springer, Heidelberg (2007)
6. Subrata, R., Zomaya, A.Y.: Location management in mobile computing. In: ACS/IEEE International Conference on Computer Systems and Applications, pp. 287–289 (2001)
7. Subrata, R., Zomaya, A.Y.: A comparison of three artificial life techniques for reporting cell planning in mobile computing. IEEE Trans. Parallel Distrib. Syst. 14(2), 142–153 (2003)
8. Taheri, J., Zomaya, A.Y.: The use of a hopfield neural network in solving the mobility management problem. In: IEEE/ACS International Conference on Pervasive Services, ICPS 2004, pp. 141–150 (July 2004)
9. Taheri, J., Zomaya, A.Y.: Realistic simulations for studying mobility management problems. Int. Journal of Wireless and Mobile Computing 1(8) (2005)
10. Taheri, J., Zomaya, A.Y.: A genetic algorithm for finding optimal location area configurations for mobility management. In: LCN 2005: Proceedings of the The IEEE Conference on Local Computer Networks 30th Anniversary, Washington, DC, USA, pp. 568–577. IEEE Computer Society, Los Alamitos (2005)
11. Wu, H.-K., Jin, M.-H., Horng, J.-T., Ke, C.-Y.: Personal paging area design based on mobile's moving behaviors. In: Twentieth Annual Joint Conference of the IEEE Computer and Communications Societies, INFOCOM 2001, vol. 1, pp. 21–30 (2001)

Adaptive Local Search for a New Military Frequency Hopping Planning Problem

I. Devarenne[1], A. Caminada[1], H. Mabed[1], and T. Defaix[2]

[1] UTBM, SET Lab, 90010 Belfort Cedex, France
[2] DGA, CELAR Lab, 44000 Rennes Cedex, France
{isabelle.devarenne,alexandre.caminada,hakim.mabed}@utbm.fr

Abstract. The military radio stations with frequency hopping propose new problems of frequency assignment which must take into account the size of the deployment, the limited resources and also new interferences constraints on transmitters. This new problem is public since the publication [1] with a set of 10 instances. The computing resources and the computing times are imposed. We tackle this problem with a method based on adaptive local search coming from the graph-coloring which manages phases of intensive research and diversified research. We developed several alternatives that are tested and compared on the scenarios.

1 Introduction

The problem of frequency assignment for the radiocommunication systems with frequency hopping is a new problem exposed in 2006 [1]. Communications must be transmitted between various groups of vehicles transporting transmitter-receiver "stations". These stations are gathered in networks and are connected by the same frequency channel. The quality of the communications depends on the various situations of interferences produced by the stations. We distinguish the "Co-vehicle" electromagnetic disturbances that come from the presence of several stations on the same vehicle ("biposte" or "triposte" disturbance for two or three transmitter-receivers on the same vehicle) and the "Co-site" disturbances generated by the presence of two close stations (for example distant of less than 100 meters). These disturbances require establishing an electromagnetic constraint of compatibility between the frequencies allocated to the networks to which the stations belong in order to avoid jamming. Of a problem made up of vehicles, stations and networks, one builds a problem of optimization made up of Co-vehicle and Co-site constraints between the networks for which it is required to assign frequencies by minimizing a function of interference.

The characteristic of this system is that the stations emit with "Frequency Hopping" (FH) using sub-bands of frequency. A communication in FH is not transmitted on only one frequency but on a list of frequency called "Frequency Plan" (FP). Each frequency is used for very a short period and during a communication the whole of the frequencies of the list is used [2]. Because of assignment of a list of frequencies to each network, compared to a problem with fixed

M. Giacobini et al. (Eds.): EvoWorkshops 2008, LNCS 4974, pp. 11–20, 2008.

frequency the problem combinatory is strongly amplified. In addition, the interferences between two networks are not expressed any more in term of difference between frequencies as in the systems at fixed frequencies, but they are the result of the whole of the frequencies used by each network. Thus, one will not define the quality of the communications by the respect of separation constraints between frequencies but by a binary level of error (noted BER for *Bit Error Rate*) which results from the number of common frequencies between the lists allotted to the networks belonging to the same disturbance. Then the level of interference on each network should not exceed a maximum threshold expressed in BER.

Because of the recent publication of this problem, only [1] made state of 3 methods applied to this new problem on private scenarios (CN-Tabu, simulated annealing and our method). We thus present the first work on the scenarios made public in 2006. In the literature there exist problems which approach it: the assignment of frequencies [3, 4] and more generally the set-T-coloring. This last problem is defined as follows: given an undirected graph $G = (V, E)$ and B a whole of needs with $B = \{b_1, \ldots, b_N | \, b_i \in \mathbb{N}\}$ corresponding to the respective needs for each node x_i, the problem is to allocate a whole of b_i colors $c(x_i) = \{c_{i,1}, \ldots, c_{i,b_i}\}$ to each node x_i while respecting the constraints inside nodes: $\forall c_{i,m}, c_{i,n} \in c(x_i), m \neq n, |c_{i,m} - c_{i,n}| \notin T_{i,i}$, and the constraints between adjacent nodes: $\forall (x_i, x_j) \in E, \forall c_{i,m} \in c(x_i), \forall c_{j,n} \in c(x_j), |c_{i,m} - c_{j,n}| \notin T_{i,j}$. The problems of set-T-coloring were very largely studied in the literature [5, 6], however they do not cover the topics of organizing the colors in sub-sets and maximizing the set size per node as it is in our problem. Looking at frequency assignment problems for other communication systems (cf. http://fap.zib.de), they often are particular cases of the graph-coloring problem [3, 8]. The differences between the problems of frequency assignment are related to the various radio systems. Only the GSM system, world standard of mobile telephony, is also equipped with the frequency hopping. Unfortunately the complexity of the management of the jump makes that there is few literature on its optimization. The majority of the publications on the frequency assignment relates to the cases without hops with methods based on population of solutions [9, 10, 11], on local search [12] as Taboo Search [13, 14, 15] or constraints programming [16, 17, 18]. [19] dealt with optimization breaking up the problem into two parts: the generation of lists of frequencies per station then the modification of these lists with a simulated annealing. But the generated interferences are calculated according to the difference between the frequencies as in a problem without hops and the lists to be allocated are not structured in sub-bands.

The state of the art on the problem we deal with is thus very thin. Major specificities of this problem are the evaluation of the FP that needs a simulator to compute the BER and the concept of sub-bands which structure the frequency sets. The benchmarks combinatory is huge ($10^{28\,400}$ for a scenario with 400 networks) and the computing resources and the computing times are imposed (P-IV with 30mn or 60mn by scenario). Our work was thus guided by the adaptation of a local search method resulting from graph-coloring to profit from experiment on a theoretical problem, requiring few resources, able to provide complete solutions and rather robust. A local search method with an adaptive mechanism for diversification and intensification phases is developed for this problem. Firstly, the formulation of the problem is described in section 2. In section 3, we present the adaptive local search method. Then, in section 4 the results on public scenarios allow to compare various alternative methodologies.

2 Problem Formalization

2.1 Constraints to Satisfy

The problem constraints can be gathered in three sets: constraints related to the networks, constraints on the frequency plan and those related to interference.

Constraints linked to the networks. There are two types of networks: networks whose FP are already allocated, noted R^{hc}, and cannot be modified, and networks whose FP must be allocated, noted R^c. Each network $r \in R = R^c \bigcup R^{hc}$ is defined by: an hierarchical level h_r representing its level of priority; the minimal size NF_r^{min} of the FP to be allocated to this network knowing that the size of a FP is the number of frequencies it uses; the maximum threshold of BER of the network, noted BER_r^{max}, which defines the maximum binary error rate authorized for the network r; and a domain of resource D_r, defining the set of frequencies available for this network. The domains are various: set of contiguous frequencies, combs of frequencies, isolated frequencies...

Constraints linked to the frequency plans. The frequency planning consists in allocating to each network the list of frequencies forming the FP. The FP is a whole of frequencies structured in intervals called "sub-bands". A plan gathers a maximum number SB_{max} of sub-bands. A sub-band sb is made up of the frequencies of an interval $[f_{min}; f_{max}]$ sampled with the step δ_f and f_{min} as initial frequency. The sub-bands of the same FP cannot overlap and their number is limited. The FP pdf_r of the network r is defined by:

$$\begin{cases} pdf_r \in \mathcal{P}(S) \text{ such that } \forall sb \in pdf_r\,,\, sb \subset D_r \\ \forall (sb_1,\, sb_2) \in pdf_r^2,\ (max(sb_1) < min(sb_2)) \text{ or } (max(sb_2) < min(sb_1)) \\ |pdf_r|_{sb} \leq SB_{max} \text{ and } |pdf_r|_f \geq NF_r^{min} \end{cases} \quad (1)$$

with $r \in R$ one network and pdf_r the frequency plan of the network r; S, the set of sub-bands sb and $\mathcal{P}(S)$ the set of parts of S; $min(sb)$, $max(sb)$ the initial and final frequencies of sub-band sb; $|pdf_r|_{sb}$ the number of sub-bands of the frequency plan pdf_r; $|pdf_r|_f$ the number of frequencies of the frequency plan pdf_r of the network r; NF_r^{min} the minimum number of frequency the network r must have. Within the framework of the public problems we treated, the maximum number of sub-bands SB_{max} is fixed at 10, the step of a sub-band δ_f can take three values (1, 2 and 4) and the whole of the domains associated with the various networks is included in the global domain, noted D_g, of 2000 frequencies.

Constraints linked to interferences. The level of interference between several FP is calculated starting from the elementary values of interference between the frequencies of each plan. Then the binary error rate of one network will be the maximum value of BER of the network as a receiver for all the disturbances which it undergoes. To identify the scramblers, we define as a link the set of stations of various networks that disturb each other in saturated traffic condition (each transmitter is transmitting). The reference value of BER for each network is then the worst case among its whole links. The level of interference of one network, noted $BER^c(R)$, is thus:

$$BER^c(r) = \max_{l(r_r;r_{b1}[,r_{b2}])\in L} \left(BER_l\left(pdf_{r_r}; pdf_{r_{b1}}[, pdf_{r_{b2}}]\right)\right) \quad (2)$$

with $BER_l(pdf_{r_r}; pdf_{r_{b1}}[, pdf_{r_{b2}}])$, the BER of the link having pdf_{r_r} as receiver and disturbed by $pdf_{r_{b1}}$ and $pdf_{r_{b2}}$. It is calculated by the sum of the elementary BER generated by each triplet of frequencies, balanced by the product of the sizes of all FP. It is thus an average BER. For each network r, BER_r^{max} defines the level of radio quality to respect. The radio quality level $BER^c(R)$ of the network r should not exceed the value threshold BER_r^{max}, then:

$$\forall r \in R, BER^c(r) \leq BER_r^{max} \quad (3)$$

2.2 Objectives to Optimize

The fitness function uses two types of information to estimate the quality of the solutions: interferences and the frequency reuse.

The interferences. The first problem to deal with consists in minimizing the interferences resulting from all FP. It includes two objectives: to minimize the greatest difference between the BER of a network and the BER value threshold; and to minimize the sum of these differences for all the networks balanced by their hierarchy. These two objectives are aggregate in a function using two weights α and α' that are given input data for each scenario. Thus the function which includes the criteria of interferences is formulated by:

$$\begin{aligned} F_{BER}(S) &= \alpha \times \max_{r\in R} \left(BER^c(r) - BER_r^{max}\right)^+ \\ &+ \alpha' \times \sum_{r\in R} \left(h_r \times \left(BER^c(r) - BER_r^{max}\right)^+\right) \end{aligned} \quad (4)$$

The frequency reuse. In addition to the measurement of the interferences, the solution quality is also measured by the frequency reuse ratio. FP of big size are favored via three weighted criteria (β, γ and γ' are given input data for each scenario).

1. Minimal size of FP: the objective is to minimize the number of networks, defined by $F_{size}(S)$, whose FP does not respect the imposed minimal number of frequency used.

$$F_{size}(S) = \beta \times \left|\{\forall r \in R,\ |pdf_r|_f < NF_r^{min}\}\right| \quad (5)$$

2. Size of the smallest FP: the objective is to maximize the size of the smallest FP defined by the function $F_{min}(S)$.

$$F_{min}(S) = \gamma \times \min_{r \in R}(|pdf_r|_f) \tag{6}$$

3. The frequency reuse rate: the objective is to maximize the balanced sum of the sizes of all the FP defined by the function $F_{sum}(S)$. The hierarchy of the networks h_r is taken into account.

$$F_{sum}(S) = \gamma' \times \sum_{r \in R}(h_r \times |pdf_r|_f) \tag{7}$$

The fitness. Finally the assignment of FP in frequency hopping consists in assigning a list of frequency to each network of the deployment by respecting the constraints of a maximum number of sub-bands SB_{max} and non-covering of the sub-bands of the same FP; on the one hand by minimizing the interference between the FP $F_{BER}(S)$ and the number of networks having a too small FP $F_{size}(S)$; and in addition by maximizing the size of the smallest FP $F_{min}(S)$ and the frequency reuse rate $F_{sum}(S)$. The fitness function to minimize is:

$$F(S) = F_{BER}(S) + F_{size}(S) - (F_{min}(S) + F_{sum}(S)) \tag{8}$$

NB: This function does not have physical reality; a multi-objective approach would be more suitable but is not on the current specifications of the problem.

3 Adaptive Local Search

The scanty means imposed in the benchmark, the complexity of the problem and the combinatory of the scenarios do not make it possible to use an exact method or a population based method as starting algorithm even if they could be relevant in the further study. The local search having proven its reliability in set-T-coloring, we defined a local search method with neighborhood extension and restriction control. It is based on two complementary mechanisms: a detection of loop and a Taboo list which are employed respectively to extend and restrict the neighborhood of the current solution. The mechanism of loop detection uses a history of research in order to diversify this one by extending the set of the candidate neighborhood solutions in case of search blocking. The Taboo list limits the set of the candidate neighborhood solutions of the current solution to prevent repetitive choices of variables at the next iterations. In both cases, one iteration is defined by successive operations whose finality is to modify the list of frequency which constitutes the FP of a network: the choice of the network; the choice of the sub-band; and the choice of the modification to apply to the sub-band.

The mechanisms of loop detection and Taboo list are only applied to the network choice. The choices of the sub-band, randomized, and on the frequencies, the best ones, do not depend on these two mechanisms. The neighborhood that we used issues from graph-coloring [7]. It is the neighborhood based on the graph nodes (here the networks) which modifies the colors (here the frequencies) of the nodes in conflict (here the networks whose at least one constraint is unsatisfied). This structure is the most used in literature of graph-coloring. The extension or the restriction of the lists of candidate neighbors will act on the selection of the networks among the more in conflict, all those in conflict or those without conflict. The motivations and the principles are explained hereafter.

Network weighting. In order to select the network to be modified, a weight function $P_c(R)$ is employed for each network r estimating the quality of the associated FP. This weight shows the influence of each FP on the evaluation of the total solution according to the level of BER reached by the network and its deficit in frequencies compared to the required minimal width. The formula (9) used to define this weight takes into account the parameters of the fitness function of the solution according to the notations defined previously.

$$\begin{aligned} P_c(r) = h_r \times \big(&\alpha' \times (BER^c(r) - BER_r^{max})^+ \\ &+ \beta \times (NF_r^{min} - |pdf_r|_f)^+ \\ &+ \gamma' \times |pdf_r|_f\big) \end{aligned} \tag{9}$$

Restriction of candidate networks with Taboo list. Initially, we defined a local search based on the modification of the FP of worse quality. This first method was used with and without Taboo list on the choice of the network. The network of the most important weight is selected. In the event of equality, the network is randomly selected among the networks of the most important weight. This *deterministic* choice of the network is carried out with each iteration. However, the purely deterministic character may involve premature convergence towards a local optimum. The option with Taboo list forces the algorithm to choose the network not taboo of the most important weight. The selected network becomes taboo for one T duration randomly taken in the interval DT defined in equation (10) as for a dynamic Taboo list [6].

$$T = rand(DT) \text{ with } DT = \left[0.5 \times \frac{\sqrt{|R|}}{2}; 1.5 \times \frac{\sqrt{|R|}}{2}\right] \tag{10}$$

Extension of candidate networks with loop detection. Up to now, we have a local search algorithm without or with Taboo list according to a model very frequently used in literature: the algorithm works on the variables which contribute more to the degradation of the solution performance. A second method based on diversification in the choice of the network is added by the use of a mechanism of loop detection. The objective of the loop detection is to detect the occurrences of choice of networks during the iterations. It is based on the list of the last visited networks that represents a temporal window updated after each iteration. The observation of the repetition of the choice of a network is characteristic of a blocking of research in spite of the taboo status of certain variables.

The loop detection allows the method to react by modifying the behavior of the local search during research. The choice of the network will not be made any more in a *deterministic* way according to the higher weight but in a *diversifying* way by a random choice among the set of networks not taboos of non null weight. The parameters of the loop detection are identical to those applied in graph-coloring [20]. When the number of visits of a network r is higher than a value threshold representing a number of authorized occurrences *nbOcc* during the m last iterations, then a loop of repetition is detected and one iteration of *diversifying* type is carried out. The memory size m of the last iterations observed is defined according to the number of variables of the problem. Here, the full number of networks (target and out-targets) of the problem is used, $m = \frac{|R|}{2}$. The number of authorized occurrences *nbOcc* is defined by a percentage of the memory size fixed empirically at 5%: $nbOcc = 0.05 \times m$.

The method and its variants. The general operation of the method is the following. At each iteration the loop detection is evaluated: for each variable there is a checking of the number of occurrence compared to the threshold. Two types of iterations can then be used; they differ by the selection of the network to modify. If the algorithm does not turn round (the occurrences are lower than the threshold), a deterministic iteration is applied: the choice of network of the highest weight is selected. If the algorithm turns round (the occurrences are higher than the threshold for at least a variable), a diversifying iteration is applied: the network is randomly selected among all. The method is known as adaptive since it dynamically adapts the list of the candidate neighbors according to the state of research. If we combine the Taboo list principle with this general operation, only the networks not taboos can be selected whatever the type of iteration. With the Taboo list, two options are considered for the management of the taboo status: all the selected networks become taboos or only the networks detected in loop become taboos. To summarize, we thus have five concurrent methods tested in the following section to manage the iterative process: deterministic local search based on the networks weight; local search with Taboo list; local search with detection of loop; and local search with combination of loop detection and Taboo list of all the nodes or only nodes in loop.

4 Results and Analysis on Public Scenarios

This section compares the five methods of local search. All the methods have the same global architecture but differ by the management of the variable access. First of all the table 1 presents the evaluation of the best solutions obtained on only one run for the 10 public scenarios. The first two methods in column do not use a mechanism of loop detection; they are methods based on the use of iterations of the local search driven by the resolution of conflicts. The selected network is always that of higher weight (deterministic step). The difference between the two methods rests on the use or not of a Taboo list defined dynamically according to equation (10). The three other methods are based on the mechanism of loop detection with the combination of determinist and diversifying methods: without Taboo list on the networks (4th column), with a Taboo list on all the networks having started a loop (5th column) and with a Taboo list on all the visited networks (6th column).

Table 1. Comparison of five local search algorithms

problems	higher conflict		loop detection		
	without	with	without	with taboo	
	taboo	taboo	taboo	networks in loop	all networks
PUB01	-10 426	-10 827	-13 764	**-13 892**	-12 111
PUB02	-928	-961	-1 353	-1 436	**-1 447**
PUB03	-8 110	-9 516	-13 932	-13 952	**-14 574**
PUB04	-15 843	-17 515	**-27 866**	-27 837	-27 786
PUB05	-7 309	-9 654	-10 720	**-12 144**	-10 912
PUB06	515 417	95 906	71 738	96 314	**70 360**
PUB07	-23 554	-53 777	-478 471	-476 693	**-484 297**
PUB08	-194 435	-336 741	**-887 354**	-869 292	-878 640
PUB09	1 667 459	962 109	876 409	755 753	**684 699**
PUB10	-5 672	-7 541	**-10 731**	-10 624	-10 680

A first comparison can be made between both methods without mechanism of loop detection. For all the scenarios, the addition of the dynamic Taboo list improves the preceding results; the use of medium-term diversification by restriction of the vicinity is thus effective. Then, a comparison can be made with the detection of loop. The use of loop detection without Taboo list improves considerably the results on all the scenarios. The loop detection to diversify by extension of the vicinity is thus more effective than the Taboo list. Lastly, combining the Taboo list with the loop detection improves the best results for 7 cases out of 10. The best run for each example (in fat) was obtained with one of the three methods using the loop detection mechanism.

The preceding results relate to only one run. The table 2 presents final results for all the methods with loop detection with the average scores on 5 runs. The dark cells of the table correspond to the best results. Thus, it is observed that the first method (without Taboo list) obtains the best result three times, it is classified second for two problems and last for the five others. The second method is classified first five times, second three times and last for two scenarios. The last method is the best for two cases, second for five others and three times last. All in all, on five runs the second method obtains the best performances. The adaptive combination of the deterministic and diversifying iterations done dynamically via the loop detection gave the best performances. In particular it

Table 2. Combination of loop detection and Taboo list (on 5 runs)

	loop detection		
	without	with taboo	
problems	taboo	networks in loop	all networks
PUB01	-13 981	-14 359	-14 828
PUB02	-1 538	-1 507	-1 543
PUB03	-13 466	-13 905	-13 625
PUB04	-32 485	-32 839	-32 512
PUB05	-9 669	-10 025	-9 562
PUB06	240 937	141 303	145 587
PUB07	-474 565	-474 072	-472 030
PUB08	-863 941	-828 613	-819 019
PUB09	895 716	810 793	826 828
PUB10	-10 139	-9 207	-10 017

has been better than an intensive method with Taboo list what shows than the taboo status is sometime not enough efficient to diversify the research.

5 Conclusion and Perspectives

This paper presented studies on an adaptive local search implemented for the problem of assignment of lists of frequencies for military networks of radiocommunications with frequency hopping. The assignment of lists of frequencies with frequency hopping is a new optimization problem with constraints whose objective is to allocate to each network a frequency plan which is a structured list while minimizing a fitness which incorporates several criteria. The ideal models of set-T-coloring are closest to our problem however this one has major differences: the interferences are not represented into frequency constraints spacing as in set-T-coloring but as average BER values; the frequency plan must be organized in non-overlapping sub-bands of the spectrum; and the fitness is made up of aggregate and balanced criteria which combine radio quality and plan size data. We explained the general outline of an adaptive local search method to deal with this problem. A procedure of loop detection on the networks and adaptive choice of the intensifying or diversifying local search framed it. We defined a function to weight the networks in order to classify them according to their conflict. We also carried out a version of the method with and without Taboo list to evaluate the contribution of this combination in this applicative context. The results obtained by various alternatives of the method were provided on public scenarios which are available. The adaptive method based on the combination of two local searches by the loop detection gave better scores that an intensive method with Taboo list. In addition the association of loop detection with a Taboo status on the networks detected in loop is the most powerful option. We must now develop other methods of research in collaboration with other teams for better evaluating this problem. Also a multi-objective approach would be very interesting to identify the compromise solutions independently of the weights on the criteria which are currently given as scenario inputs.

References

[1] Defaix, T.: SFH-FP Slow Frequency Hopping - Frequency Planning. In: Septième Congrès de la Société ROADEF 2006, Lille, France (2006)
[2] Milstein, L.B., Simon, M.K.: Spread spectrum communications. In: The mobile communications handbook, pp. 152–165. IEEE Press, Los Alamitos (1996)
[3] Aardal, K., van Hoesel, C., Koster, A., et al.: Models and Solution Techniques for Frequency Assignment Problems. ZIB-report, Berlin, pp. 01–40 (2001)
[4] Koster, A.M.C.A.: Frequency Assignment: Models and Algorithms. PHDThesis Universiteit Maastricht (1999)
[5] Comellas, F., Ozón, J.: Graph Coloring Algorithms for Assignment Problems in Radio Networks. In: Alspector, J., et al. (eds.) Applic. of Neural Nets, pp. 49–56 (1995)
[6] Dorne, R., Hao, J.: Tabu Search for graph coloring, T-coloring and Set T-colorings. In: Osman, I., et al. (eds.) Theory and Applications. Theory and Applications, Kluwer, Dordrecht (1998)
[7] Avanthay, C., Hertz, A., Zufferey, N.: A variable neighborhood search for graph coloring. European Journal of Operational Research 151, 379–388 (2003)
[8] Eisenblätter, A., Grötschel, M., Koster, A.: Frequency Planning and Ramifications of Coloring. Discussiones Mathematicae, Graph Theory 22, 51–88 (2002)
[9] Hao, J.-K., Dorne, R.: Study of Genetic Search for the Frequency Assignment Problem. In: Alliot, J.-M., Ronald, E., Lutton, E., Schoenauer, M., Snyers, D. (eds.) AE 1995. LNCS, vol. 1063, pp. 333–344. Springer, Heidelberg (1996)
[10] Renaud, D., Caminada, A.: Evolutionary Methods and Operators for Frequency Assignment Problem. SpeedUp Journal 11(2), 27–32
[11] Weinberg, B., Bachelet, V., Talbi, G.: A co-evolutionist meta-heuristic for the FAP. In: Frequency Assignment Workshop, London, England (2000)
[12] Palpant, M., Artigues, C., Michelon, P.: A Large Neighborhood Search method for solving the frequency assignment problem. LIA report 385 (2003)
[13] Hao, J.-K., Dorne, R., Galinier, P.: Tabu search for Frequency Assignment in Mobile Radio Networks. Journal of Heuristics 4(1), 47–62 (1998)
[14] Montemanni, R., Moon, J., Smith, D.H.: An Improved Tabu search Algorithm for the Fixed-Spectrum FAP. IEEE Trans. on VT 52(4), 891–901 (2003)
[15] Montemanni, R., Smith, D.H.: A Tabu search Algorithm with a dynamic tabu list for the FAP. Technical Report, University of Glamorgan (2001)
[16] Dupont, A., Vasquez, M., Habet, D.: Consistent Neighbourhood in a Tabu search. In: Workshop on Constraint Programming techniques, Nantes (2005)
[17] Schulz, M.: Solving Frequency Assignment Problem with Constraint Programming. Technische Universität Berlin, Institut für Mathematik (2003)
[18] Yokoo, M., Hirayama, K.: Frequency Assignment for Cellular Mobile Systems Using Constraint Satisfaction Techniques. In: Proc. of IEEE VTC Spring (2000)
[19] Moon, J., Hughes, L., Smith, D.: Assignment of Frequency Lists in Frequency Hopping Networks. IEEE Trans. on Vehicular Technology 54(3), 1147–1159 (2005)
[20] Devarenne, I., Caminada, A., Mabed, H.: Analysis of Adaptive Local Search for the Graph Coloring Problem. In: 6th MIC 2005, Vienne, Autriche (2005)

SS vs PBIL to Solve a Real-World Frequency Assignment Problem in GSM Networks

José M. Chaves-González, Miguel A. Vega-Rodríguez,
David Domínguez-González, Juan A. Gómez-Pulido, and Juan M. Sánchez-Pérez

Univ. Extremadura. Dept. Technologies of Computers and Communications,
Escuela Politécnica. Campus Universitario s/n. 10071. Cáceres, Spain
{jm,mavega}@unex.es, cap.ddominguez@iberia.es,
{jangomez,sanperez}@unex.es

Abstract. In this paper we study two different meta-heuristics to solve a real-word frequency assignment problem (FAP) in GSM networks. We have used a precise mathematical formulation in which the frequency plans are evaluated using accurate interference information coming from a real GSM network. We have developed an improved version of the scatter search (SS) algorithm in order to solve this problem. After accurately tuning this algorithm, it has been compared with a version fixed for the FAP problem of the population-based incremental learning (PBIL) algorithm. The results show that SS obtains better frequency plannings than PBIL for all the experiments performed.

Keywords: FAP, Frequency Planning, SS, PBIL, real-world GSM network.

1 Introduction

In this paper we compare two different metaheuristics (with two variants for each one) to discover which one is able to obtain better frequency plannings when solving a realistic-size real-world frequency assignment problem (FAP). The FAP problem is an NP-hard problem, so its resolution using metaheuristic algorithms has proved to be particularly effective. We have evaluated the algorithms PBIL (population-based incremental learning) and SS (scatter search) on a real-world GSM network with 2612 transceivers which currently operates in a quite large U.S. city (Denver city). Figure 1.a shows the topology of the GSM instance used.

GSM (global system for mobile) is the most successful mobile communication technology nowadays. In fact, by mid 2006 GSM services are in use by more than 1.8 billion subscribers [1] across 210 countries, representing approximately 77% of the world's cellular market. One of the most relevant and significant problems that it can be found in the GSM technology is the frequency assignment problem (FAP), because frequency planning is a very important and critical task for current GSM operators.

It is true that many mathematical models have been proposed since the late sixties [2] to solve the FAP problem but this work is focussed on concepts and models which are relevant for current GSM frequency planning [3]. For these reasons, we separate

M. Giacobini et al. (Eds.): EvoWorkshops 2008, LNCS 4974, pp. 21–30, 2008.

ourselves from existing results, since our problem is far different from those reported in the literature with similar names (which are benchmarking-like problems). We use realistic and accurate interference information from a real-world GSM network.

The two most significant elements in the FAP problem are the transceivers (TRXs) which give support to the communication and the frequencies which make possible the communication. Each TRX has to have assigned a concrete frequency in the most optimum way. The problem is that there are not enough frequencies (there are a few dozens) to give support to each transceiver (there are thousands of them) without causing interferences. It is completely necessary to repeat frequencies in different TRXs, so, it is necessary a good planning to minimize the number of interferences.

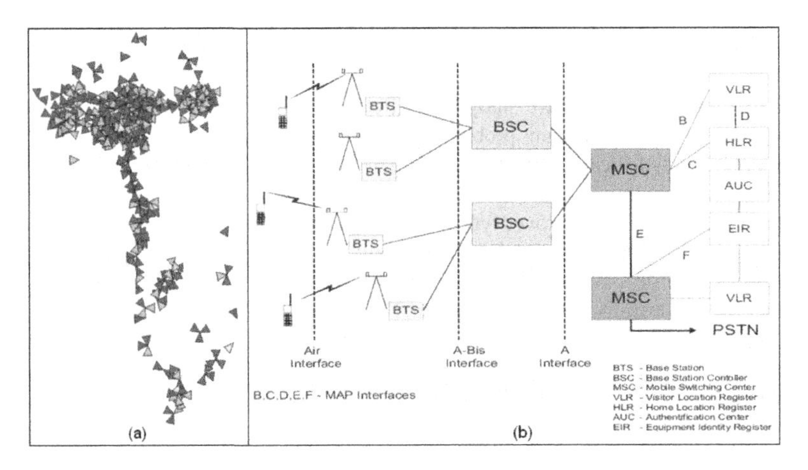

Fig. 1. (a) Topology of the GSM instance used. (b) Scheme for the GSM network architecture

The rest of the paper is structured as follows: In section 2 we present the background and the mathematical formulation of the frequency planning in GSM networks. Section 3 describes the algorithms used in this study (PBIL and SS). After that, the experimental evaluation of the algorithms is explained in section 4. Finally, conclusions and future work of the research are discussed in the last section.

2 The FAP Problem in GSM Networks

In the following subsections we firstly give a brief description of the GSM architecture; then, we give details on the frequency planning task applied to GSM networks; and finally, we introduce the mathematical formulation of the problem.

2.1 The GSM System

The Global System for Mobile communication (GSM) [4] is an open, digital cellular technology used for transmitting mobile voice and data services. A general scheme of the GSM network architecture [5] can be seen in figure 1.b.

As we can see in fig. 1.b, GSM system consists of many different components. The two most relevant ones which refer to frequency planning are the antennas or, as they are more known, base transceiver stations (BTSs) and the transceivers (or TRXs). Essentially, a BTS is a set of TRXs (grouped in sectors). Figure 1.a displays the network topology of the instance used for our work. In that figure, each triangle represents a sectorized antenna where several TRXs operate. The instance we use is quite large (it covers the city of Denver, with more than 500,000 inhabitants) and the GSM network includes 2612 TRXs, grouped in 711 sectors, distributed in 334 BTSs.

In GSM, one TRX is shared by up to eight users in TDMA (*Time Division Multiple Access*) mode. The main role of a TRX is to provide conversion between the digital traffic data on the network side and radio communication between the mobile terminal and the GSM network. The site where a BTS is installed is usually organized in sectors (of several TRXs) and the area, where each sector operates, defines a cell.

2.2 The FAP Problem Applied to the GSM Networks

The frequency planning is the last stage in the design of a GSM network. FAP lies in the assignment of a channel (or a frequency) to every TRX [3]. The optimization problem comes up because the usable radio spectrum is usually very scarce and, consequently, frequencies have to be reused for many TRXs in the network. However, the multiple use of a same frequency may cause interferences that can reduce the quality of service (QoS) down to unsatisfactory levels. In fact, significant interferences will occur if the same or adjacent channels are used in near overlapping cells. The problem is that computing this level of interference is a difficult task which depends on various factors (channels, radio signals, environment...). To quantify the interferences produced among the TRXs of a GSM network it is used what it is called the *interference matrix*. Each element *(i,j)* of this matrix contains two types of interferences: the *co-channel interference*, which represents the degradation of the network quality if the cells *i* and *j* operate on the same frequency; and the *adjacent-channel interference*, which occurs when two TRXs operate on adjacent channels (e.g., one TRX operates on channel *f* and the other on channel *f+1* or *f–1*). Therefore, an accurate interference matrix is an essential requirement for frequency planning because the ultimate goal of any frequency assignment algorithm will be to minimize the sum of all the interferences.

2.3 Mathematical Formulation

Let $T = \{t_1, t_2, \ldots, t_n\}$ be a set of n transceivers (TRXs), and let $F_i = \{f_{i1}, \ldots, f_{ik}\} \subset N$ be the set of valid frequencies that can be assigned to a transceiver $t_i \in T$, $i = 1, \ldots, n$. Note that k, which is the cardinality of F_i, is not necessarily the same for all the transceivers. Furthermore, let $S = \{s_1, s_2, \ldots, s_m\}$ be a set of given sectors (or cells) of cardinality m. Each transceiver $t_i \in T$ is installed in exactly one of the m sectors. Besides, we denote the sector in which a transceiver t_i is installed by $s(t_i) \in S$. Finally, let a matrix $M = \{(\mu_{ij}, \sigma_{ij})\}_{mxm}$, called the interference matrix where the two elements μ_{ij} and σ_{ij} of a matrix entry $M(i,j) = (\mu_{ij}, \sigma_{ij})$ are numerical values greater or equal than zero. In fact, μ_{ij} represents the mean and σ_{ij} the standard deviation of a Gaussian probability distribution describing the carrier-to-interference ratio (C/I) [6] when sectors i and j operate on a same frequency. The higher the mean value is, the lower the interference will be, and thus the better the communication quality. Note that the interference matrix is defined at sector (cell) level. We can establish that a solution to the problem is obtained by assigning to each transceiver $t_i \in T$ one of the frequencies from F_i. We will denote a solution (or frequency plan) by $p \in F_1 \times F_2 \times \ldots \times F_n$, where $p(t_i) \in F_i$ is the frequency assigned to the transceiver t_i. The objective, or the plan solution, will be to find a solution p that minimizes the following cost function [5]:

$$C(p) = \sum_{t \in T} \sum_{u \in T, u \neq t} C_{sig}(p,t,u) \tag{1}$$

In order to define the function $C_{sig}(p,t,u)$, let s_t and s_u be the sectors in which the transceivers t and u are installed, which are $s_t = s(t)$ and $s_u = s(u)$ respectively. Moreover, let $\mu_{s_t s_u}$ and $\sigma_{s_t s_u}$ be the two elements of the corresponding matrix entry $M(s_t, s_u)$ of the interference matrix with respect to sectors s_t and s_u. Then, $Csig(p,t,u) =$

$$\begin{cases} K & \text{if } s_t = s_u, |p(t) - p(u)| < 2 \\ C_{co}(\mu_{s_t s_u}, \sigma_{s_t s_u}) & \text{if } s_t \neq s_u, \mu_{s_t s_u} > 0, |p(t) - p(u)| = 0 \\ C_{adj}(\mu_{s_t s_u}, \sigma_{s_t s_u}) & \text{if } s_t \neq s_u, \mu_{s_t s_u} > 0, |p(t) - p(u)| = 1 \\ 0 & \text{otherwise} \end{cases} \tag{2}$$

$K >> 0$ is a very large constant defined by the network designer so as to make it undesirable allocating the same or adjacent frequencies to transceivers serving the same area (e.g., installed in the same sector). $C_{co}(\mu, \sigma)$ represents the cost due to *co-channel interferences*, whereas $C_{adj}(\mu, \sigma)$ is the cost in the case of *adjacent-channel interferences*. For further explanation about the $C_{co}(\mu, \sigma)$ and $C_{adj}(\mu, \sigma)$ costs, consult the references [5, 7].

3 Algorithms

This section presents the two algorithms used in this work for solving the proposed FAP: the SS algorithm (section 3.1) and the PBIL algorithm (section 3.2).

3.1 Scatter Search

Scatter Search (SS) [8, 9] works with a quite small set of solutions (called RefSet, which includes around 10 solutions). The RefSet encodes the frequency plans which solve the FAP problem and it is divided into quality solutions (the best frequency plans for the FAP problem) and diverse solutions (the most different ones). A brief description of the algorithm can be seen in figure 2.

According to figure 2, the algorithm starts with the generation of the population using a *Diversification Generation Method* which creates random individuals, so that all TRXs included for each individual are assigned with one of its random valid frequencies. Then (in the same method of initialization), an *Improvement Method* fixed to the FAP problem is applied to each individual to try to improve it. This improvement method will be applied again (line 10 in figure 2) to try to improve a frequency planning obtained as a result of the combination method. If this planning is better than the worst one contained in the *RefSet*, the later will be replaced, and the *RefSet* will be ordered again according to the fitness function used for the FAP problem [5]. Finally, when there are no subsets left to examine, an iteration of the algorithm has been completed. Then, the *b/2* best solutions are saved in the *RefSet* and a new population is generated to select the *b/2* most diverse solutions. With this new *RefSet* the algorithm is applied from the line 3 which appears in figure 2.

Algorithm 1

```
1: Start with the population P = ∅ and then Initialize(P) with PSize elements
2: Knowing x as a solution, create the reference set RefSet = {x^1,..., x^b} with the
   b/2 best solutions and the b/2 most diverse solutions of P
3: EvaluateSolutions (RefSet) and OrderSolutionsByFitness (RefSet)
4: NewSolution = TRUE
5: while (NewSolution) do
6:    NewSolution = FALSE
7:    Use a Subset Generation Method to create all possible subsets from the RefSet
8:    while (there are subsets no examined) do
9:       Select a subset and label it as examined
10:      Apply the Solution Combination Method to the solutions in the subset and
         use the Improvement Method for each solution obtained in the combination.
         Knowing x as the improved solution:
11:      if (f(x) < f(x^b) AND x ∉ RefSet) then
12:         Set x^b = x and OrderSolutionsByFitness (RefSet) again
13:         NewSolution = TRUE
14:      end if
15:   end while
16: end while
```

Fig. 2. Pseudocode for Scatter Search

3.2 Population-Based Incremental Learning

PBIL [10, 11] is a method that combines the genetic algorithms with the competitive learning (typical in artificial neural networks) for function optimization. The method is based on the evolution of a probability distribution, *P*, which is represented by a matrix with dimensions of *M* sectors by *N* frequencies. Solutions are represented as binary $M \times N$ matrices, where $S_{ij} = 1$ means that frequency *j* has been assigned to sector *i*. This representation avoids co-channel interferences within the same sector, that is, the highest-cost interferences, being not possible to assign the same frequency to two or more different TRXs in a sector. The algorithm appears in fig. 3.

Therefore, PBIL updates the probability distribution, *P*, from its initial uniform distribution (line 1 in figure 3) by using in each iteration the best individual obtained (lines 3-7) to solve the FAP problem. The update occurs in line 8. After that, in order to introduce some diversity a mutation operation is done over the probability distribution matrix *P* (lines 9-10). In conclusion, P_{ij} is the probability that the frequency *j* is assigned to the sector *i*.

```
Algorithm 2
1: P ← InitProbMatrix (each position = 0.5)
2: while (NumGenerations) do
3:    while (not enough samples generated) do
4:       sample_ij      ← GenerateSampleAccordingP()
5:       evaluation_ij  ← Evaluate(sample_ij)
6:    end while
7:    Find the best sample MAX ← FindSampleWithMaximumEvaluation()
8:    Update the probability matrix P, position by position, using the sample MAX
      and the learning rate LR: P_ij ← P_ij * (1.0 - LR) + MAX_ij * (LR)
9:    Mutate the probability matrix P, position by position, using the mutation
      probability MutP and the mutation amount MutA following the next statements:
      if (random (0,1] < MutP) then
10:      P_ij ← P_ij * (1.0 – MutA) + random (0.0 or 1.0) * (MutA)
11:   end if
12: end while
```

Fig. 3. Pseudocode for PBIL

4 Experimental Evaluation

As we said in the Introduction section, we have used a real-world instance to perform our experiments. We considered that this was much more representative than using theoretical instances, because our requirements have come directly imported from the industry (not like benchmarks generated in a computer by sampling random variables) and this makes more practical the experiments performed. In fact, due to our real approach (section 2) our solutions do not consider only the computation of high performance frequency plans, but also the prediction of QoS, which is very important for the industry. Indeed, both the definition of the interference matrix and the subsequent computations to obtain the cost values are motivated by real-world GSM networks since they are related to the computation of the BER (Bit Error Rate) performance of Gaussian Minimum Shift Keying (GMSK), the modulation scheme used for GSM [12]. In particular, we have used a real GSM network which is currently operating in a U.S. 400 km^2 city with more than 500,000 people (fig. 1.a shows the GSM network topology), so its solution is of great practical interest.

In order to fairly compare the results given by all the experiments and to make possible to do all the experiments possible (some test would take a lot of hours without time limitation), we have established a common limit of 30 minutes for all the executions. Also, 30 independent runs have been performed for statistical purposes.

Besides, according to the mathematical formulation of the problem (section 2.3), the results are given in function of the cost that a frequency plan is able to obtain. The smaller the value of the cost is, the better the frequency plan (and the result) will be. Table 1 summarizes the results which will be explained in the following subsections.

4.1 PBIL and PBIL-LS

The present work is the continuation of a previous one where we performed the adjustment of the different parameters of PBIL to solve the FAP problem [7]. In fact, we have performed a complete study through the different variations of the PBIL algorithm to check which version is better when solving the FAP problem. The variations we studied were the following:

- *PBIL-NegativeLR*: The probability matrix is moved towards the best individual (using the learning rate) and also away from the worst individual (using the negative learning rate) in each generation.
- *PBIL-M-Equitable*: The probability matrix is moved equally in the direction of each of the M selected individuals (M best samples) in each generation.
- *PBIL-M-Relative*: The probability matrix is moved according to the relative evaluations (fitness functions) of the M best individuals in each generation.
- *PBIL-M-Consensus*: The probability matrix is moved only in the positions in which there is a consensus in all of the M best individuals in each generation.
- *PBIL-Different*: In each generation, the probability matrix is only moved towards the bits in the best individual which are different than those in the worst individual.
- *PBIL-Complement*: The probability matrix is moved towards the complement of the lowest evaluation individual (the worst sample) in each generation.

As it can be consulted in table 1, the reference result obtained in our previous work (see ref. [7]) when the FAP problem is solved with the optimum configuration of the standard version of PBIL is 239459 (cost units) in average. This value is used as the reference for all the experiments which are summarized in figures 4 and 5.

The figure 4 shows the results obtained in the experiments made with the *PBIL-NegativeLR* version (*negative-learning rate* parameter is the one to adjust), whereas the figure 5 represents the results obtained in the most relevant experiments made with the versions *PBIL-M-Equitable*, *PBIL-M-Relative* and *PBIL-M-Consensus* (here the parameter to adjust is the *number of samples, M*).

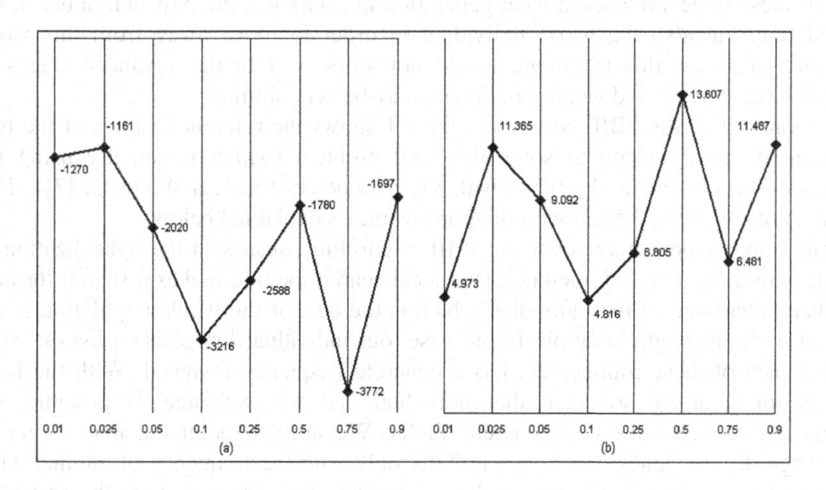

Fig. 4. (a) Means and (b) standard deviations for different negative-learning values

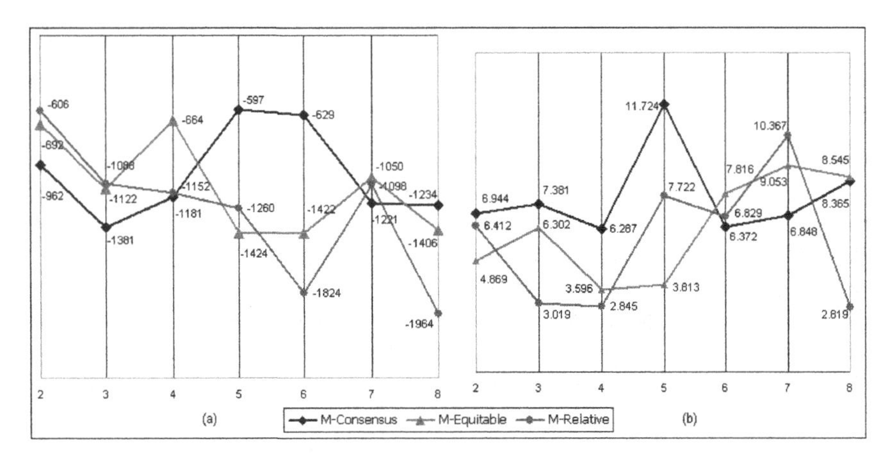

Fig. 5. (a) Means and (b) standard deviations for different M configurations

With regard to the two last versions of PBIL, *PBIL-Different* and *PBIL-Complement*, they do not require the adjustment of any parameter. These two versions are configured with the same fixed parameters used in standard PBIL. In this case, the results obtained were the following:

- For *PBIL-Different*: a mean of -2110 with 11.061 of standard deviation.
- For *PBIL-Complement*: a mean of -2161 with 4.141 of standard deviation.

Analyzing these results, we can conclude that none of the versions of PBIL studied improve the results obtained by the *standard PBIL* when they are applied to solve the FAP problem, although all of them are recommended in the literature [10, 11]. The worst versions of PBIL are *NegativeLR*, *Different* and *Complement*. We think this is due to these three versions in each generation not only use the best individual in the population, but also the worst individual in order to move away from this worst sample. However, this technique would not work well in the advanced stages of search since the best and worst individuals may be very similar.

In conclusion, the PBIL column on table 1 shows the reference values of the best version of the algorithm to solve the FAP problem (which is actually using the standard version of the algorithm with the parameters fixed as shown in [7]). This result is the one to be beaten with the improvements explained below.

The improvement made over the PBIL algorithm consists in the hybridization of PBIL with a *local search* method. This local search method makes a search through the neighbourhood of the individual which is the best of the whole population at the end of each algorithm iteration. In our case, our individual represents a list of TRXs where each of these transceivers has a concrete frequency assigned. With the local search improvement we scan the individual and we exchange, if possible, the frequencies which have two concrete TRXs. We have to point out that we try to exchange the frequency of a concrete TRX only with the frequency of another TRX which occupies a higher position in the list. So, when we are evaluating the first TRX (in the position 0), there are 2611 possibilities of exchange, when we evaluate the TRX in the position 1 there are 2610, and so on, until the last but one TRX, which can only exchange its frequency with the last one. Moreover, we have adapted the local search method to the FAP problem, so there are some constraints we have to fulfil before doing an exchange:

1. The frequencies of the TRXs to exchange have to be different.
2. The sectors of the TRXs to exchange have to be different.
3. The exchange is not valid if it causes co-channel interferences or adjacent channel interferences (see section 2.3) within any of the sectors involved in the exchange. These interferences have a very high cost (K, see equation (2)).

Only when all these three conditions are satisfied, the exchange is done, and then, it is evaluated if the exchange improves the original solution. If this is the case, the exchange is effective, but if not, the exchange is discarded and the local search continues with another TRX.

Results obtained are shown in the column PBIL-LS in table 1. As a conclusion, we can say that this hybridization (with a local search) is an important component for the success of PBIL. We think local search is key because it can correct the drastic changes in plans caused, sometimes, during the learning and mutation stages of PBIL.

4.2 SS and SS*

These experiments are divided in two parts. In the first one, we fixed the SS algorithm to solve the FAP problem (the results are presented in the column SS of table 1). The final parameters established for SS were: *PSize* = 20, *b* = 4, *Improvement Method Steps* = 4. In a second part, using these same configuration parameters, we developed a variation of the algorithm taking in consideration the good effects caused by the local search developed for PBIL (the results are presented in the column SS* of table 1). In particular, we have changed the *Improvement Method* taking into account the same restrictions of the section 4.1. As conclusion, the obtained results are quite better. In fact, with this change, we obtain the best results in the study (*113891 cost units of mean* versus the 120389 cost units obtained with the standard SS). Therefore, we can conclude that the improvement made in the *Improvement Method* of SS using a fixed *local search method adapted to the FAP problem* is very significant.

Table 1. Empirical results (in cost units) of PBIL, PBIL-LS, SS and SS* for 30 executions

	PBIL	PBIL-LS	SS	SS*
Best	239290	162810	118067	110858
Average (μ)	239459	163394	120389	113891
Std.(σ)	131.50	346.78	1053.2	1316.6

5 Conclusions and Future Work

In this paper we present and compare two different meta-heuristics to solve the FAP problem in a real-world GSM network composed of 2612 transceivers. We have improved the results obtained in a previous work which used standard PBIL to solve the problem [7], but moreover we have carefully tuned the SS algorithm and it has been compared as well with the fixed PBIL algorithm. The results obtained in the experiments show that the SS algorithm clearly outperforms the PBIL approach. But, moreover, it has been clearly proved that the usage of fixed local search methods for both algorithms is a very significant improvement (see table 1). Future work includes the evaluation of the algorithms using additional real-world instances and extending the mathematical model to deal with more advanced issues in GSM frequency planning (e.g., separation constraints, frequency hopping, etc.).

Acknowledgments. This work has been partially funded by the Spanish Ministry of Education and Science and FEDER under contract TIN2005-08818-C04-03 (the OPLINK project). José M. Chaves-González is supported by the research grant PRE06003 from Junta de Extremadura (Spain).

References

1. Wireless Intelligence (2006), http://www.wirelessintelligence.com
2. Aardal, K.I., van Hoesen, S.P.M., Koster, A.M.C.A., Mannino, C., Sassano, A.: Models and Solution Techniques for Frequency Assignment Problems. 4OR 1(4), 261–317 (2003)
3. Eisenblätter, A.: Frequency Assignment in GSM Networks: Models, Heuristics, and Lower Bounds. PhD thesis, Technische Universität Berlin (2001)
4. Mouly, M., Paulet, M.B.: The GSM System for Mobile Communications. Telecom Publishing (1992)
5. Luna, F., Blum, C., Alba, E., Nebro, A.J.: ACO vs EAs for Solving a Real-World Frequency Assignment Problem in GSM Networks. In: GECCO 2007, London, UK, pp. 94–101 (2007)
6. Walke, B.H.: Mobile Radio Networks: Networking, Protocols and Traffic Performance. Wiley, Chichester (2002)
7. Domínguez-González, D., Chaves-González, J.M., Vega-Rodríguez, M.A., Gómez-Pulido, J.A., Sánchez-Pérez, J.M.: Using PBIL for Solving a Real-World Frequency Assignment Problem in GSM Networks. In: Neves, J., Santos, M.F., Machado, J.M. (eds.) EPIA 2007. LNCS (LNAI), vol. 4874, pp. 207–218. Springer, Heidelberg (2007)
8. Martí, R., Laguna, M., Glover, F.: Principles of Scatter Search. European Journal of Operational Research 169, 359–372 (2006)
9. Laguna, M., Hossell, K.P., Martí, R.: Scatter Search: Methodology and Implementation in C. Kluwer Academic Publishers, Norwell (2002)
10. Baluja, S.: Population-based Incremental Learning: A Method for Integrating Genetic Search based Function Optimization and Competitive Learning. Technical Report CMU-CS-94-163, Carnegie Mellon University (1994)
11. Baluja, S., Caruana, R.: Removing the Genetics from the Standard Genetic Algorithm. In: Twelfth Int. Conference on Machine Learning, San Mateo, CA, USA, pp. 38–46 (1995)
12. Simon, M.K., Alouini, M.-S.: Digital Communication over Fading Channels: A Unified Approach to Performance Analysis. Wiley, Chichester (2005)

Reconstruction of Networks from Their Betweenness Centrality*

Francesc Comellas and Juan Paz-Sánchez

Departament de Matemàtica Aplicada IV, Universitat Politècnica de Catalunya
Avda. Canal Olímpic s/n, 08860 Castelldefels, Catalonia, Spain
{comellas,juan}@ma4.upc.edu

Abstract. In this paper we study the reconstruction of a network topology from the values of its betweenness centrality, a measure of the influence of each of its nodes in the dissemination of information over the network. We consider a simple metaheuristic, simulated annealing, as the combinatorial optimization method to generate the network from the values of the betweenness centrality. We compare the performance of this technique when reconstructing different categories of networks –random, regular, small-world, scale-free and clustered–. We show that the method allows an exact reconstruction of small networks and leads to good topological approximations in the case of networks with larger orders. The method can be used to generate a quasi-optimal topology for a communication network from a list with the values of the maximum allowable traffic for each node.

1 Introduction

In recent years there has been a growing interest in the study of complex networks, related to transportation and communication systems (WWW, Internet, power grid, etc.), see [1,2]. Many of these networks are large with a number of nodes very often in the thousands. To store the topological details of the network requires knowing the list of adjacencies and, although usually the networks are sparse, this means the use of a large amount of memory. In contrast, many invariants of the network (degree sequence, eccentricity, spectrum, betweenness, etc.) contain important information with significantly less memory use. Therefore it would be of interest to reconstruct, even partially, a network from one (or more) of these invariants. Another related problem is the construction of a new network from a list of desired values of some relevant parameter associated to its nodes. One useful case would be the generation a topology for a quasi-optimal communication network from the values of the maximum allowable traffic for each node. In [3], Ipsen and Mikhailov use simulated annealing with an elaborated cost function based on the spectral density to perform such

* Research supported by the Ministerio de Educación y Ciencia, Spain, and the European Regional Development Fund under project TEC2005-03575 and by the Catalan Research Council under project 2005SGR00256.

M. Giacobini et al. (Eds.): EvoWorkshops 2008, LNCS 4974, pp. 31–37, 2008.

a reconstruction from the values of the Laplacian spectrum. Here we propose a reconstruction of a network topology from the values of its (vertex) betweenness centrality, a measure of the influence of each of its nodes in the dissemination of information over the network. The use of a simple cost function, together with the information provided implicitly by the knowledge of the betweenness centrality, drives the simulated annealing optimization method towards a good network reconstruction. The method is probabilistic, i.e. it contain a random component, and as a consequence we can not guarantee that the algorithm will find an optimal reconstruction, but we show that the final networks match the originals in their main topological properties.

In the next section, we introduce the mathematical notation and concepts necessary for this study, including a short description of simulated annealing, the combinatorial optimization technique considered here. Our main results are presented in Section 3.

2 The Betweenness Centrality of a Network and Its Reconstruction

We model a network as a graph $G = G(V, E)$, with vertex set V (order $n = |V|$) and edge set E.

Vertex betweenness or betweenness centrality (BC) was first proposed by Freeman [4] in 1977 in the context of social networks and has been considered more recently as an important parameter in the study of networks associated to complex systems [5,2]. BC is usually defined as the fraction of shortest paths between all vertex pairs that go through a given vertex. To be more precise, if $\sigma_{uv}(w)$ denotes the number of shortest paths (geodetic paths) from vertex u to vertex v that go through w, and σ_{uv} is the total number of geodetic paths from u to v, then we define $b_w(u,v) = \sigma_{uv}(w)/\sigma_{uv}$ and the betweenness centrality of vertex w is $b_w = \sum_{u,v \neq w} b_w(u,v)$. The normalized betweenness centrality of vertex w is defined as $\beta_w = \frac{1}{(n-1)(n-2)} \sum_{u,v \neq w} b_w(u,v)$, see Fig. 1.

In this paper, when we refer to betweenness centrality or BC we mean the set of values $\{\beta_1, \beta_2, \ldots, \beta_n\}$. The average normalized betweenness of a graph of order n is $\overline{\beta} = (\sum_{u \in V} \beta_u)/n$ and it is related to its average distance $\overline{l}$ as $\overline{l} = (n-2)\overline{\beta} + 1$, see [6].

Here, we study the reconstruction of graphs from their BC. Note that the number of different graphs of a given order n is large even for relatively small

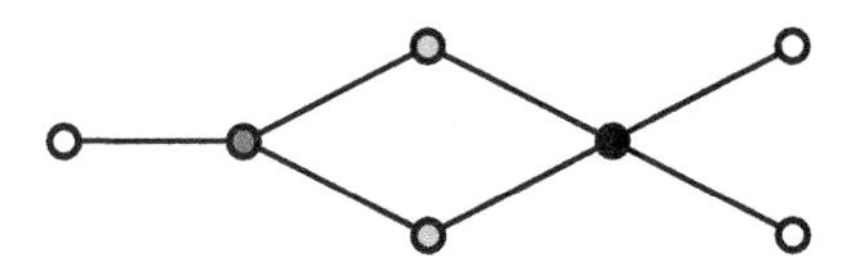

Fig. 1. The vertices of this graph have been colored according to their normalized betweenness centrality value: white 0, light grey 1/5, dark grey 11/30, black 19/30

orders. For example, for $n = 40$ there are approximately 10^{186} graphs. It makes no sense to check all of these graphs to find one with a matching BC, even in an approximate way. We are in the classical situation where combinatorial optimization algorithms (simulated annealing, genetic algorithms, tabu search, ant colony based systems, etc.) are useful, see [7].

For this initial study, we have considered as optimization method a standard version of simulated annealing (SA) [8]. As it is known, this method is inspired in the analogy made between the states of a physical system, e.g. a liquid, and the configurations of a system in a combinatorial optimization problem. A controlled heating/cooling process of the liquid (annealing) results in a true crystal (a minimum energy state) and avoids reaching a disordered glassy state. In the analogy, a change that decreases the cost of a function, ϵ, which measures the quality of a graph topology (see below), is always accepted, whereas if the cost increases, the change is accepted with a certain probability $e^{-\Delta\epsilon/T}$. (T is a control parameter known as temperature because of the analogy.) At a given temperature, a number of attempts N, large enough to obtain a good statistical set of trials, is performed and thereafter the temperature decreased. This process is repeated and the system is gradually cooled until it is stopped according to some criteria (time, number of changes accepted, etc.) In pseudo-code the SA algorithm can be written as follows:

1. Generate an initial random graph. Fix the initial value of T and T_{min}.
2. Repeat N times.
   (a) Modify the graph topology and find new cost.
   (b) If better, accept it as current solution.
   (c) If worse, accept only if $e^{-\Delta\epsilon/T} > rand()$
3. Lower T and repeat 2 until $T < T_{min}$ or other stop criterion.

In our case we will reconstruct a given reference graph G_0 from its BC, $\{\beta_1^0, \beta_2^0, \ldots, \beta_n^0\}$. To perform a reconstruction we generate an initial random connected graph with n vertices (each vertex $v \in V(G)$ has random degree δ_v, $1 \leq \delta_v \leq n-1$). During the simulated annealing process, a typical graph modification consists of reconnecting all the edges of one vertex chosen at random. This reconnection is performed by deleting all the edges of this vertex and introducing r, $1 \leq r \leq n-1$, new random edges avoiding duplicate connections and ensuring that the new modified graph is also connected. To decide if the changes should be accepted, we need a measure (cost function) of the "distance" of a given graph G_t with BC $\{\beta_1^t, \beta_2^t, \ldots, \beta_n^t\}$ to the reference graph G_0. We introduce a simple distance function based on the quadratic difference of the BC, $\epsilon = \sum_{i=1}^{n}(\beta_i^0 - \beta_i^t)^2$. This function, suggested by the least squares method, is a natural choice in some multivariable optimization problems. On the other hand, we have tested other related functions assigning weights to the BC elements, but they are more complex and their efficiency is similar.

The main problem with the reconstruction of a graph is to relate the final graph with the reference graph. The use of a vertex graph invariant in reconstructing a network might be hampered by the degeneracy of almost all known graph invariants, and in our case two or more topologically distinct vertices might have identical betweenness values. Moreover, the reconstructed graph can be isomorphic to the original graph but with permuted vertices or non-isomorphic with some topological similarity that might not be manifest. Although this is an important question when reconstructing a graph from its spectrum, in our case and as each value of the BC is directly associated to a vertex, the problem only appears when the BC contains several entries with the same value.

As in [3], we check graph similarity using the singular value decomposition of the adjacency matrices of the reference and final graphs . We recall that a matrix A can be decomposed into two matrices U and V and a diagonal singular value matrix Σ which satisfy $A = U\Sigma V^T$ and $\Sigma = U^T AV$. For any two graphs G_1 and G_2 with adjacency matrices A_1 and A_2, consider the function $F = F(A_1, A_2) = U_1 \Sigma_2 V_1^T = U_1 U_2^T A_2 V_2 V_1^T$ which is constructed from the singular vectors of G_1 and G_2. If the two graphs are isomorphic and their adjacency matrices only differ because of a different ordering of the vertices, it will happen that $A_1 = F(A_1, A_2)$. However, if the two graphs are not isomorphic, F will have real values not far from the values of A_1. Therefore, it is possible to define $\Delta = A_1 - F$ and use the norm $\delta = \sqrt{\sum_{i,j} \Delta_{ij}^2 / n}$ to measure similarity between the graphs.

We note that two isomorphic graphs have the same BC, which is independent of the labeling of the vertices, but there also exist non-isomorphic graphs (topologically different) with the same BC, which we call *isobet* graphs. For $n \leq 5$ there are no connected isobet graphs. For $n = 6$ there exist two pairs, there are 15 pairs for $n = 7$, etc. The number of isobet graphs increases rapidly with the order of the graph, but the fraction is very small. Hence, two graphs with the same BC would indeed be isomorphic with a high probability.

To know if two graphs are isomorphic is a difficult problem. It has been proved to belong to the class NP but it is thought not to be an NP-complete problem [9]. There is no known efficient (polynomial time) algorithm to solve this problem. Schmidt and Druffel [10] propose the method which we have implemented in our study. Their algorithm is not guaranteed to run in polynomial time, but has been shown to perform efficiently for a large class of graphs. Two isomorphic graphs should have the same exact degree distribution. After checking this property, the Schmidt and Druffel algorithm uses information from the distance matrices of the graphs to establish an initial vertex partition. Then, the distance matrix information is applied in a backtracking procedure to reduce the search for possible mappings between the vertices of the two graphs. The algorithm returns this mapping if the original and reconstructed graphs are indeed isomorphic.

3 Results and Conclusion

The SA algorithm was implemented in C++ (Xcode) and executed on an Apple Xserver G5 with dual PowerPC processors at 2.3 GHz. The parameters considered for the SA are $T_0 = 1.0$, $N = 2000$, $T_{min} = 0.000001$ and a geometric cooling rate $T_{k+1} = 0.9T_k$.

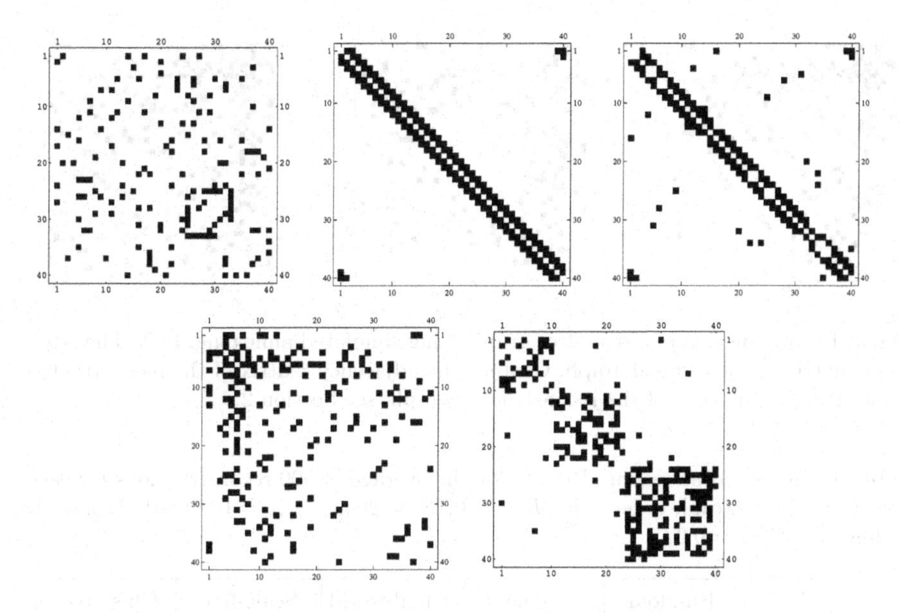

Fig. 2. Adjacency matrices of the reference graphs: random, regular (circulant), Watts-Strogatz small-world, scale-free and clustered. All graphs haver order 40.

The main study was performed as follows:

We generate one sample graph of order 40 for each of the categories considered: random, regular (circulant), Watts-Strogatz small-world [11], scale-free [1], and clustered. Fig. 2 shows a graphic representation of the adjacency matrices of these reference graphs. For each reference graph, we compute the betweenness centrality and use it to reconstruct the graph with the simulated annealing method. To be fair in the comparisons, we fix the reconstruction time for each graph to be 900 seconds. After this time, we compute the main topological parameters (diameter, average distance, degree distribution, clustering) for the best graph obtained and we check the similarity of its adjacency matrix with the original graph. Each test is repeated 500 times and the results are averaged. Fig. 3 shows a typical reconstruction.

In Table 1, we present a set of results for this method. We can see that the reconstruction gives acceptable results in all cases, but provides better approximations for graphs with some randomness in their structure, and such that their vertices have different betweenness centrality values. This is the case, obviously, of random graphs and also scale-free graphs.

We also tested the algorithm using graphs with small orders (up to 12 vertices) and in all cases we were able to obtain an exact reconstruction of the graph.

The results show that a simple metaheuristic method, simulated annealing, with an also simple cost function, reconstructs small graphs exactly from their betweenness centrality and obtains a topologically good approximation for larger

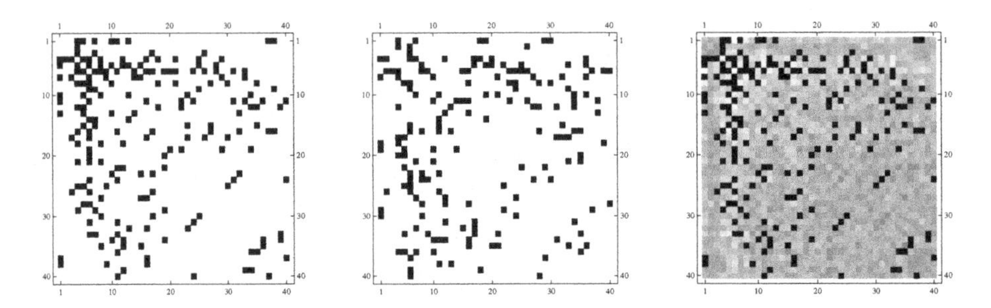

Fig. 3. Reconstruction of a scale-free graph using simulated annealing. Left: The adjacency matrix of the original graph. Center: The adjacency matrix of the reconstructed graph. Right: Matrix F of the reconstructed graph, see Section 2.

Table 1. Simulated annealing. Results for the average of 500 reconstructions for each reference graph. Graph order= 40, $T_0 = 1.0$, $N = 2000$, $T_{min} = 0.000001$, geometric cooling rate $T_{k+1} = 0.9T_k$.

		Random		Circulant		Small-world		Scale-free		Clustered	
		Ref.	*Recns.*	Ref.	*Recns.*	Ref.	*Recns.*	Ref.	*Recns.*	Ref.	*Recns.*
Diameter	avg.	6	5.870	10	8.770	6	6.687	4	4.475	5	5.271
Avg. Dist.	avg.	2.89	2.780	5.38	4.670	3.31	3.028	2.32	2.340	2.65	2.578
Degrees	min.	1	1.000	4	1.999	3	1.000	1	1.11	2	3.75
	avg.	3.8	3.950	4	2.570	4	3.690	4.95	5.096	6.3	4.464
	max.	8	10.000	4	3.870	5	6.190	17	15.760	13	14.650
Clustering	avg.	0.2	0.195	0.5	0.030	0.32	0.122	0.26	0.262	0.37	0.200
Norm. BC	avg.	0.050	0.047	0.115	0.097	0.061	0.053	0.035	0.035	0.043	0.042
δ	avg.		0.026		0.074		0.029		0.024		0.082

graphs. (We have tested graphs with up to 2000 nodes and 20000 [12].) The method works without modification in the related problem of the construction of a new network from a a list of desired values for the maximum allowable traffic for each node.

Our extensive tests show that the cost function considered, the quadratic difference of the BC, is a good choice for SA, and the method is a nice alternative to the reconstruction from the Laplacian spectrum as it is easier to implement and results in a faster algorithm, allowing reconstructions of similar quality.

Further work is planned to evaluate and compare the performance of other combinatorial optimization methods, like ant colony optimization [13], multi-agent systems [14], tabu search [7,15] and genetic algorithms.

References

1. Barabasi, A.-L., Bonabeau, E.: Scale-free networks. Scientific American 288(5), 50–59 (2003)
2. Newman, M.E.J.: The structure and function of complex networks. SIAM Review 45, 167–256 (2003)
3. Ipsen, M., Mikhailov, A.S.: Evolutionary reconstruction of networks. Phys. Rev. E. 66, 46109 (2002)
4. Freeman, L.C.: A set of measures of centrality based upon betweenness. Sociometry 40, 35–41 (1977)
5. Goh, K.-I., Oh, E., Jeong, H., Kahng, B., Kim, D.: Classification of scale-free networks. Proc. Natl. Acad. Sci. USA 99, 12583–12588 (2002)
6. Comellas, F., Gago, S.: Synchronizability of complex networks. J. Phys. A: Math. Theor. 40, 4483–4492 (2007)
7. Aarts, E., Lenstra, J.K. (eds.): Local Search in Combinatorial Optimization. John Wiley & Sons Ltd, New York (1997)
8. Kirkpatrick, S., Gelatt, C.D., Vecchi, M.P.: Optimization by simulated annealing. Science 220, 671–680 (1983)
9. Garey, M.R., Johnson, D.S.: Computers and Intractability: A Guide to the Theory of NP-Completeness. W.H. Freeman, New York (1979)
10. Schmidt, D.C., Druffel, L.E.: A fast backtracking algorithm to test directed graphs for isomorphism using distance matrices. Journal of the ACM 23, 433–445 (1976)
11. Watts, D.J., Strogatz, S.H.: Collective dynamics of 'small-world' networks. Nature 393, 440–442 (1998)
12. Paz-Sanchez, J.: Reconstrucció de grafs a partir del grau d'intermediació (betweenness) dels seus vèrtexs. PFC (Master Thesis) (in Catalan) (July 2007)
13. Dorigo, M., Stützle, T.: Ant colony optimization. MIT Press, Cambridge (2004)
14. Comellas, F., Sapena, E.: A multiagent algorithm for graph partitioning. In: Rothlauf, F., Branke, J., Cagnoni, S., Costa, E., Cotta, C., Drechsler, R., Lutton, E., Machado, P., Moore, J.H., Romero, J., Smith, G.D., Squillero, G., Takagi, H. (eds.) EvoWorkshops 2006. LNCS, vol. 3907, pp. 279–285. Springer, Heidelberg (2006)
15. Glover, F.: Future paths for integer programming and links to artificial intelligence. Comput. & Ops. Res. 13, 533–549 (1986)

A Self-learning Optimization Technique for Topology Design of Computer Networks

Angan Das and Ranga Vemuri

Department of Electrical and Computer Engineering
University of Cincinnati
Cincinnati, Ohio 45221-0030, USA
{dasan,ranga}@ececs.uc.edu

Abstract. Topology design of computer networks is a constrained optimization problem for which exact solution approaches do not scale well. This paper introduces a self-learning, non-greedy optimization technique for network topology design. It generates new solutions based on the merit of the preceding ones. This is achieved by maintaining a solution library for all the variables. Based on certain heuristics, the library is updated after each set of generated solutions. The algorithm has been applied to a MPLS-based IP network design problem. The network consists of a set of Label Edge Routers (LERs) routing the total traffic through a set of Label Switching Routers (LSRs) and interconnecting links. The design task consists of — 1) assignment of user terminals to LERs; 2) placement of LERs; and 3) selection of the actually installed LSRs and their links, while distributing the traffic over the network. Results show that our techniques attain the optimal solution, as given by GNU solver - *lp_solve*, effectively with minimum computational burden.

Keywords: Topology design, MPLS networks, Self-learning search, Constrained optimization.

1 Introduction

A computer network, in simple terms, is composed of multiple communicating entities interconnected over a wired or wireless medium to share data and other resources. Topology design in the networking domain refers to the layout, either virtual or actual, of connected devices in a network. Design of economic and realistic topologies along with network planning has always been an evergreen research topic [1]. In fact, the problem of optimal design of a network meeting all constraints and optimizing certain objectives has been an important area not only in computer networks, but several other real world applications such as telecommunications, sewage systems, oil and gas lines, road traffic and so on.

In this regard, the complexity, cost, and short turn around time of any network system design necessitates the development of automated design tools. All such tools are basically built on top of some optimization technique. Genetic algorithms (GAs) have shown some promise in this context. [2], [3] and [4] used GAs for the topology design of MPLS networks, mesh-based radio access networks, and local area networks

M. Giacobini et al. (Eds.): EvoWorkshops 2008, LNCS 4974, pp. 38–51, 2008.

respectively. Unfortunately, all of them suffer from appreciable computational burden. Other works include the topology generator IGen [5], based of some well-known design heuristics. [6] proposes a simulated evolution algorithm where the overall cost function is based on fuzzy logic. In [7], a memetic algorithm is applied to packet switched networks. The authors in [8] used a branch-and-bound based technique for MPLS networks. The method, though exact, is too time consuming. [9] demonstrated the use of Integer Linear programming for telecommunication networks. Here, the number of variables grows exponentially with the dimension of the design network, rendering the method unscalable.

In an attempt to alleviate all of the above drawbacks, we propose a self-learning or adaptive algorithm for the design of computer networks. It draws inspiration partly from the concepts underlined in compact GAs [10] and the Univariate Marginal Distribution Algorithm (UMDA) [11]. The main heuristic behind our self-learning approach is that the future set of solutions is dependent on the merit of the present solution. The method is heuristic-based, non-greedy, and scalable. Moreover, the technique adopted is generic, and hence may be applied to any constrained optimization problem in general.

The problem of designing the backbone topology for a computer network is network-specific to some extent. Among several options, we have chosen the Multi-Protocol-Label-Switching (MPLS) based network as our design problem. This owes to the following reasons. Firstly, MPLS is considered to be one of the most predominant networking technologies as a IP protocol suite, demanding substantial attention [1]. Secondly, MPLS network design is a general problem whereby the technique can be minimally tweaked and applied to other networks as well. Thirdly, MPLS requires the design approach to be scalable, whereby it should solve for a growing traffic volume and increasing number of users. Finally, MPLS does not use any specific technology process and does not affect the other layers.

The rest of the paper is organized as follows. Section 2 describes the typical backbone topology and functionality of a MPLS-based network. Section 3 formulates the design problem. Section 4 introduces the self-learning optimization technique and section 5 outlines the application of this technique to the MPLS network design problem. Section 6 reports the results. Section 7 finally concludes the work.

2 MPLS Topology and Functionality

The schematic of a typical MPLS-based network is shown in Fig. 1. It consists of a MPLS core and a MPLS edge. The core contains Label Switching Routers (LSRs) or transit nodes. The edge houses Label Edge Routers (LERs) or access nodes. A LER is basically an interface between user terminals and LSRs. It classifies incoming packets into Forwarding Equivalence Classes based on network layer or other control information. A packet is transmitted from a user terminal to a source LER, after which it is received by an ingress LSR. This LSR affixes a label to the packet based on its destination, VPN membership or other criteria. Tables are built associating the labels with different routes through the LSRs. The idea is to steer IP traffic onto a variety of routes instead of a single one discovered by an interior gateway protocol. This helps to avoid congestion or failures, or to enable guaranteed service level. Therefore, all incoming packets from different sources, but with the same label use the same path - termed as

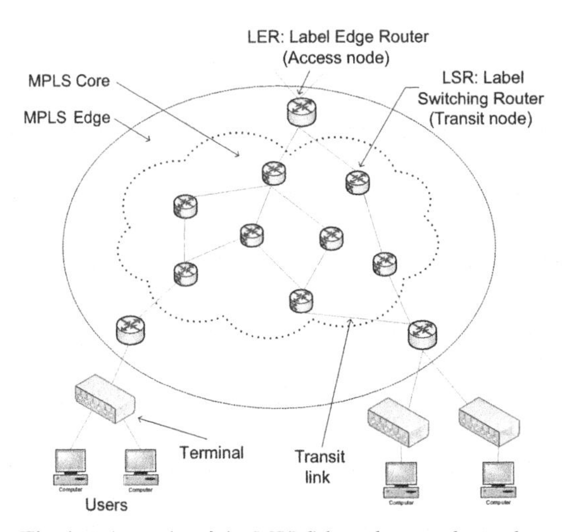

Fig. 1. Schematic of the MPLS-based network topology

Label Switched Path (LSP) or tunnel. Hence, packets destined for the same endpoint can use various LSPs to get there. Finally, an egress LSR receives the packet, removes the label and delivers the packet to the destination LER.

3 Problem Formulation

MPLS network topology design can be formulated as a constrained combinatorial optimization problem. Given a set of bandwidth (BW) and capacity constraints, a set of nodes and their interconnecting links need to placed such that the installation cost is minimized, while traffic demands are satisfied. We decompose the problem into three sub-problems such that the individual solutions combine to give the overall design.

3.1 Assignment of Terminals to Access Nodes (ATAN)

This is the first part of the design problem. Here, we assume that the location of terminals and access nodes are fixed and given. The objective is to assign or map the terminals to the access nodes in a way that the connection cost is minimized. The constraints involve a many to one correspondence between the set of terminals and access nodes. Further, the total capacity of the connected terminals is to be accommodated by the corresponding access node BW. Hence, all access nodes are needed. Let the set of terminals and set of access nodes be denoted by $\Gamma = \{1, 2, ..., M\}$ and $\Lambda = \{1, 2, ..., N\}$ respectively. The variable $x_{ij} = 1$ denotes that terminal i $(i \in \Gamma)$ is connected to access node j $(j \in \Lambda)$; $x_{ij} = 0$ otherwise. The corresponding connection cost is given by α_{ij}. $\beta = [\beta_i : i \in \Gamma]$ gives the capacity of each terminal, while $\gamma = [\gamma_j : j \in \Lambda]$ denotes the BW of an access node. Table 1 gives the optimization problem formulated.

Table 1. Constrained Optimization Problem Formulation

TYPE	VARIABLES	OBJECTIVE	CONSTRAINTS
ATAN	$x_{ij} : x_{ij} \in \{0,1\} \quad \forall i \in \Gamma, \forall j \in \Lambda$	Minimize $C = \sum_{j=1}^{N}\sum_{i=1}^{M} \alpha_{ij} x_{ij}$	(1) $\sum_{j=1}^{N} x_{ij} = 1, \ \forall i \in \Gamma$ (2) $\sum_{i=1}^{M} \beta_i x_{ij} \leq \gamma_j, \ \forall j \in \Lambda$
LAN	$x_{ij'} : x_{ij'} \in \{0,1\}$ $y_{j'} : y_{j'} \in \{0,1\}$ $\forall i \in \Gamma, \quad \forall j' \in \Lambda'$	Minimize $C = \sum_{j'=1}^{N'}\sum_{i=1}^{M} c_{ij'} x_{ij'} +$ $\sum_{j'=1}^{N'} \lambda_{j'} y_{j'}$	(1) $\sum_{j'=1}^{N'} x_{ij'} = 1, \ \forall i \in \Gamma$ (2) $\sum_{i=1}^{M} \beta_i x_{ij'} \leq \gamma_{j'} y_{j'}, \ \forall j' \in \Lambda'$
TNPC	$v_i : v_i \in \{0,1\}$ $y_j : y_j \in \{0,1\}$ $u_j : u_j \in \{0,1\}$ $z_{ji} : z_{ji} \in \{0,1\}$ x_{dp} $\forall i \in \Omega, \forall j \in E, \forall d \in \Delta, \forall p \in P_d$	Minimize $C = \sum_{i=1}^{Q} \omega_i v_i$ $+ \sum_{j \in E} \eta_j y_j + \sum_{j \in E} \zeta_j u_j$	(1) $\sum_{p \in P_d} x_{dp} = h_d, \ \forall d \in \Delta$ (2) $\sum_{j \in E} y_j z_{ji} \leq v_i V_i, \ \forall i \in \Omega$ (3) $u_j \leq y_j C_j$, where $u_j = \sum_{d \in \Delta}\sum_{p \in P_d} \omega_{jdp} x_{dp}, \ \forall j \in E$

3.2 Location of Access Nodes (LAN)

This is basically a superset of the ATAN problem. Here, a set of possible access node locations, instead of the exact locations, is provided. The total cost function now includes both the link cost and the node installation cost. The objective is to select a particular subset of the possible locations and the pertinent mapping such that the cost is minimized. Hence, all access nodes are not needed. The constraints are similar to that of ATAN. The parameters, apart from ATAN, include a set of possible access nodes locations: $\Lambda' = \{1, 2, ..., N'\}$, and their associated fixed installation cost given by $\lambda = [\lambda_{j'} : j' \in \Lambda']$. The variable $y_{j'} = 1$ if we decide to install the node j'; $y_{j'} = 0$ otherwise. Table 1 gives the optimization problem.

3.3 Transit Node Placement and Connectivity (TNPC)

TNPC problem deals with routing of the total load/demand over the network. Each demand, from a source-destination access node, needs to be distributed among a set of paths through the transit nodes and interconnecting transit links. The cost function now comprises of the transit node installation cost, link connection cost and link capacity cost. Therefore, keeping the cost to a minimum, the objectives are to optimally obtain the: 1) Location of transit nodes; 2) Selection of transit links; 3) Distribution of each demand among the admissible paths. Links are possible between any pair of a given set of transit node positions. Regarding constraints, they ensure that — 1) The total demand is catered, (2) Nodes do not handle more links than they can (3) Links do not allow demand flow through them beyond their capacities.

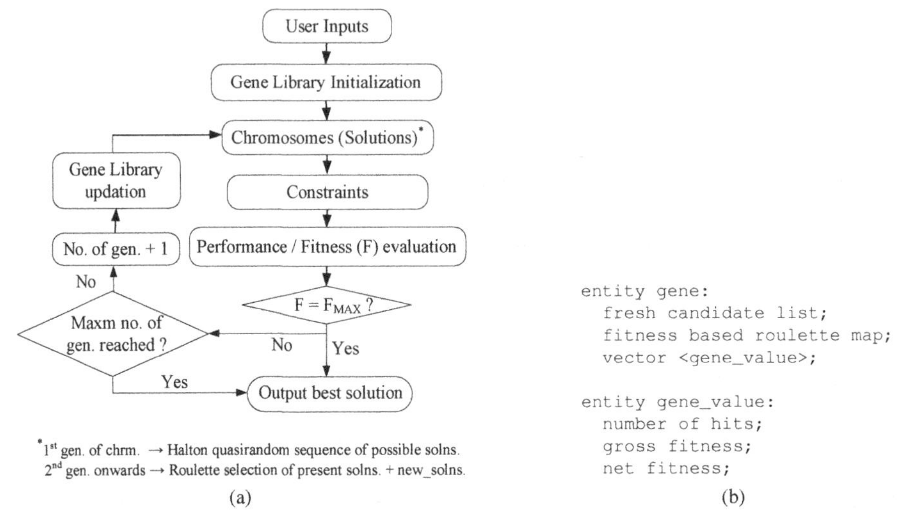

Fig. 2. (a) Overall flow: Self-learning optimization technique; (b) Components of the gene library

Let $\Lambda = \{1, 2, ..., N\}$ denote the set of access nodes and $\Omega = \{1, 2, ..., Q\}$ the set of possible transit node locations. For each such location $i (i \in \Omega)$, let ω_i be the fixed installation cost and V_i the maximum node degree. Also, variable $v_i = 1$ if the node is actually placed; $v_i = 0$ otherwise. Let $E = \{e_{kk} : k \in \Omega\}$ give the set of all transit links. For transit link $j (j \in E)$, let η_j be the fixed installation cost, ζ_j the 'per unit capacity' cost, variable u_j the capacity, and C_j the upper bound for the capacity. Variable $y_j = 1$ if link is actually provided; $y_j = 0$ otherwise. The variable $z_{ji} = 1$ if link j is connected to transit node i; $z_{ji} = 0$ otherwise. Also, a set of demands $\Delta = \{1, 2, ..., D\}$ is given. For each demand d ($d \in \Delta$), the volume (h) and the admissible paths (P_d) are given. Variable x_{dp} denotes the fractional demand in path $p (p \in P_d)$ for demand d. Finally, variable $w_{jdp} = 1$ if path p for demand d passes through link j; $w_{jdp} = 0$ otherwise. The required optimization formulation is given in Table 1.

4 Self-learning Optimization Technique

4.1 Overview

We propose a self-learning search technique to find the global optimum of a constrained combinatorial optimization problem. The self-learning nature implies that the the appropriateness of the present solution decides on how well its individual variable values qualify to be prospective candidates for future set of solutions. Though not a GA in the true sense, the method does use some concepts and terms of a GA [12].

The general flow of the algorithm is shown in Fig. 2(a). The user provides all the various run parameters and constants. Like a GA, we maintain a generation of chromosomes, i.e. a set of solutions, throughout the run. A library is formed that contains the solution set for the individual genes, along with a fitness metric denoting the closeness of the solutions to the global optimum. The library is updated after each set of solutions generated as per the algorithm given in Alg. 1. After the first generation, successive generations are produced by populating the genes with appropriate values either from the roulette map of the gene in its library (described in Section 4.5) or from new solution points, if any. The handling of constraints and fitness evaluation of each chromosome is specific to the problem at hand. The run continues either till the maximum fitness is reached, or for the allowable number of generations. The best solution is produced as the output of optimization. The individual aspects are outlined below.

4.2 Halton-Sequence Based First Generation

Any combinatorial optimization approach requires selection of permissible values for each of the variables undergoing optimization. In the first generation, we assign Halton-sequence based random values to all the variables. A Halton multidimensional quasi-random sequence generates uniformly distributed points in the permissible search space that appear to be random, but are actually constructed deterministically [13]. Compared to a general pseudorandom sequence, the Halton random sequence does a better job in filling up the whole search space evenly, avoiding otherwise large gaps. This leads to evenly spread initial solutions.

4.3 Unity Generation Gap

In our approach, barring the best solution, a new generation replaces its previous generation completely. This condition, like generational GAs, is referred to as a generation gap ($ggap : ggap \in [0, 1]$) of unity. $ggap$ is chosen unity because over here, we do not carry over traits through mechanisms like crossover and mutation. We make use of the credibility of each gene, rather than that of any substring pattern.

4.4 Individual Gene Values Responsible for Chromosome Fitness

Fitness is a metric for performance that symbolizes the quality of a solution point [12]. Each gene in a solution set contributes to the overall fitness of the solution. Now, in a CAD optimization framework, we are unaware of their exact individual contributions. Hence we divide the total chromosome fitness (F_c) equally among the total number of genes (G). Again, the fitness contribution of each of the genes (say g-th gene) F_g in turn owes to the value actually used for gene (say its i-th value). It is denoted as $(F_g)_i$. This $(F_g)_i$ adds towards the $Gross\_F$ calculation for that particular value (described in Section 4.5).

4.5 Gene Library: Content and Characteristics

The gene library stores certain aspects of each individual gene, that gets updated after each generation. These aspects are shown in Fig. 2(b) and are explained as follows:

Fresh candidate list: Throughout the run, we keep track of the unexplored solutions for each gene. These are stored in the gene's fresh candidate list. The list gradually decreases, as we explore more and more of the unvisited search space for the gene.

Algorithm 1. Gene library update after each generation

Input: Gene library in gen-(n), Chromosomes of gen-(n)
Output: Updated gene library — Required for formation of gen-(n+1)
procedure Gene library update
forall *chromosomes* $\in$ *gen-n* **do**
  F_c = `fitness` (chromosome);
  forall *genes (say g-th gene)* $\in$ *chromosome* **do**
    $F_g = F_c/G$ // $G \rightarrow$ No. of genes in chromosome
    For the particular value of g-th gene used (say i-th value)
    $(F_g)_i = F_g$
    if *i-th value is used for the first time for the gene* **then**
      Remove *i-th* value from fresh candidate list of the gene
      $(F_g)_i = (F_g)_i * \lambda * gen\_no$ // $\lambda \rightarrow constant(< 1)$
    Gross fitness and count
    $(Gross\_F_g)_i = \frac{(Gross\_F_g)_i * (N_g)_i + (F_g)_i}{(N_g)_i + 1}$
    $(N_g)_i = (N_g)_i + 1$
Net fitness
forall *genes (i.e. each of the g-th genes)* **do**
  forall *values of the gene (i.e. each of the i-th values used)* **do**
    $(Net\_F_g)_i = \alpha * (Gross\_F_g)_i + \beta * (N_g)_i$ // where $\alpha, \beta \rightarrow constants(< 1);\ \alpha + \beta = 1$

Fitness based roulette map: The already visited solutions for each gene are ranked based on their net fitness and stored in the form of a roulette map. Roulette map and its associated selection procedure is a strategy whereby better candidates have higher chances of getting selected [12].

Characteristics of a gene's value: A gene or variable may possess any permissible value. A sub-library for the values is formed and maintained that contains the following information:

- *Number of hits* (N_g)*:* This quantity signifies the total number of times that the value has been visited and thereby used for the gene.
- *Gross fitness* $(Gross\_F_g)$*:* Each time a value is used for a gene, it is necessary to award it appropriately. The average contribution of a gene's value towards the overall solution (whenever it is used) contributes towards the gross fitness for that value.
- *Net fitness* (Net_F_g)*:* The quality of a value is judged not only by its gross fitness, but also by the number of hits. For example, if a solution visited thrice has comparable gross fitness to that visited once, the former gets a much higher priority. Therefore, we assign some importance or weightage to both the gross fitness and number of hits. Denoting them by α and β respectively, the net fitness is the weighted sum of the individual parameters.

4.6 Gene Library Update

After each generation, for each of the genes, the library entries for all the values used in chromosome formation are updated with necessary information acquired from the present solutions. The updates are:

- *Fresh candidate list:* If a value is chosen, then it no longer remains fresh. Hence, if it is present in the fresh candidate list for the gene at the onset of a generation, it is subsequently erased from the list at the end of the generation run.
- *Number of hits:* The value's occurrence count increments by unity each time the value is chosen.
- *Fitness:* The new average gross fitness comes from its old gross fitness and its share from the total solution fitness, for all solutions where it is used. The net fitness is then obtained through the weighted sum approach. The roulette map for each gene is subsequently formed with these net fitness values.

4.7 New Solution Exploration Rate ($nser$)

During formation of chromosomes, apart from selecting the already explored solutions through the roulette map strategy, the algorithm also needs to visit new solution points. Hence, we introduce a fixed constant called 'new solution exploration rate' ($nser, nser < 1$). For each succeeding generation (after the 1st gen.), if the fresh candidate list of a gene is non-empty, that gene is populated with the new entries, for a randomly chosen $nser\%$ of the chromosomes. It is to be noted that the new entries are also chosen in a Halton way. Thereby, not all new entries are guaranteed to be chosen in any particular generation. The constant is somewhat similar to the harmonic memory considering rate in harmonic search technique [14]. Also, the value of $nser$ is proportional to the extent of the search space i.e. range of permissible values.

4.8 Generation-dependent Fitness of a New Solution for a Gene

In case of a newly chosen solution for a gene i.e. if the value previously belonged to the gene's fresh candidate list, it is included into the library with a fitness proportional to the generation number (λ constant). This ensures that as the generations progress, new values for genes get included with higher fitness. This gives them some chance to get selected amidst some very fit solutions already placed in the roulette maps.

4.9 Constraints and Penalty Function

The constrained optimization problem is converted to an unconstrained one by including the constraints inside the cost function. This necessitates a penalty function $P(X)$ for solution X. Here, since we are unaware of the feasible proportion of the search space, we use the widely known exterior additive penalty method [15]. If $\text{cost}_{\text{orig}}$ and cost_{mod} are the original and modified (owing to penalty) costs respectively,

$$\text{Constraint:} \quad \phi(X) \leq 0 \tag{1}$$

$$\text{Penalty:} \quad P(X) = \delta\{max[0,\ \phi(X)]\}^{\gamma} \tag{2}$$

$$\text{Cost:} \quad \text{cost}_{\text{mod}} = \text{cost}_{\text{orig}} + P(X) \qquad \text{where}\delta, \gamma \rightarrow \text{constants} \tag{3}$$

5 Application of Optimization Technique to Network Design Problem

The optimization technique developed above has been applied to the MPLS-based network topology design problem. The specifics pertaining to each sub-problem are explained in the following sections.

5.1 ATAN Problem

Chromosome Encoding and Gene Library: With M, N, Γ and Λ carrying the same meanings as in Section 3.1, the chromosome is an M-bit vector of integers, $X = [x_1, x_2, ..., x_M]$. For the $i-$th bit, if $X[i] = j$, the index i indicates the terminal number, and j denotes the access node number ($j = 1 \rightarrow N$). For each gene $i : i \in \Gamma$, the permissible solution space comprises of all the access nodes (Λ). This is also the initial fresh candidate list for each gene and the maximum size of its roulette map list.
Fitness Evaluation: The obtained cost (C_{obtained}) for this problem is assumed to be the Euclidian distance between the terminals and their corresponding access nodes. Since the exact solution to this problem can be known, hence the cost function (C) is the difference between the optimal cost (C_{optimal}) and C_{obtained}. The normalized cost gives the fitness (F). The higher the fitness, the better the solution, with unity fitness representing the optimal solution. For problems where the optimal solution is not known beforehand, this definition can be easily modified to that of a conventional GA where the saturation of cost over many generations denotes optimality. Anyways, quantitatively, if L signifies the location of a terminal, we have:

$$C_{\text{obtained}} = \sum_{i=1}^{M} |L\{X(i)\} - L\{i\}|, \quad C = |C_{\text{obtained}} - C_{\text{optimal}}|, \quad F = \frac{1}{1+C}$$

Constraints: The only constraint here is the BW constraint. Denoting the maximum BW of an access node as b and the terminal load on it as $\psi(X)$ for solution X, the constraint $\phi(X)$, as defined in Eqn.(1), for the access node j is given by $\phi(X)_j = \psi(X)_j - b_j$. The penalties owing to BW violations for all the access nodes are added up and the modified cost is calculated as per Eqn.(3).

5.2 LAN Problem

The chromosome encoding scheme is the same as that of the ATAN problem, i.e. an M-bit vector of integers, $X = [x_1, x_2, ..., x_M]$. The variable denoting whether the j-th access node is installed or not, is implicitly encoded in X. Hence if the terminal load on j, $\psi(X)_j$, is zero, it implies that the node is not installed. Therefore, the cost includes the fixed installation cost of only those access nodes for which $\psi(X)_j > 0$. The connection cost and BW penalty functions are similar to that of ATAN problem. Also, here too, the cost function and fitness denotes its closeness to the known optimal solution.

5.3 TNPC Problem

The sequential steps needed to solve the TNPC problem, along with their underlying assumptions, are:

- *Selection of source $\rightarrow$ destination path:* The path from a source to destination access node involves traveling through a certain set of transit nodes and links. Here, the paths are determined by successively selecting node points where one node lies within a certain nearness of its predecessor node.

- *Distribution of load into various paths:* The load distribution among the participating paths are assumed to be integral multiples of the total load divided by the total paths for the load. For example, for 4 possible paths, each path is assumed to carry $0, 25, 50, 75$ or 100 % of the load. All such combinations are formed and only feasible ones among them are considered.

Chromosome Encoding and Gene Library: Once the paths are identified, a chromosome represents the load distribution among the different paths. The number of genes is equal to the number of given source-destination traffic loads. The permissible value set for each gene comprises of a feasible path distribution combination for that particular load. The other variables are implicitly encoded as explained:

- *Link presence/absence:* If a link falls within any of the paths having load share $\neq 0$, it is present.
- *Transit node presence/absence:* If a link is present, its end nodes are present due to obvious reasons.
- *Link capacity:* The summation of all the net loads passing through a link gives the link's capacity.

Fitness Evaluation: The transit nodes and links that are actually installed contribute to the installation cost and the related connection cost. Likewise, the link capacity cost is calculated from the link load. The cost and fitness calculations are similar to that of ATAN or LAN.

Constraints: The demand flow constraint is automatically taken care of by considering only the feasible demand distributions. With the symbols carrying meanings as described in Section 3.3, the constraint $\phi(X)$ for solution set X as defined in Eqn.(1) is given as:

- *Degree of transit node:* For installed transit node i, with n links connected to it, $\phi(X)_i = n(X)_i - V_i$.
- *Link capacity:* For installed link j having capacity u, $\phi(X)_j = u(X)_j - C_j$.

The cost function is accordingly modified, following Eqn.(3).

6 Experiments and Results

The algorithm has been coded in C++. The program was run on a Sun Workstation having 750 MHz UltraSPARC-3 processor and 2GB RAM, with Solaris 10 OS. The multidimensional Halton sequences have been generated using the sequence generator in [13]. All the problems required tuning of the algorithm constants over several runs. α and β are set to 0.95 and 0.05 respectively. The new solution scaling factor (λ) is 0.01. $nser$ is set to 20%. δ and γ are varied between 1 and 2. The results are reported for the best case out of 20 runs. We use a modest size of 100 chromosomes for all the generations.

Three test designs with increasing difficulty are considered for each problem. But owing to space constraints, we show the topology and fitness curves for only one design in each problem. As shown in Table 2, the network $T(X)A(Y)$ for ATAN signifies X terminals (T) and Y access nodes (A); $T(X)P(Y)$ for LAN stands for X terminals and Y possible access nodes (P); and $A(X)Tr(Y)$ for TNPC symbolizes X access nodes and Y transit nodes (Tr). The position of the terminals and nodes are randomly generated on a $100X100$ grid as shown in Figs. 3-5(a). Now, a prior-known optimal solution always ensures proof of concept. Here, though the exact optimal solutions to the problems could be computed theoretically, they are practically impossible even for small-sized problems. So we use the GNU MILP solver *lp_solve* [16] to solve the ILP-formulated design problems already provided in Table 1. The solutions thus obtained are considered to be optimal. The individual designs and results are discussed below.

6.1 ATAN Problem

As shown in Fig. 3(a), access nodes are represented by $a[b]$ where a: node #, and b: BW; and terminals by $c < d >$ where c: terminal #, and d: capacity. Table 2 gives the optimal cost obtained using *lp_solve* for all the three test cases. Using our technique, the optimal topology generated for design $T(50)A(10)$ is given in Fig. 3(a). It matches the optimal solution. The capacity of the terminals is varied between 2 and 5 and access node BW between 10 and 30. It took 455 generations and a total computation time of 6.4 secs for the algorithm to converge. The corresponding average and best fitness curves are given in Fig. 3(b). The progressively increasing nature of the curves signify the movement of the search towards the optimum.

At the end of the run, the library produced was examined. It was observed that certain genes had promising fitter values compared to its other values. For example, the 29-th and 32-nd terminals had the 10-th access node; 20-th and 18-th terminals contained the 3-rd access node and so on. On the other hand, some terminals such as the 6-th terminal had comparable strengths for all the access nodes.

6.2 LAN Problem

The setup for LAN is similar to that of ATAN. The terminal capacities lie between 5 and 10, and the access node BWs range between 20 and 40. The access node installation cost ranges from $50 - 200$. For access nodes, the format $a[b, c]$ signifies node a having bandwidth b and installation cost c. The test design detailed here is $T(20)P(10)$. The

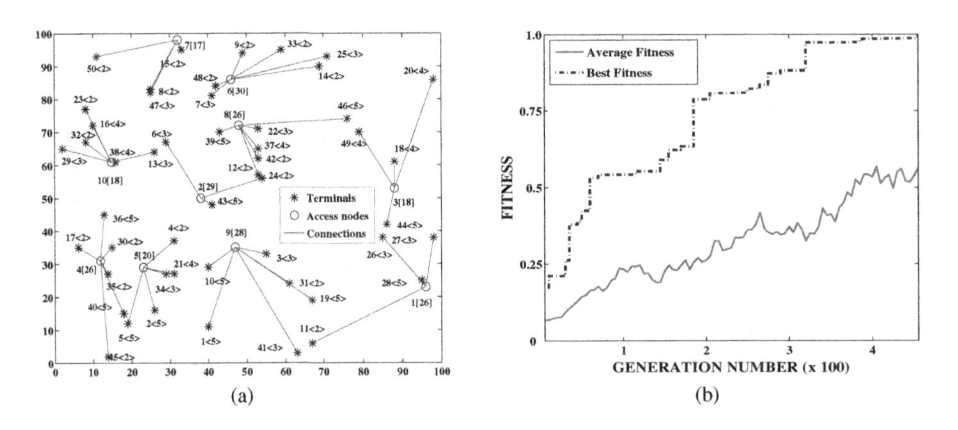

Fig. 3. ATAN $T(50)A(10)$ design →(a) Optimal topology design obtained, (b) Fitness curves

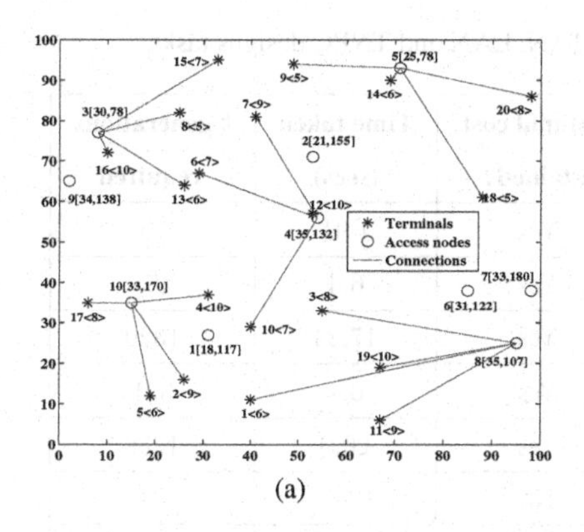

(a)

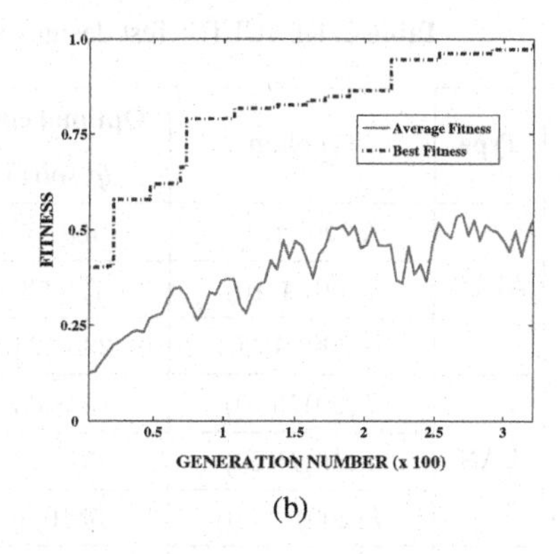

(b)

Fig. 4. LAN $T(20)P(10)$ design → (a) Optimal topology design obtained, (b) Fitness curves

optimal topology is obtained in the 321-st generation after 6.1 secs run time, and has a cost of 1048.63. It matches the optimal case. The topology thus generated is shown in Fig. 4(a). It is seen that nodes $1, 2, 6, 7$ and 9 are unutilized. Also, terminals like 15 and 8 showed greater strength for the 3-rd access node. Fig. 4(b) shows the gradually increasing average and best fitness curves.

6.3 TNPC Problem

The test design, $A(6)Tr(12)$, shown in Fig. 5(a), has source-destination demands between $(1-2)$, $(3-5)$, and $(4-6)$ pairs of access nodes. Also, $a < b - c >$ signifies the a-th transit node with b node-degree and c node installation cost. The load traffic, link capacity cost, link maximum capacity, node installation cost and maximum node degree are generated randomly within appropriate ranges. The allowable paths for the different loads are chosen considering node hops varying between 2 and 4. The path for $(1-2)$ load have transit node based path options: $\{7, 8\}$, $\{1, 8\}$, $\{9, 10\}$ and $\{2, 4, 12\}$. For load on $(3-5)$, the paths are $\{2, 7, 1\}$, $\{5, 4\}$, $\{5, 8\}$ and $\{4, 8\}$. Finally, load on

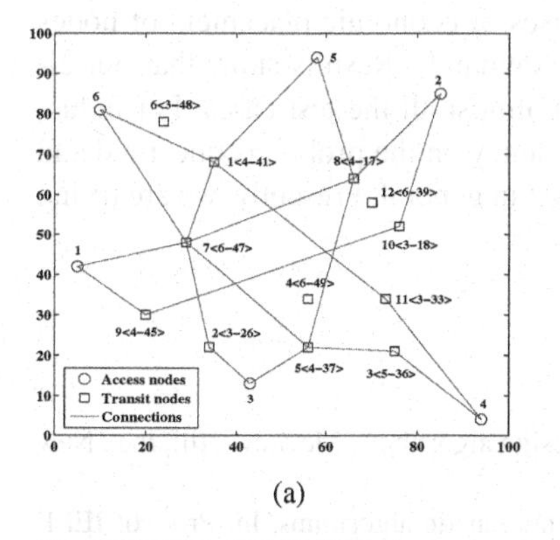

(a)

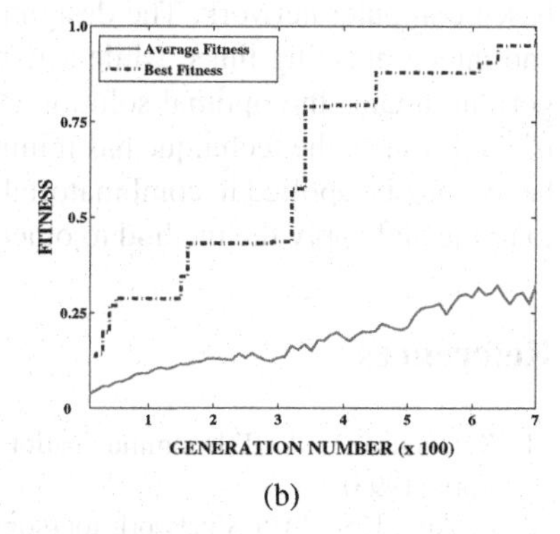

(b)

Fig. 5. TNPC $A(6)Tr(12)$ design → (a) Optimal topology design obtained, (b) Fitness curves

Table 2. RESULTS: Test designs for the ATAN, LAN and TNPC designs tasks

Type	Problem	Optimal cost (*lp_solve*)	Optimal cost attained?	Time taken (secs)	# generations required
ATAN	$T(20)A(4)$	468.87	Yes	1.01	75
	$T(50)A(10)$	720.53	Yes	6.4	455
	$T(200)A(20)$	2361.24	Yes	17.24	1320
LAN	$T(20)P(10)$	1048.63	Yes	6.1	321
	$T(50)P(20)$	2512.12	Yes	20.9	1252
	$T(200)P(50)$	5510.5	No	–	–
TNPC	$A(6)Tr(12)$	520.50	Yes	19.75	700
	$A(20)Tr(50)$	2711.34	Yes	43.7	1650
	$A(30)Tr(100)$	3871.05	Yes	79.2	3200

$(4-6)$ has path options comprising of $\{3,4,6\}$, $\{11,1\}$ and $\{3,5,7\}$. The load sharing is assumed to be equally divided among the optimized paths.

The optimal topology, given in Fig. 5(a), is obtained in the 700-th generation and has a cost of 520.50, matching the calculated optimal. All the demands needed two paths sharing equal load. The paths for the $(1-2)$, $(3-5)$, and $(4-6)$ demands are through the transit node sets $\{7,8\}$ and $\{9,10\}$, $\{2,7,1\}$ and $\{5,8\}$, $\{3,5,7\}$ and $\{11,1\}$ respectively. The transit nodes 4, 6 and 12 are not needed. The obtained fitness curves are provided in Fig. 5(b).

7 Conclusion

We have introduced a self-learning search technique for constrained combinatorial optimization problems. The relative merit of previous solutions guide the selection for future solutions. The techniques have been applied to the topology design of an MPLS-based computer network. The design task comprises of economic placement of nodes and interconnecting links, while satisfying traffic demands. Results show that our algorithm attains the optimal solution efficiently in almost all the test cases. For industrial relevance, the technique has minimum dependency on the problem concerned and hence may be applied to combinatorial optimization in general. Presently, we are trying to refine and apply the method to other domains.

References

1. Kershenbaum, A.: Telecommunications network design algorithms. McGraw-Hill, Inc., New York (1993)
2. El-Alfy, E.S.: MPLS network topology design using genetic algorithms. In: Proc. of IEEE Intl. Conf. on Computer Systems and Applications, pp. 1059–1065 (March 2006)

3. Ghosh, S., Ghosh, P., Basu, K., Das, S.K.: GaMa: An evolutionary algorithmic approach for the design of mesh-based radio access networks. In: Proc. of IEEE Conf. on Local Computer Networks (November 2005)
4. Elbaum, R., Sidi, M.: Topological design of local-area networks using genetic algorithms. IEEE/ACM Transactions on Networking 4(5) (1996)
5. Quoitin, B.: Topology generation based on network design heuristics. In: Proc. of CoNEXT, pp. 278–279 (2005)
6. Youssef, H., Sait, S.M., Khan, S.A.: Topology design of switched enterprise networks using a fuzzy simulated evolution algorithm. Engineering Applications of Artificial Intelligence 15(3), 327–340 (2002)
7. Runggeratigul, S.: A memetic algorithm for communication network design taking into consideration an existing network. Applied Optimization - Metaheuristics: computer decision-making, 615–626 (2004)
8. Pioro, M., Myslek, A., Juttner, A., Harmatos, J., Szentesi, A.: Topological design of MPLS networks. In: Proc. of IEEE Globecom, vol. 1, pp. 12–16 (November 2001)
9. Arabas, J., Kozdrowski, S.: Applying an evolutionary algorithm to telecommunication network design. IEEE Transactions on Evolutionary Computation 5(4), 309–322 (2001)
10. Harik, G.R., Lobo, F.G., Goldberg, D.E.: The compact genetic algorithm. IEEE Transactions on Evolutionary Computation 3(4), 287–297 (1999)
11. Muhlenbein, H.: The equation for response to selection and its use for prediction. Evolutionary Computation 5(3), 303–346 (1997)
12. Goldberg, D.E.: Genetic Algorithms in Search, Optimization and Machine Learning. Add.-Wesley Prof., London (1989)
13. Burkardt, J.: Halton random sequence generator, `http://people.scs.fsu.edu/~burkardt/cpp_src/halton/halton.html`
14. Geem, Z.W., Kim, J.H., Loganathan, G.V.: A new heuristic optimization algorithm: Harmony search. Simulation, the Society for Modeling and Simulation International 76(2), 60–68 (2001)
15. Yeniay, O.: Penalty function methods for constrained optimization with genetic algorithms. Mathematical and Computational Applications 10(1), 45–56 (2005)
16. GNU_Public_License: Integer linear programming (ILP) solver ‘lp_solve’, `http://lpsolve.sourceforge.net/5.5/`

A Comparative Study of Fuzzy Inference Systems, Neural Networks and Adaptive Neuro Fuzzy Inference Systems for Portscan Detection

M. Zubair Shafiq[1], Muddassar Farooq[1], and Syed Ali Khayam[2]

[1] Next Generation Intelligent Networks Research Center (nexGIN RC)
National University of Computer & Emerging Sciences (NUCES)
Islamabad, Pakistan
{zubair.shafiq, muddassar.farooq}@nexginrc.org
[2] NUST Institute of Information Technology (NIIT)
National University of Sciences & Technology (NUST)
Rawalpindi, Pakistan
khayam@niit.edu.pk

Abstract. Worms spread by scanning for vulnerable hosts across the Internet. In this paper we report a comparative study of three classification schemes for automated portscan detection. These schemes include a simple Fuzzy Inference System (FIS) that uses classical inductive learning, a Neural Network that uses back propagation algorithm and an Adaptive Neuro Fuzzy Inference System (ANFIS) that also employs back propagation algorithm. We carry out an unbiased evaluation of these schemes using an endpoint based traffic dataset. Our results show that ANFIS (though more complex) successfully combines the benefits of the classical FIS and Neural Network to achieve the best classification accuracy.

Keywords: Adaptive Neuro Fuzzy Inference System (ANFIS), Information Theoretic Features, Neural Networks, Portscan Detection.

1 Introduction

The number of vulnerable hosts on the Internet is increasing due to an increasing number of novice users using it [1]. An attacker can hack these vulnerable machines and use them as a potential army (zombies) for launching a Denial of Service (DoS) attack on any target host. It is not only the number of machines that is of concern, but also the time interval in which an attacker can gain access to the vulnerable machines [1]. Attackers usually accomplish the objective of infecting large number of machines in as little time as possible through portscans. A portscan is an attempt by the attacker to find out open (possibly vulnerable) ports on a victim machine. The attacker decides to invade the victim, through a vulnerable port, on the basis of the response to a portscan. These portscans are usually very fast and hence an attacker can take control of a major proportion of the vulnerable machines on the Internet in a small amount of time. The compromised machines can be used to launch DoS attacks. It is to be noted that

M. Giacobini et al. (Eds.): EvoWorkshops 2008, LNCS 4974, pp. 52–61, 2008.

some portscans are stealthy and slow as well. In this study, however, we only deal with fast (i.e., high-rate) portscans.

In the first phase of a DoS attack, the attacker deploys the DoS tools on the the victim machines or zombies. Worms provide an effective method to deploy DoS tools on the vulnerable hosts on the Internet automatically. Worms use random or semi-random portscans to find out vulnerable hosts across the Internet. In the second phase of DoS attack, the infected systems are used to launch a DoS attack at a specific target machine. This attack will be highly distributed due to the possible geographic spreading pattern of the hosts across the web. As a result, DoS attack will turn into, a more disruptive and difficult to counter, *Distributed Denial of Service* (DDoS) attack. Thus an attacker with the help of an intelligently written worm can very quickly gain control over millions of vulnerable systems on the Internet.

Fortunately, many existing solutions can detect and block this serious security threat. One of the most popular solution is 'firewall'. The problem with firewalls and many other similar tools is that they require a manual setting of a number of security levels (ranging from low and medium to high) which are not comprehendible to a novice user. If a user sets high security levels, this often results in the disruption of a user's activities by a high frequency of annoying pop-us and notices. This leads the user to select very low security levels, which practically makes a firewall ineffective.

The characteristics of the traffic generated by portscans is usually different from that generated by a normal user activity. This is because state-of-the-art worms spread on the principle of 'infecting maximum number of hosts in least possible time' [2]. Therefore, using certain characteristics of normal traffic of a user, we can train a classifier which will distinguish between normal traffic (due to activities of a normal user) and malicious traffic (due to portscans by a worm). Such a classifier employs some features extracted from the users' traffic to detect malicious activity.

In this paper we use two information theoretic features, namely *entropy* and *KL-divergence* of port usage, to model the network traffic behavior of normal user applications. We carry out a comparative study of the following three classifiers for the problem of automated portscan detection: 1) fuzzy rule-based system, 2) neural network, and 3) adaptive neuro fuzzy system.

Organization of the Paper. In the next section we provide a brief overview of three classifiers. In Section 3, we present the traffic features used in this paper. We describe the traffic test bed in Section 4. We will discuss the performance evaluation parameters in Section 5 and then analyze the results obtained from the experiments in Section 6.

2 A Review of Classification Schemes

In this section we present a brief overview of three classification algorithms. First is the classical inductive fuzzy rule learning classifier. Other two are neural network and ANFIS, which are bio-inspired classification algorithms.

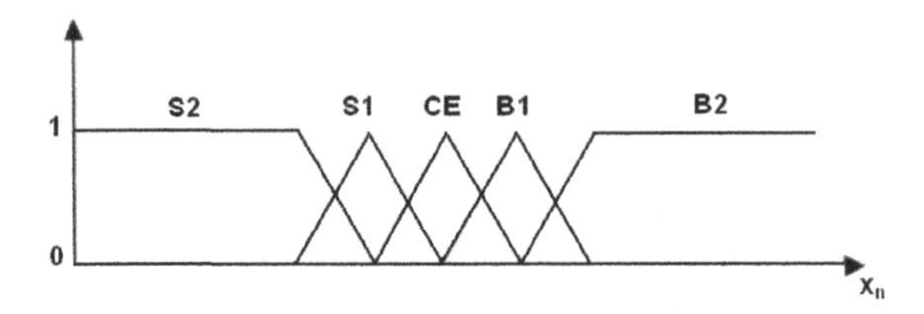

Fig. 1. Input Membership functions

2.1 Classical Inductive Fuzzy Rule Learning Classifier

In [7], the authors have given a general method to generate fuzzy rules from numerical examples for the Mamdani-type fuzzy system. In this study, 50 examples were used in the rule generation phase. This method consists of five steps:

1. In the first step, the input and output spaces are divided into fuzzy regions. This step produces initial membership functions. We divided the input space in such a manner that we get five membership functions for each input. These are denoted as S2 (Small 2), S1 (Small 1), CE (Center), B1 (Big 1) and B2 (Big 2). For symmetry, the membership functions are chosen to be isosceles triangles. Figure 1 shows the input membership functions. The output of the fuzzy classifier has only two membership functions, benign and malicious, because of its Boolean classification nature. So, the design of output membership functions does not follow procedure defined in [7].
2. The second step involves the initial generation of the fuzzy rules from given data pairs. Given the ith numerical example data (Xi_1, Xi_2, Xi_j, ... , Yi), the degree of membership for each input is calculated. It is possible that the input parameter has non-zero degree of membership (DOM) for two membership functions. In this case the membership function with a maximum DOM (DOM_{max}) is chosen. A rule is formulated as:
   `IF X1 is 'a' AND X2 is 'b' AND ... then y is 'c',`
   where a,b and c are the respective membership functions for each input.
3. A degree is assigned to each rule. For rule i, degree is defined as,
   $\mathtt{degree(i)} = \mathtt{DOM}_{max(1)} \times \mathtt{DOM}_{max(2)} \times \mathtt{DOM}_{max(i)} \times \ldots \times \mathtt{DOM}_{max(y)}$
4. After performing the third step, we will get a populated rule base. It is possible that more than one rule, with similar inputs, may have different outputs. This situation represents the conflict amongst the rules. In order to resolve this conflict, the rule with a maximum degree will be chosen.
5. We use centroid defuzzification technique to get a crisp output. This defuzzification technique was chosen because the output produced by it includes the balanced effect of all the inputs. The formula for calculating the centroid is given by:

$$F^{-1}_{Center\_Of\_Gravity}(\bar{A}) = \frac{\int_x \mu\bar{A}(x)xdx}{\int_x \mu\bar{A}(x)dx}$$

2.2 Neural Network Classifier with Back Propagation Learning Algorithm

Neural networks are bio-inspired paradigm which map concepts from the biological nervous systems. They consist of a network of neurons which are boolean in nature. The connections and the weight between these connections are crucial to the performance of the network. The neural network used in this study consisted of two neurons in its output layer to identify the traffic behavior as either benign or malicious. The network is a two-layer log-sigmoid/log-sigmoid network. The log-sigmoid transfer function was selected because its output range is suitable for learning to output Boolean values. The first hidden layer had 10 neurons. This number was chosen after pilot studies using different number of neurons. The standard back propagation learning algorithm was utilized by a batch training scheme in which weights are updated after a complete batch of training examples. 50 training examples were used in the training phase. More details can be found in [8].

2.3 Adaptive Neuro Fuzzy Inference System (ANFIS)

Adaptive Neuro Fuzzy Inference System (ANFIS) is a fuzzy rule based classifier in which the rules are learnt from examples that use a standard back propagation algorithm. Note that this algorithm is also used in neural network training. However, ANFIS is far more complex than the simple Mamdani-type fuzzy rule based system as explained in Section 2.1. ANFIS uses Sugeno-type fuzzy system. The subtractive clustering was used to divide the rule space. Five triangular membership functions were chosen for all inputs and output similar to the fuzzy system in Section 2.1. 50 training examples were chosen in the training phase. An interested reader can find the details of ANFIS in [9].

3 Traffic Feature Modeling Using Information Theoretic Measures

We employ information theoretic measures [6] to compare the probability distributions of a pre-sampled benign traffic profile and run-time traffic profile. It is assumed that the malicious traffic is not present while sampling the benign traffic profile. We have chosen *entropy* and *Kullback-Leibler (KL) divergence* [6] of port usage as tools to compare the benign traffic profile (collected prior to classification) with a run-time traffic profile. These measures have been used previously for network anomaly detection [3,4].

We are interested in outgoing unicast traffic to detect the malware residing on the system, which tries to propagate through portscan. We have calculated the entropy and KL using both source and destination port information. The source and destination port information is collected at a session level, where a session is defined as the bidirectional communication (mostly involving multiple packets) between two IP addresses. Entropy and KL are calculated in a time window of

15 seconds in our study. However, qualitatively similar results are obtained for other sizes of time window.

Entropy gives the spread of a probably distribution. In order to calculate entropy in a time window, let p_i be the number of times source port i was used and q_i be the number of times destination port i was used in a time window. Let p_n be the aggregate frequencies of source ports used in a particular time window n. Similarly, let q_n be the aggregate frequencies of destination ports used in a particular time window n. Mathematically, $p_n = \sum_{i=0}^{65,535} p_i$ and $q_n = \sum_{i=0}^{65,535} q_i$. The source and destination port entropies are defined as:

$$H_{source} = - \sum_{i=0}^{65,535} \frac{p_i}{p_n} \log_2 \frac{p_i}{p_n} \tag{1}$$

$$H_{destination} = - \sum_{i=0}^{65,535} \frac{q_i}{q_n} \log_2 \frac{q_i}{q_n} \tag{2}$$

KL divergence gives the distance between two probability distributions. In this case, it will give the difference between distributions of traffic in a particular session window and benign traffic. For source and destination port KL, let $p_i^{'}$ be the frequency of source port i in the benign traffic profile and $q_i^{'}$ be the frequency of destination port i in the benign traffic profile. Moreover, p and q represent the aggregate frequency of source and destination ports in the benign profiles. Mathematically, $p = \sum_{i=0}^{65,535} p_i^{'}$ and $q = \sum_{i=0}^{65,535} q_i^{'}$. The source and destination port KL are defined as:

$$D_{source} = \sum_{i=0}^{65,535} \frac{p_i}{p_n} \log_2 \frac{p_i/p_n}{p_i^{'}/p} \tag{3}$$

$$D_{destination} = \sum_{i=0}^{65,535} \frac{q_i}{q_n} \log_2 \frac{q_i/q_n}{q_i^{'}/q} \tag{4}$$

Since we are focusing on fast portscans, we invoke classifier only if the number of sessions per time window exceeds the `SessionThreshold`. The `SessionThreshold` in this study was set to 15 sessions per time window. This corresponds to and average rate of one session per second. The worms with the session rates lower than `SessionThreshold` are ignored. The value of `SessionThreshold` is justified since the motive of a worm is to infect large number of machines in as little time as possible.

In order to produce suitable inputs for the classification systems, the means of respective information theoretic measures are used in Equations (1), (2), (3) and (4) that are calculated from the benign traffic profile. These parameters are labeled as H_{benign_source}, $H_{benign\_destination}$, D_{benign_source} and $D_{benign\_destination}$. The differences between the parameters, calculated from run-time traffic profile, and means of respective parameters calculated from benign traffic profile were

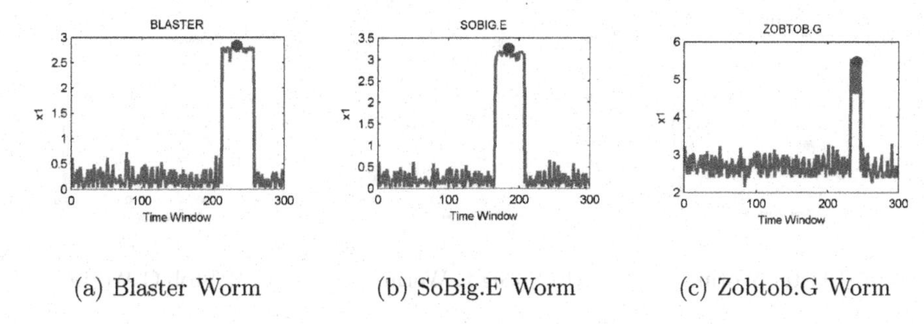

(a) Blaster Worm (b) SoBig.E Worm (c) Zobtob.G Worm

Fig. 2. Source Port Entropy for different worms

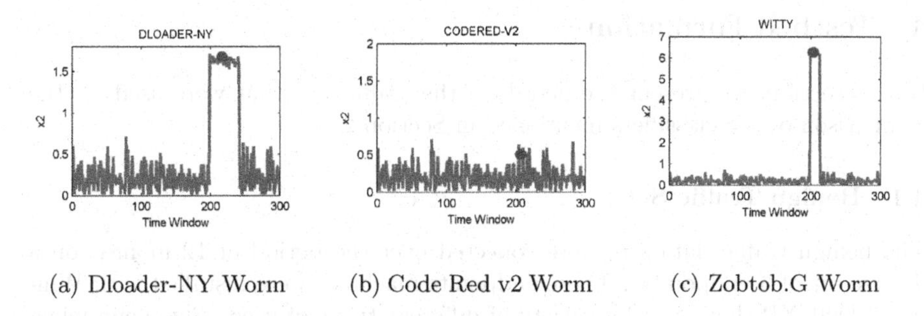

(a) Dloader-NY Worm (b) Code Red v2 Worm (c) Zobtob.G Worm

Fig. 3. Destination Port Entropy for different worms

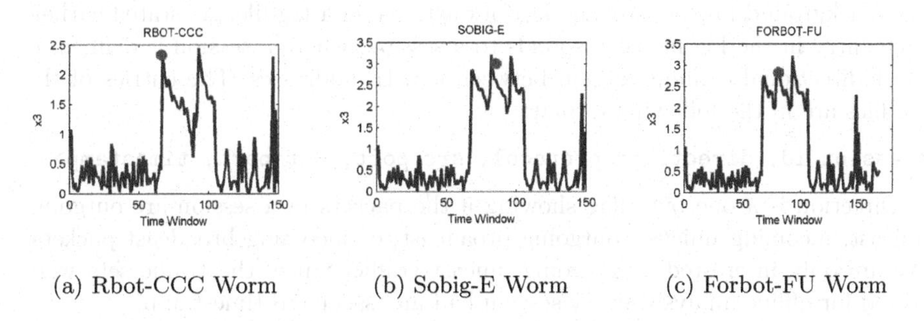

(a) Rbot-CCC Worm (b) Sobig-E Worm (c) Forbot-FU Worm

Fig. 4. Source Port KL for different worms

used as inputs to the classifiers. For simplicity absolute value of the difference is considered. Mathematically we can represent this as:

$$x_1 = |H_{source} - \mu_{Hbenign\_source}|$$
$$x_2 = |H_{destination} - \mu_{Hbenign\_destination}|$$
$$x_3 = |D_{source} - \mu_{Dbenign\_source}|$$
$$x_4 = |D_{destination} - \mu_{Dbenign\_destination}|$$

Figures 2, 3, 4 and 5 show a plot of these parameters for different worms. Note that the circle represents the middle of the infection period.

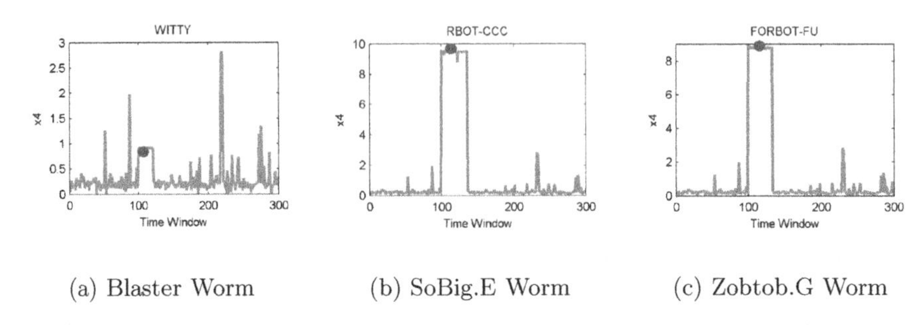

(a) Blaster Worm (b) SoBig.E Worm (c) Zobtob.G Worm

Fig. 5. Destination Port KL for different worms

4 Testbed Formation

In this section, we present the details of the traffic sets that were used for the comparison of the classifiers mentioned in Section 2.

4.1 Benign Traffic Set

The benign traffic data-sets were collected over the period of 12 months on a diverse set of 5 endpoints[1]. These endpoints machines were installed with Windows 2000/XP that served a variety of different types of users. Some endpoints were also shared by multiple users. Characteristics of the benign traffic set are tabulated in Table 1. Traffic data-sets were collected using `argus`, which runs as a background process storing network activity in a log file. As stated earlier, each entry in the log file corresponds to a session, where a session is defined as the bidirectional communication between two IP addresses. The entries of the log files are in the following format:

```
<session id, direction, protocol, src port, dst port, timestamp>
```

Direction is a one byte flag showing if the packets in a session are outgoing unicast, incoming unicast, outgoing broadcast, or incoming broadcast packets. We are only interested in outgoing unicast traffic. Since, the traffic sets were stored for offline analysis, every session had an associated timestamp.

4.2 Worm Set

Malicious traffic sets were collected by infecting virtual machines with different self-propagating malicious codes. Note that `CodeRedv2` and `Witty` were simulated. The traffic sets were collected using the same method as explained before in Section 4.1. These malicious traffic sets were embedded into benign traffic sets to form a test for evaluation of the designed system. The details of the malware used in this study are given in Table 2.

[1] "An endpoint is an individual computer system or device that acts as a network client and serves as a workstation or personal computing device." [5]

Table 1. Statistics of Bengin set used in this Study

Endpoint ID	Endpoint Type	Mean Session Rate (/sec)
1	Office	0.22
2	Home	1.92
3	Univ	0.19
4	Univ	0.28
5	Univ	0.52

Table 2. Statistics of Worm set used in this Study

Worm Name	Release Date	Ports Used
Blaster	Aug 2003	TCP 135,4444, UDP 69
Dloader-NY	Jul 2005	TCP 135,139
Forbot-FU	Sep 2005	TCP 445
Rbot.CCC	Aug 2005	TCP 139,445
CodeRedv2	Jul 2004	TCP 80
Witty	Mar 2004	UDP 4000
SoBig.E	Jun 2003	TCP 135, UDP 53
Zobtob.G	Jun 2003	TCP 135, 445, UDP 137

4.3 Formation of Infected Traffic Sets

Due to the university policies and the user reservations, we were not able to infect operational endpoints with worms. Therefore, we resorted to the following offline traffic mixing approach. We inserted T minutes of malicious traffic data of each worm in the benign profile of each endpoint at a random time instance. Specifically, for a given endpoint's benign profile, we first generated a random infection time t_I (with millisecond accuracy) between the endpoint's first and last session times. Given n worm sessions starting at times $t_1, ..., t_n$, where $t_n \leq T$, we created a special infected profile of each host with these sessions appearing at times $t_I + t_1, ..., t_I + t_n$. Thus in most of the cases once a worms traffic was completely inserted into a benign profile, the resultant profile contained interleaved benign and worm sessions starting at t_I and ending at $t_I + t_n$. For all worms except Witty, we used $T = 15$ minutes and to simulate the worstcase behavior of Witty, we inserted only $20,000$ scan packets (approximately 1 minute) in the infected profiles.

5 Receiver Operating Characteristic (ROC) Performance Parameters

In ROC analysis, the 2x2 confusion matrix for performance evaluation of boolean classifiers gives four possible outcomes [10]. These outcomes are True Positive (TP), False Positive (FP), True Negative (TN) and False Negative (FN). The metrics considered for the evaluation of our fuzzy based classifier are False Positive Rate ($fprate$), and True Positive Rate ($tprate$). These metrics are defined as, $fprate = \frac{FP}{FP+TN}$ and $tprate = \frac{TP}{TP+FN}$.

Table 3. Results on the Infected dataset

	Inductive FIS [7]	Neural Network [8]	ANFIS [9]
Endpoint ID - 1			
tp rate	0.888	0.735	0.885
fp rate	0.037	0.062	0.043
Endpoint ID - 2			
tp rate	0.857	0.728	0.861
fp rate	0.163	0.191	0.155
Endpoint ID - 3			
tp rate	0.900	0.774	0.893
fp rate	0.100	0.116	0.121
Endpoint ID - 4			
tp rate	0.865	0.731	0.829
fp rate	0.096	0.104	0.088
Endpoint ID - 5			
tp rate	0.954	0.790	0.968
fp rate	0.160	0.211	0.114

6 Classification Results

The ROC performance evaluation parameters for all the endpoints used in this study are tabulated in Table 3. It is interesting to note that the worst results for all the classifiers were reported for the endpoint 2. Endpoint 2 was a home based endpoint with a relatively high session rate because of several multimedia and video streaming applications that were running on this endpoint. High $fprate$ is because of the high volume and the bursty nature of multimedia traffic which resembles the traffic produced by the portscan activities of worms. Note that the results of endpoint deployed at the university and office are very similar.

It is clear from the results tabulated in Table 3 that ANFIS outperforms rest of the classifiers. The results of Inductive FRBS are better than those of Neural network which shows the promise of fuzzy based schemes for modeling and representation of network traffic features. Neural network is unable to cater for the inherent fuzziness in the users' traffic patterns. ANFIS combines the advantages of the back propagation learning algorithm with the fuzzy based classifier, and as a result, we get the best detection accuracy.

7 Conclusion and Future Research

Our findings from this study are that the users' behaviors are not very crisp because of the pseudo-random nature of the users' network traffic. To cater for the inherent fuzziness in modeling of the user's network traffic trends, a fuzzy rule based system is a better approach. But the generation of fuzzy rules and membership functions is a problem. One solution is to use classical inductive fuzzy rule learning for fixed membership functions which is an empirical yet an effective way to generate fuzzy rules. Another solution is to use a neural network with a more sophisticated back propagation learning algorithm which is suitable for use with numerical data pairs for training. But a neural network based solution is unable to cater for the inherent fuzziness in the users' traffic patterns. This leads us to the following question: Is a combined approach i.e. ANFIS, which uses fuzzy rule based system along with the sophisticated back propagation learning algorithm, more effective than both of the standalone approaches?. The answer to this question is positive as per our experimental study. ANFIS is able to effectively combine the benefits from both approaches. The experimental results, for portscan detection, of ANFIS are better than both classical inductive fuzzy rule base system and neural network based classifier using the standard back propagation algorithm.

In future we wish to run the experiments on a more diverse and extensive traffic set consisting of more diverse endpoints. We also wish to extend our malware traffic repository. Further, we also plan to evaluate the performance of various machine learning algorithms such as support vector machines (SVMs) and other bio-inspired schemes such as Artificial Immune Systems (AIS).

Acknowledgements. This work is supported by the National ICT R&D Fund, Ministry of Information Technology, Government of Pakistan. The information, data, comments, and views detailed herein may not necessarily reflect the endorsements of views of the National ICT R&D Fund.

References

1. Staniford, S., Paxson, V., Weaver, N.: How to Own the Internet in Your Spare Time. In: Usenix Security Symposium (2002)
2. Moore, D., Shannon, C., Voelker, G.M., Savage, S.: Internet Quarantine: Requirements for Containing Self-Propagating Code. In: IEEE Infocom (2003)
3. Gu, Y., McCullum, A., Towsley, D.: Detecting anomalies in network traffic using maximum entropy estimation. In: ACM/Usenix IMC (2005)
4. Lakhina, A., Crovella, M., Diot, C.: Mining anomalies using traffic feature distributions. In: ACM Sigcomm (2005)
5. Endpoint Security, http://www.endpointsecurity.org
6. Cover, T.M., Thomas, J.A.: Elements of Information Theory. Wiley-Interscience, Chichester (1991)
7. Wang, L.-X., Mendel, J.M.: Generating Fuzzy Rules by Learning from Examples. IEEE Transactions on Systems, Man, and Cybernetics 6(22), 1414–1427 (1992)
8. Bishop, C.M.: Neural Networks for Pattern Recognition. Oxford University Press, Oxford (1995)
9. Jang, J.-S.R.: ANFIS: Adaptive-Network-Based Fuzzy Inference System. IEEE Transactions on System, Man and Cybernetics (23), 665–685 (1993)
10. T. Fawcett.: ROC Graphs: Notes and Practical Considerations for Researchers, Technical report (HPL-2003-4), HP Laboratories, Palo Alto, CA, 2003-4, USA (2003)

Evolutionary Single-Position Automated Trading

Antonia Azzini and Andrea G.B. Tettamanzi

University of Milan
Information Technology Department - Crema, Italy
{azzini,tettamanzi}@dti.unimi.it

Abstract. Automated Trading is the activity of buying and selling financial instruments for the purpose of gaining a profit, through the use of automated trading rules. This work presents an evolutionary approach for the design and optimization of artificial neural networks to the discovery of profitable automated trading rules. Experimental results indicate that, despite its simplicity, both in terms of input data and in terms of trading strategy, such an approach to automated trading may yield significant returns.

1 Introduction

Trading is the activity of buying and selling financial instruments for the purpose of gaining a profit [9]. Usually, such operations are carried out by traders by making bids and offers, by using orders to convey their bids and offers to the brokers or, more interesting, by developing automated trading systems that arrange their trades.

Several works have been carried out in the literature by considering market simulators for program trading. Recently, Brabazon and O'Neill [4] explained that, in program trading, the goal is usually to uncover and eliminate anomalies between financial derivatives and the underlying financial assets which make up those derivatives. Trading rules are widely used by practitioners as an effective mean to mechanize aspects of their reasoning about stock price trends. However, due to their simplicity, individual rules are susceptible of poor behavior in specific types of adverse market conditions. Naive combinations of rules are not very effective in mitigating the weaknesses of component rules [14].

As pointed out in [14], recent developments in the automation of exchanges and stock trading mechanisms have generated substantial interest and activity within the machine learning community. In particular, techniques based on artificial neural networks (ANNs) [3,2] and evolutionary algorithms (EAs) [4,7,8] have been investigated, in which the use of genetic algorithms and genetic programs has proven capable of discovering profitable trading rules. The major advantages of the evolutionary algorithms over conventional methods mainly regard their conceptual and computational simplicity, their applicability to broad classes of problems,their potential to hybridize with other methods and their capability of self-optimization. Evolutionary algorithms can be easily extended to include other types of information such as technical and macroeconomic data as well as

M. Giacobini et al. (Eds.): EvoWorkshops 2008, LNCS 4974, pp. 62–72, 2008.

past prices: some genetic algorithms become useful to discover technical trading rules [1] or to find optimal parameter-values for trading agents [5].

Different kinds of evolutionary techniques can be used to solve global optimization problems, and several studies carried out in the literature demonstrate how these algorithms are a more integrated way of optimizing classifier systems since they do not require any expert knowledge of the problem. Much research has been undertaken on the combination of EAs and ANNs, giving rise to so-called evolutionary artificial neural network models (EANNs) [15,16]. EANNs have become well-established approaches for different financial problems in which the analytical solutions are difficult to obtain, and they are helpful in finding the optimal ANN design for a given financial problem. Some works focused on investigating the relationship between neural network optimization for financial trading and the efficient market hypothesis [12]; others examined the relationships between economic agents' risk attitude and the profitability of stock trading decisions [10]; others focused on modeling the mutual dependencies among financial instruments [3].

In this work, an EANN approach, performing a joint evolution of neural network weights and topology, is applied to providing trading signals to an automated trading agent. This approach has already been validated on different benchmark problems, and then tested on several real-world problems [3,2], also comparing the performances of the EANN with random approaches.

2 Single-Position Automated Trading Problem

A so called Single-position automated day-trading is the problem of finding an automated trading rule for opening and closing a single position within a trading day. In such a scenario either short or long positions are considered, and the entry strategy is during the opening auction at market price. A profit-taking strategy is also defined in this work, by waiting until market close unless a stop-loss strategy is triggered. A trading simulator is used to evaluate the performance of a trading agent. While for the purpose of designing a profitable trading rule R, i.e., solving any static automated day trading problem, the more information is available the better, whatever the trading problem addressed, for the purpose of evaluating a given trading rule the quantity and granularity of quote information required varies depending on the problem. For instance, for one problem, daily open, high, low, and close data might be enough, while for another tick-by-tick data would be required. Generally, the annualized log-return of rule R when applied to time series X of length N is

$$r(R, X) = \frac{Y}{N} \sum_{i=1}^{N} r(R, X, i), \tag{1}$$

where $r(R, X, i)$ is the log-return generated by the rule R on the ith day of time series X, and Y is the number of market days in a year. In this work, the log-return depends on a fixed *take-profit* return r_{TP}, an algorithm's parameter corresponding to the maximum performance that can be achieved by the automated trading simulation. For each day, a position is defined through two different operation thresholds θ_{buy} and θ_{sell}, that correspond, respectively, to the maximum value of a network output for buying and to the minimum value for selling. In this automated trader application these thresholds are set, respectively, to 0.34 and 0.66, i.e., in such a way as to partition the output range into three equal-sized segments. If the output obtained from the evolutionary approach is less than θ_{buy}, a long position is opened, while if the output is more than θ_{sell} a short position is opened. No position is taken in between.

The trading rules defined in this approach use past information to determine the best current trading action, and return a buy/sell signal at current day depending on the information of the previous day. Such a rule is then used to define the log-returns of the financial trading on the current day. In the trading simulation the log-returns obtained are then used, together with the risk free rate r_f and the downside risk DSR to calculate the *Sortino ratio* SR_d [13], an extension of the Sharpe ratio [11], defined by:

$$SR_d(R, X) = \frac{r(R, X) - r_f}{DSR_{r_f}(R, X)}. \tag{2}$$

SR_d represents the measure of the risk-adjusted returns of the simulation, and it will be used in the evolutionary process in order to evaluate the fitness of an individual trading agent.

At the beginning of the trading day, the trader sends to the trading simulator the order decoded from the output of the neural network. During the trading day, if a position is open, the position is closed as soon as the desired profit (indicated by a log-return r_{TP}) is attained (in practice, this could be obtained by placing a reverse limit order at the same time as the position-opening order); if a position is still open when the market close, it is automatically closed in the closing auction at market price.

3 EANN Approach

The approach implemented in this work defines a population of traders, the individuals, encoded through neural network representations. The evolutionary algorithm [3,2] considered in this approach evolves traders population by using the joint optimization of structure and weights of the NNs, and it takes advantage of the backpropagation algorithm (BP) as a specialized decoder.

Such a simultaneous evolution underlines that an EA allows all aspects of a NN design to be taken into account at once, without requiring any expert knowledge of the problem, and can overcome possible drawbacks of each single technique combining their advantages. In this evolutionary algorithm only a specific subset of NN, Multi-Layer Perceptron (MLP), is considered for neural encoding, due to the simple structure that it needs to be represented. Indeed, MLPs are feedforward NNs with a layer of input neurons, a layer of one or more output neurons and zero or more "hidden" (i.e., internal) layers of neurons in between; neurons in a layer can take inputs from the previous layer only.

Table 1. Individual Representation

Element	Description
l	Length of the topology string, corresponding to the number of layers.
topology	String of integer values that represent the number of neurons in each layer.
$\mathbf{W}^{(0)}$	Weights matrix of the input layer neurons of the network.
$\mathbf{Var}^{(0)}$	Variance matrix of the input layer neurons of the network.
$\mathbf{W}^{(i)}$	Weights matrix for the ith layer, $i = 1, \ldots, l$.
$\mathbf{Var}^{(i)}$	Variance matrix for the ith layer, $i = 1, \ldots, l$.
b_{ij}	Bias of the jth neuron in the ith layer.
$Var(b_{ij})$	Variance of the bias of the jth neuron in the ith layer.

Each individual is encoded in a structure in which basic information are maintained as illustrated in Table 1.

The values of all these parameters are affected by the genetic operators during evolution, in order to perform incremental (adding hidden neurons or hidden layers) and decremental (pruning hidden neurons or hidden layers) learning.

As described in detail in [3,2], all individuals have no pre-established topology, the population is initialized with different hidden layer sizes and different number of neurons for each individual by means of two exponential distributions, in order to maintain diversity between all the individuals in the new population. Such dimensions are not bounded in advance, even though the fitness function may penalize large networks.

In the evolutionary process, after population definition and network initialization, the genetic operators are applied to each network until termination conditions are not satisfied, with the following steps: *(i)* select from the population (of size n) $\lfloor n/2 \rfloor$ individuals by truncation, *(ii)* for all individuals of the population mutate the weights and the topology of the offspring, train the resulting network, calculate fitness on the test set (see Section 3.1), save the best individual, *(iii)* save statistic information about the entire evolutionary process.

In the truncation selection, elitism is also considered, allowing the survival of the best individual unchanged into the next generation and the solutions to get better over time. In each new generation, a new population has to be created, and the first half of such new population corresponds to the best parents that have been selected with the truncation operator, while the second part is defined by creating offspring from the previously selected parents.

Weights mutation perturbs the weights of the neurons before performing any structural mutation and applying BP to train the network.

Then, *Topology mutation* is implemented with four types of mutation by considering neurons and layer addition and elimination. The elimination of a neuron is carried out only if the contribution of that neuron is negligible with respect to the overall network output, while the addition and the elimination of a layer and the insertion of a neuron are applied with three independent probabilities p^{+}_{layer}, p^{-}_{layer}, and p^{+}_{neuron}.

3.1 Fitness Function

As in the previous approaches [3,2], the fitness function used in this work depends on the cost of each individual. So, the convention that the best fitness corresponds to the lowest fitness is adopted, defining the objective of each task as a cost minimization problem. The fitness function also works as a controller and selector, because it penalizes large networks.

In this application, the fitness function depends on the risk-adjusted return obtained by the considered trader, and is calculated, at the end of the training and evaluation process, by

$$f = \lambda kc + (1 - \lambda) * e^{-SR_d}, \tag{3}$$

where λ corresponds to the desired tradeoff between network cost and accuracy, and it has been set experimentally to 0.2 to place more emphasis on accuracy, since the NN cost increase is checked also by the entire evolutionary algorithm. This parameter also represents the measure of the correlation between the fitness value and the measure of the risk-adjusted return considered e^{-SR_d}. k is a scaling constant set experimentally to 10^{-6}, and c models the computational cost of a neural network, proportional to the number of hidden neurons and synapses of the neural network. It is important to emphasize that the Sortino ratio used in the fitness function, defines "risk" as the risk of loss.

Following the commonly accepted practice of machine learning, the problem data will be partitioned into training, test and validation sets, used, respectively for network training, to stop learning avoiding overfitting, and to to test the generalization capabilities of a network. The fitness is calculated according to Equation 3 over the test set.

3.2 Setting of the Data

The input values of the dataset are defined by considering the quotes of the daily historical prices and 24 different technical indicators for the same financial instrument, that correspond to the most popular indicators used in technical analysis. These indicators also summarize important features of the time series of the financial instrument considered, and they represent useful statistics and technical information that otherwise should be calculated by each individual of the population, during the evolutionary process, increasing the computational cost of the entire algorithm. The list of all the inputs of a neural network is shown in Table 2, and a detailed discussion about all these technical indicators can be easily found in the literature [6].

Generally, technical indicators can be directly incorporated as model inputs, or, alternatively, they can be preprocessed to produce an input by taking ratios or through the use of rules. The last case is a combinatorial problem and traditional modeling methods can provide an infinite number of possibilities, in some cases problematic. This suggests that an evolutionary algorithm in which the model structure and model inputs are not defined a priori will have potential for generating trading operations drawn from individual technical indicators [4].

Table 2. Input Technical Indicators

Index	Input Tech. Ind.	Description	Index	Input Tech. Ind.	Description
1	Open(i)	Opening value	13	EMA20(i)	20-day Exp. Mov. Avg
2	High(i)	High value	14	EMA50(i)	50-day Exp. Mov. Avg
3	Low(i)	Low value	15	EMA100(i)	100-day Exp. Mov.Avg
4	Close(i)	Closing value	16	EMA200(i)	200-day Exp.Mov. Avg
5	MA5(i)	5-day Mov. Avg	17	MACD(i)	Mov. Avg Conv./Div.
6	MA10(i)	10-day Mov. Avg	18	SIGNAL(i)	Exp. Mov. Avg on MACD
7	MA20(i)	20-day Mov. Avg	19	Momentum(i)	Rate of price change
8	MA50(i)	50-day Mov. Avg	20	ROC(i)	Rate Of Change
9	MA100(i)	100-day Mov. Avg	21	K(i)	Stochastic oscillator K
10	MA200(i)	200-day Mov. Avg	22	D(i)	Stochastic oscillator
11	EMA5(i)	5-day Exp. Mov. Avg	23	RSI(i)	Relative Strength Index
12	EMA10(i)	10-day Exp. Mov. Avg	24	Close($i-1$)	Closing val. on ($i-1$)

One target value is then defined for each day i of the time series considered with a value between 0 and 1, corresponding to the operation that should be carried out by the automated trader on that day. Three possible operations are taken into account, and correspond to buy, sell and no operations.

4 Experiments and Results

The database for this automated trader application was created by considering all daily quotes and the 24 technical indicators described in Table 2 for the ordinary stock of the Italian car maker FIAT, which is traded at the Borsa Italiana stock exchange. All the daily data are related to the period from the 31th of March, 2003, through the 1st of December, 2006. Training, test and validation sets have been created by considering, respectively, the 66, 27 and 7% of all available data. All time series of the three datasets are preprocessed by considering a normal distribution with mean 0 and standard deviation set to 1, although no future knowledge has been considered in the dataset definition.

For each run of the evolutionary algorithm, up to 800,000 network evaluations (i.e., simulations of the network on the whole training set) have been allowed, including those performed by BP. The risk free rate r_f has been set for all runs to 0.0344 (i.e., the current ECB discount rate of 3.5%), while different settings have been also considered for the log-return values of the take profit, for both target output and trading simulator settings, in order to define the combination that better suits to the results obtained by the evolutionary process to provide the best automated trader. A round of experiments, summarized in Table 3, was aimed at determining whether using a different take-profit log-return for generating the target in the learning dataset used by BP than the one used for simulating and evaluating the trading rules could bring about any improvement of the results. For this reason the best individual found for each setting is saved, and the Sortino ratio and the correlated log-returns on the validation set are reported. The conclusion is that runs in which the take profit used to construct the target is greater than the actual target used by the strategy usually lead to more profitable trading rules, although exceptions can be found: as a matter of

Table 3. A comparison of validation Sortino ratio and log-return for different combinations of target and simulation take profits

Target Take Profit	Validation Set	Simulation Take Profit								
		0.0046			0.006			0.008		
		worst	avg	best	worst	avg	best	worst	avg	best
0.0046	SR_d	-0.6210	0.2149	1.2355	-0.3877	-0.0027	0.5294	-0.3605	0.0050	0.5607
	Log-Return	-0.3336	0.1564	0.6769	-0.1973	0.0628	0.3192	-0.1770	0.0225	0.3424
0.006	SR_d	-0.1288	-0.0303	0.0712	-0.0620	0.4035	1.2289	-0.0043	0.3863	1.2432
	Log-Return	0	0.0218	0.0737	0.0363	0.2559	0.6768	0.0320	0.3041	0.6773
0.008	SR_d	-0.0620	0.2137	0.7859	-0.0620	0.2803	0.9327	0.0395	0.3683	1.2237
	Log-Return	0	0.1484	0.4531	0	0.1850	0.5312	0.0562	0.2311	0.6695

fact the best combination turned out to be a simulation take profit of 0.008 and a target take profit of 0.006.

4.1 Tenfold Cross-Validation

To validate the generalization capabilities of the automated trading rules found by the approach, a tenfold cross-validation has been carried out as follows.

The training and test sets have been merged together into a set covering the period from March 31, 2003 to August 30, 2006, representing the 93% of the entire period, since no validation set has been considered in the merging process. That set has been divided into 10 equal-sized intervals. Each interval has been used in turn as the validation set, while the remaining 9 intervals have been used to create a training set (consisting of 6 intervals) and a test set (3 intervals). The 10 (training, test, and validation) datasets thus obtained have been used to perform 10 runs each of the evolutionary algorithm, for a total of 100 runs. The best performing take-profit settings found during the previous round of experiments have been used for all 100 runs, namely a target take profit of 0.006 and a simulation take profit of 0.008. Table 4 reports the average and standard deviation of the best fitness obtained during each run; the best individual found in each run was applied to its relevant validation set, and the resulting Sortino ratio and log-return are reported in the table.

It can be observed that the results are consistent for all ten dataset used in the cross-validation, with Sortino ratios often greater than one, meaning that the expected return outweighs the risk of the strategy. This is a positive indication of the generalization capabilities of this approach.

4.2 Discussion

The evolutionary approach presented in this work is elementary and minimalistic in two respects. First, the data considered each day by the trading agent to make a decision about its action is restricted to the open, low, high, and close quotes of the last day, plus the close quote of the previous day; any visibility of the rest of the past time series is filtered through a small number of popular and quite standard technical indicators. Second, the trading strategy an agent can follow is among the simplest and most accessible even to the unsophisticated

Table 4. 10-Fold validation: experimental results for different settings of the algorithm parameters

Setting	Parameter Setting			Take Profit Log-Return =0.008			
	p^+_{layer}	p^-_{layer}	p^+_{neuron}	f_{avg}	Std. Dev.	Log-Return	SR_d
1	0.05	0.05	0.05	0.1518	0.0101	0.7863	1.4326
	0.05	0.1	0.05	0.1491	0.0267	4840	0.8105
	0.1	0.1	0.05	0.1598	0.0243	0.2012	0.2899
	0.1	0.2	0.05	0.1502	0.0190	0.7304	1.3235
	0.2	0.05	0.05	0.1494	0.0256	0.7399	1.3307
	0.2	0.1	0.2	0.1645	0.0115	0.9002	1.6834
	0.2	0.2	0.2	0.1532	0.0129	0.3186	0.5037
2	0.05	0.05	0.05	0.1892	0.0251	0.9920	1.8991
	0.05	0.1	0.05	0.1768	0.0293	0.6362	1.1229
	0.1	0.1	0.05	0.1786	0.0342	0.9726	1.8707
	0.1	0.2	0.05	0.1749	0.0274	0.7652	1.3965
	0.2	0.05	0.05	0.1765	0.0188	0.7391	1.3340
	0.2	0.1	0.2	0.1864	0.0210	0.5362	0.9355
	0.2	0.2	0.2	0.1799	0.0306	0.7099	1.2751
3	0.05	0.05	0.05	0.1780	0.0290	0.6655	1.2008
	0.05	0.1	0.05	0.1790	0.1003	1.0537	2.0371
	0.1	0.1	0.05	0.1880	0.1364	0.8727	1.6569
	0.1	0.2	0.05	0.1858	0.1683	0.3767	0.6347
	0.2	0.05	0.05	0.1894	0.1363	0.5434	0.9474
	0.2	0.1	0.2	0.1845	0.1013	0.6544	1.1784
	0.2	0.2	0.2	0.1840	0.1092	0.6882	1.2551
4	0.05	0.05	0.05	0.1909	0.0315	0.7800	1.4484
	0.05	0.1	0.05	0.2026	0.0234	0.8400	1.5745
	0.1	0.1	0.05	0.1866	0.0300	0.9303	1.7543
	0.1	0.2	0.05	0.1831	0.0293	0.8194	1.5384
	0.2	0.05	0.05	0.2011	0.0379	1.0497	2.0417
	0.2	0.1	0.2	0.2212	0.0283	0.5146	0.8842
	0.2	0.2	0.2	0.1923	0.0340	0.9758	1.8647
5	0.05	0.05	0.05	0.2520	0.0412	0.6825	1.2430
	0.05	0.1	0.05	0.2237	0.0245	0.5403	0.9552
	0.1	0.1	0.05	0.2213	0.0327	0.4932	0.8545
	0.1	0.2	0.05	0.2169	0.0331	0.4748	0.8201
	0.2	0.05	0.05	0.2295	0.0416	0.5796	1.0335
	0.2	0.1	0.2	0.2364	0.0316	0.4449	0.7644
	0.2	0.2	0.2	0.2200	0.0287	0.5799	1.0251
6	0.05	0.05	0.05	0.1932	0.0478	0.3892	0.6407
	0.05	0.1	0.05	0.2183	0.0339	0.8521	1.5979
	0.1	0.1	0.05	0.2303	0.0312	0.6858	1.2407
	0.1	0.2	0.05	0.2094	0.0444	0.6375	1.1418
	0.2	0.05	0.05	0.2168	0.0268	0.6776	1.2254
	0.2	0.1	0.2	0.2320	0.0445	0.8312	1.5614
	0.2	0.2	0.2	0.2186	0.0495	0.3634	0.6671
7	0.05	0.05	0.05	0.2020	0.0268	0.5196	0.8740
	0.05	0.1	0.05	0.2171	0.0227	0.6727	1.1781
	0.1	0.1	0.05	0.2081	0.0184	0.9178	1.7002
	0.1	0.2	0.05	0.2042	0.0381	0.6905	1.2214
	0.2	0.05	0.05	0.2050	0.0375	0.6653	1.1619
	0.2	0.1	0.2	0.2187	0.0235	0.8449	1.5530
	0.2	0.2	0.2	0.2153	0.0353	0.8321	1.5207
8	0.05	0.05	0.05	0.2109	0.0370	0.3534	0.5823
	0.05	0.1	0.05	0.2018	0.0460	0.6068	1.0832
	0.1	0.1	0.05	0.1845	0.0369	0.8938	1.6860
	0.1	0.2	0.05	0.1956	0.0338	0.5846	1.0402
	0.2	0.05	0.05	0.2172	0.0300	0.4137	0.6968
	0.2	0.1	0.2	0.1856	0.0365	0.5516	0.9617
	0.2	0.2	0.2	0.1888	0.0366	0.8130	1.5059

Table 4. *(continued)*

9	0.05	0.05	0.05	0.2010	0.0263	0.9420	1.7953
	0.05	0.1	0.05	0.1997	0.0252	0.2538	0.4051
	0.1	0.1	0.05	0.2007	0.0312	0.7444	1.3792
	0.1	0.2	0.05	0.2300	0.0373	0.8998	1.6987
	0.2	0.05	0.05	0.2170	0.0429	0.7192	1.3175
	0.2	0.1	0.2	0.2252	0.0248	0.9606	1.8470
	0.2	0.2	0.2	0.1930	0.0441	0.9813	1.8860
10	0.05	0.05	0.05	0.2161	0.0168	0.4443	0.7558
	0.05	0.1	0.05	0.2017	0.0312	0.8144	1.5233
	0.1	0.1	0.05	0.2154	0.0333	0.8133	1.5007
	0.1	0.2	0.05	0.2138	0.0424	0.9079	1.7118
	0.2	0.05	0.05	0.2079	0.0230	0.6604	1.1749
	0.2	0.1	0.2	0.2063	0.0288	0.6148	1.0804
	0.2	0.2	0.2	0.2113	0.0323	0.7083	1.2787

individual trader; its practical implementation does not even require particular kinds of information-technology infrastructures, as it could very well be enacted by placing a couple of orders with a broker on the phone before the market opens; there is no need to monitor the market and react in a timely manner. Nonetheless, the results clearly indicate that, despite its simplicity, such an approach may yield, if carried out carefully, significant returns, in the face of a risk that is, to be sure, probably higher than the one the average investor would be eager to take, but all in all proportionate with the returns expected. By simulating individual trading rules, we observed that the signal given by the neural network is correct most of the times; when it is not, the loss (or draw-down) tends to be quite severe — which is the main reason for the low Sortino ratios exhibited by all the evolved rules. Table 5 shows the history of a simulation of a trading strategy combining by majority vote the seven best trading rules discovered when using a target take-profit of 0.006 and a simulation take-profit of 0.008 on the validation set. Only days in which the majority took a position are shown.

Table 5. A simulation of the strategy dictated by the majority of the best trading rules (one for each of the 7 parameter settings of Table 4) obtained by running the algorithm with a combination of target take-profit of 0.006 and a simulation take-profit of 0.008. The action in the Action column refers to the (short)-selling or buying of FIAT. The r column shows the log-return realized by the strategy for each day. Only days in which the strategy took a position are shown.

Date	Output of the best ANN found with each parameter setting							Action	r
	1	2	3	4	5	6	7		
09/08/2006	0.37289	0.30866	0.36189	0.19916	0.11058	0.34688	0.23212	Buy FIAT	-0.00087
09/11/2006	0.37288	0.17984	0.24793	0.036435	0.42759	0.25281	0.088654	Buy FIAT	0.00605
09/12/2006	0.37287	0.13193	0.22707	0.025921	0.17347	0.19494	0.079639	Buy FIAT	0.008
09/13/2006	0.37287	0.00572	0.14981	-0.0524	0.11109	0.16788	0.073001	Buy FIAT	0.008
09/14/2006	0.33293	0.19561	0.38577	0.030935	0.47942	0.39281	0.095375	Buy FIAT	0.008
09/26/2006	0.37292	0.33825	0.39093	0.18484	0.35414	0.29081	0.095379	Buy FIAT	0.008
10/18/2006	0.37287	-0.06376	0.086698	-0.07738	0.42974	-0.09006	0.069613	Buy FIAT	0.008
10/19/2006	0.37287	-0.12859	-0.08951	-0.08207	0.28453	-0.07239	0.0692	Buy FIAT	0.008
11/02/2006	0.37287	0.33471	0.42151	0.25141	0.09108	0.36179	0.07453	Buy FIAT	0.008
11/03/2006	0.37287	0.064807	0.39822	-0.0423	0.089449	0.18745	0.069529	Buy FIAT	0.008
11/06/2006	0.37287	0.22927	0.42214	0.06885	0.26745	0.3251	0.07085	Buy FIAT	0.008
11/07/2006	0.37287	0.21738	0.42549	0.095818	0.33432	0.39917	0.071313	Buy FIAT	0.008

This case shows how with FIAT instruments few operations are enough to gain a profit from the strategy. Indeed, the compounded return is really attractive: the annualized log-return of the above strategy, obtained with target take-profit of 0.006 and actual take-profit of 0.008, corresponds to a percentage annual return of 39.87%, whereas the downside risk is almost zero (0.003).

5 Conclusion and Future Work

An application of a neuro-genetic algorithm to the optimization of very simple trading agents for static, single-position, intraday trading has been described. The approach has been validated on the trading of stock of Italian car maker FIAT. Experimental results indicate that, despite its simplicity, both in terms of input data and in terms of trading strategy, such an approach to automated trading may yield significant returns. Furthermore, this opens up the opportunity for many extensions, improvements, and sophistications both on the side of input data and indicators, and of the technicalities of the trading strategy.

References

1. Allen, F., Karjalainen, R.: Using genetic algorithms to find technical trading rules. Journal of Financial Economics 51, 245 (1999)
2. Azzini, A., Tettamanzi, A.: A neural evolutionary approach to financial modeling. In: Proceedings of the Genetic and Evolutionary Computation Conference, GECCO 2006, vol. 2, pp. 1605–1612. Morgan Kaufmann, San Francisco (2006)
3. Azzini, A., Tettamanzi, A.: Neuro-genetic single position day trading. In: Workshop Italiano di Vita Artificiale e Computazione Evolutiva, WIVACE 2007 (2007)
4. Brabazon, A., O'Neill, M.: Biologically Inspired Algorithms for Financial Modelling. Springer, Heidelberg (2006)
5. Cliff, D.: Explorations in evolutionary design of online auction market mechanisms. Electronic Commerce Research and Applications 2, 162 (2003)
6. Colby, R.: The Encyclopedia Of Technical Market Indicators, 2nd edn. McGraw-Hill, New York (2002)
7. Dempster, M., Jones, C.: A real-time adaptive trading system using genetic programming. Quantitative Finance 1(4), 397–413 (2001)
8. Dempster, M., Jones, C., Romahi, Y., Thompson, G.: Computational learning techniques for intraday fx trading using popular technical indicators. IEEE Transactions on Neural Networks 12(4) (2001)
9. Harris, L.: Trading and Exchanges, Market Microstructure for Practitioners. Oxford University Press, New York (2003)
10. Hayward, S.: Evolutionary artificial neural network optimisation in financial engineering. In: Proceedings of the Fourth International Conference on Hybrid Intelligent Systems, HIS 2004, IEEE Computer Society Press, Los Alamitos (2004)
11. Sharpe, W.: The Sharpe ratio. Journal of Portfolio Management 1, 49–58 (1994)
12. Skabar, A., Cloete, I.: Neural networks, financial trading and the efficient markets hypothesis. In: Proceedings of the Australian Computer Science Conference 2002, vol. 4, pp. 241–249. Australian Computer Science Inc. (2002)

13. Sortino, F., der Meer, R.V.: Downside risk – capturing what's stake in investment situations. Journal of Portfolio Management 17, 27–31 (1991)
14. Subramanian, H., Ramamoorthy, S., Stone, P., Kuipers, B.: Designing safe, profitable automated stock trading agents using evolutionary algorithms. In: Proceedings of the Genetic and Evolutionary Computation Conference, GECCO 2006, vol. 1, pp. 1777–1784. Morgan Kaufmann, San Francisco (2006)
15. Yao, X.: Evolutionary Optimization. Kluwer Academic Publishers, Norwell, Massachusetts (2002)
16. Yao, X., Liu, Y.: A new evolutionary system for evolving artificial neural networks. IEEE Transactions on Neural Networks 8(3), 694–713 (1997)

Genetic Programming in Statistical Arbitrage

Philip Saks and Dietmar Maringer

Centre for Computational Finance and Economic Agents, University of Essex

Abstract. This paper employs genetic programming to discover statistical arbitrage strategies on the banking sector in the Euro Stoxx universe. Binary decision rules are evolved using two different representations. The first is the classical single tree approach, while the second is a dual tree structure where evaluation is contingent on the current market position. Hence, buy and sell rules are co-evolved. Both methods are capable of discovering significant statistical arbitrage strategies.

1 Introduction

During the last decades, the *Efficient Market Hypothesis* (EMH) has been put to trial, especially with the emergence of behavioral finance and agent-based computational economics. The rise of the above mentioned fields provides a theoretical justification for attacking the EMH, which in turn stimulates the empirical forecasting literature.

A basic premise for efficiency is the existence of *homo economicus*, that the markets consists of *homogeneous rational* agents, driven by utility maximization. However, cognitive psychology has revealed that people are far from rational, instead they rely on *heuristics* in decision making to simplify a given problem [15]. This is both useful and necessary in everyday life, but in certain situations it can lead to biases such as *overconfidence*, *base-rate neglect*, *sample-size neglect*, *gamblers fallacy*, *conservatism* and *aversion to ambiguity* [3]. However, what is more important is that these biases manifest themselves on an aggregate level in the markets as momentum and mean-reversion effects [12, 7]. Accepting the existence of *heterogeneous* agents, have pronounced effects on a theoretical level. In such a scenario the market clearing price cannot be determined formally, since agents need to form expectations about other agents' expectations etc. This leads to an "infinite regress in subjectivity", where no agent irrespective of reasoning powers, can form expectations by deductive means, thus perfect rationality is not well-defined. Instead, investors are forced to hypothesize expectational models, where the only means of verification is to observe the models' performance in practice [2]. In such a world it is indeed sensible to develop expectational models beyond traditional equilibrium analysis which we seek to accomplish using *genetic programming* (GP).

The majority of existing applications of GP in financial forecasting have for some reason focused on foreign exchange. Here, the general consensus is that GP can discover profitable trading rules at high frequencies in presence of transaction costs [14, 8, 6]. For the stock market results are mixed. [1] do not outperform

M. Giacobini et al. (Eds.): EvoWorkshops 2008, LNCS 4974, pp. 73–82, 2008.

the buy-and-hold strategy on daily S&P500 data, while [5] do on a monthly frequency. Besides changing the frequency, they reduce the grammar and consider a cooperative co-evolution scheme, where buy and sell rules are evaluated separately.

In this paper, we consider genetic programming for statistical arbitrage. Arbitrage in the traditional sense is concerned with identifying situations where a self-funding is generated that will provide only non-negative cash flows at any point in time. Obviously, such portfolios are possible only in out-of-equilibrium situations. Statistical arbitrage is a wider concept where, again, self-funding portfolios are sought where one can expect non-negative pay outs at any point in time. Here, however, one accepts negative pay-outs with a small probability as long as the expected positive payouts are high enough and the probability of losses is small enough; ideally this shortfall probability converges to zero. In practice, such a situation can occur when price processes are closely linked. In the classical story of Royal Dutch and Shell [3], the pair of stocks are cointegrated since they are fundamentally linked via their merger in 1907. In most cases, however, such links are not as obvious, but that does not eliminate the possibility that such relationships might exist and can be detected by statistical analysis. Since it can be argued that these stocks are exposed to many of the same risk factors and should therefore have similar behavior, this paper considers stocks within the same industry sector.

We shall construct so-called arbitrage portfolios, where the proceedings from short selling some stocks are used to initiate long positions in other stocks. This scheme has several advantages, firstly, it is self-financing and second the profits made from this strategy are in excess of the risk-free rate and virtually uncorrelated with the market index. Furthermore, by modeling the relationships between stocks, we are not trying to predict the future, rather we focus the attention in a direction where more stable patterns should exist.

The rest of the paper is organized as follows. Section 2 provides evidence of significant clustering between sectors within the Euro Stoxx universe. Section 3 introduces the data, model and framework. Section 4 presents results under the assumptions of frictionless trading. Finally, conclusions are drawn and pointers are given to future research in Section 5.

2 Clustering of Financial Data

Previously we hypothesized that stocks within the same industry sector are exposed to many of the same risk factors and should therefore have similar behavior. In order to clarify this, we investigate the majority of stocks in the Euro Stoxx 600 index. The data is gathered from Bloomberg and includes information such as company name, ticker symbol, industry sector and industry group. In addition hereto, we obtain the adjusted closing prices in the period from 21-Jan-2002 to 26-Jun-2007. Since the index composition is changing over time, we only consider stocks where data exists for the last two years for both price and volume series. Taking this into account, the universe comprises of a total of 477 stocks.

The notion that stocks have similar behavior needs to be specified in order to conduct a proper analysis. An obvious measure for price data is the correlation of returns, where a higher correlation implies stronger similarity. To test the hypothesis that stocks within the same sector tend to be clustered together we employ the *k-means algorithm* to construct statistical clusters. Thus, if the statistical clustering is independent of the fundamental clustering, dictated by the industry sectors, then we must reject our hypothesis. Let, S and F be two stochastic variables, which describe the statistical and fundamental clusters.[1] The maximum number of clusters is denoted by the integers k_s and k_f, respectively. Define s_i (f_i) as the statistical (fundamental) cluster asset i belongs to, and $I_{\{s_i=j\}}$ $(I_{\{f_i=m\}})$ be a binary indicator which is 1 if this statistical (fundamental) cluster is equal to j (m) and 0 otherwise. Then, for a universe of N stocks,

$$V_{j,m} = \sum_{i=1}^{N} I_{\{s_i=j\}} \cdot I_{\{f_i=m\}} \quad \forall \quad j = 1, 2, \ldots, k_s \quad m = 1, 2, \ldots, k_f \qquad (1)$$

is the number of stocks that, at the same time, belong both to statistical cluster j and the fundamental factor m. The hypothesis of independence can be tested via a χ^2-statistic for contingency tables. Setting $k_s = k_f = 10$, we obtain a test statistic $\chi^2 = 1314.1$ where the critical value is $\chi^2_{0.05}(81) = 103.0$, thus strongly rejecting the null hypothesis of independence.

The analysis above proves that there are significant clustering within the sectors, and in the following we shall focus on the actual statistical arbitrage application.

3 The Framework

As mentioned previously, the objective is to develop a trading strategy for statistical arbitrage based on price and volume information, and in the following we elaborate on data, preprocessing and model construction.

3.1 Data

The data comprises Volume-Weighted Average Prices (VWAP) and volume, sampled on an hourly frequency for the banks in the Euro Stoxx 600 index. It covers the time period from 01-Apr-2003 to 29-Jun-2007, corresponding to a total of 8648 observations. Again, we only consider stocks for which we have enough data, which limits the portfolio to 30 assets.

When analyzing high frequency data it is important to take intraday effects into account, e.g., the intraday volume is higher after open and before close than during the middle of the day [11]. In the context of trading rule induction it is important to remove this bias, which is basically a proxy for the time of day, and prohibits sensible conditioning on intraday volume.

[1] The statistical clusters are obtained by using k-means on the correlation matrix, i.e., the data comprises of N observations in N dimensional space.

3.2 Preprocessing

The return series for each stock is standardized with respect to its volatility, estimated using simple exponential smoothing. Likewise, a volume indicator is constructed that removes the intraday bias, and measures the extent to which the level is lower or higher than expected. Specifically, we take the logarithm of the ratio between the realized and expected volume, where this ratio has been hard limited in the range between 0.2 and 5.

Since we are interested in cross-sectional relationships between stocks, rather than their direction, we subtract the cross-sectional average from the normalized returns and volume series for each stock. Based on these series, we calculate the moving averages over the last 8, 40 and 80 periods, corresponding to one day, a week and two weeks on the hourly frequency. All indicators are scaled in the range between 0 and 1.

3.3 Model

There are two approaches for modeling trading rules, either as decision trees where market positions or actions are represented in the terminal nodes [16], or as a single rule where the conditioning is exogenous to the program [6].

We consider the latter approach in the context of a binary decision problem, which corresponds to long and short positions. As mentioned previously, we are interested in arbitrage portfolios, where the purchase of stocks is financed by short selling others. Naturally, a precondition for this to be achieved is that not all the forecasts across the 30 stocks are the same, e.g., if the trading rule take a bullish view across the board, then short-selling opportunities have not been identified and proper arbitrage portfolios cannot be constructed. In this case we do not hold any stocks. However, when forecasts facilitate portfolio construction, this is done on a volatility adjusted basis. Let $o_t^i \in \{-1, 1\}$ denote the forecast on stock i at time t, then the holding is given,

$$h_t^i = o_t^i \frac{\frac{1}{\sigma_t^i}}{\sum_{j=1}^{n} \frac{1}{\sigma_t^j} I_{\{o_t^j = o_t^i\}}} \tag{2}$$

where σ_t is the volatility, n is the number of stocks in the universe, and $I_{\{o_t^j = o_t^i\}}$ is an indicator variable that ensures that forecasts are normalized correctly, i.e., it discriminates between long and short positions.

We employ two different methods for solving the binary decision problem. The first uses a standard single tree structure, while the second follows [4] and considers a dual structure in conjunction with cooperative co-evolution. In both methods, the trees return boolean values. For the dual structure, program evaluation is contingent on the current market position for that particular stock, i.e., the first tree dictates the long entry, while the second enters a short position. More formally, let $b_t^{j,i} \in \{0, 1\}$ be the the truth value for tree j on stock i at time t, then the forecast is given as, **if** $o_{t-1}^i < 0$ **then** $o_t^i = 2 \cdot b_t^{1,i} - 1$ **else** $o_t^i = -2 \cdot b_t^{2,i} + 1$ **end**.

In both settings, the programs are constructed from the same grammar which is fairly restricted. It consists of numeric comparators ($<$, $>$), boolean operators (`AND`, `OR`, `XOR`, `NOT`) and *if-then-else* statements (`ITE`). Furthermore, we have introduced a special function `BTWN`, that takes three arguments and evaluates if the first is between the second and third. The terminals comprises the six indicators and numerical constants ranging from 0 to 1. The parsimonious grammar reduces the risk of overfitting, and enhances interpretability of the evolved solutions.

3.4 Objective Function

The choice of objective function is essential in evolutionary computation. It has previously been found that a risk-adjusted measure improves out-of-sample performance, relative to an absolute return measure [6]. In this context the Sharpe Ratio is an obvious candidate. However, using this measure it is possible to evolve strategies that do extremely well only on a subset of the in-sample data and mediocre on the remainder.

Instead we employ the *t-statistic* of the linear fit between cumulated returns and time, since it maximizes the slope while minimizing the deviation from the ideal straight line performance graph.

3.5 Parameter Settings

In the following experiments we consider a population size of 250 individuals, initialized using the *ramped half-and-half* method. It evolves for a maximum of 51 generations, but is stopped after 15 generations if no new *best-so-far* individual has been found. We use normal tournament selection with a size of 5, and the crossover and mutation probabilities are 0.9 and 0.1, respectively. Moreover, the probability of selecting a function node during reproduction is 0.9, and the programs are constrained to a maximum complexity of 50 nodes.

The data is split into a training and test set. The former contains 6000 samples and covers the period from 01-Apr-2003 to 10-Mar-2006, and the latter has 2647 samples in the period from 13-Mar-2006 to 29-Jun-2007.

4 Empirical Results

In this section we assume the absence of market impact, i.e., that it is possible to execute on the realized VWAP. Trading on VWAP differs from a traditional market order, where a trade is executed at the current observed price. Contrary, the VWAP is a backward looking measure, and it is therefore not possible to trade on the observed VWAP at time t. Instead the execution occurs gradually between t and $t+1$, resulting in the VWAP at $t+1$. In summary, a trading decision is formed based on the VWAP at time t, the entry price is observed at time $t+1$ and the one period return is evaluated at $t+2$.

We perform 10 experiments using both the single and dual tree method, according to the settings outlined in Section 3. For each experiment, the *best-so-far* individual is evaluated on the training and test set. Figure 1 shows the

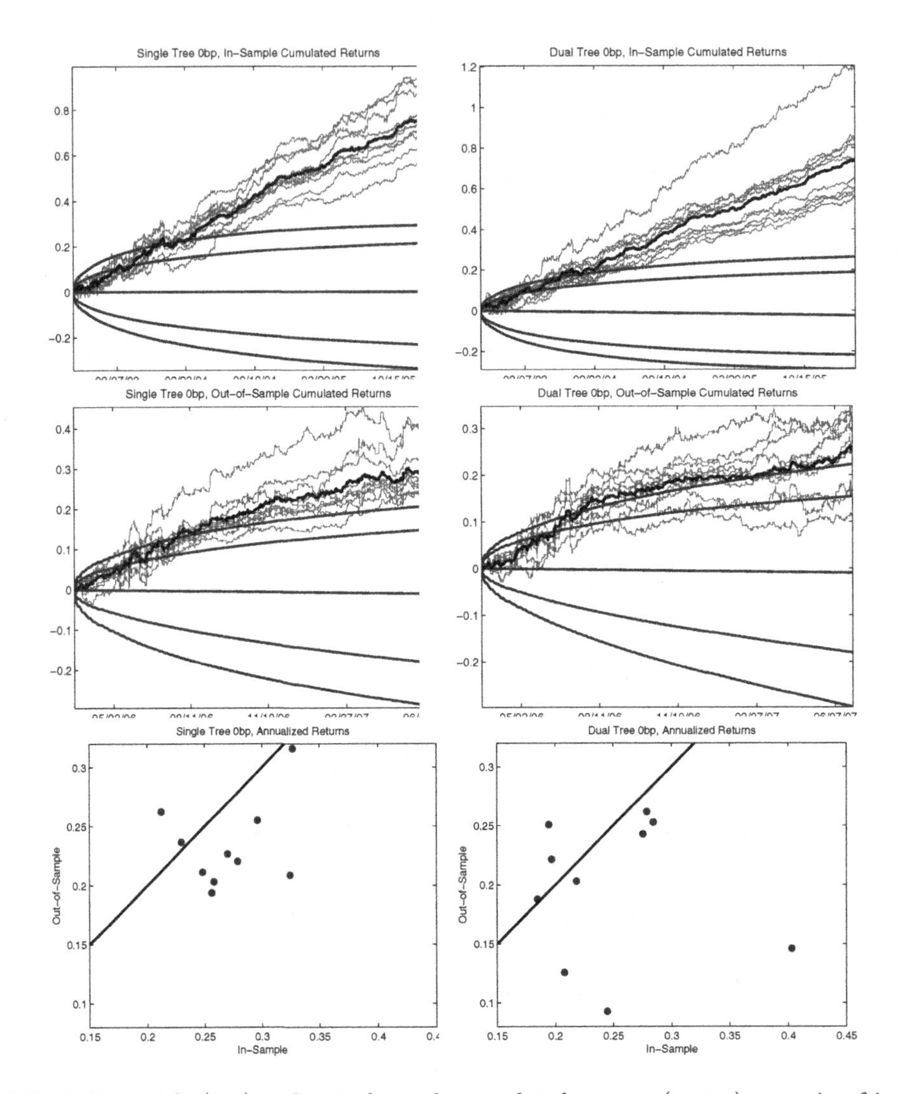

Fig. 1. In-sample (top) and out-of-sample cumulated returns (center) assuming frictionless trading. The thick lines are the average performances and the 95% and 99% confidence intervals are constructed using the stationary bootstrap procedure. Annualized in-sample versus out-of-sample returns with the 45°-line (bottom). The left and right column are the single and dual tree results, respectively.

performance for the single and dual tree method. Judging from the in-sample results, the *t-statistic* measure works as intended, and all experiments have steady increasing cumulative returns. The average annualized in-sample returns are 27.0% and 24.9%, for the single and dual tree method, respectively. As a proxy for generalization, it is instructive to consider the *shrinkage* which is defined,

$$\psi = \frac{X_{\text{train}} - X_{\text{test}}}{X_{\text{train}}} \tag{3}$$

where X is an arbitrary performance measure [6]. Both methods generalize extremely well out-of-sample, and the shrinkage is only 0.12 and 0.16, corresponding to annualized returns of 23.3% and 19.9%.

In order to asses the significance of these results, confidence intervals are constructed using the *stationary bootstrap* method, which is a superior alternative to well known *block bootstrap* procedure [17]. Instead of using a fixed block size, it varies probabilistically according to a geometric distribution.[2] Thus, sampling with replacement is performed from the holdings, and statistics are gathered from 500 bootstrap runs. The 99% upper confidence limits are 16.1% and 17.7% for the single and dual tree method, respectively. For the single tree method all ten experiments exceed the limit, while for the dual tree method it is only 7. From a risk adjusted perspective, there is also significant performance. The average annualized out-of-sample Sharpe Ratio is 2.83 and 2.60, where the 99% confidence limits are 2.17 and 2.41, for the single and dual tree method, respectively.

Positive out-of-sample returns need not imply market inefficiency. Traditionally, this is investigated by comparing the trading strategy to the buy-and-hold strategy. This, however, is not a suitable benchmark for statistical arbitrage strategies. Firstly, because a statistical arbitrage is self-financing and the buy-and-hold is not. This could be addressed by considering the returns of the buy-and-hold in excess of the risk-free rate, but this is a naive approach contingent on a specific equilibrium model.[3] Instead, we employ a special statistical test for statistical arbitrage strategies where this is not the case [10]. Hence, it circumvents the joint hypothesis problem, that abnormal returns need not imply market inefficiency, but can be due to misspecification of a given equilibrium model [9]. The constant mean version of the test assumes that the discounted[4] incremental profits satisfy,

$$\Delta v_i = \mu + \sigma i^{\lambda} z_i \qquad i = 1, 2, \ldots, n \tag{4}$$

where $z_i \sim N(0,1)$. The joint hypothesis, H1 : $\mu > 0$ and H2 : $\lambda < 0$ determines the presence of statistical arbitrage.

In order to make some general inferences, we consider an aggregation or bagging of the evolved trading strategies in the following.[5] Table 1 contains the test

[2] With the probability parameter $p = 0.01$, blocks with an expected length of 100 samples are generated.

[3] For the interested reader, the benchmark of an equally weighted portfolio of the banking stocks in excess of the risk free rate obtains an annualized return of 4.47% and has a Sharpe Ratio of 0.35 in the out-of-sample period.

[4] As a discount rate we employ the 1-month LIBOR rate for the Eurozone.

[5] All 10 evolved strategies are aggregated, but one could employ various schemes to improve out-of-sample performance. Generally, this requires the use of additional validation sets, but since data is limited this is problematic. Moreover, aggregating all strategies is clearly the conservative approach and is therefore preferred.

Table 1. Statistical arbitrage test results

	μ ($\cdot 10^{-4}$)	σ	λ	H1	H2	H1+H2 ($\cdot 10^{-6}$)
Single Tree	1.1321	0.0018	-0.0471	0.0000	0.0000	3.75
Dual Tree	0.9256	0.0020	-0.0922	0.0000	0.0000	4.78

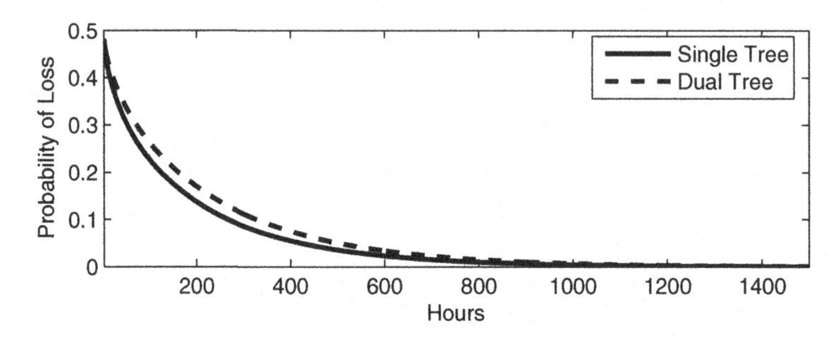

Fig. 2. Probability of loss for the bagged statistical arbitrage strategies as a function of trading time

results for the bagged models, where p-values for the joint hypothesis is obtained via the *Bonferroni* inequality. Both methods discover highly significant statistical arbitrage dynamics at all usual levels of significance.[6] Figure 2 depicts the probability of a loss as a function of trading time, and after 422 and 502 hours it is less than 0.05 for the single and dual tree, respectively. This demonstrates the essence of statistical arbitrage, that a riskless profit is earned in the limit.

5 Conclusion

In this paper genetic programming is employed to evolve trading strategies for statistical arbitrage. This is motivated by the fact that stocks within the same industry sector should be exposed to the same risk factors and should therefore have similar behavior. This certainly applies to the Euro Stoxx universe, where we find evidence of significant clustering.

Traditionally there has been a gap between financial academia and the industry. This also applies to statistical arbitrage, an increasingly popular investment style in practice, but to the authors' knowledge little formal research has been undertaken within this field. This paper addresses this imbalance and aims to narrow this gap. We consider two different representations for the trading rules. The first is a traditional single tree structure, while the second is a dual tree structure in which evaluation is contingent on the current market position. Hence, buy and sell rules are co-evolved. Both methods have substantial market timing and discovers significant statistical arbitrage strategies. However, in a

[6] For the individual strategies, 9/10 and 7/10 have significant performance on a 0.01 level of significance, for the single and dual tree method, respectively.

frictionless environment we cannot conclude that this violates the efficient market hypothesis, since "prices reflect information to the point where the marginal benefits of acting on information (the profits to be made) do not exceed the marginal costs" [13].

Having assumed a frictionless environment it is important to asses the implications of such an assumption. Naturally, were these strategies to be implemented in practice a cost would be associated with trading, but it is not a serious as one might think. When closing prices or point estimates of the price are considered, trading is associated with crossing the bid-ask spread, and thereby incurring a cost ranging from 2bp to as much as 40bp depending on the liquidity and the price level of the stock. In addition to the bid-ask spread, *slippage* occurs if the order size is large relative to the depth of the market. This paper, however, does not use closing prices, but instead proprietary VWAP's. The advantage of using these series is that algorithmic trading actually permits these prices to be obtained within 1bp for moderate order sizes.[7] Naturally, the assumption of a complete frictionless environment is an idealization. However, preliminary research suggest that positive returns can be generated under realistic market impact, but more work is needed.

References

[1] Allen, F., Karjalainen, R.: Using genetic algorithms to find technical trading rules. Journal of Financial Economics 51, 245–271 (1999)

[2] Arthur, B., Holland, J.H., LeBaron, B., Palmer, R., Tayler, P.: Asset pricing under endogenous expectations in an artificial stock market, Technical report, Santa Fe Institute (1996)

[3] Barberis, N., Thaler, R.: Handbook of the Economics of Finance, pp. 1052–1090. Elsevier Science, Amsterdam (2003)

[4] Becker, L.A., Seshadri, M.: Cooperative coevolution of technical trading rules, Technical report, Department of Computer Science, Worcester Polytechnic Institute (2003a)

[5] Becker, L.A., Seshadri, M.: GP–evolved technical trading rules can outperform buy and hold, Technical report, Department of Computer Science, Worcester Polytechnic Institute (2003b)

[6] Bhattacharyya, S., Pictet, O.V., Zumbach, G.: Knowledge–intensive genetic discovery in foreign exchange markets. IEEE Transactions on Evolutionary Computation 6(2), 169–181 (2002)

[7] Bondt, W.F.M.D., Thaler, R.: Does the stock market overreact. The Journal of Finance 40(3), 793–805 (1985)

[8] Dempster, M.A.H., Jones, C.M.: A real-time adaptive trading system using genetic programming. Quantitative Finance 1, 397–413 (2001)

[9] Fama, E.F.: Market efficiency, long-term returns, and behavioral finance. Journal of Financial Economics 49, 283–306 (1998)

[10] Hogan, S., Jarrow, R., Teo, M., Warachaka, M.: Testing market efficiency using statistical arbitrage with applications to momentum and value strategies. Journal of Financial Economics 73, 525–565 (2004)

[7] Lehman Brothers Equity Quantitative Analytics, London.

[11] Jain, P.C., Joh, G.-H.: The dependence between hourly prices and trading volume. Journal of Financial Economics 23(3), 269–283 (1988)
[12] Jegadeesh, N., Titman, S.: Returns to buying winners and selling losers: Implications for stock market efficiency. Journal of Finance 48(1), 65–91 (1993)
[13] Jensen, M.C.: Some anomalous evidence regarding market efficiency. Journal of Financial Economics 6, 95–101 (1978)
[14] Jonsson, H., Madjidi, P., Nordahl, M.G.: Evolution of trading rules for the FX market or how to make money out of GP, Technical report, Institute of Theoretical Physics, Chalmers University of Technology (1997)
[15] Kahneman, D., Tversky, A.: Judgment under uncertainty: Heuristics and biases. Science 185(4157), 1124–1131 (1974)
[16] Li, J.: FGP: a genetic programming based tool for financial forecasting, PhD thesis, University of Essex (2001)
[17] Politis, D.N., Romano, J.P.: The stationary bootstrap. Jounal of the American Statistical Association 89(428), 1303–1313 (1994)

Evolutionary System for Generating Investment Strategies

Rafał Dreżewski and Jan Sepielak

Department of Computer Science
AGH University of Science and Technology, Kraków, Poland
drezew@agh.edu.pl

Abstract. The complexity of generating investment strategies problems makes it hard (or even impossible), in most cases, to use traditional techniques and to find the strict solution. In the paper the evolutionary system for generating investment strategies is presented. The algorithms used in the system (evolutionary algorithm, co-evolutionary algorithm, and agent-based co-evolutionary algorithm) are verified and compared on the basis of the results coming from experiments carried out with the use of real-life stock data.

1 Introduction

Investing on the stock market requires analyzing of the great number of strategies (which security should be chosen, when it should be bought or sold). Accurate analysis is important during predicting and choosing the optimal investment strategy. It plays an important role in a future success. Majority of the investment decisions are based on present and historical data. The trend anticipation depends on many assumptions, parameters and conditions. Consideration of so many assumptions, combinations of parameters and their values leads to the comparison of the great number of graphs. The evaluation of parameters of many securities is difficult and time consuming for the investor and the analyst. As a result, the investor or the analyst is able to analyze only the small subset of the possible strategies, so the optimal investment strategy is usually not found [9].

The set of the strategies which consists of indicator function is infinite because the complexity of the strategy can be unlimited. Formulas of the given strategy are functions of hundreds (or thousands) of parameters. Complexity of the problem makes it impossible to use direct search methods and instead of it a heuristic approach must be used. For example, it is possible to apply here *evolutionary algorithms* because there exist many solutions to the problem and finding optimal solutions is not necessary—suboptimal solutions are usually sufficient for an investor.

Evolutionary algorithms are optimization and search techniques, which are based on the Darwinian model of evolutionary processes [2]. One of the branches of evolutionary algorithms are *co-evolutionary algorithms* [7]. The general difference between them is the way in which the fitness of the individual is evaluated. In the case of evolutionary algorithms the fitness of the individual depends only on how "good" is the solution of the given problem encoded within its genotype. In the case of co-evolutionary algorithms

M. Giacobini et al. (Eds.): EvoWorkshops 2008, LNCS 4974, pp. 83–92, 2008.

the fitness of the individual depends on the values of other individuals' fitness. The value of fitness is usually based on the results of tournaments, in which the given individual and some other individuals from the population are engaged. Co-evolutionary algorithms are generally applicable in the cases in which it is difficult or even impossible to formulate explicit fitness function. Co-evolutionary interactions between individuals have also other positive effects on the population—for example maintaining population diversity, which is one of the biggest problems in the case of "classical" evolutionary algorithms.

Agent-based (co-)evolutionary algorithms are the result of research on decentralized models of evolutionary computations. The basic idea of such approach is the realization of evolutionary processes in multi-agent system, which leads to very interesting class of systems: (co-)evolutionary multi-agent systems—(Co)EMAS [3]. Such systems have some features which radically differ them from "classical" evolutionary algorithms. The most important of them are the following: synchronization constraints of the computations are relaxed because the evolutionary processes are decentralized (individuals are agents), there exist the possibility of constructing hybrid systems using many different computational intelligence techniques within single, coherent agent architecture, there are possibilities of introducing new evolutionary and social mechanisms, which were hard or even impossible to introduce in the case of classical evolutionary algorithms. (Co)EMAS systems have been already applied to multi-modal optimization and multi-objective optimization problems. Another area of applications is the modeling and simulation of social and economical phenomena.

In the paper the component system for generating investment strategies is presented. In the system three algorithms were implemented: "classical" evolutionary algorithm, co-evolutionary algorithm, and agent-based co-evolutionary algorithm. These algorithms were compared during the series of experiments, which results conclude the paper.

2 Previous Research on Evolutionary Algorithms for Generating Investment Strategies

During last years there can be observed the growing interest in applying biologically inspired algorithms to economic and financial problems. Below, only selected applications of evolutionary algorithms in systems supporting investment decision making are presented.

S. K. Kassicieh, T. L. Paez and G. Vora used the genetic algorithm for supporting the investment decisions making [5]. Their algorithm operated on historical stock data. The tasks of the algorithm included selecting company to invest in. The time series of the considered companies were given. In their system some logical operations were carried out on the data. The genetic algorithm was used to determine, which logical operators should be applied in a given situation.

O. V. Pictet, M. M. Dacorogna, R. D. Dave, B. Chopard, R. Schirru and M. Tomassini ([6]) presented the genetic algorithm for the automatic generation of trade models represented by financial indicators. Three algorithms were implemented: genetic algorithm (it converged to local minima and had the poor capability of generalization), genetic algorithm with fitness sharing technique developed by X. Yin and N. Germay [10] (it explored the search space more effectively and had more ability to find diverse optima), and genetic algorithm with fitness sharing technique developed by authors themselves in order to prevent the concentration of the individuals around "peaks" of fitness function (it had the best capability of generalization). The proposed algorithms selected parameters for indicators and combined them to create new, more complex ones.

F. Allen and R. Karjalainen ([1]) used genetic algorithm for finding trading rules for S&P 500 index. Their algorithm could select the structures and parameters for rules. Each rule was composed of a function organized into a tree and a returned value (signal), which indicated whether stocks should be bought or sold at a given price. Components of the rules were the following: functions which operated on historical data, numerical or logical constants, logical functions which allowed for combining individual blocks in order to build more complicated rules. Function in the root always returned logical value, which ensured the correctness of the strategy. Fitness measure was based on excess return from the buy-and-hold strategy, however the return did not excess the transaction cost.

3 Evolutionary System for Generating Investment Strategies

In this section the architecture of the component system for generating investment strategies is presented. Also, the three algorithms (evolutionary algorithm, co-evolutionary algorithm, and agent-based co-evolutionary algorithm) used as computational components are described.

3.1 The Architecture of the System

Fig. 1 shows the system's architecture model. The system has the following basic components:

- *DataSource*—supplies the data to the strategy generator. Historical sessions' stock data are used.

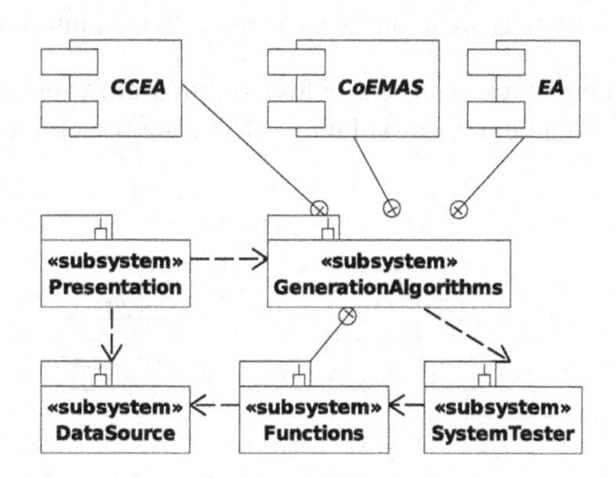

Fig. 1. The architecture of the system used in experiments

- *Functions*—contains all classes, which are necessary for creating formulas of strategies. It allows to carry out basic operations on formulas such as: initialization, exchange of single functions, adding new functions or removing existing ones. Formulas can be tested on data and, in such a case, the results will be returned.
- *SystemTester*—allows to test strategies. It can prepare reports concerning the transactions and containing the information about the gained profit. It is used by the generation algorithms to estimate the fitness.
- *GenerationAlgorithms*—contains implementation of three algorithms which generate strategies: evolutionary (EA), co-evolutionary (CCEA) and co-evolutionary multi-agent (CoEMAS). This component includes the mutation and recombination operators and also the fitness estimation mechanisms.
- *Presentation*—contains the definitions of GUI forms, which are used to monitor the generation algorithms and present the results.

3.2 The Algorithms

In all three algorithms the strategy is a pair of formulas. First formula indicates when one should enter the market and the second indicates when one should exit the market. Each formula, which is a part of strategy can by represented as a tree, which leaves and nodes are functions performing some operations. Such tree has a root and any quantity of child nodes. The function placed in the root always returns logical value. Fig. 2 shows the tree of the formula, which can be symbolically written in the following way: $STE(WillR(20),30) > 10.0$.

Formula tree is represented in memory as a tree. The root node is an object which contains the references to the functions and the references to parameters. These parameters are also the same objects as the root. The leaves of the tree are the objects which do not contain parameters. When formula is executed recursive calls occur. In the beginning, the root requires values of all parameters needed to invoke its function. Then the control flows to objects of the parameters. The parameters objects behave in the same way as the root object. Leaves do not contain parameters, so they can return the value required by the parent node.

Functions (which formulas are composed of) were divided into four categories:

- Functions returning stock data, e.g. Close (returns close prices), Open (returns opening prices), Volume (returns volumes). There are 6 such functions.

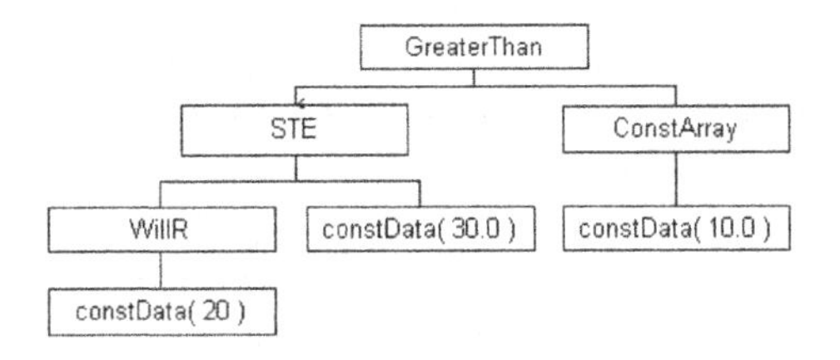

Fig. 2. Tree of exemplary formula

- Mathematical functions, e.g. Sin (sine function), Add (the addition), Pow (the power function). There are 40 such functions.
- Tool functions, e.g. Cross (finds a cross point of two functions), STE (calculates standard error), Outside (finds outside days). There are 33 of them.
- Indicator functions, e.g. AD (calculates the Accumulation/Distribution indicator), MOV (calculates diverse moving averages), WillR (calculates Williams' %R indicator). There are 14 of them.

There are 93 functions altogether. These functions accept the following types of parameters: constants (integer, float or enum), array of constant float values, and values returned by other function (array of logical values or array of float values). Logical constant does not exist because it was not needed for formula building.

Evolutionary Algorithm (EA). In the evolutionary algorithm genetic programming was used. The genotype of individuals simultaneously contains formula for entering and exiting the market. Estimation of the fitness of the strategies is carried out on the loaded historical data. Two tables of logical values are created as a result of execution of the formulas. The first one relates to the purchase action (entering the market) and the second one relates to the sale action (exiting the market). Profit computing algorithm processes tables and determines when a purchase and a sale occur. Entering the market occurs when a system is outside the market and there is the value of "true" in the enter table. Exiting the market occurs when the system is on the market and there is value of "false" in the exit table. In the other cases no operation is performed. The profit/loss from the given transaction is estimated when exiting the market and it is accumulated. The cost of each transaction is included—the commission is calculated by subtracting a certain constant from the transaction value.

Apart from the profit/loss there are also other criteria, which are included in the fitness estimation. The first one is the formula complexity—too complex formulas can slow down the computations and increase the memory usage. The complexity of the formulas is determined through the summing up of all component functions. The second criteria is the length of the transaction—it depends on the preference of the investor.

Three kinds of recombination operator are used: *returned value* recombination, *arguments* recombination, and *function* recombination. *Returned value* recombination is performed when there are two functions with different arguments but the same returned values within the formula tree. These functions are exchanged between individuals (functions are moved with their arguments). *Arguments* recombination occurs when there are two functions with the same arguments within the parents. These arguments are exchanged between individuals during the recombination. *Function* recombination can take place when two functions have the same arguments and the same returned values.

Two types of mutation were used: *function arguments* mutation and *function* mutation. In *function argument* mutation the argument of the function must be a constant. This constant is exchanged for the other one coming from the allowed range. There are three variants of the *function* mutation:

1. Traverse on functions. If a given function should be mutated, a list of the functions which take the same argument is found. If such functions exist, the exchange is performed. If there are no such functions, mutation is not carried out.

2. A given function can be exchanged for the other, completely new one, regardless of arguments—only returned types must match. Arguments of such function are created in the random manner.
3. Similarly as in (2), but, if it is possible, parameters of the replaced function are copied to a new one.

The tournament reproduction ([2]) was used in the EA algorithm. After creating the offspring, it was added to the population of parents (the reinsertion mechanism). From such enlarged population the new base population was chosen also with the use of the tournament mechanism.

Co-Evolutionary Algorithm (CCEA). While developing the co-evolutionary algorithm for generating investment strategies, co-operative approach proposed by M. A. Potter and K. A. De Jong ([8]) was used. There are two species in the implemented algorithm: individuals representing entering the market strategies and individuals representing exiting the market strategies. Interaction between these species relies on co-operation. During the fitness estimation process individuals are selected into pairs, which form the complete solution. In the first generation, for each evaluated individual from the first species a partner for co-operation from the second species is chosen randomly. For the complete solution created in this way, the fitness is computed and assigned to the individual that is being evaluated. In the next generations, the best individual from the opposite species is chosen for the evaluated individual. The pairs of individuals gaining the best profit are the result of the algorithm.

Co-Evolutionary Multi-Agent System (CoEMAS). Co-evolutionary multi-agent system used in the experiments is the agent-based realization of the co-evolutionary algorithm. Its general principles of functioning are in accordance with the general model of co-evolution in multi-agent system [3]. The CoEMAS system is composed of the environment (which include computational nodes—islands—connected with paths) and agents, which can migrate within the environment. The selection mechanisms is based on the resources, which are defined in the system. The general rule is such, that in the each time step the environment gives more resources to "better" agents and less resources to "worse" agents. The agents use the resources for every activity, like migration, reproduction, and so on. Each time step, individuals lose some constant amount of the possessed resources (the agents can live more than one generation), which is given back to the environment. The agents make all their decisions independently—especially those concerning reproduction and migration. They can also communicate with each other and observe the environment.

In the CoEMAS algorithm realized in the system for generating investment strategies each co-evolutionary algorithm (CCEA) is an agent which is located on one of the islands and independently carries out the computations. The population of each co-evolutionary algorithm also consists of the agents (there are two species of agents within each population, like in the case of CCEA).

The method of the reinsertion is different than the one used in the previous algorithms, and now works on the basis of resources. The agent which has the greater amount of resource wins the tournament. Also, on the basis of the amount of the possessed resource each individual decides whether it is ready for the reproduction. During the recombination parents give the offspring certain quantity of their resources. The possibility of migration of the agent-individual from one population to another was added as well.

4 The Experiments

In order to examine the generalization capabilities of the system and compare the proposed algorithms, strategies which earn the largest profit per year for random stocks were sought. An attempt was made to determine, which algorithm generalizes in a best way and what quantity of stocks should be used for strategies generation so that the system would not overfit. It is also interesting which algorithm generates the best strategies and has the smallest convergence.

4.1 Plan of the Experiments

The presented results of the experiments comparing the quality of the generated strategies and convergence properties of the considered algorithms are average values from 30 runs of the algorithms. Each algorithm was run for 500 generations on the data of 10 randomly chosen stocks. The session stock data came from the WIG index ([4]) and the period of 5 years was chosen (from 2001-09-29 to 2006-09-29). The size of the population in all the algorithms was equal to about 40 individuals (CoEMAS approach uses variable-size populations).

All experiments were made with the use of optimal values of the parameters. These values were found during consecutive experiments. The algorithms were run 10 times for each parameter value coming from the established range and average results were computed—on this basis the set of best parameters' values was chosen.

While examining the generalization capabilities, each algorithm generated the solution for n random stocks (stage 1). Next, different n stocks were chosen ten times at random and the profit was calculated using the best strategy obtained in the stage 1. Then, the average of these profits was counted. These calculations were carried out four times for $n = 2, 5, 7, 10$.

Like in the first type of experiments (when the quality of the solutions and the convergence properties were compared), populations had similar sizes in the case of all three algorithms. All experiments were carried out on the machine with one AMD Sempron 2600+ processor.

4.2 Results

Fig. 3 shows the average fitness (from 30 experiments) of the best individuals for each generation. Presented results show that the evolutionary algorithm achieved the best results. The co-evolutionary algorithm achieved a little bit worse results. The quality of the solution generated by the CoEMAS was close to that of the CCEA.

Fig. 4 presents the plot of the convergence. The convergence is the phenomenon of losing by the evolutionary algorithm the ability to search the solution spaces before finding a solution which would be a global optimum. It is manifested by the occurrence

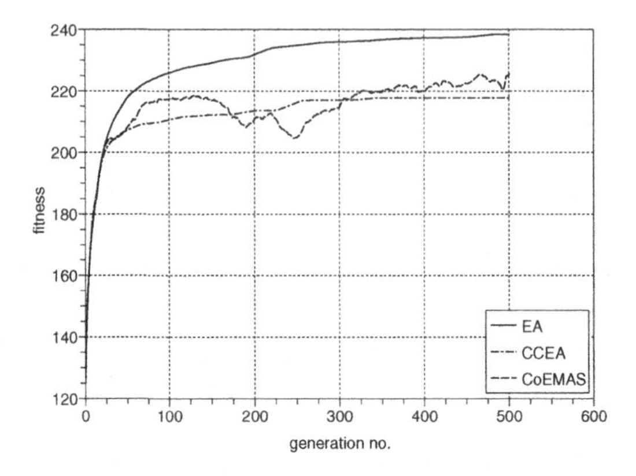

Fig. 3. Fitness of the best individuals (average values from 30 experiments)

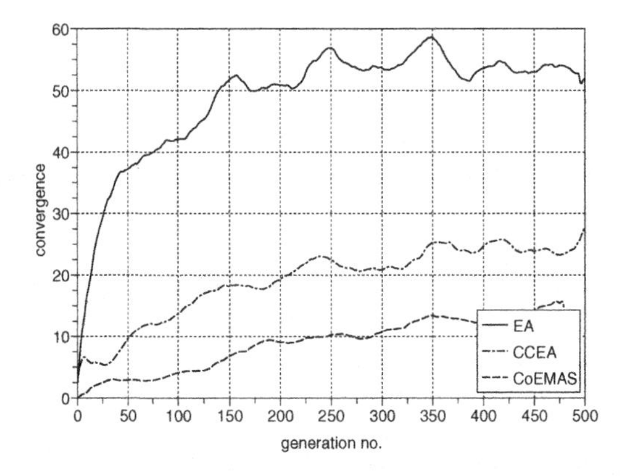

Fig. 4. Convergence values for three compared algorithms (average from 30 experiments)

of the pairs composed of identical individuals in the population. In the case of convergence the evolutionary algorithm had the worst results. Large convergence occurred already at the beginning and from 200 generation it was in the range from 50% to 60%. The co-evolutionary algorithm appeared to be much better. CoEMAS had the smallest convergence. For the last two algorithms convergence grew slowly from the beginning, but even in the end of the experiment it did not rise much.

In the table 1 the results of experiments, which goal was to investigate the capability of generalization of all compared algorithms are presented. The results show that while generating a strategy, at least 7 stocks should be used, so that the strategy could be used on any stock and earn profits in any situation (see tab. 1). If there are more stocks used during the strategy generation, the profit will be greater in the case of random stocks. For random stocks, when there was 3 or 5 of them in the group, the profits are varied and unstable. For this reason it is difficult to compare implemented algorithms with *buy and hold* strategy. It is not so, when the number of stocks in the group is 7 or 10. In the case of the random stocks, buy and hold strategy was always better (on average 2.67

Table 1. Generalization capabilities of the algorithms

Algorithm	No. of stocks in the group	Profit (%) per year (stocks from the group)	Profit (%) per year for buy & hold strategy (stocks from the group)	Profit (%) per year (random stocks)	Profit (%) per year for buy & hold strategy (random stocks)
EA	3	133.24	86.24	8.79	34.81
	5	89.15	28.85	1.8	33.62
	7	68.32	21.43	13.74	23.72
	10	87.57	24.29	20.81	26.07
CCEA	3	115.63	46.12	3.29	21.07
	5	64.82	8.83	7.39	31.36
	7	85.75	45.24	14.69	24.64
	10	69.06	26.37	20.75	25.14
CoEMAS	3	87.39	12.64	3.97	30.28
	5	66.97	7.64	-1.93	20.72
	7	86.93	39.48	19.01	34.47
	10	46.67	18.67	24.6	24.98

times) than the strategies generated by all three evolutionary algorithms, but for the stocks from the learning set generated strategies were always better (on average 1.45 times) than buy and hold strategy.

5 Summary and Conclusions

Generating investment strategies is generally very hard problem because there exist many assumptions, parameters, conditions and objectives which should be taken into consideration. In the case of such problems finding the strict solution is impossible in most cases and sub-optimal solution is usually quite sufficient for the decision maker. In such cases some (meta-)heuristic algorithms like biologically inspired techniques and methods can be used. In this paper the system for generating investment strategies, which uses three types of evolutionary algorithms was presented. The system can generate strategies with the use of "classical" evolutionary algorithm, co-evolutionary algorithm, and agent-based co-evolutionary algorithm. These algorithms were verified and compared with the use of real-life data coming from the WIG index.

The presented results show that evolutionary algorithm generated the individual (strategy) with the best fitness, the second was agent-based co-evolutionary algorithm, and the third co-evolutionary algorithm. When the population diversity (convergence) is taken into consideration, the results are quite opposite: the best was CoEMAS, the second CCEA, and the worst results were reported in the case of EA. Such observations generally confirm that co-evolutionary and agent-based co-evolutionary algorithms maintain population diversity much better than "classical" evolutionary algorithms. This can lead to stronger abilities of the population to "escape" from the local minima in the case of highly multi-modal problems.

When we consider the generalization capabilities (profit gained from 7 and 10 random stocks during one year) of the strategies generated with the use of each evolutionary algorithm, the best results were obtained by CoEMAS (21.8% profit on the average), the second was CCEA (on the average 17.7%), and the worst results were obtained in the case of EA (17.3% on the average). Implemented algorithms provide better results than buy and hold strategy for stocks from the learning set and worse results in the case of the random stocks.

The future research could concentrate on additional verification of the proposed algorithms, and on the implementation and testing of other co-evolutionary mechanisms—especially in the case of the most promising technique (CoEMAS).

References

1. Allen, F., Karjalainen, R.: Using genetic algorithms to find technical trading rules. Journal of Financial Economics 51(2), 245–271 (1999)
2. Bäck, T., Fogel, D., Michalewicz, Z. (eds.): Handbook of Evolutionary Computation. IOP Publishing and Oxford University Press, Oxford (1997)
3. Dreżewski, R.: A model of co-evolution in multi-agent system. In: Mařík, V., Müller, J.P., Pěchouček, M. (eds.) CEEMAS 2003. LNCS (LNAI), vol. 2691, Springer, Heidelberg (2003)
4. Historical stock data, http://www.parkiet.com/dane/dane_atxt.jsp
5. Kassicieh, S.K., Paez, T.L., Vora, G.: Investment decisions using genetic algorithms. In: Proceedings of the 30th Hawaii International Conference on System Sciences, IEEE Computer Society, Los Alamitos (1997)
6. Pictet, O.V., et al.: Genetic algorithms with collective sharing for robust optimization in financial applications. Technical Report OVP.1995-02-06, Olsen & Associates (1995)
7. Paredis, J.: Coevolutionary algorithms. In: Bäck, T., Fogel, D., Michalewicz, Z. (eds.) Handbook of Evolutionary Computation, 1st supplement, IOP Publishing and Oxford University Press, Oxford (1998)
8. Potter, M.A., De Jong, K.A.: Cooperative coevolution: An architecture for evolving coadapted subcomponents. Evolutionary Computation 8(1), 1–29 (2000)
9. Tertitski, L.M., Goder, A.G.: Method and system for visual analysis of investment strategies. US Patent 6493681 (December 2002)
10. Yin, X.: A fast genetic algorithm with sharing scheme using cluster analysis methods in multimodal function optimization. In: Forrest, S. (ed.) Proceedings of the Fifth International Conference on Genetic Algorithms, Morgan Kaufmann, San Francisco (1993)

Horizontal Generalization Properties of Fuzzy Rule-Based Trading Models

Célia da Costa Pereira and Andrea G.B. Tettamanzi

Università degli Studi di Milano
Dipartimento di Tecnologie dell'Informazione
via Bramante 65, I-26013 Crema, Italy
pereira@dti.unimi.it, andrea.tettamanzi@unimi.it

Abstract. We investigate the generalization properties of a data-mining approach to single-position day trading which uses an evolutionary algorithm to construct fuzzy predictive models of financial instruments. The models, expressed as fuzzy rule bases, take a number of popular technical indicators on day t as inputs and produce a trading signal for day $t+1$ based on a dataset of past observations of which actions would have been most profitable.

The approach has been applied to trading several financial instruments (large-cap stocks and indices), in order to study the *horizontal*, i.e., cross-market, generalization capabilities of the models.

Keywords: Data Mining, Modeling, Trading, Evolutionary Algorithms.

1 Introduction

Single-position automated day-trading problems (ADTPs) involve finding an automated trading rule for opening and closing a single position within a trading day. They are a neglected subclass of the more general automated intraday trading problems, which involve finding profitable automated technical trading rules that open and close positions within a trading day.

An important distinction that may be drawn is the one between static and dynamic trading problems. A *static* problem is when the entry and exit strategies are decided before or on market open and do not change thereafter. A *dynamic* problem allows making entry and exit decisions as market action unfolds.

Dynamic problems have been the object of much research, and different flavors of evolutionary algorithms (EAs) have been applied to the discovery and/or the optimization of dynamic trading rules (cf., e.g., [3].

Static problems are technically easier to approach, as the only information that has to be taken into account is information available before market open. This does not mean, however, that they are easier *to solve* than their dynamic counterparts.

This paper focuses on the generalization properties of the solutions to a class of static single-position automated day-trading problems found by means of a data-mining approach which uses an EA to construct a fuzzy predictive model of a

M. Giacobini et al. (Eds.): EvoWorkshops 2008, LNCS 4974, pp. 93–102, 2008.

financial instrument. The model takes the values of a number of popular technical indicators computed on day t as inputs to produce a *go short*, *do nothing*, *go long* trading signal for day $t+1$ based on a dataset of past observation of which actions would have been most profitable.

2 Evaluating Trading Rules

Informally, we may think of a trading rule R as some sort of decision rule which, given a time series $X = \{x_t\}_{t=1,2,\ldots,N}$ of prices of a given financial instrument, for each time period t returns some sort of trading signal or order.

Following the financial literature on investment evaluation [2], the criteria for evaluating the performance of trading rules, no matter for what type of trading problem, should be measures of risk-adjusted investment returns. The reason these are good metrics is that, in addition to the profits, consistency is rewarded, while volatile patterns are not.

While the Sharpe ratio [7] is probably the most popular measure of risk-adjusted returns for mutual funds and other types of investments, it has been criticized for treating positive excess returns, i.e., windfall profits, the same way as it treats negative returns; however, traders, just like investors, do not regard windfall profits as something to avoid as unexpected losses. A variation of the Sharpe ratio which acknowledges this fact is Sortino ratio [8], which may be defined as

$$\mathrm{SR}_d(R;X) = \frac{r(R;X) - r_f}{\mathrm{DSR}_{r_f}(R;X)}, \tag{1}$$

where $r(R;X)$ is the annualized average log-return of rule R applied to time series X, r_f is the risk-free rate r_f, assumed to be constant during the timespan covered by X, and $\mathrm{DSR}_\theta(R;X) = \sqrt{\frac{Y}{N}\sum_{t=1}^{N} \min\{0, \theta - r(R;X,t)\}^2}$ is called the downside risk [4,8] of rule R on X. Unlike the Sharpe ratio, the Sortino ratio adjusts the expected return for the risk of falling short of the risk-free return; positive deviations from the least acceptable return θ are not taken into account to calculate risk.

3 The Trading Problem

We focus on a particular class of static ADTP, whereby the trading strategy allows taking both long and short positions at market during the opening auction, a position is closed as soon as a pre-defined profit r_{TP} has been reached, or otherwise at market during the closing auction as a means of preventing losses beyond the daily volatility of an instrument.

Such problems make up the simplest class of problems when it comes to rule evaluation: all is required is open, high, low, and close quotes for each day, since a position is opened at market open, if the rule so commands, and closed either with a fixed profit or at market close.

A trading rule for this static problem has just to provide a ternary decision: *go short*, *do nothing*, or *go long*.

Given time series $X = \{x_t^O, x_t^H, x_t^L, x_t^C\}_{t=1,\ldots,N}$, of daily open, high, low, and close quotes, the log-return generated by rule R in the tth day of time series X is

$$r(R; X, t) = \begin{cases} r_{TP} & \text{if signal is } \textit{go long} \text{ or } \textit{short} \text{ and } \bar{r} > r_{TP}, \\ s \ln \frac{x_t^C}{x_t^O} & \text{if signal is } \textit{go long} \text{ or } \textit{short} \text{ and } \bar{r} \leq r_{TP}, \\ 0 & \text{otherwise,} \end{cases} \quad (2)$$

where $\bar{r} = \ln \frac{x_t^H}{x_t^O}$ for a long position, and $\bar{r} = \ln \frac{x_t^O}{x_t^L}$ for a short one.

This problem is therefore among the most complex single-position day-trading problems whose solutions one can evaluate when disposing only of open, high, low, and close quotes for each day. The reason we chose to focus on such problem is indeed that while such kind of quotes are freely available on the Internet for a wide variety of securities and indices, more detailed data can in general only be obtained for a fee.

We approach this problem by evolving trading rules that incorporate fuzzy logic. The adoption of fuzzy logic is useful in two respects: first of all, by recognizing that concept definitions may not always be crisp, it allows the rules to have what is called an *interpolative behavior*, i.e., gradual transitions between decisions and their conditions; secondly, fuzzy logic provides for linguistic variables and values, which make rules more natural to understand for an expert.

4 The Approach

Data mining is a process aimed at discovering meaningful correlations, patterns, and trends between large amounts of data collected in a dataset. A model is determined by observing past behavior of a financial instrument and extracting the relevant variables and correlations between the data and the dependent variable. We describe below a data-mining approach based on the use of EAs, which recognize patterns within a dataset, by learning models represented by sets of fuzzy rules.

4.1 Fuzzy Models

A model is described through a set of fuzzy rules, made by one or more antecedent clauses ("IF ...")and a consequent clause ("THEN ..."). Clauses are represented by a pair of indices referring respectively to a variable and to one of its fuzzy sub-domains, i.e., a membership function.

Using fuzzy rules makes it possible to get homogenous predictions for different clusters without imposing a traditional partition based on crisp thresholds, that often do not fit the data, particularly in financial applications. Fuzzy decision rules are useful in approximating non-linear functions because they have a good interpolative power and are intuitive and easily intelligible at the same time.

Their characteristics allow the model to give an effective representation of the reality and simultaneously avoid the "black-box" effect of, e.g., neural networks.

The intelligibility of the model is useful for a trader, because understanding the rules helps the user to judge if a model can be trusted.

4.2 The Evolutionary Algorithm

The described approach incorporates an EA for the design and optimization of fuzzy rule-based systems originally developed to learn fuzzy controllers [9,6], then adapted for data mining, [1].

A model is a rule base, whose rules comprise up to four antecedent and one consequent clause each. Input and output variables are partitioned into up to 16 distinct linguistic values each, described by as many membership functions. Membership functions for input variables are trapezoidal, while membership functions for the output variable are triangular.

Models are encoded in three main blocks:

1. a set of trapezoidal membership functions for each input variable; a trapezoid is represented by four fixed-point numbers, each fitting into a byte;
2. a set of symmetric triangular membership functions, represented as an area-center pair, for the output variable;
3. a set of rules, where a rule is represented as a list of up to four antecedent clauses (the IF part) and one consequent clause (the THEN part); a clause is represented by a pair of indices, referring respectively to a variable and to one of its membership functions.

An island-based distributed EA is used to evolve models. The sequential algorithm executed on every island is a standard generational replacement, elitist EA. Crossover and mutation are never applied to the best individual in the population.

The recombination operator is designed to preserve the syntactic legality of models. A new model is obtained by combining the pieces of two parent models. Each rule of the offspring model can be inherited from one of the parent models with probability 1/2. When inherited, a rule takes with it to the offspring model all the referred domains with their membership functions. Other domains can be inherited from the parents, even if they are not used in the rule set of the child model, to increase the size of the offspring so that their size is roughly the average of its parents' sizes.

Like recombination, mutation produces only legal models, by applying small changes to the various syntactic parts of a fuzzy rulebase.

Migration is responsible for the diffusion of genetic material between populations residing on different islands. At each generation, with a small probability (the migration rate), a copy of the best individual of an island is sent to all connected islands and as many of the worst individuals as the number of connected islands are replaced with an equal number of immigrants.

A detailed description of the algorithm and of its genetic operators can be found in [6].

4.3 The Data

In principle, the modeling problem we want to solve requires finding a function which, for a given day t, takes the past history of time series X up to t and produces a trading signal *go short*, *do nothing*, or *go long*, for the next day.

Instead of considering all the available past data, we try to take advantage of *technical analysis*, an impressive body of expertise used everyday by practitioners in the financial markets, which is about summarizing important information of the past history of a financial time series into few relevant statistics. The idea is then to reduce the dimensionality of the search space by limiting the inputs of the models we look for to a collection of the most popular and time-honored technical analysis statistics and indicators.

For lack of space, we cannot give here mathematical definitions for the indicators used, and we refer the interested reader to specialized publications [5].

After a careful scrutiny of the most popular technical indicators, we concluded that more data were needed if we wanted an EA to discover meaningful models expressed in the form of fuzzy IF-THEN rules. Combinations of statistics and technical indicators are required that mimic the reasonings analysts and traders carry out when they are looking at a technical chart, comparing indicators with current price, checking for crossings of different graphs, and so on.

Combinations may take the form of differences between indicators that are pure numbers or that have a fixed range, or of ratios of indicators such as prices and moving average, that are expressed in the unit of measure of a currency. Following the use of economists, we consider the natural logarithm of such ratios, and we define the following notation: given two prices x and y, we define

$$x : y \equiv \ln \frac{x}{y}. \tag{3}$$

Eventually, we came up with the following combinations:

- all possible combinations of the Open (O), High (H), Low (L), Close (C), and previous-day Close (P) prices: $O : P$, $H : P$, $L : P$, $C : P$, $H : O$, $C : O$, $O : L$, $H : L$, $H : C$, $C : L$;
- close price compared to simple and exponential moving averages, $C : \text{SMA}_n$, $C : \text{EMA}_n$, $n \in \{5, 10, 20, 50, 100, 200\}$;
- the daily changes of the close price compared to simple and exponential moving averages, $\Delta(C : \text{SMA}_n)$, $\Delta(C : \text{EMA}_n)$, where $\Delta(x) \equiv x(t) - x(t-1)$;
- the MACD histogram, i.e., MACD – signal, and the daily change thereof, $\Delta(\text{Histogram})$;
- Fast stochastic oscillator minus slow stochastic oscillator, $\%K - \%D$, and the daily change thereof, $\Delta(\%K - \%D)$.

The full list of the statistics, technical indicators, and their combinations used as model inputs is given in Table 1.

4.4 Fitness

Modeling can be thought of as an optimization problem, where we wish to find the model M^* which maximizes some criterion which measures its accuracy in

Table 1. The independent variables of the dataset

Name	Formula	Explanation
Open	x_t^O	the opening price on day t
High	x_t^H	the highest price on day t
Low	x_t^L	the lowest price on day t
Close	x_t^C	the closing price on day t
Volume	x_t^V	the volume traded on day t
O:P	$x_t^O : x_{t-1}^C$	opening price on day t vs. previous-day closing price
H:P	$x_t^H : x_{t-1}^C$	high on day t vs. previous-day closing price
L:P	$x_t^L : x_{t-1}^C$	low on day t vs. previous-day closing price
C:P	$x_t^C : x_{t-1}^C$	close on day t vs. previous-day closing price
H:O	$x_t^H : x_t^O$	high on day t vs. same-day opening price
C:O	$x_t^C : x_t^O$	closing on day t vs. same-day opening price
O:L	$x_t^O : x_t^L$	opening price on day t vs. same-day lowest price
H:L	$x_t^H : x_t^L$	high on day t vs. same-day low
H:C	$x_t^H : x_t^C$	high on day t vs. same-day closing price
C:L	$x_t^C : x_t^L$	closing price on day t vs. same-day low
dVolume	$x_t^V : x_{t-1}^V$	change in volume traded on day t
C:MAn	$x_t^C : \mathrm{SMA}_n(t)$	n-day simple moving averages, for $n \in \{5, 10, 20, 50, 100, 200\}$.
dC:MAn	$\Delta(x_t^C : \mathrm{SMA}_n(t))$	daily change of the above
C:EMAn	$x_t^C : \mathrm{EMA}_n(t)$	n-day exponential moving averages, for $n \in \{5, 10, 20, 50, 100, 200\}$.
dC:EMAn	$\Delta(x_t^C : \mathrm{EMA}_n(t))$	daily change of the above
MACD	$\mathrm{MACD}(t)$	Moving average convergence/divergence on day t
Signal	$\mathrm{signal}(t)$	MACD signal line on day t
Histogram	$\mathrm{MACD}(t) - \mathrm{signal}(t)$	MACD histogram on day t
dHistogram	$\Delta(\mathrm{MACD}(t) - \mathrm{signal}(t))$	daily change of the above
ROC	$\mathrm{ROC}_{12}(t)$	rate of change on day t
K	$\%K_{14}(t)$	fast stochastic oscillator on day t
D	$\%D_{14}(t)$	slow stochastic oscillator on day t
K:D	$\%K_{14}(t) - \%D_{14}(t)$	fast vs. slow stochastic oscillator
dK:D	$\Delta(\%K_{14}(t) - \%D_{14}(t))$	daily change of the above
RSI	$\mathrm{RSI}_{14}(t)$	relative strength index on day t
MFI	$\mathrm{MFI}_{14}(t)$	money-flow index on day t
AccDist	$\Delta(\mathrm{AccDist}(t))$	The change of the accumulation/distribution index on day t
OBV	$\Delta(\mathrm{OBV}(t))$	The change of on-balance volume on day t
PrevClose	x_{t-1}^C	closing price on day $t-1$

predicting $y_i = x_{im}$ for all records $i = 1, \ldots, N$ in the training dataset. The most natural criteria for measuring model accuracy are the mean absolute error and the mean square error.

One big problem with using such criteria is that the dataset must be *balanced*, i.e., an equal number of representative for each possible value of the predictive attribute y_i must be present, otherwise the underrepresented classes will end up being modeled with lesser accuracy. In other words, the optimal model would be very good at predicting representatives of highly represented classes, and quite poor at predicting individuals from other classes.

To solve this problem, we divide the range $[y_{\min}, y_{\max}]$ of the predictive variable into 256 bins. The bth bin, X_b, contains all the indices i such that

$$1 + \lfloor 255 \frac{y_i - y_{\min}}{y_{\max} - y_{\min}} \rfloor = b. \tag{4}$$

For each bin $b = 1, \ldots, 256$, it computes the mean absolute error for that bin

$$\text{err}_b(M) = \frac{1}{\|X_b\|} \sum_{i \in X_b} |y_i - M(x_{i1}, \ldots, x_{i,m-1})|, \tag{5}$$

then the total absolute error (TAE) as an integral of the histogram of the absolute errors for all the bins, $\text{tae}(M) = \sum_{b:\|X_b\| \neq 0} \text{err}_b(M)$. Now, the mean absolute error for every bin in the above summation counts just the same no matter how many records in the dataset belong to that bin. In other words, the level of representation of each bin (which, roughly speaking, corresponds to a class) has been factored out by the calculation of $\text{err}_b(M)$. What we want from a model is that it is accurate in predicting all classes, independently of their cardinality.

The fitness used by the EA is given by $f(M) = \frac{1}{\text{tae}(M)+1}$, in such a way that a greater fitness corresponds to a more accurate model.

5 Experiments

A desirable property for models is their capability of generalizing, i.e., correctly predicting other data than those used to discover them. There are two dimensions of generalization that might be of interest here:

1. a *vertical* dimension, which has to do with being able to correctly model the behavior of the financial instrument used for learning for a timespan into the future;
2. a *horizontal* dimension, which has to do with being able to correctly model the behavior of other financial instruments than the one used for learning: here we might be interested in applying the model to similar instruments (i.e., same sector, same market, same asset class) or to instruments taxonomically further away.

We have tested our approach with the specific aim of assessing its *horizontal* generalization properties. The reason why this type of generalization is desirable is that it would allow the user to trade "young" financial instruments, for which too few data are available, by using models trained on similar, but "older" financial instruments.

5.1 Experimental Protocol

The following financial instruments have been used for the experiments:

- the Dow Jones Industrial Average index (DJI);
- the Nikkei 225 index (N225);
- the common stock of Italian oil company ENI, listed since June 18, 2001 on the Milan stock exchange;
- the common stock of world's leading logistics group Deutsche Post World Net (DPW), listed since November 20, 2000 on the XETRA stock exchange;
- the common stock of Intel Co. (INTC), listed on the NASDAQ.

For all the instruments considered, three datasets of different length have been generated, in an attempt to gain some clues on how much historical data is needed to obtain a reliable model:

- a "long-term" dataset, generated from the historical series of prices since January 1, 2002 till December 31, 2006, consisting of 1,064 records, of which 958 are used for training and the most recent 106 are used for testing;
- a "medium-term" dataset, generated from the historical series of prices since January 1, 2004 till December 31, 2006, consisting of 561 records, of which 505 are used for training and the most recent 56 are used for testing;
- a "short-term" dataset, generated from the historical series of prices since January 1, 2005 till December 31, 2006, consisting of 304 records, of which 274 are used for training and the most recent 30 are used for testing;

Table 2. Summary of experimental results. Minimum, average, and maximum values are over the best models produced in ten independent runs of the island-based EA, when applied to the corresponding validation set.

Performance Measure	Dataset								
	Long-Term			Medium-Term			Short-Term		
	min	avg	max	min	avg	max	min	avg	max
Dow Jones Industrial Average									
Fitness	0.3297	0.3394	0.3484	0.3188	0.3327	0.3457	0.3183	0.3398	0.3671
Return*	0.1618	0.2303	0.4017	0.1062	0.2280	0.5503	0.0996	0.3225	0.5416
Sortino Ratio	1.5380	2.5572	4.7616	0.7642	2.7949	6.5557	0.7215	4.0799	6.9968
Nikkei 225									
Fitness	0.3211	0.3414	0.3651	0.3241	0.3418	0.3575	0.3205	0.3351	0.3529
Return*	−0.1467	−0.0006	0.2119	−0.1118	0.0006	0.1436	−0.1063	−0.0161	0.1040
Sortino Ratio	−1.9181	0.0782	4.1485	−1.5070	0.0253	2.1311	−1.9135	−0.1033	3.2197
ENI Stock									
Fitness	0.2459	0.3268	0.3500	0.2475	0.2907	0.3425	0.2402	0.2949	0.3277
Return*	−0.1389	0.0122	0.2120	−0.0856	0.0248	0.1547	−0.1936	−0.0372	0.2643
Sortino Ratio	−2.3274	−0.2751	3.0867	−2.4578	−0.1799	2.4460	−2.8959	−0.9655	3.2188
Deutsche Post World Net Stock									
Fitness	0.3182	0.3306	0.3451	0.3200	0.3342	0.3506	0.3118	0.3299	0.3403
Return*	−0.0607	0.0476	0.2646	−0.0246	0.0547	0.2480	0.0117	0.1169	0.2820
Sortino Ratio	−15.8114	−2.3809	10.5500	−15.8114	−0.1780	12.7425	−10.2067	0.0920	4.6700
Intel Co. Stock									
Fitness	0.2490	0.3050	0.3443	0.2433	0.2838	0.3658	0.2394	0.2665	0.3333
Return*	0.0247	0.1015	0.1669	0.0131	0.2254	0.4292	−0.0244	0.1252	0.3632
Sortino Ratio	−0.2467	0.8624	3.2520	−0.4569	2.9042	6.1129	−15.8114	−0.7107	3.4903

*) Annualized logarithmic return.

The validation dataset, in all cases, consists of records corresponding to the first half of 2007, which require a historical series starting from March 17, 2006 (200 market days before January 2, 2007) to be generated, due to the 200-day moving averages and their changes that need to be computed.

5.2 Results

For each combination of instrument and dataset, ten runs of the EA with four islands of size 100 connected according to a ring topology and with a standard parameter setting have been performed. Each run lasted as many generations as required to reach convergence, defined as no improvement for 100 consecutive generations. The results are summarized in Table 2.

A superior performance of models evolved against the short-term dataset can be noticed. That is an indication that market conditions change over time and a profitable trading model one year ago may not be profitable today. Furthermore, while very profitable models for the DJIA are found on average, performance is much less consistent on the other four instruments, probably due to the specific volatility patterns of the instruments considered.

6 Horizontal Generalization

In order to study *horizontal* generalization, the best performing models (in terms of their Sortino ratio) for each instrument have been applied to the other four. The results of this experiment are reported in Table 3.

From those results, we can draw the following conclusions: it appears that the DJI model has interesting generalization capabilities and performs well on N225, DPW, and ENI, but, surprisingly enough, fails on one of its components, namely

Table 3. Results of applying to an instrument models trained on another instrument. The instruments used for training the models (with an indication of the relevant trainin set) are in the rows; the instruments to which the models have been applied are in the columns.

Performance Measure	Validation Instrument				
	DJI	N225	DPW	ENI	INTC
Dow Jones Industrial Average, Short-Term, Run #1					
Annualized Log-Return	0.5416	0.1195	0.2768	0.0630	−0.0428
Sortino Ratio	6.9968	2.5756	4.4929	0.4528	−1.2073
Nikkei 225, Long-Term, Run #6					
Annualized Log-Return	0.2620	0.2119	0.1183	0.2653	0.0627
Sortino Ratio	3.6495	4.1485	0.6594	3.8662	0.1410
Deutsche Post World Net Stock, Medium-Term, Run #2					
Annualized Log-Return	0	0.0150	0.0688	0.1034	0.0011
Sortino Ratio	−15.8114	20.9006	12.7425	1.9863	−15.5462
ENI Stock, Short-Term, Run #3					
Annualized Log-Return	0.0135	0.1815	0.3626	0.2643	0.1601
Sortino Ratio	−11.6329	3.7711	4.3744	3.2188	1.1145
Intel Co. Stock, Medium-Term, Run #10					
Annualized Log-Return	0	−0.0207	0.2492	0.1177	0.1968
Sortino Ratio	−15.8114	−1.4857	2.4810	0.9493	6.1129

INTC; the other model trained on an index, N225, extends satisfactorily to all other instruments; models trained on stocks, instead, show poor generalization capabilities when it comes to modeling the two indices, but, with one exception, extend quite well to the other instruments of the same type.

7 Conclusions

An experimental test of the generalization capabilities of a fuzzy-evolutionary data-mining approach to static ADTPs has been performed. The results demonstrate that the idea of using high-performance models discovered for a financial instrument for trading other somehow related financial instruments is feasible, although with a grain of salt. As a matter of fact, evidence has been gathered that models trained on indices tend to perform well, with rare exceptions, on other indices and stocks, whereas models trained on individual stocks tend to perform well on other stocks, but poorly on indices.

Future work will involve, besides examining a larger data set with more runs, evolving the models on a heterogenous set of securities to boost the robustness and generalization capabilities of the evolved rules.

References

1. Beretta, M., Tettamanzi, A.: Learning fuzzy classifiers with evolutionary algorithms. In: Pasi, G., Bonarini, A., Masulli, F. (eds.) Soft Computing Applications, pp. 1–10. Physica Verlag, Heidelberg (2003)
2. Bodie, Z., Kane, A., Marcus, A.J.: Investments, 7th edn. McGraw-Hill, New York (2006)
3. Brabazon, A., O'Neill, M.: Biologically Inspired Algorithms for Financial Modelling. Springer, Berlin (2006)
4. Harlow, H.V.: Asset allocation in a downside-risk framework. Financial Analysts Journal, 30–40 (September/October 1991)
5. Kirkpatrick, C.D., Dahlquist, J.R.: Technical Analysis: The Complete Resource for Financial Market Technicians. FT Press, Upper Saddle River (2006)
6. Tettamanzi, A., Poluzzi, R., Rizzotto, G.G.: An evolutionary algorithm for fuzzy controller synthesis and optimization based on SGS-Thomson's W.A.R.P. fuzzy processor. In: Zadeh, L.A., Sanchez, E., Shibata, T. (eds.) Genetic algorithms and fuzzy logic systems: Soft computing perspectives, World Scientific, Singapore (1996)
7. Sharpe, W.F.: The Sharpe ratio. Journal of Portfolio Management 21(1), 49–58 (1994)
8. Sortino, F.A., van der Meer, R.: Downside risk — capturing what's at stake in investment situations. Journal of Portfolio Management 17, 27–31 (1991)
9. Tettamanzi, A.: An evolutionary algorithm for fuzzy controller synthesis and optimization. In: IEEE International Conference on Systems, Man and Cybernetics. IEEE Systems, Man, and Cybernetics Society, vol. 5/5, pp. 4021–4026 (1995)

Particle Swarm Optimization for Tackling Continuous Review Inventory Models

K.E. Parsopoulos[1,3], K. Skouri[2], and M.N. Vrahatis[1,3]

[1] Computational Intelligence Laboratory (CI Lab), Department of Mathematics, University of Patras, GR–26110 Patras, Greece
{kostasp,vrahatis}@math.upatras.gr
[2] Department of Mathematics, University of Ioannina, GR–45110 Ioannina, Greece
kskouri@uoi.gr
[3] University of Patras Artificial Intelligence Research Center (UPAIRC), University of Patras, GR–26110 Patras, Greece

Abstract. We propose an alternative algorithm for solving continuous review inventory model problems for deteriorating items over a finite horizon. Our interest focuses on the case of time–dependent demand and backlogging rates, limited or infinite warehouse capacity and taking into account the time value of money. The algorithm is based on Particle Swarm Optimization and it is capable of computing the number of replenishment cycles as well as the corresponding shortage and replenishment instances concurrently, thereby alleviating the heavy computational burden posed by the analytical solution of the problem through the Kuhn–Tucker approach. The proposed technique does not require any gradient information but cost function values solely, while a penalty function is employed to address the cases of limited warehouse capacity. Experiments are conducted on models proposed in the relative literature, justifying the usefulness of the algorithm.

1 Introduction

Inventory maintenance of deteriorating items is a major concern in the supply chain of business organizations, since many products undergo decay or deterioration over time. Deterioration and demand rates play a crucial role in such problems. For this purpose, relative models have been proposed in the literature for different deterioration rates [1,2,3]. Also, a constant demand rate is usually valid in the mature stage of a product's life cycle, while it can be linearly approximated in the growth and/or end stage of the life cycle. Such models with linearly time varying demand were studied initially in [4,5], while most recent trends are reported in [6,7].

Another important issue of inventory systems is the management of unsatisfied demand. Often, complete backlogging of unsatisfied demand is assumed. However, in practice, there are customers who are willing to wait and receive their orders at the end of shortage period, while others are not. To this extent, considerable attention has been paid in the last few years to inventory

M. Giacobini et al. (Eds.): EvoWorkshops 2008, LNCS 4974, pp. 103–112, 2008.

models with partial backlogging, where the backlogging rate can be modeled taking into account the behavior of customers [8, 9, 10]. In addition, the effects of inflation and time value of money are vital in practical environments especially in developing countries. Recently, Chern *et al.* [11] studied an inventory model for deteriorating items with time varying demand and partial backlogging, taking into account the time value of money. This model can be considered as a generalization of older models. Basic assumption of the model is the unlimited storage capacity. However, this assumption does not often hold in practice.

Particle Swarm Optimization (PSO) was introduced in 1995 as a stochastic population–based algorithm for numerical optimization by Eberhart and Kennedy [12, 13]. It belongs to the class of *swarm intelligence* algorithms, whose dynamics are based on principles that govern socially organized groups of individuals [14]. Up–to–date, PSO has received a lot of attention from researchers due to its efficiency in solving different problems in science and engineering [15,14,16,17].

This paper is devoted to the investigation of the efficiency of PSO on solving an extended version of the model of Chern *et al.* [11], where limited storage is also considered. The detection of replenishment cycles, replenishment instances and replenishment orders is required, while warehouse capacity constraints can be present. The underlying optimization problem is mixed–integer with the solutions having variable length as well as posing constraints on the magnitude of their components, since the ordering of time instances must be preserved. For this purpose, hard bounding constraints are posed on the search points, while a penalty function is employed to tackle capacity constraints. The workings of the proposed approach are illustrated on three test problems considered in [11,9].

The rest of the paper is organized as follows: Section 2 contains the necessary background information on the considered inventory models and PSO. Section 3 describes the proposed approach, while experimental results are reported in Section 4. The paper concludes in Section 5.

2 Background Information

In the following subsections we describe the basic concepts of the continuous review inventory model proposed in [11] as well as the PSO algorithm.

2.1 The Considered Review Inventory Model

The model under investigation is an extension of the model of Chern *et al.* [11], assuming that the storage can also have limited capacity. The selection of this model was based on its generality due to the time varying demand, deterioration and backlogging rates. Thus, it can be considered to include different previously proposed models as special cases. The assumptions under which the model is developed are:

Table 1. Notation used for the parameters of the model

Param.	Description
n	Number of replenishment cycles during the planning horizon.
s_i	Time at which shortage starts during the i–th cycle, $i = 1, 2, \ldots, n$.
t_i	Time at which the i–th replenishment is made, $i = 1, 2, \ldots, n$.
r	Discount rate.
i_1	Internal inflation rate, which is varied by the company operation status.
i_2	External inflation rate, which is varied by the social economical situation.
r_1	$r - i_1$, discount rate minus the internal inflation rate.
r_2	$r - i_2$, discount rate minus the external inflation rate.
c_0	Internal fixed purchasing cost per order.
c_p	External variable purchasing cost per unit.
c_{h_1}	Internal inventory holding cost per unit and per unit of time.
c_{h_2}	External inventory holding cost per unit and per unit of time.
c_{b_1}	Internal backlogging cost per unit and per unit of time.
c_{b_2}	External backlogging cost per unit and per unit of time.
c_{l_1}	Internal cost of lost sales per unit and per unit of time.
c_{l_2}	External cost of lost sales per unit and per unit of time.
W	Storage area or volume.

1. The planning horizon is finite and equal to H time units. The initial and final inventory levels during the planning horizon are both set to zero.
2. Replenishment is instantaneous (replenishment rate is infinite).
3. The lead–time is zero.
4. The on hand inventory deteriorates at time varying deterioration rate $\theta(t)$.
5. The demand rate at time $t \in [0, H]$, is a continuous function $f(t)$.
6. The system allows for shortages in all cycles, and each cycle starts with shortages.
7. Shortages are backlogged at a rate $\beta(x)$, which is a non–increasing function of the waiting time x up to the next replenishment, with $0 \leqslant \beta(x) \leqslant 1$ and $\beta(0) = 1$.

The notation that will be used hereafter is reported in Table 1, along with the descriptions of the parameters. Let

$$\delta(t) = \int_0^t \theta(u)du,$$

then the total cost of the inventory system during the planning horizon H, as defined by Chern *et al.* [11], is:

$$
\begin{aligned}
TC(n, s_i, t_i) = & \sum_{i=1}^{n} c_0 e^{-r_1 t_i} \\
& + \sum_{i=1}^{n} c_p e^{-r_2 t_i} \left(\int_{s_{i-1}}^{t_i} \beta\left(t_i - t\right) f(t) dt + \int_{t_i}^{s_i} e^{\delta(t) - \delta(t_i)} f(t) dt \right) \\
& + \sum_{i=1}^{n} \sum_{j=1}^{2} c_{h_j} \int_{t_i}^{s_i} e^{-r_j t} \int_{t}^{s_i} e^{\delta(u) - \delta(t)} f(u) du dt \\
& + \sum_{i=1}^{n} \sum_{j=1}^{2} \frac{c_{b_j}}{r_j} \int_{s_{i-1}}^{t_i} \left(e^{-r_j t} - e^{-r_j t_i}\right) \beta(t_i - t) f(t) dt \\
& + \sum_{i=1}^{n} \sum_{j=1}^{2} c_{l_j} \int_{s_{i-1}}^{t_i} e^{-r_j t} \left[1 - \beta\left(t_i - t\right)\right] f(t) dt, \qquad (1)
\end{aligned}
$$

subject to $s_0 = 0$, $s_{i-1} < t_i \leqslant s_i$, and $s_n = H$. Considering additionally the capacity constraints, we end up with the following constrained, mixed–integer minimization problem:

$$
\begin{aligned}
& \min_{n, t_i, s_i} TC(n, t_i, s_i) \\
\text{s.t.} \quad & \int_{t_i}^{s_i} e^{\delta(u) - \delta(t_i)} f(u) du \leqslant W, \qquad (2) \\
& s_0 = 0, \ s_n = H, \ s_{i-1} < t_i \leqslant s_i, \ i = 1, 2, \ldots, n.
\end{aligned}
$$

Ignoring the constraints, $s_{i-1} < t_i \leqslant s_i$, $i = 1, 2, \ldots, n$, and for given n, the application of the classical Kuhn–Tucker approach can find the optimal solution after solving 2^n nonlinear systems of equations with $2n$ up to $3n$ variables [11,9]. Clearly, the computational cost for solving the problem for unknown n using the Kuhn–Tucker approach is heavy. For this purpose, we propose a technique for concurrent computation of n, t_i, and s_i (which are used for determining the size of the replenishment order). The approach is based on the application of the PSO algorithm, which is described in the next section.

2.2 Particle Swarm Optimization

PSO employs a population of search points that probe the search space simultaneously. The population is called a *swarm*, while the search points are called the *particles*. The particles are initialized randomly in the search space and move with an adaptive velocity within it. Also, each particle has a memory where it stores its best experience during the search, i.e., the best position it has ever visited in the search space. An iteration of the algorithm corresponds to an update of the positions of all particles. The update for each particle is performed by computing the new velocity of the particle, taking into account both its own experience as well as the experience of other particles. These particles are said to constitute its *neighborhood.*

Let $S \subset \mathbb{R}^D$ be a D–dimensional search space and $F : S \to \mathbb{R}$ be the objective function (without loss of generality only the minimization case is considered). A swarm is a set of N particles, $\mathbb{S} = \{x_1, x_2, \ldots, x_N\}$, each of which, is a D–dimensional search point, $x_i = (x_{i1}, x_{i2} \ldots, x_{iD})^\top \in S$, $i = 1, \ldots, N$, and it has an adaptive velocity, $v_i = (v_{i1}, v_{i2}, \ldots, v_{iD})^\top$. Also, each particle x_i remembers the best position, $b_i = (b_{i1}, b_{i2}, \ldots, b_{iD})^\top \in S$, it has ever visited.

A neighborhood, NB_i, is defined for each particle x_i, $i = 1, 2, \ldots, N$. There are several different neighborhood schemes (also called *topologies*) presented in the literature [18, 19]. Most of them are defined based on the indices of the particles rather than their actual positions in S. The most common scheme is the *ring topology*, where the particles are assumed to be organized on a ring, communicating with their immediate neighbors. Under this topology, a neighborhood of radius q of x_i is defined as the set $NB_i^q = \{x_{i-q}, \ldots, x_i, \ldots, x_{i+q}\}$, where x_1 follows immediately after x_N. We denote with g_i the index of the best particle in NB_i, i.e., the particle that has visited the best position in S in terms of its function value, $F(b_{g_i}) \leqslant F(b_j)$, for all j such that $x_j \in NB_i$.

Let t to be the iteration counter. Then, the swarm is updated using the equations [20],

$$v_{ij}(t+1) = \chi \Big[v_{ij}(t) + c_1 R_1 \big(b_{ij}(t) - x_{ij}(t) \big) + c_2 R_2 \big(b_{g_i,j}(t) - x_{ij}(t) \big) \Big], \quad (3)$$

$$x_{ij}(t+1) = x_{ij}(t) + v_{ij}(t+1), \quad (4)$$

where $i = 1, 2, \ldots, N$, $j = 1, 2, \ldots, D$. The parameter χ is called *the constriction coefficient* and it is used to constrain the magnitude of the velocities during the search. The positive constants c_1 and c_2 are referred to as the *cognitive* and *social* parameter, respectively; while R_1, R_2 are random variables uniformly distributed in $[0, 1]$. Default values for χ, c_1 and c_2 are determined in the theoretical analysis of Clerc and Kennedy [20]. The best positions of the particles are updated at each iteration according to the relation:

$$b_i(t+1) = \begin{cases} x_i(t+1), & \text{if } F(x_i(t+1)) < F(b_i(t)), \\ b_i(t), & \text{otherwise,} \end{cases} \qquad i = 1, 2, \ldots, N.$$

The particles are usually constrained to move strictly in the search space, posing explicit bounds on each component of the particles.

3 The Proposed Approach

In the proposed approach, PSO is used to determine both the number of replenishment cycles, n, as well as the corresponding solution, $(s_0, t_1, s_1, \ldots, t_n, s_n)$, concurrently. Since the first and the last component of a possible solution vector are known *a priori* for a specific problem instance ($s_0 = 0$, $s_n = H$), it is sufficient to determine only the remaining $(2n-1)$ solution components, t_1, s_1, ..., t_n, which will be called the *time components* hereafter. However, in our approach, n is also a variable, rising questions regarding the encoding of the variable length PSO particles. For this purpose, we consider a fixed, user–defined maximum number of cycles, $n_{\max} \geqslant n$, which is used to fixate the particles' dimension to $D = 2n_{\max}$. Then, the i–th particle of the swarm is defined as a D–dimensional vector of the form

$$x_i = (x_{i1}, x_{i2}, \ldots, x_{iD}) = (n, t_1, s_1, \ldots, t_n, s_n, \ldots, t_{n_{\max}}),$$

in order to facilitate arithmetic operations, while only the first $2n$ components that correspond to n, t_1, s_1, ..., t_n, are used for the evaluation of x with the cost function $TC(x)$, i.e., $TC(x) = TC(n, s_0, t_1, s_1, \ldots, t_n, s_n)$. The rest of the components are simply ignored. Alternatively, one could incorporate special operators for updating particles of variable length. Such operators have been introduced in the literature for PSO [21]. However, their use in our case requires special handling regarding the ordering of the particle's components, due to constrictions posed by the problem on the time components. Therefore, the simpler solution of assuming a reasonable maximum number of cycles and defining particles of fixed and equal dimensionality, was adopted.

The values of n must be integers and lie within the range $[1, n_{\max}]$. The restriction of the corresponding particle component, x_{i1}, to integer values would require the use of special operators for the particle's update. In order to retain the simplicity and straightforward applicability of the algorithm, we allowed x_{i1} to assume real values in the range $[0.6, n_{\max}]$ in the particle's update procedure, while, for the function evaluation, we round its value to the nearest integer. Such rounding approaches have been shown to work efficiently also in different problems with PSO [22]. The initialization of x_{i1} for each particle is performed randomly and uniformly within $[0.6, n_{\max}]$.

The rest of the components of the particle (i.e., the time components) must lie within the range $[0, H]$, preserving the ordering $t_1 \leqslant s_1 \leqslant \cdots \leqslant t_{n_{\max}}$. For this purpose, the j–th component, x_{ij}, is constrained within the range

$$x_{i,j-1} \leqslant x_{ij} \leqslant x_{i,j+1}, \qquad j = 2, 3, \ldots, D-1, \tag{5}$$

in the particle update procedure at each iteration. Clearly, this restriction fosters the danger of biasing the components of the particles towards H, if one of the preceding time components assumes a large value close to H. If this effect takes place in the initialization phase, then it can be detrimental for the algorithm's performance, since it will inhibit the initialization of particles in specific parts of the search space. In order to avoid such an effect, we initialize each time component of x_i randomly in equidistant intervals within the range $[0, H]$, i.e.,

$$x_{ij}^{\text{initial}} = (j - 2 + \text{rand})\Delta, \qquad j = 2, 3, \ldots, D,$$

where $\Delta = H/(D-1)$ and "rand" is a random variable uniformly distributed in $[0, 1]$.

In the cases where constraints on the capacity of the warehouse were considered, the following penalty function was used:

$$TC_{\text{pen}}(n, s_i, t_i) = TC(n, s_i, t_i) + \sum_{k=1}^{K} \frac{TC(n, s_i, t_i)}{n}, \tag{6}$$

where K is the number of violated constraints in Eq. (2), $0 \leqslant K \leqslant n$. Thus, for each violated constraint, a fixed portion of the cost function is added to the actual cost function value.

4 Experimental Analysis

The proposed approach was applied on the following test problems, denoted as TP1, TP2, and TP3, respectively:

TEST PROBLEM 1 [9]. This problem is based on a simplified version of the model described by Eq. (1), with demand rate $f(t) = 20+2t$, $\beta(x) = e^{-\alpha x}$, $r_1 = r_2 = 0$, $c_0 = 100$, $c_p = 0.2$, $c_{b_2} = 1.5$, $c_{l_2} = 0.5$, $c_{h_2} = 55$, $c_{h_1} = c_{b_1} = c_{l_1} = 0$ and $\theta(t) = 0.01$. The problem was considered for three different levels of the parameter α, namely $\alpha = 0.08$, 0.05, and 0.02.

TEST PROBLEM 2 [11]. In this problem, the shortages are completely backlogged, i.e., $\beta(x) = 1$ for all t, the demand rate is $f(t) = 200 + 50t$, and the parameters assume the values: $H = 10$, $c_0 = 80$, $c_{h_1} = 0.2$, $c_{h_2} = 0.4$, $c_{b_1} = 0.5$, $c_{b_2} = 0.4$, $c_p = 9$, $r = 0.2$, $i_1 = 0.08$, $i_2 = 0.09$, and $\theta(t) = 0.01$.

TEST PROBLEM 3 [11]. In this problem, the shortages are also completely backlogged, i.e., $\beta(x) = 1$ for all t, the demand rate is $f(t) = 200 + 50t - 3t^2$, and the parameters assume the values: $H = 10$, $c_0 = 80$, $c_{h_1} = 0.2$, $c_{h_2} = 0.4$, $c_{b_1} = 0.8$, $c_{b_2} = 0.6$, $c_p = 15$, $r = 0.2$, $i_1 = 0.08$, $i_2 = 0.1$, and $\theta(t) = 0.01$.

Regarding the parameters of PSO, the typical values $\chi = 0.729$, $c_1 = c_2 = 2.05$, derived from the theoretical analysis of Clerc and Kennedy [20] were used. The neighborhood radius was equal to 1 for all particles, while the swarm size was set to $N = 100$ in all experiments, and the algorithm was terminated after a maximum number of 15000 iterations, in all test problems. The maximum number of replenishments, which is used for the determination of the particles' dimension, was equal to $n_{\max} = 20$ for all test problems, resulting in 40–dimensional optimization problems (recall that $D = 2n_{\max} = 40$). The maximum inventory size in the constrained cases of TP1 was equal to $W = 90$, while for TP2 and TP3 it was $W = 300$. For each test problem and case, 50 independent experiments were performed to derive statistics regarding the performance of the proposed approach.

The first component, x_{i1}, of the particle, which corresponds to n (it is not a time component), is initialized randomly and uniformly within $[0.6, n_{\max}]$. The obtained results are reported in Table 2. More specifically, the first column of the table specifies the test problem, while the second and third columns specify the

Table 2. The obtained results in terms of the required number of iterations

Problem	α	W	n^*	TC^*	Suc.	Mean	St.D.	Min	Max
1	0.08	∞	3	685.888	50/50	223.96	66.47	106	421
	0.08	90	3	688.354	50/50	2985.30	2364.64	444	11568
	0.05	∞	3	687.686	50/50	205.84	44.36	98	305
	0.05	90	3	690.702	50/50	1676.90	971.32	466	5130
	0.02	∞	3	689.405	50/50	219.66	58.69	111	432
	0.02	90	3	693.010	50/50	988.74	557.28	339	2826
2		∞	1	21078.04	50/50	8.20	1.88	5	13
		300	1	21078.04	50/50	46.48	28.01	10	133
3		∞	1	29990.68	50/50	7.64	1.86	5	13
		300	1	29990.68	50/50	28.30	19.29	9	86

corresponding value of the parameter α (applicable only to TP1) and the value of W. Infinite warehouse capacity in the unconstrained cases is denoted as "∞". In the rest of the columns, the detected optimal number of replenishment cycles, n^*, is reported per case, along with the corresponding value, TC^*, of the cost function. Also, the number of experiments where the algorithm was successful, i.e., it detected the optimal solution, is reported, along with the mean, standard deviation, minimum and maximum number of iterations required to obtain the solution. Since PSO is a stochastic algorithm, the obtained solution at each experiment for the same problem and case is expected to vary slightly. One of the obtained optimal solutions for each case is reported in Table 3.

Table 3. The obtained solutions rounded up to 6 decimal digits

Problem	α	W	n^*	s_0	t_1	s_1	t_2	s_2	t_3	s_3
1	0.08	∞	3	0.0	1.518444	4.546430	5.776002	8.463058	9.536687	12.0
	0.08	90	3	0.0	1.644925	4.854017	6.164020	8.711135	9.873837	12.0
	0.05	∞	3	0.0	1.470722	4.529375	5.732727	8.448895	9.506381	12.0
	0.05	90	3	0.0	1.609384	4.881638	6.175431	8.721005	9.873838	12.0
	0.02	∞	3	0.0	1.434863	4.523818	5.702551	8.447381	9.487877	12.0
	0.02	90	3	0.0	1.575821	4.907751	6.179812	8.724783	9.873839	12.0
2		∞	1	0.0	10.0	10.0				
		300	1	0.0	10.0	10.0				
3		∞	1	0.0	10.0	10.0				
		300	1	0.0	10.0	10.0				

It is clear that the imposition of constraints in TP1 increases its difficulty significantly, as it is revealed by the increased mean number of iterations required by the algorithm. Nevertheless the algorithm was successful in all cases, detecting both n^* and the corresponding solution without any user intervention. The hard constraints posed on the particles do not prevent PSO from detecting the optimal value, although the bounds change continuously for each particle and iteration, in order to preserve the ordering of the time components of the particles. The penalty function defined in Eq. (6) was adequate to prevent PSO from converging to unfeasible solutions in the cases with constrained warehouse capacity, without any assumptions needed regarding the feasibility of the initial population. Thus, all solutions reported in Table 3 were feasible.

The same observations can be made also for TP2 and TP3. However, in these cases, the reported solutions coincide for the unconstrained and constrained warehouse capacity, since the optimal solution corresponds to a single cycle, and time components lie exactly on the bounds of the time horizon. Overall, the considered test problems were addressed efficiently, rendering the proposed approach a useful alternative for solving continuous review inventory models of the considered type.

5 Conclusions

A major concept in supply chain is the maintenance of inventories of deteriorating items. Such problems are usually addressed through analytical approaches, based on the theory of Kuhn–Tucker. However, the corresponding computational cost is high and the problems reported in the literature usually do not take into account the limited warehouse capacity.

We proposed an alternative approach for solving such problems through PSO. The proposed approach computes the number of replenishment cycles as well as the corresponding shortage and replenishment instances, concurrently, without the need of gradient information. Experiments conducted on an extension of a recently proposed model indicate that the proposed approach can tackle the problem efficiently. Future work will consider further test problems as well as the development of specialized operators that can incorporate model information in the PSO update schemes.

References

1. Ghare, P.M., Shrader, G.F.: A model for exponentially decying inventories. J. Ind. Eng. 14, 238–243 (1963)
2. Covert, R.P., Philip, G.C.: An EOQ model for items with Weibull distribution deterioration. Am. Inst. Ind. Eng. Trans. 5, 323–326 (1973)
3. Tadikamalla, P.R.: An EOQ inventory model for items with Gamma distribution. AIIE Trans. 5, 100–103 (1978)
4. Resh, M., Friedman, M., Barbosa, L.C.: On a general solution of the deterministic lot size problem with time–proportional demand. Oper. Res. 24, 718–725 (1976)

5. Donaldson, W.A.: Inventory replenishment policy for a linear trend in demand: An analytical solution. Oper. Res. Quart. 28, 663–670 (1977)
6. Goyal, S.K., Giri, B.C.: Recent trends in modeling of deteriorating inventory. Eur. J. Oper. Res. 134, 1–16 (2001)
7. Raafat, F.: Survey of literature on continuously deteriorating inventory model. J. Oper. Res. Soc. 42, 27–37 (1991)
8. Chang, H.J., Dye, C.Y.: An EOQ model for deteriorating items with time varying demand and partial backlogging. J. Oper. Res. Soc. 50, 1176–1182 (1999)
9. Skouri, K., Papachristos, S.: A continuous review inventory model, with deteriorating items, time–varying demand, linear replenishment cost, partially time–varying backlogging. Applied Mathematical Modelling 26, 603–617 (2002)
10. Teng, J.T., Chang, H.J., Dye, C.Y., Hung, C.H.: An optimal replenishment policy for deteriorating items with time–varying demand and partial backlogging. Oper. Res. Let. 30, 387–393 (2002)
11. Chern, M.S., Yang, H.L., Teng, J.T., Papachristos, S.: Partial backlogging inventory lot–size models for deteriorating items with fluctuating demand under inflation. European Journal of Operational Research (in press 2007)
12. Eberhart, R.C., Kennedy, J.: A new optimizer using particle swarm theory. In: Proc. 6th Symp. Micro Machine and Human Science, pp. 39–43 (1995)
13. Kennedy, J., Eberhart, R.C.: Particle swarm optimization. In: Proc. IEEE Int. Conf. Neural Networks, vol. IV, pp. 1942–1948 (1995)
14. Kennedy, J., Eberhart, R.C.: Swarm Intelligence. Morgan Kaufmann, San Francisco (2001)
15. Engelbrecht, A.P.: Fundamentals of Computational Swarm Intelligence. Wiley, Chichester (2006)
16. Parsopoulos, K.E., Vrahatis, M.N.: Recent approaches to global optimization problems through particle swarm optimization. Natural Computing 1(2–3), 235–306 (2002)
17. Parsopoulos, K.E., Vrahatis, M.N.: On the computation of all global minimizers through particle swarm optimization. IEEE Transactions on Evolutionary Computation 8(3), 211–224 (2004)
18. Kennedy, J.: Small worlds and mega–minds: Effects of neighborhood topology on particle swarm performance. In: Proc. IEEE Congr. Evol. Comput., Washington, D.C., USA, pp. 1931–1938. IEEE Press, Los Alamitos (1999)
19. Suganthan, P.N.: Particle swarm optimizer with neighborhood operator. In: Proc. IEEE Congr. Evol. Comput., Washington, D.C., USA, pp. 1958–1961 (1999)
20. Clerc, M., Kennedy, J.: The particle swarm–explosion, stability, and convergence in a multidimensional complex space. IEEE Trans. Evol. Comput. 6(1), 58–73 (2002)
21. Ó Neill, M., Leahy, F., Brabazon, A.: Grammatical swarm: A variable–length particle swarm algorithm. In: Nedjah, N., Macedo Mourelle, L. (eds.) Studies in Computational Intelligence, vol. 26, pp. 59–74. Springer, Heidelberg (2006)
22. Laskari, E.C., Parsopoulos, K.E., Vrahatis, M.N.: Particle swarm optimization for integer programming. In: Proceedings of the IEEE 2002 Congress on Evolutionary Computation, Hawaii (HI), USA, pp. 1582–1587. IEEE Press, Los Alamitos (2002)

Option Model Calibration Using a Bacterial Foraging Optimization Algorithm

Jing Dang[1,2], Anthony Brabazon[1], Michael O'Neill[1], and David Edelman[2]

[1] Natural Computing Research and Applications Group
University College Dublin, Ireland
jing.dang@ucd.ie, anthony.brabazon@ucd.ie, m.oneill@ucd.ie
[2] School of Business, University College Dublin, Ireland
davide@ucd.ie

Abstract. The Bacterial Foraging Optimization (BFO) algorithm is a biologically inspired computation technique which is based on mimicking the foraging behavior of E.coli bacteria. This paper illustrates how a BFO algorithm can be constructed and applied to solve parameter estimation of a EGARCH-M model which is then used for calibration of a volatility option pricing model. The results from the algorithm are shown to be robust and extendable, suggesting the potential of applying the BFO for financial modeling.

1 Introduction

This paper illustrates the financial application of a biologically-inspired computation technique (see [1] for a general introduction to biologically inspired algorithms), the Bacterial Foraging Optimization (BFO) algorithm introduced by Passino [14] in 2002, which models the foraging behavior of *Escherichia coli* bacteria present in our intestines.

The algorithm has been developed and applied to solve various real-world problems [7,8,10,19], in a number of application domains. Mishra [11] shows that BFO can converge to the global optimum faster than the canonical genetic algorithm. Kim [8] suggests that the BFO could be applied to find solutions for difficult engineering design problems. In this paper, we examine the potential of applying BFO algorithm - with and without swarming effect - within the financial domain. We employ BFO to estimate the parameters of the EGARCH-M model which is a nonlinear problem. The results are used to price the volatility options.

The paper is organized as follows: Section 2 provides a concise overview of the BFO algorithm. Section 3 gives background information of volatility option pricing and section 4 outlines the experimental methodology adopted. Section 5 provides the results of these experiments followed by conclusions in Section 6.

2 The Bacterial Foraging Optimization (BFO) algorithm

Natural selection tends to eliminate animals with poor foraging strategies and favors the propagation of genes of those animals that have successful foraging

M. Giacobini et al. (Eds.): EvoWorkshops 2008, LNCS 4974, pp. 113–122, 2008.

strategies, since they are more likely to enjoy reproductive success. After many generations, poor foraging strategies are either eliminated or shaped into good ones. This activity of foraging led researchers to use it as optimization process: animals search for nutrients to maximize the energy obtained per unit time spent foraging, in the face of constraints presented by its own physiology. The *E.coli* bacteria present in our intestines also undertake a foraging strategy [14].

Algorithm 1. BFO algorithm

Randomly distribute initial values for $\theta^i, i = 1, 2, ..., S$ across the optimization domain. Compute the initial cost function value for each bacterium i as J^i, and the initial total cost with swarming effect as J^i_{sw}.

for *Elimination-dispersal loop* **do**
  for *Reproduction loop* **do**
    for *Chemotaxis loop* **do**
      for *Bacterium i* **do**
        Tumble: Generate a random vector $\phi \in \mathcal{R}^D$ as a unit length random direction
        Move: Let $\theta^{new} = \theta^i + c\phi$ and compute corresponding J^{new}. Let $J^{new}_{sw} = J^{new} + J_{cc}(\theta^{new}, \theta)$
        Swim: Let m=0
        while $m < N_s$ **do**
          let m=m+1
          if $J^{new}_{sw} < J^i_{sw}$ **then**
            Let $\theta^i = \theta^{new}$, compute corresponding J^i and J^i_{sw}
            Let $\theta^{new} = \theta^i + c\phi$ and compute corresponding $J(\theta^{new})$.
            Let $J^{new}_{sw} = J^{new} + J_{cc}(\theta^{new}, \theta)$
          else
            let $m = N_s$
          end
        end
      end
    end
    Sort bacteria in order of ascending cost J_{sw}The $S_r = S/2$ bacteria with the highest J value die and other S_r bacteria with the best value split Update value of J and J_{sw} accordingly.
  end
  Eliminate and disperse the bacteria to random locations on the optimization domain with probability p_{ed}. Update corresponding J and J_{sw}.
end

Here, the objective is to find the minimum of $J(\theta), \theta \in \mathcal{R}^D$, where we do not have the gradient information $\nabla J(\theta)$. Suppose θ is the position of the bacterium and $J(\theta)$ represents a nutrient profile, i.e.,$J(\theta) < 0, J(\theta) = 0$ and $J(\theta) > 0$ represent the presence of nutrients, a neutral medium and noxious substances respectively. The bacterium will try to move towards increasing concentrations of nutrients (i.e. find lower values of J), search for ways out of neutral media and avoid noxious substances (away from positions where $J > 0$). It implements a type of biased random walk.

Bacteria can also engage in a form of chemically-mediated 'social communication' during their search process. Let J^i denote the actual cost (or the nutrient surface) at the position of the ith bacterium θ^i. A bacterium that has uncovered good sources of nutrients during its search can release a chemical signal which attracts other bacteria to converge (or swarm) to its current location. This process is mediated by the release of a 'repellent signal' to ensure that the bacteria do not get too close to each other. Including this mechanism in our optimization algorithm, the problem becomes the minimization of $J^i_{sw} = J^i + J_{cc}(\theta_i, \theta)$, which represents the time-varying total cost value for bacterium i. The mathematical swarming (cell-cell signalling) function can be represented by:

$$J_{cc}(\theta^i, \theta) = \begin{cases} -M\left(\sum_{k=1}^{S} e^{-W_a\|\theta^i-\theta^k)\|^2} - \sum_{k=1}^{S} e^{-W_r\|\theta^i-\theta^k\|^2}\right) & \text{With swarming} \\ 0 & \text{No swarming} \end{cases}$$

where $\|.\|$ is the Euclidean norm, W_a and W_r are measures of the width of the attractant and repellent signals respectively, M measures the magnitude of the cell-cell signalling effect. These parameter values are chosen according to the nutrient profile used.

During the lifetime of *E.coli* bacteria, they undergo different stages such as chemotaxis, reproduction and elimination-dispersal. A description of each of these is given below. The details of the algorithm are presented in Algorithm 1.

2.1 Chemotaxis

Chemotaxis is the tendency of a bacterium to move toward distant sources of nutrients. In this process, the bacterium alternates between tumbling (changing direction) and swimming behaviors. Here, a *tumble* is represented by a unit walk with random direction $\phi \in \mathcal{R}^D$ (i.e. $\phi = \frac{\Delta}{\sqrt{\Delta^T \Delta}}$, where Δ is a vector with each element a random number on [-1,1]), a *swim* is indicated as movement in the same direction as the previous tumble. After one step move, the new position of the ith bacterium can be represented as $\theta^{new} = \theta^i + c\phi$, where $\theta^i \in \mathcal{R}^D$ indicates the position of the ith bacterium across the optimization domain. c is the chemotactic step size taken in the direction of ϕ. In this paper, we consider a fixed step size c for all bacteria.

If at θ^{new}, the total cost J^{new}_{sw} is better (lower) than the cost at θ^i, another swimming step is taken, and is continued as long as it continues to reduce the cost, but only up to a maximum number of steps, N_s. This means that the bacterium will tend to keep moving if it is headed in the direction of an increasingly favorable environment.

2.2 Reproduction

After N_c chemotactic steps, a reproduction step is taken. In reproduction, the least healthy bacteria die and the other S_r healthiest bacteria each split into two bacteria, which are then placed in the same location. The health of the bacteria are measured by J_{sw}, higher cost represents that the bacterium did not get as many nutrients during its life of foraging (hence is not as "healthy") and thus unlikely to reproduce.

2.3 Elimination - Dispersal

Let N_{ed} be the number of elimination-dispersal steps. The elimination-dispersal step happens after N_{re} reproduction steps. In elimination-dispersal, individual bacterium is stochastically selected for elimination from the population and is replaced by a new bacterium located at a random new location within the optimization domain, according to a preset probability p_{ed}. This mimics the real-world process whereby bacteria can be dispersed to new locations, for example via wind dispersal.

3 Estimation of Volatility Option Pricing Model

Volatility is a measure of how much a stock can move over a specific amount of time. In 1993, the first measure of volatility in the overall market - the S&P 500 Volatility Index (VIX) was created. It is a widely disseminated benchmark index commonly referred to as the market's "fear gauge" and serves as a proxy for investor sentiment - rising when investors are anxious or uncertain about the market and falling during times of confidence or complacency.

Options are financial instruments that convey the right, but not the obligation, to engage in a future transaction on some underlying assets. In February 2006, options on the S&P500 volatility index (VIX Options) began trading on the Chicago Board of Exchange (CBOE), which is the first product on market volatility to be listed on an regulated securities exchange. VIX options offer investors the ability to make trades based on their view of future direction or movement of the VIX, and option buyers have the advantage of limited risk. VIX options also offer the opportunity to hedge volatility risk of a portfolio, distinct from price risk.

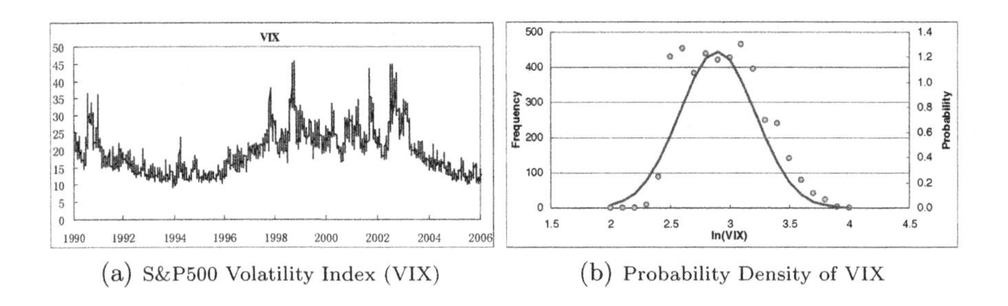

(a) S&P500 Volatility Index (VIX) (b) Probability Density of VIX

Fig. 1. Empirical Results of the VIX

Detemple and Osakwe [3] provide analytic pricing formulae for European volatility options[1] assuming that volatility follows a mean-reverting log process (MRLP). We consider MRLP to be a reasonable assumption with small misidentification error, as can be seen from Fig.1, the volatility tends to revert to some long-running average (mean-reversion properties) and the probability density function of the volatility based on unconditional distribution of VIX could be approximated by a lognormal distribution. The closed from expression for the European call option c_t written on the volatility V, with strike price K and maturity T is [3]:

$$c_t(V_t, K, \tau) = e^{-\lambda\tau}[V_t^{\phi_\tau} \exp(\frac{\alpha}{\lambda}(1-\phi_\tau) + \frac{1}{2}a_\tau^2)\mathrm{N}(d_\tau + a_\tau) - K\mathrm{N}(d_\tau)] \quad (1)$$

$$\text{where } d_\tau = \frac{1}{a_\tau}[\phi_\tau \ln(V_t) - \ln(K) + A_\tau]$$

$$\phi_\tau = \mathrm{e}^{-\lambda(\tau)}$$

$$A_\tau = \frac{\alpha}{\lambda}(1-\phi_\tau)$$

$$a_\tau = \frac{\alpha}{\sqrt{2\lambda}}(1-\phi_\tau^2)^{\frac{1}{2}}$$

and where $\tau = T - t$ is the time-to-maturity, t is the current time, V_t is the volatility of the underlying asset at time t. The above model is a function of parameters α, λ, and σ. α/λ denotes a long run mean for log (V), and $exp\left((\alpha + \frac{1}{4}\sigma^2)/\lambda\right)\sqrt{285}$ denotes a long run mean annualized volatility (based on 285 days). These parameters for the option pricing model can be calculated as below by estimating the corresponding EGARCH-M (the exponential GARCH[2] in mean, first proposed by Nelson [13]) model and then taking the limit [4,12]:

$$\alpha = \frac{K}{2} + \frac{A_1}{\sqrt{2\pi}}\,, \quad \lambda = 1 - G_1\,, \quad \sigma = \frac{1}{2}\sqrt{(\frac{\pi-2}{\pi}){A_1}^2 + {L_1}^2} \quad (2)$$

The EGARCH-M model has estimation issues such as choice of starting values and choice of the optimization routine [15]. We employ BFO to optimize the EGARCH-M model parameters: K, G_1, A_1 and L_1, the details are described in the following section.

4 Experimental Approach

Due to the existence of noise in the newly-traded volatility option data, we calibrate the MRLP option pricing model by estimating the corresponding discrete

[1] A European call option on an asset V_t with maturity date T and strike price K is defined as a contingent claim with payoff at time T given by max $[V_T - K, 0]$.

[2] GARCH models are popular econometric modeling methods, having been initially specified by Engle [5] and Bollerslev [2], they are specifically designed to model and forecast changes in variance, or volatility per se.

time EGARCH-M model and then taking the limit. The EGARCH-M model is an asymmetric model designed to capture the leverage effect, or negative correlation, between asset returns and volatility. The EGARCH-M (1,1) model considered in this paper is set up as follows:

Conditional mean model:

$$y_t = C - \frac{1}{2}\sigma_t^2 + \varepsilon_t \tag{3}$$

where $\varepsilon_t = \sigma_t z_t$, $z_t \sim N(0,1)$ and y_t represents the log returns of S&P 500

Conditional variance model:

$$\log\sigma_t^2 = K + G_1\log\sigma_{t-1}^2 + A_1(|z_{t-1}|) + L_1 z_{t-1} \tag{4}$$

where $z_{t-1} = \frac{|\varepsilon_{t-1}|}{\sigma_{t-1}}$

The left-hand side of equation 4 is the log value of the conditional variance. This implies that the leverage effect is exponential, rather than quadratic, and the forecasts of the conditional variance are guaranteed to be nonnegative. In equation 3, the coefficient of σ_t^2 is fixed at -0.5, and the constant C is assumed to be 0.0005, hence the parameters to be estimated are those appeared in equation 4, namely, K, G_1, A_1, and L_1. Where K is the conditional variance constant, G_1 (GARCH term) is the coefficients related to lagged conditional variances, A_1 (ARCH term) is the coefficients related to lagged innovations, L_1 is the leverage coefficients for asymmetric EGARCH-M(1,1) model.

The EGARCH-M model can be estimated by maximum likelihood estimation (MLE). The idea behind maximum likelihood parameter estimation is to determine the parameters that maximize the probability (likelihood) of the sample data. From a statistical point of view, the method of maximum likelihood is considered to be more robust and yields estimators with good statistical properties. Although the methodology for maximum likelihood estimation is simple, the implementation is mathematically intense. For the EGARCH-M models specified in equations 3 and 4, the objective is to maximize the log likelihood function (LLF) as follows:

$$LLF = -\frac{1}{2}\sum_{t=1}^{T}[\log(2\pi\sigma_t^2) + \frac{\varepsilon_t^2}{\sigma_t^2}] \tag{5}$$

The residuals ε_t and the conditional variances σ_t^2 are inferred by recursive substitution based on equations 3 and 4, given the observed log return series, the current parameter values and the starting value of $z_1 \sim N(0,1)$ and $\sigma_1^2 = exp(K)$. Since minimizing the negative log-likelihood ($-LLF$) is the same as maximizing the log-likelihood(LLF), we use $-LLF$ as our nutrient function (the objective function). And the goal is to minimize the $-LLF$ value, by optimizing parameters K, G_1, A_1, L_1 within the search domain.

The EGARCH-M model is fitted to the return series of S&P 500 daily index using the BFO algorithm (a modified version of Passino's original Matlab code [20]). The S&P 500 (Ticker SPX) equity index is obtained from CBOE, with data drawn from 02/01/1990 to 30/12/2006, giving a total of 4084 daily observations.

5 Results

The estimated parameters are constrained within [-1, 1]. During the searching process for optimal parameter value, if it breaches the lower/upper bound, its value is set to be -1 or 1. In each run of the BFO, we use parameter values specified in Table 1, they are chosen based on trial and error experimentation.

Fig.2 depicts the evolution of the objective function value, measured using negative maximum likelihood ($-LLF$), as a function of the iteration number for a random single run of the algorithm for BFO with, and for comparison purposes without, the swarming effect. Obviously, the BFO algorithm containing swarming effect has quicker convergence than BFO without this mechanism, though the accuracy of the final results are not too different. Figs. 3(a), 3(b), 3(c) and 3(d) depict the evolution of the parameters K, G_1, A_1 and L_1 as a function of the iteration number for a random single run of the BFO algorithm with the swarming effect. In the early iterations BFO mainly performs global search for the optimum value and displays quick convergence. Later in the run, the focus switches to local optimal search.

Running both BFO with and without swarming effect over 30 trials, we obtain the results shown in Table 2. Where 'Optimum' provides the best results over 30 runs for the objective value $-LLF$ and relevant parameters K, G_1, A_1, L_1. The best results averaged over 30 runs and the standard deviation of the best results over 30 runs are reported in the 'Mean' and 'S.D.' respectively. In order to provide a benchmark for the accuracy of the results obtained by BFO, a Matlab (Ver.7.0.1(R14)) optimizer *fmincon*was used. This optimizer uses sequential quadratic programming (SQP) methods, which closely mimic Newton's method for constrained optimization.

From Table 2, the BFO algorithm with swarming outperforms that without swarming effect. The objective value (the minimal $-LLF$) of -14242.27 is close

Parameter Values and Definition	
D=4	Search space dimension
S=50	Bacteria population size
N_c=20	No. of chemotactic steps
N_s=4	No. of swimming steps
N_{re}=4	No. of reproduction steps
N_{ed}=2	No. of elimination-dispersal
p_{ed}=0.25	Prob. for elimination-dispersal
c=0.007	Chemotactic step size
M=10	Magnitude of swarming effect
W_a=0.2	Coefficients of attractant effect
W_r=10	Coefficients of repellent effect

Table 1. BFO Parameters

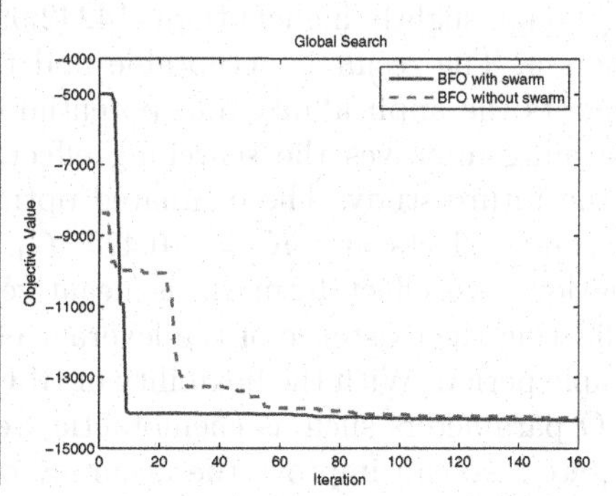

Fig. 2. Objective value vs. Iteration

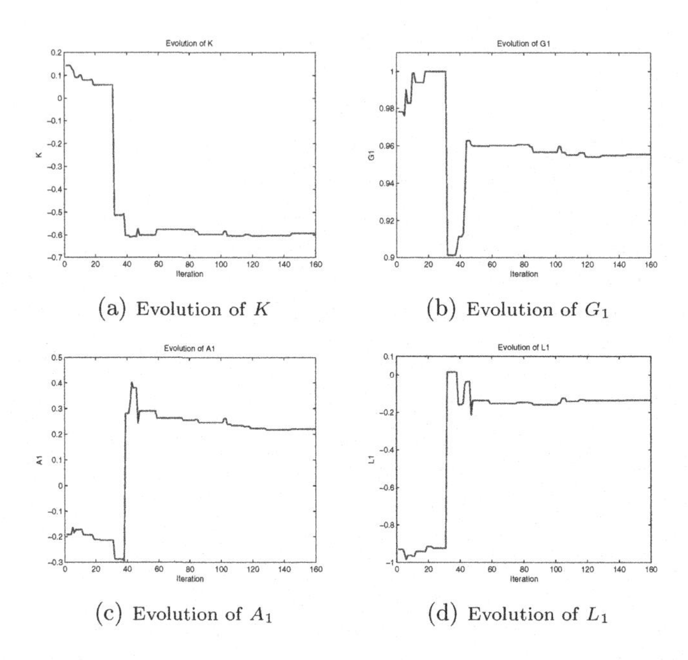

(a) Evolution of K

(b) Evolution of G_1

(c) Evolution of A_1

(d) Evolution of L_1

Fig. 3. Evolution of parameters as a function of the iteration number

Table 2. Results of BFO with and without swarming effect

Method		Output	$-LLF(Obj)$	K	G_1	A_1	L_1
BFO	With Swarming	Optimum	-14242.27	-0.474	0.965	0.182	-0.121
		Mean	-14064.26	-0.685	0.956	0.382	-0.086
		(S.D.)	(206.69)	(0.219)	(0.020)	(0.230)	(0.183)
	No Swarming	Optimum	-14241.72	-0.539	0.960	0.198	-0.125
		Mean	-13841.24	-0.726	0.945	0.302	-0.119
		(S.D.)	(1072.5)	(0.212)	(0.030)	(0.224)	(0.164)
Matlab Optimizer		Optima	-14242.99	-0.491	0.963	0.178	-0.114

to (though slightly higher) than -14242.99 obtained in Matlab using the *fmincon* function. The result is acceptable and the standard deviation is small, which suggests the applicability and potential of BFO in finance. Also, the bacterial swarming improves the searching effect, which could be further investigated in the future study. The estimated optimal parameter values using BFO with swarming effects are: $K = -0.474$, $G_1 = 0.965$, $A_1 = 0.182$, $L_1 = -0.121$. The leverage effect term L_1 is negative and statistically different from zero, indicating the existence of the leverage effect in future stock returns during the sample period. With the flexibility of BFO, it is believed that by further evolving BFO parameters such as chemotactic step size c, number of chemotactic steps N_c, etc., we can improve the accuracy of the results, however, there is always trade off between accuracy (achieved by adding complexity to the algorithm) and convergence speed.

Based on the above results and equation 2, the resulting stochastic volatility option pricing model parameters are: $\alpha = -0.164$, $\lambda = 0.035$ and $\sigma = 0.082$. The negative α implied mean reversion with a long run mean for log (V) of $\alpha/\lambda = -4.707$, and a long run mean annualized volatility (based on 285 days) of $exp\left((\alpha + \frac{1}{4}\sigma^2)/\lambda\right)\sqrt{285} = 0.160$ percent. The speed of reversion λ, is small, indicating strong autocorrelation in volatility which in turn implies volatility clustering. These are consistent with the empirical results found from Fig.1.

Furthermore, based on the estimated parameters of the volatility option pricing model, hedgers can manage the risk/volatility of their existing investment/portfolio. Traders can also use the generated theoretical volatility option prices as a trading guide to make arbitrage/speculating profits.

6 Conclusion and Future Work

In this paper, we applied the proposed Bacterial Foraging Optimization (BFO) algorithm for parameter estimation of a EGARCH-M model, the results of which can be used to price volatility options. Compared to traditional parameter estimation methods, BFO provides satisfactory results for this non-linear parameter estimation problem, indicating proof of concept of the potential utility of applying this algorithm to the domain of finance. The results also indicate that the BFO algorithm with a social swarming effect provides quicker convergence and more stable results compared with the BFO algorithm without the swarming mechanism.

In future work, we intend to extend the application of the BFO algorithm to harder higher-dimensional and dynamic optimization problems in finance. These complexities are widespread in financial modeling and pose serious problems for traditional, gradient-based statistic computing methods which might only give local solutions. Hence, it is important that the utility of population-based approaches such as the BFO algorithm is tested in the finance domain. It is noted that there is of yet, still a limited literature on the application of the BFO algorithm for real-world problems. Hence, it is necessary to undertake comprehensive scaling and dynamic environment benchmarking of the BFO algorithm and examine the contribution of BFO compared to other global optimium searching algorithms.

References

1. Brabazon, A., O'Neill, M.: Biologically-inspired Algorithms for Financial Modelling. Springer, Berlin (2006)
2. Bollerslev, T.: Generalized Autoregressive Conditional Heteroskedasticity. Journal of Econometrics 31(3), 307–327 (1986)
3. Detemple, J.-B., Osakwe, C.: The Valuation of Volatility Options. European Finance Review 4(1), 21–50 (2000)

4. Duan, J.-C.: Augmented GARCH (p,q) Process and its Diffusion Limit. Journal of Econometrics 79(1), 97–127 (1997)
5. Engle, R.-F.: Autoregressive conditional heteroskedasticity with estimates of the variance of U.K. inflation. Econometrica 50(4), 987–1008 (1982)
6. Hentschel, L.: All in the Family: Nesting Symmetric and Asymmetric GARCH Models. Journal of Financial Economics 39(1), 71–104 (1995)
7. Kim, D.-H., Cho, J.-H.: Intelligent Control of AVR System Using GA-BF. In: Khosla, R., Howlett, R.J., Jain, L.C. (eds.) KES 2005. LNCS (LNAI), vol. 3684, pp. 854–859. Springer, Heidelberg (2005)
8. Kim, D.-H., Abraham, A., Cho, J.-H.: A Hybrid Genetic Algorithm and Bacterial Foraging Approach for Global Optimization. Information Sciences 177(18), 3918–3937 (2007)
9. Li, M.-S., Tang, W.-J., et al.: Bacterial Foraging Algorithm with Varying Population for Optimal Power Flow. In: Giacobini, M. (ed.) EvoWorkshops 2007. LNCS, vol. 4448, pp. 32–41. Springer, Heidelberg (2007)
10. Mishra, S.: A Hybrid Least Square-Fuzzy Bacterial Foraging Strategy for Harmonic Estimation. IEEE Transactions on Evolutionary Computation 9(1), 61–73 (2005)
11. Mishra, S., Bhende, C.-N.: Bacterial Foraging Technique-Based Optimized Active Power Filter for Load Compensation. IEEE Transactions on Power Delivery 22(1), 457–465 (2007)
12. Nelson, D.-B.: ARCH Models as Diffusion Approximations. Journal of Econometrics 45(1–2), 7–38 (1990)
13. Nelson, D.-B.: Conditional Heteroskedasticity in Asset Returns: A New Approach. Econometrica 59(2), 347–370 (1991)
14. Passino, K.-M.: Biomimicry of bacterial foraging for distributed optimization and control. Control Systems Magazine, IEEE 22(3), 52–67 (2002)
15. St. Pierre, E.F.: Estimating EGARCH-M models: Science or art? The Quarterly Review of Economics and Finance 38(2), 167–180 (1998)
16. Tang, W.-J., Wu, Q.-H., Saunders, J.-R.: A Novel Model for Bacterial Foraging in Varying Environments. In: Gavrilova, M.L., Gervasi, O., Kumar, V., Tan, C.J.K., Taniar, D., Laganá, A., Mun, Y., Choo, H. (eds.) ICCSA 2006. LNCS, vol. 3980, pp. 556–565. Springer, Heidelberg (2006)
17. Tang, W.-J., Wu, Q.-H., Saunders, J.-R.: Bacterial Foraging Algorithm for Dynamic Environments. IEEE Congress on Evolutionary Computation, 1324–1330 (July 2006)
18. Tang, W.-J., Wu, Q.-H., Saunders, J.-R.: Individual-Based Modeling of Bacterial Foraging with Quorum Sensing in a Time-Varying Environment. In: Marchiori, E., Moore, J.H., Rajapakse, J.C. (eds.) EvoBIO 2007. LNCS, vol. 4447, pp. 280–290. Springer, Heidelberg (2007)
19. Ulagammai, M., Venkatesh, P., et al.: Application of Bacterial Foraging Technique Trained Artificial and Wavelet Neural Networks in Load Forecasting. Neurocomputing 70(16-18), 2659–2667 (2007)
20. Original matlab codes of Passino can be obtained from, `http://www.ece.osu.edu/~passino/ICbook/ic_index.html`

A SOM and GP Tool for Reducing the Dimensionality of a Financial Distress Prediction Problem

E. Alfaro-Cid[1], A.M. Mora[2], J.J. Merelo[2], A.I. Esparcia-Alcázar[1], and K. Sharman[1]

[1] Instituto Tecnológico de Informática,
Universidad Politécnica de Valencia, Spain
{evalfaro,aesparcia,ken}@iti.upv.es
[2] Departamento de Arquitectura y Tecnología de Computadores
Universidad de Granada, Spain
{amorag,jmerelo}@geneura.ugr.es

Abstract. In order to build prediction models that can be applied to an extensive number of practical cases we need simple models which require the minimum amount of data. The Kohonen's self organizing map (SOM) are usually used to find unknown relationships between a set of variables that describe a problem, and to identify those with higher significance. In this work we have used genetic programming (GP) to produce models that can predict if a company is going to have book losses in the future. In addition, the analysis of the resulting GP trees provides information about the relevance of certain variables when solving the prediction model. This analysis in combination with the conclusions yielded using a SOM have allowed us to reduce significantly the number of variables used to solve the book losses prediction problem while improving the error rates obtained.

1 Introduction

The financial distress prediction is one of the most interesting topics to research from a company manager point of view. To begin with, it is a difficult problem to solve because there are lots of variables to take into account and moreover, there are many relationships (visible or hidden) between them which makes it hard to advance this situation. If we look at the history of this subject, it is seen that the path of the studies moves from complex, accurate and difficult to implement approaches like univariate and statistical analysis [1,2]; to more generic (less restrictive) tools, like the application of artificial neural networks (ANN) [3,4,5,6] and genetic programming (GP) [7,8,9,10].

The problem of using statistical techniques is the requirement of the existence of a functional relation among dependent and independent variables, to get good results. In addition, these methods generalize badly because they are

M. Giacobini et al. (Eds.): EvoWorkshops 2008, LNCS 4974, pp. 123–132, 2008.

very sensitive to exceptions. On the other hand, soft-computing techniques are more flexible.

In this paper we use GP to predict the book losses of a company from a data set of more than 400 companies. The database used includes not only financial data from the companies but also general information that can be relevant when predicting failure. The same database has previously been used for the prediction of bankruptcy [11,12]. One of the factors that was found to be the key in the successful application of GP to the bankruptcy prediction problem was the reduction of the number of variables. Usually bankruptcy prediction models are based on 7 or 8 economic ratios. In that work, in order to handle the amount of data in the database and in order to ensure that the GP generated models which were understandable, the prediction was done in two steps. To start with, GP was run with all the available variables and then used to identify which data were relevant for solving the problem. This could be done since GP creates analytical models as a final result. Then the proper prediction models were evolved using only those variables that had been identified as important in the first stage. The results in [11,12] showed that reducing the number of variables not only simplifies the GP classifiers structures but also improves the classification rates.

In this study a self-organizing map (SOM) [13] is used to reduce the dimensionality of the book losses prediction problem. The SOM was introduced by Teuvo Kohonen in 1982. It is a non-supervised neural network that tries to imitate the self-organization done in the sensory cortex of the human brain, where neighbouring neurons are activated by similar stimulus. SOMs are able to show non-linear relationships among variables. The main property of the SOM is that it makes a linear projection from a high dimensional data space (one dimension per variable) on a regular, low dimensional (usually 2D) grid of neurons. Also, in changing environments, as new examples become available, they are able to adapt their output. This method is usually used as a clustering/ classification tool, or used to find unknown relationships between data.

In [14] a Kohonen's SOM was used for surveying the financial status of Spanish companies. From the map the authors infer which are the most relevant variables so that a fast diagnostic on the status of a company can be reached. Thus, we have decided to consider some of the conclusions yielded in that work to choose a set of variables which may be significant to predict the book losses for a company.

The prediction of book losses is interesting since there is a direct relationship between continued book losses and legal bankruptcy [15]. Moreover book losses usually happen at a stage prior to insolvency so predicting book losses gives the management of a company more time to react and find a solution to the problem (if there is one).

The remainder of this paper is organized as follows: section 2 describes the dataset used to make the study. The methods used to process the samples are introduced in section 3. Section 4 shows the results yielded by these methods, and the related conclusions are reported in section 5.

2 Dataset and Problem Description

The data used in this work were extracted from the Infotel database[1], and it is a set composed by data from about 470 companies. 170 of these companies had continuous book losses during the years 2001 - 2003, and the remaining 300 companies presented a good financial health. There are available data of these companies from years 1998, 1999 and 2000 to perform the prediction.

Table 1 shows the independent variables, their description and type. As can be seen, the variables can take values from different numerical ranges: real, integer and binary. Also, some of the non-financial data take categorical values; these are the size of the company, the type of company and the auditor's opinion. Usually, company size is a real variable but in this case the companies are grouped in three separated categories according to their size. Each categorical variable can take 3 different values. To work with them they have been transformed into 3 binary variables each. For example, if we have the variable *size*, with 3 possible values, '1', '2' or '3', we can create three new variables *size1*, *size2* and *size3*. Each one will have a value of '1' if the old value of *size*, was '1', '2' or '3', respectively, and a value of '0' otherwise. Therefore, after this modification the available data set for each company includes 37 independent variables: 18 real, 7 integer and 12 binary variables.

3 Methodology

In this section we briefly describe the GP framework that we have used to predict book losses. Basically, the GP algorithm must find a structure (a function) which can, once supplied with the relevant data from the company, decide if this company is going to have book losses or not. In short, it is a binary classification problem.

Classification. The classification works as follows. Let $X = \{x_0, \ldots, x_N\}$ be the vector comprising the data of the company undergoing classification. Let $f(X)$ be the function defined by an individual GP tree structure. We can apply X as the input to the GP tree and calculate the output $f(X)$. Once the numerical value of $f(X)$ has been calculated, it will give us the classification result according to:

$$f(X) > 0, \; X \in L \tag{1}$$

$$f(X) \leq 0, \; X \in \overline{L} \tag{2}$$

where L represents the class to which companies with book losses belong and $\overline{L}$ represents the class to which healthy companies belong.

The task of GP is to find the function $f(X)$.

Fitness Evaluation. Since there is an unbalance in the database in the sense that only 170 companies have book losses versus 300 healthy companies, we

[1] Bought from `http://infotel.es`

Table 1. Independent Variables

Financial Variables	Description	Type
Debt Structure	Long-Term Liabilities /Current Liabilities	Real
Debt Cost	Interest Cost/Total Liabilities	Real
Leverage	Liabilities/Equity	Real
Cash Ratio	Cash Equivalent /Current Liabilities	Real
Working Capital	Working Capital/ Total Assets	Real
Debt Ratio	Total Assets/Total Liabilities	Real
Operating Income Margin	Operating Income/Net Sales	Real
Debt Paying Ability	Operating Cash Flow/Total Liabilities	Real
Return on Operating Assets	Operating Income/Average Operating Assets	Real
Return on Equity	Net Income/Average Total Equity	Real
Return on Assets	Net Income/Average Total Assets	Real
Asset Turnover	Net Sales/Average Total Assets	Real
Receivable Turnover	Net Sales/Average Receivables	Real
Stock Turnover	Cost of Sales/Average Inventory	Real
Current Ratio	Current Assets/Current Liabilities	Real
Acid Test	(Cash Equivalent + Marketable Securities + Net receivables) /Current Liabilities	Real
Non-financial Variables	**Description**	**Type**
Size	Small/Medium/Large	Categorical
Type of company		Categorical
Auditor's opinion		Categorical
Audited	If the company has been audited	Binary
Delay	If the company has submitted its annual accounts on time	Binary
Linked in a group	If the company is part of a group holding	Binary
Number of partners		Integer
Number of employees		Integer
Age of the company		Integer
Number of changes of location		Integer
Number of judicial incidences	Last year	Integer
Historic number of judicial incidences	Since the company was created	Integer
Historic number of serious incidences	Such as strikes, accidents...	Integer
Amount of money spent on judicial incidences	Last year	Real
Historic amount of money spent on judicial incidences	Since the company was created	Real

Table 2. GP parameters

Initialization method	Ramped half and half
Replacement operator	Generational with elitism (0.2%)
Selection operator	Tournament selection
Tournament group size	7
Cloning rate	0.1
Crossover operator	Bias tree crossover
Internal node selection rate	0.9
Crossover rate	0.4
Mutation rate	0.1
Tree maximum initial depth	7
Tree maximum depth	18
Population size	1000
Termination criterion	250 generations

have modified the cost associated to misclassifying the positive and the negative classes to compensate for the imbalanced ratio of the two classes [16]. For example, if the imbalance ratio is 1:10 in favor of the negative class, the penalty for misclassifying a positive example should be 10 times greater. Basically, it rewards the correct classification of examples from the small class over the correct classification of examples from the over-sized class. It is a simple but efficient solution.

Therefore, the fitness function to maximize is:

$$Fitness = \sum_{i=1}^{n} u_i \tag{3}$$

where

$$u_i = \begin{cases} 0 & : \text{ incorrect classification} \\ \frac{n_h}{n_l} & : \text{ company with losses classified correctly} \\ 1 & : \text{ healthy company classified correctly} \end{cases} \tag{4}$$

n_l is the number of companies with book losses and n_h is the number of healthy companies.

GP implementation. The GP implementation used is based on ECJ [2], a research evolutionary computation system developed at George Mason University's Evolutionary Computation Laboratory (ECLab). Table 2 shows the main parameters used during evolution.

As a method of bloat control we have included a new crossover operator, bloat-control-crossover, that occurs with a probability of 0.4. This crossover operator implements a bloat control approach described in [17] and inspired in the "prune and plant" strategy used in agriculture. It is used mainly for fruit trees and it consist of pruning some branches of trees and planting them in order to grow

[2] `http://cs.gmu.edu/~{}eclab/projects/ecj`

Table 3. Function set

Functions	Number of arguments	Arguments type	Return type
+, -, *, /	2	real	real
If $arg_1 \leq arg_2$ then arg_3 else arg_4	4	real	real
If arg_1 then arg_2 else arg_2	3	arg_1 is a boolean arg_2, arg_3 are real	real
If $arg_1 \leq int$ then arg_2 else arg_2 (*int* is randomly chosen)	3	arg_1 is an integer arg_2, arg_3 are real	real

new trees. The idea is that the worst tree in a population will be substituted by branches "pruned" from one of the best trees and "planted" in its place. This way the offspring trees will be of smaller size than the ancestors, effectively reducing bloat.

Strong Typing. Strongly typed GP (STGP) [18] is an enhanced version of GP that enforces data-type constraints, since standard GP is not designed to handle a mixture of data types. In STGP, each function node has a return-type, and each of its arguments also have assigned types. STGP permits crossover and mutation of trees only with the constraint that the return type of each node matches the corresponding argument type in the node's parent.

A STGP has been implemented in order to ensure that in the resulting classifying models the functions operate on appropriate data types so that the final model has a physical meaning. That is, the objective is to avoid results that operate on data which are not compatible, for instance, models which add up the liabilities and the age of a company.

The terminal set used consists of 38 terminals: the independent variables from Table 1 plus Koza's ephemeral random constant.

Table 3 shows the function set used and the chosen typing.

4 Experiments and Results

The mean and standard deviation of 30 runs were calculated. Table 4 shows the results obtained using all the available variables for prediction.

The results show the difficulty of predicting book losses, since the same GP strategy obtained better classification rates when using the database for predicting bankruptcy [11,12].

Table 4. Average results for the prediction of book losses using all variables

Training error	Testing error	Overall error
22.58 ± 2.18	34.32 ± 2.11	25.92 ± 1.85

Table 5. Variables used more frequently in the GP trees

Variable	Type
Debt Cost	Real
Leverage	Real
Operating Income Margin	Real
Stock Turnover	Real
Return on Operating Assets	Real
Return on Equity	Real

Table 6. Average results for the prediction of book losses using the variables used more frequently by GP

Training error	Testing error	Overall error
23.61 ± 1.53	35.85 ± 1.21	27.09 ± 1.25

As a first approach we used GP to reduce the number of independent variables, since this method yielded good results in previous work [11,12]. We analyzed the final tree structures the GP algorithm converged to in the previous experimental runs and we observed that there are 6 variables that were used more frequently than others (in more than 70% of the final trees). We then ran a second set of experiments considering only these variables. In Table 5 the variables used in the second set of experiments are presented.

Table 6 shows the numerical results obtained when solving the prediction problem using the reduced set of variables. This method of reducing the number of variables yielded good results for the bankruptcy prediction problem [11,12] but it has not worked well in this case. Both the average training and testing errors have increased.

Given that the results obtained using GP for reducing the variables are not good we have used SOM to fulfill the task. In [14] three clusters were identified:

- *Warm spot*: large companies (*size3* set to 1), which have been audited, with *type of company1* set to 1, with a favorable auditor's opinion (*auditor opinion1* set to 1), high acid test, high current ratio, delay in reporting the annual accounts, high leverage and belonging to a group. There are more failed companies in this cluster than successful ones, so the cluster probably corresponds to old members of company conglomerates, with a big size, which are conveniently closed without incurring in big losses
- *Hot spot*: successful companies, with small size (*size1* set to 1) and most economic indicators in a healthy shape.
- *Small spot*: most companies with losses with *type of company3* set to 1 and no other distinctive value.

These variables are summarized in Table 7.

Table 7. Variables identified as relevant by the SOM

Variable	Type
Size1	Binary
Size3	Binary
Type of company1	Binary
Type of company3	Binary
Audited	Binary
Auditor opinion1	Binary
Delay	Binary
Linked in a group	Binary
Leverage	Real
Acid Test	Real
Current Ratio	Real

Table 8. Average results for the prediction of book losses using the variables identified as relevant by the SOM

Training error	Testing error	Overall error
21.94 ± 1.39	32.79 ± 1.33	25.03 ± 1.18

Thus, a third set of experiments has been run where the book losses prediction problem has been solved using the variables identified by the SOM as relevant plus two other variables that the GP algorithm used in more than 80% of the final trees: *return on equity* and *debt cost.*

Table 8 shows the error rates obtained in this third set of experiments, smaller than when using all the variables. Kruskal-Wallis tests have been used to compare the results. When comparing the latter results with those obtained using all variables, no significant differences were found between training error rates. However, differences in the testing error rate were significant to level 95%. If the comparison is made between the results obtained using the SOM variables and those obtained using GP, the differences are significant for both training and testing test with confidence levels of 99%.

5 Conclusions

In this study we present an improved method for prediction of financial distress using GP and SOMs. It is based on work carried out in [11,12], where the importance of variable selection when applying GP to the financial failure problem was emphasized. We have merged our results with those obtained by [14] employing Kohonen's Self-Organizing Maps on the same database. As a result, we have got a set of variables that are significant for the book losses prediction problem.

Considering this set of variables and the previously tested GP approach, the classification (prediction) rates have improved.

So, reducing the dimensionality of the prediction of financial distress problem can not only simplify the understanding of the resulting classifiers, but also improve the prediction error rates.

On the other hand, the analysis of the resulting GP trees, which in previous work provided enough information for reducing the number of variables, has not yielded satisfactory results in this case, given that the error rates increased.

However, when the variables used more frequently in the resulting GP trees have been combined with those variables identified by the SOM as more relevant, the error rates have decreased.

Thus, the application of SOM to the analysis of variables has allowed an improvement in the prediction rates and a reduction of the number of variables from 37 to 13, nearly three times smaller.

Acknowledgements

This work has been supported by project TIN2007-68083-C02 (Spanish Ministry of Education and Culture, NOHNES Project).

References

1. Beaver, W.H.: Financial ratios as predictors of failures. Empirical research in accounting: Selected studies. Journal of Accounting Research 5(Suppl.), 71–111 (1966)
2. Altman, E.I.: The success of business failure prediction models. An international survey. Journal of Banking, Accounting and Finance 8, 171–198 (1984)
3. Castillo, P.A., la Torre, J.M.D., Merelo, J.J., Román, I.: Forecasting business failure. A comparison of neural networks and logistic regression for spanish companies. In: Proceedings of the 24th European Accounting Association, Athens, Greece (2001)
4. Charalambous, C., Charitou, A., Kaourou, F.: Application of feature extractive algorithm to bankruptcy prediction neural networks. In: Proceedings of the IEEE-INNS-ENNS International Joint Conference (IJCNN 2000), vol. 5 (2000)
5. Kaski, S., Sinkkonen, J., Peltonen, J.: Bankruptcy analysis with self-organizing maps in learning metrics. IEEE Trans. Neural Networks 12(4), 936ss (2001)
6. Kiviluoto, K.: Predicting bankruptcies with the self-organizing map. Neurocomputing 21(1-3), 191–201 (1998)
7. Brabazon, A., O'Neill, M.: Biologically inspired algorithms for finantial modelling. Springer, Berlin (2006)
8. Lee, W.-C.: Genetic programming decision tree for bankruptcy prediction. In: Proceedings of the 2006 Joint Conference on Information Sciences, JCIS 2006, Kaohsiung, October 8-11, 2006, Atlantis Press (2006)
9. Lensberg, T., Eilifsen, A., McKee, T.E.: Bankruptcy theory development and classification via genetic programming. European Journal of Operational Research 169, 677–697 (2006)

10. Salcedo-Sanz, S., Fernández-Villacañas, J.L., Segovia-Vargas, M.J., Bousoño-Calzón, C.: Genetic programming for the prediction of insolvency in non-life insurance companies. Computers and Operations Research 32, 749–765 (2005)
11. Alfaro-Cid, E., Sharman, K., Esparcia-Alcázar, A.: A genetic programming approach for bankruptcy prediction using a highly unbalanced database. In: Giacobini, M. (ed.) EvoWorkshops 2007. LNCS, vol. 4448, pp. 169–178. Springer, Heidelberg (2007)
12. Alfaro-Cid, E., Cuesta-Cañada, A., Sharman, K., Esparcia-Alcázar, A.I.: Strong Typing, Variable Reduction and Bloat Control for Solving the Bankruptcy Prediction Problem Using Genetic Programming. In: Natural Computing in Computational Economics and Finance. Studies in Computational Intelligence Series, Springer, Heidelberg (to appear)
13. Kohonen, T.: The Self-Organizing Maps. Springer, Heidelberg (2001)
14. Mora, A.M., Laredo, J.L.J., Castillo, P.A., Merelo, J.J.: Predicting financial distress: A case study using self-organizing maps. In: Sandoval, F., Prieto, A.G., Cabestany, J., Graña, M. (eds.) IWANN 2007. LNCS, vol. 4507, pp. 765–772. Springer, Heidelberg (2007)
15. Román, I., Gómez, M.E., la Torre, J.M.D., Merelo, J.J., Mora, A.M.: Predicting financial distress: Relationship between continued losses and legal bankrupcy. In: Proceedings of the 27th Annual Congress European Accounting Association, Dublin, Ireland (2006)
16. Japkowicz, N., Stephen, S.: The class imbalance problem: a systematic study. Intelligent Data Analysis 6(5), 429–449 (2002)
17. Fernández de Vega, F., Rubio del Solar, M., Fernández Martínez, A.: Implementación de algoritmos evolutivos para un entorno de distribución epidémica. In: Arenas, M.G., et al (eds.) Actas del IV Congreso Español de Metaheurísticas, Algoritmos Evolutivos y Bioinspirados (MAEB 2005), Granada, Spain, September 2005, pp. 57–62 (2005)
18. Montana, D.J.: Strongly typed genetic programming. Evolutionary Computation 3(2), 199–230 (1995)

Quantum-Inspired Evolutionary Algorithms for Financial Data Analysis

Kai Fan[1,2], Anthony Brabazon[1], Conall O'Sullivan[2], and Michael O'Neill[1]

[1] Natural Computing Research and Applications Group
University College Dublin, Ireland
kai.fan@ucd.ie, anthony.brabazon@ucd.ie, michael.oneill@ul.ie
[2] School of Business, University College Dublin, Ireland
conall.osullivan@ucd.ie

Abstract. This paper describes a real-valued quantum-inspired evolutionary algorithm (QIEA), a new computational approach which bears similarity with estimation of distribution algorithms (EDAs). The study assesses the performance of the QIEA on a series of benchmark problems and compares the results with those from a canonical genetic algorithm. Furthermore, we apply QIEA to a finance problem, namely non-linear principal component analysis of implied volatilities. The results from the algorithm are shown to be robust and they suggest potential for useful application of the QIEA to high-dimensional optimization problems in finance.

1 Introduction

A wide-variety of biologically-inspired algorithms have been applied for financial modelling [1] in recent years. One interesting avenue of this research has been the hybridisation of quantum-inspired concepts with evolutionary algorithms [11,8,13] producing a family of algorithms known as quantum-inspired evolutionary algorithms (QIEA). A claimed benefit of these algorithms is that because they use a quantum representation, they can maintain a good balance between exploration and exploitation. It is also suggested that they offer computational efficiencies as use of a quantum representation can allow the use of smaller population sizes than typical evolutionary algorithms. As yet, apart from [4,5,6], there have been no studies applying these algorithms in the finance domain. This paper extends these proof of concept studies in two important ways. First, it explores the utility of a real-valued QIEA by applying the algorithm to a series of benchmark problems and comparing the results with those produced by a canonical GA. Then it applies the methodology to undertake a non-linear principal component analysis (NLPCA). The NLPCA is used to determine the non-linear principal components that drive the variations in the implied volatility smile for financial options.

The next section provides a short introduction to the real-valued QIEA and outlines the benchmark tests undertaken. This is followed by a description of the experimental approach adopted in the NLCPA, followed by the results and finally, the conclusions of this paper.

M. Giacobini et al. (Eds.): EvoWorkshops 2008, LNCS 4974, pp. 133–143, 2008.

2 Quantum-Inspired Evolutionary Algorithm

The real-valued quantum-inspired evolutionary algorithm applied in this paper is described in [5,6] and readers are referred there for more details. A short overview of the algorithm is provided below.

Quantum mechanics is an extension of classical mechanics which models behaviours of natural systems that are observed particularly at very short time or distance scales. In the initial literature which introduced the QIEA, a binary representation was adopted, wherein each quantum chromosome was restricted to consist of a series of 0s and 1s. The methodology was modified to include real-valued vectors by da Cruz et al., [3]. As with binary-representation QIEA, real-valued QIEA maintains a distinction between a quantum population and an observed population of, in this case, real-valued solution vectors. However the quantum individuals have a different form to those in binary-representation QIEA. The quantum population $Q(t)$ is comprised of N quantum individuals ($q_i : i = 1, 2, 3, \ldots, N$), where each individual is comprised of G genes ($g_{ij} : j = 1, 2, 3, \ldots, G$). Each of these genes consist of a pair of values $g_{ij} = (p_{ij}, \sigma_{ij})$ where $p_{ij}, \sigma_{ij} \in \Re$ represent the mean and the width of a square pulse. Representing a gene in this manner has a parallel with the quantum concept of superposition of states as a gene is specified by a range of possible values, rather than by a single unique value.

The da Cruz et al algorithm does periodically sample from a distribution to get a "classical" population, which can be regarded as a wave-function (quantum state) collapsing to a classical state upon observation. It is also noted that this bears similarity to the operation of a number of estimation of distribution algorithms (EDAs).

Algorithm. The real-valued QIEA algorithm is as follows

Algorithm. Real-valued Quantum-inspired Genetic Algorithm

```
Set t=0;
Initalise Q(t) (the quantum chromosome);

while t < max_t do
    Create the PDFs (and corresponding CDFs) for each gene locus using the quantum
    individual;
    Create a temporary population, denoted E(T), of K real-valued solution vectors through
    a series of 'observations' via the CDFs;
    if t=0 then
        C(t)=E(t);
        Note: the population C(t) is maintained between iterations of the algorithm;
    else
        E(t)=Outcome of crossover between E(t-1) and C(t-1);
        Evaluate E(t);
        C(t)= K best individuals from E(t) ∪ C(t-1);
    end
    With the N best individuals from C(t);
    Q(t+1)=Output of translate operation on Q(t);
    Q(t+1)=Output of resize operation on Q(t+1);
    t=t+1;
end
```

Initialising the Quantum Population. At the start of the algorithm, each quantum gene is initialised by randomly selecting a value from within the range of allowable values for that dimension. For example, if the known allowable values for dimension j are $[-75, 75]$ then q_{ij} (dimension j in quantum chromosome i) is initially determined by randomly selecting a value from this range (say) -50. The corresponding width value will be 150. Hence, $q_{ij} = (-50, 150)$. The square pulse need not be entirely within the allowable range for a dimension when it is initially created as the algorithm will automatically adjust for this as it executes. The height of the pulse arising from a gene j in chromosome i is calculated using

$$h_{ij} = \frac{1/\sigma_{ij}}{N} \tag{1}$$

where N is the number of individuals in the quantum population. This equation ensures that the probability density functions (PDFs) used to generate the observed individual solution vectors will have a total area equal to one.

Observing the Quantum Chromosomes. In order to generate a population of real-valued solution vectors, a series of observations must be undertaken using the population of quantum chromosomes (individuals). A pseudo-interference process between the quantum individuals is simulated by summing up the square pulses for each individual gene across all members of the quantum population. This generates a separate PDF (just the sum of the square pulses) for each gene and eq. 1 ensures that the area under this PDF is one. Hence, the PDF for gene j on iteration t is

$$PDF_j(t) = \sum_i^j g_{ij} \tag{2}$$

where g_{ij} is the squared pulse of the jth gene of the ith quantum individual (of N). To use this information to obtain an observation, the PDF is first converted into its corresponding Cumulative Distribution Function (CDF)

$$CDF_j(x) = \int_{L_j}^{U_j} PDF_j(x)dx \tag{3}$$

where U_j and L_j are the upper and lower limits of the probability distribution. By generating a random number r from (0,1) following a specific distribution, the CDF can be used to obtain an observation of a real number x, where $x = CDF^{-1}(r)$. Once these have been calculated, the observation process is iterated to create a temporary population with K members, denoted by E(t).

Updating the Quantum Chromosomes. In this study we adjust the quantum probability amplitude with specified operators by comparing each successive generation's best fitness function so that the quantum chromosome can produce more promising individuals with higher probability in the next generation, i.e. if the best fitness function has improved (disimproved) we shrink (enlarge) the width in order to improve the local (global) search.

Table 1. Benchmark functions

f	Function	Mathematical representation	Range	$f(x_i^*)$	x_i^*
f_1	DeJong(Sphere)	$f(x) = \sum_{i=1}^{p} {x_i}^2$	$-5 \leq x_i \leq 5$	0	0
f_2	Rosenbrock	$f(x) = \sum_{i=1}^{p-1} 100(x_{i+1} - {x_i}^2)^2 + (1 - x_i)^2$	$-50 \leq x_i \leq 50$	0	1
f_3	Rastrigin	$f(x) = 10p + \sum_{i=1}^{p}({x_i}^2 - 10cos(2\pi x_i))$	$-100 \leq x_i \leq 100$	0	0
f_4	Griewangk	$f(x) = \sum_{i=1}^{p} \frac{{x_i}^2}{4000} - \prod_{i=1}^{p} \cos\left(\frac{x_i}{\sqrt{i}}\right) + 1$	$-600 \leq x_i \leq 600$	0	0

Table 2. Parameters setting in QIEA and GA

QIEA	Population=50	Observation=200	Shrinkage=0.005	Enlargement=30
GA	Population=50	Generations=200	Mutation=0.005	Crossover=0.75

3 Benchmark Test

Four major static benchmark functions are chosen to test the ability of QIEA to find a global minimum. The results are compared to those of a canonical GA. Details of the benchmark functions are shown in Table 1.

In order to make a fair comparison between QIEA and GA, we call the evaluation function 10000 times for both algorithms. The parameters used for QIEA and GA, selected from sensitivity test, are shown in Table 2.

Table 3. DeJong results

[-5,5]	Best	Mean	S.D.	Time(s)
Dimension : 5				
QIEA	0.0021	0.0012	0.0007	27.96
GA	0.0001	0.0000	0.0002	23.58
Dimension : 10				
QIEA	0.0378	0.0090	0.0216	28.28
GA	0.0002	0.0000	0.0002	39.74
Dimension : 50				
QIEA	3.427	0.289	2.562	29.21
GA	3.511	0.787	2.304	103.00
Dimension : 100				
QIEA	24.953	2.077	21.434	31.86
GA	54.778	6.809	39.564	190.63

Table 4. Rosenbrock results

[-50,50]	Best	Mean	S.D.	Time(s)
Dimension : 5				
QIEA	68.9335	44.1549	14.1392	29.36
GA	578.2585	1141.90	4.2406	24.34
Dimension : 10				
QIEA	786.7354	362.0550	80.5617	30.66
GA	939.3839	1334.200	10.8133	45.14
Dimension : 50				
QIEA	8.751e+5	2.245e+5	3.901e+5	31.04
GA	2.089e+6	7.242e+5	9.400e+5	125.34
Dimension : 100				
QIEA	3.50e+7	8.47e+6	1.88e+7	35.87
GA	1.56e+8	2.86e+7	1.07e+8	228.07

Table 5. Rastrigin results

[-100,100]	Best	Mean	S.D.	Time(s)
Dimension : 5				
QIEA	18.687	4.579	7.285	30.61
GA	15.777	6.016	9.186	21.21
Dimension : 10				
QIEA	47.485	10.104	25.542	30.97
GA	32.161	5.858	19.622	37.52
Dimension : 50				
QIEA	1479.3	190.5	866.9	37.84
GA	1763.8	276.6	1286.4	107.18
Dimension : 100				
QIEA	10599.0	858.7	8244.5	38.13
GA	22845.0	2594.9	18547.0	194.28

Table 6. Griewangk results

[-600,600]	Best	Mean	S.D.	Time(s)
Dimension : 5				
QIEA	0.398	0.092	0.182	44.64
GA	0.355	0.167	0.105	25.95
Dimension : 10				
QIEA	0.803	0.078	0.638	45.18
GA	0.397	0.107	0.214	25.95
Dimension : 50				
QIEA	10.06	1.35	5.65	47.92
GA	12.57	2.21	7.34	127.72
Dimension : 100				
QIEA	173.36	21.76	131.43	55.13
GA	204.37	24.80	163.53	232.33

The results of benchmark tests are shown in Tables 3 to 6, where both QIEA and GA are run 30 times. The first column lists the minimal (overall best) objective value found during the 30 runs. The second and third column lists the mean and standard deviation for the best value found in each of the 30 runs. The *Time* column shows the total processing time taken for 30 runs.

The results indicate that QIEA performs better (relative to GA) as the search space becomes more complex and the dimensionality of the search space becomes larger. It is also notable that the algorithm's efficiency vs the GA increases as the problem becomes larger. These results suggest the interesting potential of QIEA as an optimising algorithm in hard, high-dimensional problems.

4 Experimental Approach

In this section we explain the importance of the implied volatility smile in option trading and how QIEA will be applied for non-linear principal component analysis.

4.1 Implied Volatility Smile

Implied volatilities are frequently used in the financial markets to quote the prices of options. Option traders and brokers monitor movements in volatility smiles closely. As option prices change over time the implied volatility smile (for various maturities) also changes.

If we stack the implied volatility smile (for one particular maturity) according to the time the IVS data was recorded, a time series of panel data, with highly correlated entries, results. Implied volatilities at different strike prices are highly correlated because as the volatility of the asset rises all implied volatilities rise, yet some may rise more than others. However the economic forces of no-arbitrage (no free-lunch) ensures that the implied volatilities cannot get too detached from one another because if they did this represents a riskless trading opportunity for savvy investors, who sell the more expensive option (with the higher implied volatility) and hedge it with cheaper options (with lower implied volatilities).

4.2 Non-linear Principal Component Analysis

Suppose $X \in M^{m,n}$ is a panel data set that contains correlated data points along the columns, evaluated at different points in time along the rows. Given that X consists of correlated data points, the variation in X can be decomposed into a small number r of orthogonal principal components with $r < n$, resulting in a reduction of the dimension of the problem with only a small loss in information. The principal components from standard PCA are linear combinations (along the rows) of the original data set. If it is suspected that the data set contains non-linearities, a common procedure is to "linearise" the data set using suitable transformations prior to analysis. This approach has the advantage that it retains the simplicity of the underlying principal component analysis (PCA)

Table 7. Mapping functions employed for non-linear principal component analysis

$g_1(X)$	$g_2(X)$	$g_3(X)$	$g_4(X)$	$g_5(X)$	$g_6(X)$	$g_7(X)$	$g_8(X)$	$g_9(X)$	$g_{10}(X)$
$4X(1\text{-}X)$	$1\text{-}1.4X_t^2+0.3X_{t-1}$	$0.25X_{t-1}+\varepsilon$	$\exp(X)$	$\sin(X)$	$\cos(X)$	X_{t-1}	X_{t-2}	$X_t\text{-}X_{t-1}$	$X_t\text{-}X_{t-2}$

whilst gaining the ability to cope with non-linear data. To do this we construct a modified data set X_{NL} from the original data set X:

$$X_{NL} = G(X), \tag{4}$$

where G is a function consisting of n individual mapping functions from linear to non-linear space:

$$G = w_1 g_1(X) + w_2 g_2(X) + \cdots + w_n g_n(X), \tag{5}$$

and where $g_i(X)$ is an individual non-linear mapping function of X and w_i is the weight on the function g_i. There are an infinite number of mapping functions $g_i(X)$ to choose from and in this paper we consider a small number of mapping functions we think are important given the domain knowledge of the problem under consideration. There are a total of ten functions chosen in this study, including time-series models, and they are given in Table 7.

The previous evidence for PCA applied to implied volatility smiles ([10,12,7]) suggests that changes in the implied volatility smile are driven predominantly by three factors. The first factor is a level factor which controls the overall height of the implied volatility smile. The second and third factors are slope and curvature factors across the strike price axis. However options and the implied volatilities associated with options are multi-dimensional non-linear instruments and standard PCA may neglect some of non-linear subleties inherent in option implied volatilities. This is the reason NLPCA is applied to the IVS in this paper.

5 Results

5.1 Data

The data used in this study are option implied volatilities across 11 different strikes and a number of different maturities on the FTSE 100 index. The data consists of end-of-day settlement option implied volatilities from the 26th of March 2004 till the 17th of March 2006 consisting of 500 trading days. FTSE 100 index options are European style options and the underlying asset is the FTSE 100 performance index. To price options on this index one must adjust the index by extracting the present value of future cash dividend payments before each options expiration date. The annualised dividend yield of the FTSE 100 index was downloaded from Datastream. The one-month LIBOR rate was used as the risk-free rate where the LIBOR rate was converted into a continuously compounded rate. The forward price used in the option calculations is then simply $F_t = S_0 e^{(r-q)t}$ where S_0 is the current index price level, F_t is the price for the forward contract maturing at time t, r is the continuously compounded risk-free rate and q is the continuously compounded dividend yield.

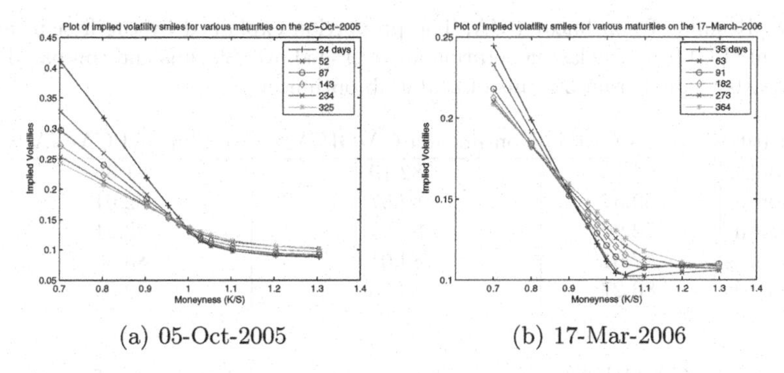

(a) 05-Oct-2005 (b) 17-Mar-2006

Fig. 1. Implied Volatility Smiles on two different dates

We have interpolated implied volatilities on a fixed grid of moneyness and maturity for all the days in the data sample. For each day t in the sample we define the implied volatility smile at a fixed moneyness n_m and maturity τ_j by

$$IVS(t) = \{I_t(1, \tau_j), \ldots, I_t(n_m, \tau_j)\}\,.$$

We then stack these implied volatility smiles over time to form the data matrix $X = \{IVS(1), \ldots, IVS(500)\}'$. Non-linear principal component analysis (NLPCA) is conducted on the implied volatility smile matrix for maturities ranging from 2 to 6 months.

5.2 Result Analysis

The first three principal components from linear PCA explain up to approximately 96% of the variation in the level of the implied volatility smile, depending on the maturity of the IVS considered. As 96% in PCA analysis may be overfitting, we would rather target the first principal component than the first three components and this why the objective function in the NLPCA was chosen to be proportion of variation explained by the first principal component.

The analysis of the eigenfactors from standard PCA for the implied volatility smiles of each maturity shows that the first factor has a positive effect on all implied volatilities. This eigenfactor can be interpreted as a level or a volatility factor. An increase in this factor causes the whole IVS to increase and causes all options to become more expensive since options are increasing functions of volatility. The second factor has a negative effect for implied volatilities with $K < S$, e.g. out-of-the-money puts, and a positive effect for implied volatilities with $K > S$, e.g. out-of-the-money calls. This factor can be interpreted as a skew factor and increase in this factor causes out-of-the money calls to become more expensive relative to out-of-the-money puts. The third factor has a positive effect for implied volatilities with $K < S$ and $K > S$ e.g. out-of-the-money calls and puts, and a negative effect for implied volatilities that are close to the money with $K \approx S$. This factor can be interpreted as a curvature factor an an increase in this factor causes out-of-the money calls and puts to become more expensive relative to near-the-money calls and puts.

Table 8. Results of non-linear PCA. The proportion explained by the first principal component (PC) from the last generation are averaged over 30 runs and compared with the parameter values from 30 runs of a Matlab optimiser.

Maturity	Linear PCA(%)	Non-linear PCA-QIEA(%)	Non-linear PCA-GA(%)
2 months	64.15	82.19	81.35
3 months	69.57	82.67	82.01
4 months	72.90	83.94	83.21
5 months	77.01	84.01	83.93
6 months	80.27	83.23	82.39

In our NLPCA-QIEA analysis, the weights on the mapping functions are optimised by using a quantum-inspired evolutionary algorithm to maximise the objective function which is the proportion of variation in the data explained by the first principal component. The weights are also optimised using the GA Matlab toolbox developed by Andrew Chipperfield. Fig. 2 depicts the evolution of the objective function versus the generation number. The parameter settings in the QIEA are given in Table 2. NLPCA is more efficient than linear PCA especially for the options with shorter times-to-maturity. For example, for the 2 month IVS the 1st principal component from NLPCA explains approximately 82% of the variation of the data versus only 64% for standard PCA. However the outperformance of NLPCA is to be expected given the extra degrees of freedom involved since it uses four non-linear functions that first operate on the data before PCA is applied. It is interesting to note that for the two month IVS the first component from NLPCA with ten non-linear functions explains 82% of the variation whilst the first *three* components from linear PCA explain up to 96% of the variation in the data. Although 96% is a higher level of explanatory power this is more than likely overfitting historical data at the expense of poor out-of-sample performance. If we forecast the evolution of the IVS out-of-sample using the techniques in this paper, a parsimonious procedure would be to include a more general set of time series models in the set of non-linear functions and use these to forecast the first factor from NLPCA and reconstruct future IVS's from the weights derived from historical analysis. This would be more parsimonious than fitting a separate time series model to three linear principal components and then reconstructing the future IVS as would have to be done in linear PCA. Thus, at least for shorter term options, the NLPCA method can explain 82% of the variation in the data with one linear combination of non-linear functions of the data versus approximately 64% for linear PCA. Thus rather than increasing the number of principal components in the analysis we have shown that another route is to use non-linear principal components to achieve a statistical significant increase in explanatory power.

It is interesting to note the weights on various functions that were derived in the NLPCA. The f_2 (Hénon function) captures a curvature effect (mentioned earlier) due to the squaring of the data, and a time series effect due to the dependence on past values. The weight on this function is close to one implying that the function depending on the curvature of the current IVS and the past

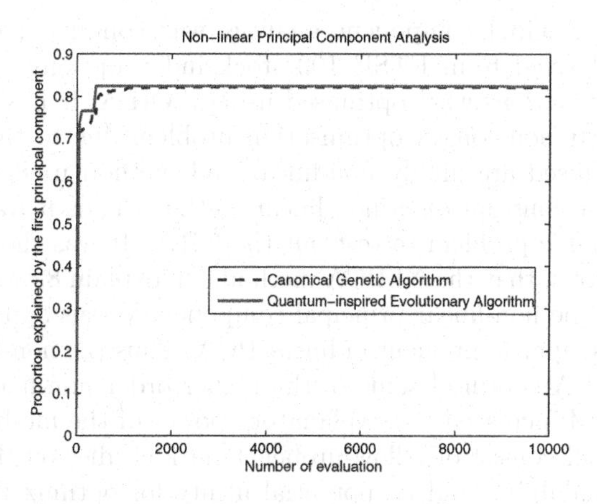

Fig. 2. Global search

Table 9. Weights on mapping functions. These weights on mapping functions from the last generation are averaged over 30 runs.

Maturity	f_1	f_2	f_3	f_4	f_5	f_6	f_7	f_8	f_9	f_{10}
2 months	0.124	0.925	0.328	0.024	0.036	0.897	0.106	0.152	0.194	0.375
3 months	0.121	0.901	0.371	0.016	0.047	0.923	0.097	0.133	0.189	0.342
4 months	0.151	0.917	0.368	0.020	0.068	0.925	0.072	0.067	0.169	0.274
5 months	0.141	0.917	0.324	0.016	0.059	0.918	0.095	0.102	0.185	0.293
6 months	0.137	0.882	0.363	0.019	0.066	0.903	0.074	0.073	0.168	0.247

level of the IVS is very important for explaining the variation in the IVS over time. The f_3 (auto regressive function) is capturing serial correlation in the daily movements of the IVS (something that cannot be done under linear PCA). Thus there is positive serial correlation in the data and this represents a possible trading strategy.

The weight on $cos(X)$ is approximately 0.9, which means this function contribute much during the mapping process. Also, both QIEA and GA can find the global optima. As they are sensitive to the parameters, i.e. crossover and mutation rate in GA, enlargement, shrinkage, and resize factor in QIEA, further analysis would need to be conducted for sensitivity test, other transformation functions and methods, such as Fourier transformation.

6 Conclusions

The results of benchmark tests suggest the potential of QIEA for application to high-dimensional problems. This is particularly interesting for financial applications which often require optimization in complex and high-dimensional environments.

A non-linear principal component analysis was conducted on the implied volatility smile derived from FTSE 100 stock index options. The weights on these non-linear functions were optimised using a QIEA. It has potential to be a highly non-linear non-convex optimisation problem due to the fact that the options data analysed are highly non-linear and method used to describe the variation in the options data is a non-linear method. Thus it was thought that this was a reasonable problem to test out the QIEA. It was shown, at least for shorter term options, that the NLPCA method can explain 82% of the variation in the data with one non-linear principal component versus approximately 64% for one linear principal component in linear PCA. Thus the non-linear functions used in the NLPCA captured some of the higher order non-linear factors that affect the data and increased the explanatory power of the method.

Future work will consist of follow-up benchmark studies on the QIEA to examine both its scalability and its potential utility for optimization in dynamic environments. In the context of the examination of the implied volatility smile, future work consists of expanding the number of non-linear functions being considered with a focus on including a larger number of time series models. This would be very useful in predicting the IVS out-of-sample and in constructing options trading strategies. Future work could also look at multi-objective NLPCA where the proportion of variation explained by the first factor is maximised followed by the proportion of variation explained by the second factor, etc. Also it would be useful to relax the restriction on the parameters of the non-linear functions used in NLPCA and allow the QIEA to find optimal values for these parameters. All of these extensions will result in very high-dimensional optimisation problems where the use of evolutionary algorithms such as the QIEA may be essential.

References

1. Brabazon, A., O'Neill, M.: Biologically-inspired Algorithms for Financial Modelling. Springer, Berlin (2006)
2. da Cruz, A., Barbosa, C., Pacheco, M., Vellasco, M.: Quantum-Inspired Evolutionary Algorithms and Its Application to Numerical Optimization Problems. In: Pal, N.R., Kasabov, N., Mudi, R.K., Pal, S., Parui, S.K. (eds.) ICONIP 2004. LNCS, vol. 3316, pp. 212–217. Springer, Heidelberg (2004)
3. da Cruz, A., Vellasco, M., Pacheco, M.: Quantum-inspired evolutionary algorithm for numerical optimization. In: Proceedings of the 2006 IEEE Congress on Evolutionary Computation (CEC 2006), Vancouver, 16-21 July, pp. 9180–9187. IEEE Press, Los Alamitos (2006)
4. Fan, K., Brabazon, A., O'Sullivan, C., O'Neill, M.: Quantum-Inspired Evolutionary Algorithms for Calibration of the VG Option Pricing Model. In: Giacobini, M. (ed.) EvoWorkshops 2007. LNCS, vol. 4448, pp. 189–198. Springer, Heidelberg (2007)
5. Fan, K., Brabazon, A., O'Sullivan, C., O'Neill, M.: Option Pricing Model Calibration using a Real-valued Quantum-inspired Evolutionary Algorithm. In: GECCO 2007, pp. 1983–1990. ACM Press, New York (2007)

6. Fan, K., O'Sullivan, C., Brabazon, A., O'Neill, M.: Testing a Quantum-inspired Evolutionary Algorithm by Applying It to Non-linear Principal Component Analysis of the Implied Volatility Smile. In: Natural Computing in Computational Finance, Springer, Heidelberg (in press, 2008)
7. Fengler, M., Härdle, W., Schmidt, P.: Common factors governing VDAX movements and the maximum loss. Jounal of Financial Markets and Portfolio Management 1, 16–19 (2002)
8. Han, K.-H., Kim, J.-H.: Quantum-inspired evolutionary algorithm for a class of combinatorial optimization. IEEE Transactions on Evolutionary Computation 6(6), 580–593 (2002)
9. Han, K.-H., Kim, J.-H.: On setting the parameters of quantum-inspired evolutionary algorithm for practical applications. In: Proceedings of IEEE Congress on Evolutionary Computing (CEC 2003), August 8–December 12, 2003, pp. 178–184. IEEE Press, Los Alamitos (2003)
10. Heynen, R., Kemma, K., Vorst, T.: Analysis of the term structure of implied volatilities. Journal of Financial and Quantitative Analysis 29, 31–56 (1994)
11. Narayanan, A., Moore, M.: Quantum-inspired genetic algorithms. In: Proceedings of IEEE International Conference on Evolutionary Computation, May 1996, pp. 61–66. IEEE Press, Los Alamitos (1996)
12. Skiadopoulos, G., Hodges, S., Clewlow, L.: The Dynamics of the S&P 500 Implied Volatility Surface. Review of Derivatives Research 3, 263–282 (1999)
13. Yang, S., Wang, M., Jiao, L.: A novel quantum evolutionary algorithm and its application. In: Proceedings of IEEE Congress on Evolutionary Computation 2004 (CEC 2004), June 19-23, 2004, pp. 820–826. IEEE Press, Los Alamitos (2004)

Analysis of Reconfigurable Logic Blocks for Evolvable Digital Architectures

Lukas Sekanina and Petr Mikusek

Faculty of Information Technology, Brno University of Technology
Božetěchova 2, 612 66 Brno, Czech Republic
sekanina@fit.vutbr.cz, imikusek@fit.vutbr.cz

Abstract. In this paper we propose three small instances of a reconfigurable circuit and analyze their properties using the brute force method and evolutionary algorithm. Although proposed circuits are very similar, significant differences were demonstrated, namely in the number of unique designs they can implement, the sensitiveness of functions to the inversions in the configuration bitstream and the average number of generations needed to find a target function. These findings are quite unintuitive. Once important (sensitive) bits of the reconfigurable circuit are identified, evolutionary algorithm can incorporate this knowledge. We believe that the proposed type of analysis can help those designers who develop new reconfigurable circuits for evolvable hardware applications.

1 Introduction

One of possible approaches to building adaptive hardware is to combine reconfigurable hardware with search algorithms. In the field of evolvable hardware, the evolutionary algorithm is used to find a suitable configuration of a reconfigurable device [1, 2].

In the area of digital circuits, application-specific reconfigurable circuits and field programmable gate arrays (FPGA) can be considered as the most popular reconfigurable platforms for evolvable hardware. In general, the reconfigurable digital circuit consists of an array of programmable logic elements, programmable interconnects and programmable I/O ports. The function of programmable logic elements and their interconnection (i.e. the circuit functionality) is defined using a configuration bitstream. The configuration bitstream is stored in a configuration register (or memory) whose bits directly control the configurable switches and multiplexers of the platform.

When the evolvable system is completely implemented on a single chip, a part of the chip is devoted for evolving designs and another part is used to implement the evolutionary algorithm. In these systems, evolutionary algorithm usually directly operates with the configuration register, i.e. the chromosome is considered as a candidate configuration. A kind of internal reconfiguration has to be employed (for example, ICAP in Xilinx Virtex II+ families [3]). Another option is to configure the reconfigurable device externally, for example, from a PC where the evolutionary algorithm is implemented [4]. The quality of evolved solutions

M. Giacobini et al. (Eds.): EvoWorkshops 2008, LNCS 4974, pp. 144–153, 2008.

depends on the evolutionary algorithm as well as the reconfigurable device. As this paper primarily deals with reconfigurable hardware in evolvable digital architectures, we will focus our attention only on the reconfigurable circuit.

When one is building a new reconfigurable ASIC, the reconfigurable circuit can be designed exactly according to requirements of a given application. Designer can choose the optimal type and count of configurable logic elements, suitable interconnecting network as well as configuration subsystem (organization of the configuration memory, the style of reconfiguration etc.).

When one is building a reconfigurable device with the FPGA, there are two options. (1) Evolution can work at the level of logic blocks available in the FPGA. In other words, it operates directly with the configuration bitstream of the FPGA [4, 3]. This solution requires the knowledge of the internal structure of the FPGA and the configuration bitstream. It is usually very efficient in terms of resources; however, it can be slow. (2) A new reconfigurable circuit is created on the top of an FPGA [5]. Using this method, sometimes called Virtual Reconfigurable Circuit (VRC), a very efficient reconfigurable device can be created for a given application. However, its implementation cost can be significant, as everything must be implemented using resources available in the FPGA.

In both cases, designer has to come up with a suitable configurable logic blocks and configurable interconnections with respect to the target application. Designer has to define the organization of the configuration register (memory) in order to maximize the efficiency of evolutionary algorithm.

The goal of this paper is to demonstrate how these design choices can influence the class of functions which will be implementable in the particular reconfigurable circuit and the efficiency of a search algorithm in the space of possible configurations. In particular, we will be interested in those architectures in which the reconfiguration subsystem is implemented using multiplexers, i.e. the function of a configurable element as well as the interconnection is determined using multiplexers. This is typical for reconfigurable ASICs as well as for VRCs. In this study we will consider a simple reconfigurable circuit with four inputs and two outputs and a 32-bit configuration register. As the number of possible configurations is relatively small (2^{32}), we can analyze its behavior by brute force to see how its structure influences the number of implementable designs, the efficiency of the search algorithm and the effect of mutations in the configuration bitstream. Results of the analysis can be exploited for designing new evolvable systems for real world-applications in which the small reconfigurable circuit can represent a single reconfigurable logic block of a complex reconfigurable system.

2 Proposed Model of a Reconfigurable Circuit

In order to perform the analysis of a typical reconfigurable unit observable in current evolvable hardware systems, we propose to investigate the structure and properties of a small instance of VRC. The VRC is used to implement a combinational logic circuit of four inputs (x_0, x_1, x_2, x_3) and two outputs (y_0, y_1) whose

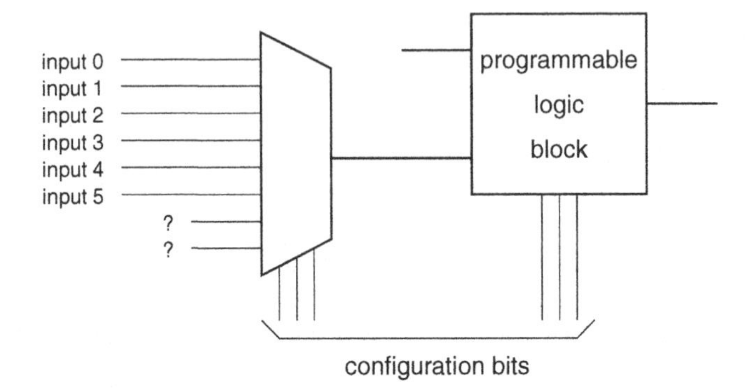

Fig. 1. A non-optimized use of multiplexers

function is defined using 32 bits. The VRC contains four logic blocks. Obviously, such circuit can be implemented using two 16-bit look-up tables. However, that type of implementation would not allow us to estimate the behavior of larger instances of VRCs which exhibit many similar features with the proposed architecture.

Because it is supposed that the chromosome directly represents the configuration bits of the VRC, all possible bit combinations should represent valid circuits. Moreover, in order to make the evolution efficient, the implementation of EA easier and the utilizations of hardware resources economical, it is desirable to perfectly utilize all possible combinations of groups of bits. Consider the example given in Figure 1. The 8-input multiplexer effectively uses only six inputs; the last two inputs has to be connected somewhere. Anyway, three bits must be included in the chromosome to control the multiplexer's selector. Then, the two out of eight combinations are not used effectively, which can turn the search algorithm to a wrong part of the search space.

For comparisons, we propose three architectures of VRC, labeled as cfg4f, cfg8f and cfg16f. They have the same number of inputs, outputs, configurable blocks and the size of configuration register. They differ in the number of functions supported by configurable blocks and the reconfiguration options.

2.1 Reconfigurable Circuit cfg4f

Figure 2 shows the reconfigurable circuit cfg4f which consists of four programmable elements B0–B3 and three stages of multiplexers. Each of configurable blocks can implement four different functions. The first stage of multiplexers selects a primary input which will be connected to blocks B0 and B1. As there are only six possible input points for the second stage of multiplexers (four primary inputs and the outputs of B0 and B1), the 8-input multiplexers can not be utilized perfectly. Hence, the output of block B0 and B1 is connected to the multiplexers twice. From the point of evolutionary design, the probability that a connection is made between block B0/B1 and the second stage of multiplexers is higher than for the primary inputs and the second stage of multiplexers. The

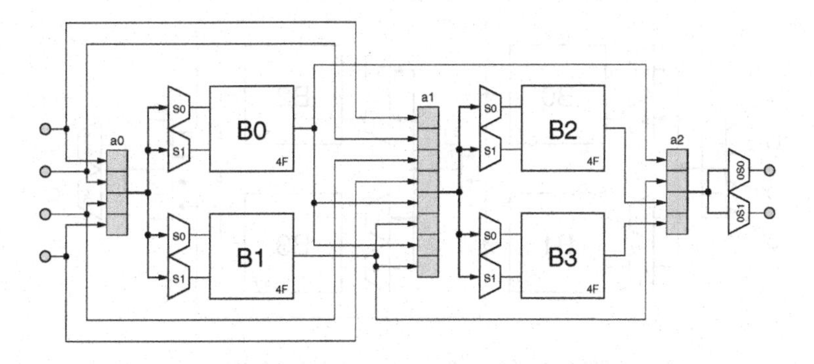

Fig. 2. Reconfigurable circuit cfg4f

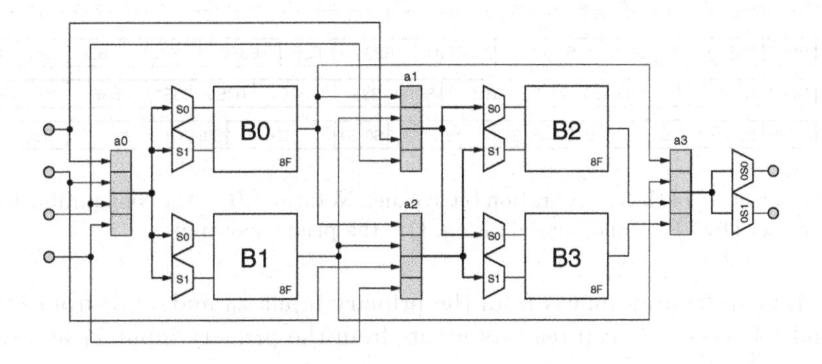

Fig. 3. Reconfigurable circuit cfg8f

primary outputs can be connected either to blocks B0, B1, B2 or B3 using the third stage of multiplexers. Selection bits of the third stage of multiplexers are perfectly utilized. For next comparisons, we will consider three variants of cfg4f which differ in the function sets supported in configurable blocks:

- cfg4f(xornot) utilizes functions NAND(0), NOR(1), XOR(2) and NOT(3)
- cfg4f(xoror) utilizes functions NAND(0), NOR(1), XOR(2) and OR(3)
- cfg4f(xorxnor) utilizes functions NAND(0), NOR(1), XOR(2) and XNOR(3)

2.2 Reconfigurable Circuit cfg8f

Figure 3 shows reconfigurable circuit cfg8f which employs configurable blocks with eight functions (NOR (0), x AND $\bar{y}$ (1), $\bar{x}$ AND y (2), AND (3), OR (4), $\bar{x}$ OR y (5), x OR $\bar{y}$ (6), NAND (7)). Because three bits of the configuration bitstream are devoted to the selection of a function, fewer bits can be used to define the interconnects.

The first and third stage of multiplexers is identical with cfg4f. As the second stage uses 4-input multiplexers, not all six possible points (primary inputs and the outputs of blocks B0 and B1) can be connected to block B2 and B3. Hence,

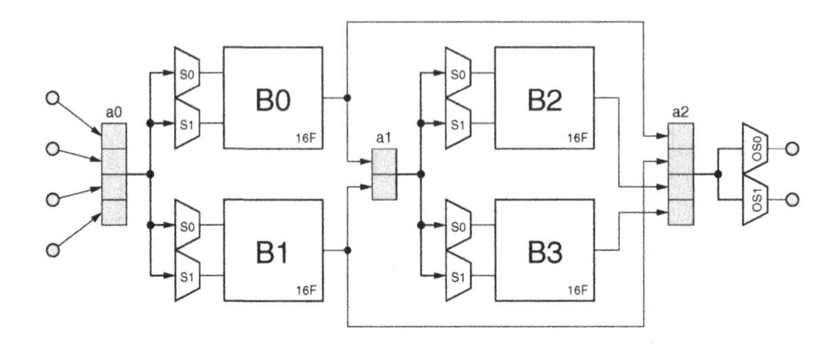

Fig. 4. Reconfigurable circuit cfg16f

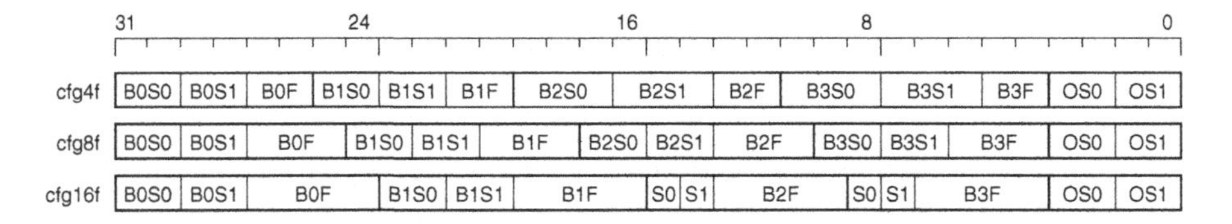

Fig. 5. Structure of the configuration bitstreams. Notation: B_x – the configurable block, S_x – the selector, F – function definition, O – the primary output.

block B2 can read its inputs from the primary input x_0 and x_2 or from blocks B0 and B1. Block B3 can read its inputs from the primary input x_1 and x_3 or from blocks B0 and B1.

2.3 Reconfigurable Circuit cfg16f

Similarly to cfg8f, also cfg16f restricts the interconnection options. Blocks B2 and B3 can be connected only with blocks B0 or B1. On the other hand, this allows the use of a full repertoire of possible logic functions over two logic variables in all configurable blocks. Figure 4 shows architecture of cfg16f. Finally, the structure of configuration bitstreams of all circuits is given in Figure 5.

3 Experimental Evaluation

In order to analyze the behavior of proposed reconfigurable circuits a well-optimized software simulator was created. As a single configuration can be evaluated in approx. 100 ns, it is possible to test all configurations in less than 8 minutes on a common PC. The number of different configurations is $|C| = 2^{32}$. The theoretical number of possible logic behaviors is $|F| = 2^{n_o.2^{n_i}}$, where n_i is the number of primary inputs and n_o is the number of primary outputs, i.e. $2^{2.2^4} = 2^{32}$ in our case. However, the number of logic functions which can be implemented in the reconfigurable circuit is much lower because two and more different configurations quite often represent the same logic behavior. This is typical for all reconfigurable devices used for evolvable hardware.

Table 1. Characterization of reconfigurable circuits in terms of the number of different (unique) logic functions which can be implemented and the number of different implementations of a single logic function (occurrence)

Circuit	cfg4f (xornot)	cfg4f (xoror)	cfg4f (xorxnor)	cfg8f	cfg16f
Unique designs	57,837	119,502	104,468	178,764	57,712
Avr. occurence	74,260	35,941	41,113	24,026	74,421
Max. occurence	86,994,432	86,926,336	95,109,120	113,224,704	308,514,048
Min. occurence	256	128	256	64	256
Designs with min. occ.	13,272	1,512	29,160	29,952	15,864

3.1 Achievable Functions

First series of experiments is devoted to characterizing proposed reconfigurable circuits in terms of the number of unique designs (logic functions) and the occurrence of some specific designs. Table 1 shows that the number of unique designs (in the space of 2^{32} possible designs) is quite small. Circuit cfg8f provides the highest number of unique designs (178,764). We can observe how significantly the number of unique designs decreases when only one of functions in configurable blocks is changed from the two-input OR to the single input NOT. Some functions are very frequent on all reconfigurable circuits, for example $y_0 = y_1 = 0$, $y_0 = y_1 = 1$, or $y_0 = y_1 = x_k$. No function exists which can be implemented uniquely; the minimum number of occurrences of a function is 64. There are only 8,888 different functions which can be implemented on all five variants of the circuit.

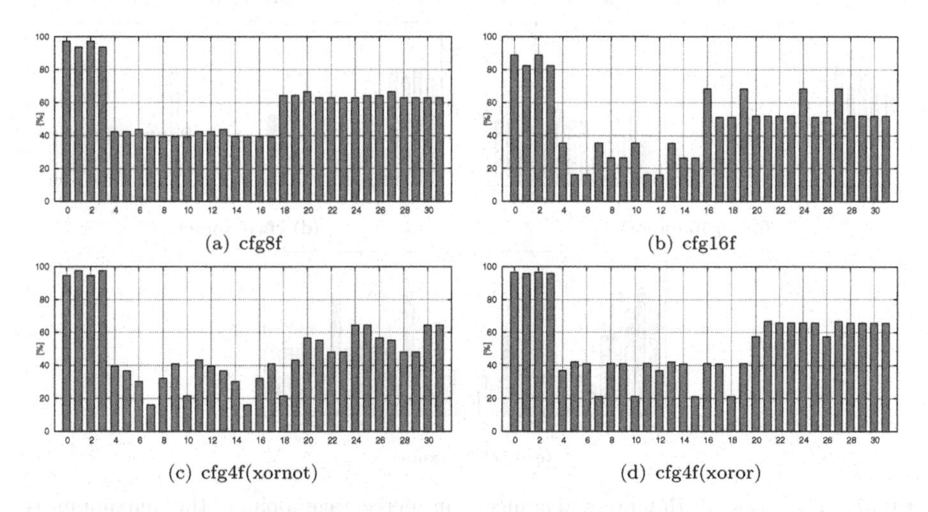

(a) cfg8f (b) cfg16f

(c) cfg4f(xornot) (d) cfg4f(xoror)

Fig. 6. The number of changes in logic functions when jth bit is inverted (calculated for all configurations)

3.2 The Sensitivity of Functions to Inversions

As mutation is usually implemented using the inversion of a particular bit, it is important to analyze to what extent the circuit function is sensitive to bit inversions of the configuration bitstream. For every configuration c_i, we calculated the corresponding logic function $f_i(c_i)$. Then, for every single independent inversion of the configuration bit $j, j \in \{1, \ldots, 32\}$, we checked whether the new function $f_i(c_i)^{(j)}$ is different from $f_i(c_i)$. Figure 6 shows how many times (in percentage points) the logic function is changed when jth bit of the configuration c_i is inverted ($i \in \{1, \ldots, 2^{32}\}$). A general observation is that independently of the reconfigurable circuit and its configuration, we can see that the logic function is changed in more than 90 % cases when bit 0, 1, 2 or 3 are inverted. The circuit function is also very sensitive to other four bits in cfg16f. Other bits do not seem to be so important. Results are not shown for cfg4f(xorxnor) because they are indistinguishable from cfg4f(xoror) in Figure 6.

Figure 7 shows the results of the same experiment; however, the y-axis does not give the number of changes in logic functions. It displays the sum of Hamming distances (in percentage points, the maximum is $2^{32}.2^{n_o.2^{n_i}}$) between truth tables of original logic functions and truth tables of logic functions obtained using the inversion of jth bit. Thus, we can see how significant the inversions are for

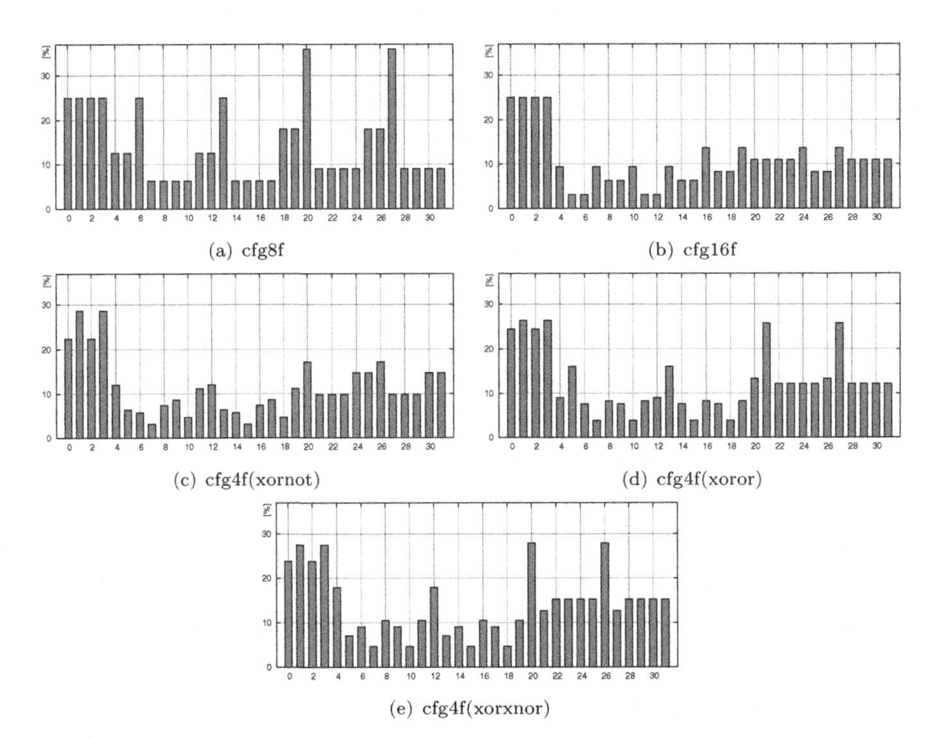

(a) cfg8f (b) cfg16f

(c) cfg4f(xornot) (d) cfg4f(xoror)

(e) cfg4f(xorxnor)

Fig. 7. The sum of Hamming distances (in percentage points, the maximum is $2^{32}.2^{n_o.2^{n_i}}$) between truth tables of original logic functions and truth tables of logic functions obtained using the inversion of jth bit

Table 2. The average percentages of changes in logic functions and Hamming distances of truth tables over all configurations and independent bit inversions

Circuit	cfg4f (xornot)	cfg4f (xoror)	cfg4f (xorxnor)	cfg8f	cfg16f
Average changes [%]	49.27	53.97	54.30	57.72	48.32
Average Hamming distance [%]	11.62	12.70	14.16	14.55	10.94

a particular bit of the configuration. For example, while logic functions strongly depend on bits 20 and 27, the importance of bits 0, 1, 2 and 3 is less significant for cfg8f in comparison to Fig. 6. The graphs obtained using the Hamming distance are not as uniform as in the previous case.

Table 2 summarizes average values of percentage points from Fig. 6 and Fig. 7. Both metrics suggest that the cfg8f architecture is the most sensitive to the (independent) inversions of the configuration bits. These results are also correlated with the number of unique design which can be implemented in a given reconfigurable circuit (see Table 1).

3.3 Circuit Evolution on Proposed Architectures

In order to evaluate proposed reconfigurable circuits for purposes of the evolutionary circuit design, a suitable target problem has to be chosen. As we have already mentioned, there are 8,888 functions which can be implemented on all five architectures. The goal of this experiment is to calculate the average number of generations that are needed to find all these functions on all architectures. A simple evolutionary algorithm was utilized which directly operates at the level of configuration bitstream of a particular reconfigurable circuit. It utilizes the population of five individuals. A new population is formed using (1+4) evolutionary strategy. A single bit is always inverted during mutation. The probability of selection is identical (uniform) for all bits. The fitness value is determined as the number of output bits which are correctly calculated by a candidate circuit for all possible input combinations. The maximum fitness is 32. Evolutionary algorithm was executed 200 times for every target function. A single run is stopped when a perfect fitness value is obtained or 50,000 generations are exhausted. Results are given in the first part of Table 3 which shows the average number of generations for a single function, calculated as the average number of generations from all runs and for all 8,888 target functions. We can observe that the best value is obtained for the cfg16f (797) circuit and the worst one is for the cfg8f circuit (2,734). This result corresponds with the number of unique designs. If more unique designs exist in the search space then it is more difficult to find a particular one.

3.4 Exploiting the Knowledge of Reconfigurable Architectures

Because we have recognized that logic functions are (in average) more sensitive to some configuration bits than to some others, we can speculate whether

Table 3. Summary of results for evolutionary design of 8,888 functions

Circuit	cfg4f (xornot)	cfg4f (xoror)	cfg4f (xorxnor)	cfg8f	cfg16f
Average generations (uniform mut.)	1,115	1,929	1,553	2,734	797
Average generations (nonuniform mut.)	1,009	1,728	1,480	2,710	813
Speedup	1.11	1.17	1.05	1.01	0.98

Table 4. Masks for the bits selected for nonuniform mutation

Circuit	sensitive	sensitive-inverted	average
cfg4f(xornot)	0xc410000f	0x3beffff0	0x38e80800
cfg4f(xoror)	0x0820202f	0xf7dfdfd0	0x33c01010
cfg4f(xorxnor)	0x0410101f	0xfbefefe0	0x0be80100
cfg8f	0x0810204f	0xf7efdfb0	0xf0001830
cfg16f	0x0909000f	0xf6f6fff0	0x00f02490

Table 5. Summary of results for evolutionary design of 8,888 functions when non-uniform mutation is used on selected bits

Circuit	cfg4f (xornot)	cfg4f (xoror)	cfg4f (xorxnor)	cfg8f	cfg16f
Avr. genenerations (nonunif. mut. 1/4)	1,970	3,151	2,669	3,691	1,105
Speedup	0.57	0.61	0.58	0.74	0.72
Avr. genenerations (average bits)	1,524	2,445	2,065	3,545	1,021
Speedup	0.73	0.78	0.75	0.77	0.78

a higher/lower mutation probability of these sensitive bits can improve the convergence of the evolutionary algorithm. For each reconfigurable circuit we have chosen eight the most sensitive bits and mutated them with the probability 4-times higher than other bits. This is called the nonuniform mutation in Table 3. The position of selected bits is given with respect to Figure 5 in Table 4 (column "sensitive").

We used the same evolutionary algorithm as in the previous section. Table 3 shows a small speedup of convergence (1–17 %) for four out of five investigated architectures. Therefore, it seems that our selection of sensitive bits is good.

In order to validate that the bits, which we have identified as sensitive, are really more important than other bits, two additional experiments were performed on all reconfigurable circuits. Firstly, we repeated the previous experiment, however, decreased the probability of mutation of eight the most sensitive configuration bits four times in comparison to other bits. Secondly, the previous experiment was repeated, but the probability of mutation was increased four times for 8 really average sensitive bits (their position is given as a mask in Table 4, column "average"). In both cases the speedup is much smaller than 1. That means that more generations are needed in average to find a solution and that this approach is not useful.

4 Conclusions

Although proposed circuits are very similar, significant differences were demonstrated, namely in the number of unique designs they can implement, the sensitiveness of functions to the inversions in the configuration bitstream and the average number of generations needed to find a target function. These findings are quite *unintuitive*. We believe that the proposed type of analysis can help those designers who develop new reconfigurable circuits for evolvable hardware applications. Once important (sensitive) bits of the reconfigurable circuit are identified, evolutionary algorithm can incorporate this knowledge. Additional knowledge can be included to the evolutionary algorithm and circuit architecture from the target domain. Typically, only a specific subset of all possible functions is evolved using the reconfigurable device. In this paper, we assumed that all possible functions belong to the application domain and will be evolved. Further research is needed to identify a suitable probability of mutation for the sensitive bits of a particular reconfigurable circuit and other parameters of the evolutionary algorithm.

Acknowledgements. This work was supported by the Grant Agency of the Czech Republic No. 102/06/0599 and the Research Plan No. MSM 0021630528.

References

[1] Higuchi, T., et al.: Evolving Hardware with Genetic Learning: A First Step Towards Building a Darwin Machine. In: Proc. of the 2nd International Conference on Simulated Adaptive Behaviour, pp. 417–424. MIT Press, Cambridge (1993)
[2] de Garis, H.: Evolvable hardware – genetic programming of a darwin. In: Int. Conf. on Artificial Neural Networks and Genetic Algorithms, Innsbruck, Springer, Heidelberg (1993)
[3] Upegui, A., Sanchez, E.: Evolving hardware with self-reconfigurable connectivity in Xilinx FPGAs. In: The 1st NASA/ESA Conference on Adaptive Hardware and Systems (AHS-2006), pp. 153–160. IEEE Computer Society, Los Alamitos (2006)
[4] Thompson, A., Layzell, P., Zebulum, S.: Explorations in Design Space: Unconventional Electronics Design Through Artificial Evolution. IEEE Transactions on Evolutionary Computation 3(3), 167–196 (1999)
[5] Sekanina, L.: Virtual Reconfigurable Circuits for Real-World Applications of Evolvable Hardware. In: Tyrrell, A.M., Haddow, P.C., Torresen, J. (eds.) ICES 2003. LNCS, vol. 2606, pp. 186–197. Springer, Heidelberg (2003)

Analogue Circuit Control through Gene Expression

Kester Clegg and Susan Stepney

Dept. of Computer Science, University of York
{kester,susan}@cs.york.ac.uk

Abstract. Software configurable analogue arrays offer an intriguing platform for automated design by evolutionary algorithms. Like previous evolvable hardware experiments, these platforms are subject to noise during physical interaction with their environment. We report preliminary results of an evolutionary system that uses concepts from gene expression to both discover and decide when to deploy analogue circuits. The output of a circuit is used to trigger its reconfiguration to meet changing conditions. We examine the issues of noise during our evolutionary runs, show how this was overcome and illustrate our system with a simple proof-of-concept task that shows how the same mechanism of control works for progressive developmental stages (canalisation) or adaptable control (homoeostasis).

1 Background and Motivation

We present a system for the automated discovery and deployment of software configured analogue circuits. High level circuit components (for example, analogue filters) are modelled as 'genes'. Genes are 'expressed' in response to circuit output; as conditions change new genes are expressed, triggering the deployment of a new circuit. Gene characteristics are defined in a genome representation discovered using evolutionary algorithms.

Evolutionary computation has a proven record as a technique for search-based optimisation [1,2,3]. The method has also been used to automate design discovery [4,5]. To date most work in the field focuses on the *performance* of evolutionary search. Fewer researchers are questioning the precepts on which stochastic, population-based algorithms are founded, whether their evolutionary model corresponds to modern biological thinking [6] or the role of developmental processes in evolution [7]. The latter is particularly important with regard to how gene expression explores the functional search space [8].

Thompson's ground-breaking work in the mid-1990s established that search algorithms could be used to discover novel hardware configurations that lie outside conventional engineering knowledge [9]. Thompson concentrated on *innovation* rather than rapid discovery, and was careful to design his experiments so that the search could exploit physical characteristics not normally incorporated into an engineer's design [10]. Following on from this, further examples using evolutionary algorithms in software configured physical media have come from

M. Giacobini et al. (Eds.): EvoWorkshops 2008, LNCS 4974, pp. 154–163, 2008.

Miller and Harding [11,12], and in both real and simulated physical environments from Stoica and others [13].

The problem of analogue circuit design has been met using evolutionary computation by John Koza and his colleagues [14]. In Koza's work, the evolutionary design process and fitness evaluations are run as software simulations. However, physical simulation by software imposes limits on what search algorithms can find. If one wants to expose the algorithms to resources outside conventional design knowledge, it is necessary to allow software to interact with the richer physics provided by real hardware. An alternative example using analogue circuit simulation and evolutionary computation comes from Mattiussi and Floreano [15,16]. They propose a generic model to represent genetic regulatory networks, metabolic networks, neural networks and analogue electronic circuits as 'analogue networks'. The representation and variable strength of links between components are found using evolutionary algorithms, and like Koza's discoveries, the network designs are claimed to be human-competitive [16].

Most representations in evolutionary computation evolve a single solution tested against a static problem, often with an almost one-to-one correspondence between the genotype and phenotype. This rigid translation contrasts with one of the most remarkable qualities of natural organisms: the ability of their encoding to permit adaptability in different conditions, particularly during developmental stages, yet retain a large degree of homoeostasis or canalisation. The eventual form of a natural organism is taken from a multitude of potential 'solutions' within its DNA, each of which differs and only one of which is expressed. We have tried to increase the distance between digital genotypes and physical, decoded phenotypes. In our architecture, the indirect translation gives genes different roles according to their context of expression. The model of gene expression drew inspiration from the descriptions of evolutionary developmental processes by Carroll, Wolpert and others [8,17]. Details of the model and how it maps to this biology can be found in [18], with more details on the hardware in [19].

2 Platform and System Description

Switched-capacitor based arrays allow the implementation of software configurable analogue circuits. Sometimes called Field Programmable Analogue Arrays (FPAA),[1] these platforms implement an analogue circuit by downloading a configuration bitstream onto an integrated circuit (IC). A typical FPAA application can be thought of as a set of analogue circuits, with a hosted application controlling when to reconfigure to a new circuit. Circuits on the Anadigm AN221E04 FPAA are composed of Configurable Analogue Modules (CAMs) that can contain filters, multipliers, integrators, differentiators, etc. or even signal or power sources. They are configured by setting options, floating point parameters and clock speeds. An application hosted on a PC is able use the Anadigm API to both create circuits and control the reconfiguration process.

[1] Alternative names for these ICs include Field Programmable Transistor Arrays (FPTA) or Dynamically Programmable Analogue Signal Processors (*dp*ASP).

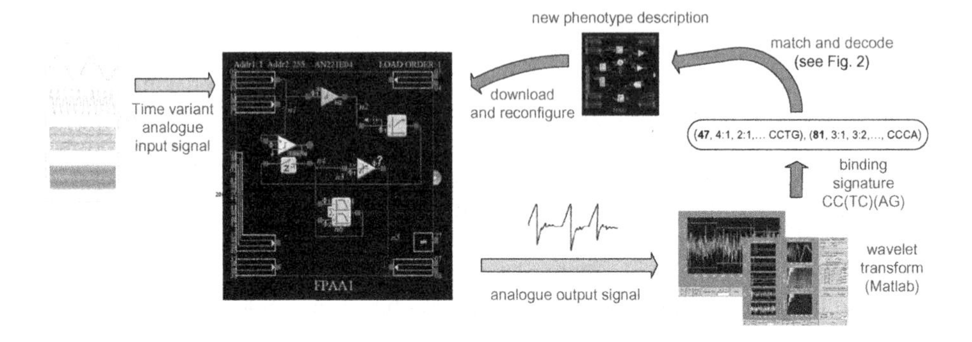

Fig. 1. System overview: analogue outputs are converted into digital binding signatures. Genes with binding sites that match the binding signature are expressed in the next circuit reconfiguration.

Our genome circuit specifications are represented using an adapted form of Cartesian Genetic Programming (CGP) [20] in a 4+1 evolutionary scheme. The genotype in CGP is a list of integers that encode the function and connection of each node in a directed graph. Nodes correspond to genes in our genome. Each gene encodes a CAM ID, its parameters, connections to other CAMs and its binding site. The binding site for each gene is a string representing some combination of the 4 bases (ACGT) in DNA. Reconfiguration is triggered by circuit output changing and generating a new 'binding signature', which is matched against gene binding sites in the genome. A gene that matches is expressed.

Due to the risk of exceeding the on-chip resources if all genes are expressed, we limit each genome to 3 genes giving 7 possible circuits (plus the 'empty' circuit which is ignored). In our experiments, the number of possible circuits is more than twice the number that can be deployed, ensuring there a degree of redundancy in the genome. As the wiring specification of the full genome does not match the wiring of partially expressed subsets of genes, the decoding process gives highly context dependent results: being expressed does not guarantee a gene then plays a functional part in a circuit, as it may have lost input or output connections depending on which other genes were also expressed [19].

The binding process provides feedback from the functional domain to the genome. It converts the circuit's analogue output into a binding signature by taking a wavelet transform of the output, normalising the coefficients and thresholding to get a binary valued matrix, with 4 rows corresponding to the 4 DNA bases, and a task dependent number of columns. More than one value in a column represents a 'wildcard' position that can match several bases, e.g. A{AC}GT is a signature of length 4 with 2 wildcards in position 2. If a column has no values above the threshold it is ignored. The grid is read column by column to produce a binding signature of fixed bases and wildcards which is continuously matched against the binding sites (see Fig. 1). As many positions are read as required to match the binding site lengths.

3 Task Description

The prototype was set a task with two primary objectives a) to evolve reliable reconfiguration b) to configure the IC in stages according to different fitness criteria. To meet these objectives, it was decided to input a series of frequencies (1kHz, 5kHz and 10kHz) in steps of fixed duration . The fitness criteria were to minimise power output from the chip on the 1kHz and 10kHz steps, and maximise power output from the 5kHz step.

Each phenotype starts against the output of a 'bare wire' (i.e. a straight connection between input and output), so that for the population to move to higher fitness a reconfiguration must take place. A frequency step is input for 5 seconds to allow the system to settle, after which the wavelet transform is converted to a binding signature and checked against each gene binding site. Each match is added to the list of genes that will be expressed to make up the next circuit. Reconfiguration takes place while input continues for a further 5 seconds, allowing the system to settle again. Finally, the Fourier transform and power reading for that circuit is taken and converted into a fitness score. The input is then changed to the next frequency step and the process repeated, or a new test is started.

The fitness scores are based on a formula that moves the selection process in the direction of circuits that maximise power in the middle frequency band:

$$Fitness = (B + R) - (A + C) \tag{1}$$

where B, A and C are the power readings at mid, lower and upper frequencies respectively. R is the reconfiguration bonus. As mentioned previously, the phenotype must initially respond to the output of a bare wire that passes the input signal through unaltered (except for a slight reduction in voltage). If no reconfiguration occurs, the power readings return negative fitness values. Due to noise in the system and a result of elitism in a 4+1 evolutionary scheme, phenotypes that do nothing (i.e. fail to reconfigure) can get selected above those that reconfigure badly if power readings are used exclusively as the basis of selection. As this imposes an unacceptable delay for the search process to 'get started', we penalise any phenotype that fails to reconfigure from the initial setting. Conversely phenotypes that reconfigure — even to a bad circuit — have their fitness scores augmented by a small bonus (usually less than 10% of final fitness score). On runs for random sequence frequency steps (see §6), more reconfigurations are required so bonuses are accordingly reduced.

4 Preliminary Runs

4.1 Noise

There are two principal areas for noise to affect testing phenotypes. The first is during wavelet transform, where wavelet coefficients measure similarity of the wavelet to a signal as it is time shifted at different scales. Part of our binding

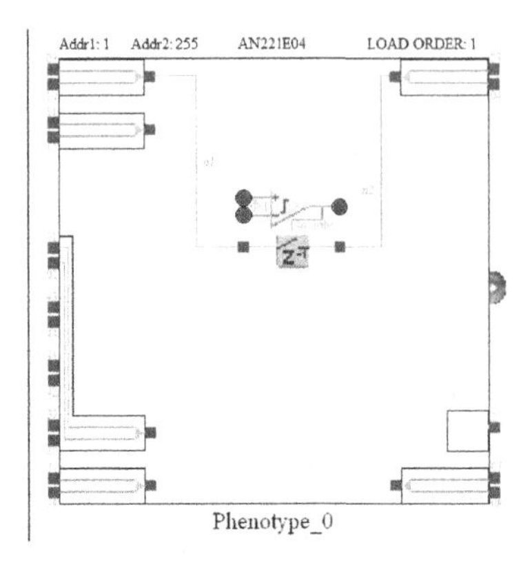

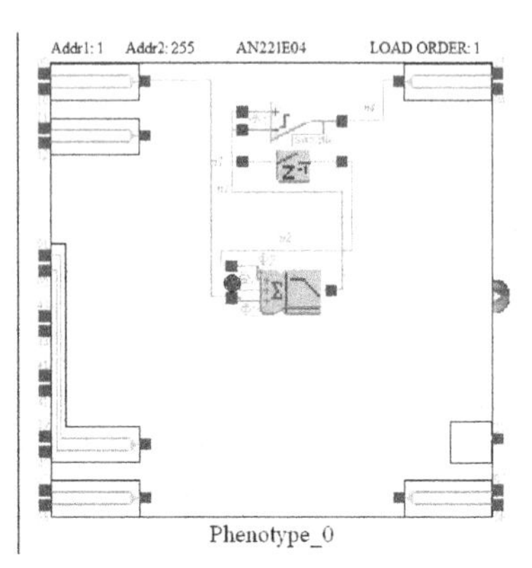

Fig. 2. The circuits deployed by an unreliable champion phenotype. Left circuit (HoldVoltageControlled) was downloaded at 1kHz, the circuit on the right (with SumFilter added) was downloaded at 5kHz and 10kHz.

process converts the wavelet coefficients into a four-row grid of binary values determined by a threshold. The more noise present in a signal, particularly a very weak signal, the more likely it will have coefficients scaled over the threshold. This translates into columns with more than one value — giving 'wildcards' for that binding position — meaning that more genes will be expressed as a result.

During evolutionary runs, no attempt is made to ensure gene expressions make valid circuits and it is possible that circuits may be broken (no output) or the signal output may be severely reduced or otherwise made noisy. In these cases, as coefficients are scaled, the noise introduced into the binding process results in many wildcards being produced for all positions. An unexpected effect of this is that a particular circuit can be downloaded by a 'lucky' binding sequence (i.e. one brought about by random noise due to a previous stage's bad circuit). The new circuit may get a high fitness score. However, the binding that caused the good circuit to download might never be repeated. Elitism results in the evolutionary process being 'conned' into keeping and mutating the poor solution for many generations. The mutations never make successive phenotypes improve above the lucky fitness score of the original phenotype, as the binding that allowed the good circuit to be downloaded occurred by chance, leaving the evolutionary process stranded on a false peak of high fitness.

A second source of noise in the system comes from the Fourier transforms at each frequency step. These readings are susceptible to fluctuations due to variations in heat and interference from surrounding electrical and computer appliances. Power readings of the same circuit configuration will therefore always vary from one test to another. These fluctuations are small but can vary by as much as 40%, although the figure is generally closer to 10% depending on the CAMs involved.

Aside from these examples, a more serious source of fluctuations was noticed after the first week of preliminary runs. Despite the degree of expected noise, some champion phenotypes would not reproduce anything close to the scores they achieved during the evolutionary run. For example, a phenotype that had scored 4130 during one run was re-tested five times at the end of the run. It scored successively -1067, -1036, 300, 4096, -1993. On closer inspection, it was noticed that despite the huge variation in fitness scores, the phenotype was reconfiguring using the same circuits at each frequency step. Two circuits were used (see Fig. 2) by the phenotype. The tests were repeated with the circuits loaded separately to observe their behaviour at each step. Neither circuit produced the high power rating in any of the frequency steps. In fact the circuits seemed inert and produced no output at all. Each of the circuits was then downloaded while input signals were being put through the chip. At first nothing happened, but then at random the switch in one circuit latched producing a high power output. Once downloaded the circuit reversed the switch at high frequency so that no output signal was produced at that frequency and no reconfiguration was required to produce a high fitness score. The reason for this behaviour lies in the fact that CAMs such as those shown in Fig. 2 exhibit hysteresis. Certain CAM behaviour depends on the input to a comparator (GainSwitch, GainPolarity, etc.). If the two inputs of the comparator are connected together then even small amounts of noise are going to lead to the CAM behaviour 'flipping'.[2]

Rather than exclude such CAMs from the pool of 'primitives' that evolution could select from, we decided to test each phenotype 5 times, taking the median result for our fitness scores. This strategy, although increasing the length of each run fivefold, had the effect of reducing the worst variability due to noise or unexpected CAM behaviour such as hysteresis. The results can be seen in Fig 3, where the heavy lines are the champion phenotype scores, while the light lines are the best of a generation. Runs that tested each phenotype once show best of generation scores varying considerably from the current champion. Taking the median of 5 tests gives best of generation scores much closer to the current

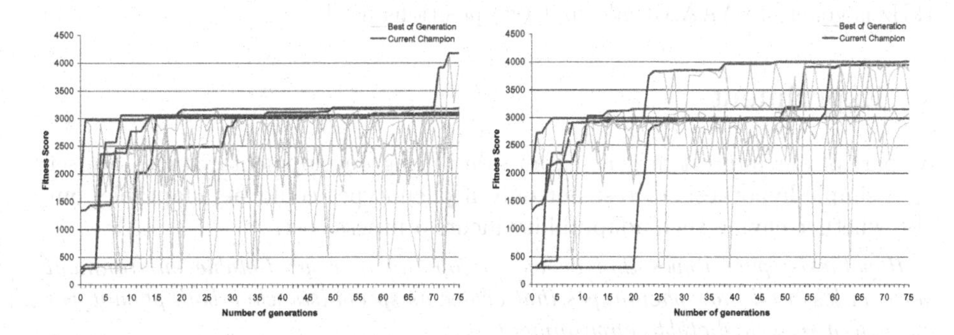

Fig. 3. Five runs using a binding signature length of two: one test per phenotype (left) and median of 5 tests per phenotype (right). The charts show current champion score (bold lines) against the best of each generation. Taking the median results in more stable phenotypes.

[2] Our thanks to Dave Lovell of Anadigm for this explanation.

champion, and produces final solutions that seem more stable with respect to the forms of noise mentioned.

4.2 Evolutionary Parameters

Tests were performed over 75 generations using a 4+1 evolutionary scheme. In this scheme, the best of previous generations is cloned, mutated and the selection process repeated. In our scheme, it was necessary to introduce variable mutation. Variable mutation is based on the numbers in each generation, so that the first clone receives no mutations per gene, the second one mutation per gene, the third two mutations and so on. This gave a maximum of 4 mutations per gene — enough to move poor phenotypes some distance from their parents, but also allow gradual mutations of good phenotypes. This can be verified from the selection history, where early progress is often made by the most heavily mutated phenotypes, but later stages rely on small improvements to good solutions.

Initial tests tried up to 500 generations, but further improvement was rare after 50 or 60 generations. This may be because the tests were too easy for the phenotypes to solve, as it became clear that there were many ways of achieving similar levels of high fitness. Another possibility is that relatively few mutations were required to reach the higher levels of fitness once an initial solution had been found.

In addition to reconfiguration bonuses, further 'bootstrapping' assistance was provided by allowing genes to have 'seeded' binding sites. Thus for a bare wire configuration, the binding signature produces a set of signatures that run through the frequencies (if no reconfiguration occurs) from AA, C/G, to TT. Having binding sites of length 2 gave a good chance of matches being found, but to speed up the search we seeded initial generations with binding site bases known for signatures with a bare wire configuration at the initial frequency step. For example, the first generation in a step-up specialist environment might have all binding sites seeded with A or C bases, as the bare wire output signature for 1kHz is generally AAAA (only first two positions used).

5 Experiment

As part of our interest in gene expression during developmental stages, we set up a simple hypothesis to test not only if phenotypes could be adaptable, but also whether having such adaptability incurs a fitness cost.

Hypothesis: phenotypes that evolve to cope an in unpredictable environment perform less well than phenotypes that evolve as specialists when both phenotypes are placed in a predictable environment.

To test this, the three frequency steps were considered developmental stages during which different behaviour was required from the phenotype. In specialist environments the steps in frequency either increased or decreased, in non-specialist environments all possible transitions occurred. In all environments the fitness test remained the same: maximise power in the 5kHz band, minimise it elsewhere. As a phenotype uses circuit output to trigger a reconfiguration, the

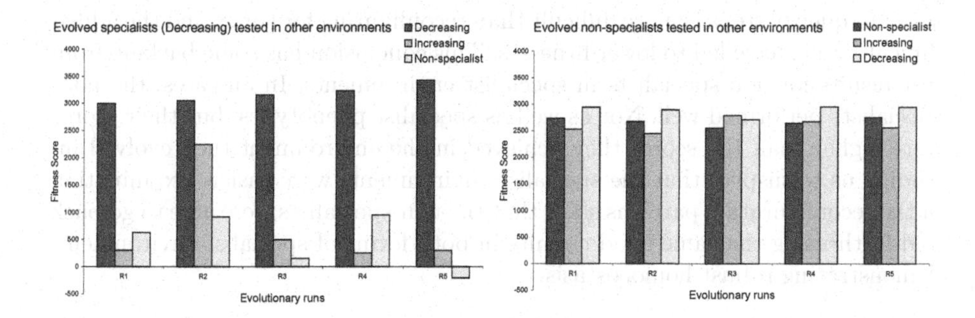

Fig. 4. Results of specialists (left) and non-specialists (right) tested across all environments. Dark bars show phenotype scores in their own environment.

current configuration is crucial to how the phenotype configures the next step. For example, a specialist phenotype may have configured to a high pass filter with gain suitable for high fitness during a 5kHz stage. On stepping up to 10kHz, this filter produces a binding signature that the phenotype can use to configure to the next circuit. However, if the same step occurs as part of a sequence that runs from 10kHz to 5kHz and back to 10kHz, the previously unknown step from 10kHz to 5kHz may cause the phenotype to configure the chip to another circuit. The step from 5kHz to 10kHz now results in a different output signature and the previously used genes for the good circuit no longer match. Specialists evolving in predictable environments can therefore both profit and suffer from previously encountered steps in a non-specialist environment. To counter this potential bias, the non-specialist environment starts with a bare wire configuration at 5kHz and tries to avoid known steps where possible.

6 Results

Each of the specialist and non-specialist phenotypes were evolved over 75 generations. The champion phenotype was then tested in environments it had not evolved in. The tests used the average of 2 results (each the taking median of 5 scores, as during evolutionary runs). The results, shown in Fig. 4, show poor capability for specialists in their 'opposite' specialist environment (where the frequency steps are reversed) and in the non-specialist environment. The reason seems largely the result of a strategy employed by phenotypes in the final stage of specialist environments. Circuits that reconfigure to 'broken' last circuits (at 1kHz or 10kHz stages) help their fitness scores and have nothing to lose from reconfiguring to a circuit that no longer uses circuit input or one that produces no output. But the same phenotypes suffered badly if they needed to recover from that stage in the non-specialist environment, and generally failed to reconfigure.

The runs for non-specialist phenotypes show a slightly reduced fitness in their environment compared to specialists. This would seem to back the hypothesis given in §5, however it should be borne in mind that reconfiguring to manage seven frequency steps is more difficult than reconfiguring for three, and the added difficulty may have led to lower fitnesses. This conclusion has some backing from test results for non-specialists in specialist environments. In all cases, the non-specialists performed well. Not as well as specialist phenotypes, but their scores were higher than the scores they achieved in the environment they evolved in, leading us to suspect that the specialist environments were easier. Examination of the reconfiguration patterns show that the non-specialists were able to go back and forth using the same good circuits in both forms of specialist environment, demonstrating robust homoeostasis.

7 Conclusions and Future Work

Our system has demonstrated the potential of using a feedback mechanism as a means of reconfiguring solutions specified by an evolved genome representation. The representation is capable of maintaining multiple solutions and a large degree of redundancy. This redundancy has the potential to be expressed if conditions change, something which could be valuable where evolution is carried out *in situ* (see Stoica et al in [13]). The downsides to the technique are that the total number of potential gene expressions quickly rises as genome length increases. In such cases, it may be impossible to predict what solutions may be expressed if the conditions used to trigger reconfiguration change unexpectedly.

Biological control through stages of development relies on gene expression triggered by the context of transcription factors present in the nucleus of the cell at that moment in time. Homoeostatic control relies on gene expression controlled in a similar fashion, but with the option of reversibility. In our experiment, we wanted to see if our system could devise mechanisms of control of either type. The specialist phenotypes show in many cases the 'blindness' of their future proofing after evolutionary runs have left them stranded up specialist peaks of perfection. Non-specialists cannot afford the luxury of such high fitnesses, as it means they won't remain adaptable in the face of unexpected changes in the environment.

Our next plans are to take our system and investigate the effects of scaling upwards. Long genomes, with many thousands of potential solutions, require an equally wide set of reconfiguration triggers to allow phenotypes to explore functional search spaces. We want to understand the role of developmental stages in these large evolutionary searches, and by scaling up see whether our system can evolve adaptability in more complex environments.

References

1. Whitley, D.: An overview of evolutionary algorithms: practical issues and common pitfalls. Information and Software Technology 43(14), 817–831 (2001)
2. Coley, D.: An Introduction to Genetic Algorithms for Scientists and Engineers. World Scientific Publishing, Singapore (1999)
3. Coello, C., Lamont, G., van Veldhuizen, D.: Evolutionary Algorithms for Solving Multi-objective Problems. Springer, Heidelberg (2006)

4. Koza, J.R., Keane, M.A., Streeter, M.J., Mydlowec, W., Yu, J., Lanza, G.: Genetic Programming IV: Routine Human-Competitive Machine Intelligence. Kluwer Academic Publishers, Dordrecht (2003)
5. Streeter, M., Keane, M., Koza, J.: Routine human-competitive automatic synthesis using genetic programming of both the topology and sizing for five post-2000 patented analog and mixed analog-digital circuits. In: 2003 Southwest Symposium on Mixed-Signal Design, pp. 5–10. IEEE Circuits and Systems Society (2003)
6. Banzhaf, W., Beslon, G., Christensen, S., Foster, J., Kepes, F., Lefort, V., Miller, J.F., RAdman, M., Ramsden, J.J.: Guidelines: From artificial evolution to computational evolution: a research agenda. Nature Reviews Genetics 7(9), 729–735 (2006)
7. Kumar, S., Bentley, P. (eds.): On Growth, Form and Computers. Elsevier, Amsterdam (2003)
8. Carroll, S.: Endless Forms Most Beautiful: The New Science of Evo Devo and the Making of the Animal Kingdom. Weidenfeld & Nicolson (2006)
9. Thompson, A.: Silicon evolution. In: Genetic Programming 1996: Proceedings of the First Annual Conference, pp. 444–452. MIT Press, Cambridge (1996)
10. Thompson, A.: An Evolved Circuit, Intrinsic in Silicon, Entwined with Physics. In: Higuchi, T., Iwata, M., Weixin, L. (eds.) ICES 1996. LNCS, vol. 1259, pp. 390–405. Springer, Heidelberg (1997)
11. Harding, S., Miller, J.: Evolution in materio: Initial experiments with liquid crystal. In: Evolvable Hardware, p. 298. IEEE Computer Society, Los Alamitos (2004)
12. Miller, J.F., Downing, K.: Evolution in materio: Looking beyond the silicon box. [13], 167–176
13. Proc. of NASA/DoD Conference on Evolvable Hardware, IEEE Computer Society, Los Alamitos (2002)
14. Koza, J.R., Jones, L., Keane, M., Streeter, M.: Towards industrial strength automated design of analog electrical circuits by means of genetic programming. In: Genetic Programming Theory and Practice II, pp. 121–138. Kluwer, Dordrecht (2004)
15. Mattiussi, C.: Evolutionary Synthesis of Analog Networks. PhD thesis, EPFL, Lausanne (2005)
16. Mattiussi, C., Marbach, D., Dürr, P., Floreano, D.: The Age of Analog Networks. AI Magazine (to appear, 2007)
17. Wolpert, L.: Relationships Between Development And Evolution. In: [7], ch. 2, pp. 47–62
18. Clegg, K., Stepney, S., Clarke, T.: Using feedback to regulate gene expression in a developmental control architecture. In: Lipson, H. (ed.) GECCO, pp. 966–973. ACM, New York (2007)
19. Clegg, K., Stepney, S., Clarke, T.: Evolutionary Search Applied to Reconfigurable Analogue Control. In: Field-Programmable Logic and Applications: FPL 2007, IEEE Press, Amsterdam (2007)
20. Miller, J.F., Thomson, P.: Cartesian Genetic Programming. In: Poli, R., Banzhaf, W., Langdon, W.B., Miller, J., Nordin, P., Fogarty, T.C. (eds.) EuroGP 2000. LNCS, vol. 1802, pp. 121–132. Springer, Heidelberg (2000)

Discovering Several Robot Behaviors through Speciation

Leonardo Trujillo[1], Gustavo Olague[1],
Evelyne Lutton[2], and Francisco Fernández de Vega[3]

[1] EvoVisión Project, CICESE Research Center, Ensenada, B.C., México
trujillo@cicese.mx, olague@cicese.mx
[2] APIS Team, INRIA-Futurs, Parc Orsay Université 4, ORSAY Cedex, France
evelyne.lutton@inria.fr
[3] Grupo de Evolución Artificial, Universidad de Extremadura, Mérida, Spain
fcofdez@unex.es

Abstract. This contribution studies speciation from the standpoint of evolutionary robotics (ER). A common approach to ER is to design a robot's control system using neuro-evolution during training. An extension to this methodology is presented here, where speciation is incorporated to the evolution process in order to obtain a varied set of solutions for a robotics problem using a single algorithmic run. Although speciation is common in evolutionary computation, it has been less explored in behavior-based robotics. When employed, speciation usually relies on a distance measure that allows different individuals to be compared. The distance measure is normally computed in objective or phenotypic space. However, the speciation process presented here is intended to produce several distinct robot behaviors; hence, speciation is sought in behavioral space. Thence, individual neurocontrollers are described using *behavior signatures*, which represent the traversed path of the robot within the training environment and are encoded using a character string. With this representation, *behavior signatures* are compared using the normalized Levenshtein distance metric (N-GLD). Results indicate that speciation in behavioral space does indeed allow the ER system to obtain several navigation strategies for a common experimental setup. This is illustrated by comparing the best individual from each species with those obtained using the Neuro-Evolution of Augmenting Topologies (NEAT) method which speciates neural networks in topological space.

1 Introduction

Evolutionary Robotics (ER) [1] can be seen as an extension to behavior-based robotics (BBR) [2,3]. In classic BBR behaviors are hand-designed by a human expert. On the other hand, in ER the sensory-motor mappings that control the way in which a robot interacts with its surroundings emerge from an artificial evolutionary process. Consequently, ER encourages robot behaviors to emerge from complex interactions between: 1) the autonomous agent; 2) the control mechanism; and 3) the physical environment. ER employs evolutionary computation (EC) methods in the design process of artificial neural networks (ANN) that provide the control mechanism for an autonomous robot. When using ER techniques, most researchers are only interested in finding a single solution for the problem at hand, e.g. a navigation strategy. However, using evolution

M. Giacobini et al. (Eds.): EvoWorkshops 2008, LNCS 4974, pp. 164–174, 2008.

to find a single *super* individual can have several disadvantages [4]. For instance, a large amount of computational effort is not exploited because only one solution from the population is used. Moreover, populations can converge prematurely and solutions may become overfitted to the training problem instance. A workaround to the previous shortcomings is to employ diversity preservation methods. For instance, speciation allows individuals to compete within their own species instead of the entire population. In this way, novel but perhaps less apt solutions can still propagate their genetic material and populations can stay in a more heterogeneous state. Therefore, through speciation a diverse set of solutions could conceivably be obtained from a single evolutionary run, even when all the individuals are trained using the same environment in an ER system.

Outline of the Proposed Approach. This work introduces a behavior-based speciation method, where the behavior exhibited by each neurocontroller is described by what are called *behavior signatures* which allow different behaviors to be compared. Behavior signatures are given as character strings that contain the path followed by the robot within a topological, or graph-based, representation of the environment. Hence, a string similarity measure can be used to compare behavior signatures, in this case the normalized Levenshtein distance metric (N-GLD) is proposed [5]. Thence, speciation can be carried out in behavioral space, a more natural approach for BBR than using objective or phenotypic space to speciate, both of which are more prevalent in other EC problem domains; see Figure 1a. The technique promotes the emergence of distinct robot behaviors, each following a different navigation strategy within the same training environment. The speciation strategy is incorporated within the Neuro-Evolution of Augmenting Topologies (NEAT) method [6]. NEAT adds ANN complexity in an incremental manner and evolves the topology and connection weights concurrently.

This paper proceeds as follows: Section 2 describes the basic concept of speciation and reviews related work. Section 3 introduces the proposed speciation method. Implementation details and the evolutionary setup are discussed in Section 4. Experimental results are presented in Section 5. Finally, in Section 6 concluding remarks are given.

2 Speciation and Diversity Preservation

When a multimodal space exists, it may be desirable to find as many solutions as possible. To achieve this goal, a common approach within EC is to incorporate a speciation mechanism within the evolutionary algorithm (EA) of choice.

In formal terms, a speciating algorithm applies a mechanism $\mathcal{D}$ that maximizes the diversity of individuals within a population $\mathcal{P}$, and also maintains a high mean population fitness $\mathcal{F}(\mathcal{P})$. Therefore, an idealized mechanism would imply that,

$$\mathcal{D}(\mathcal{P}) \longrightarrow \max\{H(\mathcal{P})\} \wedge \max\{\mathcal{F}(\mathcal{P})\} \ , \tag{1}$$

where $H(\mathcal{P})$ is the entropy of population $\mathcal{P}$. Some speciation methods are of general use like fitness sharing [7], while others are domain specific such as symbiosis [8].

Related Work. Speciating methods can be grouped into two classes. The first contains techniques that perform problem decomposition and specialization, what some

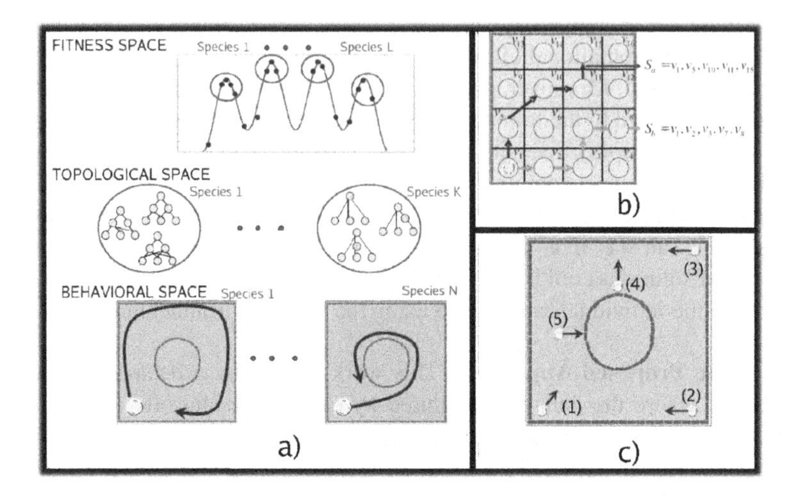

Fig. 1. a) The top row shows the basic niching technique carried out in fitness space. Next, speciation based on topological similarities between ANNs (NEAT). Finally, the proposed behavior-based speciation. **b)** Two sample behavior signatures generated in the topological map representation. Each node is labeled, an the path consists of the string of visited nodes by the robot. **c)** Training environment used: (1) represents the initial position for behavior signature generation; (2 - 5) each of the starting positions and headings for the four training epochs. The topological representation of this environment is the same as in **b)** using a 4×4 grid.

researchers call "evolutionary divide and conquer" [9]. Some examples include the work by Moriarty & Mikkulainen [8] and Dunn *et al.* [9]. The second group performs speciation in order to find problem solutions that "perform different versions of basically the same job" [10]. In other words, each individual represents a complete standalone solution, and each species contains solutions with distinctive properties. Relevant examples include the work by Hocaoğlu & Sanderson [11] and Stanley & Miikkulainen [6]. Hocaoğlu & Sanderson evolve alternative paths in 2D and 3D environments. However, their problem formulation is given in terms of a deliberative control mechanism, as apposed to the BBR approach of ER. Stanley & Miikkulainen introduce the NEAT method, a specialized GA that uses speciation to obtain alternative ANNs. NEAT evolves both the topology and connection weights, thus carrying out incremental learning of network complexity. NEAT has shown an ability to solve *hard* problems, hence, it is used as the basis for the proposed ER system.

Neuro-Evolution of Augmenting Topologies. The NEAT method introduces several advantages when compared with other neuro-evolution systems. For instance, the encoding used allows for crossover operations to be carried out between networks with different topologies. NEAT simulates *incremental learning* by starting from an initial topology and incrementally adding new nodes and synapses. Finally, NEAT protects topological innovations with speciation. To the authors knowledge, NEAT has not been used in BBR problems, marking the present work as the first such instance.

Speciation in NEAT. Speciation in NEAT groups ANN using a measure of topological similarity. The method defines a similarity measure between two ANN chromosomes using the number of disjoint genes, excess genes, and connection weight differences; where genes can represent a network node or a synaptic weight. Therefore, a measure of similarity δ_{NEAT} is given by

$$\delta_{NEAT} = \frac{c_1 \cdot G + c_2 \cdot D}{N} + c_3 \cdot \overline{W} , \tag{2}$$

where G is the number of excess genes, D the number of disjoint genes, $\overline{W}$ the average weight difference of matching genes, c_x are weight coefficients normally set to $c_1 = c_2 = 1$ and $c_3 = 0.4$, and N a normalization factor; for details see [6]. Thus, given a similarity threshold δ_t a new individual a is added to the first species B where its distance δ_{NEAT} to a randomly selected species member $b \in B$ is $\delta_{NEAT}(a, b) < \delta_t$. If no such species is found, then a new species A is created for a. Explicit fitness sharing is used within each species. The adjusted fitness f'_i for the individual i is calculated according to its distance δ to every other individual j in the population,

$$f'_i = \frac{f_i}{\sum_{j=1}^{n} sh(\delta(i,j))} , \tag{3}$$

where function $sh(\delta(i, j))$ is set to 0 if $\delta(i, j) \geq \delta_t$ and 1 otherwise.

Limitations of Topological Speciation. The goal behind speciation is to produce a functionally diverse set of solutions. Building complexity with varying topologies is less interesting if different species do not exhibit an appreciable difference in their functional response. The speciation mechanism proposed by NEAT can only guarantee a diverse set of network topologies not a diverse set of functional solutions. This can be understood with the concept of competing conventions [12], because two ANNs can produce the same functional response even when they are topologically different. In the present work, it is hypothesized that if an appropriate comparative measure can be defined then species will develop in different regions of behavior space, see Figure 1.

3 Behavior-Based Speciation

In order to be able to speciate in behavior space an appropriate behavior representation is necessary along with a proper comparative measure. This work presents a behavior representation that employs *signatures* expressed as character strings. Thus, similarity measures are taken from string comparison techniques.

Behaviors and Neurocontrollers. The distinction between a behavior and an individual in the evolutionary process must be stressed because they do not represent the same concept. An individual represents a particular neurocontroller x, while a behavior is a navigation strategy a induced by the sensory-motor mapping of x within an environment $\mathcal{E}$, written as $x \overset{\mathcal{E}}{\rightsquigarrow} a$. Moreover, due to competing conventions a many-to-one relationship should be assumed between individuals and behaviors. Consequently, let two individuals x and y induce behaviors $x \overset{\mathcal{E}}{\rightsquigarrow} a$ and $y \overset{\mathcal{E}}{\rightsquigarrow} a$ respectively. The notation implies that the underlying navigation strategy a is shared by both x and y. Also, it is assumed that each individual neurocontroller x induces one and only one behavior within $\mathcal{E}$. Furthermore, a behavior is considered to be a subjective concept, while its corresponding signature S_a represents an objective characterization of a. It can be said that S_a is obtained by way of an interpretation process denoted by ψ.

Definition 1. *Let x represent an individual neurocontroller and a the behavior induced by x within environment $\mathcal{E}$, written as $x \overset{\mathcal{E}}{\rightsquigarrow} a$. Then, the* ***behavior signature*** *S_a represents a description of behavior a, obtained through a* ***behavior interpretation process*** *ψ, written as $\psi(a) \hookrightarrow S_a$.*

Indeed, making measurements of specific attributes of a behavior is common, however the same cannot be trivially done for the behavior itself. The reason for this is that ψ is an attempt to interpret a behavior as if it had concrete existence, when in fact it represents an abstract concept. In this work, ψ is such that S_a represents the traversed path of the robot within $\mathcal{E}$. Note that the proposed speciation method works under the assumption that each behavior a is characterized by one and only one signature S_a.

Figure 1b gives a graphical representation of the proposed behavior signatures. The environment is represented using a topological map $\mathcal{M} = (V, E)$ where V is the set of nodes in $\mathcal{M}$ and E the set of edges. A neurocontroller x, starting from an initial node $v_1 \in V$, will guide the robot across the map generating a path S, represented by the sequence of nodes visited by the robot $S = v_i, ..., v_j, ..., v_n$. In order to obtain a signature S, a controller x navigates the robot for 4000 cycles, and the position of the robot is updated every 10 cycles. If at a given update cycle t, the node v^t that the robot occupies is different from the node it occupied at the previous update cycle v^{t-1}, then v^t is added to S. To avoid having the same initial nodes in all behavior signatures, which could influence the similarity measure, nodes are added to S only after an initial stabilizing time period of 500 cycles. The stabilizing time eliminates nodes from S that all behavior signatures would have as their leading characters due to the shared starting position and not due to any meaningful similarity. Because S is a character string a string similarity measure $\delta(S_a, S_b)$ can be applied to compare different signatures. Therefore, $\delta(S_a, S_b)$ defines a distance between behaviors a and b.

N-GLD: Normalized Levenshtein Distance. Before describing the N-GLD metric some preliminary definitions must be established. The alphabet is Σ, Σ^* is the set of strings over Σ, and $\lambda \notin \Sigma$ is the null string. Here, $\Sigma = V$ and Σ^* is the set of possible paths in $\mathcal{M}$. A string $S \in \Sigma^*$ is expressed as $S = s_1, s_2 ... s_n$, where $s_i \in \Sigma$ is the *ith* symbol of S, and $|S| = n$ the size of the string (the null string has $|\lambda| = 0$). The Generalized Levenshtein Distance (GLD), also known as the edit distance, compares strings by various edit operations, commonly using the deletion, insertion, and substitution of individual symbols [5]. If $v, u \in \Sigma$, an elementary edit operation is defined as a pair $(v, u) \neq (\lambda, \lambda)$, and is written as $v \rightarrow u$, where $|v|, |u| \in \{0, 1\}$. The operations $\lambda \rightarrow v$, $v \rightarrow u$, and $u \rightarrow \lambda$, represent insertions, substitutions and deletions respectively. It is possible to define the *edit transformation* $T_{S_a,S_b} = T_1, T_2 ... T_l$ as a sequence of edit operations that transforms S_a into S_b. If a weight function $\gamma(v \rightarrow u) \geq 0$ assigns a non-negative weight to each edit operation, then the total weight of T_{S_a,S_b} is $\gamma(T_{S_a,S_b}) = \sum_{i=1}^{l} \gamma(T_i)$, and the GLD is defined as,

$$GLD(S_a, S_b) = \min \{\gamma(T_{S_a,S_b})\} . \quad (4)$$

The GLD is a metric over Σ^* if :

1. $\forall\, v, u \in \Sigma \cup \{\lambda\}$, $\gamma(v \rightarrow v) = 0$.
2. $\gamma(v \rightarrow u) > 0$ if $(v \neq u) \wedge [\gamma(v \rightarrow u) = \gamma(u \rightarrow v)]$.

In order to account for the common situation in which $|S_a| \neq |S_b|$, a normalized version of GLD is required. Yuijian and Bo [5] define the normalized GLD δ_{N-GLD} for two strings $S_a, S_b \in \Sigma^*$ as

$$\delta_{N-GLD}(S_a, S_b) = \frac{2 \cdot GLD(S_a, S_b)}{\alpha(|S_a| + |S_b|) + GLD(S_a, S_b)} , \quad (5)$$

where $\alpha = \max \{\gamma(v \rightarrow \lambda), \gamma(\lambda \rightarrow u),\ v, u \in \Sigma\}$, and $\delta_{N-GLD}(\lambda, \lambda) = 0$.

It was shown in [5] that the δ_{N-GLD} has the following properties:

1. It satisfies $0 \leq \delta_{N-GLD}(S_a, S_b) \leq 1$.
2. $\delta_{N-GLD}(S_a, S_b) = 0$ if and only if $S_a = S_b$.
3. It is symmetric, because $\delta_{N-GLD}(S_a, S_b) = \delta_{N-GLD}(S_b, S_a)$.
4. It satisfies the triangle inequality, thence, it is a metric over Σ^* if, $\forall v \in \Sigma, \gamma(v \rightarrow \lambda) = \gamma(\lambda \rightarrow v) = \alpha$, and γ is a metric over the set of elementary operations. In [5] the following weight function is suggested, and is used in the present work: $\gamma(v, v) = 0, \gamma(v, u) = 1$, and $\gamma(v, \lambda) = \gamma(\lambda, u) = 1\ \forall v, u \in \Sigma$.

Species Behaviors. Before presenting the experimental setup, another domain specific concept is defined that will facilitate further discussion of the proposed method.

Definition 2. *A population $\mathcal{P} = \{x_1, x_2 ... x_j ... x_N\}$ of N neurocontrollers x, can be divided into M different species R_k with $k = 1 ... M$, such that*

$$\mathcal{P} = \textstyle\bigcup_{k=1}^{M} R_k \ where\ R_k \cap R_l = \varnothing\ for\ k \neq l . \quad (6)$$

Furthermore, let $f(x)$ represent the fitness value of neurocontroller x within environment $\mathcal{E}$. Then, the ***species behaviors*** *of population $\mathcal{P}$ within $\mathcal{E}$ is given by the multiset $\mathcal{B} = \{a^1, ... a^i, ... a^L\}$ of L behaviors, such that $\forall\ a^i \in \mathcal{B}$ if $x \overset{\mathcal{E}}{\rightsquigarrow} a^i$ and $x \in R_k$ then*

$$f(x) > \sup \{f(y) |\ \forall\, y \in R_k,\ y \neq x\} \wedge f(x) > h , \quad (7)$$

where h is called the behavior threshold which is set empirically.

Therefore, every $a^i \in \mathcal{B}$ is induced by one and only one neurocontroller $x \in \mathcal{P}$, and every such neurocontroller is the super-individual of its corresponding species. Given

Definition 2, it is possible to observe that *species behaviors* are contingent on the environment $\mathcal{E}$ that the neurocontrollers interact with. In the general ER framework, $\mathcal{E}$ refers to the training environment employed. An ER system that produces a large $\mathcal{B}$ is said to have found several super-individuals. However, it cannot be assumed that these behaviors represent distinctively different navigation strategies. Therefore, an objective evaluation must be performed in order to determine which of the members of $\mathcal{B}$ do indeed represent "different versions of basically the same job".

4 Implementation of the ER System

This section first describes the Kephera robot, outlines the ER algorithm and gives details on the training environment and fitness function employed.

The Kephera Robot and Simulator. The Kephera is very common within the ER community, it possesses a simple structure and control mechanism that makes it ideal to test novel methods. The Kephera has two DC motors act_1 and act_2 as actuators, and eight infrared proximity sensors $I_1, I_2, ..., I_8$. Evolving neurocontrollers on-line on a real Kephera robot can be quite cumbersome and problematic [1]. Therefore, much of the ER research is conducted on a simulated environment. Robot and environment simulation in the present work is done on the freeware Kephera Simulator version 2.0

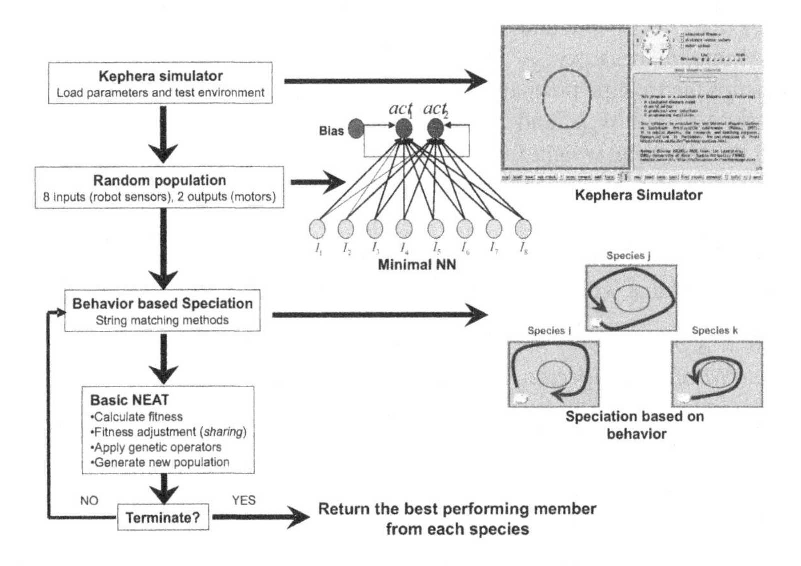

Fig. 2. An overview of the ER system used to evolve alternatives behaviors. First, the Kephera Simulator loads all the algorithm parameters and acts as the interface with the user. Next, the minimal topology of the ANNs used for control. Followed by the neuro-evolutionary system, beginning with the behavior-based speciation process that groups ANNs according to their behavior signatures at each generation. The last steps are basic GA processes, with special genetic operators used by the NEAT method. Finally, a representative Neurocontroller from each species is obtained, the set of *Species Behaviors* $\mathcal{B}$.

[13]. The simulator gives a satisfactory modeling of the physical properties of a real Kephera robot, and the ability to write any kind of control algorithms in C or C++.

The ER System for Behavior-Based Speciation. Figure 2 is a high-level view of the algorithm, based on the NEAT [1] [6] method and employing the Kephera simulator [13]. The ER system is integrated into the Kephera Simulator where the robot parameters, EA, and training environment are loaded. The initial population contains an homogeneous collection of ANN topologies. The minimal topology is a fully connected ANN with 8 input neurons (one for each sensor) and 2 output neurons (for act_1 and act_2) with randomly assigned weights. This is followed by the basic NEAT method which is a straightforward generational GA with fitness proportional selection. The only additional mechanism is that related to speciation and fitness adjustment. The basic process of speciation in NEAT was described in Section 2. The main difference with the proposed behavior-based speciation is the use of behavior signatures with a similarity measure based on the N-GLD metric. Signatures are obtained for each ANN placing the robot in node v_1 at a 45° heading, see Figure 1c.

Training Environment. The training environment is very similar to the one used in [1] shown in Figure 1c. It is simple, basically a square room with a "big" obstacle in the middle. In spite of this, the environment offers a multimodal landscape in behavioral space where different navigation strategies are possible.

Fitness Evaluation. The type of behavior that simulated evolution *should* be searching for is one where the robot navigates around the environment exhibiting the following properties: 1) the robot moves forward in a straight line; 2) the robot moves as fast as possible; and 3) the robot avoids collisions. For these properties to emerge, fitness is assigned as in [14], where for an individual neurocontroller x,

$$f(x) = \frac{1}{N \cdot M} \sum_{j=1}^{M} \sum_{k=1}^{N} V_i (1 - \sqrt{\Delta v_k})(1 - \varphi_k) , \qquad (8)$$

where V_k is the sum of the two motor speeds at time step k. Δv_k is the absolute difference between the two motors. φ is the normalized activation value of the infrared sensor with the highest activation. Moreover, M is the number of test runs, or epochs, and N the total number of time steps or cycles within an environment during an epoch j. The number of epochs is set to $M = 4$ with the initial position and heading of the robot for each epoch shown in Figure 1c, while the number of cycles per epoch is $N = 3000$. The fitness function $f(x)$ is maximized with better performance.

5 Experimental Results

This section describes the results of the proposed speciation method and how it compares with the NEAT method. The parameters employed by each method are the following: number of runs = 6; population size = 100; generations = 50; crossover rate = 0.75;

[1] Source code downloaded from the web site of the Neural Networks Research Group of the University of Texas at Austin: http://www.cs.utexas.edu/ nn/

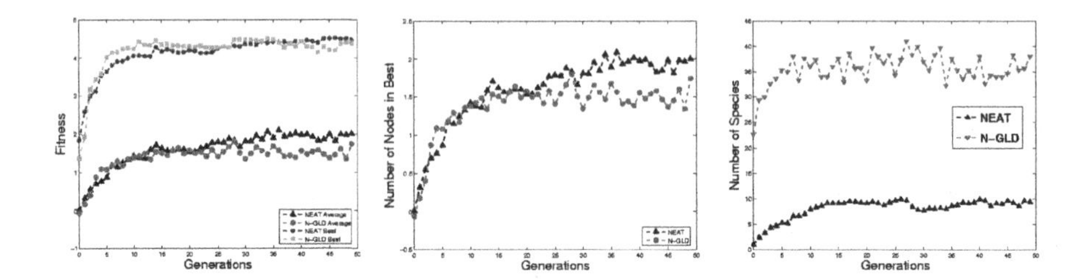

Fig. 3. Performance plots that compare NEAT and the proposed behavior speciation method with the N-GLD metric. The plots, from left-to-right, are: a) average population and best individual fitness; b) number of nodes in best solution; and c) number of species.

compatibility threshold (used to assign species membership): $\delta_{N-GLD} = 0.4$, $\delta_{NEAT} = 3$; behavior threshold (used to identify *species behaviors*): $h = 3.7$. Both methods share all the run-time parameters except for the compatibility threshold δ_t, which was set experimentally for N-GLD and as in [6] for NEAT.

Figure 3 presents three comparative performance curves. All graphs are plotted relative to the number of generations and represent averages over the total number of runs. The plots, from left-to-right, are: a) average population and best individual fitness; b) number of nodes in best solution; and c) number of species. In the first graph, performance is mostly equivalent between both methods. NEAT performs slightly better in average performance which suggests that solutions are better fitted, or overfitted, to the training environment. In the second, the number of nodes of the best individual shows that the NEAT method produces more complex individuals. However, these larger individuals do not yield a higher fitness. With regards to the number of species, the NEAT method produces a lower number of species than does the N-GLD measure. Therefore, the proposed speciation method keeps a more diverse set of solutions.

Additionally, Figure 4 presents two sets of species behaviors $\mathcal{B}$, one each for the NEAT topological measure and the N-GLD measure. Each of the six runs produced a corresponding $\mathcal{B}$, however only one of them is shown due to the length constraints of

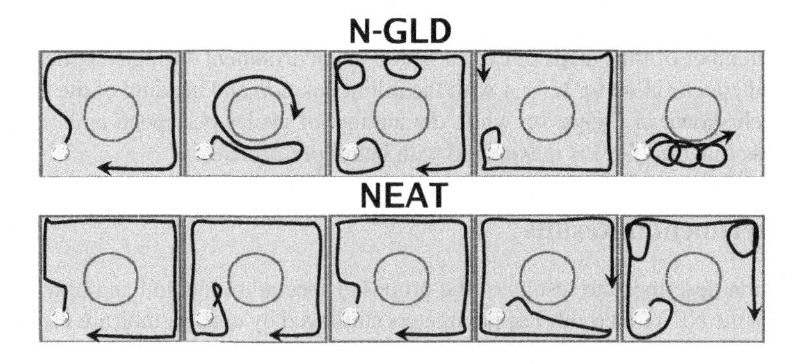

Fig. 4. The set of species behaviors $\mathcal{B}$ found for each of the compared speciation methods: Top row, behavior-based speciation with N-GLD metric; bottom row NEAT's topological speciation

the paper. Nevertheless, the $\mathcal{B}$ shown for each are highly representative, and any further discussion derived from these behaviors generalizes well to the other sets of results. The behavior-based speciation with G-NLD produced *species behaviors* that are all unique. Each have a different manner in which they perform navigation within the environment. Therefore, every behavior represents a qualitatively different solution from the rest. On the other hand, NEAT's topology-based speciation fails to obtain the same degree of diversity. In this case, only one of the species behaviors is different from the rest. Therefore, most NEAT species converge to very similar navigation strategies. In sum, behavior-based speciation was indeed able to find solutions that "perform different versions of basically the same job", while topological speciation fails in this task.

6 Conclusions

In an ER system, obtaining several behaviors could provide a better characterization of the space of possible solutions because the same tasks can usually be performed using different behaviors. Thence, a system that is capable of obtaining several solutions from a single evolving population is of interest. The present work describes a novel behavior-based speciation method that encourages several navigation strategies to evolve concurrently within a single evolutionary process. Behaviors are compared using their *signatures*, which represent a traversed path across the training environment. A similarity measure employing the string edit distance is proposed, the N-GLD metric. This measure is incorporated into the NEAT method, substituting, and subsequently compared with, NEAT's topology-based similarity measure. Results indicate that the EA was able to produce several different navigation strategies using the proposed behavior-based speciation; the same could not be achieved using NEAT's similarity measure. This work presents the first instance within ER literature where various navigation strategies are evolved concurrently, thus providing several strategies from which the end user can choose from. Finally, future work should focus on how to relax the two main assumptions made within the proposed speciation method, namely: 1) that each neurocontroller induces one and only one behavior; and, 2) that each behavior can be instantiated by one and only one signature.

Acknowledgements. Research funded by the Ministerio de Educación y Ciencia (project Oplink - TIN2005-08818-C04), the LAFMI project, and the Junta de Extremadura Spain. First author supported by scholarship 174785 from CONACyT México.

References

1. Nolfi, S., Floreano, D.: Evolutionary Robotics: The Biology, Intelligence, and Technology of Self-Organizing Machines. Bradford Book (2004)
2. Brooks, R.A.: A robust layered control system for a mobile robot. IEEE Journal of Robotics and Automation 2(1), 14–23 (1986)
3. Brooks, R.A.: Intelligence without representation. Artif. Intell. 47(1-3), 139–159 (1991)
4. Yao, X.: Evolving artificial neural networks. PIEEE: Proceedings of the IEEE 87(9), 1423–1447 (1999)

5. Yujian, L., Bo, L.: A normalized levenshtein distance metric. IEEE Trans. Pattern Analysis and Machine Intelligence 29(6), 1091–1095 (2007)
6. Stanley, K.O., Miikkulainen, R.: Evolving neural networks through augmenting topologies. Evolutionary Computation 10(2), 99–127 (2002)
7. Goldberg, D.E., Richardson, J.: Genetic algorithms with sharing for multimodal function optimization. In: Proceedings of the Second International Conference on Genetic Algorithms and their application, pp. 41–49. Lawrence Erlbaum Associates, Inc, Mahwah (1987)
8. Moriarty, D.E., Mikkulainen, R.: Efficient reinforcement learning through symbiotic evolution. Machine Learning 22(1-3), 11–32 (1996)
9. Dunn, E., Olague, G., Lutton, E.: Parisian camera placement for vision metrology. Pattern Recogn. Lett. 27(11), 1209–1219 (2006)
10. Darwen, P.J., Yao, X.: Speciation as automatic categorical modularization. IEEE Trans. Evolutionary Computation 1(2), 101–108 (1997)
11. Hocaoğlu, C., Sanderson, A.C.: Planning multiple paths with evolutionary speciation. IEEE Trans. Evolutionary Computation 5(3), 169–191 (2001)
12. Montana, D.J., Davis, L.: Training feedforward neural networks using genetic algorithms. In: Sridharan, S. (ed.) Proceedings of the Eleventh International Joint Conference on Artificial Intelligence, pp. 762–767. Morgan Kaufmann, San Francisco (1989)
13. Michel, O.: Khepera Simulator v2 User Manual, University of Nice-Sophia, Antipolis (1996)
14. Miglino, O., Lund, H.H., Nolfi, S.: Evolving mobile robots in simulated and real environments. Artificial Life 2(4), 417–434 (1995)

Architecture Performance Prediction Using Evolutionary Artificial Neural Networks

P.A. Castillo[1], A.M. Mora[1], J.J. Merelo[1], J.L.J. Laredo[1], M. Moreto[2], F.J. Cazorla[3], M. Valero[2,3], and S.A. McKee[4]

[1] Architecture and Computer Technology Department
University of Granada
{pedro,amorag,jmerelo,juanlu}@geneura.ugr.es
[2] Computer Architecture Department
Technical University of Catalonia
HiPEAC European Network of Excellence
{mateo,mmoreto}@ac.upc.edu
[3] Barcelona Supercomputing Center
{francisco.cazorla,mateo.valero}@bsc.es
[4] Cornell University
sam@csl.cornell.edu

Abstract. The design of computer architectures requires the setting of multiple parameters on which the final performance depends. The number of possible combinations make an extremely huge search space. A way of setting such parameters is simulating all the architecture configurations using benchmarks. However, simulation is a slow solution since evaluating a single point of the search space can take hours. In this work we propose using artificial neural networks to predict the configurations performance instead of simulating all them. A prior model proposed by Ypek et al. [1] uses multilayer perceptron (MLP) and statistical analysis of the search space to minimize the number of training samples needed. In this paper we use evolutionary MLP and a random sampling of the space, which reduces the need to compute the performance of parameter settings in advance. Results show a high accuracy of the estimations and a simplification in the method to select the configurations we have to simulate to optimize the MLP.

1 Introduction

Designing a computer architecture needs a huge number of parameters to be calibrated. Each parameter can take different values which could impact in the architecture performance.

Usually, simulation techniques are used to evaluate different settings, searching for either the best combination of values or a promising niche within the search space. Although the improvement in simulators, search space size makes simulation times too high [1]. Even small search spaces can be impracticable when simulating [2,3,4]. That is why using a system that predicts performance without actually running the simulator would save a lot of time in researching

M. Giacobini et al. (Eds.): EvoWorkshops 2008, LNCS 4974, pp. 175–183, 2008.

new hardware configurations, giving a range or a set of parameters that can then be simulated for an effective test of performance.

This paper extends Ypek's work [1], who proposed using artificial neural networks (ANN) for architecture performance (instructions per cycle, IPC) prediction. In order to optimize the ANN the training and validation patterns are sampled using *Active learning* [5].

In this paper we intend to simplify the sampling method of the parameter space, using random selection. We propose to focus the effort on the ANN optimization using GProp [6,7,8,9], an evolutionary method for the design and optimization of neural networks.

The experimentation process consists in randomly selecting 1% of the search space configurations. Those simulated points are used to train the MLP, and this is used afterwards to predict the rest of architecture configuration performance. Since the MLP is a fast method, a big amount of configurations can be evaluated in a shorter time. Furthermore, once the configurations with best IPC are found, the designer can focus the study on that zone of the search space.

The rest of this paper is structured as follows: In section 2 related work is analysed. Section 3 describes the problem of exploring architectural design spaces. In section 4 the GProp algorithm is introduced. Section 5 describes the experiments and presents the results obtained, followed by a brief conclusion in section 6.

2 Related Work

There are some recent works tackling the computer architecture design problem, mainly under two approaches: analytic and simulation methods.

Within the analytic approaches, Karkhanis and Smith [10] proposed a superscalar microprocessor model which yields 87 – 95% of accuracy in estimations. Yi et al. [11] studied parameter priority using fractional factorial design. By focusing on the most important parameters, the number of simulations required to explore a large design space can be reduced.

Other researchers (Chow and Ding [12] and Cai et al. [13]) proposed using principal components analysis to identify the most important parameters and their correlations for processor design. Eeckhout et al. [14] and Phansalkar et al. [15] used similar methods for workload and benchmark composition.

Muttreja et al. [16] developed high-level models to estimate performance and energy consumption. They simulated several embedded benchmarks with 1.3% error. Lee and Brooks [17] used regression for predicting performance and power consumption. However, their approach is not easy to apply and it requires some statistical knowledge.

The alternative to analytic methods is simulation [4]. Oskin et al. [18] developed a hybrid simulator to model instruction and data streams, Rapaca et al. [19] used another hybrid simulator and instructions code to infer information that is used to estimate statistics for other application code. Other authors, such as Wunderlich et al. [20] modeled minimal instruction stream to achieve results within desired confidence intervals. Haskins and Skadron [21] sampled application code to create a cache and branch predictor state.

Ypek et al. [1] developed accurate predictive design-space models simulating sampled points and using the results to train an ANN. Their methods yielded a high accuracy but the design space sampling method is rather complex.

In this work, we intend to simplify the sampling method (using a random selection method that simulates less architecture configurations) and to improve performance approximation results using an evolutionary method for ANN design.

3 The Problem

Computer architects have to deal with several types of parameters that define a design: quantitative parameters (i.e. cache size), selections (i.e. cache associativity), numerical values (i.e. frequency) and logic values (i.e. core configuration). The encoding and the way these values are used to train and to exploit an ANN can influence the model accuracy.

In this work, we study the memory system and the CPU design problems. These are defined by a set of parameters (see [22] for details). We use the benchmark suite SPEC CPU 2000 [23] which is composed by a wide range of applications. Following prior work [1], we use bzip2, crafty, gcc, mcf, vortex, twolf, art, mgrid, applu, mesa, equake and swim. They cover a wide spectrum of the total set of benchmarking programs.

Table 1 shows parameters in the memory hierarchy study. Core frequency is 4GHz. The L2 bus runs at core frequency and the front-side bus is 64 bits. The cross product of all parameter values requires 23040 simulations per benchmark.

Table 2 shows parameters in the microprocessor study. We use core frequencies of 2GHz and 4GHz, and calculate cache and SDRAM latencies and branch misprediction penalties based on these. We use 11- and 20-cycle minimum latencies for branch misprediction penalties in the 2GHz and 4GHz cases, respectively. For register files, we choose two of the four sizes in Table 2 based on ROB size (e.g., a 96 entry ROB makes little sense with 112 integer/fp registers). When choosing the number of functional units, we choose two sizes from Table 2 based on issue width. The number of load, store and branch units is the same as the number of floating point units. SDRAM latency is 100ns, and we simulate a 64-bit front-side bus at 800MHz. Taking into account these parameters and their values, the microprocessor study requires 20736 simulations per benchmark.

4 The Method

We propose using GProp, an algorithm that evolves an MLP population. This method searches for the best network structure and initial weights, while minimizing the error rate. It makes use of the capabilities of two types of algorithms: the ability of evolutionary algorithms (EA) [24,25] to find a solution close to the

Table 1. Parameter values in memory system study

Variable Parameters	Values
L1 DCache Size	8, 16, 32, 64 KB
L1 DCache Block Size	32, 64 B
L1 DCache Associativity	1, 2, 4, 8 Way
L1 Write Policy	WT, WB
L2 Cache Size	256, 512, 1024, 2048 KB
L2 Cache Block Size	64, 128 B
L2 Cache Associativity	1, 2, 4, 8, 16 Way
L2 Bus Width	8, 16, 32 B
Front Side Bus Frequency	0.533, 0.18, 1.4 GHz
Fixed Parameters	**Value**
Frequency	4 GHz
Fetch/Issue/Commit Width	4
LD/ST Units	2/2
ROB Size	128 Entries
Register File	96 Integer / 96 FP
LSQ Entries	48/48
SDRAM 100 ns	64 bit FSB
L1 ICache	32 KB / 2 Cycles
Branch Predictor	Tournament (21264)

global optimum, and the ability of the quick-propagation algorithm [26] to tune it and to reach the nearest local minimum by means of local search from the solution found by the EA.

The complete description of the method and the results obtained using classification problems have been presented elsewhere [6,7,8,9]. The designed method uses an elitist [27] algorithm.

In GProp, an individual is a data structure representing a complete MLP with two hidden layers, which implies the use of specific operators. Five variation operators are used to change MLPs: mutation, crossover, addition and elimination of hidden units, and quick-propagation training applied as operator.

The genetic operators act directly upon the ANN object, but only initial weights and the learning constant are subject to evolution, not the weights obtained after training. In order to compute fitness, a clone of the MLP is created, and thus, the initial weights remain unchanged in the original MLP.

The fitness function of an individual (MLP) is given by the mean squared error obtained on the validation process that follows training. In the case of two individuals showing an identical classification error, the one with the hidden layer containing the least number of neurons would be considered the best (the aim being small networks with a high generalization ability).

To present the data to the MLP, cardinal and continuous parameters are encoded as a real number in the [0,1] range, normalizing with minimax scaling via minimum and maximum values over the design space. For nominal parameters

Table 2. Parameter values in the processor study

Variable Parameters	Values
Fetch/Commit Width	4, 6, 8 Instructions
Frequency	2, 4 GHz (affects Cache/DRAM/Branch Misprediction Latencies)
Max Branches	8, 32
Branch Predictor	1K, 2K, 4K Entries (21264)
Branch Target Buffer	1K, 2K, Sets (2 way)
ALUs/FPUs	2/1, 4/2, 3/1, 6/3, 4/2, 8/4 (2 choices per Issue Width)
ROB Size	96, 128, 160
Register File	64, 80, 96, 112 (2 choices per ROB Size)
LD/ST Queue	16/16, 24/24, 32/32
L1 ICache	8, 32 KB
L1 DCache	8, 32 KB
L2 Cache	256, 1024 KB
Fixed Parameters	**Value**
L1 DCache Associativity	1, 2 Way (depends on L1 DCache Size)
L1 DCache Block Size	32 B
L1 DCache Write Policy	WB
L1 ICache Associativity	1, 2 Way (depends on L1 ICache Size)
L1 ICache Block Size	32 B
L2 Cache Associativity	4, 8 Way (depends on L2 Cache Size)
L2 Cache Block Size	64 B
L2 Cache Write Policy	WB
Replacement Policies	LRU
L2 Bus	32B/Core Frequency
FSB	64 bits / 800 MHz
SDRAM	100 ns

we allocate an input unit for each parameter setting, making the input corresponding to the desired setting 1 and those corresponding to other settings 0. Boolean parameters are represented as single inputs with 0/1 values. Target value (IPC) for model training is encoded like inputs. Normalized IPC predictions are scaled back to the actual range. Following the method presented in [1], when reporting error rates, we perform calculations based on not normalized values.

5 Experiments and Results

The following experiments have been carried out: We have searched and optimized an MLP to predict the IPC values for the Memory System and CPU problems. The MLP is trained using the 1% of the total points (architecture configurations), and afterwards it predicts the IPC values for the whole design space. We choose this percentage as proposed in [1].

Then, the best configuration for each one of the benchmarking applications (either for Memory System and CPU problems) is found and the best MLP is used to predict the IPC for those architecture settings.

We conducted our experiments on a bi-processor AMD AthlonXP with 1.66GHz and 1GB RAM. The evolutionary method and the later exploitation of the obtained MLPs consume about nine minutes, while the phase of approaching the whole design space takes less than a second.

Tables 3 (a) and (b) show the results obtained training an MLP using intelligent sampling [1] and those obtained using GProp with random sampling after 30 independent runs (mean squared error and standard deviation are reported).

Table 3. Mean squared error and standard deviation for the Memory System (a) and the CPU (b) problems. Only a 1% of the design space has been simulated to train the MLPs. The table shows the results obtained by Ypek et al. [1] and with the GProp method.

Application	Ypek et al.	GProp
applu	3.11 ± 2.74	4.27 ± 1.08
art	6.63 ± 5.23	4.11 ± 0.45
bzip2	1.95 ± 1.84	1.62 ± 0.08
crafty	2.16 ± 2.10	2.96 ± 0.47
equake	2.32 ± 3.28	2.42 ± 0.35
gcc	3.69 ± 4.02	1.77 ± 0.16
mcf	4.61 ± 5.60	1.46 ± 0.10
mesa	2.85 ± 4.27	13.75 ± 4.22
mgrid	4.96 ± 6.12	4.34 ± 2.47
swim	0.66 ± 0.52	0.83 ± 0.11
twolf	4.13 ± 6.23	1.52 ± 0.22
vortex	5.53 ± 4.63	8.91 ± 0.59

(a) Memory system study

Application	Ypek et al.	GProp
applu	1.94 ± 1.45	4.83 ± 0.64
art	2.41 ± 1.91	1.09 ± 0.19
bzip2	1.30 ± 0.95	2.25 ± 0.23
crafty	2.65 ± 2.03	4.21 ± 0.50
equake	1.80 ± 1.39	3.03 ± 0.42
gcc	1.88 ± 1.48	2.39 ± 0.24
mcf	1.67 ± 1.38	1.05 ± 0.17
mesa	2.57 ± 1.96	8.38 ± 1.28
mgrid	1.39 ± 1.13	3.08 ± 0.58
swim	2.65 ± 2.05	1.72 ± 0.28
twolf	4.85 ± 4.76	1.32 ± 0.17
vortex	2.90 ± 2.17	6.01 ± 1.36

(b) CPU study

Although GProp trains the MLP with a random 1% from the whole possible configurations, results are comparable and even better than those obtained using *Active Learning* for pattern sampling. Furthermore, GProp shows its robustness with the low standard deviations reported versus those reported in [1] (Ypek column in the table).

Tables 4 (a) and (b) show the best simulated configuration IPC and the prediction obtained using GProp for that configuration. The MLP yields a good prediction concerning the IPC value for the best setting (obtained by simulation). Furthermore, we observe from experimentation that MLP predicts the best settings within the same niche in the design space. In this experiment, Ypek et al. [1] only report the value for the Memory system problem in the bzip2 application. The best setting yields an IPC of 1.09, very close to the optimum and to the value obtained using GProp.

Table 4. Best simulated configuration and the prediction obtained using GProp for the Memory System (a) and the CPU (b) problems. First column show the benchmarking applications, the second one the IPC of the best configuration after simulating the whole search space. The third column shows the prediction obtained using GProp for that configuration (mean squared error and standard deviation).

Application	IPC Best Simulated Configuration	IPC GProp Predicted Configuration
applu	1.79	1.74 ± 0.01
art	1.56	1.48 ± 0.01
bzip2	1.10	1.077 ± 0.002
crafty	1.33	1.29 ± 0.01
equake	1.17	1.15 ± 0.01
gcc	1.05	1.036 ± 0.003
mcf	0.47	0.444 ± 0.004
mesa	1.82	1.81 ± 0.01
mgrid	1.55	1.52 ± 0.02
swim	0.77	0.755 ± 0.002
twolf	0.90	0.889 ± 0.001
vortex	1.71	1.67 ± 0.01

(a) Memory system study

Application	IPC Best Simulated Configuration	IPC GProp Predicted Configuration
applu	2.25	2.15 ± 0.03
art	0.53	0.502 ± 0.001
bzip2	1.48	1.40 ± 0.03
crafty	1.76	1.65 ± 0.02
equake	1.66	1.56 ± 0.01
gcc	1.29	1.20 ± 0.01
mcf	0.58	0.54 ± 0.01
mesa	3.04	2.88 ± 0.08
mgrid	1.73	1.68 ± 0.02
swim	0.95	0.917 ± 0.004
twolf	1.01	0.97 ± 0.01
vortex	2.48	2.29 ± 0.07

(b) CPU study

6 Conclusions and Future Work

This work tackles the computer architecture design using the benchmark problems proposed in [1]. We have shown how an ANN can shape a wide search space from the knowledge of a small and random portion. Thus, the experiments just use a randomly chosen 1% of all the possible design settings; this implies that by randomly choosing 1% of possible parameter settings to simulate, we can obtain a good representation of the architecture performance function.

We propose using GProp, a method that evolves an MLP population to obtain a model that predicts the IPC value. The designed MLP predicts any architecture parameter configuration performance with a small error rate.

Furthermore, the proposed method uses a simple random pattern sampling mechanism for the training set. Results obtained are comparable to those presented by other authors, with a low standard deviation (algorithm robustness) as an improvement over them.

We have demonstrated that randomly selecting a small configurations set, it is possible to make accurate predictions. Moreover, our proposal is able to explore a wide search space far from the current simulation methods capabilities.

As future work, we plan the automatic exploitation of the promising settings that the MLP has discovered within the search space applying evolutionary techniques.

Acknowledgements

This work has been supported by the Spanish MICYT projects TIN2007-68083-C02-01, TIN2004-07739, TIN2007-60625 and grant AP-2005-3318, the Junta de Andalucia CICE project P06-TIC-02025 and the Granada University PIUGR 9/11/06 project.

References

1. Ipek, E., McKee, S.A., de Supinski, B.R., Schulz, M., Caruana, R.: Efficiently Exploring Architectural Design Spaces via Predictive Modeling. In: ASPLOS 2006, pp. 195–206 (2006)
2. Martonosi, M., Skadron, K.: NSF computer performance evaluation workshop (2001), http://www.princeton.edu/mrm/nsf_sim_final.pdf
3. Jacob, B.: A case for studying DRAM issues at the system level. IEEE Micro 23(4), 44–56 (2003)
4. Davis, J., Laudon, J., Olukotun, K.: Maximizing CMP throughput with mediocre cores. In: Proc. IEEE/ACM International Conference on Parallel Architectures and Compilation Techniques, pp. 51–62 (2005)
5. SaarTsechansky, M., Provost, F.: Active learning for class probability estimation and ranking. In: Proc. 17th International Joint Conference on Artificial Intelligence, pp. 911–920 (2001)
6. Castillo, P.A., Carpio, J., Merelo, J.J., Rivas, V., Romero, G., Prieto, A.: Evolving Multilayer Perceptrons. Neural Processing Letters 12(2), 115–127 (2000)
7. Castillo, P.A., Merelo, J.J., Rivas, V., Romero, G., Prieto, A.: G-Prop: Global Optimization of Multilayer Perceptrons using GAs. Neurocomputing 35(1-4), 149–163 (2000)
8. Castillo, P., Arenas, M., Merelo, J.J., Rivas, V., Romero, G.: Optimisation of Multilayer Perceptrons Using a Distributed Evolutionary Algorithm with SOAP. In: Guervós, J.J.M., Adamidis, P.A., Beyer, H.-G., Fernández-Villacañas, J.-L., Schwefel, H.-P. (eds.) PPSN 2002. LNCS, vol. 2439, pp. 676–685. Springer, Heidelberg (2002)
9. Castillo, P., Merelo, J., Romero, G., Prieto, A., Rojas, I.: Statistical Analysis of the Parameters of a Neuro-Genetic Algorithm. IEEE Transactions on Neural Networks 13(6), 1374–1394 (2002)
10. Karkhanis, T., Smith, J.: A 1st-order superscalar processor model. In: Proc. 31st IEEE/ACM International Symposium on Computer Architecture, pp. 338–349 (2004)
11. Yi, J., Lilja, D., Hawkins, D.: A statistically-rigorous approach for improving simulation methodology. In: Proc. 9th IEEE Symposium on High Performance Computer Architecture, pp. 281–291 (2003)
12. Chow, K., Ding, J.: Multivariate analysis of Pentium Pro processor. In: Proceedings of Intel Software Developers Conference Track 1, Portland, Oregon, USA, October 27-29, 1997, pp. 84–104 (1997)
13. Cai, G., Chow, K., Nakanishi, T., Hall, J., Barany, M.: Multivariate prower/performance analysis for high performance mobile microprocessor design. In: Power Driven Microarchitecture Workshop (ISCA 1998), Barcelona (1998)

14. Eeckhout, L., Bell Jr, R., Stougie, B., DelBosschere, K., John, L.: Control flow modeling in statistical simulation for accurate and efficient processor design studies. In: Proc. 31st IEEE/ACM International Symposium on Computer Architecture, pp. 350–336 (2004)
15. Phansalkar, A., Josi, A., Eeckhout, L., John, L.: Measuring program similarity: Experiments with SPEC CPU benchmark suites. In: Proc. IEEE International Symposium on Performance Analysis of Systems and Software, pp. 10–20 (2005)
16. Muttreja, A., Raghunathan, A., Ravi, S., Jha, N.: Automated energy/performance macromodeling of embedded software. In: Proc. 41st ACM/IEEE Design Automation Conference, pp. 99–102 (2004)
17. Lee, B., Brooks, D.: Accurate and efficient regression modeling for microarchitectural performance and power prediction. In: Proc. 12th ACM Symposium on Architectural Support for Programmming Languages and Operating Systems (ASPLOS-XII), San Jose, California, USA, pp. 185–194. ACM Press, New York (2006)
18. Oskin, M., Chong, F., Farrens, M.: HLS: Combining statistical and symbolic simulation to guide microprocessor design. In: Computer Architecture, 2000. Proc. 27th IEEE/ACM International Symposium on Computer Architecture (SIGARCH Comput. Archit. News), pp. 71–82. ACM Press, New York (2000)
19. Rapaka, V., Marculescu, D.: Pre-characterization free, efficient power/performance analysis of embedded and general purpose software applications. In: Proc. ACM/IEEE Design, Automation and Test in Europe Conference and Exposition, pp. 10504–10509 (2003)
20. Wunderlich, R., Wenish, T., Falsafi, B., Hoe, J.: SMARTS: Accelerating microarchitecture simulation via rigorous statistical sampling. In: Proc. 30th IEEE/ACM International Symposium on Computer Architecture (ISCA), San Diego, California, USA, June 9-11, 2003, vol. 8, pp. 84–95. IEEE Computer Society Press, Los Alamitos (2003)
21. Haskins, J., Skadron, K.: Minimal subset evaluation: Rapid warm-up for simulated hardware state. In: Proceedings of the International Conference on Computer Design: VLSI in Computers and Processors, September 23-26, 2001, p. 32. IEEE Computer Society Press, Washington (2001)
22. Renau, J.: SESC (2007), `http://sesc.sourceforge.net/index.html`
23. SPEC: Standard Performance Evaluation Corporation. SPEC CPU benchmark suite (2000), `http://specbench.org/osg/cpu2000`
24. Goldberg, D.: Zen and the art of genetic algorithms. In: Procs. of the 6th International Conference on Genetic Algorithms, ICGA 1995, pp. 80–85 (1995)
25. Michalewicz, Z.: Genetic Algorithms + Data Structures = Evolution Programs, 3rd Extended edn., Springer, Heidelberg (1996)
26. Fahlman, S.: Faster-Learning Variations on Back-Propagation: An Empirical Study. In: Proceedings of the 1988 Connectionist Models Summer School, Morgan Kaufmann, San Francisco (1988)
27. Whitley, D.: The GENITOR Algorithm and Selection Presure: Why rank-based allocation of reproductive trials is best. In: Schaffer, J.D. (ed.) Procc of The 3th Int. Conf. on Genetic Algorithms, pp. 116–121. Morgan Kaufmann, San Francisco (1989)

Evolving a Vision-Driven Robot Controller for Real-World Indoor Navigation

Paweł Gajda and Krzysztof Krawiec

Institute of Computing Science
Poznan University of Technology, Poznań, Poland

Abstract. In this paper, we use genetic programming (GP) to evolve a vision-driven robot controller capable of navigating in a real-world environment. To this aim, we extract visual primitives from the video stream provided by a camera mounted on the robot and let them to be interpreted by a GP individual. The response of GP expressions is then used to control robot's servos. Thanks to the primitive-based approach, evolutionary process is less constrained in the process of synthesizing image features. Experiments concerning navigation in indoor environment indicate that the evolved controller performs quite well despite very limited human intervention in the design phase.

1 Introduction and Related Work

Using genetic programming (GP) to evolve robot controllers, whether for real or for virtual environments, may be traced back to the very beginning of GP [8,11]. Typical robotic tasks successfully solved by means of GP include wall following [7], mapping sensor readings to robot locations [4], and learning an obstacle avoidance strategy [6], to mention a few representative contributions. Also, much research has been done on evolving cooperative behaviors of robots, for instance to control a team of robots playing soccer [1]. An extensive review on evolving controllers for real robots may be found in [14].

In most of the aforementioned contributions, the evolved GP programs usually process scalar data coming from distance sensors (or virtual distance sensors in case of simulation). In this paper, we are particularly interested in evolving a *vision-driven* real-world robot controller. Past research within this area includes several contributions. In [3], Ebner used genetic programming to evolve edge detectors for robotic vision. Graae, Nordin, and Nordahl [5] evolved a stereoscopic vision system for a humanoid robot using GP. Langdon and Nordin used machine code GP to evolve hand-eye coordination for a humanoid robot [9]. Seok, Lee, and Zhang applied a variant of linear GP and FPGA hardware to evolve the behavior of locating light sources and avoiding obstacles [13]. Wolff and Nording made use of visual feedback for evolving gait controllers of a bipedal robot [15]. There are also other reports on GP-based robotic vision in an virtual environment, which is not considered in this paper (e.g., in [2] an OpenGL framework is used to simulate the visual environment of a physical robot solving the task of line following).

M. Giacobini et al. (Eds.): EvoWorkshops 2008, LNCS 4974, pp. 184–193, 2008.

Fig. 1. The PPRK robot with the mounted CMUCam2+ camera

This paper demonstrates the possibility of evolutionary learning of a high-level behavioral pattern directly from low-level visual sensory input. In our approach, we avoid defining a fixed repertoire of high-level visual features to be used by the learner. Rather than that, we feed a low-level, though non-raster, visual data into GP learners. Next, using an appropriately defined fitness function, we entice the evolving learners to navigate in a real-world indoor environment. As our robot's perception is not tuned to the particular task or visual target, the evolutionary learning has to build up an appropriate and effective higher-level representation of visual patterns. An experimental evaluation on a real-world robotic platform in an indoor environment shows the ability of our approach to solve the task of approaching a visual target.

2 The Hardware Platform

We used Palm Pilot Robot Kit (PPRK) as the hardware platform for our approach. PPRK is a small, three-wheeled, autonomous robot designed by the Robotics Institute at the Carnegie Mellon University and produced by Acroname [12]. As its name suggests, it may be controlled by a handheld computer (Palm Pilot or PocketPC), but it is also equipped with a built-in controller called *BrainStem* that is able to store and execute simple behavioral patterns. BrainStem is based on PIC18C252 processor clocked at 40MHz.

PPRK uses *holonomic* drive, composed of three equidistant wheels arranged in a circle and propelled by three independent servos. Each of its 'omni-wheels', thanks to built-in rolls, may move freely in directions that do not lay in its plane of rotation; this happens, for instance, when the wheel is dragged by the other wheels. This kind of drive allows for arbitrary rotation and translation, so that the total number of robot's degrees of freedom (3) is equal to the number of controllable degrees of freedom. For instance, applying the same potential v to all three servos causes the robot to rotate in place. Applying potentials v, $-v$, and 0 to servos #1, #2, and #3, respectively, makes the robot move perpendicularly to the section connecting wheels #1 and #2, dragging the wheel #3 behind it.

We mounted a CMUCam2+ board on robot's chassis (see Fig. 1). The CMUCam2+ card, also provided by Acroname, uses Omnivision's CMOS OV6620 camera controlled by SX52 microcontroller. Though it implements some simple image processing and feature tracking functionalities, they have not been used in our setup, and the digitized video stream was directly sent to the controlling computer. The three infrared proximity sensors included in PPRK were also inactive in our experiments.

The camera board acquires video data at resolution of 87×143 pixels and is able to perform some elementary image analysis at 25-50 frames per second. Unfortunately, it communicates with other modules (including the BrainStem) via standard serial interface, which limits the maximum data transfer rate to 115200 bits per second. This seriously reduces the number of frames that can be sent to the image analysis module in real time. This is why, in the following experiments the actual number of frames processed per seconds amounts to approximately 1, though the evolved GP expressions could easily handle two orders of magnitude higher frame rates.

In contrast to most of related research, we do not extract any predefined high-level image features to help evolution to learn the desired robot's behavior. Rather than that, we rely on GP-based processing of low-level visual primitives, an approach described in the following section.

3 Using GP Trees to Process Visual Primitives

Our GP-based approach to visual processing, originally proposed in [16] and later extended in [17], has the following rationale. The volume of raw raster data is usually too large to make it direct subject to evolutionary learning. To reduce these data, one commonly uses a predefined set of visual features extracted from the training images; the evolutionary process works then with such features only. There is, however, a significant risk that the features pre-selected by the human are not the best ones to cope with the particular visual task. Also, the sole process of defining and implementing such features may be time-consuming and difficult.

To keep the amount of visual training data within reasonable limits on one hand and avoid arbitrary pre-selection of visual features on the other, our approach relies on *visual primitives* (VP). We define the visual primitive as a local salient feature extracted from an image location characterized by a prominent gradient. In the beginning of processing, each VP is described by three scalars called hereafter *attributes*; these include two spatial coordinates (x and y) and orientation of the local gradient vector. The complete set P of VPs is usually much more compact than the original image s in terms of information content, yet it well preserves the overall sketch of the visual input.

The right-hand part of Figure 2 presents the VPs extracted from an exemplary frame of the video sequence used in the following experiment, with each primitive depicted as a short section. Note, however, that the individuals described in the following learn from training examples that are technically *sets* of VPs, each of them described by a triple of numbers. In other words, learners do not explicitly 'perceive' the image as a two-dimensional raster.

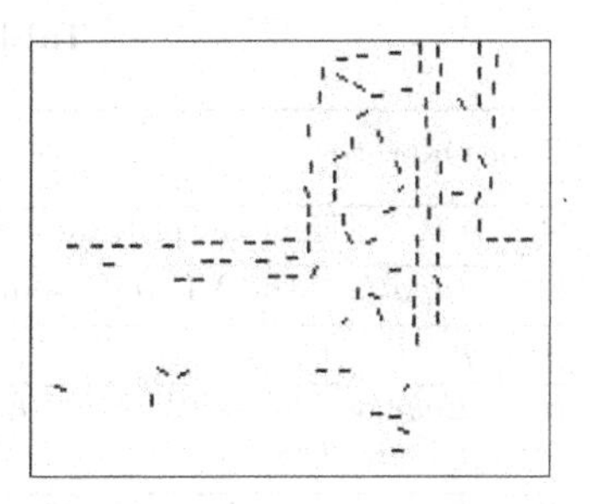

Fig. 2. An exemplary input image (left) and the corresponding visual primitives (right)

Each learner L is a GP expression written in a form of a tree, with nodes representing *functions* that process sets of VPs. The feed of image data for the tree has been implemented by introducing a special terminal function (named *ImageNode*) that fetches the set of primitives P derived from the input image s. The consecutive internal nodes process the primitives, all the way up to the root node. We use strongly-typed GP, so child node's output type must match parent node's input type. The list of types includes numerical scalars, sets of VPs, attribute labels, binary arithmetic relations, and aggregators.

The functions, presented in Table 1, may be divided into (a) selectors, which select some VPs based on their attributes, (b) iterators, which process VPs one by one, (c) grouping operators, which group VPs based on their attributes and features, e.g., spatial proximity, and (d) scalar arithmetic functions. In addition,there is a group of functions that compute simple set operations in the domain of VPs, like set union (*SetUnion*), set difference (*SetMinus*), or symmetric difference (*SetMinusSym*). Implementation of most functions is straightforward. For instance, the *SelectorMin* function applied to a set of primitives S and attribute label p_y selects from S the VP (or VPs) with the minimal value of attribute (coordinate) y. The *ForEach* function iterates over the set of elements returned by its left child node, and passes each of them through the subtree rooted in its right child node, finally grouping the results into one set. The semantics of the remaining functions may be decoded from their mnemonics.

It is worth emphasizing that in this process, VPs and sets of VPs are used interchangeably. Technically, a set may contain both VPs and other nested sets of VPs. Therefore, the processing carried out by an individual-learner L applied to the input image s boils down to building a hierarchy of VP sets derived from s. Each invoked tree node creates a new set of VPs built upon the elements (VPs or sets of VPs) provided by its child node(s). When needed, we recursively compute an attribute value of a VP set as an average of its elements.

To control robot's actuators (servos), our trees must compute real-valued responses at the root node. To this aim, we employ an advanced feature of our approach, namely its ability to create new attributes, apart from the pre-defined ones (x, y, and orientation). The new attribute may be computed from and attached to VPs or VP sets. Technically, this is implemented by the *AddAttribute*

Table 1. The GP operators

Type	Operator
$\Re$	ERC – Ephemeral Random Constant
Ω	*Input*() – the VP representation P of the input image s
A	p_x, p_y, p_o,
R	*Equals*, *Equals5Percent*, *Equals10Percent*, *Equals20Percent*, *LessThan*, *GreaterThan*
G	*Sum*, *Mean*, *Product*, *Median*, *Min*, *Max*, *Range*
$\Re$	$+(\Re,\Re)$, $-(\Re,\Re)$, $*(\Re,\Re)$, $/(\Re,\Re)$, $sin(\Re)$, $cos(\Re)$, $abs(\Re)$, $sqrt(\Re)$, $sgn(\Re)$, $ln(\Re)$, *AttributeValue*(Ω,A)
Ω	*SetIntersection*(Ω,Ω), *SetUnion*(Ω,Ω), *SetMinus*(Ω,Ω), *SetMinusSym*(Ω,Ω), *SelectorMax*(Ω,A), *SelectorMin*(Ω,A), *SelectorCompare*$(\Omega,A,R,\Re)$, *SelectorCompareAggreg*(Ω,A,R,G), *CreatePair*(Ω,Ω), *ForEach*(Ω,Ω), *ForEachCreatePair*(Ω,Ω,Ω), *Ungroup*(Ω), *GroupHierarchyCount*$(\Omega,\Re)$, *GroupHierarchyDistance*$(\Omega,\Re)$, *GroupProximity*$(\Omega,\Re)$, *GroupOrientationMulti*$(\Omega,\Re)$, *AddAttribute*$(\Omega,\Re)$, *AddAttributeToEach*$(\Omega,\Re)$

and *AddAttributeToEach* functions (see Table 1). For instance, the *AddAttribute* function takes the VP set S returned by its left child node and passes it through its right child subtree. Due to syntactic constraints imposed by strong typing, the right child subtree is forced to return a scalar value computed using ERCs, scalar functions (e.g., +, -, *, /, *abs*), and the values of existing attributes fetched from S using, among others, the *AttributeValue* function. The computed value is attached as a new attribute to S, which is subsequently returned by the *AddAttribute* function (the VP hierarchy in S remains therefore unchanged). The *AddAttributeToEach* function operates similarly, however, it repeats the steps listed here recursively for each element of S.

Given this functionality, we expect evolution to elaborate individuals that define a new attribute, different from p_x, p_y, and p_o, which should be attached to the set of VPs returned by the root node as the final response of the tree. The returned value (essentially a very specific image feature) is subsequently evaluated by the fitness function which compares it to the desired value.

4 The Experiment

4.1 The Task

Our task consists in navigating the robot in a small (3.2×3.2m) indoor environment, in varying lighting conditions. More specifically, the robot has to find and approach a 15×15cm diamond-shaped marker placed close to the floor on one of the room's walls. To make the task non-trivial, we placed some other objects in the environment; these included a chair and a cupboard. Other artifacts often visible in robot's field of view include power outlets and floor-wall boundaries.

4.2 The Training Procedure

From an evolutionary perspective, the most desirable approach to evolve our controllers would be to directly bind the evolutionary process to the real-world and estimate the fitness of each controller (individual) by downloading it to the physical robot and letting it navigate in our environment. Such an procedure would be obviously implausible. An alternative approach of evaluating individuals in a virtual simulator has been often criticized as being far from perfect in terms of fidelity to the real-world. Because our objective was to evolve a controller that operates in real-world, we took another way.

To devise a computationally feasible experimental setup without abandoning the real-world data, we came up with the following teacher-driven approach. The training data has been collected by guiding the robot by a human operator, starting from random initial locations and ending close to the target marker. For each such experiment, we recorded the sequence of video frames together with the synchronized values of potentials applied to the servos. Specifically, servo values have been recorded at one second delay with respect to the video stream to compensate for operator's reaction time.

When preparing to data acquisition, we found out that direct controlling of the holonomic drive is cumbersome and non-intuitive for humans. Therefore, we designed a more handy set of controls, composed of 4 buttons: *Forwards*, *Backwards*, *Left*, and *Right*. A single click on a button increases the intensity of particular action; e.g., each click on the *Forwards* button increases the speed of forward movement; after that, a couple of clicks on the *Backwards* button are needed to stop the robot.

The buttons determine the values of two intermediate variables: linear speed and angular speed. The values of these variables are subsequently mapped to voltages to be applied to particular servos. Therefore, each video frame is accompanied by three desired effector values.

Figure 3 shows the typical training frames selected from one of the recording sessions. After a couple of such sessions that a few hours in total (during which the lighting conditions changed significantly), the total number of collected frames amounted to 734. As using such a number of frames for evolutionary learning would be prohibitive from the viewpoint of computational burden (even when using visual primitives instead of raw raster data), we included only 30 representative frames in the final training set. These 30 frames were subject to extraction of visual primitives (see Section 3). In this process, we created VPs only for image locations for which the response of the gradient exceeded 150 (on the scale 0..255), and enforced a lower limit on their mutual distance (5 pixels). The number of primitives extracted from one image (i.e., $|P|$) was not allowed to exceed 300. To speed up calculation, this procedure has been carried out only once, prior to the evolutionary run, and the resulting VPs have been cached in the memory.

As the complete controller of our robot has to provide three output (effector) values, one for each wheel, we carried out three separate evolutionary runs. The

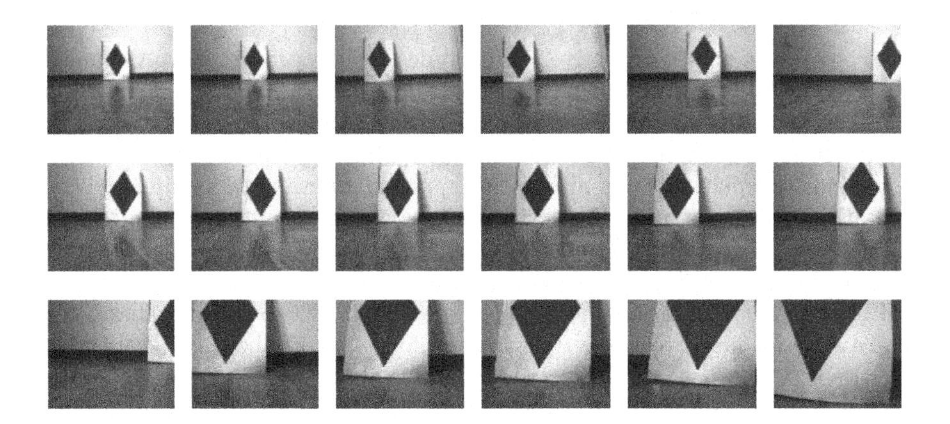

Fig. 3. Selected frames from the training set

final controller has been built by combining the best-of-run individuals from these runs.

In each of those runs, a tree is expected to return a set of visual primitives with a new attribute attached to it (cf. end of Section 3). The value of that attribute, $r(t)$, computed by the individual for the frame $\#t$, is subsequently interpreted as the voltage to be applied to the corresponding servo. In the training phase, these values were compared to the desired values $d(t)$ provided by the human controller using fitness function that aggregates the SSDs of the individual's responses $r(t)$ with respect to the desired values $d(t)$ over the entire set of 30 training frames:

$$\sum_{t=1}^{30}(r(t) - d(t))^2.$$

In each of the evolutionary runs, the following parameter values have been used: population size: 5000, number of generations: 100, tournament selection with tournament size 3, probability of crossover: 0.8, probability of mutation: 0.2, maximal tree depth: 6. The remaining parameters were set to their default values as provided in the ECJ software package [10]. Evolving controller for each servo took about 10 hours on a Pentium PC computer with 1GHz processor.

4.3 Testing the Evolved Controller

The best evolved controller, shown partially in Figure 4 (left-wheel tree only), has been subsequently used for real-time control of the robot in the testing phase. Thirty testing sessions have been carried out. Among them, 14 ended successfully, with the robot reaching the marker. Figure 5 presents a typical correct trajectory traversed by the robot. In the remaining cases, at some stage of approaching the marker the robot usually executed an extensive turn, probably distracted by spurious visual primitives resulting from image noise. Having lost the marker

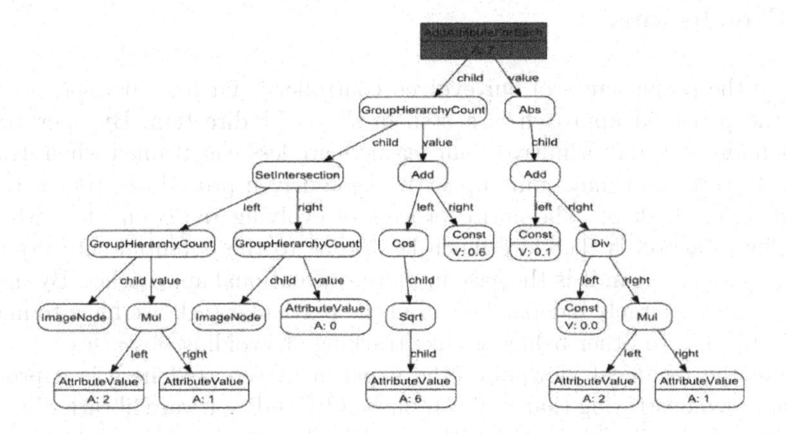

Fig. 4. The best controller evolved for the left wheel

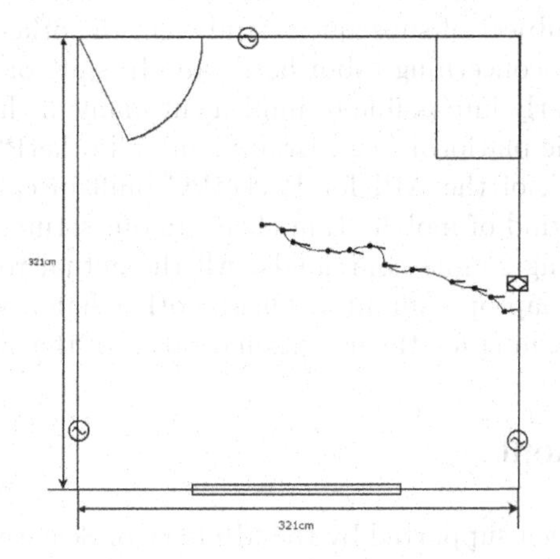

Fig. 5. A typical correct trajectory of the robot

from its field of view, it then started to move randomly. In some cases, it was able to redirect its camera towards the marker and continue the journey. Though this deficiency in robot's behavior could be circumvented, e.g., by introducing an extra heuristic, such intervention was beyond the subject of this study.

In the presence of other objects, like power sockets or chairs, the robot was usually able to navigate correctly towards the marker. However, in about 40% of trials, it erroneously targeted at the distracting objects. On the other hand, we have observed many times that a navigation error caused by one of the trees was corrected by the others. In other words, the controllers of particular servos seemed to cooperate well, despite the fact that their training was carried out independently.

5 Conclusions

Though the performance of our evolved controller is far from perfect, we think that the proposed approach is a step in the right direction. By operating in the domain of visual primitives, our learners are less constrained when defining visual features and may come up with vision-driven procedures that a human would never think of. The entire process of evolving the complete path from stimulus to action is therefore much less prone to the potential subjectivity of human designer than it is the case in more conventional approaches. By making visual learning largely independent from the particular task, we hope to make it applicable also to other behaviors like tracking or avoiding obstacles.

From the technical viewpoint, the experiment reported here is a proof-of-concept, demonstrating that well-established EC/GP software libraries like ECJ may be quite easily integrated with a real-world robotic systems. Although, as explained earlier, the learning process does not take place on-line here, an extension of our experimental setup to on-line learning is straightforward and will be most likely the subject of subsequent study. On the other hand, we have also learned hard lessons concerning robot hardware. In spite of our expectations, it turned out to be nearly impossible to implement many of the required functionalities on the robotic platform (e.g., BrainStem or PocketPC), due to technical difficulties (e.g., lack of the API for PocketPC built-in camera). Thus, in the end, the robot is a kind of mobile 'thin client' in our setup, merely carrying the camera and executing simple commands. All the actual robot's intelligence is hosted by a nearby laptop computer. On the other hand, we appreciated such setup as less constraining for the evolutionary part of the system.

Acknowledgement

This research has been supported by the Ministry of Science and Higher Education grant # N N519 3505 33.

References

1. Ciesielski, V., Wilson, P.: Developing a team of soccer playing robots by genetic programming. In: McKay, B., Tsujimura, Y., Sarker, R., Namatame, A., Yao, X., Gen, M. (eds.) Proceedings of The Third Australia-Japan Joint Workshop on Intelligent and Evolutionary Systems. School of Computer Science Australian Defence Force Academy, Canberra, Australia, November 22-25, 1999, pp. 101–108 (1999)
2. Dupuis, J.-F., Parizeau, M.: Evolving a vision-based line-following robot controller. In: The 3rd Canadian Conference on Computer and Robot Vision (CRV 2006), p. 75. IEEE Computer Society Press, Los Alamitos (2006)
3. Ebner, M.: On the evolution of edge detectors for robot vision using genetic programming. In: Groß, H.-M. (ed.) Workshop SOAVE 1997 - Selbstorganisation von Adaptivem Verhalten, VDI Reihe 8 Nr. 663, Düsseldorf, pp. 127–134. VDI Verlag (1997)

4. Ebner, M.: Evolving an environment model for robot localization. In: Langdon, W.B., Fogarty, T.C., Nordin, P., Poli, R. (eds.) EuroGP 1999. LNCS, vol. 1598, pp. 184–192. Springer, Heidelberg (1999)
5. Graae, C.T.M., Nordin, P., Nordahl, M.: Stereoscopic vision for a humanoid robot using genetic programming. In: Oates, M.J., Lanzi, P.L., Li, Y., Cagnoni, S., Corne, D.W., Fogarty, T.C., Poli, R., Smith, G.D. (eds.) EvoWorkshops 2000. LNCS, vol. 1803, pp. 12–21. Springer, Heidelberg (2000)
6. Harding, S., Miller, J.F.: Evolution of robot controller using cartesian genetic programming. In: Keijzer, M., Tettamanzi, A.G.B., Collet, P., van Hemert, J.I., Tomassini, M. (eds.) EuroGP 2005. LNCS, vol. 3447, pp. 62–73. Springer, Heidelberg (2005)
7. Koza, J.R.: Evolution of a subsumption architecture that performs a wall following task for an autonomous mobile robot via genetic programming. In: Hanson, S.J., Petsche, T., Rivest, R.L., Kearns, M. (eds.) Computational Learning Theory and Natural Learning Systems, June 1994, vol. 2, pp. 321–346. MIT Press, Cambridge (1994)
8. Koza, J.R., Rice, J.P.: Automatic programming of robots using genetic programming. In: Proceedings of Tenth National Conference on Artificial Intelligence, pp. 194–201. AAAI Press/MIT Press (1992)
9. Langdon, W.B., Nordin, P.: Evolving hand-eye coordination for a humanoid robot with machine code genetic programming. In: Miller, J., Tomassini, M., Lanzi, P.L., Ryan, C., Tetamanzi, A.G.B., Langdon, W.B. (eds.) EuroGP 2001. LNCS, vol. 2038, pp. 313–324. Springer, Heidelberg (2001)
10. Luke, S.: ECJ 15: A Java evolutionary computation (2006), `http://cs.gmu.edu/~library.eclab/projects/ecj/`
11. Nordin, P., Banzhaf, W.: Genetic programming controlling a miniature robot. In: Siegel, E.V., Koza, J.R. (eds.) Working Notes for the AAAI Symposium on Genetic Programming, November 10–12, 1995, pp. 61–67. MIT Press, Cambridge (1995)
12. Reshko, G., Mason, M.T., Nourbakhsh, I.R.: Rapid prototyping of small robots. Technical Report CMU-RI-TR-02-11 (2002)
13. Seok, H.-S., Lee, K.-J., Zhang, B.-T.: An on-line learning method for object-locating robots using genetic programming on evolvable hardware. In: Sugisaka, M., Tanaka, H. (eds.) Proceedings of the Fifth International Symposium on Artificial Life and Robotics, Oita, Japan, January 26-28, 2000, vol. 1, pp. 321–324 (2000)
14. Walker, J., Garrett, S., Wilson, M.: Evolving controllers for real robots: A survey of the literature. Adaptive Behavior 11(3), 179–203 (2003)
15. Wolff, K., Nordin, P.: Evolution of efficient gait with humanoids using visual feedback. In: Proceedings of the 2nd IEEE-RAS International Conference on Humanoid Robots, pp. 99–106. Institute of Electrical and Electronics Engineers, Inc (2001)
16. Krawiec, K.: Learning High-Level Visual Concepts Using Attributed Primitives and Genetic Programming. In: Rothlauf, F., Branke, J., Cagnoni, S., Costa, E., Cotta, C., Drechsler, R., Lutton, E., Machado, P., Moore, J.H., Romero, J., Smith, G.D., Squillero, G., Takagi, H. (eds.) EvoWorkshops 2006. LNCS, vol. 3907, pp. 515–519. Springer, Heidelberg (2006)
17. Krawiec, K.: Generative Learning of Visual Concepts using Multiobjective Genetic Programming. Pattern Recognition Letters 28(16), 2385–2400 (2007)

Evolving an Automatic Defect Classification Tool

Assaf Glazer[1,2] and Moshe Sipper[1]

[1] Dept. of Computer Science, Ben-Gurion University, Beer-Sheva, Israel
www.moshesipper.com
[2] Applied Materials, Inc., Rehovot, Israel*
assaf_glazer@amat.com

Abstract. Automatic Defect Classification (ADC) is a well-developed technology for inspection and measurement of defects on patterned wafers in the semiconductors industry. The poor training data and its high dimensionality in the feature space render the defect-classification task hard to solve. In addition, the continuously changing environment—comprising both new and obsolescent defect types encountered during an imaging machine's lifetime—require constant human intervention, limiting the technology's effectiveness. In this paper we design an evolutionary classification tool, based on genetic algorithms (GAs), to replace the manual bottleneck and the limited human optimization capabilities. We show that our GA-based models attain significantly better classification performance, coupled with lower complexity, with respect to the human-based model and a heavy random search model.

1 Introduction

Traditional classification approaches suffer from a problem of poor generalization on image classification tasks. In this paper we focus on the image classification task for the semiconductors industry. During the production process within a fab, which is a customer's wafer fabrication facility, we would like to automatically find and characterize defects, and determine their sources. Classified defects may be fixed, thus increasing the yield of the wafer production process. The motivation for this paper is the industry's demand for better classification results with higher throughput (wafers produced per hour), and the desire for an automated process with minimal human intervention. Poor data, and a deceptive environment in the fab where the classification problem itself varies over time, renders the ADC task hard to solve. Our primary goal is to provide a solution for the above challenges by using genetic-algorithm techniques, providing an evolutionary classification tool that automatically optimizes itself across a machine's lifetime and adapts to the changing environment inside the fab.

* This research was supported in part by the IMG4 consortium (4th Generation Imaging Machines) of the Israeli Ministry of Industry, Trade, and Employment, and also by the Lynne and William Frankel Center for Computer Science.

M. Giacobini et al. (Eds.): EvoWorkshops 2008, LNCS 4974, pp. 194–203, 2008.

Different models of Automatic Defect Classification (ADC) tools already exist in different fabs, using a wide range of classification models. Most of the ADC models are based on *machine learning* [3] techniques, which is a broad subfield of *artificial intelligence.* In our research we would like to optimize a given ADC classifier, which is based on a radial basis function neural network (RBFN) with Gaussian radial basis function kernels, by applying a genetic algorithm to select kernels that are used within the RBFN hidden units. The research has taken place in the SEM division, Applied Materials, Inc. (AMAT) and Ben-Gurion University. We use the SEMVision classification tool, which is based on a RBFN, as a benchmark for our research.

In Section 2 we introduce the defect classification problem. In Section 3 we provide background on *radial basis function neural networks* (RBFN). In Section 4, we present a reference model for the SEMVision ADC tool we intend to use later. In Section 5 we describe the basic evolutionary model. An additional model of heavy random search is defined for comparison with our model. In Section 6 we introduce the enhanced evolutionary model for the defect classification problem. Finally, Section 7 summarizes the results obtained in our research and suggests some directions for future research.

2 The Defect Classification Problem

The goal of the defect classification process is quite simple: given an image, classify the defect type found in the image. The need for an image classification tool arises in various fields; in our research we focus on the defect classification process in the semiconductors industry. Semiconductor wafer manufacturers invest much of their time in isolating the causes of yield-impacting defects during the lithographic printing and processing of integrated circuits on wafers. Automatic Defect Classification (ADC) automates the slow manual process of defect review and classification during optical microscopy and scanning electron microscopy (SEM). The ADC machine is a key step in the identification of the root cause of manufacturing problems. A fast, accurate, and reliable ADC tool is required. However, though the problem definition is simple, its solution involves many challenges. During the fab's lifetime new defects appear, and old defects become obsolete; the ADC model has to adapt itself to this changing environment. Additional challenges, such as poor data with inaccurate classified samples and high throughput demands from the customers, require a fast automated and reliable ADC tool.

The automatic defect classification (ADC) problem can be defined as a subset selection problem. Given a training set TR of pre-classified images and a pre-classified validation set VL, we would like to find an optimal subset $S \subseteq TR$, which maximizes the classification rate of the given ADC classifier. Each exemplar in the training set S is translated into a hidden unit in the classifier. For example, if we use a Gaussian kernel function in a *RBFN*, then $\forall t \in S$ we generate a hidden unit of a Gaussian function with a *mean* $\mu = t$ (in *RBFN* models we usually define a fixed *variance* σ for all hidden units). Eventually, the size of the given subset S is equal to the number of units inside the RBFN hidden layer. We briefly discuss the RBFN hidden units in Section 3. In addition, we keep an independent test set $TEST$ of pre-classified images to estimate the classification rate that can be achieved in a real-time application with the optimized classifier. The test set $TEST$ can also allow us to gauge the quality of the achieved solution on real field data.

We view each of our customers' fabrication factories as a 'greenhouse' for images of defects. Each greenhouse generates defect prototypes with some common characteristics at a specific time and place, and we have to fit and optimize the ADC tool to the current fab with the current process and the current time stamp. As a result, the objective function can change along the ADC machine's lifetime by modifying the input data sets TR, VL, and $TEST$ during the classification process. Therefore, our method should be based on an *anytime algorithm*, which generates the best solution within the scope of available data that have been explored up to the allowed time. Thus, we can change the data sets during the ADC's lifetime and the algorithm can fit itself to the new environment.

Usually, the classification rate is determined as the *accuracy* of the classified result, i.e., the percentage of the defects classified correctly, or as the *purity* of the classified result. *Purity* is the exactness of the classification, i.e., the fraction of defects classified correctly with respect to the number of total classified images, not including 'Unknown Defect' classification results. We can also use a hybrid definition involving both *accuracy* and *purity* results.

3 Radial Basis Function Neural Networks (RBFNs)

As shown in Figure 1, a RBFN is a three-layer, feed-forward network, with each layer fully connected to the next layer. The RBFN consists of an input layer L_0, a hidden layer L_1, and an output layer L_2. The k-dimensional input vector enters the RBFN through the k units of input layer L_0, and passes through weighted edges $w_{i,j}$ to the hidden units in hidden layer L_1. A kernel threshold function is activated in each of the hidden units, passing the activation result through additional weighted edges $w_{j,l}$ to the output layer L_2. The hidden layer consists of a set of radial basis functions. Associated with each hidden-layer node is a parameter vector c_i called a center. For each hidden unit calculate the Euclidean distance r between c_i and the input vector, and pass the result to the kernel function. Different kernel functions can be chosen as the network activation function, such as *Gaussian*, *Multiquadric*, and *Thin plate spline* functions [10].

Training an RBFN consists of three main stages [3]: the **first** stage is *network initialization*, which defines the selected features for the input layer and determines the number of centers and the radii c_i of the kernel functions in the hidden layer. The **second** stage is *obtaining the weights for the output layer*: once we have determined the network-initialization parameters we can seek the weights for the output layer using least-squares error function minimization, similar to single-layer networks, e.g., *QR decomposition*. The **third** stage is *iterative optimization*, where several methods can be used for RBFN optimization [3].

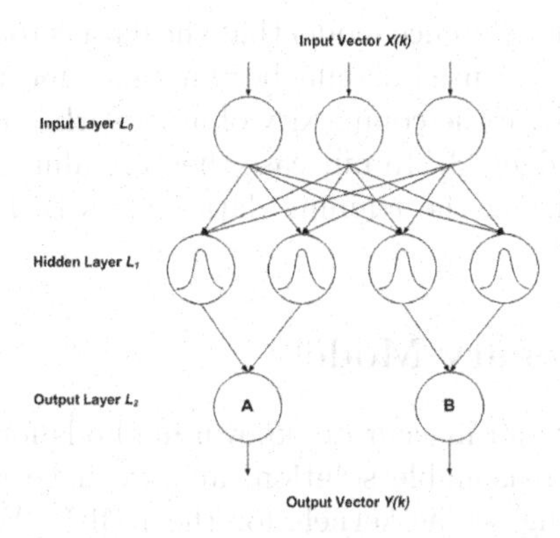

Fig. 1. Radial basis function neural network (RBFN)

Genetic algorithms can be used for different optimization tasks for RBFNs, including kernel-function optimization [1, 4, 7, 8, 9, 11], architecture optimization [5, 6, 11], and training-set optimization [6]. The evolutionary process is especially advantageous when the search space is large and the objective function is non-differentiable, e.g., subset selection for training-set optimization. We also find the GA solution to be advantageous when a deceptive objective function is involved, e.g., as caused due to a changing environment during the ADC machine's lifetime.

4 Reference Model for SEMVision ADC Tool

In order to understand the problem we introduce a Matlab-based model, using the *netlab toolbox*, defined as a reference for the SEMVision ADC model. The data set we used has 2613 individuals from 9 different defect classes, each one represented by a vector with 76 features. The data set is separated into a training set of size 605 and a validation set of size 2008. We ran a RBFN with *Gaussian kernel* functions. We used a principal components analysis (PCA) algorithm for the feature selection part. The *k-means* algorithm was used to select the centers c_i for each of the *Gaussian kernels* in the hidden units, and we used a fixed width for the Gaussian variance. After setting the centers of the kernels we optimized the network with an *EM (expectation maximization) algorithm* [2] using a *least-squares* error function.

The results showed no convergence. We tried a *cross-validation* technique in order to optimize the number of features and kernels with no success. A balancing method also proved unsuccessful in avoiding a trivial solution. If we use the whole training set, as the SEMVision ADC does, the classification accuracy reaches 85.7, a much higher performance than in the reference model. We can see that traditional *Machine Learning* algorithms for kernel selection, such as *cross-validation* for selecting the number of kernels, and *k-means* for clustering, attain low performance compared to the SEMVision ADC.

We conclude from the reference model that the data is too poor. The literature suggests that the data we used should be ten times more abundant than the network complexity, where the complexity of a network is measured as the total number of connected edges [3]. In our case they are almost equal, and we have no other choice than using the training data set, as is, for the centers in the hidden neurons.

5 Basic Evolutionary Model

When the data is poor and inaccurate, known methods for RBFN optimization do not converge to a reasonable solution, and we have no other choice than using the whole training set as kernels for the RBFN. When the data is too poor for using a clustering method, using real kernels in the classifier—which represent real defects—can reduce the risk of overfitting. However, though using the entire training set as kernels for the RBFN achieves reasonable results, this method suffers from a stability problem and demands frequent skilled Customer Engineer (CE) support, since every classified sample has an immediate impact on the classified product of an image with the same characteristics. Erroneous flyers in the training set can dramatically reduce the accuracy of the classifier, and the changing environment requires constant CE support. Moreover, the sampling of images for the training set is done manually by humans, who can hardly optimize the selected samples for the ADC tool with no computational support. Perhaps we can attain better accuracy by using a subset from the training data set.

In the model we use we would like to optimize the subset of kernels from the training set for our ADC tool such that we attain better results with less computational complexity: the fewer the kernels in the ADC, the less complex is the ADC model, and the higher the throughput and generalization of our model.

The problem of subset selection from a training set is hard to solve. Under the assumption that the data set is too poor for learning by using known optimization methods, this problem of subset selection is *NP-Hard* by a simple reduction from the *subset sum problem.*

The GA. We introduce a genetic algorithm for this optimization, with classifier accuracy being the fitness. The GA is defined as follows:

- *Initialization.* Each generation comprises a population set π of 50 individual genomes. We use random initialization with uniform distribution to construct the first generation. Each bit in the genome is set to either 0 or 1 with probability 0.5.
- *Genome representation.* Let $TR = \{t_1, ..., t_m\}$ be a training set of m classified images. We use a binary encoding for the genome representation, such that each genome represents a subset of the initial training set, e.g., the genome $S = [01101001...]$ represents the subset $S \subset TR, S = \{t_2, t_3, t_5, t_8, ...\}$. The genome length is equal to the size of the training set.
- *Fitness function.* In each generation we calculate the fitness function f_{fit} per individual. Each genome represents a subset S of the initial training set TR. The fitness function is defined as the accuracy rate of the ADC classifier.

- *Selection, crossover, mutation.* We use the entire population π as the mating set in order to maintain diversity. In our case the difference in fitness between individuals is small, and the use of the *fitness-proportionate selection* method might reduce selection pressure. Therefore, we use the *rank selection* mechanism. We apply an *elitism selection* mechanism (of one individual per generation), thus maintaining the best individual in the population set π for the next generation. *Single-point crossover* is performed with crossover rate $p_{cross} = 0.8$. Each individual is subjected to mutation. This operation inverts each bit in the genome with a probability of 0.05.
- *Termination criterion.* Though this is an anytime algorithm, for empirical reasons, we stop the algorithm after 50 generations.

To compare with our GA we run a *heavy random* stress test for kernel selection. Although random selection of kernels from the training set is apparently a naive choice, sometimes it is more justifiable than the more sophisticated techniques. The reason is that, especially in small sample size problems, intricate algorithms can easily lead to overfitting [6].

The Experiment. We have already seen that we can attain better performance on the same data set by using the SEMVision ADC, than by using the reference model. Therefore, in order to reduce overall run time, we allowed ourselves to use a smaller data set than that used for the reference model above.

The data consists of 1905 individuals from 5 different classes. Each of the individuals represents a defect image with 76 features. The data is separated into a training set of size 498, a validation set of size 946, and a test set of size 461. In addition to the genetic algorithm, we ran a *heavy random* stress test for kernel selection. The application runs on an XP OS with Xeon Dual-Core computer and 2Gb RAM.

Results. If we use the whole training set, as the CE currently does, classification accuracy reaches 0.870. With our GA we attained 0.897—with less than half the number of kernels, i.e., achieving a 0.297 improvement with half the complexity. The *heavy random* stress test shows poor results of 0.883 with almost no convergence over 50 generations, compared to the evolutionary model. Moreover, we left the *heavy random* to run for 150 generations, and no significant improvement was found. The total GA runtime is approximately 60 hours. Three separate runs were made for this basic model, showing almost identical behavior.

An improvement from 87.0% to 89.7% is much more impressive than, say, from 77.0% to 79.7%, for instance. Moreover, using 193 kernels found by the GA instead of 498 used by the manual method (CE), is a major improvement in terms of throughput, since the ADC buildup time and its classification processing time is exponential with the number of kernels inside the RBFN.

The total runtime of approximately 60 hours can be reduced dramatically with several optimization methods, such as parallel fitness evaluation, an efficient way to modify the ADC tool to learn a new training set without a new reconstruction on each evaluation, parsing elimination, and hardware improvement. We must remember that this optimization algorithm is an off-line problem, where the computational efforts can take place during the computer's idle time. Moreover, monitoring and optimizing the classification process in spaces of 2-3 day is good enough for most of the fabs.

6 Enhanced Evolutionary Model

Given our experience with the basic evolutionary model, and given the characteristics of the classification problem—poor and inaccurate data in a changing environment—and the requirement for an accurate, fast, and automatic classification tool, we formulate the following guidelines for a solution, in order to attain even better results:

- **Anytime algorithm.** *genetic algorithms* can be considered as a form of *anytime algorithm*. Our model must provide the best solution it can find by using the data that have been explored until the current moment.
- **Optimization of training set.** Our model should find the optimized subset from the given training set.
- **Generic model.** We treat the SEMVision ADC model as a black box. This way our optimization model can be used with different types of classification methods.
- **Reduced complexity.** The use of a subset, instead of the whole training set, reduces the complexity, improves throughput, and increases generalization.
- **Robust solution with no human intervention.** Our only requirement from the customer is a daily classification of a constant number of samples, which is a reasonable demand.

The Enhanced GA. The input for our model consists of a training set TR, a validation set VL, and a test data set $TEST$. We assume the data set can increase and change during the fab's lifetime. The main *genetic engine* module is responsible for the entire problem optimization. The input for the *genetic engine* module is the data sets, and the output is an optimal subset $S \subseteq TR$, which maximizes the classification rate of the given ADC classifier—as measured on the test data set $TEST$. The *genetic engine* module comprises two separate GAs working in parallel: the *defect controller* module is responsible for defining the problem we are trying to solve. The output of the *defect controller* module is a *hint* array with the relevant samples of the training set. The *ADC enhancer* module is responsible for optimizing the current problem. The *hint* array, produced by the *defect controller*, is part of the input for the *ADC enhancer*. Basically, we run two separate GAs in parallel—*defect controller* and *ADC enhancer*—the latter constantly receiving hints from the former, which help improve its performance.

The output of the *defect controller* GA module is a *hint* array with the relevant samples of the training set. If the training set consists of n elements, the output of the *defect controller* module is a *hint* array of size n with the fraction of the kernels' relevances inside the given training set TR. The *hint* array can dynamically change during the evolutionary process. The *ADC Enhancer* GA module uses the *hint* array to achieve better problem optimization. The output of the *ADC Enhancer* module is an optimal subset $S \subseteq TR$, which maximizes the classification rate on the validation data set VL. The use of the *hint* array in the *ADC Enhancer* module is through the mutation operation, as shown in Figure 2.

```
Mutate(gene, hint)

parameter(s): gene – the genome
parameter(s): hint – the hint array
output: the mutated genome

Initialization :
i ← 0
while i < hint.length
   do { if random < mutation rate
        then { if random < hint(i)
               then gene(i) ← 1
               else gene(i) ← 0
        i ← i + 1 }
return (gene)
```

Fig. 2. Pseudocode of the mutation operation

For example, if one of the samples inside the training set becomes obsolete, then its fraction will drop in the *hint* array, and the flipping probability for the equivalent bit inside the mutated genome will decrease as well. This way we can also explore new and old defects in the classification process.

The *defect controller* module dynamically produces the *hint* array in the following manner: a subset with the *best* result during a pre-defined window of time is maintained, i.e., the last fixed number of generations backward. We can define *best* results in two different ways:

1. Classification rate above a predefined value.
2. Classification rate in top of predefined fixed percentage of the results from the fixed number of generations backward.

For each of the kernels in the training set we calculate the fraction of usage inside the *best* subsets from the relevant generations. The *hint* itself is an array with the fraction of kernels usage in the *best* subsets. The higher the fraction the higher the relevance of the kernel in the current problem.

We should note that both *defect controller*, and *ADC enhancer* modules can only use the validation set VL during the evolutionary process, while the output of the *genetic engine* module is measured on the output of *ADC enhancer* module with an independent test set $TEST$. This way we assure that the output of our model estimates the classification rate that can be achieved in a real-time application.

In addition to the above two modules, we propose to apply a *fitness weight decay* process to the samples in the data sets. According to the images' creation time tag we weight the data sets, such that an old sample will have lower effect upon the overall classification rate. This way we can use the *fitness weight decay* process to attain better results for the current classified images, and we can also explore obsolete defects in the data sets. Unfortunately, our classifier doesn't generate time tags for the classified images. Therefore, we cannot implement *fitness weight decay*, and we leave it for future research.

The Experiment. We used the same data set as in the basic experiment in Section 5, with the same evolutionary parameters for both the *defect controller* and the *ADC optimizer* modules.

Results. Five separate runs were made for the enhanced model, showing almost identical behavior. Using the enhanced model we attained a classification rate of 0.901 after 15 generations, while with the basic model we obtained 0.897 after 50 generations. Both models attained significantly better results than the manual process, with half the complexity—a dramatic improvement in throughput terms.

As opposed to the *ADC enhancer* module, it is important to keep the *defect controller* module as objective as we can. Our goal in this module is to monitor the changing environment inside the fab, and not to achieve an optimized solution. For this reason we always look a fixed predefined number of generations backward, and we don't use the *hint* array during its own evolutionary process.

Both the reduced complexity, the automatic control process, and the higher accuracy attained, are practical benefits of our enhanced model. The major breakthrough of the enhanced model is its ability to independently fit itself to the changing environment inside the fab—with a deceptive environment of poor and inaccurate information—achieving a high classification rate, an increased throughput, and better generalization. Obsolete defects can be isolated using the *hint* array, and the input data set can dynamically change during the machine's lifetime.

By using the *hint* array we can not only identify obsolete and relevant samples in the data sets, we can also find an obsolete class of defects: when we find all the samples for a specific class of defects irrelevant we can eliminate it from the data sets.

7 Concluding Remarks and Future Research

Genetic algorithms are useful for numerous real-life classification challenges. When the search space is large, when the objective function is inaccurate and changes over time, when the data is impoverished, and when the problem we are trying to solve has subset-selection characteristics, a GA may be the answer. In our problem we find all the above elements. Finding an optimal subset out of a group of 498 elements is a hard problem with subset-selection characteristics. Some of our samples in the training set have been falsely classified so we have a deceptive objective function. When the defects we are trying to classify change during the fab life-time our objective function varies and we have to adapt to the new environment. In the experiments we performed, the genetic algorithm has proved itself as the best solution to date, as shown in Table 1.

Table 1. A comparison between the different optimization methods: 1) the manual method, where the CE optimizes the training set by hand, is the currently used method; 2) the heavy random stress test; 3) the basic evolutionary model; and 4) the enhanced evolutionary model. The best classification rate is shown in the first row, the number of kernels, which reflects the RBFN complexity and throughput, is shown in the second row, and the number of generations for the method to converge to an optimal solution is shown in the third row. We should remember that only the enhanced method can adapt itself to a changing environment.

	Manual	Random	Basic	Enhanced
Accuracy	0.870	0.883	0.897	0.910
Kernels	498	196	193	198
Convergence	Manually	50	47	15

In the future, when we can obtain a time stamp for each classified sample, we will be able to test the proposed *fitness weight decay* process. In addition, we hope that we can gather more data and test our new and obsolete defects exploration model.

References

1. David Sanchez, A.V.: Searching for a solution to the automatic RBF network design problem. Neurocomputing 42, 147–170 (2002)
2. Bilmes, J.: A gentle tutorial on the EM algorithm and its application to parameter estimation for Gaussian mixture and hidden Markov models (1997)
3. Bishop, C.: Neural Networks for Pattern Recognition. Oxford University Press, Oxford (1995)
4. Buchtala, O., Klimek, M., Sick, B.: Evolutionary optimization of radial basis function classifiers for data mining applications. IEEE Transactions on Systems, Man and Cybernetics, Part B 35, 928–947 (2005)
5. Camacho, F., Manrique, D., Rodriguez-Paton, A.: Designing radial basis function networks with genetic algorithms. In: IASTED International conference artificial intelligence and soft computing, September 2004, vol. 451(8), pp. 398–403 (2004)
6. Kuncheva, L.I.: Initializing of an RBF network by a genetic algorithm. Neurocomputing 14, 273–288 (1997)
7. Kuo, L.E., Melsheimer, S.S.: Using genetic algorithms to estimate the optimum width parameter in radial basis function networks. In: American Control Conference, vol. 2, pp. 1368–1372 (July 1994)
8. Maillard, E.P., Gueriot, D.: RBF neural network, basis functions and genetic algorithm. In: Neural Networks International Conference, vol. 4, pp. 2187–2192 (June 1997)
9. Wai Mak, M., Wai Cho, K.: Genetic evolution of radial basis function centers for pattern classification, `citeseer.ist.psu.edu/8,1322.html`
10. Specht, D.F.: Probabilistic neural networks. Neural networks 3, 109–118 (1990)
11. Whitehead, B.A., Choate, T.D.: Cooperative-competitive genetic evolution of radial basis function centers and widths for time series prediction. IEEE Transactions on Neural Networks 7, 869–880 (1996)

Deterministic Test Pattern Generator Design

Gregor Papa[1], Tomasz Garbolino[2], and Franc Novak[1]

[1] Jožef Stefan Institute, SI-1000 Ljubljana, Slovenia
[2] Silesian University of Technology, PL-44100 Gliwice, Poland
gregor.papa@ijs.si, tomasz.garbolino@polsl.pl, franc.novak@ijs.si

Abstract. This paper presents a deterministic test pattern generator structure design based on genetic algorithm. The test pattern generator is composed of a linear register and a non-linear combinational function. This is very suitable solution for on-line built-in self-test implementations where functional units are tested in their idle cycles. In contrast to conventional approaches our multi-objective approach reduces the gate count of built-in self-test structure by concurrent optimization of multiple parameters that influence the final solution.

1 Introduction

To cope with the problem of growing complexity of modern integrated circuits and increasing testing demands, boundary-scan approach has been developed and is widely adopted in practice [10,15]. Limited number of input/output pins represents a bottleneck in testing of complex embedded cores where transfers of large amounts of test patterns and test results between the automatic test equipment (ATE) and the unit-under-test (UUT) are required. One of the alternative solutions is to implement a built-in self-test (BIST) of the UUT with on-chip test pattern generator (TPG) and on-chip output response analysis logic. In this way, communication with external ATE is reduced to test initiation and transfer of test results. One of the disadvantages of BIST implementation is the area overhead, which typically results in performance penalties due to longer signal routing paths. Therefore, minimization of the BIST logic is one of the commonly addressed problems in practice.

In the past different TPG approaches have been proposed. They can be classified as ROM-based deterministic, algorithmic, exhaustive and pseudo-random. The first approach, where deterministic patterns are stored in a ROM and a counter is used for their addressing, is limited to small test pattern sets. Algorithmic TPG are mostly used for testing regular structures such as RAMs. Exhaustive TPG is counter-based approach which is not able to generate specific sequence of test vectors. With some modifications, however, counter-based solutions are able to generate deterministic test patterns. Pseudo-random TPG is most commonly applied technique in practice, where Linear Feedback Shift Register (LFSR) or Cellular Automata (CA) are employed to generate pseudo-random test patterns. In order to decrease the complexity of a TPG, designers

M. Giacobini et al. (Eds.): EvoWorkshops 2008, LNCS 4974, pp. 204–213, 2008.

usually try to embed deterministic test patterns into the vector sequence generated by some linear register. Such embedding can be done either by re-seeding a TPG or modifying its feedback function. There are also solutions that modify or transform the vector sequence produced by a LFSR, such that it contains deterministic test patterns. Most proposed LFSR structures are based on D-type flip-flops. In recent years, LFSR composed of D-type and T-type flip-flops is gaining popularity due to its low area overhead and high operating speed [7,8].

The motivation for this work came with the need of deterministic TPG for an on-line BIST structure - composed of idle functional units and registers, where functional units and registers that are not used for the computations during individual time slots are organized into a structure that is continuously tested in parallel with normal system operation. Normally, pseudo-random test vectors can be employed for such on-line self-test. In mission critical systems, where low fault latency is required, TPGs that generate deterministic test sequence are needed. In this paper an approach for the generation of deterministic TPG logic based on a LFSR, composed of D-type and T-type flip-flops, is described. The approach employs a genetic algorithm (GA) to find an acceptable practical solution in a large space of possible LFSR implementations.

The rest of the paper is organized as follows: in Section 2 we describe a structure of the proposed TPG, and give an example of area minimization through the modification of TPG structure and its test vectors; in Section 3 we describe the implemented GA and its operators; in Section 4 we describe the whole optimization process and evaluate it; and in Section 5 we list our conclusions.

2 Structure of the Proposed TPG

Typically, a TPG is initialized with a given deterministic seed and run until the desired fault coverage is achieved. The test application time using an LFSR is significantly larger than what is required for applying the test set generated using a deterministic TPG. This is due to the fact that vector set generated by a LFSR includes, besides useful vectors, also many other vectors that do not contribute to the fault coverage. In our approach, the goal is to develop a TPG that would generate only the required test vectors (i.e., no non-useful vectors).

The structure of the proposed n bit test pattern generator is composed of a Multiple-Input Signature Register (MISR) and modification logic. The MISR has a form of a ring that is composed of n flip-flops with either active high or active low inputs. Each flip-flop (either T-type or D-type) can also have inverter on their input (denoted as $\overline{\text{D}}$ or $\overline{\text{T}}$). Thus, the register may have one of $4n$ different structures. The inputs of the MISR are controlled by the modification logic, while the outputs of the MISR are fed back to the modification logic which is a simple combinational logic and acts like a decoder. The modification logic allows that in the subsequent clock cycles the contents of the MISR assume the values specified by the target test pattern set. Hence MISR and the modification logic are application specific: they are synthesized according to the required test pattern set.

Particularly important parameter in the case of deterministic test pattern generators is the area overhead, which is influenced by:

- the structure of each MISR stage,
- the order of the test patterns in a test sequence,
- the bit-order of the test patterns.

The first property influences the complexity of both the MISR and the modification logic while the remaining two impact the area of the modification logic only. These relationships are illustrated by the following examples.

Initial structure and test vectors. Having the set of six 3-bit vectors (see Figure 1a), the resulting structure of the TPG consists of D-type flip-flops in all stages of the MISR. Since all the flip-flops are scannable and have asynchronous reset, the total area of the TPG manufactured in AMS 0.35 μm technology is 1821 μm^2.

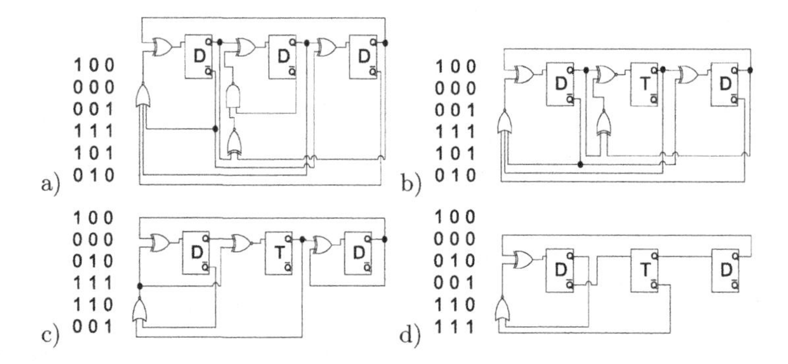

Fig. 1. TPG structure: a) initial configuration, b) modified flip-flop type, c) permutated columns, and d) permutated test patterns

Flip-flop type replacement. Replacing the second flip-flop with a T-type flip-flop, the new configuration of the TPG is presented in Figure 1b. While there is no T type flip-flop with inverted input in the standard cell library of the AMS 0.35 μm technology, the negation is implemented by replacing the XOR gate with an XNOR. The total area of the TPG is 1784 μm^2.

Column permutation. Permutation of columns of the test pattern sequence (i.e., by simultaneous switching of bits in all test patterns) further decreases the area of the TPG. If we permute the second and the third column in the test sequence (as illustrated in Figure 1c), the TPG is simplified to the structure with the area of 1657 μm^2.

Vectors permutation. Further we can permute test patterns in the test sequence. Exchanging the test patterns 4 and 6 in the test sequence simplifies the structure (Figure 1d) to the area of 1421 μm^2.

The above examples indicate that a suitable change of the MISR structure, the order of the test patterns in a test sequence, and the bit-order of the test patterns may result in a substantial area reduction of the TPG. The solution space is very broad: for an n-bit TPG producing the sequence of m test patterns there are $4nm!n!$ possible solutions. Therefore, effective optimization algorithms are required to find an acceptable practical solution.

3 Genetic Algorithm

Well-known population-based evolutionary approach, implemented through GA [1,9], was used for optimization because of its intrinsic parallelism that allows searching within a broad database of solutions in the search space simultaneously, climbing many peaks in parallel. There is always some risk of converging to a local optimum, but efficient results of various research work obtained in other optimization problem areas [11,12,13,14] encouraged us to consider GA approach as one of the possible alternatives in TPG synthesis optimization.

Our GA implementation for this multi-parameter and multi-objective [4,16] problem optimizes multiple parameters within multiple design aspects (type of flip-flops, order of patterns in test sequence, and bit-order of a test pattern).

3.1 TPG Structure Encoding

Three different chromosomes were used to encode the parameters of the TPG to be optimized. Each solution is presented by three chromosomes, which do not interact with each other. With three chromosomes we concurrently optimized the structure of the TPG, the order of the test patterns, and the bit order of test patterns. The first chromosome, which encodes the structure of n-bit TPG, looks like

$$C_1 = t_1 i_1 t_2 i_2 \ldots t_n i_n, \tag{1}$$

where t_j $(j = 1, 2, \ldots, n)$ represents the type of the flip-flop (either D or T) and i_j $(j = 1, 2, \ldots, n)$ represents the presence of the inverter on the output of the j-th flip-flop.

The second and third chromosome, which encode the order of the test patterns, and the bit order of test patterns, look like

$$C_2 = a_1 a_2 \ldots a_m, \tag{2}$$

where m is the number of test vectors and a_j $(j = 1, 2, \ldots, m)$ is the label number of the test vector from the vector list, and

$$C_3 = b_1 b_2 \ldots b_k, \tag{3}$$

where k is the number of flip-flops in the structure and b_j $(j = 1, 2, \ldots, k)$ is the label number of the bit order of test patterns.

3.2 Population Initialization

The initial population consisted of N chromosomes, of each type, that represent the initial structure. To ensure versatile population some chromosomes were mirrored. The values on the left side (beginning) of the structure chromosome were mirrored to the right side (ending); either type of registers or inverter presence or both values were mirrored. In case of other two, i.e. order of test patterns and their bit order, chromosomes, their initial reproduction included mirroring of orders between the beginning and the ending positions. Mirroring was used as an alternative to random generating of initial solutions.

3.3 GA Operators

In the selection process the elitism strategy was applied through the substitution of the least-fit chromosomes with the equal number of the best-ranked chromosomes.

Crossover. In a two-point crossover scheme, chromosome mates were chosen randomly and with a probability p_c all values between two randomly chosen positions were swapped, which led to the two new solutions that replaced their original solutions (steady-state type of the genetic algorithm). Figure 2(top) shows the example of crossover with crossover points on positions 1 and 4.

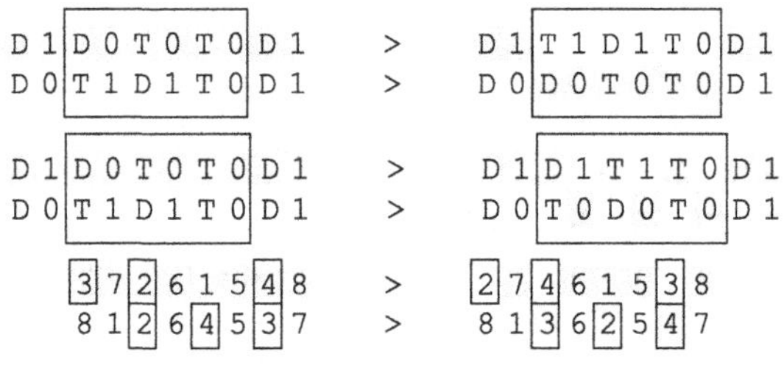

D 1 | D 0 T 0 T 0 | D 1 > D 1 | T 1 D 1 T 0 | D 1
D 0 | T 1 D 1 T 0 | D 1 > D 0 | D 0 T 0 T 0 | D 1

D 1 | D 0 T 0 T 0 | D 1 > D 1 | D 1 T 1 T 0 | D 1
D 0 | T 1 D 1 T 0 | D 1 > D 0 | T 0 D 0 T 0 | D 1

3 7 2 6 1 5 4 8 > 2 7 4 6 1 5 3 8
8 1 2 6 4 5 3 7 > 8 1 3 6 2 5 4 7

Fig. 2. Crossover: flip-flop types and inverter presence (top), inverter presence only (middle), pattern and bit order position (bottom)

In the first chromosome, register type and inverter presence are considered as one indivisible block (i.e., two values for one position in the chromosome). Moreover, with some probability p_r only the values of inverters within the crossover points were swapped, as presented in Figure 2(middle).

The crossover in case of test patterns order and bit-order of the test patterns was performed with the interchange of positions that store the ordered numbers within the range; for an example within the range [2, 4], see Figure 2(bottom).

Mutation. In the mutation process each value of the chromosome mutated with a probability p_m. To prevent a random walk through the GA search space, p_m had to be chosen to be somewhat low. Three different types of mutation were applied (see Figure 3 for details):

```
D 1[D]0[T]0 T 0 D 1   >   D 1[T]0[D]0 T 0 D 1
D[1]D 0 T[0]T 0 D 1   >   D[0]D 0 T[1]T 0 D 1
    3 7[2]6 1[5]4 8   >     3 7[5]6 1[2]4 8
```

Fig. 3. Mutation: flip-flop types (top), inverter presence (middle), pattern orders and bit orders (bottom)

- D/T type change, where only flip-flop types were changed,
- inverter change, where inverter presences were changed,
- order change, where pattern order and bit order were changed.

Furthermore, the variable mutation probability p_m was decreasing linearly with each new population. Since each new population generally was more fit than the previous one, we overcome a possible disruptive effect of mutation at the late stages of the optimization, and speed up the convergence of the GA to the optimal solution in the final optimization stages.

3.4 Fitness Evaluation

After the recombination operators modified the solutions, the whole new population was ready to be evaluated. The external evaluation tool was used to evaluate each new chromosome created by the GA. Here two results were obtained for each solution:

- cost of the flip-flops in the register, and
- cost of the combinational next state function.

The first one represents the cost of all flip-flops that build the MISR, while the second one represents the cost of the modification logic that executes the next state function.

Evaluation tool. On the basis of the equations for the register's next-state, values of the outputs of the modification logic for each vector but last in the test sequence can be derived. In that way ON-set and OFF-set of the modification logic are defined. Further, Espresso software [5] was used for Boolean minimization of the modification logic and its approximate cost estimation. This software takes a two-level representation of a two-valued (or multiple-valued) Boolean function as input, and produces a minimal equivalent representation (number of equivalent gates).

3.5 Termination and Solutions

GA operated repetitively; when a certain number of populations had been generated and evaluated, the system was assumed to be in a non-converging state. The solutions on the Pareto front were taken to be the final implementation solution candidates. They were further evaluated by the experienced engineer to choose the one (or more) that best fits the given problem.

4 Results

The optimization procedure sets the initial TPG structure through the desired sequence of test patterns. Then the GA tries to optimize the circuit (make new configuration) while checking the allowed TPG structure and using the external evaluation tool. The evaluation tool calculates the cost of a given structure through the input test patterns and TPG configuration. After a number of iterations the best structure is chosen and implemented.

There are two sets of GA parameters used in our experiments; (1) for first three circuits: number of generations is 50, population size is 10, probability of crossover is 0.8, and probability of mutation is 0.01, (2) for the next three circuits: number of generations is 100, population size is 50, probability of crossover is 0.7, and probability of mutation is 0.05. These parameters were set on the basis of a short pre-experimentation tests, and ensure quick enough TPG structure optimization.

Table 1 presents the results of the evaluation of the optimization process with the ISCAS test-benchmark combinational circuits. These simple combinatorial circuits are used to benchmark various test pattern generation systems. The ISCAS benchmark suite has been introduced in simple netlist format at the International Symposium of Circuits and Systems in 1985 (ISCAS '85). The 1989 ISCAS symposium introduced a set of sequential circuits, similar to the 1985 circuits, but with some additional flip-flop elements.

The used test sets were transformed by the input reduction procedure proposed in [3]. The test pattern width (denotes the number of the inputs) and the number of test patterns (number of different input test vectors to cover all possible faults) are presented in the second and the third column, respectively, for each benchmark. The next two columns present the total cost (number of equivalent gates) of the modification logic reported by Espresso software for the initial and optimized TPG structure. The last column shows the achieved improvement.

No report on total execution time is included. It was measured in minutes, but since this is off-line and one-time optimization procedure, optimization effectiveness was considered more important as optimization time.

Table 1. Results of modification logic size

circuit	test pattern width	number of test patterns	initial TPG	optimized TPG	improvement in %
c432	36	27	348	280	19.5
c499	41	52	312	164	47.4
c880	60	16	536	402	25.0
c1355	41	84	584	488	16.4
c1908	33	106	2077	1840	11.4
c6288	11	12	74	49	33.8

Table 2. Comparison of *complexity* in [6] and of *area_per_bit* in [2] with GA approach

	complexity		*area_per_bit*	
circuit	[6]	GA approach	[2]	GA approach
c432	0.33	0.29	0.22	0.11
c499	0.13	0.08	0.21	0.10
c880	0.38	0.35	0.19	0.29
c1355	0.19	0.14	0.20	0.09
c1908	0.29	0.53	0.19	0.32
c6288	–	0.44	0.24	0.54

As stated before, the bit-order of the test patterns and the order of the test patterns in a test sequence influence the area of the modification logic. In this respect it may be interesting to compare our results with the results of column matching algorithm [6]. Both approaches use MISR of similar complexity, but the main differences are in the design of the modification logic. The second and the third column of Table 2 show the results of the comparison of these two approaches for the same benchmark circuits.

The *complexity* values in Table 2 are expressed in terms of a total cost reported by Espresso software per bit of the produced test pattern:

$$complexity = \frac{total\_cost}{test\_pattern\_width * number\_of\_test\_patterns}. \quad (4)$$

We need to apply such a measure because in our experiments we used different test pattern sets than those reported in [6].

The presented comparison indicates that the proposed approach has a higher potential to provide solutions of TPG generating deterministic test patterns than column matching. There is also a big difference in testing time. In column matching solution all deterministic test patterns are embedded in a long test sequence composed of 5000 test vectors, which contains a lot of patterns not contributing to the fault coverage in the CUT. On the other hand, the GA based solution produces all deterministic test patterns as a one short test sequence that does not contain any superfluous vectors.

The fourth and the fifth column of Table 2 present the area of TPG logic for AMS 0.35 μm technology for the implementations reported in [2] and the GA based solutions. The area is expressed in terms of equivalent two input NAND gates. Like for *complexity* we need to apply a specific measure of the area overhead of the TPGs due to the fact that different deterministic patterns sets have been used for TPG synthesis. The proposed measure is expressed by the following equation:

$$area\_per\_bit = \frac{area}{test\_pattern\_width * number\_of\_test\_patterns}. \quad (5)$$

Shown experimental results indicate that for some benchmarks the proposed TPG and the GA optimization procedure provide solutions with lower area overhead than the TPG presented in [2] while for some other benchmarks the TPG in [2] are better. This may be due to the fact that we used Espresso software as a fast evaluation tool in the TPG optimization process and Synopsys software as a tool for synthesizing the final solution. Applying Synopsys software as both the evaluation tool and the final synthesis tool is likely to improve the results.

These evaluations and results illustrate the advantages of the proposed approach in comparison with the existing solutions. However, one should be aware that the employed benchmark circuits are relatively small. Realistic assessment of techniques for automatic deterministic test pattern generation requires more complex circuits. Since such examples are not reported in the referred papers, we performed GA optimization approach on some larger benchmark circuits. While the results regarding the *complexity* and the *area_per_bit* are in average comparable to the GA examples reported above, the computation time for larger circuits considerably increases and may represent a bottleneck in practical implementations. For example, the computation time for circuit s38417 was 140 times larger than for c880.

5 Conclusion

Pseudo random pattern generators provide reasonable fault coverage for different circuits-under-test. However, if a TPG fails to provide the desired fault coverage within the given test length, application specific deterministic TPGs are employed. Deterministic TPGs are by default more complex since they employ additional logic to prevent the generation of non-useful test patterns. Area overhead is one of the important issues of the design of deterministic TPGs. In this paper, a design of a new type of deterministic TPG is presented based on a feedback shift register composed of D- and T-type flip-flops and inverters. It is also equipped with a modification logic that can invert any bit in any pattern generated by the register. The search for the optimal structure of the TPG is performed by a genetic algorithm and some illustrative case studies were performed on ISCAS test-benchmark circuits. Promising initial results have been obtained on small and medium benchmark circuits, while the computation time for larger circuits considerably increases and may represent a bottleneck in practical implementations.

References

1. Bäck, T.: Evolutionary Algorithms in Theory and Practice. Oxford University Press, Oxford (1996)
2. Bellos, M., Kagaris, D., Nikolos, D.: Test set embedding based on phase shifters. In: Bondavalli, A., Thévenod-Fosse, P. (eds.) EDCC-4 2002. LNCS, vol. 2485, pp. 90–101. Springer, Heidelberg (2002)
3. Chen, C.-A., Gupta, K.: Efficient BIST TPG design and test set compaction via input reduction. IEEE Transactions on Computer-Aided Design of Integrated Circuits and Systems 17, 692–705 (1998)
4. Coello, C.C.: A comprehensive survey of evolutionary-based multiobjective optimization techniques. Knowledge and Information Systems 1(3), 269–308 (1999)

5. Espresso UC Berkeley (1988), `http://www-cad.eecs.berkeley.edu:80/software/software.html`
6. Fišer, P., Hlavička, J.: Column matching based BIST design method. In: Proceedings IEEE European Test Workshop, Corfu, Greece, pp. 15–16 (2002)
7. Garbolino, T., Hlawiczka, A.: A new LFSR with D and T flip flops as an effective test pattern generator for VLSI circuits. In: Hlavicka, J., Maehle, E., Pataricza, A. (eds.) EDDC 1999. LNCS, vol. 1667, pp. 321–338. Springer, Heidelberg (1999)
8. Garbolino, T., Hlawiczka, A., Kristof, A.: Fast and low area TPG based on T type flip flops can be easily integrated to the scan path. In: Proceedings IEEE European Test Workshop, Cascais, pp. 161–166 (2000)
9. Goldberg, D.: Genetic Algorithms in Search, Optimization, and Machine Learning. Addison-Wesley, Reading (1989)
10. Khalil, M., Robach, C., Novak, F.: Diagnosis strategies for hardware or software systems. Journal of electronic testing: Theory and Applications 18, 241–251 (2002)
11.Ageroušić-Seljak, B.: Timetable construction using general heuristic techniques. Journal of Electrical Engineering 53, 61–69 (2002)
12. Oblak, K., Korošec, P., Kosel, F., Šilc, J.: Multi-parameter numerical optimization of selected thin-walled machine elements using a stigmergic optimization algorithm. Thin-walled structures 45(12), 991–1001 (2007)
13. Papa, G., Koroušić-Seljak, B.: An artificial intelligence approach to the efficiency improvement of a universal motor. Engineering applications of artificial intelligence 18(1), 47–55 (2005)
14. Papa, G., Šilc, J.: Automatic large-scale integrated circuit synthesis using allocation-based scheduling algorithm. Microprocessors and Microsystems 26, 139–147 (2002)
15. Parker, K.: The boundary-scan handbook, 3rd edn. Kluwer Academic Publishers, Dordrecht (2003)
16. Zitzler, E., Deb, K., Thiele, L.: Comparison of multiobjective evolutionary algorithms: Empirical results. Evolutionary Computation 8(2), 173–195 (2000)

An Evolutionary Methodology for Test Generation for Peripheral Cores Via Dynamic FSM Extraction

D. Ravotto, E. Sanchez, M. Schillaci, and G. Squillero

Dipartimento di Automatica e Informatica, Politecnico di Torino, Italy
{danilo.ravotto,edgar.sanchez,massimiliano.schillaci,
giovanni.squillero}@polito.it

Abstract. Traditional test generation methodologies for peripheral cores are performed by a skilled test engineer, leading to long generation times. In this paper a test generation methodology based on an evolutionary tool which exploits high level metrics is presented. To strengthen the correlation between high-level coverage and the gate-level fault coverage, in the case of peripheral cores, the FSMs embedded in the system are identified and then dynamically extracted via simulation, while transition coverage is used as a measure of how much the system is exercised. The results obtained by the evolutionary tool outperform those obtained by a skilled engineer on the same benchmark.

Keywords: Peripheral testing, test generation, μGP^3, evolutionary methods, approximate methods.

1 Introduction

A system-on-chip (SoC) can integrate into a single device one or more processor cores with standard peripheral memory and application-oriented logic modules. This high integration of many components leads to an increased complexity of the test process since it decreases the accessibility of each functional module into the chip. Thus, the ever increasing usage of such devices demands for cheap testing methodologies.

The *Software-based Self-test* (SBST), whereby a program is executed on the processor core to extract information about the functioning of the processor or other SoC modules and provide it to the external test equipment [2] meets this demands since: it allows cheap at-speed testing of the SoC; it is relatively fast and flexible; it has very limited, if any, requirements in terms of additional hardware for the test; it is applicable even when the structure of a core is not known, or can not be modified. Even though SBST is currently being increasingly employed, the real challenge of software-based testing techniques is to generate effective test programs.

Many SBST techniques have been developed for the test of microprocessor cores; traditional methodologies resort to functional approaches based on exciting specific functions and resources of the processor [1]. New techniques, instead, differ on the basis of the kind of description they start from: in some cases only the information coming from the processor functional descriptions are required [3]; other simulation-based approaches require a pre-synthesis RT-level description [4] or the gate-level description [5].

M. Giacobini et al. (Eds.): EvoWorkshops 2008, LNCS 4974, pp. 214–223, 2008.

Simulation-based strategies are heavily time consuming, thus, the use of RT-level descriptions to drive the generation of test sets is preferable to allow much faster evaluation. Relying on high-level models not only helps the user of the SoC to perform more simulations increasing the confidence in the generated tests, but is also of value to the manufacturer allowing early generation of a significant part of the final test set. Whereas the correlation between RT-level code coverage metrics (CCM) and gate-level fault coverage is not guaranteed in the general case, several RT-level based methodologies maximize the CCMs to obtain a good degree of confidence on the quality of the generated test set.

This paper describes the application of an evolutionary algorithm in test set generation process for different types of peripheral cores embedded in a SoC. Furthermore the generation process is fully automated and requires a very low human effort. The generation process is driven by the transition coverage on the peripheral's *finite state machine* (FSM) and by the RT-level Code Coverage Metrics (CCMs). Exploiting the correlation between high-level and low-level metrics, during the generation process only logic simulation is performed allowing the reduction of the generation time. The results are finally validated running a gate-level fault simulation.

Results show that the combination of the FSM transition coverage and CCMs can effectively guide the test block generation and a high fault coverage can be achieved. Moreover, we show that the new approach makes the test generation process more robust, improving the relationship between high- and low-level metrics.

The rest of the paper is organized as follows: section 2 recalls some background concepts in peripheral testing; section 3 outlines the methodology adopted for the generation of test sets and details the evolutionary tool. Section 4 introduces the experimental setup, describing the case study and presents the experimental results. Finally, section 5 draws some conclusions.

2 Peripheral Testing

2.1 Basics

A typical SoC is composed of a microprocessor core, some peripheral components, memory modules, and possibly customized cores. An external ATE is supposed to be available for test application: its purpose is to load a test program in the memory, start execution, and interact with the peripherals applying data to the input ports and collecting values from the outputs while the program is running.

To make effective use of the test setup both the test programs and the peripheral input/output data have to be specified; therefore, a complete set for testing peripheral cores is composed of some test blocks [9], defined as basic test units composed of two parts: a configuration and a functional part. The configuration part includes a program fragment that defines the configuration modes used by the peripheral, and the functional part contains one or more program fragments that exercise the peripheral functionalities as well as the data set or stimuli set provided/read by the ATE.

Researchers have long sought high-level methodologies to generate high quality test sets; this is possible only if a correlation between high-level metrics and gate-level fault coverage exists. Differently from the general case, where the correlation is vague, in the case of peripheral cores this correlation actually exists. It is not complete but, as experimentally shown in [10], suitable for test set generation.

Therefore, an automatic methodology for the generation of test sets for peripheral cores that uses a high-level model of the peripheral in the generation phase is an interesting solution to overcome new testing issues on SoCs.

As mentioned in [9], traditional code coverage metrics suitable for guiding the development of the test sets for peripheral cores are: *Statement coverage* (SC), *Branch coverage* (BC), *Condition coverage* (CC), *Expression coverage* (EC), *Toggle coverage* (TC). Maximizing all the coverage metrics allows to better exercise the peripheral core. It is not possible to accept a single coverage metric as the most reliable and complete one [6]; thus different metrics must be exploited in order to guarantee better performance of the test sets [7].

2.2 Previous Work

An attempt to provide effective solutions for peripheral test set generation is presented in [9]; the process is performed by hand and mainly relies on the experience of a test engineer, who maximizes sequentially the various coverage metrics, generating one or more test blocks for every metric. This process is repeated until sufficiently high coverage values are obtained for all the chosen metrics.

In [8] a pseudo-exhaustive approach to generate functional programs for peripheral testing was presented. The proposed method generates a functional program for each possible operation mode of the peripheral core in order to generate control sequences which would place the peripheral in all possible functional modes. The pseudo-exhaustive approach produces a large number of functional programs, since one has to be written for every operation mode.

In [13] the authors describe a generic and systematic flow of SBST application on two communication peripheral cores. The methodology achieves high fault coverage but needs a deep knowledge of the peripheral core leading to long test development time with a high human effort.

In [10] the peripheral test set generation has been automated using an evolutionary algorithm, called μGP^3 The test block generation was supported by the construction of couples of templates: one for program and the other for data generation. The evolutionary algorithm is used to optimize parameter values, leaving the structure of the test block fixed. The obtained results compare favorably with respect to the manually generated [9].

In [12] an improved version of the evolutionary algorithm has been described, able to optimize both the structure and the parameters. The same results as [10] are obtained with no need of the rigid templates used previously, reducing significantly the required generation time.

3 Proposed Approach

As stated before, traditional CCMs extracted at the RT-level do not, in general, show a tight correlation with gate-level fault coverage. Furthermore, the RT-level descriptions use, especially in the case of complex cores, many modules that interact among each other in order to perform the core functionalities. The traditional CCMs do not consider these interactions and only aim at maximizing the coverage metrics in each module. After the synthesis process, at the gate level, the distinction between modules of a core is less clear and therefore it is important to consider the interactions to enforce a correlation between high-level metrics and low level ones.

One way to model a system is to represent it with a FSM. Coverage of all the possible transitions in the machine ensures thoroughly exercising the system functions. Additionally, the use of FSM transition coverage has the additional advantage that it makes the interactions between functional modules in the peripheral explicit. Figure 3 sketches the proposed methodology.

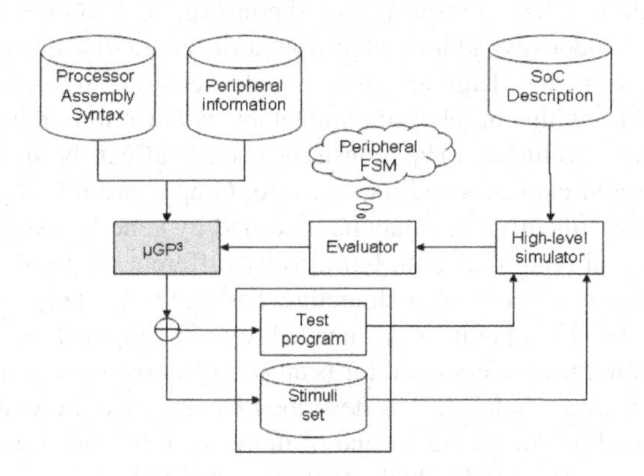

Fig. 1. Evolutionary generation loop

The evolutionary approach generates test blocks starting from information about the peripheral core and the processor assembly syntax only. Every new test block generated is evaluated using a high-level simulator. The evaluation stage assigns a fitness to every individual. The procedure ends when a time limit is elapsed or when a steady state is detected, that is, a predefined number of test blocks are generated without any improvement of the coverage metrics. At the end of the evolutionary run a single test block is provided as output.

The sketched procedure is iteratively repeated to generate a complete test set. In the steps following the first one, the evaluation phase is modified in order to only take into account the additional coverage provided by the new test blocks. The rationale for this methodology is that in general it is not possible to completely solve the problem with one single test block. The end result of the process is a set of test blocks that cumulatively maximize the targeted coverage metrics.

3.1 Evolutionary Tool

For the automatic generation of the test blocks an evolutionary tool named μGP^3 [14] has been employed. μGP^3 is a general-purpose approach to evolutionary computation, derived from a previous version specifically aimed at test program generation.

The tool is developed following the rules of software engineering and was implemented in C++. All input/output, except for the individuals to evaluate, is performed using XML with XSLT. The use of XML with XSLT for all input and output allows the use of standard tools, such as browsers, for inspection of the constraint library, the populations and the configuration options.

The current version of the μGP^3 comprises about 50,000 lines of C++ code, 113 classes, 149 header files and 170 C++ files.

Evolution Unit. μGP^3 bases its evolutionary process on the concept of constrained tagged graph, that is a directed graph every element of which may own one or more tags, and that in addition has to respect a set of constraints. A tag is a name-value pair whose purpose is to convey additional information about the element to which it belongs, such as its name. Tags are used to add semantic information to graphs, augmenting the nodes with a number of parameters, and also to uniquely identify each element during the evolution. The constraints may affect both the information contained in the graph elements and its structure. Graphs are initially generated in a random fashion; subsequently, they may be modified by genetic operators, such as the classical mutation and recombination, but also by different operators, as required by the specific application. The tool architecture has been specially thought for easy addition of new genetic operators as needed by the application. The activation probability and strength for every operator is an endogenous parameter.

The genotype of every individual is described by one or more constrained tagged graphs, each of which is composed by one or more sections. Sections allow to define a global structure for the individuals that closely follows the structure of any candidate solution for the problem.

Constraints. The purpose of the constraints is to limit the possible productions of the evolutionary tool, and also provide them with semantic value.

The constraints are provided through a user-defined library that provides the genotype-phenotype mapping for the generated individuals, describes their possible structure and to define which values the existing parameters (if any) can take. Constraint definition is left to the user to increase the generality of the tool.

The constraints are divided in sections, every section of the constraints matching a corresponding section in the individuals. Every section may also be composed of subsections and, finally, the subsections are composed of macros.

Constraint definition is flexible enough to allow the definition of complex entities, such as the test blocks described above, as individuals. Different sections in the constraints, and correspondingly in the individual, can map to different entities.

In this specific case the constraints define three sections: a program configuration part, a program execution part and a data part or stimuli set. The first two are composed of assembly code, the third is written as part of a VHDL testbench. Though syntactically different, the three parts are interdependent in order to obtain good solutions.

Fitness. Individual fitnesses are computed by means of an external evaluator: this may be any program able to provide the evolutionary core with proper feedback.

The fitness of an individual is represented by a sequence of floating point numbers optionally followed by a comment string. This is currently used in a prioritized fashion: one fitness A is considered greater than another fitness B if the n-th component of A is greater than the n-th component of B and all previous components (if any) are equal; if all components are equal then the two fitnesses are considered equal.

Evolutionary Scheme. The evolutionary tool is currently configured to cultivate all individuals in a single panmictic population, although it can be configured to use an island model. The population is ordered by fitness. Choice of the individuals for reproduction is performed by means of a tournament selection; the tournament size τ is also endogenous. The population size μ is set at the beginning of a run, and the tool employs a variation on the plus ($\mu+\lambda$) strategy: a configurable number λ of genetic operators are applied on the population. Since different operators may produce different number of offspring the number of individuals added to the population is variable. All new unique individuals are then evaluated, and the population resulting from the union of old and new individuals is sorted by decreasing fitness. Finally, only the first μ individuals are kept.

To promote diversity, the individuals genetically equal to already existing ones, called clones, may have their fitness scaled by a fixed value in the range [0.0,1.0].

The possible termination conditions for the evolutionary run are: a target fitness value is achieved by the best individual; no fitness increase is registered for a predefined number of generations; a maximum number of generations is reached.

At the end of every generation the internal state of the algorithm is saved in a XML file for subsequent analysis and for providing a minimal tolerance to system crashes.

3.2 Evaluator

The proposed approach is based on modeling the entire system as a FSM which is dynamically constructed during the test generation process. Thus, differently from other approaches, the FSM extraction is fully automated, and requires minimum human effort: the approach only requires the designer to identify the state registers in the RT-level code; every global state in the peripheral represents a possible configuration of values of all the state registers. Thus, whenever a state register in any module changes its value, also the global state of the peripheral is affected.

Given the dynamic nature of the FSM construction, it is not possible to assume known the maximum number of reachable states, not to mention the possible transitions. For this reason it is impossible to determine the transition coverage with respect to the entire FSM.

As experimentally demonstrated [6], maximizing more than one metric usually leads to better quality tests. Thereby, the simulation-based method proposed here exploits the FSM transition coverage, that enforce a maximum interaction between peripheral modules, and all the available CCMs to thoroughly exercise the peripheral functionalities.

The implemented evaluator collects the output of the simulation and dynamically explores the FSM; it assesses the quality of the test block considering the transition coverage on the FSM and the CCMs.

The fitness fed back to the evolutionary tool is composed of many parts: the FSM transition coverage followed by all the others CCMs (SC, BC, CC, EC, TC). As we mentioned before the metrics are considered in order of importance. In this way it is possible, during the generation process, to select more thoroughly those test blocks that are able to better excite the peripheral.

4 Experimental Analysis

4.1 Test Case

The benchmark is a purposely designed SoC which includes a Motorola 6809 microprocessor, a Universal Asynchronous Receive and Transmit (UART), a Peripheral Interface Adapter (PIA), a Video display unit (VDU) and a RAM memory core. The system derives from one available on an open source site [11]. The methodology is used to test the UART, the PIA and the VDU in the targeted SoC.

The peripherals are described at RT-level in VHDL code and are composed of different modules. The SoC was synthesized using a generic home-developed library.

Table 1. Implementation characteristics

Description	measure	PIA	VDU	UART
RT-level	statements	149	153	383
	branches	134	66	182
	condition	75	24	73
	expression	0	9	54
	toggle	77	199	203
Gate level	Gates	1,016	1,321	2,247
	Faults	1,938	2334	4,054

Table 1 shows details of the targeted peripherals, including information at high and low level. Rows labeled with RT-level present CCM information while the remaining rows illustrate the number of gates counted on the synthesized devices and the number of collapsed faults for the stuck-at model, respectively.

At the end of the generation process, some gate-level fault simulation were performed only to validate the proposed methodology; the gate-level fault coverage figures reported in the following sections target the single stuck-at fault model.

4.2 Experimental Results

All the reported experiments have been performed on a PC with an Athlon XP3000 processor, 1GB of RAM, running Linux.

The algorithm parameters for the evolutionary experiments are the same both when targeting only the CCMs, and when the number of transitions in the FSM is also taken into account: for the PIA and the VDU experiments, μ=50 and λ=70; and as the UART is more complex than the PIA the evolutionary parameters were set to perform a lower number of simulations: μ was set to 30 and λ to 40.

In order to provide the reader with a reference value, we recall that the fault coverage obtained by the manual approach presented in [9] is 80.96% for the UART and 89.78% for the PIA.

Table 2 summarizes the results obtained for the targeted peripherals, reporting the number of FSM transitions covered, the high-level CCMs and the stuck-at fault coverage (FC) in percentage. The reader should note that the value of traditional CCMs are expressed as absolute values (instead of percentages).

Table 2. Results for considered peripherals

	PIA	VDU	UART
FSM Transition	115	191,022	142
Statement	149	153	383
Branch	129	66	180
Condition	68	23	72
Expression	0	9	51
Toggle	77	191	203
FC(%)	**91.4**	**90.8**	**91.28**

For every peripheral considered the methodology is able to reach a good value of gate-level fault coverage. In the case of the VDU the number of transition is very high; this is due to the state registers that hold the current position on the screen.

To experimentally demonstrate that the use of the FSM transition coverage is essential to strengthen the correlation between high an low level metrics 100 experiments on the UART are performed, using both the evolutionary approach presented in [12] and the generation process detailed above.

Table 3. Comparison between the two methodologies

		FSM	SC	BC	CC	EC	TC	FC
[12]	Average	NA	381.8	178.7	70.7	50.7	201.3	84.8
	std.dev.	NA	0.36	0.39	0.30	0.32	0.40	6.37
New methodology	Average	141.0	382.2	179.3	71.8	50.8	202.2	90.9
	std.dev.	1.49	0.28	0.33	0.22	0.24	0.36	1.10

Table 3 reports a comparison between the results of the experiments performed following the methodology presented in [12] and the current one; the table illustrates the average and standard deviation of the different CCMs and of the stuck-at fault coverage (FC). In all cases the CCMs are very near to the absolute maximum, and both methodologies lead to small standard deviations on the considered metrics. In the first case, however, the standard deviation in the fault coverage of each test set is relatively high. Although the methodology obtains good results, it is not as robust as desirable, and the obtained solution may not exhibit the expected quality.

Using the new methodology the average fault coverage is increased by more than 6% and, more importantly, the standard deviation of the fault coverage is dramatically reduced. This clearly shows that the robustness of the methodology is increased, and solutions of consistent quality can be obtained.

Table 4. Overall comparison

	FC	TGEN	TAPP	Size
[12]	90.7	5.1	28,842	1,953/72
New Methodology	91.3	2.2	32,762	2,345/87

Table 4 synthetically reports a comparison between the two methodologies in the case of the UART, highlighting the obtained fault coverage (FC) in percentage, the average generation time (TGEN) expressed in hours, the average application time (TAPP) in clock cycles, and the average size of the test sets, reported as program bytes and data bytes. The results clearly show that the new methodology outperforms the previous one in terms of fault coverage and generation time. The latter, in particular, is less than a half with respect to the previous methodology, highlighting the efficiency of the new approach.

Other approaches [8][13] to peripheral test are not directly comparable with our methodology since they are referred to different devices, although their complexity and the results are similar to the devices analyzed here. Furthermore, our methodology only needs RT-level simulation and does not need the time-expensive fault-simulations.

5 Conclusions

In this paper a successful application of the evolutionary tool for the generation of sets of test blocks for different types of peripheral modules in SoCs driven by the FSM transition coverage and the high-level CCM has been described.

The evolutionary tool is able to generate test blocks where the relation between high-level coverage metrics and low level one is much stronger; this better relation has been experimentally demonstrated with a experimental analysis where many test blocks are generated and evaluated.

The experimental results on different type of peripheral cores, communication peripherals and VDU controller, show the effectiveness of the proposed methodology.

Acknowledgment. The authors thank Alessandro Aimo, Luca Motta and Alessandro Salomone for their help in designing and implementing the µGP^3, and Alberto Cerato for performing most of the experiments.

References

1. Thatte, S., Abraham, J.: Test Generation for Microprocessors. IEEE Transactions on Computers C-29, 429–441 (1980)
2. Kranitis, N., Paschalis, A., Gizopoulos, D., Xenoulis, G.: Software-based self-testing of embedded processors. IEEE Transactions on Computers 54(4), 461–475 (2005)
3. Corno, F., Cumani, G., Sonza Reorda, M., Squillero, G.: Fully Automatic Test Program Generation for Microprocessor Cores. In: DATE2003: IEEE Design, Automation & Test in Europe, pp. 1006–1011 (2003)

4. Cheng, A., Parashkevov, A., Lim, C.C.: A Software Test Program Generator for Verifying System-on-Chip. In: 10th IEEE International High Level Design Validation and Test Workshop 2005 (HLDVT 2005), pp. 79–86 (2005)
5. Corno, F., Cumani, G., Sonza Reorda, M., Squillero, G.: An RT-level Fault Model with High Gate Level Correlation. In: HLDVT2000: IEEE International High Level Design Validation and Test Workshop (2000)
6. Chien-Nan, J.L., Chen-Yi, C., Jing-Yang, J., Ming-Chih, L., Hsing-Ming, J.: A novel approach for functional coverage measurement in HDL Circuits and Systems. In: ISCAS 2000: The 2000 IEEE International Symposium on Circuits and Systems, pp. 217–220 (2000)
7. Sanchez, E., Sonza Reorda, M., Squillero, G.: Test Program Generation From High-level Microprocessor Descriptions. In: Sonza Reorda, M., Violante, M., Peng, Z. (eds.) Test and validation of hardware/software systems starting from system-level descriptions, p. 179, pp. 83–106. Springer, Heidelberg (2004)
8. Jayaraman, K., Vedula, V.M., Abraham, J.A.: Native Mode Functional Self-test Generation for System-on-Chip. In: IEEE International Symposium on Quality Electronic Design (ISQED 2002), pp. 280–285 (2002)
9. Sanchez, E., Veiras Bolzani, L., Sonza Reorda, M.: A Software-Based methodology for the generation of peripheral test sets Based on high-level descriptions. In: SBCCI 2007, Symposium on Integrated Circuits, pp. 348–353 (2007)
10. Bolzani, L., Sanchez, E., Schillaci, M., Squillero, G.: An Automated Methodology for Cogeneration of Test Blocks for Peripheral Cores. In: IOLTS 2007: International On-Line Testing Symposium, pp. 265–270 (2007)
11. http://www.opencores.org/
12. Bolzani, L., Sanchez, E., Schillaci, M., Squillero, G.: Co-Evolution of Test Programs and Stimuli Vectors for Testing of Embedded Peripheral Cores. In: CEC 2007. IEEE Congress on Evolutionary Computation, pp. 3474–3481 (2007)
13. Apostolakis, A., Psarakis, M., Gizopoulos, D., Paschalis, A.: A functional Self-Test Approach for Peripheral Cores in Processor-Based SoCs. In: IOLTS 2007: IEEE International On-Line Testing Symposium (2007)
14. http://ugp3.sourceforge.net/

Exploiting MOEA to Automatically Geneate Test Programs for Path-Delay Faults in Microprocessors

P. Bernardi[1], K. Christou[2], M. Grosso[1], M.K. Michael[2], E. Sánchez[1], and M. Sonza Reorda[1]

[1] Politecnico di Torino – Dipartimento di Automatica e Informatica - Torino, Italy
{paolo.bernardi,michelangelo.grosso,edgar.sanchez, matteo.sonzareorda}@polito.it
[2] University of Cyprus – Department of Electrical and Computer Engineering - Nicosia, Cyprus
{christou,mmichael}@ucy.ac.cy

Abstract. This paper presents an innovative approach for the generation of test programs detecting path-delay faults in microprocessors. The proposed method takes advantage of the multiobjective implementation of a previously devised evolutionary algorithm and exploits both gate- and RT-level descriptions of the processor: the former is used to build Binary Decision Diagrams (BDDs) for deriving fault excitation conditions; the latter is used for the automatic generation of test programs able to excite and propagate fault effects, based on a fast RTL simulation. Experiments on an 8-bit microcontroller show that the proposed method is able to generate suitable test programs more efficiently compared to existing approaches.

Keywords: MOEA, path-delay testing, microprocessor, BDD.

1 Introduction

In order to guarantee product quality for today's microprocessor cores, traditional stuck-at tests are no longer sufficient and more complex fault models have to be considered when devising test strategies. At-speed delay fault testing, in particular, has been widely addressed by academia and is becoming common practice in industry [1]-[4]. Among all existing delay fault models, the path-delay fault model is considered the most accurate since it can detect both lumped and distributed delays [3][5], but also the most challenging, due to the enormous number of faults (paths).

Delay test has been approached adopting different strategies, purely relying on an external tester or applying structural self-testing methodologies such as *Built-In Self-Test* (*BIST*), or exploiting the execution of suitable self-test programs. The latter strategy is usually referred to as *Software-Based Self-Test* (*SBST*) and is generally more affordable, as it exploits the processor instructions in the normal mode of operation; it can be used in stand-alone modules as well as when the processors are deeply embedded in a System on Chip (SoC) and their accessibility is reduced.

Regarding test generation addressing path-delay faults, several techniques exist for enhanced full-scan circuits, based on either structural ATPG tools [6][7] or function-based tools using *Binary Decision Diagrams* (*BDDs*) [8]-[10] and Boolean-SAT

M. Giacobini et al. (Eds.): EvoWorkshops 2008, LNCS 4974, pp. 224–234, 2008.

[11][12] implementations. Some work on software-based test generation has been done exploiting deterministic techniques [13]-[15]. Evolutionary algorithms have been successfully exploited for the automatic generation of program sets for verification, test [16], and diagnosis [17] for processors described at different levels of abstraction. In most cases, the evolutionary algorithm faces the test set generation as a single-objective optimization problem, e.g., resorting to a multi-run strategy. However, hardware optimization techniques belong to a real-world classification of problems that usually require the simultaneous optimization of many objectives. Therefore, hardware optimization problems could be addressed resorting to multiobjective optimizers. *Multiobjective Evolutionary Algorithms* (*MOEA*s) were initially introduced in 1985, by the implementation of the first evolutionary algorithm dealing with multiobjective optimization problems [18]. Roughly speaking, MOEAs produce a set of potentially optimal solutions, rather than an unique solution, that represents a subset of the Pareto optimal set.

This paper presents an innovative approach for the automatic generation of path-delay functional test programs for microprocessors exploiting both gate- and RT-level descriptions. The former is used to select the set of critical paths to be considered and to obtain path excitation requirements based on BDD analysis; the latter is used for effectively identifying the test programs able to reproduce the conditions activating the targeted fault (*excitation*), and to make the fault effect(s) visible on the processor outputs (*propagation*). For automatically generating test programs, the new implementation of an evolutionary algorithm addressing multiobjective optimization is employed. The main advantage introduced is the improvement in the flow performances compared to other approaches based only on gate-level simulation [19].

The organization of this paper is as follows: Section 2 provides the needed background; Section 3 details the proposed methodology; Section 4 presents the case of study. Finally, in Section 5 some conclusions are drawn.

2 Background

2.1 Software-Based Path Delay Testing

A path-delay fault occurs when a defect in a circuit causes the cumulative delay of a combinational path to exceed some specified duration [5][20]. The combinational path begins at a primary input or a clocked flip-flop (*startpoint*), includes a connected chain of gates, and ends at a primary output or a clocked flip-flop (*endpoint*) (Fig. 1). The specified time duration can be the duration of the clock period (or phase), or the vector period. The propagation delay is the time that an event (i.e., a transition) takes to traverse the path. For each combinational path in a circuit, there are two path-delay faults, corresponding to rising and falling transitions on the startpoint. Signals that compose the path and feed the traversed gates are called *on-path signals*; signals that are not on the path but feed the gates on the path are called *off-path signals*.

In order to examine the timing operation of a circuit we should examine signal transitions: delay tests consist of vector pairs ($V_1 \rightarrow V_2$) to be applied on the inputs feeding the path (*a*, *b*, *c*, *d* and *e* in Fig. 1), so that an input transition on the startpoint propagates to the endpoint.

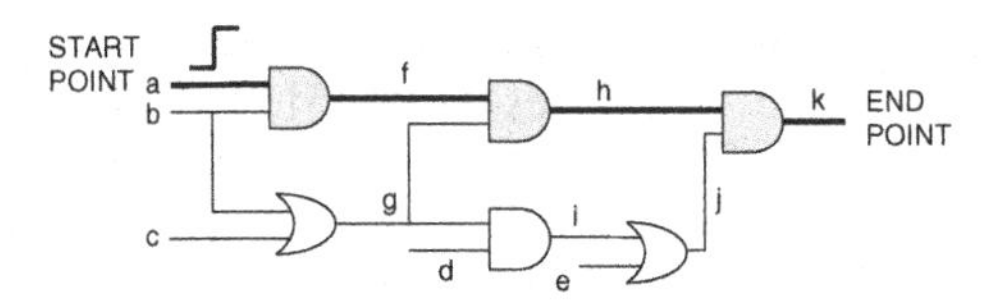

Fig. 1. Example of a path-delay fault: *on-path* signals indicated by thick lines (*a, f, h, k*); *b, g* and *i* are *off-path* signals

Path-delay test application can be performed resorting to suitable scan-chains or by employing functional techniques. In scan-based test methodologies, the patterns are serially loaded into the scan chains (at reduced speed if necessary). Consequently, the two test vectors are applied in succession with a defined timing and the test results are shifted out through the scan chains, thus achieving full observability. In the case of Software-Based path-delay testing the test vectors $V_1 \rightarrow V_2$ reach the targeted path inputs during the normal at-speed circuit operations, hence depending on the sequence of data feed (instructions in case of processors) and allowing continuous application of test vectors. When targeting microprocessors, a test program must be made to ensure that the excitement conditions of the targeted path-delay fault are met in a consecutive pair of clock cycles, and that the fault effect(s) propagate to suitable observable points (e.g., output ports).

If a test can be applied in the normal operations of a circuit, we refer to it as a *functional test.* A path is *functionally testable* if there exists a functional test for that path. Otherwise, the path is *functionally untestable* [14]. Functionally untestable faults never determine the performance in normal operations of the circuit, and if detected during testing may lead to overkill (i.e., discarding functioning chips). On the other hand, defects on functional testable paths may degrade the circuit performance when path-delay faults occur. Software-based testing concentrates on the latter class, intrinsically avoiding over-testing redundant paths.

2.2 Exploiting Gate- and RT Level Descriptions for Path-Delay Testing

Commonly adopted solutions for path-delay test generation in sequential circuits are mostly based on the analysis of gate-level descriptions. Addressing a fault list provided by *timing analysis* tools, test patterns for path excitation are calculated. At this phase it is seldom possible to assess whether the faults are functionally testable. The test patterns correspond to two consecutive vectors to be applied at speed to the inputs of the combinational circuit partition including the selected path. From this point forward, they will be referred as V_1 and V_2.

When dealing with functional test (in the absence of scan structures) V_1 and V_2 are *functionally justifiable* iff they can be consecutively reproduced on the memory elements and primary inputs feeding the path by a sequence of instructions and data. In this case, the processor RT-level description may be employed to establish whether an instruction sequence is able to apply V1 and V2 to the selected combinational part. Since the observation of flip-flop values is required, only, it is possible to relate each considered flip-flop in the gate-level description to a signal in the RTL one.

2.3 BDDs for Structural Path Delay Fault Tests

Rather than devising a specific couple of vectors V_1 and V_2 that excite a specific fault, through BDD analysis of the gate-level netlist it is possible to derive a wider set of requirements for the combinational subcircuit inputs to excite the path it contains.

A reduced ordered Binary Decision Diagram (referred to as a BDD here) is a canonical graphical representation of a Boolean function [10]. BDDs have been widely used in test generation, for various fault models. For the case of path-delay faults in enhanced scan designs [8][9][23], given one (or more) fault(s) a Boolean function can be formulated whose solution space is all the possible pairs of test vectors that can detect the fault(s). This function is derived based on all the necessary values on on-path and off-path signals of the path-delay fault(s). The variables of the function correspond to the primary inputs of the circuit. When such a function is given by a BDD, we have a very compact (due to the suppression of variables with the x value) and implicit (non-enumerative) representation of the entire solution space. This is of high importance for several issues in test generation: untestable faults are very easily determined; hard-to-detect faults, that require a lot of time in structural-based ATPG tools, are also efficiently handled (BDD is very small since it contains a small number of cubes); fault simulation, for fault dropping, can be trivially performed on the BDD and not on the gate-level netlist. Moreover, if an input pattern is not a valid test, the BDD can be used to quickly determine how far the input pattern is from becoming a valid test (% of bits that must be changed in the input pattern). The latter is of particular importance in the proposed methodology, since it can quickly and accurately guide the evolutionary engine to generate the necessary path-delay fault tests.

2.4 Basic Concepts on MOEAs

Multiobjective evolutionary algorithms, as their single-objective counterpart, are population-based searching algorithms that mimic natural evolution. However, differently from single-objective algorithms, MOEAs exploit the population of individuals to simultaneously evolve solutions to multiple and usually conflicting goals [21][22]. The expected result from a MOEA is a set of trade-off individuals called nondominated solutions, Pareto-optimal solutions, or Pareto optimal set. For each individual into the population, a fitness vector $f_i = (x_1, x_2, \ldots x_n)$ represents the figures of merit obtained by the individual regarding to the n pursued objectives.

Pareto optimality is defined using the concepts of domination: given two individuals *A* and *B*, *A dominates B* iff *A* is at least as good as *B* in all objectives, and better in at least one. *A* is *equivalent* to *B* iff results on *A* and *B* are identical in all objectives. *A covers B* if *A* either dominates or is equivalent to *B*. Similarly, given two sets of individuals Y and Z, Y dominates Z if every individual of Z is dominated by some individual of Y. Similar definitions relative to sets of individuals can be made for equivalence and coverage concepts. Thus, the Pareto optimal set is the set of all Pareto optimal individuals, and the corresponding set of fitness vectors is the *Pareto optimal front*. Individuals belonging to the Pareto optimal set are equally important. Indeed, for the individuals belonging to the Pareto optimal set, no improvement is possible in any objective without harming at least one of the other objectives.

Different strategies have been proposed in order to properly sort individuals belonging to the population; for example: aggregation-based approaches, lexicographical ordering, target-vector approaches, criterion-based approaches, and Pareto-based approaches. Some of them do not incorporate directly the concept of optimality outlined before, whereas others not only exploit it but include additional mechanisms to guarantee the diversity of the population. One of the most popular strategies used by MOEAs is based on a ranking scheme that divides the whole population on different sets, in such a way that each set contains only non-dominated individuals, and lower ranked sets are dominated by higher ones [21]. It is interesting to highlight that in a successful experiment the highest set contains the individuals belonging to the Pareto optimal set.

3 The Proposed Approach

The proposed approach targets the automatic generation of test programs (i.e., instruction sequences) for processors addressing the path-delay fault model. This low-cost generation procedure exploits both gate- and RT-level descriptions.

Four main steps have been devised to approach the generation process:

- Path list grouping: preliminary step aimed at reducing the cost of the following generation step. The path list provided by timing analysis tools is analyzed and a set of shorter fault lists is produced, each one corresponding to a *coherent* set of critical paths in the processor netlist, i.e., a set of paths related to the same processor elements. As a matter of fact, excitation conditions for faults belonging to the same structurally coherent fault group are likely to be stressed by the same instructions. Details on this topic can be found in [19].
- Circuit subdivision and BDD analysis: Given the gate-level netlist and the addressed path list, for each path a combinational subcircuit (or *chunk*) is automatically extracted, which contains the path and, therefore, all the information needed for the analysis of its excitation conditions. A BDD is then derived that contains all the possible input vectors that bring necessary excitation values at the inputs of the path under consideration. Structurally untestable faults are removed in this phase. The BDD representation will be used in the *sequential fault excitation* step for evaluating the ability of each program to excite specific faults: the fitness function depends on the minimum hamming distance of the vectors applied from the set of vectors that can excite the path. It can be computed optimally and quickly when the set of vectors is represented by a BDD.
- Sequential fault excitation: this step aims at generating the test programs that effectively excite the considered path-delay faults. A MOEA is exploited to automatically generate instruction sequences, whose fitness is evaluated through RT-level simulation, avoiding highly expensive gate-level simulations, and relying on the already available BDDs. This step will be analyzed in detail.
- Sequential Error propagation: this step targets error propagation to the processor output ports and uses an evolutionary algorithm implementing a single-objective strategy. For this task, during the RTL simulation of the test program execution, the values of the flip-flops feeding the investigated path are analyzed at each clock cycle in order to check for the excitation conditions (both on on-path and

off-path); whenever they are met, a faulty value is forced on the path endpoint for one clock cycle (*fault injection*, [19]). From that point in time, the state of all flip-flops is saved at each clock cycle and compared to the original (fault-free) simulation: if the simulation of the already generated program on the sabotaged RTL introduces a change on the processor output ports at any time following the fault injection, the test program achieves excitation and observation of the addressed fault and is complete. Otherwise, the number of flip-flops with different contents with respect to the fault-free simulation is used as a fitness function to be maximized, until the fault effects are propagated to the outputs.

The purpose of the sequential fault excitation phase (Fig. 2) is the generation of suitable instruction sequences that excite the path-delay faults in coherent lists. This process is based on the usage of a new implementation of a well known *evolutionary algorithm* (EA), called μGP^3, able to automatically generate suitable test programs.

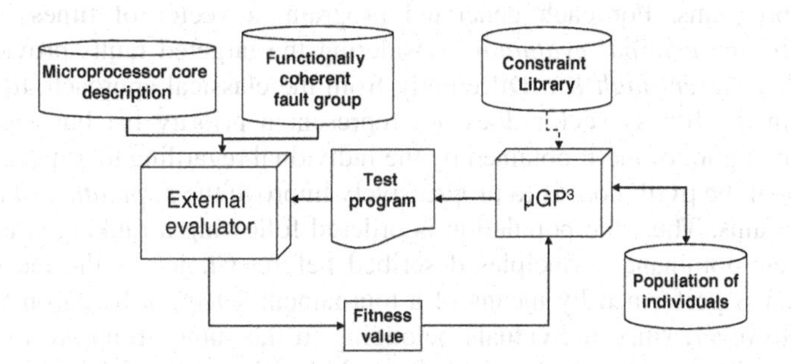

Fig. 2. Sequential fault excitation phase

Roughly speaking, an EA is a population-based optimizer that imitates the natural process of biological evolution. Following this perspective, a test program is an *individual* and the tool handles a population of individuals (i.e., a collection of assembly programs). The initial population is generated randomly, then iteratively refined mimicking the Darwinian Theory: new individuals are generated either by *mutation* (an individual is slightly modified) or by *recombination* (two or more individuals are mixed in some way); the best performing individuals are selected for survival. The process is blocked after a certain number of steps, called *generations*, or when a steady state is reached. The best individual is eventually provided as output.

Differently from the standard approach described in [1], the evolutionary tool implements a MOEA [21] able to deal with several path-delay faults at a time. In this case the main goal of the evolutionary process is not to obtain a single best program but a set of best programs able to correctly excite the targeted faults. The main idea behind the MOEA implementation of μGP^3 is to simultaneously optimize a complete functionally coherent group. As mentioned before, faults belonging to the same structurally coherent fault group are probably excited by similar test programs. Thus, the MOEA will evolve a population of individuals working on a specific portion of the processor core rather than a single program focusing on a unique fault.

μGP^3 bases its evolutionary process on a constrained tagged graph, which is a directed graph whose elements may own one or more tags, and that in addition has to respect a set of constraints. The constraints may affect both the information contained in the graph elements and its structure. Graphs are initially generated in a random fashion; subsequently, they may be modified by genetic operators (e.g., the classical mutation and recombination, but also by different operators, as required; the tool architecture has been specially thought for easy addition of new genetic operators).

The purpose of the constraints is to limit the possible productions of the evolutionary tool, and also provide them with semantic value. The constraints are provided through a user-defined library that provides the genotype-phenotype mapping for the generated individuals, describes their possible structure and defines which values the existing parameters (if any) can take. Constraint definition is left to the user to increase the generality of the tool; it is flexible enough to allow the definition of complex entities to easily describe a wide range of processor *instruction sets architecture* (*ISA*).

The evolutionary core reads the *constraint library* in order to adequately generate assembly programs. For each generated program, a vector of fitness values are computed by the *external evaluator* considering the targeted faults provided by the *functionally coherent fault list*. Differently from the classical approach, the sequence of values in the fitness vector does not represent a priority list but each of them describes the figure of merit obtained by the individual regarding to a specific fault.

The task of the μGP^3 core is to progressively improve the *population of individuals* or test programs. Thus, the population is ordered following a ranking strategy based on the Pareto-dominance principles described before. Choice of the individuals for reproduction is performed by means of a tournament selection based on the ranking position. However, since individuals belonging to the same group are by definition non-dominated ones, the selection is performed resorting to the *delta entropy value* of the individual [16]. The purpose of the entropy value is not to rank a population in absolute terms, but to detect whether the amount of genetic diversity in a set of individuals is increasing or decreasing. The tournament size τ is also endogenous. The population size μ is set at the beginning of a run, and the tool employs a variation on the plus ($\mu+\lambda$) strategy: a configurable number λ of genetic operators are applied on the population. Since different operators may produce different number of offspring, the number of individuals added to the population is variable; the activation probability and strength for every operator is an endogenous parameter. All new unique individuals are then evaluated, and the population resulting from the union of old and new individuals is ordered resorting to the ranking approach described previously. Clearly, if a new individual dominates the complete population, a new individuals set is created and it is placed at the top of the rank list. Finally, only the first μ individuals are kept.

In order to customize this architecture to the specific goal we address here, we use the BDD-based *fitness function* described above, which is effective in guiding the algorithm towards the solution, and can be computed in reasonable times.

In this case, the evaluation of the generated test programs (or instruction sequences) is performed on the RT-level microprocessor core description by means of a logic simulation: during the simulation, at each clock cycle the vectors feeding the path are passed to the fitness function, and the maximum value obtained during the program run identifies the program's fitness.

4 Experimental Data

The proposed flow has been preliminary evaluated on a description of an 8051 microcontroller, addressing non-robust path-delay testing. The processor reads the test programs from an external memory and its output ports are directly accessible.

The critical timing analysis of the synthesized architecture has been performed utilizing the Synopsys PrimeTime suite ver. X-2005.12. The 92,430 worst paths were selected. This data is related to an in-house developed library. For each path, a combinational subcircuit is automatically extracted from the circuit and the BDD representation is generated and used to remove structurally untestable faults.

The set of structurally testable paths contains 10,394 faults. They have been automatically divided in classes depending on their structural coherence, using a simple tool based on set covering principles, and obtaining 96 coherent fault lists, each one including an average of about 108 faults.

The sequential fault excitation step has been performed resorting to the new MOEA implementation of μGP^3 [24], which also includes a new operator called *local-scan mutation*, whose purpose is the generation of a reduced set of individuals in the neighborhood of the selected parent by performing slight mutations to only one determined parameter. In this case the fitness evaluator comprised a commercial logic simulator (Mentor Graphics ModelSim v.6.2h) and an ad-hoc C-language software monitor implemented in the simulator environment. The evolutionary experiment has been set up with the aim of performing a multi-objective optimization. The initial population is composed of 300 random individuals; the population size is 100 and at each generation 80 genetic operators are applied. For each of the coherent fault lists, the evolutionary experiment was set up in the following manner:

1) the first 20 faults in the list are initially considered (in order not to slow excessively the simulation, not all faults in the list are addressed together) and the EA is started, evaluating the excitation fitness (20 paths implies 20 fitness values)
2) whenever a test program fitness hits 100% for one of the inspected faults, that fault is removed from the experiment and replaced from a new one from the same list (fault dropping strategy). The obtained test program is saved.
3) if the algorithm does not improve the fitness for a set number of generations (10 in this case), the 20 paths are replaced with the following 20 in the list.

The process continues until all paths in the coherent fault list have been considered. This phase took about 110 hours for the whole fault list.

The error propagation step took about 35 hours. The fitness has been evaluated resorting to the ModelSim simulator running a script performing fault injection and to an ad-hoc tool elaborating the simulation dump. The majority of the test program set achieves test observability without modification; for the ones whose fault effects are still not propagated, the EA modifies the original test program maximizing the observability fitness, making sure that the excitation conditions are still met.

The obtained coverage values (Table I) are comparable to the ones obtained using other approaches [14][19]. It must be noted that not-covered faults include functionally untestable ones, which do not determine the circuit performances and cannot be tested functionally. The required time computation compares favorably with the time required in [19]. The experiments run on an Intel E6400 @2.13 GHz.

In order to detail the behavior of the approach, the following pictures describe the evolution of an experiment targeting one coherent fault list that contains 84 faults. Fig. 3 shows the first 300 steps of the evolutionary process: the continuous dark line represents the average of the 20 considered fitness values (mean value on the population), while vertical bars indicate the maximum fitness obtained at each step. For this coherent fault list, the final coverage is 50%. It is important to notice that whenever excitation is found for a fault (e.g., step 28), the average fitness falls down due to the fault dropping strategy. Similarly, this average value undergoes a big

Table 1. Excitation and propagation figures on the case study

	# of faults
Complete path set	92,430
Structurally Justified paths	10,394
Excited path-delay faults	2,731
Propagated faults (before error prop.)	1,536
Propagated faults (final)	2,489

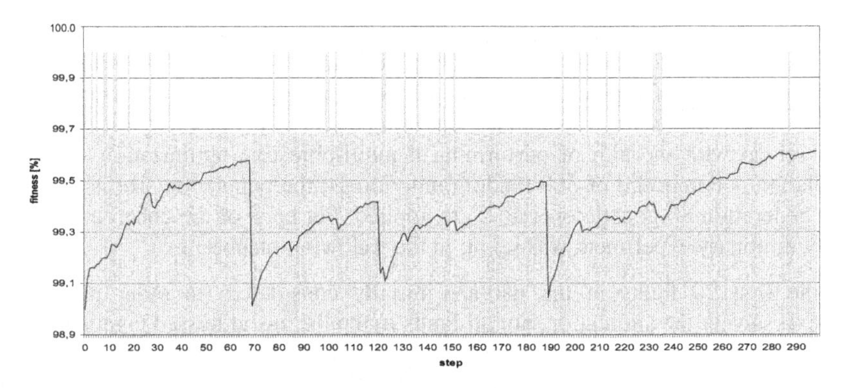

Fig. 3. Fitness behavior on a coherent path list, average and maximum values

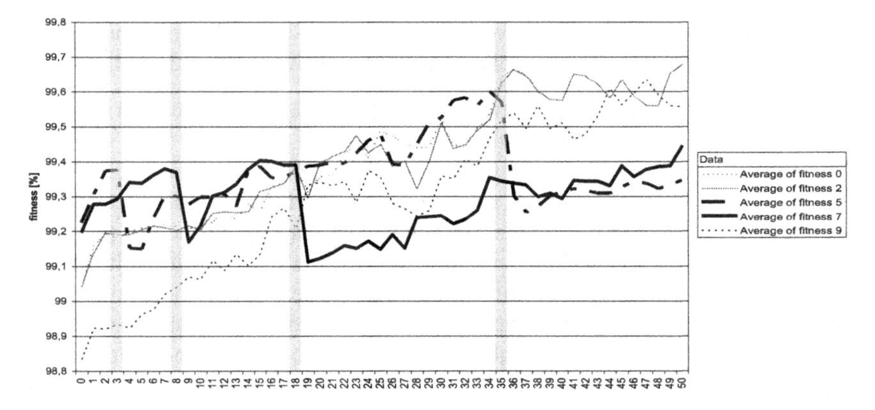

Fig. 4. Trajectories of 5 fitness values during the first 50 steps

depression each time the steady state is reached and all targeted faults are replaced (steps 68, 118 and 189). Nevertheless, the average fitness tends to increase along the experiment. Fig. 4 shows the first 50 steps of the same experiment; in this case, 5 out of the 20 evaluated fitness values are shown (average values on the population). Fitness 5 and 7 show that when a 100% is found the fitness value decreases, due to the substitution of the path-delay fault under inspection; however, the other fitness values seem not to be considerably affected by the replacement mechanism. It is also interesting to note that fitness 9 is continuously increased without finding a 100%. Finally, fitness 0 and 2 describe a very similar trajectory during the first 50 step, thus demonstrating the advantage of evolving coherent fault lists in the same experiment.

5 Conclusions

We presented an innovative approach to fully-automatic generation of path-delay test programs for microprocessors exploiting a MOEA.

Preliminary experimental results show that this methodology allows reducing the test generation time, by concentrating on suitably classified structurally coherent fault lists and avoiding computation-intensive gate-level simulations. The employed evolutionary algorithm takes advantage of the introduced BDD-based fitness evaluation functions for directing the test programs generation flow towards optimal solutions. The obtained coverage results are comparable to manual/deterministic approaches in literature.

Acknowledgments. The authors thank Alessandro Salomone, Massimiliano Schillaci, Giovanni Squillero and Sonia Drappero for their invaluable help in designing the new MOEA implementation of μGP3.

References

1. Mak, T.M., et al.: New challenges in delay testing of nanometer, multigigahertz designs. IEEE Design & Test of Computers 21(3), 241–248 (2004)
2. Lin, C.J., Reddy, S.M.: On Delay Fault Testing in Logic Circuits. IEEE Trans. on CAD 6(5), 694–703 (1987)
3. Chakraborty, T.J., et al.: Delay fault models and test generation for random logic sequential circuits. In: ACM/IEEE Design Automation Conference, pp. 165–172 (1992)
4. Kim, K.S., Mitra, S., Ryan, P.G.: Delay defect characteristics and testing strategies. IEEE Design & Test of Computers 20(5), 8–16 (2003)
5. Krstic, A., Cheng, K.-T.: Delay Fault Testing for VLSI circuits. Kluwer Academic Publishers, Dordrecht (1998)
6. Fuchs, K., Pabst, M., Roessel, T.: RESIST: A Recursive Test Pattern Generation Algorithm. IEEE Trans. on CAD 13(12), 1550–1561 (1994)
7. Tafertshofer, P., Ganz, A., Antreich, K.J.: IGRAINE–An Implication GRaph-bAsed engINE for Fast Implication, Justification, and Propagation. IEEE Trans. on CAD 19(8), 907–927 (2000)
8. Bhattacharya, D., et al.: Test Pattern Generation for Path Delay Faults using Binary Decision Diagrams. IEEE Trans. on Computers 44(3), 434–447 (1995)

9. Michael, M.K., Tragoudas, S.: Functions-based Compact TestPattern Generation for Path Delay Faults. IEEE Trans. on VLSI 13(8), 996–1001 (2005)
10. Bryant, R.: Graph-based algorithms for Boolean function manipulation. IEEE Trans. on Computers C-35(8), 677–691 (1986)
11. Cheng, C.A., Gupta, S.K.: Test generation for path delay faults based on satisfiability. In: IEEE Design Automation Conference (1996)
12. Yang, K., Cheng, K.T., Wang, L.C.: TranGen: A SAT-Based ATPG for Path-Oriented Transition Faults. In: ASP-DAC, pp. 92–97 (2004)
13. Singh, V., Inoue, M., Saluja, K.K., Fujiwara, H.: Instruction-Based Delay Fault Self-Testing of Processor Cores. In: IEEE International Conference on VLSI Design, pp. 933–938 (2004)
14. Lai, W.-C., Krstic, A., Cheng, K.-T.: Test Program Synthesis for Path Delay Faults in Microprocessor Cores. In: IEEE International Test Conference, pp. 1080–1089 (2000)
15. Gurumurthy, S., et al.: Automatic Generation of Instructions to Robustly Test Delay Defects in Processors. In: IEEE European Test Symposium, pp. 173–178 (2007)
16. Corno, F., et al.: Evolving Assembly Programs: How Games Help Microprocessor Validation. IEEE Trans. on Evolutionary Computation 9, 695–706 (2005)
17. Sanchez, E., Schillaci, M., Sonza Reorda, M., Squillero, G.: An Enhanced Technique for the Automatic Generation of Effective Diagnosis-oriented Test Programs for Processors. In: IEEE Design, Automation and Test in Europe, pp. 1–6 (2007)
18. Schaffer, J.D.: Multiple Objective Optimization with Vector Evaluated Genetic Algorithms. In: Int'l Conf. on Genetic Algorithms and Their Applications, pp. 93–100 (1985)
19. Bernardi, P., et al.: On the Automatic Generation of Test Programs for Path-Delay Faults in Microprocessor Cores. In: IEEE European Test Symposium, pp. 179–184 (2007)
20. Bushnell, M.L., Agrawal, V.D.: Essentials of Electronic Testing for Digital, Memory & Mixed-Signal VLSI Circuits. Kluwer Academic Publishers, Dordrecht (2000)
21. CoelloCoello, C.A., Van Veldhuizenand, D.A., Lamont, G.B.: Evolutionary Algorithms for Solving Multi-Objective Problems. Kluwer Academic Publishers, Dordrecht (2002)
22. Huband, S., et al.: A Review of Multiobjective Test Problems and a Scalable Test Problem Toolkit. IEEE Trans. on Evolutionary Computation 10(5), 477–506 (2006)
23. Padmanaban, S., Tragoudas, S.: Efficient Identification of (Critical) Testable Path Delay Faults Using Decisions Diagrams. IEEE Trans. on CAD 24(1), 77–87 (2005)
24. MicroGP++, http://ugp3.sourceforge.net

Evolutionary Object Detection by Means of Naïve Bayes Models Estimation

Xavier Baró[1] and Jordi Vitrià[1,2]

[1] Computer Vision Center, Edifici O, Campus UAB, Bellaterra, Barcelona
[2] Dept. Matemàtica Aplicada i Anàlisi, UB, Gran Via 585, 08007 Barcelona, Spain
{xbaro,jordi}@cvc.uab.cat

Abstract. This paper describes an object detection approach based on the use of Evolutionary Algorithms based on Probability Models (EAPM). First a parametric object detection schema is defined, and formulated as an optimization problem. The new problem is faced using a new EAPM based on Naïve Bayes Models estimation is used to find good features. The result is an evolutionary visual feature selector that is embedded into the Adaboost algorithm in order to build a robust detector. The final system is tested over different object detection problems obtaining very promising results.

1 Introduction

The detection and classification of objects in images that have been acquired in unconstrained environments is a challenging problem because objects can occur under different poses, lighting conditions, backgrounds and clutter. This variation in the object appearance makes unfeasible the design of handcrafted methods for object detection. Although this problem has been the subject of research from the early beginning of the computer vision field, it has not been until the recent past years that researchers have developed generic object recognition systems for a broad class of real world objects. The key point for this achievement has been the use of a machine learning framework that makes use of very large sets of sample images to learn robust models: Given a training set of n pairs $(\mathbf{x}_i, y_i)$, where $\mathbf{x}_i$ is the ith image and y_i is the category of the object present in $\mathbf{x}_i$, we would like to learn a model, $h(\mathbf{x}_i) = y_i$ that maps images to object categories.

One of the most extended approaches to object detection is based on *local methods*, which model an object as a collection of local visual features or "patches". Thus an image $\mathbf{x}_i$ can be considered to be a vector $(\mathbf{x}_{i,1}, \ldots, \mathbf{x}_{i,m})$ of m patches. Each patch $\mathbf{x}_{i,j}$ has a feature-vector representation $F(\mathbf{x}_{i,j}) \in \Re^d$; this vector might represent various features of the appearance of a patch, as well as features of its relative location and scale. We can choose from a wide variety of features, such as the fragments-based representation approach of Ullman [1], the gradient orientation-based SIFT [2], or some forms of geometric invariant descriptors. One of the most successfully used *a priori* image feature, at least for a

M. Giacobini et al. (Eds.): EvoWorkshops 2008, LNCS 4974, pp. 235–244, 2008.

broad class of visual objects, is known as Haar feature. These features, which are related to the wavelet decomposition, were originally proposed in the framework of object detection by Viola and Jones [3] in their face detection algorithm.

The two-dimensional Haar decomposition of a square image with n^2 pixels consists of n^2 wavelet coefficients, each of which corresponds to a distinct Haar wavelet. The first such wavelet is the mean pixel intensity value of the whole image; the rest of the wavelets are computed as the difference in mean intensity values of horizontally, vertically, or diagonally adjacent squares.

Haar features present an interesting property in the context of object recognition: they can be computed very fast using the *integral image*. Integral image at location of x, y contains the sum of the pixel values above and left of x, y inclusive.

Viola and Jones [3] used the Adaboost algorithm [4] to learn a real time face detector with a very high classification performance. In their proposal, weak classifiers are threshold-based classification rules on the values of several Haar features of the image. In order to select the best features, they performed an exhaustive search on the whole set of Haar features for a window. Finally, they built a cascade of strong classifiers to achieve a very low level of false positives.

It is important to note that Haar features constitute an over-complete dictionary of the image and that there are more than 2^{18} different features for a small image window of 576 pixels (24x24 pixels). This fact imposes a high computational cost on the learning step of the Adaboost algorithm, which involves several rounds of exhaustive searches. From a practical point of view, the development of a high performance object detector represents, when using conventional hardware, a learning time of the order of several hundred hours.

The work of Viola and Jones was extended by Lienhart and Maydt [5], who showed that the use of a larger feature set may improve the convergence and the performance of the final classifier. The extension of the feature set was done by adding rotated versions of original Haar-like features, and thus adding a factor to the exponential relation between the feature set size and the training time.

Another natural extension of the Haar features is the dissociated dipoles proposed by Balas and Sinha in [6] in the context of computational neuroscience. As in the case of Haar-like features, they are region based comparisons, but in this case we always have only two regions, which do not have to be adjacent regions. The main reason to consider these features is because while Haar-like features are local descriptors, the dissociated dipoles maintain this ability but also includes non-local descriptors of the images. In addition, they demonstrated that human visual system can perform that type of non-local comparisons, and that these comparisons benefits recognition systems. The proposal of Balas and Sinha, in spite of it can be easily adapted to Adaboost-based detectors, has not been used in any real world object detector due to computational limitations: there are more than 2^{28} different dissociated dipoles in standard image window of 24×24 pixels, what makes the use of the Adaboost algorithm unfeasible.

Using a parameterizable feature set as Haar-like features or dissociated dipoles, the object detection problem can be reformulated as an optimization problem, where we must find the best parameters to minimize an error function (i.e. the weighted classification error in the Adaboost algorithm or the number of miss-classified samples). The new problem corresponds to search in a large and sparse solution space in order to deal with the best parameters. Nowadays, an emerging field is growing up to deal with those type of problems, the Evolutionary Algorithms Based on Probabilistic Models (EAPM) [7]. EAPMs are a new paradigm in the evolutionary computing field. This paradigm was started with the publication of a simple algorithm named *Population Based Incremental Learning* (PBIL) [8], based on a simple univariate model, where all the variables are assumed to be independent. The best individuals of each generation are used to update these variables, and finally the model is sampled to obtain a new generation. In spite of its simplicity, these algorithms demonstrated to converge to good solutions for several problems. Few years later, Schmidth et al. [9] re-introduced the genetic operators to the PBIL algorithm improving significantly its performance. In spite of this return to the origins, the PBIL algorithm introduced an interesting view on evolutionary computation: the extraction of a statistical description of the promising solutions, in terms of a probability distribution is the base of EAPMs and the new systematic way to solve hard search and optimization problems that they represent.

In the literature we can find a wide variety of EAPMs, in which the most important difference is the used probability model. Taking into account the considered interactions between variables, we can classify the models within three main types: Univariate models where no interactions are considered, bivariate models with only pair-wise interactions and finally the models that allow multiple interactions. Once the most convenient probability model is selected, different estimation and sample strategies can be used, thus, we can find different algorithms that share the same type of model. The most known and used algorithms are the UMDA [10], PBIL [8] and cGA [11] for univariate models, MIMIC [12], COMIT [13] and BMDA [14] in the case of bivariate models and finally, considering models with multiple interactions the FDA [15], BOA [16] and EBNA [17]. The use of a complex model allows to better represent the features space, but it adds complexity to the estimation and sampling stages. In [7] a comparison between different EAPMs over several optimization problems suggests that for simple functions, where there are no interaction between the variables, the performance of the univariate and bivariate models perform as well as more complex models, but when we face more complex problems, a more sophisticated probability model is required. As a general rule, more complex models are more reliable but at the expense of bigger execution times.

2 Object Detection Formulation

To formulate the problem, in this section we first define the object detection strategy and the used features. At the end, a parametric definition which can be used in optimization approaches is obtained. The object detection used in this paper is based on the Adaboost algorithm and Haar-like features of Viola and Jones in [3], but in our case we do not use combinations of features and threshold

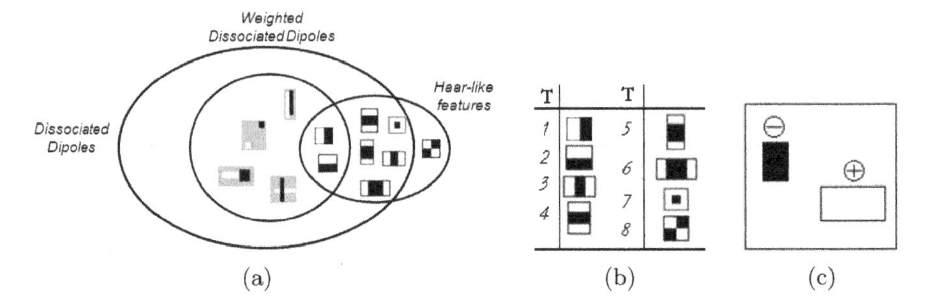

Fig. 1. *(a)* Graphical comparison between the different feature sets. *(b)* Values for the parameter T for the Haar-like features. *(c)* Excitatory and inhibitory poles.

to obtain the simple classifiers, only the sign is used instead. This type of features are called ordinal features, which have demonstrated to be robust against various intra-class variations and illumination changes [18]. In addition, we experiment with two larger feature sets than Haar-like features, which cannot be used with the classical Adaboost approach, but can be used with the Adaboost based on EAPMs. These features set are the *Dissociated Dipoles* introduced by Balas and Sinha in [6] and the *Weighted Dissociated Dipoles*, an extension of the dissociated dipoles that allows to represent most part of the Haar-like features (see Fig.1a).

The main idea of Adaboost is to maintain a weight distribution that reflects how many times each sample has been miss-classified by the previous added weak learners, and therefore, allowing to add new simple classifiers that concentrate on those samples which are systematically miss classified. At the end, the final classifier is a weighted combination of several simple classifiers. In Adaboost jargon, the simple classifier is named *Weak Classifier* and the final classifier is named *Strong Classifier*. The process of learning the best weak classifier using the weights distribution is named *Weak Learner* and is discussed in this section.

Given a training set $\langle(\mathbf{x}_1, y_1), ..., (\mathbf{x}_M, y_M)\rangle$, where $y_i \in \{-1, +1\}$ is the target value for sample $\mathbf{x}_i$, the goal of an object detection learning algorithm is to deal with the strong classifier $H(\mathbf{x}_i) = y_i$. In the boosting framework, we define a distribution $W = \{w_1, ..., w_M\}$ over the training set, where each w_i is the weight associated to the sample $\mathbf{x}_i$, and $H(\mathbf{x})$ corresponds to an additive model $H(\mathbf{x}) = \sum_t \alpha_t h_t(\mathbf{x})$ where the final decision is a combination of the decisions of several *weak classifiers* $h(\mathbf{x}) \in \{-1, +1\}$. In contrast to the strong classifier $H(\mathbf{x})$ where we expect a good prediction for any sample $\mathbf{x}_i$ in the training set, in the case of weak classifier we only expect they are better than a random decision.

Given $\mathcal{H}$ the set of all possible weak classifiers, $h^{\mathbf{s}} \in \mathcal{H}$ a certain weak classifier defined by parameters $\mathbf{s}$, W the weights distribution of the Adaboost and $\mathcal{E}(h^{\mathbf{s}}) = Pr_{i \sim W}[h^{\mathbf{s}}(\mathbf{x}_i) \neq y_i]$ the error function, the regression step consists on finding $\mathbf{s}^*$ that $\mathcal{E}(h^{\mathbf{s}^*}) \leq \mathcal{E}(h^{\mathbf{s}}) | \forall h^{\mathbf{s}^*}, h^{\mathbf{s}} \in \mathcal{H}$, where the complexity of finding $\mathbf{s}^*$ depends on the size of $\mathcal{H}$.

Using ordinal measures, Haar-like features can be parameterized by the upper-left position of one of the regions (X, Y) and their size (W, H), because the size and position of all the regions in Haar-like features is predefined for each type T (see Fig. 1b). Therefore, a Weak Classifier with ordinal Haar-like [5] features can be defined as:

$$h^{\mathbf{s}}(\mathbf{x}) \mapsto \{-1,+1\}. \\ \text{where} \quad \mathbf{s} = (X, Y, W, H, T). \tag{1}$$

To evaluate this feature, the mean intensity value of negative region (black one) is subtracted from mean value of positive region. The sign of this subtraction is used as the final class value. In the case of dissociated dipoles we can analogously parameterize the two regions as:

$$h^{\mathbf{s}}(\mathbf{x}) \mapsto \{-1,+1\}. \\ \text{where} \quad \mathbf{s} = (X_e, Y_e, W_e, H_e, X_i, Y_i, W_i, H_i). \tag{2}$$

where the subscript e refers to the excitatory pole (white in Fig. 1c) and the subscript i to the inhibitory pole (black).Finally, to deal with the weighted dissociated dipoles, we just add a weight parameter $RW \in \{1, 2\}$ to each pole:

$$h^{\mathbf{s}}(\mathbf{x}) \mapsto \{-1,+1\}. \\ \text{where} \quad \mathbf{s} = (X_e, Y_e, W_e, H_e, RW_e, X_i, Y_i, W_i, H_i, RW_i). \tag{3}$$

In addition to the mandatory parameters defined above, we can add extra parameters to improve the results. For instance, we add a polarity parameter which inverts the classification value. This parameter allows a fast step on the search process, inverting the regions. Other parameters can be added to extend the features to multichannel images, but the use of color images is out of the focus of this work.

Once the differences between the feature sets are described, we summarize the problem as finding the best instance of a set of random variables $\mathbf{s} = \{X_1, ..., X_K\}$, to minimize $\mathcal{E}(h^{\mathbf{s}}) = Pr_{i \sim W}[h^{\mathbf{s}}(\mathbf{x}_i) \neq y_i]$, where the only difference between each feature set is how to evaluate $h^{\mathbf{s}}$ and the dimension of $\mathbf{s}$.

If we analyze the random variables, it is easy to discover that there are multiple dependencies between them, for instance, regions located near the right or bottom sides of the training window cannot have large sizes. The rest of the paper is concentrated on defining an evolutionary approach based on a novel EAPM which allows to solve this optimization problem taking into account these dependencies. From this point, we can forget about the fact we are in an object detection problem and the use of Adaboost. The problem now is how to find the best parameters of a feature given a weight distribution over the samples.

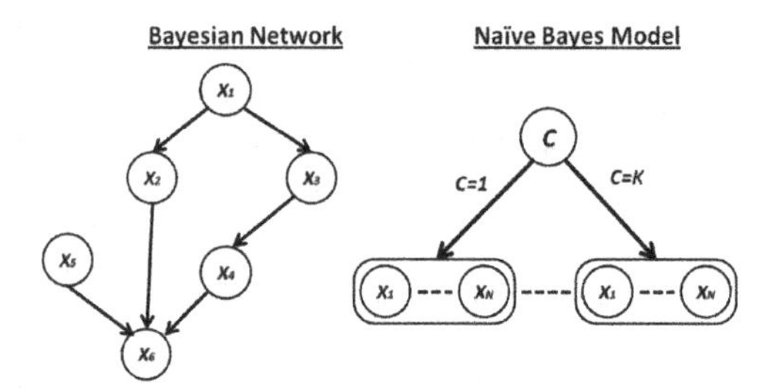

Fig. 2. Probabilistic models

3 Evolutionary Algorithm Based on Naïve Bayes Estimation

As we show before, the formulation derived from the object detection problem has multiple dependencies between the variables, and thus, we need to use a model that allows to represent all these interactions. In addition, as the optimization step must be repeated several times in order to obtain a good detector, we need a fast method. In this section we present a novel EAPM algorithm which accomplish both requirements. First we introduce the probability model, and lastly the evolutionary algorithm.

The most extended probabilistic model to capture multiple dependencies is the Bayesian network (or a belief network), represented as a probabilistic graphical model, specifically an acyclic directed graph, where each node corresponds to a variable (measured parameter, latent variable or hypothesis) and whose arcs encode the dependence between variables. Learning a Bayesian network from data is a two-fold problem: Structure learning and parameter estimation. Although there exist good methods to estimate the structure and parameters of a Bayesian network, because exact inference is $\#P$-complete and thus the existent methods are often too costly, approximate methods like Markov Chain Monte Carlo [19] and loopy belief propagation [20] must be used. The applicability of the Bayesian networks is limited by the fact that these methods have an unpredictable inference time and its convergence is difficult to diagnose.

An alternative to the Bayesian networks are the Naïve Bayes models, where with the "naive" assumption that all variables are mutually independent given a "special" variable C the model is simplified. Although the resulting model is very simple (see Fig. 2), in [21] Lowd and Domingos demonstrated from an empirical point of view that this simpler model has the same representation power than a Bayesian network. In addition to several experiments comparing both models, in this work they also propose the Naïve Bayes Estimation (NBE) algorithm to efficiently estimate Naïve Bayes Models from data. This algorithm consists of an Expectation Maximization (EM) method wrapped in an outer loop that progressively adds and prunes mixture components. Once the probability model is defined, and we have the NBE algorithm to estimate this model from a given data, the final algorithm is presented in Alg. 1.

Algorithm 1. EANBE learning algorithm

Input: Initial population P_0, evaluation function $\mathcal{E} : \mathbb{R}^D \mapsto \mathbb{R}$ (where D is the dimension of each individual $I \in P_0$ and L the number of individuals in P_0), the selection percentage P_S and the hold-out percentage P_H.
Output: The best individual I_{best} among all evaluated individuals.
$L_S \Leftarrow round(P_S \times L)$
$L_H \Leftarrow round(P_H \times L_S)$
repeat
Evaluate the population P_i using the evaluation function $\mathcal{E}$
Select the best individual I_i in the population P_i
Select the L_S best individuals using the evaluation value
if $\mathcal{E}(I_i)$ is better than $\mathcal{E}(I_{Best})$ or it is the first iteration **then**
$I_{Best} \Leftarrow I_i$
end if
Divide the selected individuals into two random subsets T and H with $L_S - L_H$ and L_H individuals respectively.
Use the NBE algorithm to learn a model M using T as the training set and H as the hold-out set.
Sample the model M to obtain a new population P_{i+1} with L individuals.
until The maximum number of generations or some other stopping criteria are achieved
return I_{Best} and $\mathcal{E}(I_{Best})$.

4 Results

To validate the proposed object detection method, we have developed the *eapmlib*, a C++ based library that implements a general framework to work with EAPMs. The learning process is done by a Gentle Adaboost version in Matlab, using MEX files to communicate with the *eapmlib*. All the source codes for the library, MEX files and Matlab files are available in an online appendix at *www.cvc.uab.cat/~xbaro/eanbe/*.

In the first part of this section we describe the data sets used to perform the experiments, and after we describe the experiments. Finally, we analyze the obtained results.

4.1 Data Sets

In order to verify the usefulness of the presented methodology we have selected different real world object detection problems, using for each one a state of the art public database to facilitate the reproduction and comparison of the results. In the databases where background images are not provided, we create them using images from the *Corel Photo Libraries*. Data sets are described in the following:

Faces: We use the MIT-CBCL face database with a random selection of 1.000 face images and 3.000 non-face images. All the images correspond to frontal faces with several illumination changes.

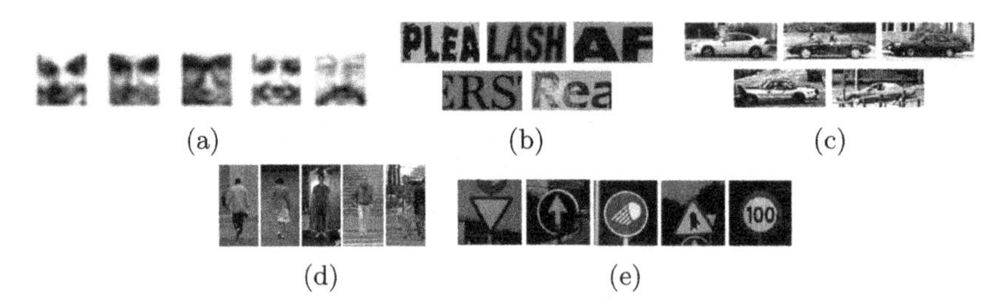

Fig. 3. Data set examples. *(a)* Faces *(b)* Text *(c)* Cars *(d)* Pedestrians *(e)* Traffic signs.

Text: We use the text location dataset from the *7th International Conference on Document Analysis and Recognition (ICDAR03)*. The images correspond to text regions over a wide set of surfaces and with several illumination changes. To obtain the training data, each text region has been split into overlapped subregions with the same size.

Cars: We use the UIUC cars database, with a total of 1.050 images containing 550 instances of lateral views of different cars in urban scenes and 500 images of background.

Pedestrians: We use the INRIA Person Dataset, with 2.924 images divided into 924 pedestrian instances and 2.000 background images.

Traffic Signs: In this case we use real images acquired in the context of a mobile mapping project provided by the Institut Cartogràfic de Catalunya[1]. The database consists on 1.000 images containing a traffic sign and 3.000 background images [22].

4.2 Measures

To compare the performance of each configuration of the system, the classification Balanced Error (BER) is used. This measure takes into account the different number of examples in each of the classes in order to face unbalanced problems. Given a samples set $\{(\mathbf{x}_i, y_i)|i = 1, ..., M\}$ with $y_i \in \{-1, 1\}$ and a certain hypothesis $h(\mathbf{x}) \mapsto \{-1, 1\}$ the balanced error is defined as:

$$BER = \frac{\frac{FN}{N^+} + \frac{FP}{N^-}}{2}. \quad (4)$$

where N^+ and N^- are the number of positive and negative samples respectively, and FP and FN the number of false positive and false negative samples.

4.3 Experiments

To validate the proposed schema and compare each selection of feature set and evolutionary algorithm, we perform a stratified 10-fold cross-validation. The population size for genetic algorithm and EANBE is set to 100 individuals, using $T = 200$ as the number of iterations in the Gentle Adaboost algorithm. In the

[1] Institut Cartogràfic de Catalunya (`www.icc.es`).

Table 1. Results obtained in the experiments

Data Set	Dissociated Dipoles		Haar-like features		Weighted Diss. Dip.	
	GA	*EANBE*	*GA*	*EANBE*	*GA*	*EANBE*
Cars	**79.84%**	79.67%	70.71%	**71.47%**	**95.95%**	95.82%
Faces	70.98%	**73.53%**	56.50%	**57.23%**	88.65%	**88.97%**
Text	55.42%	**58.98%**	48.17%	**49.82%**	**89.57%**	89.52%
Pedestrians	71.04%	**75.00%**	55.95%	**57.34%**	89.05%	**90.41%**
Traffic Signs	66.37%	**70.00%**	64.50%	**66.33%**	89.05%	**90.28%**

case of genetic algorithm, the individuals are encoded using binary gray code, while in the case of EANBE, the decimal code is used directly. The BER mean vales obtained in the experiments are shown in table 1.

Note that the results obtained using the Weighted Dissociated Dipoles outperform always the ones obtained with the smaller features sets. The results obtained by the other two feature sets, are not significantly different, but in the majority of the cases the dissociated dipoles outperform the Haar-like. Following the trend highlighted by Lienhart and Mayd in [5], the results improve with the size of the feature set.

5 Conclusions and Future Work

Although the tests must be extended to a larger set of databases, the results obtained with weighted dissociated dipoles in combination to the new evolutionary algorithm encourage to continue with this line. As a future work we want to extend this approach to multi-class learning and use a priori knowledge from the learning data in order to speed-up the selection process and obtain not only good features but meaningful features.

Acknowledgments. This work has been developed in a project in collaboration with the *Institut Cartogràfic de Catalunya* under the supervision of Maria Pla. This work was partially supported by MEC grant TIC2006-15308-C02-01 and CONSOLIDER-INGENIO 2010 (CSD2007-00018).

References

1. Ullman, S., Sali, E.: Object classification using a fragment-based representation. Biologically Motivated Computer Vision, 73–87 (2000)
2. Lowe, D.G.: Object recognition from local scale-invariant features. In: Proc. of ICCV, Corfu., pp. 1150–1157 (1999)
3. Viola, P., Jones, M.: Rapid object detection using a boosted cascade of simple features. In: Proc. of the CVPR 2001, vol. 1, pp. 511–518 (2001)
4. Freund, Y., Schapire, R.E.: Experiments with a new boosting algorithm. In: International Conference on Machine Learning, pp. 148–156 (1996)

5. Lienhart, R., Maydt, J.: An extended set of haar-like features for rapid object detection. In: Proc. of the Int. Conf. on Image Processing, pp. 900–903 (2002)
6. Balas, B., Sinha, P.: Dissociated dipoles: Image representation via non-local comparisons. In: Annual meeting of the Vision Sciences Society, Sarasota, FL (2003)
7. Blanco, R., Lozano, J.: An Empirical Comparison of Discrete Estimation of Distribution Algorithms, pp. 167–180. Kluwer Academic Publishers, Dordrecht (2001)
8. Baluja, S., Caruana, R.: Removing the genetics from the standard genetic algorithm. In: The Int. Conf. on Machine Learning 1995, pp. 38–46 (1995)
9. Schmidt, M., Kristensen, K., Jensen, T.R.: Adding genetics to the standard PBIL algorithm. In: Proceedings of the 1999 Congress on Evolutionary Computation, vol. 2, pp. 1527–1534 (1999)
10. Muhlenbein, H.: The equation for response to selection and its use for prediction. Evolutionary Computation 5(3), 303–346 (1997)
11. Harik, G.R., Lobo, F.G., Goldberg, D.E.: The compact genetic algorithm. IEEE-EC 3(4), 287 (1999)
12. de Bonet, J.S., Isbell Jr, C.L., Viola, P.: MIMIC: Finding optima by estimating probability densities. In: Mozer, M.C., Jordan, M.I., Petsche, T. (eds.) Advances in Neural Information Processing Systems, vol. 9, p. 424. MIT Press, Cambridge (1997)
13. Baluja, S., Davies, S.: Using optimal dependency-trees for combinational optimization. In: Proceedings of the Fourteenth Int. Conf. on Machine Learning, pp. 30–38. Morgan Kaufmann, San Francisco (1997)
14. Pelikan, M., Mühlenbein, H.: The bivariate marginal distribution algorithm. In: Roy, R., Furuhashi, T., Chawdhry, P.K. (eds.) Advances in Soft Computing - Engineering Design and Manufacturing, pp. 521–535. Springer, London (1999)
15. Mühlenbein, H., Mahning, T.: The factorized distribution algorithm for additively decomposed functions. In: Second Symposium on Articial Intelligence. Adaptive Systems, CIMAF 1999, La Habana, pp. 301–313 (1999)
16. Pelikan, M., Goldberg, D.E., Cantú-Paz, E.: BOA: The Bayesian optimization algorithm. In: Banzhaf, W., et al. (eds.) Proceedings of the Genetic and Evolutionary Computation Conference GECCO 1999, Orlando, FL, 13-17 1999, vol. I, pp. 525–532. Morgan Kaufmann, San Francisco (1999)
17. Etxeberria, R., Larrañaga, P.: Global optimization with bayesian networks. In: Second Symposium on Artifcial Intelligence(CIMAF 1999), Cuba, pp. 332–339 (1999)
18. Thoresz, K., Sinha, P.: Qualitative representations for recognition. Journal of Vision 1(3), 298–298 (2001)
19. Gilks, W.R., Richardson, S., Spiegelhalter, D.J.: Markov Chain Monto Carlo in Practice. Interdisciplinary Statistics Series. CRC Press, Boca Raton (1996)
20. Yedidia, J.S., Freeman, W.T., Weiss, Y.: Generalized belief propagation. In: NIPS, pp. 689–695.
21. Lowd, D., Domingos, P.: Naive bayes models for probability estimation. In: ICML 2005: Proceedings of the 22nd international conference on Machine learning, pp. 529–536. ACM Press, New York (2005)
22. Baró, X., Vitrià, J.: Traffic sign detection on greyscale images. In: Recent Advances in Artificial Intelligence Research and Development, IOS Press, Amsterdam (2004)

An Evolutionary Framework for Colorimetric Characterization of Scanners

Simone Bianco, Francesca Gasparini,
Raimondo Schettini, and Leonardo Vanneschi

Dipartimento di Informatica, Sistemistica e Comunicazione
University of Milano-Bicocca, 20126 Milan, Italy
{bianco,gasparini,schettini,vanneschi}@disco.unimib.it

Abstract. In this work we present an evolutionary framework for colorimetric characterization of scanners. The problem consists in finding a mapping from the RGB space (where points indicate how a color stimulus is produced by a given device) to their corresponding values in the CIELAB space (where points indicate how the color is perceived in standard, i.e. device independent, viewing conditions). The proposed framework is composed by two phases: in the first one we use genetic programming for assessing a characterizing polynomial; in the second one we use genetic algorithms to assess suitable coefficients of that polynomial. Experimental results are reported to confirm the effectiveness of our framework with respect to a set of methods in the state of the art.

1 Introduction

Many devices, like scanners or printers, have their own reference systems for the specification of color (device-dependent spaces). To facilitate the reproduction of colors on various devices and supports, it is often useful to employ a system of description that allows us to define the color in a univocal fashion, i.e. in a device-independent space, separating the way colors are defined from the way the various devices represent them. A point in RGB space indicates how a color stimulus is produced by a given device, while a point in a colorimetric space, such as CIELAB space, indicates how the color is perceived in standard viewing conditions. Now let us consider the function that at every point in the device dependent space associate the colorimetric value of the corresponding color. The colorimetric characterization of a scanner device means to render this function explicitly. It must take into account the peculiar characteristics of the device; consequently, every device calls for specific conversion functions.

Several different approaches to the characterization problem have appeared to date [5]. Spectral characterization approaches try to recover reflectance information from the scanner responses and to compute from these the colorimetric values [1,18]. These approaches assume that the spectral scanner sensitivity can be accurately measured or recovered mathematically. The advantage of these techniques over colorimetric ones for traditional RGB scanners using a single illuminant is not evident [17]. Among various colorimetric approaches, neural networks and polynomial regression have been widely investigated. Kang and Anderson [11], Schettini et al. [15], Vrhel and Trussel [20]

M. Giacobini et al. (Eds.): EvoWorkshops 2008, LNCS 4974, pp. 245–254, 2008.

and Cheung et al. [4] applied artificial neural networks to scanner calibration and characterization.

High-order multidimensional polynomials are often used for scanner characterization. However, since there is no clear relationship between polynomials adopted and imaging device characteristics, they must be empirically determined and defined for each device, and for a specific device, each time a change is made in any component of the system [10]. Usually, the accuracy of the characterization increases as the number of terms in the polynomial increases, however there is not a simple way to avoid data overfitting. Moreover, polynomial regression usually minimizes just the average color error and therefore it does not guarantee a uniform accuracy across the entire color gamut. Using an average color error as the functional to be minimized, the polynomial coefficients can be found by using the Least Square method (LS) or the Total Least Squares method (TLS) [9]. More powerful methods, such as Total Color Difference Minimization (TCDM) and CIELAB Least Squares minimization (LAB-LS), have been recently proposed [16] to minimize non linear functionals that take into account the average CIELAB colorimetric error.

Taking these methods and results as our point of departure, in this paper we try to improve the accuracy of the characterization by using a new evolutionary framework composed by two phases: in the first one we try to assess a polynomial using Genetic Programming (GP) [13] and in the second one we use this polynomial to solve the characterization problem by means of a Genetic Algorithm (GA) [8,7].

GP has already been applied in related, although rather different, problems. This is the case for instance of [6], where Ebner used GP to evolve an algorithm for calculating colour constancy, i.e. the ability of a device to compute colour constant descriptors of objects in view irrespective of the light illuminating the scene.

Nevertheless, to the best of our knowledge, this work represents the first attempt to use an evolutionary framework integrating GP and GAs for the characterization problem.

This paper is structured as follows: in Section 2 we introduce the problem of colorimetric characterization of scanners. In Section 3 we describe the optimization methods using in the state of the art for this problem. Section 4 presents our evolutionary framework and discusses the experimental results that we have obtained using it. Finally Section 5 concludes the paper and offers some hints for future research.

2 Scanner Characterization

The basic model that describes the response $\rho = [R,G,B]$ of a three-channel color scanner can be formulated as [16,17,18]: $\rho = \mathcal{F}(\mathbf{SIR}+\mathbf{n})$, where ρ is a 3×1 vector, $\mathbf{S}$ is the $3 \times N$ matrix formed by stacking the scanner spectral sensitivities row-wise, $\mathbf{I}$ is the $N \times N$ diagonal matrix whose elements are the samples of the scanner-illuminant spectral power distribution, $\mathbf{R}$ is the $N \times 1$ vector of the reflectance of the surface being scanned, $\mathbf{n}$ is the 3×1 noise vector and $\mathcal{F}$ is an optoelectronic conversion function representing the input-output nonlinearity that may characterize the scanner response.

Similarly, the CIEXYZ tristimulus values, denoted by a 3×1 vector $\mathbf{s}$, can be defined as: $\mathbf{s} = \mathbf{CLR}$, where $\mathbf{C}$ is the $3 \times N$ matrix of the CIEXYZ color matching functions and $\mathbf{L}$ is the $N \times N$ diagonal matrix whose elements are the samples of the viewing-illuminant spectral power distribution.

The characterization problem is to find the mapping $\mathcal{M}$ which transforms the recorded values ρ to their corresponding CIEXYZ values $\mathbf{s}$: $\mathbf{s} = \mathcal{M}(\rho)$. In this paper we address this problem using a two step procedure: first the optoelectronic conversion function $\mathcal{F}$ is estimated and $\mathcal{F}^{-1}$ is applied to the ρ data to linearize them; then a mth-order polynomial mapping $\mathbf{M}$ is applied to the linearized data $\mathcal{F}^{-1}(\rho)$ to obtain $\mathbf{s}$. The general m^{th}-order polynomial $P(R,G,B)$ with three variables can be given as: $P(R,G,B) = \sum_{i=0}^{m}\sum_{j=0}^{m}\sum_{k=0}^{m} R^i G^j B^k$, with $i+j+k \leq m$. Given the scanner response ρ, their linearized values $\mathcal{F}^{-1}(\rho)$ and the polynomial model P to use, we can calculate the polynomial expansion $\mathbf{r}$ of $\mathcal{F}^{-1}(\rho)$ as $\mathbf{r} = P\left(\mathcal{F}^{-1}(\rho)\right)$. Using the polynomial modeling, the previous equation then becomes: $\mathbf{s} = \mathbf{M}P\left(\mathcal{F}^{-1}(\rho)\right) = \mathbf{Mr}$. The first step to find the matrix $\mathbf{M}$ is to select a collection of color patches that spans the device gamut. The reflectance spectra of these N_c color patches will be denoted by $\mathbf{R}_k$ for $k \in \{1,\ldots,N_c\}$. These patches are measured using a spectrophotometer or a colorimeter which provides the device independent values: $\mathbf{s}_k = \mathbf{CLR}_k$ with $k \in \{1,\ldots,N_c\}$. Without loss of generality, $\mathbf{s}_k$ can be transformed in any colorimetric or device independent values. The same N_c patches are also acquired with the scanner to be characterized providing $\rho_k = \mathcal{F}(\mathbf{SIR}_k + \mathbf{n})$ with calculated polynomial expansions $\mathbf{r}_k$, for $k \in \{1,\ldots,N_c\}$. In equation, the characterization problem is to find the matrix $\mathbf{M}$:

$$\mathbf{M} = \arg\left(\min_{M \in \mathbb{R}^{3\times q}} \sum_{k=1}^{N_c} ||M\mathbf{r}_k - \mathcal{L}(\mathbf{s}_k)||^2\right) \tag{1}$$

where $\mathcal{L}(\cdot)$ is the transformation from CIEXYZ to the appropriate standard color space chosen and $||\cdot||$ is the error metric in the color space. $\mathbf{M}$ is a $3 \times q$ matrix, where q is the number of terms of the polynomial $P(R,G,B)$; the number of terms q is related to the order m of the polynomial by: $q = \left[\sum_{k=1}^{m}\binom{k+2}{2} + 1\right]$. Being $\mathbf{M}$ a $3 \times q$ matrix, the problem of finding $\mathbf{M}$ amounts to determine $3q$ coefficients; to have enough equations to solve for the $3q$ unknowns and to deal with a less ill-posed problem, we have to use $N_c \geq q$ different color patches. In other words, the bigger is the order of the polynomial, the greater is the number of its terms and consequently the greater is the number of different color patches we have to use.

Different functionals to be minimized can be defined to find the unknown matrix $\mathbf{M}$. Depending on the functional adopted, different optimization methods have to be considered.

3 State of the Art Scanner Characterization Methods

The easiest way to find the unknown matrix $\mathbf{M}$ in Equation (1) is to minimize the functional: $\mathcal{H}_{LS} = ||\mathbf{s} - \hat{\mathbf{s}}||_2$ with $\hat{\mathbf{s}} = \mathbf{Mr}$ using the Least Squares minimization (LS). Analytically, the matrix $\mathbf{M}$ can be easily found by: $\mathbf{M} = \mathbf{sr}^T(\mathbf{rr}^T)^{-1}$. The LS method assumes that errors are present only in the matrix $\mathbf{s}$, while the matrix $\mathbf{r}$ is assumed free of error [9].

The Total Least Squares (TLS) method [9] is a generalization of the LS method: it assumes that both the matrices $\mathbf{s}$ and $\mathbf{r}$ are affected of error. It searches for the solution $\mathbf{M}$ that minimizes the functional: $\mathcal{H}_{TLS} = ||[\mathbf{r};\mathbf{s}] - [\hat{\mathbf{r}};\hat{\mathbf{s}}]||_F \quad \text{with} \quad \hat{\mathbf{s}} = \mathbf{M}\hat{\mathbf{r}}$, where $||\cdot||_F$ is the Frobenius norm. The analytic solution for the minimization of $\mathcal{H}_{TLS}$ exists and can be found using Singular Value Decomposition.

Being the color accuracy of the characterization methods evaluated using a color error in the CIELAB color space, it has been proposed [16] Total Color Difference Minimization (TCDM) to search for the matrix $\mathbf{M}$ that minimizes the sum of the CIELAB ΔE_{94} color error between the measured and the predicted CIELAB values (respectively $\mathbf{s}_{Lab}$ and $\hat{\mathbf{s}}$) for the n patches, i.e.: $\mathcal{H}_{TCDM} = \sum_{i=1}^{n} \Delta E_{94}(\mathbf{s}_{Lab}, \hat{\mathbf{s}}) \quad \text{with} \quad \hat{\mathbf{s}} = \mathbf{M}\mathbf{r}$. The minimization of $\mathcal{H}_{TCDM}$ does not have an analytic solution, and has to be solved with an iterative method. In this paper, following [16] it has been solved using a downhill simplex method as it does not require the calculation of derivatives of the objective function [14].

The CIELAB Least Squares minimization (LAB-LS) method [16] employs a pre-processing consisting in a pth root correction of the scanner responses $\mathbf{r}$ before calculating the LS regression with the CIELAB values $\mathbf{s}_{Lab}$ of the measured patches. The LAB-LS method then minimizes: $\mathcal{H}_{LAB-LS} = ||\mathbf{s}_{Lab} - \hat{\mathbf{s}}||_2 \quad \text{with} \quad \hat{\mathbf{s}} = \mathbf{M}\mathbf{r}^{\frac{1}{p}}$. This p^{th} root correction has the aim to compensate for the cubic root relationship between the RGB scanner color space and the CIELAB color space. It can be easily found that, if the RGB data have been properly linearized, the best choice for the pth root is $p = 3m$, where m is the order of the polynomial used. Such a choice for p permits to cancel out the cubic relationship whatever is the polynomial used. Performing a LS regression, the LAB-LS method admits an analytic solution that can be found as: $\mathbf{M} = \mathbf{s}_{Lab}\left(\mathbf{r}^{1/p}\right)^T \left(\mathbf{r}^{1/p}\left(\mathbf{r}^{1/p}\right)^T\right)^{-1}$.

3.1 State of the Art Experimental Results

In this section, we report the results presented by Shen *et al.*. These results will be compared with the ones of our evolutionary framework later. These results have been obtained using the same identical values of the acquisitions as in [16,17,18], the characterization procedures LS, TLS, TCDM and LAB-LS and the full 3rd-order polynomial suggested for the considered device in [16], i.e.:

$$full3d = 1 + R + G + B + R^2 + RG + RB + G^2 + GB + B^2 + R^3 + \\ + R^2G + R^2B + G^3 + RG^2 + G^2B + B^3 + RB^2 + GB^2 + RGB$$

As in those contributions, we have used as color targets three benchmarks designed to determine the true color balance of any color rendition system: the Macbeth ColorChecker DC (MDC), the Kodak Q60 photographic standard (IT8), and the Kodak Gray Scale Q-14. The spectral reflectance values of MDC and Q14 were measured using a GretagMacbeth Spectrophotometer 7000A, and those of IT8 were measured using a GretagMacbeth Spectrolino spectrophotometer. The CIEXYZ and CIELAB values under the CIE D65 standard illuminant were then calculated from these reflectance data for scanner characterization. These three targets were scanned using Epson GT-10000+.

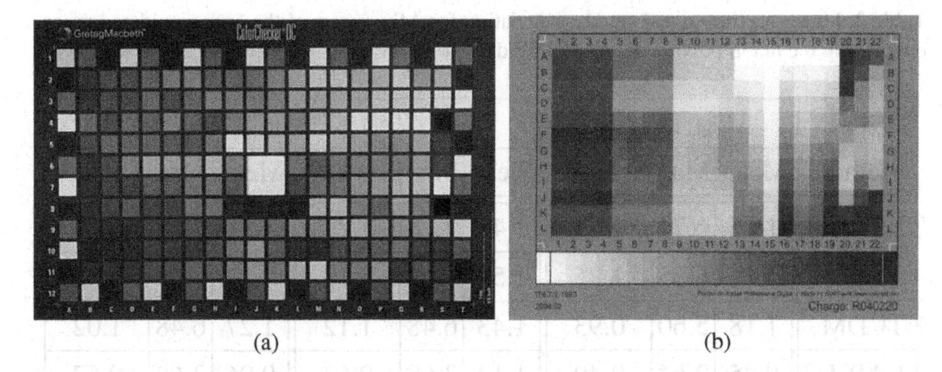

Fig. 1. The two datasets used in this paper. (a): Macbeth ColorChecker DC (MDC). (b): Kodak Q60 photographic standard (IT8).

Table 1. Mean, maximum and standard deviation for ΔE_{94} errors of the four considered procedures using the *full3d* polynomial on the Macbeth ColorChecker DC dataset

	ΔE_{94} Training			ΔE_{94} Test			ΔE_{94} Total		
Method	Mean	Max	Std Dev	Mean	Max	Std Dev	Mean	Max	Std Dev
LS	1.71	12.92	1.83	1.61	6.39	1.53	1.68	12.92	1.71
TLS	2.74	51.65	6.24	2.14	15.92	2.74	2.55	51.65	5.33
TCDM	1.53	8.52	1.52	1.55	8.30	1.51	1.54	8.52	1.52
LAB-LS	1.33	6.66	1.13	1.23	3.16	0.80	1.29	6.66	1.05

During the scanning process, all the color adjustment functions of the scanner were disabled. The RGB values of gray patches on target Q14 and their corresponding average reflectance values were used to calculate the inverse optoelectronic conversion function $\mathcal{F}$ in Equation (3). The targets MDC and IT8,reported in Figure 1, were used to evaluate the color accuracy of each characterization method.

For all the experiments, we study both the performance during the training phase and the generalization ability. We have partitioned both the Macbeth ColorChecker DC and the IT8 datasets into a training set (*Train*) and a test set (*Test*). Following [16], a loop over the lines (patches) of these two datasets has been implemented; each iteration considers three consecutive lines inserting the first two of them into *Train* and the third one into *Test*. In this way, training set and test set are interleaved parts of the whole dataset and the training set has approximately the double of the size of the test set. For all the patches of the training and test sets, we calculate the ΔE_{94} color error between the predicted and measured CIELAB values. As in [16], in this paper we report the mean, the maximum and the standard deviation for the ΔE_{94} color errors for the minimization procedures LS, TLS, TCDM and LAB-LS using the $full3d$ polynomial and for our evolutionary framework.

Table 2. Mean, maximum and standard deviation for ΔE_{94} errors of the four considered procedures using the *full3d* polynomial on the IT8 dataset

	ΔE_{94} Training			ΔE_{94} Test			ΔE_{94} Total		
Method	Mean	Max	Std Dev	Mean	Max	Std Dev	Mean	Max	Std Dev
LS	1.25	6.74	1.07	1.45	7.78	1.25	1.31	7.78	1.14
TLS	1.40	6.75	1.25	1.59	7.69	1.37	1.47	7.69	1.30
TCDM	1.18	5.60	0.93	1.43	6.48	1.12	1.27	6.48	1.02
LAB-LS	0.85	2.64	0.49	1.13	3.68	0.67	0.96	3.68	0.57

Tables 1 and 2 show the results returned by the considered procedures for Macbeth ColorChecker DC and IT8 respectively, using polynomial *full3d*.

In both these tables column one identifies the method; columns 2, 3 and 4 report the results on the training set; columns 5, 6 and 7 report the results on the test set, while columns 8, 9 and 10 report the results of the models learned on the training set executed on both the training and test set together.

As these tables clearly show, the method that returns the best results both on the training and on the test set (and also on the training and test sets considered together) and both for Macbeth ColorChecker DC and IT8 is LAB-LS, that also has lower values of the standard deviations. These results are consistent with ones presented in [16].

4 The Presented Evolutionary Framework

Our evolutionary framework is composed by two phases: in the first one we try to automatically assess the best polynomial for a given imaging device using GP. In the second one we look for the most suitable coefficients of that polynomial by means of GAs. These two phases are described in sections 4.1 and 4.2 respectively. Finally, section 4.3 presents the experimental results that we have obtained using our evolutionary framework.

4.1 Generating New Polynomials with Genetic Programming

Chosen the order m and the maximum number n of polynomial terms, the polynomials evolved by GP have been built using the set of functional (or non-terminal) symbols $F = \{Join\}$ and the set of terminal symbols: $T = \{(R^i, G^j, B^k) \mid 0 \leq i+j+k \leq m\}$. Given two terminal symbols $t_1 \in T$ and $t_2 \in T$, the $Join$ function concatenates them in a new list, i.e.: $Join(t_1, t_2) = [t_1 \;\; t_2]$; thus, the expressions evolved by GP may be seen as a list of polynomial terms, each one of the form (R^i, G^j, B^k) with $0 \leq i+j+k \leq m$. In the experiments reported here, we have set the polynomial order m to 4 and 7.

GP individuals, corresponding to different polynomials, have been evaluated using a fitness function composed by the measure of the ΔE_{94} color error on a chosen training set, weighted for the value of the Leave-one-out Cross-validation (LOOCV) [12].

In other words, the GP fitness function was: $fitness = LOOCV \cdot \text{mean}(\Delta E_{94})$. The LOOCV involves using a single observation from the original sample as the validation data, and the remaining observations as the training data. This process is repeated to allow each observation in the sample to be used exactly once as the validation data. The use of the LOOCV permits to evaluate the generalization ability without using a real test set. The results that we present in this section have been obtained and using the following set of parameters: population size $N = 200$; maximum number of tree nodes $= 2n - 1$ (where n is the maximum number of polynomial terms); maximum number of generations $= 20000$; algorithm used to initialize the population: Ramped Half-and-Half; selection algorithm: tournament, with tournament size $= 10$; crossover rate $p_c = 0.5$; mutation rate $p_r = 0.5$; presence of elitism, i.e. copy of the best unchanged individual into the next population at each generation.

4.2 Genetic Algorithms for the Characterization Problem

In this section, we present our GA based characterization procedure to asses matrix **M** using the polynomial found by GP[1]. It is an hybrid procedure based on both TCDM and LAB-LS. It uses the same pre-processing p^{th} root correction of the scanner responses calculated by LAB-LS, but then minimizes the CIELAB ΔE_{94} color error between these values $\hat{\mathbf{s}}$ and the CIELAB measured values $\mathbf{s}_{Lab}$. Furthermore, it does not use the same $p = 3m$ value of LAB-LS for pre-processing, but lets the GA look for it. This is justified by the fact that the $p = 3m$ is the best choice if the scanner responses are linear: even if the scanner responses have been linearized, some non-linearity may remain. The use of the GA to look for the best value of p permits to correct a part of the residual non-linearity, if present. The functional to be minimized is defined as: $\mathcal{H}_{GA} = \sum_{i=1}^{n} \Delta E_{94}(\mathbf{s}_{Lab}, \hat{\mathbf{s}}_p)$ with $\hat{\mathbf{s}}_p = \mathbf{M}\mathbf{r}^{\frac{1}{p}}$. This procedure inherits from the TCDM method the non existence of an analytical solution. To minimize such functional, we have used a GA with the following set of parameters: population size $N = 1000$; individuals' size $= 25$; maximum number of generations $= 100000$; selection type: tournament selection; crossover rate $p_c = 0.8$; mutation rate $p_m = 0.2$; presence of elitism. The genetic operators we have used are the standard one-point crossover and point mutation defined in [8,7]. The GAs results that we report in this work are the best ones obtained over 5 independent runs for each GA version. The choice of performing a limited number of runs (5) for a large number of generations (100000) is justified in [3], where the authors show that under a constant cost constraint, executing a small number of large runs often allows to reach better solutions faster than executing many small runs.

4.3 Experimental Results

First of all, we use GP to look for new polynomials with similar characteristics to *full3d*: thus we look for a fourth degree polynomial ($m = 4$), composed by a maximum of 20 terms ($n = 20$, the same number of terms of the *full3d*). The fitness of the polynomials

[1] We point out that we have also tested GAs using the *full3d* polynomial. The results, which are very similar to the ones presented in tables 1 and 2 are not shown in this paper to save space.

Table 3. Mean, maximum and standard deviation for ΔE_{94} errors of our evolutionary framework using the $\mathcal{P}_{(m=4,n=20)}$ polynomial found by GP on the two datasets Macbeth ColorChecker DC and IT8

Dataset	ΔE_{94} Training			ΔE_{94} Test			ΔE_{94} Total		
	Mean	Max	Std Dev	Mean	Max	Std Dev	Mean	Max	Std Dev
Macbeth ColorChecker DC	1.14	7.69	1.03	1.07	3.60	0.75	1.11	7.69	0.96
IT8	0.75	1.67	0.32	1.04	2.39	0.53	0.85	2.39	0.39

evolved by GP has been set equal to the error obtained by LAB-LS, since it is faster than the other techniques. The best polynomial obtained is:

$$\mathcal{P}_{(m=4,n=20)} = R+G+B+RG+RB+GB+R^2+G^2+R^3+RG^2+G^2B+B^3+ \\ +R^4+R^2G^2+R^2B.^2+RG^3+RG^2B+G^4+G^2B^2+GB^3 \; .$$

Table 3 reports the results obtained by the GA using the $\mathcal{P}_{(m=4,n=20)}$ polynomial for the Macbeth ColorChecker DC and on IT8 datasets. This table can be read as tables 1 and 2 except that this time the first column identifies the dataset that has been optimized by the GA. If we compare these results with the ones of Tables 1 and 2, we can remark that our evolutionary framework overcomes all the other studied methods both on training and test sets (and also on the training and test sets considered together).

Once obtained these results, the next step in our study has been to relax the constraint that the degree of the polynomial m has to be equal to 4 and that its maximum number of terms n has to be equal to 20. In other words, we have used GP imposing a maximum degree m for the polynomial equal to 7 and we have given no constraint on the maximum number of terms, except that it does not exceed the cardinality of the training set minus one. The best seventh order polynomial obtained has 28 terms and it is reported below:

$$\mathcal{P}_{(m=7,n=28)} = 1+R^2+GB+R^3+RGB+R^4+RG^2B+RB^3+G^4+G^3B+R^4G+R^2B^3+ \\ +G^4B+G^2B^3+B^5+R^4G^2+R^2G^4+R^2G^2B^2+RB^5+G^4B^2+G^3B^3+ \\ +G^2B^4+B^6+R^4G^3+G^4B^3+G^3B^4+G^2B^5+B^7 \; .$$

Once again, the fitness of the polynomials evolved by GP has been set equal to the error obtained by LAB-LS, since it is faster than the other techniques.

In Table 4 we report the results of our evolutionary framework on the Macbeth Color Checker DC and on the IT8, using the $\mathcal{P}_{(m=7,n=28)}$ polynomial found by GP.

Comparing these results with the ones of Table 3 we remark a further improvement in the performances when using this new polynomial. Furthermore, if we compare these results with the ones in tables 1 and 2, we observe that the performance improvement of our evolutionary framework compared to the other studied methods is remarkable. Considering that the results of tables 1 and 2 are the best results found in literature for this problem, we can conclude that these final results are an important, further than original, contribution.

Table 4. Mean, maximum and standard deviation for ΔE_{94} errors of our evolutionary framework using the $\mathcal{P}_{(m=7,n=28)}$ polynomial found by GP on the two datasets Macbeth ColorChecker DC and IT8

Dataset	ΔE_{94} Training			ΔE_{94} Test			ΔE_{94} Total		
	Mean	Max	Std Dev	Mean	Max	Std Dev	Mean	Max	Std Dev
Macbeth ColorChecker DC	1.02	6.63	1.05	0.91	3.46	0.74	0.98	6.63	0.91
IT8	0.64	1.70	0.37	0.89	2.53	0.49	0.72	2.53	0.41

5 Conclusions and Future Work

In this paper we have addressed the problem of scanner characterization using multidimensional polynomials and we have proposed a new evolutionary framework to solve it. Our framework is composed by two phases: the first one in which Genetic Programming (GP) is used to search for a characterization polynomial in a completely automatic way and the second one in which the problem is solved by means of a Genetic Algorithm (GA) using that polynomial. GAs are a class of optimization methods that is well-suited for the colorimetric characterization problem as they permit to minimize functionals that take into account the mean perceptual colorimetric error together with other error statistics. Experimental results have shown that our evolutionary framework globally outperforms the state of the art methods. Moreover, relaxing the constraints on the polynomial degree and number of terms, and evaluating the best polynomial with GP, the results is further improved.

Currently, we are trying to modify our GAs in order to add some constraints to our functionals, like for instance neutral axis preservation, and we are trying to improve the rendering of some specific classes (such as skin tone). In the future, we plan to develop a co-evolutionary environment to solve the characterization problem, in which GP, GAs and possibly other evolutionary techniques more closely cooperate in the construction of the final solution. In this idea, we have been inspired by the work [2,19] where Cagnoni and coworkers present a co-evolutionary environment integrating GP and GAs for functions optimization.

References

1. Berns, R.S., Shyu, M.J.: Colorimetric characterization of a desktop drum scanner using a spectral model. Journal of Electronic Imaging 4(4), 360–372 (1995)
2. Cagnoni, S., Rivero, D., Vanneschi, L.: A purely-evolutionary memetic algorithm as a first step towards symbiotic coevolution. In: Proceedings of the 2005 IEEE Congress on Evolutionary Computation (CEC 2005), Edinburgh, Scotland, pp. 1156–1163. IEEE Press, Piscataway (2005)
3. Cantu-Paz, E., Goldberg, D.E.: Are multiple runs of genetic algorithms better than one? In: Cantú-Paz, E., Foster, J.A., Deb, K., Davis, L., Roy, R., O'Reilly, U.-M., Beyer, H.-G., Kendall, G., Wilson, S.W., Harman, M., Wegener, J., Dasgupta, D., Potter, M.A., Schultz, A., Dowsland, K.A., Jonoska, N., Miller, J., Standish, R.K. (eds.) GECCO 2003. LNCS, vol. 2723, pp. 801–812. Springer, Heidelberg (2003)

4. Cheung, T.L.V., Westland, S., Connah, D.R., Ripamonti, C.: Characterization of colour cameras using neural networks and polynomial transforms. Journal of Coloration Technology 120(1), 19–25 (2004)
5. Cheung, V., Westland, S., Li, C., Hardeberg, J., Connah, D.: Characterization of trichromatic color cameras by using a new multispectral imaging technique. J. Opt. Soc. Am. A 22, 1231–1240 (2005)
6. Ebner, M.: Evolving color constancy. Pattern Recognition Letters 27(11), 1220–1229 (2006)
7. Goldberg, D.E.: Genetic Algorithms in Search, Optimization and Machine Learning. Addison-Wesley, Reading (1989)
8. Holland, J.H.: Adaptation in Natural and Artificial Systems. The University of Michigan Press, Ann Arbor, Michigan (1975)
9. Huffel, S.V., Vandewalle, J.: The total least squares problem: computational aspects and analysis, Society for industrial and applied mathematics, Philadelphia (1991)
10. Kang, H.R.: Computational coolor technology, vol. PM159. SPIE Press (2006)
11. Kang, H.R., Anderson, P.G.: Neural network application to color scanner and printer calibrations. Journal of Electronic Imaging 1(2), 125–135 (1992)
12. Kohavi, R.: A study of cross-validation and bootstrap for accuracy estimation and model selection. In: Ryan, C., et al. (eds.) Proceedings of the Fourteenth International Joint Conference on Artificial Intelligence, vol. 2, pp. 1137–1143 (1995)
13. Koza, J.R.: Genetic Programming. The MIT Press, Cambridge (1992)
14. Nelder, J.A., Mead, R.: A simplex method for function minimization. Comput. J. 7(4), 308–313 (1965)
15. Schettini, R., Barolo, B., Boldrin, E.: Colorimetric calibration of color scanners by back-propagation. Pattern Recognition Letters 16(10), 1051–1056 (1995)
16. Shen, H.-L., Mou, T.-S., Xin, J.H.: Colorimetric characterization of scanners by measures of perceptual color error. Journal of Electronic Imaging 15(4), 1–5 (2006)
17. Shen, H.-L., Xin, J.H.: Colorimetric and spectral characterization of a color scanner using local statistics. Journal of Imaging Science and Technology 48(4), 342–346 (2004)
18. Shen, H.-L., Xin, J.H.: Spectral characterization of a color scanner by adaptive estimation. Journal of the Optical Society of America A 21(7), 1125–1130 (2004)
19. Vanneschi, L., Valsecchi, A., Cagnoni, S., Mauri, G.: Heterogeneous cooperative coevolution: Strategies of integration between gp and ga. In: Keijzer, M., et al. (eds.) Proceedings of the Genetic and Evolutionary Computation Conference, GECCO 2006, vol. 1, pp. 361–368. ACM Press, New York (2006)
20. Vrhel, M.J., Trussell, H.J.: Color scanner calibration via a neural networks. In: Proceedings IEEE International Conference on Acoustics, Speech and Signal Processing, vol. 6, pp. 3465–3468 (1999)

Artificial Creatures for Object Tracking and Segmentation

Luca Mussi[1] and Stefano Cagnoni[2]

[1] Università degli Studi di Perugia, Dipartimento di Matematica e Informatica
mussi@dipmat.unipg.it
[2] Università degli Studi di Parma, Dipartimento di Ingegneria dell'Informazione
cagnoni@ce.unipr.it

Abstract. We present a study on the use of soft computing techniques for object tracking/segmentation in surveillance video clips. A number of artificial creatures, conceptually, "inhabit" our image sequences. They explore the images looking for moving objects and learn their features, to distinguish the tracked objects from other moving objects in the scene. Their behaviour is controlled by neural networks evolved by an evolutionary algorithm while the ability to learn is granted by a Self Organizing Map trained while tracking. Population performance is evaluated on both artificial and real video sequences and some results are discussed.

1 Introduction

Detection, tracking and segmentation of moving objects is a critical task in computer vision, whose role is especially important in video surveillance. There are still issues to be addressed before a really effective general-purpose tracking system can be developed. New techniques are continuously being tested. Within these, computational intelligence paradigms have been employed. While many papers, for example [1,2,3], use Artificial Neural Networks (ANNs) as global filters to identify specific pixel characteristics, and hence localize objects of interest, in recent work [4] a swarm intelligence approach has been used to achieve real-time tracking as the emerging property of collective behaviour. In [5] we have already described an approach to object tracking/segmentation based on the use of artificial creatures which "inhabit" video sequences, inspired to a method originally proposed in [6]. Each creature has a retina which allows it to see a very small portion of the current frame from the creature's position. It is also aware of the relative distance and state of other individuals in the population (also referred to as *swarm* in the following). An ANN, acting as its nervous system, decides the next move based on current input information. The creatures' behaviour is evolved through a two-step evolutionary algorithm. In a first phase individuals learn to satisfy a set of minimal behavioural requirements from a small specialised training set. This ensures that creatures perform the main functions required of any tracking/segmentation filter. Then, a population, initialised with clones of the best individual obtained in the first phase, is evolved. In this second phase, the behaviours of the individuals are differentiated as much

M. Giacobini et al. (Eds.): EvoWorkshops 2008, LNCS 4974, pp. 255–264, 2008.

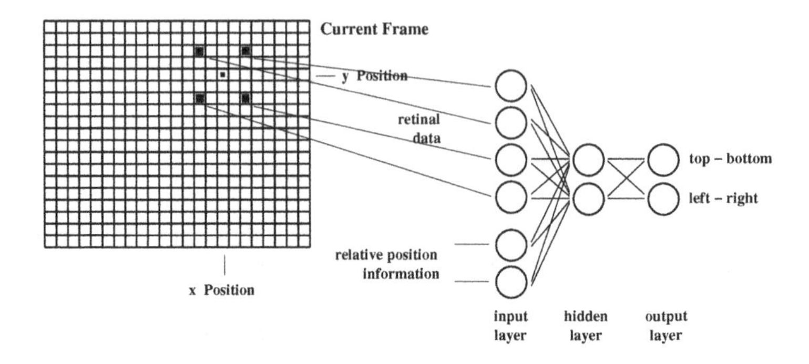

Fig. 1. Artificial creatures' architecture

as possible, even if the individuals keep the basic behaviour learned in the first phase. Tracking and segmentation are performed by the population as a whole. As shown in [5], different creatures' behaviours provide better performances than a population of clones. After these two phases, the swarm exhibits an emergent behaviour that results in finding, tracking and segmenting a moving object.

In this paper we present an improved version of the system. Keeping the same method to evolve the population, the novelty mainly stands in letting the individuals learn some features of the object they are tracking. This produces more effective tracking systems capable of solving critical situations, such as crossing targets and partial occlusions, that usually have to be faced in real-world scenarios. The paper is organised as follows. In Section 2 we describe the structure of the individuals and their inputs. In Section 3 the evolution procedure is briefly presented. In Section 4 we report results obtained with both real and artificial videos. Finally, in Section 5, we propose some future research directions.

2 System Architecture

In surveillance applications where a fixed camera is placed in an environment with stable lighting conditions, the initial image of a video sequence can be taken as a reference background. Motion occurring in the image can be identified by computing the difference between the corresponding pixel values in the current video frame and in the background. In our tracking system, a global "supervisor" keeps track of the position of all individuals. In every cycle, it sends inputs to individuals and collects outputs to update their state. While retrieving input information from the frames, it also trains a Self-Organising Map (SOM) [7], which is used to learn features of the object which is being tracked. Using the classification policy described in section 2.1, some of the visited pixels are labeled as belonging to the tracked object. When a new video frame is available, the supervisor moves the whole population over it keeping individuals' positions unchanged.

Each creature has a very simple feed-forward ANN as "brain". Through the input layer it senses environmental information while through the output layer it informs the supervisor about its next move. Each frame is considered to be toroidal. Figure 1 shows the structure of the network which controls a creature. The six input neurons are grouped into two sets. Four neurons form a minimalist retina that provides creatures with vision. The sensing units are at the corners of a 5×5-pixel square, centered on the creature's current position. This retina receives binary inputs depending on the classification criterion described later in section 2.1. The remaining two input neurons are used to provide the net with information about an individual's position with respect to the rest of the population. Their decoding differs with two different states of the population: if all individuals are still looking for a moving object, the inputs are both set to zero, otherwise they encode the relative position of the closest individual located on a tracked object with 1, if the latter has higher x (y) coordinate than the individual being considered, or -1 otherwise. The hidden layer includes only two neurons, both connected with all inputs. Their activation function is the classical sigmoid function. Finally, the output layer consists of two motor neurons connected with the hidden layer. Their purpose is to tell the supervisor how to update the creature's position in the next cycle. One neuron refers to horizontal while the other refers to vertical motion. The activation function of the output neurons is

$$o(x) = 2\,(W - 1) \cdot \mathrm{round}(s(x) - 0.5) \tag{1}$$

where W is the width of the retina (five pixels in our case) and $s(x)$ is the sigmoid function. As a result, the next position of an individual is always within a $(2\,W - 1) \times (2\,W - 1)$-pixel square centered in the present position.

2.1 Pixel Classification Policy and Self Organizing Map Training

As a first upgrade of the system described in [5] we have provided the swarm with a SOM trivially trained with the color of the pixels in the RGB color space. The criterion used to classify a pixel as part of the tracked object is the following: if the difference between the image and the background in that position is greater than a certain threshold δ, set by the user, the pixel color is checked to be close enough to those encoded in the SOM. Colors of pixels correctly classified are immediately used as input patterns for one training step of the SOM. It is also required that all sensing elements of an individual's retina satisfy the previous condition or sense no difference with the background; otherwise all units of the retina are marked as not activated. Therefore, the object membership measure of a single pixel is not based only on its color but color gradient or color discontinuities of neighbouring pixels are also taken into account, when present. Since the colors of the object of interest are likely to be sampled randomly while the creatures are inside it during tracking, the usual SOM training [8] is performed "on-line". Anyway, the decaying time of the learning rate, described by an exponential as usual, is kept very short to ensure that learning of object's features is obtained in a few frames.

During every training epoch the maximum distances between each component of the input patterns and the winning neurons are stored. In the next epoch, these values are used to decide whether an input pattern belongs to the object or not. More precisely, to classify an input as 'target' the following inequalities must be verified:

$$|x_i - w_i(\boldsymbol{x}))| <= \gamma(n)e_i \qquad \text{for all } i \tag{2}$$

where x_i is the i-th component of the input pattern, $w_i(\boldsymbol{x})$ is the corresponding weight of the best-matching (winning) neuron, $\gamma(n)$ is a tolerance factor, which depends on the frame number n, and e_i is the i-th component of the maximum distance vector computed in the previous training epoch. To make this classification criterion effective, the SOM content is randomly initialized and $\gamma(n)$ is described by the following exponential curve:

$$\gamma(n) = \alpha + (255 - \alpha)e^{-n/\tau} \tag{3}$$

In this way, at the very beginning, all colors are classified as good but, after a short time, depending on the time constant τ, $\gamma(n)$ tends toward the value α and the content of the SOM begins to be relevant for classification. The value of these parameters is a critical aspect for the effectiveness of the learning process: many experiments showed that $\alpha = 1.2$ and $\tau \simeq 8$ are good settings to achieve a proper learning in about ten frames.

3 Evolving Behaviours

The behaviour of our artificial creatures depends only on the connection weights inside their "brain". So, each individual can be described by a real vector. Evolution Strategies or Evolutionary Programming are among the most suitable and commonly used evolutionary techniques to deal with real number representations [9]. When using an entire swarm to track moving objects, besides maintaining diversity between individual behaviours to achieve better generalisation and exploration capabilities [10], we also want the swarm to have an efficient collective, swarm-type behaviour. This is obtained by letting each individual be attracted towards the swarm center while being repulsed by its neighbours [11].

In the proposed two-step evolution, an individual that satisfies a small training set which exemplifies the correct behaviour in terms of attractions to and repulsion from other individuals is firstly evolved. When such an individual is found, it is cloned to form a new population that is further evolved to maximize behaviour differentiation. This final population is then used in the tracking system. The final population size is a parameter set by the user.

3.1 Minimum Behaviour

A specific training set is used to provide attraction and repulsion as well as other basic behaviours. Figure 2 shows part of it. The inputs are specified numerically: 0 or 1 for the sensory inputs `I0-I3`, and 1 or -1 for the inputs `cloLR` and `cloTB` which specify the relative position of the closest individual. For the two outputs (left-right and top-bottom motion) we do not specify exact target values but

```
               INPUTS                                OUTPUTS
IO I1 I2 I3   cloLR cloTB      minLR   maxLR   minTB   maxTB

//Retina Activations only in one of the four corners
1  0  0  0     d-    d-          -4      -1      -4      -1
0  1  0  0     d-    d-           1       4      -4      -1
0  0  1  0     d-    d-          -4      -1       1       4
0  0  0  1     d-    d-           1       4       1       4

//Retina with no activations
0  0  0  0      1     1           1       4       1       4
0  0  0  0      1    -1           1       4      -4      -1
0  0  0  0     -1     1          -4      -1       1       4
0  0  0  0     -1    -1          -4      -1      -4      -1
```

Fig. 2. Part of the training set used in the first evolution phase. The "d-" symbol means that both 1 and -1 are allowed in that position.

minimum (minLR and minTB) and maximum (maxLR and maxTB) acceptable values. The first set of examples requires that, when only one of the four units in the retina is activated, the individual move towards the quadrant corresponding to such a unit, irrespective of other individuals' positions. The second set of examples requires that, when the retina is not activated, the individual move towards the quadrant of its nearest neighbour on the tracked object. The same behaviour must be kept also when the retina is fully activated. The supervisor keeps track of which individuals are on the tracked object. It normally moves creatures according to their outputs; however, when an individual is fully inside the tracked object, the supervisor does the opposite of what the output neurons suggest. As a result, the creature is moved away from the center of mass of the swarm. This makes it possible to relieve the net from detecting whether a creature is all inside or all outside the tracked object, which would make training much harder, particularly with only two neurons in the hidden layer. Therefore, a more computation-intensive supervisor has been preferred to larger nets. This first evolution phase is based on a classical Evolution Strategy. At every generation, a child is created for each individual by Gaussian mutation. Then, in the selection phase, round-robin tournament competitions are performed among parents and offspring: the individuals with the greatest number of "wins" survive. To compute an individual's fitness all the examples in the training set are shown to its neural network. The corresponding outputs are checked against the ranges specified by minLR, minTB, maxLR and maxTB in each example and the fitness is simply the number of output patterns within the desired ranges. The first evolutionary phase ends when an individual with a fitness value equal to the size of the training set appears or when a maximum number of generations is reached, in which case evolution is aborted.

3.2 Population Differentiation

The second phase starts as soon as an individual satisfying the minimum requirements is found. A new initial population is formed, made up of copies of the best individual found in the previous phase, and a new evolution process

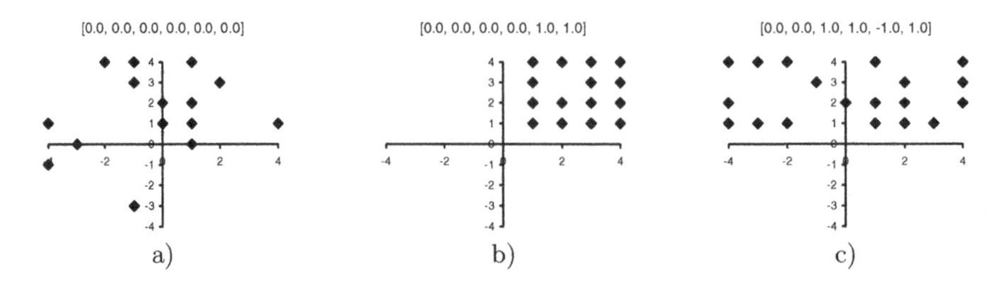

Fig. 3. Responses of each of the twenty individuals in a population to some input patterns. Notice that the y coordinate in the images increases downwards and that some outputs could be overlapped.

starts. This phase aims at differentiating behaviours within the population as much as possible. This means to make sure, for each of the possible input patterns, that the distribution of the outputs within the population is as uniform as possible. As a result, the distribution of the outputs corresponding to these inputs will also tend to be uniform within a sub-region of the output domain. Obviously, we also need to check whether the rules encoded in the training set used in the first phase are still satisfied. As before, in each generation every individual generates a child by means of Gaussian mutation. If this satisfies the minimum requirements, it replaces its parent and the fitness of the entire population is re-evaluated. If fitness does not improve, the offspring is substituted with its parent (thereby undoing the change) and the next offspring is generated and checked. The fitness function to be minimized is

$$F = \sum_{i \in I} \sum_{m \in O} \sum_{n \in O} (1 - n_i(m,n))^2 \tag{4}$$

in which I is the set of all possible input patterns, $O = \{-(W-1), \dots, (W-1)\}$ is the domain for the output of motor neurons and $n_i(m,n)$ is the number of individuals that produce the output (m,n) when i is the input pattern. This second evolutionary stage ends when a pre-set maximum number of generations is reached. The last population is saved and used for tracking.

Figure 3 shows the responses to some input patterns produced by a typical population at the end of the second phase. When all individuals are still looking for the moving object, they receive a null input. As one can see, however, despite the input being the same for everyone, each individual responds with a slightly different move. In Figure 3(a), all responses of the population to the null input are plotted. When an individual is still in the search phase and there is at least one creature on the moving object, its retina does not sense anything, but the supervisor provides him with the relative position of the closest neighbour on the object. As shown in Figure 3(b) for one of the four possible cases, every individual in this situation responds by moving towards the correct direction, but once again responses are differentiated. Finally, Figure 3(c) shows all possible responses for an individual whose bottom part of the retina is activated. In these conditions the individual is probably on the object's upper border, so it makes sense to move downward. It should be noticed that this situation was not included in our basic training set, where only activations of the corners of the retina were included among the input patterns. Nonetheless, as a result of generalization and of behaviour differentiation, the response of every individual within the population is fully appropriate for the specific situation.

4 Results

The code for the system was originally written in Java™ using JINGPaC[1]. Now, the evolutionary motor is still coded in Java, but the tracking system has been re-implemented in C++ to achieve better real-time performance.

In the experiments, in the first phase a population of 2,000 individuals, a maximum number of generations of 3,000 and tournament selection of size 5 were used. In the second evolution phase, a population of 20 individuals was evolved for 40,000 generations. We tested the new system on both artificially-created color videos and real-world color videos. Experiments were run on a PC equipped with an Intel® Core™2 Duo processor at 1.80GHz, playing the videos in real time (25fps). A two-dimensional 10×10-unit SOM was used in these experiments.

Fig. 4. On the left, a sample frame extracted from the artificially-created video used in the experiments presented here. On the right, a frame from the real-world sequence.

Figure 4 (left) shows a frame from one of the artificially-created videos: a 640x480 pixels picture of a landscape was used as a fixed background and some stylised bees, having slightly different colors, were superimposed and moved over it letting their trajectories cross more than once. Since our system tracks the first moving object that enters the scene, the bee that comes from the right-hand side of the frame was the target.

Figure 5 shows ten frame details extracted while processing this sequence: dark-gray pixels of the system output represent the segmented region of the object while white points represent visited pixels which have not been marked as part of the object. The tracked bee is approaching another one: the system, thanks to the data stored in the SOM, distinguishes between the two bees even

[1] Available at http://jingpac.sourceforge.net/

Fig. 5. Sequence of frame details with the output of the elaboration superimposed on them. As can be seen object tracking and segmentation have been correctly achieved.

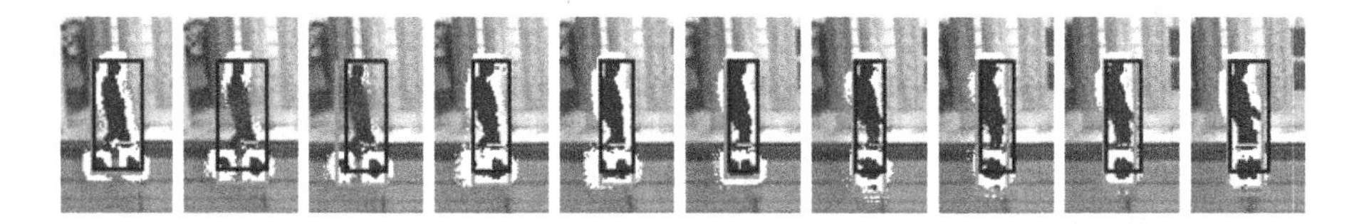

Fig. 6. Sequence of frame details from the real-world video sequence in which a person walks in front of a shop. Superimposed on each frame are the ground-truth bounding box (darker) and the bounding-box detected by the system (brighter).

if both are moving. As a result, the first bee is correctly tracked and its visible part is always almost fully segmented.

Part of the real-world videos used in the experiments belong to the CAVIAR Test Case Scenarios[2] and show a man walking inside a shopping center (Figure 4, right). The ground truth for these sequences has been hand-labeled by the CAVIAR team members and saved as XML files. This makes it possible to evaluate quantitatively the results of our system. Figure 6 shows a sequence of frame details extracted from the "man" sequence. Since this video was acquired at half-resolution PAL standard (384x288 pixels, 25 fps) and compressed using MPEG2, it is quite noisy. Furthermore, the brightness of the area is low, most people are in dark clothes and the floor surface reflects their silhouettes. Because of this, identifying the walking man accurately by means of simple backround subtraction is tricky. On each frame in Figure 6, it is once again possible to see the results of tracking superimposed onto the images. In this case a comparison can be made between the position and size of the ground-truth bounding box (darker) and the system output (brighter). For example, in the third image, the bounding box computed is slightly taller and wider than the ground truth: this is probably due to a few false detections on the man's shadow.

Considering the ground-truth bounding box, the detected bounding box, and their overlap area we have computed three different scores: the distance between

[2] Available at http://groups.inf.ed.ac.uk/vision/CAVIAR/CAVIARDATA1/

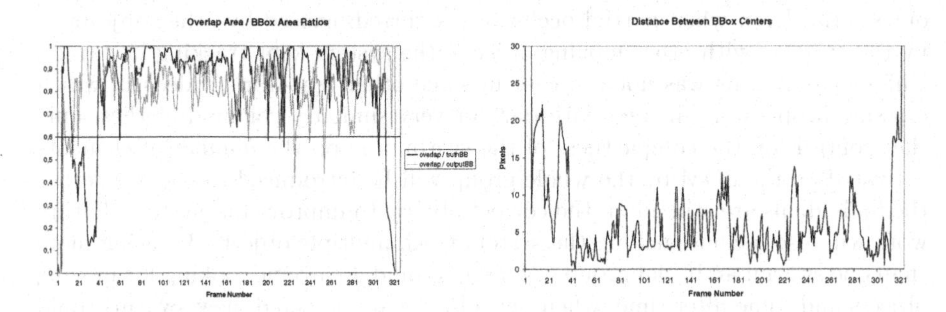

Fig. 7. Results from a sample run: on the left, percentage of the ground-truth box covered by the overlap area (dark line) and percentage of the detected box covered by the overlap area (light line); on the right, distance between box centers

box centers, the percentage of the ground-truth box overlapping the detected area, and the percentage of the detected box overlapping the ground-truth. We consider successful a detection for which both percentages are above 60%. Figure 7 shows the values obtained for these scores in a random run. The two percentage ratios are very often above the threshold: 267 correct detections out of 323 frames were recorded. Most unsuccessful detections occur at the very beginning and at the end of the sequence, when the swarm is still searching for the target and when the target enters a very dark area, respectively. However, even in most detections marked as unsuccessful in the central part of the sequence, the distance between the actual center of the object and the estimated one is low (see Figure 7, right). The same analysis was repeated ten times, showing good repeatability: the average number of valid detections was 262, with a standard deviation of about 9.5. The average mean values for the three scores over the ten run were 5 pixels, 89% and 84%, while their standard deviations were 3.25 pixels, 8.5% and 9.4%, respectively.

One of the most appreciable characteristics of this system stands in the number of pixels visited within each frame, which is very low, in particular when compared to the number of pixels belonging to the object.

Further details and videos about this project can be found at the following address: http://www.ce.unipr.it/people/mussi/projects/evolutionary-tracking/.

5 Concluding Remarks and Future Work

We have described an object tracking system based on a population of artificial creatures controlled by an ANN. As image inhabitants, they go around looking for a moving object, indicating to a global supervisor which pixels to check: if these pixels satisfy certain criteria, they are classified as part of the moving object. The criterion used for classification is partly based on background subtraction and partly on features of the object learned on-line during tracking.

The experimental results proved that our approach is effective. The first moving object entering the scene has been successfully identified and tracked by our swarm. Even when partial occlusions occurred, caused by other objects or by the crossing with other moving objects, the system kept tracking its target. Unfortunately, this was not the case in some other real-world videos in which crossing people were dressed with dark or very similarly coloured clothes. Further controls on the compactness of the swarm and on the number of elements activated on the SOM by the whole group will be introduced to check whether the individuals are still all on the correct object, to improve our system. Future work will also include making the system track multiple objects, by using more than one swarm: an initial group of foragers could go around looking for moving objects and, time after time, when new objects are detected, new swarms could be locally initialized. In this way, different objects could be detected and tracked separately. Such an architecture has already been used in [12] for a different purpose, achieving good results.

The ability to learn features of the objects being tracked offered by the SOM could help solve occasional difficult situations such as crossing targets and partial occlusions.

References

1. Cuevas, E., Zaldivar, D., Rojas, R.: LVQ color segmentation applied to face localization. In: 1st International Conference on Electrical and Electronics Engineering (ICEEE), pp. 142–146. IEEE Computer Society Press, Los Alamitos (2004)
2. Valli, G., Poli, R., Cagnoni, S., Coppini, G.: Neural networks and prior knowledge help the segmentation of medical images. Journal of Computing and Information Technology (CIT) 6(2), 117–133 (1998)
3. Kulkarni, A.D.: Artificial Neural Networks for Image Understanding. VNR Computer Library. Van Nostrand Reinhold (1994)
4. Anton-Canalis, L., Hernandez-Tejera, M., Sanchez-Nielsen, E.: Particle swarms as video sequence inhabitants for object tracking in computer vision. In: Proc. ISDA 2006, pp. 604–609. IEEE Computer Society, Los Alamitos (2006)
5. Mussi, L., Poli, R., Cagnoni, S.: Object tracking and segmentation with a population of artificial neural networks. In: Workshop Italiano di Vita Artificiale e Computazione Evolutiva (WIVACE 2007) (2007) (published on CD), `http://www.ce.unipr.it/people/mussi/index.php?page=publications`
6. Poli, R., Valli, G.: Neural inhabitants of MR and echo images segment cardiac structures. In: Proc. Computers in Cardiology, pp. 193–196. IEEE Computer Society Press, Los Alamitos (1993)
7. Kohonen, T.: Self-Organizing Maps, 3rd edn. Springer, Heidelberg (2001)
8. Haykin, S.: Neural Networks. Prentice-Hall, Englewood Cliffs (1999)
9. Eiben, A., Smith, J.: Introduction to Evolutionary Computing. Springer, Heidelberg (2003)
10. Yao, X.: Evolving artificial neural networks. Proc. of IEEE 87(9), 1423–1447 (1999)
11. Couzin, I., Krause, J., Franks, N., Levin, S.: Effective leadership and decision making in animal groups on the move. Nature 433, 513–516 (2005)
12. Olague, G., Puente, C.: The Honeybee search algorithm for three-dimensional reconstruction. In: Rothlauf, F., Branke, J., Cagnoni, S., Costa, E., Cotta, C., Drechsler, R., Lutton, E., Machado, P., Moore, J.H., Romero, J., Smith, D.G., Squillero, G., Tagaki, H. (eds.) EvoWorkshops 2006. LNCS, vol. 3907, pp. 427–437. Springer, Heidelberg (2006)

Automatic Recognition of Hand Gestures with Differential Evolution

I. De Falco[1], A. Della Cioppa[2], D. Maisto[1], U. Scafuri[1], and E. Tarantino[1]

[1] ICAR-CNR, Via P. Castellino 111, 80131 Naples, Italy
{ivanoe.defalco,domenico.maisto,umberto.scafuri, ernesto.tarantino}@na.icar.cnr.it
[2] DIIIE Lab, University of Salerno,
Via Ponte don Melillo 1, 84084 Fisciano (SA), Italy
adellacioppa@unisa.it

Abstract. Automatic recognition of hand gestures is a crucial step in facing human–computer interaction. Differential Evolution is used to perform automatic classification of hand gestures in a thirteen–class database. Performance of the resulting best individual is computed in terms of error rate on the testing set, and is compared against those of other ten classification techniques well known in literature. Results show the effectiveness and the efficiency of the approach in solving the classification task. Furthermore, the implemented tool allows to extract the most significant parameters for differentiating the collected gestures.

1 Introduction

Human-made hand gestures recognition is an interesting task in artificial intelligence and human–computer interaction, with important influences in some high–technology application areas such as virtual reality, entertainment, human factors analysis, real–time animation, etc.

A specialized software is needed to correctly distinguish among a set of hand gestures after these have been recorded by cameras, sensors or haptic gloves and digitalized. This is a typical classification task for which several tools are nowadays available. Different heuristic methods can be designed to build classifiers. Among them, Evolutionary Algorithms appear of interest.

In this paper, Differential Evolution (DE) [1], a version of an Evolutionary Algorithm [2], is considered. The aim is to evaluate DE effectiveness and efficiency in performing a centroid–based supervised classification by taking into account a database composed by human–made hand gestures, digitally recorded by haptic gloves and collected in a thirteen–class data set, extracted from Auslan database [3]. The idea is to use DE to find the positions of the class centroids, here seen as simply "class representatives" and not necessarily as "average points" of a cluster in the multidimensional space of features, such that for any class the average distance of instances belonging to that class from the relative class centroid is minimized. Error percentage for classification on testing set is computed

M. Giacobini et al. (Eds.): EvoWorkshops 2008, LNCS 4974, pp. 265–274, 2008.

on the resulting best individual. Moreover, the results are compared with those obtained by ten well–known classification techniques.

Paper structure is as follows: Section 2 outlines DE basic scheme while Section 3 illustrates the application of the system based on DE and centroids to the classification problem. Section 4 describes the Auslan sign database, reports on the transformation of the original database into another which could be directly dealt with by the DE tool, and shows the results and the comparison against ten typical classification techniques. Finally Section 5 contains conclusions.

2 Differential Evolution

Differential Evolution [1] is a stochastic, population–based optimization algorithm and uses vectors of real numbers as representations of solutions.

The seminal idea of DE is that of using vector differences for perturbing the genotype of the individuals in the population. Basically, DE generates new individuals by adding the weighted difference vector between two population members to a third member. If the resulting trial vector yields a better objective function value than a predetermined population member, the newly generated vector replaces the vector with which it was compared. By using components of existing population members to construct trial vectors, recombination efficiently shuffles information about successful combinations, enabling the search for an optimum to focus on the most promising area of solution space.

In more detail, given a minimization problem with m real parameters, DE faces it starting with a randomly initialized population $P(t = 0)$ consisting of n individuals each made up by m real values. Then, the population is updated from a generation to the next one by means of some transformation. The authors of [1] established a sensible naming–convention denoting any DE strategy with a string like *DE/x/y/z*. In it DE stands for Differential Evolution, x is a string which specifies the vector to be perturbed (*best* = the best individual in current population, *rand* = a randomly chosen one, *rand–to–best* = a random one, but the current best participates in the perturbation too), y is the number of difference vectors taken for perturbation of x (either 1 or 2), while z is the crossover method (*exp* = exponential, *bin* = binomial). We have decided to perturb a random individual by using one difference vector and by applying binomial crossover, so our strategy can be referenced as *DE/rand/1/bin*. For the generic i–th individual in the current population, this strategy consists in randomly generating three integer numbers r_1, r_2 and r_3 in $[1, n]$, differing one another and different from i. Furthermore, another integer number k in the range $[1, m]$ is randomly chosen. Then, starting from the i–th individual a new trial one i' is generated whose generic j–th component is given by:

$$x_{i',j} = x_{r_3,j} + F \cdot (x_{r_1,j} - x_{r_2,j})$$

provided that either a randomly generated real number ρ in $[0.0, 1.0]$ is lower than a value CR (parameter of the algorithm, in the same range as ρ) or the position j under account is exactly k. If neither is verified then a simple copy takes place: $x_{i',j} = x_{i,j}$. The parameter F is a real and constant factor in $[0.0, 1.0]$ which controls the magnitude of the differential variation $(x_{r_1,j} - x_{r_2,j})$.

This new trial individual i' is compared against the i–th individual in current population and, if fitter, replaces it in the next population, otherwise the old one survives and is copied into the new population. This basic scheme is repeated for a maximum number of generations g.

3 DE Applied to Classification

Encoding. We have chosen to face the classification task by using a tool in which DE is coupled with centroids mechanism (we shall hereinafter refer to it as DE–C system). Specifically, given a database with C classes and N attributes, DE–C should find the optimal positions of the C centroids in the N–dimensional space, i.e., it should determine for any centroid its N coordinates, each of which can take on, in general, real values. With these premises, the i–th individual of the population is encoded as $(\boldsymbol{p}_i{}^1, \ldots, \boldsymbol{p}_i{}^C)$, where the position of the j–th centroid $\boldsymbol{p}_i{}^j$ is constituted by N real numbers representing its N coordinates in the problem space $\boldsymbol{p}_i{}^j = \{p^j_{1,i}, \ldots, p^j_{N,i}\}$. Then, any individual in the population consists of $C \cdot N$ components, each of which is represented by a real value.

Fitness. Following the classical approach to supervised classification, the database is divided into a training and a testing set. The tool learns on the former and its performance is evaluated on the latter.

The fitness function ψ is computed as the sum on all the training set instances of the euclidean distance in the N–dimensional space between the generic instance $\boldsymbol{x}_j$ and the centroid of the known class it belongs to according to the database $(\boldsymbol{p}_i{}^{\mathbf{CL}_{\text{known}}(\boldsymbol{x}_j)})$. This sum is normalized with respect to D_{Train}, i.e., the number of instances which compose the training set. In formulae, the fitness of the i–th individual is given by:

$$\psi(i) = \frac{1}{D_{\text{Train}}} \cdot \sum_{j=1}^{D_{\text{Train}}} d\left(\boldsymbol{x}_j, \boldsymbol{p}_i{}^{\mathbf{CL}_{\text{known}}(\boldsymbol{x}_j)}\right)$$

When computing distance, any of its components in the N–dimensional space is normalized with respect to the maximal range in the dimension, and the sum of distance components is divided by N. With this choice, any distance can range within $[0.0, 1.0]$ and so can ψ. Given the chosen fitness function, the problem becomes a typical minimization problem.

Performance of a run, instead, is computed as the percentage $\%err$ of instances of testing set which are incorrectly classified by the best individual (in terms of the above fitness) achieved in the run. With this choice DE–C results can be directly compared to those provided by other classification techniques.

4 Experiments and Results

Auslan database. The gesture language of the Australian Deaf community is known as Auslan [4]. There are about 4000 signs in Auslan, each having several components that can be mixed to form a wide variety of other signs.

Quite recently a set of 95 Auslan signs were captured [5] from a native signer using a two–handed acquisition system based on high–quality position trackers and instrumented gloves. For each hand, the following variables are recorded:

- x, y, z positions, in meters, relative to a zero point set below the chin;
- *roll* expressed as a value between -0.5 and 0.5 with 0 being palm down. Positive means the palm is rolled clockwise from the perspective of the signer;
- *pitch* expressed as a value between -0.5 and 0.5 with 0 being palm flat (horizontal). Positive means the palm is pointing up;
- *yaw* expressed as a a value between -1.0 and 1.0 with 0 being palm straight ahead from the perspective of the signer. Positive means clockwise from the perspective above the signer;
- *thumb*, *forefinger*, *middle finger*, *ring finger*, *little finger*, each with values between 0 (totally flat) and 1 (totally bent).

Database samples were collected in nine different sessions over nine weeks and in each session 3 samples of 95 different gestures were gathered. Thus there are 95 different signs with 27 samples per sign, for a total of 2,565 instances.

Each sign is recorded on average in 57 time frames; for any frame, 22 parameters (11 for each hand) were captured by sensors. Nevertheless there are quite significant differences depending on the particular sign. A first difference is in the fact the representations of a same gesture may be slower or faster, and the second is in the fact that different gestures usually require different time spans, for instance the sign *no* is the fastest. Due to the above reasons, some items last about 50 frames and others about 65.

Data processing. Given the huge amount of data, the Auslan database [3] (downloadable at [6]) cannot be directly dealt with rather it must be reduced.

In first place, not all of the 95 signs seem necessary for any application related to human–computer interaction. In fact, in some cases it may be sufficient to have a small number of useful gestures that the computer must distinguish. Therefore, we have focused the attention on a subset of thirteen simple signs: *cold, drink, eat, exit, hello, hot, no, read, right, wait, write, wrong, yes* which constitute the thirteen classes of the reduced database for a total of $13 \cdot 27 = 351$ instances.

Besides, to use the classification scheme described in the previous section the need arises to identify all the variables which represent any database item. The idea in order to represent an item would be to use all of the data of each frame in their sequential order of appearance. Unfortunately this leads to a problem, since the signs have different duration in terms of number of frames. To overcome this, all recordings must be forced to have exactly the same number of frames.

Choice that has been made was to cut all the recordings at exactly the average length of all signs, i.e., 57 frames. By doing so, the longer ones have lost some of their last frames, whereas for those lasting less than 57 frames the hypothesis has been made that we can extrapolate it as if the hand stood still for the remaining frames. After doing this, the resulting number of attributes for any database item would be $57 \cdot 22 = 1,254$, which is huge for any classification tool. In particular, for DE–C any individual would consist of $1,254 \cdot 13 = 16,302$ positions, which is unmanageable and must be reduced.

A second choice has been taken to downsample the recordings by saving one frame every five, hoping that this would not affect the meaning of the gesture. This is a quite common choice in the field, and usually does not yield problems, provided that data are continuous, as this is the case. By doing so, just 11 frames are considered for any gesture.

A third choice has to be considered only the variables related to the right hand, since the vast majority of Auslan signs requires that hand only, and in any case a right–handed Auslan speaker mostly uses his right hand, so the left channel likely contains less information than the right one. This hypothesis is reinforced by the fact that a preliminary database of Auslan signs was collected by the same author by reckoning right hand movements only [5].

Finally, a fourth choice has been to consider, for any frame, only the first six parameters related to hand positions and angles, since, as it is stated by the donors of the database, measurements about fingers are much less precise [5].

With the above choices, the length for any database item reduces to $11 \cdot 6 = 66$ plus the class, and that for any DE–C individual drops down to $66 \cdot 13 = 858$.

Of course, the above hypotheses might lead to a database with different features from the original one and thus to a classification with incorrect results. For example some signs might now be confused since information important to distinguish them might have been lost, therefore the following experiments will also assess the goodness of the hypotheses made.

The reduced database is then separated into a training set and a testing one. The former consists of 234 instances, divided into thirteen classes, one for each of the thirteen signs listed above. Any class is represented in this training set by 18 instances, i.e., those recorded in the first six weeks. The testing set, instead, consists of 117 elements (9 for each class), i.e., those in the last three weeks. Thus, the training set is assigned 67% of the database instances, and the testing set the remaining 33%, which are quite usual percentages in classification.

The Experiments. As concerns the other classification techniques used for the comparison we have made reference to the Waikato Environment for Knowledge Analysis (WEKA) system release 3.4 [7] which contains a large number of such techniques, divided into groups (bayesian, function–based, lazy, meta–techniques, tree-based, rule–based, other) on the basis of the underlying working mechanism. From each such group we have chosen some among the most widely used representatives. They are: among the bayesian the Bayes Net [8], among the function-based the MultiLayer Perceptron Artificial Neural Network (MLP) [9], among the lazy IB1 [10] and KStar [11], among the meta–techniques the Bagging

Table 1. Achieved results in terms of $\%err$ and σ

	DE–C	BAYES NET	MLP ANN	IB1	KSTAR	BAG-GING	J48	NB TREE	PART	RIDOR	VFI
$\%err$	11.32	12.41	7.46	26.72	38.09	15.34	35.55	25.00	30.12	27.66	37.62
$\%err_b$	9.40	10.34	6.03	–	30.17	12.06	28.44	–	22.41	18.10	30.17
$\%err_w$	13.67	12.93	9.62	–	43.10	18.10	51.72	–	37.06	37.93	39.65
σ	1.17	0.90	1.30	–	4.43	1.73	7.07	–	4.08	4.62	2.11

[12], among the tree–based ones J48 [13] and Naive Bayes Tree (NBTree) [14], among the rule–based ones PART [15] and Ripple Down Rule (Ridor) [16] and among the others the Voting Feature Interval (VFI) [17].

On the basis of a preliminary tuning phase carried out on this and on other databases, DE–C parameters have been chosen as follows: $n = 500$, $g = 3000$, $CR = 0.01$ and $F = 0.01$. It is interesting to note that the values for CR and F are much lower than the ones classically used according to literature, which range higher than 0.5. A hypothesis about the reason for this may be that any chromosome in the population consists in this case of $N \cdot C = 66 \cdot 13 = 858$ components, so search space is very large indeed and high values for CR and F would change too many alleles and create individuals too different from the parents. This would drive the search far from the promising regions already encountered, thus creating worse individuals than those present in the current population. As a consequence, given the elitist kind of replacement strategy, only very few new individuals would be able to enter the next generation. The just hypothesized scenario seems confirmed by the evolution shown by the system for high values of CR and F: in this case fitness decrease of the best individual is not continuous, rather it takes place in steps, each of which lasts tens of generations. This kind of evolution seems similar to a random search with elitism.

Results of DE–C technique are averaged over 20 runs only differing one another for the different starting seed provided in input to the random number generator. For the other techniques, instead, some (MLP, Bagging, Ridor and PART) are based on a starting seed so that also for them 20 runs have been carried out by varying this value. Other techniques (Bayes Net, KStar, J48, VFI) do not depend on any starting seed, so 20 runs have been executed as a function of a parameter typical of the technique (*alpha* for Bayes Net, *globalBlend* for KStar, *numFolds* for J48 and *bias* for VFI). NBTree and IB1, finally, depend neither on an initial seed nor on any parameter, so only one run has been performed.

Table 1 shows the results attained by the 11 techniques on the database. Namely, for any technique the average values of $\%err$ and the related standard deviations σ are given together with the best ($\%err_b$) and the worst ($\%err_w$) values achieved in the 20 runs. Of course σ is meaningless for NBTree and IB1.

As it can be observed from the values in Table 1, DE–C is the second best technique in terms of $\%err$, very close to MLP which is the best and quite closely followed by Bayes Net. Bagging follows at a distance, while all other techniques

Table 2. Confusion matrix for the best run effected by DE–C

	cold	*drink*	*eat*	*exit*	*hello*	*hot*	*no*	*read*	*right*	*wait*	*write*	*wrong*	*yes*
cold	9												
drink		9											
eat			9										
exit				9									
hello					9								
hot			1	1		7							
no							9						
read								7	1		1		
right							1		7				1
wait										9			
write											9		
wrong							3					6	
yes							1			1			7

are quite far from the three best ones. The standard deviation σ for DE–C is not too high, meaning that, independently of the different initial populations, the final classifications achieved have similar correctness. Some techniques like J48, Ridor, KStar and Part, instead, show very different final values of $\%err$, thus sensitivity to different initial conditions.

Hence, the exploitation of DE to find positions of centroids has proven effective to face this reduced version of Auslan database: about 89 signs out of 100 are correctly recognized, on average, and 91 in the best case.

Table 2 reports the confusion matrix for the best DE–C run, i.e. that with the lower $\%err$. For any class the row shows how its 9 examples have been scattered among the 13 classes (columns). The numbers on the main diagonal represent how many examples have been correctly assigned to classes. As it can be seen, 8 gestures, like for instance *cold*, *drink* and *eat*, are always recognized by the tool, whereas some others, such as *hot* and *read*, are more difficult to distinguish. The sign *wrong* is in 3 cases confused with *no*. It is worth noting that also in the best run of MLP *read* is confused once with *write* and *yes* is confused twice with *wait*, so this might depend on the way those signs were made, perhaps slower or faster or with very different hand position or rotation with respect to the "average" movement. Thus, those items might represent *outliers*.

From an evolutionary point of view, in Fig. 1 we report the behavior of the best run in terms of lower final fitness value achieved. The left panel shows the evolution in terms of best individual fitness, average fitness and worst fitness in the population as a function of the number of generations. DE–C shows a first phase of about 700 generations in which fitness decrease is strong and almost linear, starting from 0.71 for the best and 0.75 for the average, and reaching about 0.42 for the best and 0.43 for the average. A second phase follows, lasting until about generation 1800, in which decrease is slower, and the two values tend to become closer, until they reach 0.3450 and 0.3456 respectively. From now on the decrease in fitness is slower and slower, about linear again but with a much lower slope, and those two values become more and more similar. Finally, at generation 3000 the two values are 0.3409 and 0.3410 respectively.

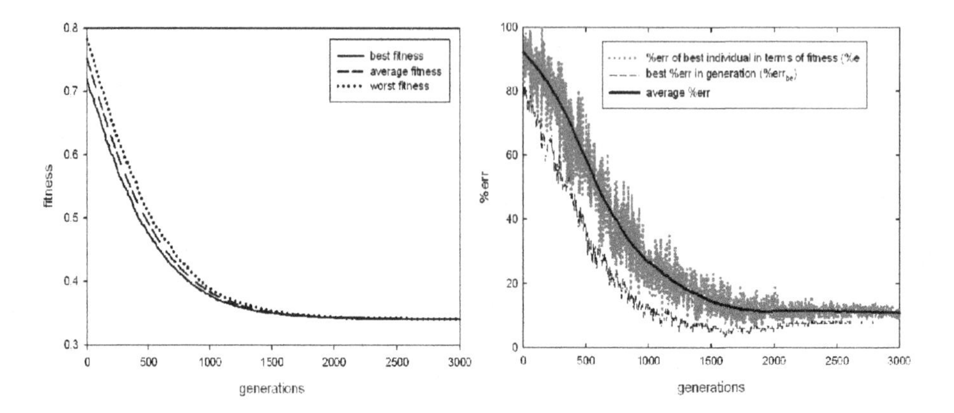

Fig. 1. Behavior of fitness (left panel) and $\%err$ (right panel) as a function of the number of generations for the best run

The right panel of Fig. 1, instead, reports the behavior of $\%err$ as a function of the generations. Namely, for any generation its average value, the lowest error value in the generation $\%err_{be}$ and the value of the individual with the best fitness value $\%err_{bf}$ are reported. It should be remarked here that $\%err_{bf}$ does not, in general, coincide with $\%err_{be}$, and is usually greater than this latter. This is due to the fact that our fitness does not take $\%err$ into account, so evolution is blind with respect to it, and it does not know which individual has the best performance on the testing set.

What described above for a specific run is actually true for all the runs carried out, and is probably a consequence of the good parameter setting chosen. This choice on the one hand allows a fast decrease in the first part of the run and, on the other hand, avoids the evolution being stuck in premature convergence as long as generations go by, as it is evidenced by the fact that best and average fitness values are different enough during the whole evolution. Preliminary experiments not reported here showed, instead, that using high values for CR and F leads to best and average fitness values to rapidly become very close.

Finally, for any attribute in database we compute the average distances d_{aver} in the features space among the 13 centroids, and the related standard deviation σ_d, as they are provided by the best solution found in the 20 runs. Table 3 reports the values of the average distances only for the attributes with $d_{aver} > (\delta + \sigma_\delta)$, where δ is the average of the d_{aver}s computed for the 66 attributes ($\delta = 0.1906$) and σ_δ is its standard deviation ($\sigma_\delta = 0.1442$). Besides, those attributes with $d_{aver} \geq 0.40$, approximately twice the value of δ, are shown in bold.

The idea is that, among the 66 attributes, the ones for which the d_{aver}s are higher are the most helpful in dividing classes and, thus, in recognizing hand gestures. Indeed, by observing the table we can see that the angles formed by the hands in the first frames during motion are very meaningful in recognizing the signs. Further, Figure 2 outlines that the frames most significant for distinguishing the gestures are those following the starting position (frames 2,3 and

Table 3. Achieved results in terms of d_{aver} and σ_d

attribute	d_{aver}	σ_d	*attribute*	d_{aver}	σ_d	*attribute*	d_{aver}	σ_d	*attribute*	d_{aver}	σ_d
$roll_1$	0.38	0.25	$roll_2$	**0.42**	**0.31**	y_3	0.36	0.24	yaw_3	**0.42**	**0.29**
yaw_1	0.33	0.31	$pitch_2$	**0.44**	**0.29**	z_3	**0.46**	**0.30**	z_4	0.34	0.25
y_2	0.34	0.22	yaw_2	0.36	0.24	$roll_3$	**0.50**	**0.38**	$roll_4$	0.38	0.30
z_2	0.32	0.24	x_3	0.35	0.27	$pitch_3$	**0.49**	**0.34**	$pitch_4$	**0.43**	**0.29**
									yaw_4	0.34	0.23

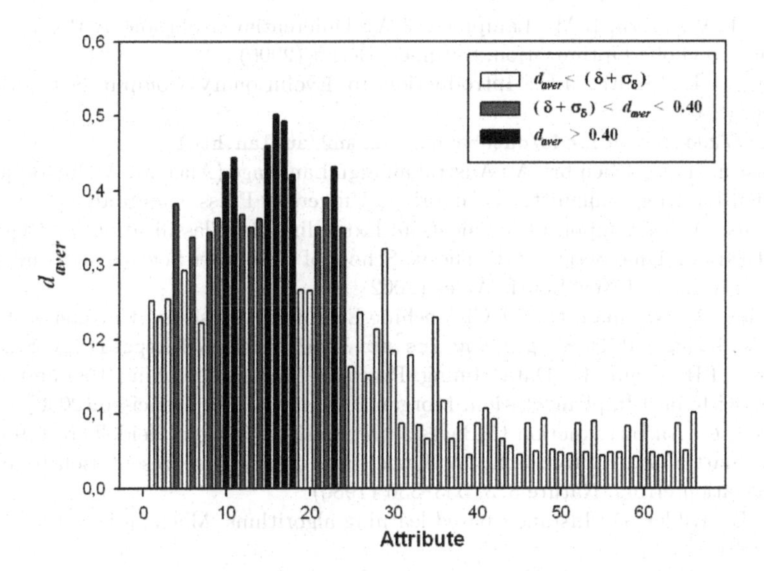

Fig. 2. Average distances d_{aver} as a function of attributes

4), and though the starting position (frame 1) is less important than its next frames, it is much more helpful than frames from 5 to 11 whose importance quickly decreases.

5 Conclusions

This paper has considered the issue of automatic recognition of hand gestures, an important topic in human–computer interaction.

By using a reduced version of Auslan, a tool has been designed and implemented in which Differential Evolution is used to find the positions of the class centroids in the search space, such that for any class the average distance of the instances belonging to that class from the relative class centroid is minimized. The classification performance is computed in terms of the error percentage on the testing set for the resulting best individual.

The experimental results prove that the tool is successful in tackling the task and is very competitive in terms of error percentage on the testing set when compared with other ten classification tools widely used in literature. In fact, only MLP shows slightly better performance. Furthermore, the implemented DE–C tool suggests that the parameters related to the angles formed by the hands in the first frames during motion are crucial in recognizing the gestures.

Future works will aim to further shed light on the efficiency of our evolutionary system in this field, and on its limitations as well.

References

1. Price, K.V., Storn, R.M., Lampinen, J.A.: Differential Evolution: A Pratical Approach to Globe Optimization. Springer, Berlin (2006)
2. Eiben, A.E., Smith, J.E.: Introduction to Evolutionary Computing. Springer, Berlin (2003)
3. `http://kdd.ics.uci.edu/databases/auslan2/auslan.html`
4. Johnston, T.A., Schembri, A.: Australian Sign Language (Auslan): An Introduction to Sign Language Linguistics. Cambridge University Press, Cambridge (2007)
5. Kadous, M.W.: Temporal Classification: Extending the Classification Paradigm to Multivariate Time Series, PhD Thesis, School of Computer Science and Engineering, University of New South Wales (2002)
6. Asucion, A., Newman, D.J.: UCI Machine Learning Repository, University of California, Irving (2007), `http://www.ics.uci.edu/~mlearn/MLRepository.html`
7. Witten, I.H., Frank, E.: Data Mining: Practical Machine Learning Tool and Technique with Java Implementation. Morgan Kaufmann, San Francisco (2000)
8. Jensen, F.: An Introduction to Bayesian Networks. Springer, Heidelberg (1996)
9. Rumelhart, D.E., Hinton, G.E., Williams, R.J.: Learning representation by back–propagation errors. Nature 323, 533–536 (1986)
10. Aha, D., Kibler, D.: Instance–based learning algorithms. Machine Learning 6, 37–66 (1991)
11. Cleary, J.G., Trigg, L.E.: K*: an instance–based learner using an entropic distance measure. In: Proceedings of the 12th International Conference on Machine Learning, pp. 108–114 (1995)
12. Breiman, L.: Bagging predictors. Machine Learning 24(2), 123–140 (1996)
13. Quinlan, R.: C4.5: Programs for Machine Learning. Morgan Kaufmann Publishers, San Mateo (1993)
14. Kohavi, R.: Scaling up the accuracy of naive-bayes classifiers: a decision tree hybrid. In: Proceedings of the Second International Conference on Knowledge Discovery and Data Mining, pp. 202–207. AAAI Press, Menlo Park (1996)
15. Frank, E., Witten, I.H.: Generating accurate rule sets without global optimization. In: Machine Learning: Proceedings of the Fifteenth International Conference, pp. 144–151. Morgan Kaufmann Publishers, San Francisco (1998)
16. Compton, P., Jansen, R.: Knowledge in context: a strategy for expert system maintenance. In: Proceedings of AI 1988, pp. 292–306. Springer, Berlin (1988)
17. Demiroz, G., Guvenir, H.A.: Classification by voting feature intervals. In: Proceedings of the 9th European Conference on Machine Learning, pp. 85–92 (1997)

Optimizing Computed Tomographic Angiography Image Segmentation Using Fitness Based Partitioning

Jeroen Eggermont[1], Rui Li[2], Ernst G.P. Bovenkamp[1], Henk Marquering[1], Michael T.M. Emmerich[2], Aad van der Lugt[3], Thomas Bäck[2], Jouke Dijkstra[1], and Johan H.C. Reiber[1]

[1] Div. of Image Processing, Dept. of Radiology C2S, Leiden University Medical Center, Leiden, The Netherlands
{J.Eggermont,E.G.P.Bovenkamp,H.A.Marquering, J.Dijkstra,J.H.C.Reiber}@lumc.nl
[2] Natural Computing Group, Leiden University, Leiden, The Netherlands
{ruili,emmerich, baeck}@liacs.nl
[3] Department of Radiology, Erasmus Medical Center, Rotterdam, The Netherlands
{A.vanderLugt}@erasmusmc.nl

Abstract. Computed Tomographic Angiography (CTA) has become a popular image modality for the evaluation of arteries and the detection of narrowings. For an objective and reproducible assessment of objects in CTA images, automated segmentation is very important. However, because of the complexity of CTA images it is not possible to find a single parameter setting that results in an optimal segmentation for each possible image of each possible patient. Therefore, we want to find optimal parameter settings for different CTA images. In this paper we investigate the use of Fitness Based Partitioning to find groups of images that require a similar parameter setting for the segmentation algorithm while at the same time evolving optimal parameter settings for these groups. The results show that Fitness Based Partitioning results in better image segmentation than the original default parameter solutions or a single parameter solution evolved for all images.

1 Introduction

Medical images often represent complex and variable structures that are not easily modeled. Moreover, they can suffer from a range of imperfections due to the applied acquisition modalities. This makes manual segmentation highly sensitive to intra- and inter-observer variability and very time consuming. Today's methods, directed at the automated recognition of certain structures in images, are applicable only over a limited range of standard situations and in some cases reach suboptimal results only.

M. Giacobini et al. (Eds.): EvoWorkshops 2008, LNCS 4974, pp. 275–284, 2008.

In [1,2] we compare Mixed-Integer Evolution Strategies (MIES) and standard Evolution Strategies (ES) for finding optimal parameter settings for the segmentation of Intravascular Ultrasound images. The results show that the parameter solutions evolved by the MIES and ES algorithms are better than the original parameter settings. However, the results also indicate that different sets of images require different parameter settings for an optimal image segmentation.

The ideal solution would be to cluster images according to their image segmentation context and optimize parameters for each individual context separately. Unfortunately the number of image segmentation contexts is not known a priori nor their characteristics. There is usually also no natural distance measure [3] to cluster images into groups that need similar parameter settings for an optimal segmentation result. Only their degree of belonging to a group, characterized by a particular set of parameters, can be measured by means of a training error for that image, after the parameters have been optimized for that group.

A possible approach for this kind of multi-level optimization problem could be cooperative coevolution (e.g., see [4,5]) in which we evolve both a set of parameter solutions and sets of images at the same time. However, this approach requires a large number of fitness evaluations which is very computationally (and thus time) intensive, since we have to do a lot of image processing, and this is therefore not attractive for our problem.

Another approach to solve this problem is Fitness Based Partitioning which we proposed and tested on artificial test problems in [6]. There the goal was to find multi-dimensional clusters by evolving combinations of uniform and normal distributions based on given data points in a multi-dimensional space. Here we apply Fitness Based Partitioning to the real-world problem of Computed Tomographic Angiography (CTA) lumen segmentation.

This paper is structured as follows: Computed Tomographic Angiography lumen detection is introduced in Section 2. Fitness Based Partitioning is discussed in Section 3. Experiments and results are presented in Section 4 followed by the conclusions and outlook in Section 5.

2 Computed Tomographic Angiography Lumen Detection

Since the introduction and increasing propagation of modern multi-slice computed tomography scanners, computed tomographic angiography has become a popular diagnostic modality in the visualization and evaluation of arteries and the detection of narrowings (stenoses). Computed Tomography is an imaging technique which results in a 3D image of the internals of an object using a series of 2D X-ray images.

We have developed a system for the quantitative analysis of Computed Tomographic Angiography images(CTA) [7] which consists of 5 steps. In the first step the vessels are segmented in the 3D image, followed in step 2 by the extraction of the vessel centerline. The third step is to construct a curved multiplanar reformatted (CMPR) image using the detected centerline (see Figure 1). The resulting 3D image stack contains 2D images perpendicular to the centerline,

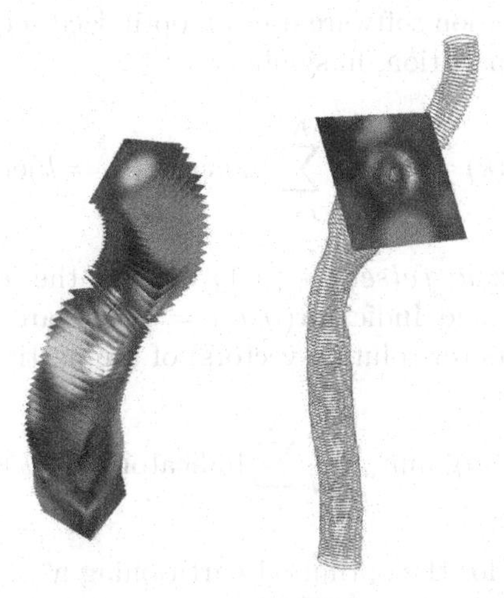

Fig. 1. A stack of CMPR images on the left with the centerline going through the center of each image and the corresponding lumen contours with a single CMPR slice on the right

and allows for the visualization of the the entire length of the vessel in a single 3D image. The fourth step is the segmentation of the lumen boundary (the part of the vessel where the blood flows) using a combination of longitudinal and transversal contour detection. It is this step that we will optimize using Mixed-Integer Evolution Strategies (MIES) and Fitness Based Partitioning. The fifth step is the quantification of the vessel morphological parameters.

3 Fitness Based Partitioning

The Fitness Based Partitioning algorithm was first tested on a set of artificial test problems in [6]. In general, our optimization problem is to find a partitioning of problem instances and for each of these partitions an optimal solution such that the optimal solution for each partition is also the optimal solution for each problem instance within that partition.

Let $\mathcal{I} = \{I_1, \ldots, I_N\}$ denote a set of images (or training instances), $\mathbf{a} \in A = \{1, \ldots, K\}^N$ an assignment of the images to one of K partitions, and $\mathbb{S}$ denote a set of control parameters for the segmentation algorithm. Then the optimization problem of finding an optimal partitioning is stated as follows:

$$\mathbf{a}^* = \arg\min_{\mathbf{a} \in A} \sum_{k=1}^{K} \mathrm{MME}_{\mathbf{a}}(k) \tag{1}$$

Here $\mathrm{MME}_{\mathbf{a}}(k)$ stands for 'minimized mean error' and denotes the average error on instances of a partition k over all training instances in that partition, provided the segmentation software uses an optimized set of control parameters for solutions on that partition, in symbols:

$$\mathrm{MME}_{\mathbf{a}}(k) = \min_{\mathbf{s} \in \mathbb{S}} \frac{1}{N} \sum_{j=1}^{N} \mathrm{Indicator}(a_j = k) \mathrm{error}_{\mathbf{s}}(I_j) \tag{2}$$

Here Indicator : $\{true, false\} \rightarrow \{0, 1\}$ denotes the indicator function with $\mathrm{Indicator}(false) = 0$ and $\mathrm{Indicator}(true) = 1$. We are also interested in the optimal parameter sets (or solution vectors) of the partitions $k = 1, \ldots, K$, i.e.

$$\mathbf{s}^*(\mathbf{a}, k) = \arg\min_{\mathbf{s} \in \mathbb{S}} \frac{1}{N} \sum_{j=1}^{N} \mathrm{Indicator}(a_j = k) \mathrm{error}_{\mathbf{s}}(I_j), \tag{3}$$

in particular in those for the optimized partitioning $\mathbf{a}^*$.

More concretely, in the case of CTA lumen segmentation we want to automatically find groups of CTA images while at the same time evolving a set of optimal parameters for detecting the lumen in the images in each of these groups.

In order to solve this multi-level optimization problem we use Fitness Based Partitioning. The top level goal is to optimize the (re-)assignment of problem instances, in our case CTA images, to partitions so that the optimal solution for each partition is also the optimal solution for each particular problem instance in that partition.

The second level optimization task is to find an optimal solution for all problem instances within a partition. For this we use Mixed-Integer Evolution Strategies (MIES), introduced in [8]. Mixed-Integer Evolution Strategies are a special type of evolution strategy that can handle mixed-integer parameters (continuous, ordinal discrete, and nominal discrete) by combining mutation operators of Evolution Strategies in the continuous domain [9], for integer programming [10], and for binary search spaces [11].

3.1 Algorithm

The detailed procedure for this 2-level optimization method is described in Algorithm 1. During the initialization phase all the problem instances (e.g., images) are distributed over the K partitions. Next a MIES algorithm MIES_k is assigned to each partition P_k.

The main loop of the Fitness Based Partitioning algorithm consists of four steps. The first step is to run each MIES_k algorithm on the problem instances in its corresponding partition P_k for G iterations. This step performs the second level optimization task.

The second step is to select the best evolved parameter solution $\mathbf{s}_k$ evolved by each MIES_k algorithm and to test it on all problem instances.

Step 3 is then to reassign all problem instances so that each problem instance I is assigned to the partition whose corresponding MIES algorithm offers the best parameter solution. This step performs the top level optimization task.

After all the problem instances have been reassigned to their "new" partitions the fourth step is to check for "empty" partitions (partitions with no problem instances). Empty partitions are not useful, since their corresponding MIES algorithms cannot optimize anything. The solution we have chosen is move half the problem instances of the largest partition to the empty partition. Additionally, we replace the population of the MIES algorithm associated with the empty partition with a copy of the population of the MIES algorithm associated with the largest partition. This effectively removes a non-useful empty partition and splits a large partition into two.

Algorithm 1. Fitness Based Partitioning

/* Initialization */
Divide the set of problem instances $\mathcal{I}$ randomly over the partitions.
Initialize the populations of the K MIES algorithms.
for T main loop iterations **do**
  /* step 1 */
  for each partition P_k **do**
    run MIES$_k$ on P_k for G iterations.
  end for
  /* step 2 */
  for each MIES$_k$ **do**
    select best individual/solution $\mathbf{s}_k$
    apply best individual/solution $\mathbf{s}_k$ to all problem instances in $\mathcal{I}$
  end for
  /* step 3 */
  for each problem instance $I \in \mathcal{I}$ **do**
    redistribute I to the partition P_k for which $\mathbf{s}_k$ offered the best solution.
  end for
  /* step 4 */
  while the smallest partition P_S is empty **do**
    copy the population of MIES$_L$ of the largest partition P_L to MIES$_S$
    divide the problem instances of P_L over P_L and P_S.
  end while
end for

4 Experiments and Results

The Fitness Based Partitioning approach as described above is tested on 9 CMPR image stacks of carotid arteries. Each CTA image stack consists of 59 to 82 images and each image consists of 32×32 pixels (16 bit signed grayscale with a spacing of 0.5mm).

To test the effect of the number of partitions, we experimented with up to 6 partitions. In case 1 partition is used the algorithm behaves like a normal single MIES algorithm since there is no need to redistribute the images to other partitions. For each data set and number of partitions we run the Fitness Based

Partitioning algorithm 10 times using different random seeds to initialize the MIES algorithms. To initialize the K partitions with images we simply divide a data set sequentially into K (almost) equally sized parts. We also experimented with other initialization techniques (e.g., random), but these gave slightly worse results. This is probably caused by the fact that two consecutive images in a stack correspond to two consecutive pieces of artery and therefore in general will require a similar parameter solution.

The MIES algorithms used for evolving an optimal parameter solution for a partition of images are programmed using the *Evolving Objects* library (EOlib) [12]. EOlib is an Open Source C++ library for evolutionary computation and is available from `http://eodev.sourceforge.net`. For the MIES algorithms in step 1 of Algorithm 1 we use a so-called plus-strategy $(\mu + \lambda)$ with $\mu = 4$ parents and $\lambda = 28$ offspring individuals. All variables have their own stepsize or mutation probability parameter which undergo self-adaptation as described in [8]. The parameters for the CTA lumen segmentation consists of 13 integer and 2 nominal discrete (Boolean) parameters.

4.1 Evaluation

In order to evaluate the fitness of a parameter solution evolved by a MIES algorithm, the lumen contour resulting from a particular parameter setting is compared to the expert contour drawn by a physician. The fitness function computes the average error $F_k(I)$ for each image I in partition P_k as:

$$F_k(I) = \sum_{p=1}^{|\text{points}|} \frac{d(C_p, E_p)}{|points|}, \tag{4}$$

where $d(C_p, E_p)$ is the Euclidean distance between the p-th point of the "evolved" contour C and the expert drawn contour E. Note that F_k corresponds to the function error in the general problem definition given in Eq. 1 to 3. Both contours have the same number of points since we resample all contours from the center of the image every 2 degrees resulting in 180 points for each contour.

The fitness of an individual parameter solution is then computed as the average minimized error of all images I in partition P_k:

$$\text{fitness} = \sum_{I \in P_k} \frac{F_k(I)}{|P_k|} \tag{5}$$

To determine the overall fitness result of our Fitness Based Partitioning algorithm we compute the average fitness of all images $I \in \mathcal{I}$ as:

$$\text{overall fitness} = \sum_{k=1}^{K} \sum_{I \in P_k} \frac{F_k(I)}{|\mathcal{I}|} \tag{6}$$

4.2 Results

The results in Tables 1 and 2 show that generally the overall average fitness improves as the number partitions gets larger.

If we look at the results on the individual datasets we find that for dataset 2 all increments in the number of partitions are improvements which are statistically significant according to the independent samples T-test with p = 0.05. For data sets 4 and 5 this is true for the difference between 3 and 4 partitions and for data sets 7 and 8 this is true for the transition from 4 to 5 partitions. For data set 6 we find that the best result is obtained with 5 rather than with 6 partitions, but this difference is not statistically significant.

Table 1. The average (plus standard deviation), minimum and maximum overall fitness values for 1 to 3 partitions. Lower values correspond to better contours.

Data Set	Number of Partitions											
	1				2				3			
	avg	s.d.	min	max	avg	s.d.	min	max	avg	s.d.	min	max
1	0.242	0.004	0.236	0.248	0.203	0.011	0.183	0.227	0.196	0.010	0.183	0.214
2	0.152	0.003	0.148	0.159	0.135	0.005	0.128	0.146	0.124	0.003	0.119	0.129
3	0.176	0.004	0.169	0.182	0.165	0.006	0.156	0.173	0.157	0.005	0.150	0.168
4	0.186	0.006	0.175	0.193	0.169	0.005	0.162	0.182	0.162	0.004	0.156	0.169
5	0.327	0.016	0.297	0.364	0.272	0.009	0.259	0.293	0.261	0.009	0.240	0.275
6	0.320	0.026	0.275	0.343	0.257	0.041	0.195	0.321	0.209	0.032	0.185	0.298
7	0.307	0.011	0.276	0.313	0.253	0.014	0.232	0.278	0.232	0.017	0.204	0.254
8	0.169	0.001	0.167	0.170	0.155	0.009	0.143	0.167	0.150	0.007	0.137	0.160
9	0.199	0.011	0.185	0.219	0.186	0.027	0.153	0.227	0.153	0.013	0.141	0.181

Table 2. The average (plus standard deviation), minimum and maximum overall fitness values for 4 to 6 partitions. Lower values correspond to better contours.

Data Set	Number of Partitions											
	4				5				6			
	avg	s.d.	min	max	avg	s.d.	min	max	avg	s.d.	min	max
1	0.184	0.007	0.177	0.203	0.175	0.006	0.166	0.184	0.171	0.005	0.159	0.178
2	0.118	0.003	0.113	0.124	0.113	0.002	0.110	0.116	0.108	0.003	0.104	0.113
3	0.151	0.004	0.144	0.160	0.149	0.005	0.141	0.156	0.144	0.004	0.140	0.152
4	0.153	0.004	0.149	0.163	0.152	0.005	0.144	0.161	0.151	0.005	0.144	0.159
5	0.233	0.009	0.215	0.249	0.229	0.012	0.213	0.253	0.223	0.012	0.209	0.249
6	0.207	0.025	0.185	0.276	0.188	0.008	0.180	0.199	0.192	0.011	0.178	0.213
7	0.228	0.010	0.215	0.250	0.210	0.010	0.194	0.234	0.197	0.010	0.184	0.218
8	0.143	0.005	0.137	0.156	0.136	0.003	0.132	0.142	0.135	0.003	0.129	0.138
9	0.146	0.010	0.134	0.170	0.140	0.006	0.135	0.156	0.135	0.006	0.130	0.150

When we examine the differences between the fitness values on all data sets for 1 and 2 partitions, we find that these are all statistically significant except for dataset 9. This indicates that our problem requires at least 2 partitions.

When we look at the final image partitioning after the algorithm has ended we see that different random seeds (and thus MIES population initializations) do not always lead to exactly the same partitions. However, analysis of the results shows that there are well defined groups of images which repeatedly end up in the same partitions. There are several reasons why we do not see all images end up in similar partitions every single run. The main reason seems to be that the partitioning process does not always stabilize for some random seeds. Other possible reasons for finding different image partitionings are that there are

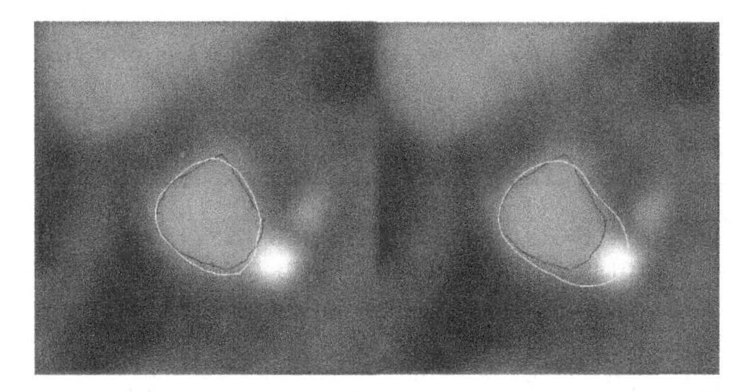

Fig. 2. Carotid lumen contours segmented using two different parameters settings. The light contour in the left image was found using parameter settings evolved for the partition to which this image was assigned. The light contour on the right was found using the parameter settings evolved for the other partition. The dark contour in both images indicates the expert drawn contour.

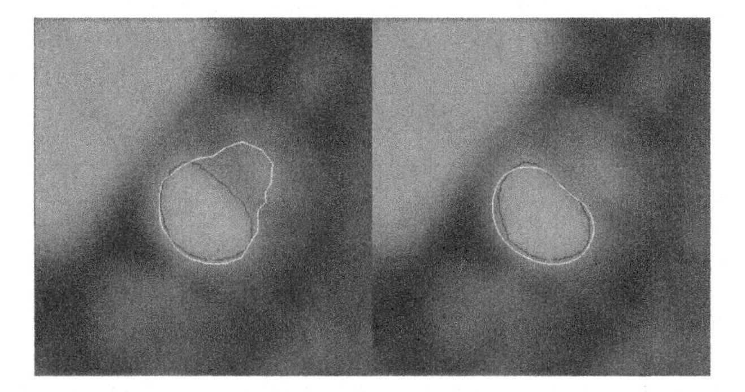

Fig. 3. Carotid lumen contours segmented using two different parameters settings. The light contour in the right image was found using parameter settings evolved for the partition to which this image was assigned. The light contour on the left was found using the parameter settings evolved for the other partition. The dark contour in both images indicates the expert drawn contour.

more image segmentation contexts than partitions or maybe there are no real distinct groups of images with respect to the image segmentation parameters. Naturally, the number of image segmentation contexts also depends on our image segmentation algorithm and how robust or sensitive it is.

In Figures 2 and 3 results are shown for 2 images from data set 2 after fitness based partitioning using 2 partitions. The light contours in the left images are found using parameter settings evolved for a partition including the image in Figure 2. The light contours in the right images are found using parameter settings evolved for the other partition, which included the image in Figure 3. As can be see both parameter settings result in contours similar to the dark expert drawn contours for the image for which they were optimized, but fail to produce satisfactory contours in the other images.

5 Conclusions and Outlook

In this paper we investigate the use of Fitness Based Partitioning in order to find sets of optimal parameters for the segmentation of the lumen in Computed Tomographic Angiography images. The purpose of Fitness Based Partitioning is to group images into partitions which require similar parameters settings while at the same time evolving optimal parameter settings for each group. Grouping images into different partitions is done, because one optimal parameter setting for each and every image is not to be expected.

The results in Tables 1 and 2 show that Fitness Based Partitioning does indeed produce sets of parameter settings which lead to better lumen segmentations when compared to one global optimal solution for all images.

Analysis of the final image partitioning results, obtained by running the algorithm with different random seeds, shows that groups of images (but not all) usually end up on the same island. However, there remains some sensitivity to the random seed used.

In the future we want to reduce this sensitivity by using larger populations which cover the search space more completely. This does, however, have a negative impact on the computation time. Another option is to make the image re-assignment method more flexible and less "greedy". We intend to extend the Fitness Based Partitioning algorithm with merge and split heuristics to automatically find an optimal number of partitions.

Once the partitions found by the Fitness Based Partition algorithm become more stable we are interested in extracting common features from these images that can act as a kind of image fingerprint, so we can automatically determine which parameter solution to use for a new image.

Acknowledgments

This research is supported by the Netherlands Organization for Scientific Research (NWO) and the Technology Foundation STW.

References

1. Bovenkamp, E., Eggermont, J., Li, R., Emmerich, M., Bäck, T., Dijkstra, J., Reiber, J.: Optimizing IVUS Lumen Segmentations using Evolutionary Algorithms. In: The 1st International Workshop on Computer Vision for IntraVascular and IntraCardiac Imaging, October 1–5, pp. 74–81 (2006)
2. Li, R., Emmerich, M., Eggermont, J., Bovenkamp, E., Bäck, T., Dijkstra, J., Reiber, J.: Mixed-Integer Optimization of Coronary Vessel Image Analysis using Evolution Strategies. In: Keijzer, M., et al. (eds.) Proceedings of the 8th annual conference on Genetic and Evolutionary Computation (GECCO), Seattle, WA, USA, pp. 1645–1652 (2006)
3. Jain, A.K., Murty, M.N., Flynn, P.J.: Data clustering: a review. ACM Comput. Surv. 31(3), 264–323 (1999)
4. Vanneschi, L., Mauri, G., Valsecchi, A., Cagnoni, S.: Heterogeneous cooperative coevolution: strategies of integration between gp and ga. In: Proceedings of Genetic and Evolutionary Computation Conference, GECCO, Seattle, Washington, USA, July 8-12, pp. 361–368 (2006)
5. Roberts, M., Claridge, E.: Cooperative coevolution of image feature construction and object detection. In: Yao, X., Burke, E.K., Lozano, J.A., Smith, J., Merelo-Guervós, J.J., Bullinaria, J.A., Rowe, J.E., Tiňo, P., Kabán, A., Schwefel, H.-P. (eds.) PPSN 2004. LNCS, vol. 3242, pp. 902–911. Springer, Heidelberg (2004)
6. Li, R., Eggermont, J., Emmerich, M., Bovenkamp, E., Bäck, T., Dijkstra, J., Reiber, J.: Towards Dynamic Fitness Based Partitioning for IntraVascular UltraSound Image Analysis. In: Giacobini, M. (ed.) EvoWorkshops 2007. LNCS, vol. 4448, pp. 388–395. Springer, Heidelberg (2007)
7. Marquering, H., Dijkstra, J., de Koning, P., Stoel, B., Reiber, J.: Towards quantitative analysis of coronary cta. Int. J. Cardiovasc Imaging 21(1), 73–84 (2005)
8. Emmerich, M., Grötzner, M., Groß, B., Schütz, M.: Mixed-integer evolution strategy for chemical plant optimization with simulators. In: Parmee, I. (ed.) Evolutionary Design and Manufacture - Selected papers from ACDM, pp. 55–67. Springer, London (2000)
9. Schwefel, H.P.: Evolution and Optimum Seeking. Sixth-Generation Computer Technology Series. Wiley, New York (1995)
10. Rudolph, G.: An evolutionary algorithm for integer programming. In: Davidor, Y., Männer, R., Schwefel, H.-P. (eds.) PPSN 1994. LNCS, vol. 866, pp. 139–148. Springer, Heidelberg (1994)
11. Bäck, T.: Evolutionary Algorithms in Theory and Practice. Oxford University Press, New York (1996)
12. Keijzer, M., Merelo, J.J., Romero, G., Schoenauer, M.: Evolving objects: a general purpose evolutionary computation library. In: EA 2001, Evolution Artificielle, 5th International Conference in Evolutionary Algorithms, pp. 231–244 (2001)

A GA-Based Feature Selection Algorithm for Remote Sensing Images

C. De Stefano, F. Fontanella, and C. Marrocco

Dipartimento di Automazione, Elettromagnetismo, Ingegneria dell'Informazione e Matematica Industriale (DAEIMI)
Università di Cassino
Via G. Di Biasio, 43 02043 Cassino (FR) – Italy
{destefano,fontanella,cristina.marrocco}@unicas.it

Abstract. We present a GA–based feature selection algorithm in which feature subsets are evaluated by means of a separability index. This index is based on a filter method, which allows to estimate statistical properties of the data, independently of the classifier used. More specifically, the defined index uses covariance matrices for evaluating how spread out the probability distributions of data are in a given n–dimensional space. The effectiveness of the approach has been tested on two satellite images and the results have been compared with those obtained without feature selection and with those obtained by using a previously developed GA–based feature selection algorithm.

1 Introduction

In the last years, the interest about the feature selection problem has been increasing. In fact, new applications dealing with huge amounts of data have been developed, such as data mining [1] and medical data processing [2]. In this kind of applications, a large number of features is usually available and the selection of an effective subset of them, i.e. a subset that allows to maximize the performance of the subsequent clustering or classification process, represents a very important task. The feature selection problem plays also a key role when set of features belonging to different domains are used: this typically happens in applications such as remote sensing [3] or handwriting recognition [4].

In a classification task in which the objects to be classified are represented by means of a set of features, the feature selection problem consists in selecting, from the whole set of available features, the subset of them providing the most discriminative power. The choice of a good feature subset is a crucial stage in any classification process since:

– If the considered set features does not include all the information needed to discriminate patterns belonging to different classes, the achievable performances may be unsatisfactory regardless the effectiveness of the learning algorithm employed.

M. Giacobini et al. (Eds.): EvoWorkshops 2008, LNCS 4974, pp. 285–294, 2008.

- The features used to describe the patterns determine the search space to be explored during the learning phase. Then, irrelevant and noisy features make the search space larger, increasing both the time and the complexity of the learning process;
- The computational cost of classification depends on the number of features used to describe the patterns. Then, reducing such number results in a significant reduction of this cost.

When the cardinality N of the candidate feature set Y is high, the problem of finding the optimal feature subset, according to a given evaluation function, becomes computationally intractable because of the resulting exponential growth of the search space, made of all the 2^N possible subsets of Y. Many heuristic algorithms have been proposed for finding near–optimal solutions [5]. Among these algorithms, greedy strategies that incrementally generate feature subsets has been proposed. Since these algorithms do not take into account complex interactions among several features, in most of the case they lead to sub–optimal solutions.

GA's, which have demonstrated to be an effective search tools for finding near–optimal solutions in complex and non–linear search spaces [6] as is the case of feature selection problems, have been widely used to solve feature selection problems [7,8]. Moreover, comparative studies have demonstrated the superiority of GA's in feature selection problems involving large numbers of features [9].

We propose a GA–based feature selection approach in which each individual, whose genotype encode the selection of a feature subset, is evaluated by means of a separability index. This index is based on a filter method which takes into account statistical properties of the input data in the subspace represented by that individual and is independent of the used classification scheme. The proposed index uses covariance matrices, which are the generalization of the variance of a scalar variable to multiple dimensions. These matrices estimate how spread out the probability distributions are in the considered $n-$dimensional space. Given a feature subset X, the proposed index estimates the separability of the classes in X by evaluating two aspects: (i) how patterns belonging to a given class are spread out around the corresponding class mean vector (the centroid); (ii) distances among class mean vectors.

The effectiveness of the proposed approach has been tested on two datasets extracted from satellite images. In both datasets image pixels are represented by feature vectors in high–dimensional spaces. These vectors describe textural measurements based on grey–level co–occurrence matrices (GLCM) [10]. The effectiveness of the selected features has been tested on a neural network classifier. The obtained results have been compared with those obtained without feature selection and with those obtained by using a previously presented approach [11].

The remainder of the paper is organized as follows: in Section 2 the problem of feature selection is described, Section 3 illustrates the GA–based method implemented, while the separability index used for subset evaluations is presented in Section 4. In Section 5 the experimental results are detailed. Finally, Section 6 is devoted to conclusions.

2 The Feature Selection Problem

Feature Selection (FS) is one of the first stages in any classification process in which data are represented by feature vectors. Its role is to reduce the number of features to be considered later in the classification stage. This task is performed by removing irrelevant and noisy features from the set of the available features. Feature selection is accomplished by reducing as much as possible the information loss due to the feature set reduction. Moreover, this selection process should not reduce classification performances.

In a classification process involving a dataset $\mathcal{D}$, in which patterns are represented by means of a set $Y = \{1, 2, \ldots, N\}$ of N features, the feature selection problem can be formulated as follows: find the subset $S \subseteq Y$ of n features which optimizes the function J. Given a generic subset $X \subseteq Y$, $J(X)$ measures how well the patterns in D, belonging to different classes, are discriminated by using the n features in X. The methods implemented by the function J can be divided in two wide class:

- *filter methods* which evaluate a feature subset independently of the classifier but are usually based on some statistical measures of distance between the patterns belonging to different classes.
- *wrapper methods* which are based on the classification results achieved by a certain classifier.

Filter methods are usually faster than wrapper ones, as the latter requires the training of the classifier used for each evaluation and this process may make this approach unfeasible when a large number of features is involved. Moreover, while filter–based evaluations are more general, as they give statistical information on the data, wrapper–based evaluations may give raise loss of generality because they depend on the specific classifier used.

Once the evaluation function $J(X)$ has been chosen, the feature selection problem becomes an optimization problem whose search space is the set of all the subsets of Y. The size of this search space is exponential (2^N). As a consequence, the exhaustive search for the optimal solution is unfeasible for those problems involving a large number of features ($N > 50$). Search strategies like branch and bound [12] have been proposed to strongly reduce the amount of evaluations, but the exponential complexity of the problem still remains. The exponential size of the search space for the feature selection problem makes appropriate the use of heuristic algorithms, for finding near–optimal solutions. Among these search algorithms, greedy search strategies are computationally advantageous but may lead to suboptimal solutions. They come in two flavours: forward selection and backward elimination. Forward selection strategies generate near–optimal feature subsets by a stepwise procedure which starts with an empty set and adds to the so far built subset the feature, among those not yet selected, that more increases the evaluation function J, this procedure is repeated until a stop criterion is not satisfied. In backward elimination, instead, the whole subset of feature is initially considered, and at each step the feature that least reduce the evaluation function is eliminated. Both procedures are optimal at each step, but they cannot discover complex interactions among several features, as is the case in most of the real world feature selection problems. Then heuristic search algorithms, like genetic algorithms and simulated annealing seems to be appropriate for finding near–optimal solutions which take into account multiple interactions among several features.

3 Genetic Algorithms for Feature Selection

The principles governing the phenomena of natural evolution have been widely studied by mathematicians and computer scientist since the end of the 50's of the last century. These studies have led to the development of a new computational paradigm named evolutionary computation [6], which during the last decades has shown to be very effective as methodology for solving optimization problems whose search space are discontinuous and very complex. In this field, genetic algorithms (GA's in the following) represent a subset of these optimization techniques, in which solutions are represented as binary strings. To these strings, operators such as selection, crossover and mutation are applied to a population of competing individuals (problem's solutions). GA's have been applied to a wide variety of both numerical and combinatorial optimization problems [7].

GA's can be easily applied to the problem of feature selection, as given a set Y having cardinality equal to N, a subset X of Y ($X \subseteq Y$) can be represented by a binary vector $\mathbf{b}$ having N elements whose i-th element is set to 1 if the i-th features is included in X, 0 otherwise. Besides the simplicity in the solution encoding, GA's are well suited for these class of problems as the search in this exponential space is very hard since interactions among features can be highly complex and strongly nonlinear. Some studies on the GA's effectiveness in solving features selection problems can be found in [7,8].

The system presented here has been implemented by using a generational GA, which starts randomly generating a population of P individuals. Each individual's chromosome is a binary vector encoding a feature subset that represents an allowed solution of the problem. The value of the i-th element is set to 1 according a given probability (called *one_prob*), which represents a parameter algorithm and is usually set to 0.1 at the aim to force the early stage of the search toward solutions having a small number of features. Afterwards, the fitness of the generated individuals is evaluated. This fitness takes into account two terms, the former measures the separability of the patterns belonging to the different classes in the problem at hand, in the feature subset encoded by the individual, while the latter terms measures the cardinality of the subset so as to favour solutions containing a smaller number of features. After this evaluation phase a new population is generated, by first copying the best e individuals of the initial population in order to implement an elitist strategy. Then $(P - e)/2$ couples of individuals are selected using the tournament method, to control loss of diversity and selection intensity. The one point crossover operator is applied to each of the selected couples, according to a given probability factor p_c. Afterwards, the mutation operator is applied. As regards this operator, two different probability factor p_0 and p_1 have been defined. These factors represent the mutation probability respectively of the 0's and 1's in the chromosome. These two different probability factors have been adopted since in a chromosome the 0's and 1's occurrences can be very different: typically in a chromosome 0's are much more than 1's. This fact is due to both the generation procedure of the individuals in the initial population and the fitness function, which favours the individuals with a smaller number of features. As a consequence, as in an individual the number of the 1's is much smaller than the that of 0's the value of p_1 is set much smaller than that of p_0. The purpose is to make, on average, the probability mutation of the 1's about equal to that of the 0's. Finally these individuals are added to the new population. The process just described is repeated for N_g generations.

4 The Separability Index

In any EC–based algorithm the design of a suitable fitness function for the problem to be solved is a crucial task. To be successful in feature selection problems, the fitness function must be able to effectively evaluate how well patterns belonging to different classes are discriminated in the subspace represented by the solution to be evaluated. The fitness function adopted is based on a well known class separability index J, which is usually adopted in Multiple Discriminant Analysis for finding a linear transformation that allows to reduce the dimension of the data.

According to so-called Fisher Criterion, the separability index J has been defined by using covariance matrices, which measure how spread out the probability distribution of the data is in the considered space. In a particular n–dimensional space, given a set of patterns belonging to different classes, the i–th class can be described by using its covariance matrix Σ_i, which is obtained only considering the patterns belonging to class i. Note that the covariance matrix Σ_i contains information about variability of data points belonging to the i–th class around their mean value μ_i.

The separability index J used for finding the most discriminative features, makes use of this dispersion concept. In particular, the classes information is condensed in two scatter matrices Σ_B and Σ_W:

$$\Sigma_W = \sum_i P(\omega_i)\Sigma_i$$

$$\Sigma_B = \sum_i P(\omega_i)(M_i - M_0)(M_i - M_0)^T$$

where $P(\omega_i)$ denotes the prior probability of the i–th class, Σ_i and M_i are respectively the covariance matrix and the mean vector of i–th class, and M_0 denotes the overall mean:

$$M_0 = \sum_i P(\omega_i)M_i \tag{1}$$

Note that Σ_W is a *within-class scatter matrix*, as it measures the spread of the classes about their means, while Σ_B is a *between-class scatter matrix*, since it measures distances between class mean vectors, i.e. centroids.

Given a feature subset X, the separability index $J(X)$, has been defined as the ratio between a *within-class scatter matrix* Σ_W and a *between-class scatter matrix* Σ_B:

$$J(X) = trace(\Sigma_W^{-1}\Sigma_B) \tag{2}$$

High values of the separability index $J(X)$ indicate that in the subspace represented by the feature subset X the class means are well separated and, at the same time, patterns appear to be not much spread out around their means values.

5 Experimental Results

The proposed approach has been tested on data represented by feature vectors in high dimensional spaces (> 200). These feature vectors describe textural measures based on the grey–level co–occurrence matrices (GLCM) [10], computed on patterns belonging to datasets extracted from Landsat Satellite images. Two datasets have been considered and for each of them 20 runs have been performed. The reported results are those obtained using the individual having the highest fitness among those obtained during the 20 performed runs. Some preliminary trials have been performed to set the basic evolutionary parameters reported in Table 1. This set of parameters has been used for all the experiments reported below.

Given an individual I, its fitness value F has been computed by applying the formula:

$$F(I) = J(I) + k\frac{N_{FT} - N_F(I)}{N_{FT}} \tag{3}$$

where $J(I)$ is the separability index, described in Section 4, computed on the subset of features represented by I, while N_{FT} is the maximum number of features available and $N_F(I)$ is the cardinality of the subset represented by I, i.e. number of bits equal to 1 in its chromosome. Finally, k is a constant value; this constant is used so as to weight the second term in the (3), which is in inverse proportion to the number of features in I. The role of this second term is essential in order to avoid an excessive increase of the number of features, as may result from the selection process because of the monotonic trend of the index. Thanks to this term individuals having a smaller number of features are favoured.

5.1 The Datasets

The first dataset (DS1 in the following) used for testing the proposed approach is the standard dataset Satimage included in the UCI dataset repository. This dataset was generated from 4–band Landsat Multi-Spectral Scanner image data. Each Landsat frame consists of four digital images of the same scene in different

Table 1. Values of the basic evolutionary parameters used in the experiments. Note that p_0 and p_1 depend on the chromosome length. For the experiments involving the first dataset (DS1) the 0's probability mutation has been set equal to 0.0047 (0.047 for the 1's), while for the second dataset (DS2) this probability has been set equal to 0.003 (0.03).

Parameter	**symbol**	**value**
Population size	$\mathcal{P}$	100
Tournament size	$\mathcal{T}$	6
elithism size	$\mathcal{E}$	5
Crossover probability	p_c	0.4
Mutation probability of 0's	p_0	$1/N_F$
Mutation probability of 1's	p_1	$10/N_F$
Number of Generations	N_g	1500

spectral bands. Two of these are in the visible region (corresponding approximately to green and red regions of the visible spectrum) and two are in the (near) infra-red. Each pixel is a 8-bit binary word, with 0 corresponding to black and 255 to white. The spatial resolution of a pixel is about 80m×80m. The patterns belong to 6 different classes, namely: red soil, cotton crop, grey soil, damp grey soil, soil with vegetation stubble and very damp grey soil.

DS1 contains 6435 patterns, organized in two sets of data: a training set (TR1 in the following) containing 4435 patterns and a test set (TS1 in the following) containing 2000 patterns. Each pattern corresponds to a 3×3 square neighbourhood of pixels and is described by considering the pixel values in the four spectral bands of each of the 9 pixels in that neighbourhood. To each pattern is assigned as label the class of the central pixel. Thus, each pattern of the dataset is represented by a feature vector of 36 integer values in the range [0,255].

For each original pattern, a new feature vector has been built by computing its Grey Level Co–occurrence Matrix (GLCM) [10] with a moving window equal to the 3×3 neighbourhood. For each of the four spectral bands, 4 GLCM in the directions $(0°, 45°, 90°, 135°)$ has been computed. As a consequence for each pattern 16 GLCM has been computed. Afterwards, for each GLCM 13 textural features have been computed. Finally, to each new feature vector the values of the 4 spectral bands and the spectral feature NDVI (Normalized Difference Vegetation Index) has been added. Then in our experiments each pattern in DS1 has been described by using a feature vectors of 213 elements (13 texture features $\times$ 4 directions $\times$ 4 spectral bands + 4 pixel values in the spectral bands + 1 NDVI). The constant k in 3 has been set to 0.2 for this dataset.

The second dataset (DS2 in the following) contains data relative to a satellite image of a residential area (city of Anzio, Italy) and was recorded by the ETM sensor on the Landsat 7 satellite, which has a ground spatial resolution of about 30m×30m and six spectral bands. Each pixel is a 8-bit binary word, with 0 corresponding to black and 255 to white. Each pattern corresponds to a 3×3 square neighbourhood of pixels and contains 54 attributes (6 spectral bands 9 pixels in the neighbourhood), resulting in a feature vector of 54 integer values

Table 2. Number of features found by GA1 and GA2

	Best ind.		**Average**		**std. dev.**	
	GA1	GA2	GA1	GA2	GA1	GA2
DS1	8	9	6.05	9.1	1.1	0.3
DS2	11	4	9.4	4.5	1.8	0.5

in the range [0,255]. The data were divided into a training set (say TR2) with 800 patterns and a test set (TS2) with 712 patterns, randomly extracted from the original Landsat 7 scene. In this scene five classes must be discriminated, namely: water, grey soil, wood, urban area, sea sand and bare soil. As for DS1, also patterns in DS2 has been described by using the GLCM textural features mentioned above. In this case, as six spectral bands are available, feature vectors of 319 elements (13 texture features $\times$ 4 directions $\times$ 6 spectral bands + 6 pixel values in the spectral bands + 1 NDVI) has been built. In this case the value of the constant k has been set to 4.0.

5.2 Comparison Findings

In order to asses the effectiveness of the implemented system, the selected features have been tested by using them to implement a simple and widely adopted neural network classifier: the Multi Layer Perceptron (MLP) trained with the Back Propagation algorithm.

The results of our GA–based method (GA2 in the following) have been compared with those obtained by another GA-based feature algorithm previously proposed in [11] (GA1 in the following), which used a different separability index for feature subset evaluations. That index was computed by using a training set Tr, containing C classes, of labelled patterns represented as feature vectors in the initial N-dimensional feature space. Given an individual I, representing a subset X, its separability index was computed as follows: first, for each class the corresponding centroid is computed in X by averaging the components of the feature vectors belonging to that class; Then, each pattern in Tr is labelled with the label of the nearest centroid (nearest neighbour rule) in X; Finally, the percentage of patterns correctly labelled is assumed as separability index of X.

Table 2 shows the number of features obtained by GA1 and GA2. As regards the results on DS1, GA1 and GA2 have found, on average, respectively 6.05 and 9.1 features. Then, on this dataset our method has found a higher number of features. However, the number of features of the best individuals (8 for GA1 and 9 for GA2) are comparable. Probably this means that though GA2 finds, on average, smaller feature subsets, it obtains good performances only when larger subsets are found. Moreover, in GA2 the number of features of the best individual is about equal to the average one. As regards the results on DS2, the average number of features found by GA1 and GA2 is equal respectively to 9.4 and 4.5. Then, in this case GA2 has been able to find smaller feature subsets than GA1. Moreover, also in this case the number of features of the GA2 best individual is lower than the average one, while for GA1 the best individual has a number

Table 3. Classification rates (expressed in percentages) of the MLP classifier on DS1 (left) and DS2 (right) on the best feature subsets found by GA1, GA2 and the 3×3–neighbourhood feature sets of the original datasets. The actual number of features used is reported in parenthesis. In the column N the number of hidden nodes of the MLP classifier is reported.

MLP results on DS1						
N	**3 × 3 neigh. (36)**		**GA1 (8)**		**GA2 (9)**	
	TR1	TS1	TR1	TS1	TR1	TS1
30	90.38	87.60	89.10	87.40	89.99	89.25
40	90.14	87.20	89.16	**88.00**	89.63	88.85
50	90.38	**87.80**	89.26	87.40	89.99	**89.50**

MLP results on DS2						
N	**3 × 3 neigh.(54)**		**GA1 (8)**		**GA2 (9)**	
	TR1	TS1	TR1	TS1	TR1	TS1
30	92.37	**74.80**	87.73	77.20	83.62	**78.65**
40	92.28	74.60	87.75	**77.60**	84.50	77.67
50	92.90	72.20	88.98	77.00	84.00	78.09

of features larger than the average. This fact seems to confirm the hypothesis stated above that GA2 obtains good performances only when a larger number of features is used. Finally, it is worth noting that GA2 standard deviations are lower than those of GA1. This fact indicates that the separability index used in GA2 is able, almost always, to find the smaller subsets providing, according to that index, the most discriminative power.

In Table 3 the recognition rates achieved by the MLP classifier on the best feature subsets found by GA1, GA2 and on the 3×3 neighbourhood feature set of the original datasets are reported. On both the databases analyzed, GA2 has achieved better results than those obtained by GA1. As regards DS1, GA2 has further improved the good rates obtained by GA1. Also for DS2 the MLP has obtained better results on the features selected by GA2 than on those selected by GA1. It is also worth noting that, for both datasets, GA2 has obtained better results than GA1 for each value of hidden nodes. This fact demonstrates that GA2, whatever the number of nodes in the hidden layer, always gives better results than GA1. Finally, on both datasets, the selected features has been able to significantly improve the performances obtained by the feature sets made of the spectral band values in the 3×3 neighbourhood of each pixel.

6 Conclusions

A new GA-based algorithm for feature selection in high dimensional feature spaces has been presented. The proposed approach uses a separability index for evaluating feature subsets. This index is based on a filter method, which estimates statistical properties of the data and is independent from the classifier used. The index is able to effectively measure both the spreading of the patterns around their mean vectors and the distances of the mean vectors of the different classes.

The proposed approach has been tested on two datasets extracted from satellite images. From these data a wide set of GLCM textural features has been computed and from this set the near–optimal subsets has been found by using the GA–based method. The results have been compared with those obtained by another GA–based method. The comparison has been done by implementing a MLP, which has been trained and tested on the best subsets found by the two methods. The accuracies on the test sets have been compared. For both the datasets, the comparison has demonstrated that our method is able to found subsets that yields better accuracies than those obtained by using the subsets found by the other method.

References

1. Martin-Bautista, M., Vila, M.A.: A survey of genetic feature selection in mining issues. In: Proc. 1999 Congress on Evolutionary Computation (CEC 1999), pp. 1314–1321 (1999)
2. Puuronen, S., Tsymbal, A., Skrypnik, I.: Advanced local feature selection in medical diagnostics. In: Proc. 13th IEEE Symp. Computer-Based Medical Systems, pp. 25–30 (2000)
3. Hung, C., Fahsi, A., Tadesse, W., Coleman, T.: A comparative study of remotely sensed data classification using principal components analysis and divergence. In: Proc. IEEE Intl. Conf. Systems, Man, and Cybernetics, pp. 2444–2449 (1997)
4. Oh, I.S., J.L.,, Suen, C.: Analysis of class separation and combination of class-dependent features for handwriting recognition. IEEE Trans. Pattern Analysis and Machine Intelligence 21, 1089–1094 (1999)
5. Guyon, I., Elisseeff, A.: An introduction to variable and feature selection. J. Mach. Learn. Res. 3, 1157–1182 (2003)
6. Goldberg, D.E.: Genetic Algorithms in Search Optimization and Machine Learning. Addison-Wesley, Reading (1989)
7. Lee, J.S., Oh, I.S., Moon, B.R.: Hybrid genetic algorithms for feature selection. IEEE Trans. Pattern Anal. Mach. Intell. 26, 1424–1437 (2004)
8. Yang, J., Honavar, V.: Feature subset selection using a genetic algorithm. IEEE Intelligent Systems 13, 44–49 (1998)
9. Kudo, M., Sklansky, J.: Comparison of algorithms that select features for pattern recognition. Pattern Recognition 33, 25–41 (2000)
10. Haralick, R.M.: Textural features for image classification. IEEE Trans. System, Man and Cybernetics 3, 610–621 (1973)
11. De Stefano, C., Fontanella, F., Marrocco, C., Schirinzi, G.: A feature selection algorithm for class discrimination improvement. In: IEEE International Geoscience and Remote Sensing Symposium (2007)
12. Yu, B., Yuan, B.: A more efficient branch and bound algorithm for feature selection. Pattern Recognition 26, 883–889 (1993)

An Evolutionary Approach for Ontology Driven Image Interpretation

Germain Forestier, Sébastien Derivaux, Cédric Wemmert, and Pierre Gançarski

LSIIT - CNRS - University Louis Pasteur - UMR 7005
Pôle API, Bd Sébastien Brant - 67412 Illkirch, France
{forestier,derivaux,wemmert,gancarski}@lsiit.u-strasbg.fr

Abstract. Image mining and interpretation is a quite complex process. In this article, we propose to model expert knowledge on objects present in an image through an ontology. This ontology will be used to drive a segmentation process by an evolutionary approach. This method uses a genetic algorithm to find segmentation parameters which allow to identify in the image the objects described by the expert in the ontology. The fitness function of the genetic algorithm uses the ontology to evaluate the segmentation. This approach does not needs examples and enables to reduce the *semantic gap* between automatic interpretation of images and expert knowledge.

1 Introduction

Automatic interpretation of images becomes a more and more complex data mining process. For example, in the field of remote sensing, the rapid evolution in terms of spatial resolution (image size) and spectral resolution (number of bands) increases the complexity of available images. Automatic analysis methods are needed to avoid a manual processing which is often costly. The most promising and the most studied approach is the *object oriented* approach which consists in identifying objects composed of several connected pixels and having an interest for the domain expert, by using a segmentation algorithm.

There exists many algorithms of segmentation like the watershed transform [1] or region growing [2]. These algorithms need a complex parametrization like the selection of thresholds or weights which are usually meaningless for the user. Thus, a difficult task for the user is to find the link between his knowledge about the objects present in the image and the appropriate parameters for the segmentation allowing to create and identifying these objects.

The use of genetic algorithm [3] is a solution to find an optimal (at least near-optimal) parameters set. They can be used to optimize a set of parameters if an evaluation function of these parameters is available. The existing methods of segmentation optimization with genetic approaches [4,5,6,7] are based on evaluation function where examples of segmented objects provided by the expert are needed. If examples are not available, it is possible to use some unsupervised criteria [5,7], which evaluate the intrinsic quality of a segmentation (e.g. region homogeneity). Nevertheless these unsupervised criteria are often insufficient to

M. Giacobini et al. (Eds.): EvoWorkshops 2008, LNCS 4974, pp. 295–304, 2008.

produce a good quality segmentation, especially for the analysis of complex images.

In this article, we propose to use domain knowledge to evaluate the quality of a segmentation. Indeed, with the oriented object approach the expert is able to express his knowledge about objects of the image. It allows a natural and intuitive description of the objects potentially present in an image. Thus, an ontology (i.e. a knowledge base) can be used to define the different objects (i.e. concepts) and their characteristics. Then, the coherence of a segmentation can be evaluated thanks to the concepts defined in the ontology. This approach does not needs examples and uses the knowledge defined in the ontology.

The outline of this paper is the following. In Section 2, we introduce the used segmentation algorithm and a description of the needed parameters. In Section 3, we present the formalization of the knowledge through an ontology. In Section 4, we study the proposed evolutionary approach used to find the set of parameters for segmentation thanks to an evaluation using the ontology. Finally, we present experimentations on the interpretation of images for Earth observation.

2 Image Segmentation

The watershed segmentation is a well-known segmentation method which considers the image to be processed as a topographic surface. In the immersion paradigm from Vincent and Soille [1], this surface is flooded from its minima thus generating different growing catchment basins. Dams are built to avoid merging water from two different catchment basins. A example of a cut of an elevation image and its minima is presented in figure 1 (a).

The segmentation result is defined by the locations of the dams (i.e. the watershed lines). In this approach, an image gradient is most often taken as the topographic surface, since object edges (i.e. watershed lines) are very probably located at pixels with high gradient values (high heterogeneity areas). To build the gradient image, each pixel is replaced by the difference between the maximal value and the minimal value of a 3×3 windows centered on the pixel. The final elevation image is obtained by combining the elevation of the different spectral bands thanks to an Euclidean norm.

The watershed has the advantage to be a completely unsupervised method without parameters. Nevertheless, it produces generally an over-segmented result, which means an image where each object (e.g. a house) is represented by several regions (e.g. the two sides of its roof). To resolve this problem, many methods can be used independently or simultaneously.

When the gradient image is computed, a threshold of the gradient [8] can be made. Every pixel having an inferior value to the threshold is set to zero. Thus, the small variations within the homogeneous region are deleted. On figure 1 (b), the line *hmin* represents a threshold, and the value beside its are considers as null. Another method consists in using the depth of the basins [9]. Let m_r be the elevation of a local minimum of the basin r and d_r be the minimal elevation when it will be separated of an another basin by a watershed. Every local minimum

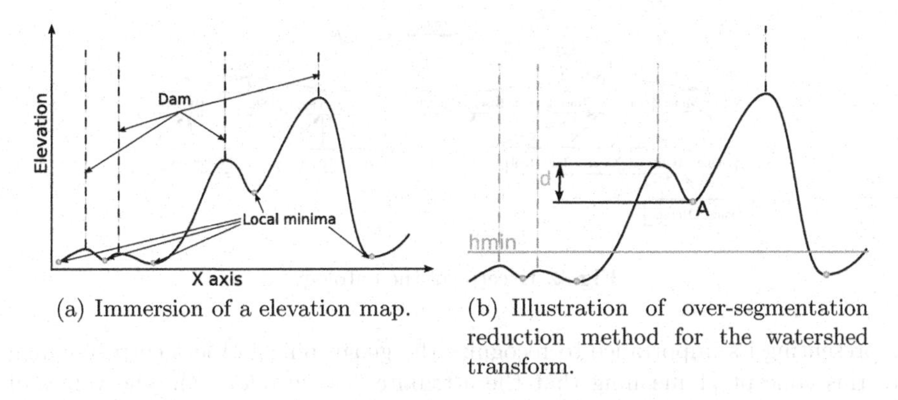

(a) Immersion of a elevation map.

(b) Illustration of over-segmentation reduction method for the watershed transform.

Fig. 1. Example of a watershed segmentation (a) and effect of over-segmentation reduction methods (b)

where $d_r - m_r < d$, with d a given threshold, will not be considered during the basin immersion step. On the figure 1 (b), the local minimum A will not be taken into account during the immersion because its dynamic, the difference between m_r and d_r, is too small. Finally, it is also possible to use a region merging technique [8]. Two regions can be separated by an heterogeneous area (implying a generation of a frontier by the watershed) but they can be spectrally similar (using the average value). To solve this problem, it is possible to use a filter which merge adjacent regions having an euclidean distance between their means lower than a threshold ft.

These different techniques can be used simultaneously to reduce over segmentation caused by the watershed and need the selection of three parameters (the level $hmin$, the threshold d and the fusion threshold ft). The optimal values of these parameters are difficult to find because the value for a given parameter depends heavily of the values selected for the other ones. Moreover, there exists a lot of local optima which increase the difficulty to find the best solution.

3 Geographical Objects Ontology

We propose here a model allowing the representation of geographic objects through an ontology and a matching process which allows to compare a region build during a segmentation and the different concepts defined in the ontology. The matching process has been fully described in [10], we remain here the principal functionalities.

3.1 Ontology Description

The proposed ontology is composed of a hierarchy of concepts (an extract is given on figure 2). In this hierarchy each node corresponds to a concept. Each concept has a label (e.g. *house*) and is defined by its attributes. Each attribute is associated to an interval of accepted values (e.g. [50; 60]) and a weight (in [0; 1])

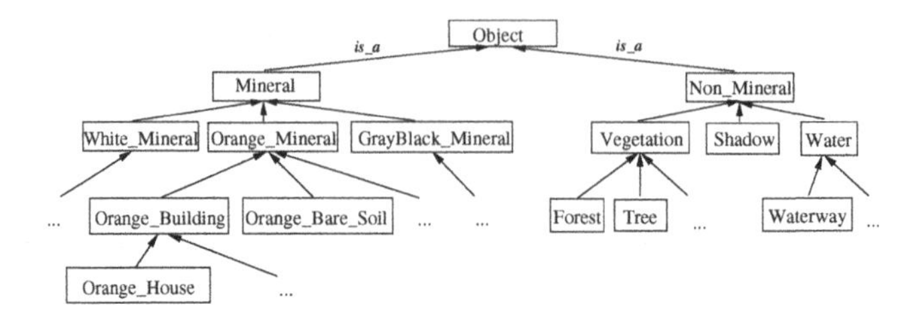

Fig. 2. Excerpt of the ontology

representing its importance to recognize the geographical object corresponding to this concept (1 meaning that the attribute is very relevant). The values of these concepts have been filled by geographers experts thanks to their knowledge about the morphology of urban objects and machine learning tools [11].

3.2 Region Matching

A matching mechanism of the region allows to evaluate the similarity between a region built during a segmentation and the concepts defined in the hierarchy of the ontology. The region matching consists in verifying the validity of feature values of a region (spectral response, size, elongation, ...) according to the properties and the constraints defined in the concepts of the ontology. The measure of matching computes the similarity between the characteristics $v_1^r \dots v_n^r$ of a region r and the attribute $a_1^k \dots a_n^k$ of a concept k is composed of a local component (dealing with the inner properties of the concept) and a global component (evaluating the pertinence in the hierarchy of concepts).

The *degree of validity* $Valid(v_i^r, a_i^k)$ evaluates the validity between the extracted characteristics v_i of a region r and the boundaries of accepted values of the attribute a_i of the concept k.

$$Valid(v_i^r, a_i^k) = \begin{cases} 1 & if \;\; v_i^r \in [min(a_i^k); max(a_i^k)] \\ \frac{v_i^r}{min(a_i^k)} & if \;\; v_i^r < min(a_i^k) \\ \frac{max(a_i^k)}{v_i^r} & if \;\; v_i^r > max(a_i^k) \end{cases}$$

The measure of *local similarity* $Sim(r, k)$ compares all the common characteristics of the region r with the attributes of the concept k. The value λ_i^k is the weight of a_i^k and represents the importance of a_i^k to identify k.

$$Sim(r, k) = \frac{\sum_{i=1}^{n} \lambda_i^k Valid(v_i^r, a_i^k)}{\sum_{i=1}^{n} \lambda_i^k}$$

The *matching score* $Score(r, k)$ evaluates the relevance of the matching between the region r and the concept k within the hierarchy of concepts. The matching score is a linear combination of local similarity measure obtained with the concepts k_j of the path starting from the root of the ontology to the studied concept k_m. This calculation integrates the proof of the concepts β_i to advantage lower concept in the hierarchy.

$$Score(r, k_m) = \frac{\sum_{j=1}^{m} \beta_j Sim(r, k_j)}{\sum_{j=1}^{m} \beta_j}$$

With such a matching process, each region produced by the segmentation can have a score which represents its suitability to the concepts formalized in the ontology.

4 Genetic Algorithm

In this section, we are interested in using an evolutionary approach to find the parameters of the segmentation algorithm by using the knowledge contained in the ontology. We start by describing the genetic algorithm and we detail the chosen evaluation function.

4.1 Description

A genetic algorithm is an optimization method. Given an evaluation function $\mathbb{F}(g)$ where g is taken in a space $\mathbb{G}$, the genetic algorithm searches the value of g where $\mathbb{F}(g)$ is maximized. Genetic algorithms are known to be effective even if $\mathbb{F}(g)$ contains many local minima. In order to consider this optimization as a learning process, it is required that the optimization performed on a learning set could be generalized to other (unlearned) datasets.

Here we consider g (the genotype in the genetic framework) as a vector containing the parameters to be tuned automatically in the watershed segmentation process. All these parameters are normalized in $[0;1]$, so here $\mathbb{G} = [0;1]^3$ as we consider 3 parameters to optimize: *hmin*, d and ft.

A genetic algorithm requires an initial population defined as a set of genotypes to perform the evolutionary process. In this process, the population evolves to obtain better and better genotypes, i.e. solutions of the optimization problem under consideration. In order to build the initial population, each genotype is randomly chosen in the space $\mathbb{G}$ except one which uses default parameters. By this way, we ensure that the final solution is as good as the default one. In our case, the default set of parameters is $\{0, 0, 0\}$, thus disabling the various over-segmentation reduction methods described previously.

Once the initial population has been defined, the algorithm relies on the following steps which represent the transition between two generations:

1. assessment of genotypes in the population.
2. selection of genotypes for crossover weighted by their score rank, as discussed in the following subsections.
3. crossover: two genotypes (p_1 and p_2) breed by combining their parameters (or genes in the genetic framework) to give a child e. For each parameter

g_i, $g_i(e)$ is computed as the value $\alpha \times g_i(p_1) + (1 - \alpha) \times g_i(p_2)$ where α is a random value between 0 and 1. We apply an elitist procedure to keep in the next generation the best solution of the current generation.

4. mutation: each parameter may be replaced by a random value with a probability $\mathbb{P}_m$. Thus we avoid the genetic algorithm to be trapped in a local minimum. As indicated previously, the best genotype of a generation is kept unchanged.

We use a mutation rate of $\mathbb{P}_m = 1\%$ and a number of generations of 15, as experiments shown that more generation do not increase the results.

4.2 Choice of the Evaluation Function

A critical point of genetic algorithm methods is how the quality of potential solutions (i.e. genotypes) is estimated, through evaluation criterion. We use here the ontology knowledge to drive the evolutionary process and find the set of parameters which allows to maximize the discovery of objects within an image. Thus, we propose to use as evaluation function, the percentage of the image which is identified by the ontology. Let $\mathcal{R}^g$ be the set of regions of a segmentation obtained with the parameters g and $\mathcal{R}^g_o$ be the set of regions identified by the ontology ($\mathcal{R}^g_o \subseteq \mathcal{R}^g$). The percentage of the surface of the image recognized by the ontology is defined as :

$$\mathbb{F}(g) = \frac{\sum_{r \in \mathcal{R}^g_o} Area(r)}{\sum_{r \in \mathcal{R}^g} Area(r)}$$

with $Area(r)$ a function returning the surface in pixels of the region r. The surface of the identified regions has been preferred to their number to evaluate the result. Indeed, a segmentation algorithm can produce many small regions which do not cover any type of concept in the ontology and thus that can not be identified by the ontology. These small regions can perturb a criteria based on the number of regions. The criterion based on the surface of the regions allows to quantify the quality of the segmentation according to the knowledge present in the ontology. The increasing of this criterion indicates that the regions built by the segmentation correspond more and more to the description of the geographical objects described in the ontology. By maximizing this criterion we build a segmentation matching with the expert knowledge about geographical objects.

5 Experimentations

The proposed method have been evaluated on an image of Strasbourg taken by the satellite Quickbird. The size of the image is 900 x 900 pixels and the spectral resolution is 8 bits for each of the four band. The figure 3 presents the image to interpret. The figure 4 (a) presents an extract of the image and the figure 4 (b) presents the segmentation of this extract by the watershed without parametrization. The figures 4 (c) (d) (e) and (f) present extracts of segmentation with the

Fig. 3. Quickbird image of Strasbourg (France). The square area used to illustrate the segmentation is outlined in white.

parameters found during the genetic evolution. We observe an amelioration of the construction of the objects, the image being better identified by the ontology with the number of generation. To validate these results we have evaluated the quality of the segmentations obtained with ground truth of three thematic classes, *house*, *vegetation* and *road*. These evaluations are done on geographic objects built and labeled manually by an expert. Three quality indexes have been used to evaluate the quality of the segmentations.

The first index used is the *recall*. It consists in considering the identification of the ontology as a classification of the image. The pixels of the objects identified by the ontology are then compared to the pixels of the objects provided by the expert:

$$recall = \frac{number\ of\ well\ labelled\ pixels}{number\ of\ expert\ pixels}$$

It takes its values in $[0; 1]$, the more it is near to 1, the more the image is well identified.

The second index used is the index of *Janssen* defined in [12]. It evaluates the concordance between the expert objects and the corresponding regions in the segmentation. For each expert object i and each corresponding region j having the biggest intersection with the object i, this index is defined as :

$$Janssen_{(i,j)} = \sqrt{\frac{Area(i \cap j)}{Area(i)} \times \frac{Area(i \cap j)}{Area(j)}}$$

It takes its value in $]0; 1]$, 1 meaning a perfect correspondence between the expert objects and the regions of the segmentation.

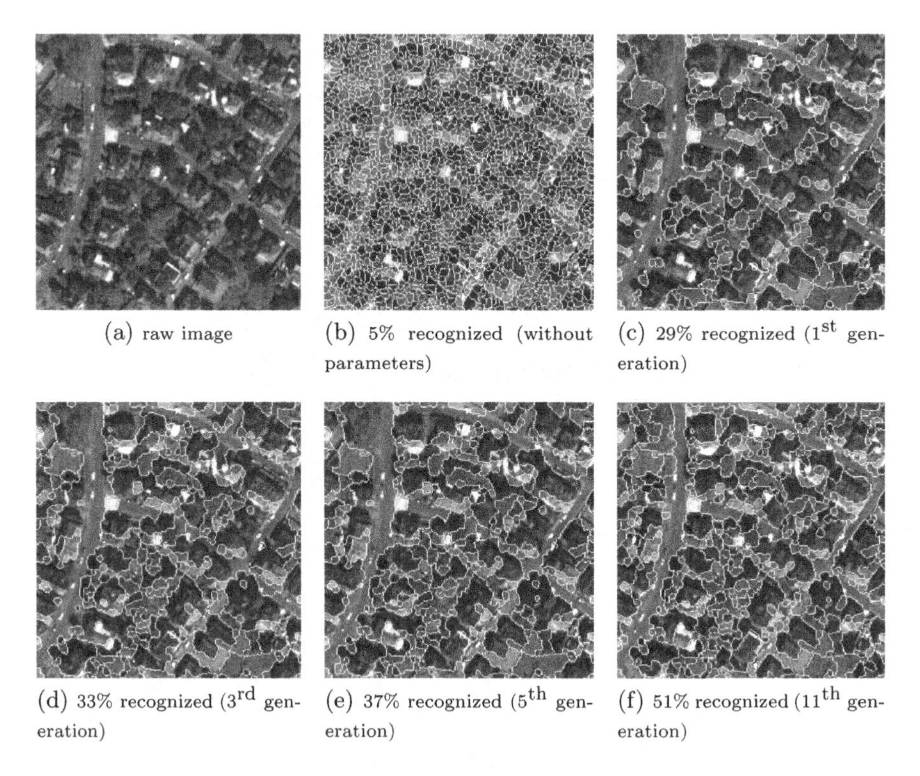

(a) raw image

(b) 5% recognized (without parameters)

(c) 29% recognized (1st generation)

(d) 33% recognized (3rd generation)

(e) 37% recognized (5th generation)

(f) 51% recognized (11th generation)

Fig. 4. Extracts of segmentations obtained at different generations during a genetic evolution. The outline of the regions is drawn in white.

The third and last index is the index of *Feitosa* defined in [7]. It also evaluates the correspondence between the expert object and the corresponding regions in the segmentation. With the same notation, it is defined as :

$$Feitosa_{(i,j)} = \frac{Area(i \setminus (i \cap j)) + Area(j \setminus (i \cap j))}{Area(i)}$$

It takes its value in $[0, (Area(i) - 1) + (Area(j) - 1)]$, the nearer from 0 it is, the more the regions correspond to the expert objects.

We have compared our evaluation criterion to these three criteria. For each criterion a mean is computed on the set of objects provided by the expert. The goal of this analysis is to check that maximizing our criterion leads to a real amelioration of the segmentation. During the evolutionary process we have evaluated each individual according to the different criteria. Thus, 200 possible parametrizations have been evaluated. The different set of parameters have been ordered according to our criterion of evaluation. The figure 5 shows the curves

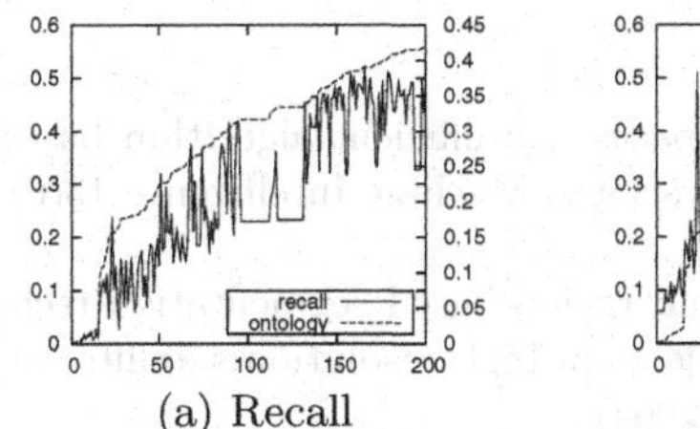

(a) Recall

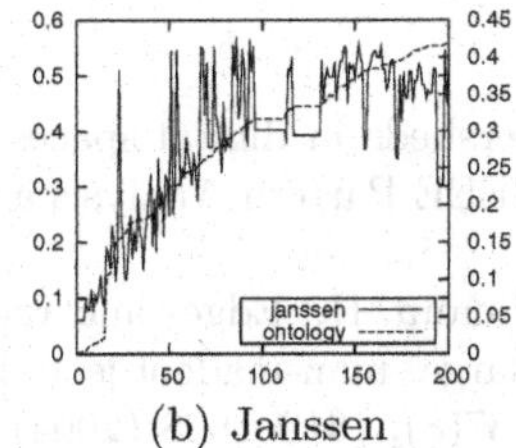

(b) Janssen

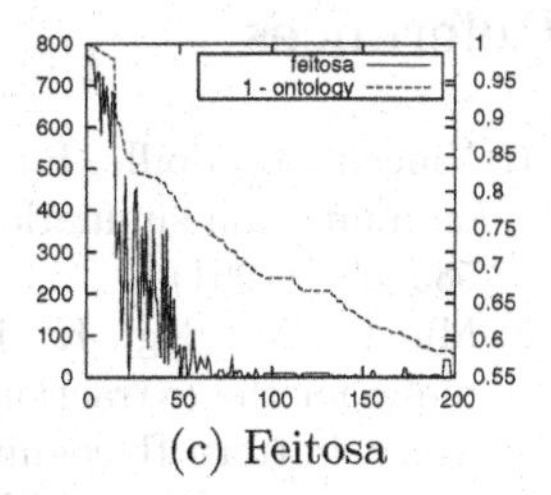

(c) Feitosa

Fig. 5. Evolution of evaluation functions for 200 individuals ordered by the criterion based on the ontology

Table 1. Results of the evaluation of the method on expert objects for 4 generations

generation	Ontology	Recall	Janssen	Feitosa
1st	29.64 %	24.98 %	0.32	25.19
3rd	33.92 %	27.83 %	0.35	17.55
5th	37.56 %	31.74 %	0.42	6.63
11th	51.91 %	49.72 %	0.48	7.10

for the three indexes. We can notice that for the three cases, the two curves seem to have the same behaviour and are highly correlated. These results show that optimizing our criterion is relevant and allows to compute segmentation of quality without forcing the expert to provide examples.

Finally, the table 1 presents values of these indexes for the segmentations with the parameters found during the generations 1, 3, 5 et 11 of the genetic evolution. After the 11th generation, the quality of the results does not increase significantly. The genetic algorithm has found the limit of the ontology recognition (approximatively 52% as shown on table 1). This limit can be explained in two ways : first many pixels in the image do not belong to the concepts given in the ontology (noise from the sensors, shadows, cars, etc.), and second the concepts as defined in the ontology do not match all objects in the image.

6 Conclusion

In this article, we have presented how an evolution algorithm could fill the semantic gap between meaningless parameters for a segmentation algorithm and knowledge of a domain expert. Results show the relevance of this approach. In the future, we want to introduce contextual knowledge like the position of the objects between each others. The evolutionary algorithm will be more complex and will check the constraints defined by this contextual knowledge.

References

1. Vincent, L., Soille, P.: Watersheds in digital spaces: An efficient algorithm based on immersion simulations. IEEE Pattern Analysis and Machine Intelligence 13(6), 583–598 (1991)
2. Mueller, M., Segl, K., Kaufmann, H.: Edge- and region-based segmentation technique for the extraction of large, man-madeobjects in high-resolution satellite imagery. Pattern Recognition 37(8), 1619–1628 (2004)
3. Goldberg, D.E.: Genetic Algorithms in Search, Optimization, and Machine Learning. Addison-Wesley Professional, Reading (1989)
4. Pignalberi, G., Cucchiara, R., Cinque, L., Levialdi, S.: Tuning range image segmentation by genetic algorithm. EURASIP Journal on Applied Signal Processing 8, 780–790 (2003)
5. Bhanu, B., Lee, S., Das, S.: Adaptive image segmentation using genetic and hybrid search methods. IEEE Transactions on Aerospace and Electronic Systems 31(4), 1268–1291 (1995)
6. Song, A., Ciesielski, V.: Fast texture segmentation using genetic programming. IEEE Congress on Evolutionary Computation 3, 2126–2133 (2003)
7. Feitosa, R.Q., Costa, G.A., Cazes, T.B., B., F.: A genetic approach for the automatic adaptation of segmentation parameters. In: International Conference on Object-based Image Analysis (2006)
8. Haris, K., Efstradiadis, S.N., Maglaveras, N., Katsaggelos, A.K.: Hybrid image segmentation using watersheds and fast region merging. IEEE Transaction On Image Processing 7(12), 1684–1699 (1998)
9. Najman, L., Schmitt, M.: Geodesic saliency of watershed contours and hierarchical segmentation. IEEE Transactions on Pattern Analysis and Machine Intelligence 18(12), 1163–1173 (1996)
10. Durand, N., Derivaux, S., Forestier, G., Wemmert, C., Gancarski, P., Boussaid, D., O., Puissant, A.: Ontology-based object recognition for remote sensing image interpretation. In: IEEE International Conference on Tools with Artificial Intelligence, Patras, Greece, pp. 472–479 (2007)
11. Sheeren, D., Puissant, A., Weber, C., Gancarski, P., Wemmert, C.: Deriving classification rules from multiple sensed urban data with data mining. In: 1rst Workshop of the EARSel Special Interest Group Urban Remote Sensing, Berlin (2006)
12. Janssen, L., Molenaar, M.: Terrain objects, their dynamics and their monitoring by the integration of gis and remote sensing. 33, 749–758 (1995)

Hybrid Genetic Algorithm Based on Gene Fragment Competition for Polyphonic Music Transcription

Gustavo Reis[1], Nuno Fonseca[1], Francisco Fernández de Vega[2], and Anibal Ferreira[3]

[1] School of Technology and Management
Polytechnic Institute of Leiria, Portugal
{gustavo.reis,nfonseca}@estg.ipleiria.pt
[2] University of Extremadura, Spain
fcofdez@unex.es
[3] University of Porto, Portugal
ajf@fe.up.pt

Abstract. This paper presents the Gene Fragment Competition concept that can be used with Hybrid Genetic Algorithms specially in signal and image processing. Memetic Algorithms have shown great success in real-life problems by adding local search operators to improve the quality of the already achieved "good" solutions during the evolutionary process. Nevertheless these traditional local search operators don't perform well in highly demanding evaluation processes. This stresses the need for a new semi-local non-exhaustive method. Our proposed approach sits as a tradeoff between classical Genetic Algorithms and traditional Memetic Algorithms, performing a quasi-global/quasi-local search by means of gene fragment evaluation and selection. The applicability of this hybrid Genetic Algorithm to the signal processing problem of Polyphonic Music Transcription is shown. The results obtained show the feasibility of the approach.

Keywords: Polyphonic Music Transcription, Evolutionary Algorithms, Genetic Algorithms, Memetic Algorithms, Intelligent Recombination Operator, Gene Fragment Competition.

1 Introduction

Although Genetic Algorithms (GAs) are very good at rapidly identifying good areas of the search space (exploration), they are often less good at refining near-optimal solutions (exploitation). For instance: when a Genetic Algorithm (GA) is applied to the "OneMax" problem[1], near-optimal solutions are quickly found but convergence to the optimal solution is slow. Therefore hybrid GAs using local

[1] The OneMax problem is a binary maximization problem, where the fitness function is simply the count of the number of genes set to "1".

M. Giacobini et al. (Eds.): EvoWorkshops 2008, LNCS 4974, pp. 305–314, 2008.

search can search more efficiently by incorporating a more systematic search in the vicinity of "good" solutions[1]. For instance: a bit-flipping hill-climber could be quickly applied within each generation for the OneMax to ensure fast convergence. Memetic Algorithms (MAs) are a class of stochastic global search heuristics in which Evolutionary Algorithms-based approaches are combined with local search techniques to improve the quality of the solutions created by evolution[1]. This means that Memetic Algorithms go a further step by combining the robustness of GAs on identifying good areas of the search space with local search for refining near-optimal solutions. Recent studies on MAs have revealed their successes on a wide variety of real world problems[1]. Particularly, they not only converge to high quality solutions, but also search more efficiently than their conventional counterparts. In diverse contexts, MAs are also commonly known as hybrid EAs, Baldwinian EAs, Lamarkian EAs, cultural algorithms and genetic local search.

But a new problem with traditional local search operators arises due to the cost of evaluation. If the calculation of the fitness function is heavy, having local search operators changing each individual several times means lots of individual evaluations thus lots of computational cost. Therefore in problems that demand high computational cost in fitness evaluation this might be a prohibitive solution. Fitness evaluation allows us to measure the whole "quality" of each individual, and in many cases, it is obtained by simply combining the values of the evaluations of each gene or gene fragment. But when the evaluation of gene fragments is possible, these fragment values are not taken on consideration during recombination.

This paper presents a Gene Fragment Competition, a different approach to recombination, that takes advantage of gene/fragment evaluation and gene/fragment selection as a way to speed up the process, especially when evaluation of individuals is a demanding computational task. The presented method arises from the work of the authors on the field of Automatic Music Transcription using Genetic Algorithms. In this kind of approach the individuals evaluation demands lots of digital signal processing based on a high number of FFTs, which results on a heavy computational cost making impracticable the use of traditional memetic algorithms, but still needing a special operator to increase significantly the performance. The rest of the paper is structured in the following way: Section 2 presents our proposed approach, Section 3 describes the Polyphonic Music Transcription problem, the Genetic Algorithm approach to the problem and how gene fragment competition is applied, Section 4 presents our experiments and results and finally Section 5 summarizes our conclusions and future work.

2 Proposed Approach

2.1 Gene Fragment Competition

In the traditional GA approach (see Fig. 1a), genetic algorithms are based on a cycle made of evaluation, selection, recombination and mutation - individuals are

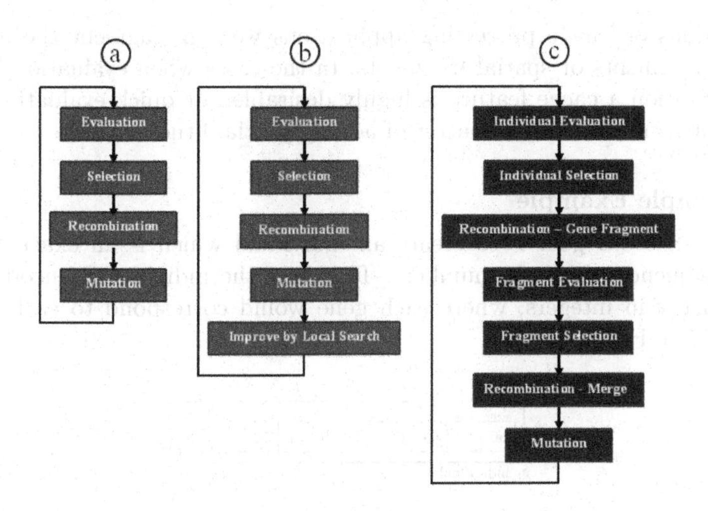

Fig. 1. Classic Genetic Algorithm approach (a) vs Traditional Memetic Algorithm approach (b) vs Gene Fragment Competition (c)

evaluated, based on their evaluation parents are selected for recombination, creating new individuals that are subject to mutation. On the other hand classical MAs apply a new local search operator in each individual just after the mutation (or even after recombination), looking for better solutions in the neighborhood of already found good solutions. Gene Fragment Competition uses a different kind of global/local search approach. Instead of using separate operations for global and local search, like the Memetic Algorithms, a different type of recombination is proposed which is responsible for a semi-global/semi-local search.

We can consider that traditional recombination operators are made of two operations: fragmentation (parent's genes are divided on two or more fragments), and merging (these gene fragments are merged to create new individuals). The main idea of the proposed method is to add two additional steps inside recombination: gene fragment evaluation and gene fragment selection. Parent genes are split on n fragments, each fragment is evaluated and then a selection method is applied to choose the best gene fragments, which will be merged to create a new born individual. To split the parents in n fragments two methods can be applied: static fragment size, on which equally sized fragments are created or dynamic fragment size where are created fragments with random sizes. For selecting gene fragments classic selection methods also apply (roulette, tournament, etc.). Although standard recombination operators breed two individuals from two parents, our method breeds only one individual from two or more parents.

As can be seen, a very important requirement must be fulfilled: it must be possible to evaluate gene fragments. If this is not possible, the method cannot be applied. This does not mean that the system must be able to evaluate individual genes, or every kind of group of genes. In some applications it simply means that some special type of fragmentation must be applied to make possible the evaluation of its fragment (see section 3.3). For instance, in signal processing applications or image processing applications we can fragment the individual in time fragments or spatial fragments. In the cases when evaluation is a complex operation a cache feature is highly desirable, for quick evaluation of gene fragments, since its only a matter of adding partial fitness values.

2.2 Simple Example

Imagine that our goal is to create an individual which is an exact copy of a target sequence of integer numbers. Therefore the individual's encoding could be an array in integers, where each gene would correspond to each sequence number (see Fig. 2).

Gene	Gene	Gene	Gene	Gene	Gene	Gene	Gene	Gene	Gene
60	62	64	65	67	69	71	72	74	76

Individual

Fig. 2. Individual's encoding on the "Find the sequence problem"

Let's consider that the fitness value is obtained by the sum of the absolute differences (Equation 1) of each individual's gene ($X(i)$) and the our target individual ($O(i)$) (see Fig. 3).

$$Fitness = \sum_{i=1}^{genes} |O(i) - X(i)| \tag{1}$$

Important note: since the best individuals are the ones closer to our target and the fitness function is measuring that distance, the best individuals are the ones who have less fitness values.

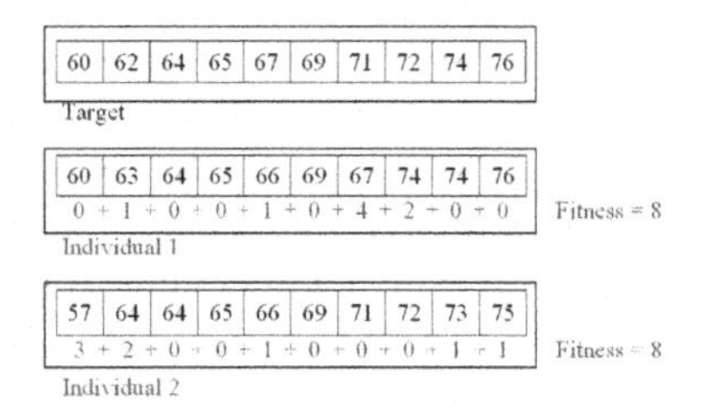

Fig. 3. Fitness values of Individual1 and Individual2. The fitness value of each individual is calculated by the sum of the absolute difference between the values of their genes and the target individual's genes.

If we want to breed a new Individual from the parents Individual1 and Individual2 with 2 random points of cut (dynamic fragment size) our intelligent recombination operator will calculate the fitness value of each fragment and then choose the fragment for the new born individual with the best fitness (see Fig. 4).

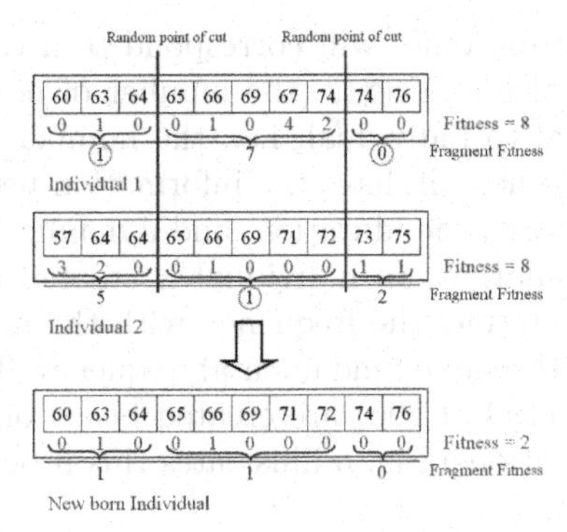

Fig. 4. Breeding of a new born individual. The best fragments of each father are inherited to the new born individual.

3 Polyphonic Music Transcription Problem

3.1 Automatic Music Transcription

Automatic Music Transcription is the process in which a computer program writes the instrument's partitures of a given song or an audio signal. Hence, automatic music transcription from polyphonic audio recordings is the automatic transcription of music in which there is more than one sound occurring at the same time: multiple notes on a single instrument (like a piano), single notes in multiple instruments, etc. (usually, only pitched musical instruments are considered). Music transcription is a very difficult problem, not only from the computational point of view but also in a musical view since it can only by addressed by the most skilled musicians.

Traditional approaches to Automatic Music Transcription try to extract the information directly from audio source signal (using frequency analysis, autocorrelation functions [2,3], and other digital signal processing techniques [4,5,6,7,8,9,10]). Nevertheless Automatic Music Transcription can be considered as a search space problem where the goal is to find the sequence of notes that best models our audio signal[11]. Instead of trying to deconstruct the audio signal, a search space approach will try to find a solution which allows the creation of a similar audio signal. Usually search space approaches are not addressed in music transcription problems due to the huge size of the search space. Nevertheless genetic algorithms[12] have proven to be an excellent tool to find solutions in extremely large search spaces, since they only need to use a very small subset of the entire search space.

3.2 Genetic Algorithm Approach

To apply genetic algorithm to the music transcription problem there are some important considerations regarding: the encoding of the individuals, the creation of the initial population, recombination and mutation and also the fitness function (as reviewed by Reis et al. [13]).

Each individual (chromosome) will correspond to a candidate solution, and is made of a sequence of note events (the number of notes will most likely be different from individual to individual, i.e.: the number of genes is not fixed). Each note, acting as a gene, will have the information needed to represent that note event (e.g. note, start time, duration, and dynamics).

For the starting population, an individual is created based on the peaks of the FFT - in each time frame, the frequency with the highest peak creates (or maintains) a note with the same fundamental frequency. The additional individuals of the population (199 of 200) are created based on the initial individual and after 10 forced mutations. Fig. 5 illustrates this process.

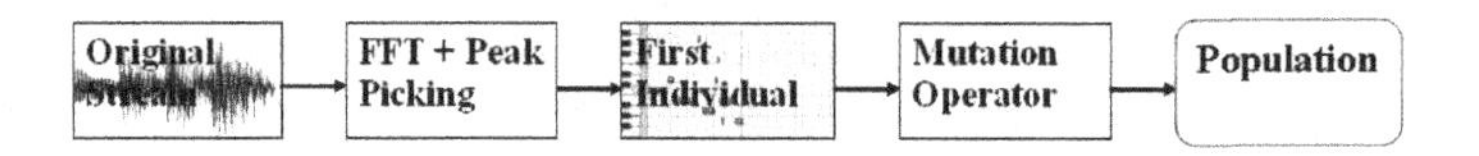

Fig. 5. Starting Population Process

There are several mutation operators: note change (±1 octave, ± half tone), start position (up to ± 0.5 second change), duration (from 50% to 150%), velocity (up to ± 16 in a scale of 128), event split (split in two with a silent between), event remove, new event (random or event duplication with different note). Besides these event mutations, there are 2 mutation operators (with lower probability) that are applied equally to all events: velocity change (up to ± 4 in a scale of 128) and duration (from 50% to 150%).

To evaluate an individual, a synthesizer is used to convert the note sequence into an audio stream, which will be compared with the original audio stream. Therefore each note has to pass through an internal synthesizer which in our case is made of a simple sampler, using piano samples.

The original stream and the synthesized one are compared in the frequency domain. The streams are segmented in time frames with 4096 samples (fs = 44.100kHz) and an overlapping of 75%. A FFT is calculated for each frame and the differences are computed (Equation 2).

$$Fitness = \sum_{t=0}^{tmax} \sum_{f=27.5Hz}^{\frac{fs}{2}} \frac{||O(t,f)| - |X(t,f)||}{f} \quad (2)$$

Table 1. Algorithm parameters

Parameters	Values
Population	100
Survivor Selection	Best 100 individuals (population size)
Crossover probability	75%
Mutation probability	1%
Note minimal duration	20 ms

The $|O(t,f)|$ is the magnitude of frequency f at time frame t in the source audio signal, and $|X(t,f)|$ is the same for each individual. The division by f acts as a frequency normalization. Fitness is computed from frame slot 0 to $tmax$, traversing all time from the beginning to the end, and from $fmin = 27,5$ Hz (corresponding to the first note of the piano keyboard) to $fmax = 22050$ Hz, which is half of the sample rate - 44100 Hz.

The GA approach is resumed on Table 1.

3.3 Applying Gene Fragment Competition to Music Transcription

To apply the proposed operator to the music transcription approach presented in [13] and summarized in section 3.2, there are some remarks. The requirement needed for gene fragment competition is that it must be possible to evaluate gene or gene fragments. In the music transcription approach presented before, each gene represents a musical note, and it is not possible to evaluate each note, especially in polyphonic parts. Nevertheless, there is a solution for that. The overall fitness of each individual is obtained by adding the FFT differences over the different time frames, which means that although it is not possible to evaluate note by note, it is possible to evaluate time frames (for instance, it is possible to evaluate the behavior of an individual on a time fragment between time=2.0s and time=4.0s). Then it is possible to map that time interval to the genes (notes) acting on that time fragment. If some notes on that fragment began before or end after the time frontiers, the note is split, and only the inside part are considered. Later, during the recombination merge phase, if a note ends on the exact same time that a similar one begins, notes are merged as one, since the algorithm considers that a previous split happened.

For global selection, the "deterministic tournament" method was used, but in fragment selection a "non-deterministic tournament" method was used as a means to preserve some biodiversity. Regarding fragment size, that in this case is measure in seconds since we consider time fragments, the value of 5 seconds was used. Increasing the fragment size should decrease the impact of the operator, and decreasing fragment size has the side effect of splitting too much the notes.

For each individual, a cache feature was implemented, that stored the fitness values for each time frame, which means that evaluating a time fragment is simply done by adding fitness values of its internal time frames.

4 Experiments and Results

To analyze the impact of the presented method, several tests were made. The proposed method was applied on our music transcription approach on an audio file with the first 30 seconds of the piano performance of the Schubert's Impromptu No.2 in E Minor. Each bank of tests was created with 1000 generations, and with at least 4 different runs[2] (the values presented correspond to the average values).

[2] The value differences between runs are very small ($< 3\%$).

Table 2. New parameters

Classic GA	
Selection	Deterministic Tournament (5 individuals)
Recombination	1-point time crossover (with eventual note split)
Gene Fragment	
Individual Selection	Deterministic Tournament (5 individuals)
Fragment Selection	Tournament (5 individuals)
Fragment size	5 seconds[3]

Table 3. Average fitness values over 1000 generations using classic GA, using static GFC and dynamic GFC

Generation	GA (1%)	Dyn GFC	Static GFC	GA (0,5)%	GA (5%)
250	1,808 E12	1,670 E12	1,718 E12	1,839 E12	1,882 E12
500	1,690 E12	1,554 E12	1,580 E12	1,708 E12	1,811 E12
750	1,621 E12	1,492 E12	1,512 E12	1,624 E12	1,765 E12
1000	1,571 E12	1,455 E12	1,464 E12	1,563 E12	1,728 E12

In the first bank of tests, there were 3 different scenarios: classic GA approach, static fragment size and dynamic fragment size of the proposed method (see Table 2). In these tests (and on the following ones), the presented values are the fitness values. Since the goal of the paper is not to evaluate our music transcription approach, presenting other results (% of transcribed notes, etc) would remove the focus from the goal of the paper.

Table 3 and left part of the Fig. 6 shows the fitness evolution (average fitness values over several different runs) over 1000 generations. Once again, it is important to recall that in our implementation, since fitness measure the error between FFT's, the lower values of fitness, means better individuals.

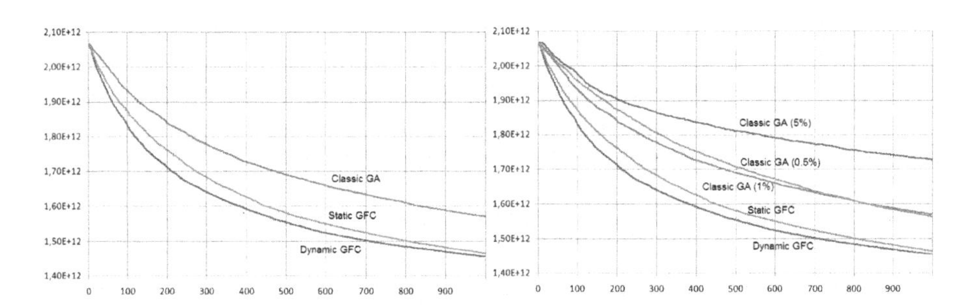

Fig. 6. Fitness values over 1000 generations using classic GA, using static fragment size gene competition and dynamic fragment size gene competition (left) and with different values of mutation probabilities in generic GA approach (right)

[3] Resulting in 6 fragments on 30 seconds audio files.

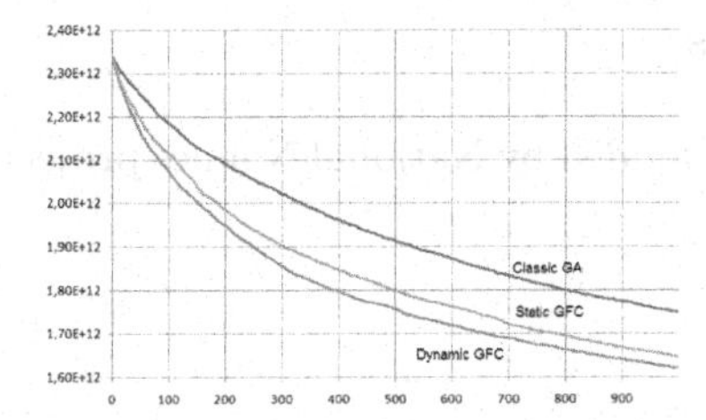

Fig. 7. Average fitness values over 1000 generations using classic GA, using static fragment size gene competition and dynamic fragment size gene competition for the Mozart's Piano Sonata

The first test, shows us that the proposed method achieves a better performance in this scenario. Nevertheless, there are other situations that needed to be tested in order to discard other hypothesis. One question that could rise is regarding the "Tournament" vs. "Deterministic Tournament" selection. A new run of tests were made using the classic GA, but changing the selection mode from "Deterministic Tournament" to "tournament". The obtained results were identical within a range $> 0.1\%$. The other question that could rise is related to mutation probabilities. Since the proposed method fragments the genes, could it be that changing the mutation probabilities of the classic GA could result in much better results? A new bank of tests was made with classic GA approach with different mutation probabilities (5%, 1% and 0.05%). Fig. 6(right) and Table 3 show the results. Using different mutation probabilities above (5%) or bellow (0.5%) didn't present significantly results.

Tests were made also with another audio file to confirm the earlier results. A 30s seconds audio file of Mozart's Piano Sonata n. 17 in B flat K570 was used and is presented in Fig. 7. Once again the proposed method presents an increase of performance. In this test the performance difference between static and dynamic fragment size also increases comparatively with the initial test.

It is important to say that in both tests by applying our operator with dynamic size, we have achieved in only 500 generations the same results the classical GA achieves in 1000 generations, which is a very significative gain in performance.

5 Conclusions and Future Work

This paper has presented a Gene Fragment Competition as a new technique for improving quality of results when applying GAs to several signal and image processing problems.

Although the proposed method presents some requirements that are not fulfilled on several GA applications (capability of evaluating fragment of genes), it is shown that at least in some scenarios can achieve an important performance increase.

In the future, additional tests must be made to have a better understanding of the impact of the operator applied on different problem applications.

Acknowledgments

This work was partially funded by National Nohnes project TIN2007-68083-C02-01 Spanish MCE.

References

1. Hart, W., Krasnogor, N., Smith, J.: Memetic Evolutionary Algorithms. In: Recent Advances in Memetic Algorithms, Springer, Heidelberg (2004)
2. Klapuri, A.P.: Qualitative and quantitative aspects in the design of periodicity estimation algorithms. In: Proceedings of the European Signal Processing Conference (2000)
3. Klapuri, A.P.: Automatic music transcription as we know it today. Journal of New Music Research 33(3), 269–282 (2004)
4. Marolt, M.: On finding melodic lines in audio recordings (2004)
5. Dixon, S.: On the computer recognition of solo piano music (2000)
6. Bello, J.P.: Towards the automated analysis of simple polyphonic music: A knowledge-based approach. PhD thesis, University of London, London, UK (2003)
7. Walmsley, P., Godsill, S., Rayner, P.: Bayesian graphical models for polyphonic pitch tracking (1999)
8. Walmsley, P., Godsill, S., Rayner, P.: Polyphonic pitch tracking using joint bayesian estimation of multiple frame parameters (1999)
9. Goto, M.: A robust predominant-f0 estimation method for real-time detection of melody and bass lines in cd recordings.
10. Gómez, E., Klaupuri, A., Meudic, B.: Melody description and extraction in the context of music content processing. Journal of New Music Research 32(1) (2003)
11. Lu, D.: Automatic music transcription using genetic algorithms and electronic synthesis. Computer Science Undergraduate Research, University of Rochester, New York, USA
12. Holland, J.H.: Adaptation in Natural and Artificial Systems: An Introductory Analysis with Applications to Biology, Control, and Artificial Intelligence. The MIT Press, Cambridge (1992)
13. Reis, G., Fonseca, N., Fernandez, F.: Genetic algorithm approach to polyphonic music transcription. In: Proceedings of WISP 2007 IEEE International Symposium on Intelligent Signal Processing, pp. 321–326 (2007)

Classification of Seafloor Habitats Using Genetic Programming

Sara Silva[1] and Yao-Ting Tseng[2]

[1] CISUC, Department of Informatics Engineering, University of Coimbra, Polo II, P-3030 Coimbra, Portugal
sara@dei.uc.pt

[2] Centre for Marine Science and Technology, Curtin University of Technology, GPO Box U1987, Perth, WA 6845, Australia
y.tseng@cmst.curtin.edu.au

Abstract. In this paper we use Genetic Programming for the classification of different seafloor habitats, based on the acoustic backscatter data from an echo sounder. By dividing the multiple-class problem into several two-class problems, we were able to produce nearly perfect results, providing total discrimination between most of the seafloor types used in this study. We discuss the quality of these results and provide ideas to further improve the classification performance.

Keywords: genetic programming, classification, acoustic characterization, seafloor habitats.

1 Introduction

Genetic Programming (GP) can solve complex problems by evolving computer programs using Darwinian evolution and Mendelian genetics as sources of inspiration [1, 2]. Most GP systems represent the programs as trees. Tree-based GP has not been often used for multiclass classification tasks, although some studies have already been developed on this subject [3, 4, 5].

The aim of this work is to provide a better understanding of the acoustic backscatter from marine macro-benthos (MMB), including mainly seagrass, algae, and other marine organisms living on the seafloor. Since these organisms live on or around their substrates, the understanding of the acoustic backscatter from their substrates is also essential.

The analysis of the acoustic backscattered signals of MMB and related substrates has been studied with a variety of different approaches [6, 7, 8, 9]. One of them [9] has been the target for improvement in a first work using GP [10], where GP was able to provide an improved discrimination between the different seafloor habitats. The initial motivation to use GP for this task came from a study on diesel engine diagnosis [11].

Here we develop a tree-based GP system to tackle this problem, testing the pairwise separability of the different classes involved in the study, and ultimately

M. Giacobini et al. (Eds.): EvoWorkshops 2008, LNCS 4974, pp. 315–324, 2008.

dividing the 5-class problem into several 2-class problems, whose solutions can be joined to provide a perfect discrimination of most of the seafloor habitats [12].

The next section describes the data used in this study, how it was acquired and prepared for being used. Section 3 describes the GP system used, its main parameters and the fitness function developed for this particular problem. Section 4 describes the results achieved, and how they were combined to build the final solution. Section 5 discusses the quality and usefulness of the proposed solution, suggesting future developments of this work. Finally, Sect. 6 concludes this study.

2 The Data

This section describes the data collection process, the removal of incomplete data and definition of representative data sets, and the statistical preprocessing suffered by the data before being used by the GP system.

2.1 Data Acquisition

The acoustic backscattered signals were collected from Cockburn Sound Western Australia on the 10th of August 2004 by a SIMRAD EQ60 single beam echo sounder. The data collection was made on two sites of shallow coastal waters where the water depths were less than 6 meters. In site 1, the seafloor habitats are predominantly sand, seagrass 1 (*Posidonia sinuosa*), and seagrass 2 (*Posidonia australis*). On the other hand, site 2 mainly consists of sand, reef and macro algae with canopy heights much higher than both of the seagrasses in site 1. Along with the collection of the acoustic data, synchronized tridimensional (3D) still images were also taken simultaneously.

Figure 1 shows an echo sounder transmitting a signal to the seafloor. The sound is backscattered from the seafloor to the sea surface, and back to the seafloor, several times. The echo sounder receives several returns for each sample. Figure 2 represents a typical sample of acoustic backscatter collected from an

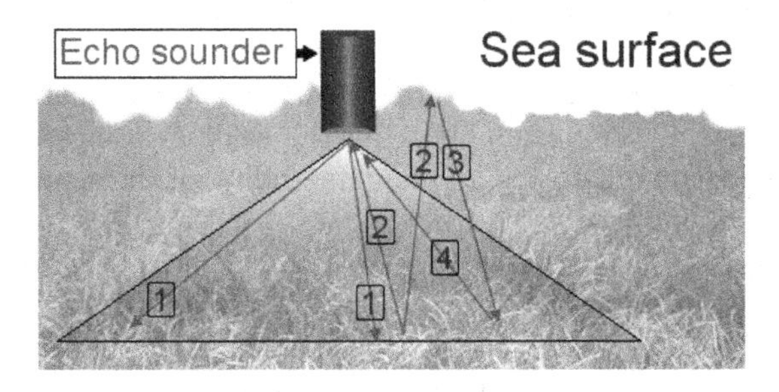

Fig. 1. Sound transmitted from the echo sounder to the seafloor (1), echo from the seafloor producing the first bottom return (2), echo from the sea surface to the seafloor (3), and again from the seafloor to the sea surface and to the echo sounder, producing the second bottom return (4)

echo sounder, showing several echo returns. Although SIMRAD EQ60 provides both 38 and 200kHz sampling ability, only the 200kHz signals have been used in our study, due to its higher resolution of 25μs sampling rate. The volume backscatter coefficient (in decibel scale) was used due to its ready availability.

2.2 Defining Data Sets

Due to the cost and technical limitations of this study, only 1232 samples of both echoes and still images were acquired. The MMB and the related substrates were roughly classified into five classes: sand, bare reef, macro algae, seagrass 1 (*P. sinuosa*), and seagrass 2 (*P. australis*) according to the visual interpretation of the 3D still images. From now on, we will refer to the seagrasses simply as sinuosa and australis. After further examination of the images and the echoes, 689 samples were rejected for not being fully intact, and some others were discarded for containing mixed habitat types. In the end, 300 samples were used as pure representatives of the five classes, unevenly distributed like this: 81 (sand), 10 (reef), 8 (algae), 21 (sinuosa), 180 (australis). Each sample consisting of several bottom returns was then truncated to contain only the first bottom return (see Fig. 2), represented by a 100-point sequence that is believed to fully describe the interactions between the transmitted sound and the respective targets.

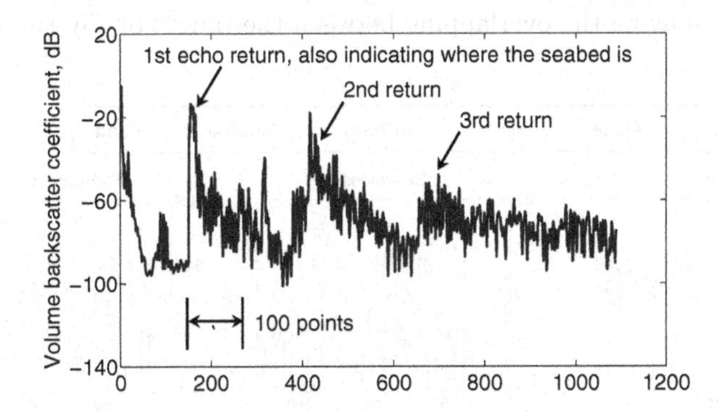

Fig. 2. A typical sample of the acoustic backscatter collected from an echo sounder. Several echo returns are present.

2.3 Statistical Preprocessing

Before being given to the GP system, each sample suffers a major transformation, one that may well determine the success or failure of the GP learning. Each of the 100-point sequences ($S = \{p_1, p_2, ..., p_n\}, n = 100$) is reduced to only seven statistical features ($F = \{x_1, x_2, ..., x_7\}$):

Kurtosis: $x_1 = \frac{\frac{1}{n}\sum_{i=1}^{n} p_i^4}{x_4^2}$
Maximum: $x_2 = \max p_i, i = 1, 2, ..., n$
Mean: $x_3 = \frac{1}{n}\sum_{i=1}^{n} p_i$

Second-order Moment from Origin: $x_4 = \frac{1}{n}\sum_{i=1}^{n} p_i^2$
Skewness: $x_5 = \frac{\frac{1}{n}\sum_{i=1}^{n} p_i^3}{x_4^{3/2}}$
Standard Deviation: $x_6 = (\frac{1}{n-1}\sum_{i=1}^{n}(p_i - x_3)^2)^{\frac{1}{2}}$
Minimum: $x_7 = \min p_i, i = 1, 2, ..., n$

This particular set of statistical features was based on the set by [11]. After calculating the different statistical features for all the samples, the 300-element vectors obtained for each feature are normalized. Each fitness case of the GP system is a 7-tuple with all the statistical features of the sample $(x_1, x_2, ..., x_7)$, along with an identifier of the class to which it belongs.

By themselves, these seven features do not seem to have the ability to discriminate between our several classes. Figure 3 shows the plotting of three of these features (Mean, Skewness, Standard deviation) for the five classes involved. The plots produced by the remaining features are not very different from the ones shown. The samples of the least represented classes were randomly replicated until each class contained 180 points in the plot (x axis), for visualization purposes only. For an easier recognition of the ranges occupied by each classes, consecutive points are connected by lines. Sand is the class that seems to be more easily separated from the rest, but still none of the single features is able to do that. GP is expected to be able to combine the single features into a compound feature that will avoid the overlapping between the ranges of any two classes.

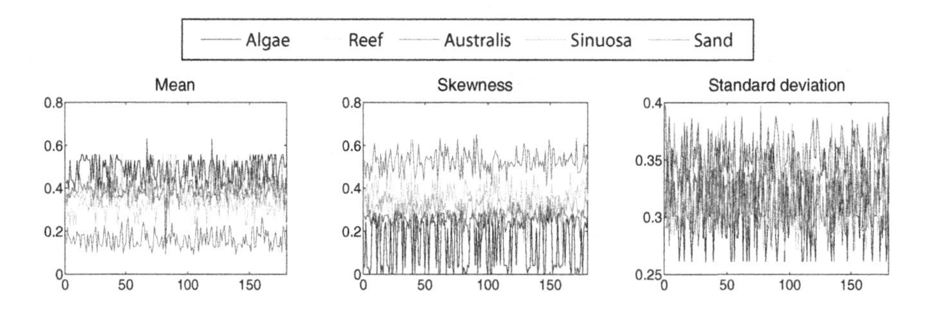

Fig. 3. Plotting of three statistical features for the five classes involved in the study

3 The GP System

This section describes the main parameters of the GP system used, as well as the fitness function developed for this work.

3.1 Parameters

The GP system used was an adaptation of GPLAB, a GP toolbox for MATLAB, version 2.0 [13]. The function set used was $\{+, -, \times, \div\}$, protected as in [1], and the terminal set contained only the seven variables $\{x_1, x_2, x_3, x_4, x_5, x_6, x_7\}$. A population of 500 individuals was initialized with the Ramped Half-and-Half procedure [1] with an initial maximum tree depth of 6. The Heavy Dynamic Limit on tree depth [14] was used to avoid excessive code growth, without imposing any static upper limit. The crossover and mutation rates were 0.5 each. A reproduction rate of 0.1 was used. Selection for reproduction was performed using a Lexicographic Parsimony Pressure tournament [15] of size 50. Selection for survival performed a total replacement of the parents with their offspring, using no elitism.

3.2 Fitness Function

The fitness function simply calculates the percentage of points in the plot (like in Fig. 3) that fall within the range of more than one class. Minimizing this percentage is our final goal, a simple and clear objective that allows the GP system complete freedom to devise any possible discrimination strategy, as long as it reaches its purpose. The evolution stops when it converges to a solution with null fitness.

Figure 4 shows the pseudocode for the fitness function. Because the available samples are unequally distributed between the classes, we give more weight to the points of the under-represented classes, such that each class contributes equally to the calculation of the fitness value. So, a point from class algae (8 samples) weights 22.5 times more than a point from class australis (180 samples).

```
nclasses = number of classes
nsamples_c = number of samples in class c
min_c = minimum value plotted in class c
max_c = maximum value plotted in class c

overlapped = 0
for c = 1 to nclasses
   for s = 1 to nsamples_c
      value_s = value plotted for sample s
      for nc = 1 to nclasses, nc <> c
         if value_s between min_nc and max_nc
            overlapped = overlapped + 1/nsamples_c

fitness = 100 * overlapped / nclasses
```

Fig. 4. Pseudocode of the fitness function

4 Results

Preliminary tests were performed in terms of pairwise separability of the classes involved, concluding that some pairs are easily separable (like all pairs involving sand) and others seem to be impossible to separate (like the pair of both seagrasses). A strategy of divide and conquer was adopted, where the 5-class problem was divided into several 2-class problems. The first step was to separate sand from the remaining classes; then, with sand left out, reef was separated from the remaining classes; algae was then separated from the seagrasses; finally an attempt was made at separating both seagrasses. What follows is the list of solutions obtained by GP at each step of this process. All except the last (where fitness was 69.7) are perfect solutions. The dashed lines in Fig. 5 represent the thresholds allowing the class discrimination.

Separating Sand

$$\frac{x_3^2}{x_2} \tag{1}$$

Separating Reef

$$\begin{aligned}
&\frac{x_6(x_5 - x_7 - x_2x_6)}{x_3 - x_4x_6} \times \\
&\times \left[1 + x_4x_5(x_3 - x_2) + x_3x_7(x_3 - x_4) - \frac{x_3^3x_7}{x_4x_5} - \right. \\
&\quad - \frac{x_2(x_3x_7 - x_5) + x_3(x_3x_7 + x_5)}{x_4} + \frac{x_3^2}{x_5} + \\
&\quad \left. + \frac{x_3x_5(x_3 - x_2) + x_4x_5^2(x_2 - x_3)}{x_4(x_1x_3x_4x_5 - x_4x_5(x_3 + 1) + x_3)}\right]^{-1}
\end{aligned} \tag{2}$$

Separating Algae

$$\begin{aligned}
&\frac{x_4x_5(x_1 + x_4)}{x_4 + x_5} \times \frac{x_6 + (x_3 + x_7)(2x_2 + x_5)}{2x_2 + x_5} \times \\
&\times \left[\frac{x_6(x_2 + 2x_7)(x_6 + x_2x_5 + x_5^2(x_2 + x_5 + 1))}{x_6(x_2 + 2x_7) + x_5^2(x_2 + x_5)(x_2 + 2x_7)} \right. \\
&\quad \left. + \frac{x_5x_7(x_2 + x_5)(x_1 + x_3 - x_7)}{x_6(x_2 + 2x_7) + x_5^2(x_2 + x_5)(x_2 + 2x_7)}\right]^{-1}
\end{aligned} \tag{3}$$

Separating Seagrasses

$$\begin{aligned}
&x_2 + x_4 + x_5 - \frac{2(x_6(x_3 - 1) + 2x_2 - x_1)(x_2 - x_1)}{x_4x_5x_6} + \\
&+ \frac{x_5x_7}{x_3(x_1 + x_3 + x_4)(x_6 - 2x_5)} - \frac{x_2(x_3 + x_6)}{x_1x_5} + \\
&+ \frac{x_1x_4x_6(x_5(x_6 - x_2) - x_3 + x_4)(x_6 - 2x_5)(x_7 + x_1x_3)}{(x_1 + x_2 - x_5)(x_2x_7 + x_1^2(x_6 - 2x_5)(x_7 + x_1x_3))}
\end{aligned} \tag{4}$$

The results of the divide and conquer strategy can now be joined together to produce a candidate solution for the original 5-class problem. This solution is

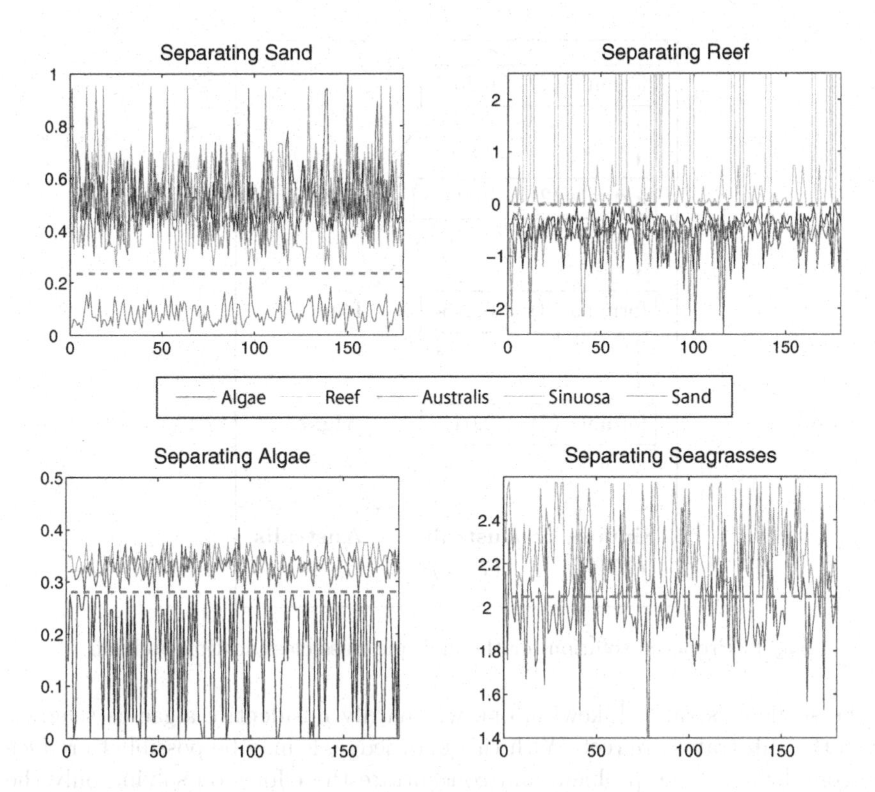

Fig. 5. Plots produced by the compound features found for the several sub-problems. The dashed lines represent the thresholds specified in Fig. 6.

a binary tree where each node compares a compound feature with a threshold value, in order to determine if a class is already identified or if other nodes need to be visited. The threshold values are the dashed lines plotted in Fig. 5. Figure 6 shows the proposed solution.

5 Discussion and Future Work

In the results presented in Sect. 4, only the seagrasses could not be completely separated from each other. The reasons for this limitation may be inherent to the data, that may not contain enough information to perform this discrimination, or to the large removal of information caused by the adoption of the statistical features as sole representatives of the samples. Using different statistical features, or ones that concentrate on smaller areas of the 100-point sequences, may allow us to finally distinguish between these two similar classes.

From the marine science point of view, it may not even be that important to achieve a solution that discriminates among all the classes. For example, it is known that live seagrasses only exist on sand, and algae only grow on reef. So the chances of erroneously identifying seagrass as algae are very low when most

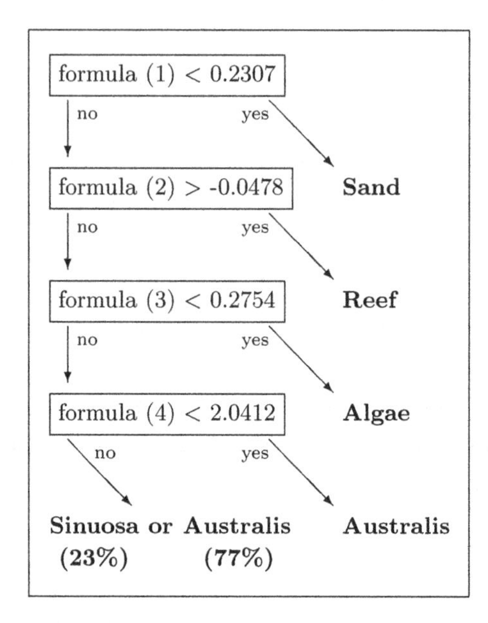

Fig. 6. Proposed solution for the multiple class classification problem

of the seafloor is sand. Likewise, one will hardly misidentify algae as seagrass when the substrates are reef. With this knowledge, it may be possible to reduce the complexity of the problem and concentrate the efforts on solving only the most important and practical issues.

In spite of its apparent quality, the solution proposed must be used with caution. Because the available data samples were highly unbalanced between the different classes, it was not possible to perform any cross-validation of the results. The solution may be biased toward some outliers that may be present and may represent a high proportion of data in the most under-represented classes. Overfitting may be occurring, a phenomenon that is already hinted by the large size of most of the compound features found.

It should also be noted that we have only used pure habitats types to derive the proposed solution. If we had used mixed types, finding compound features to completely separate between classes would not only be virtually impossible, but would also not make much sense in practical terms. The fact is that the proposed solution is not appropriate for dealing with mixed types at all. Since the divide and conquer strategy seems to be a promising way of dealing with the inadequacy of tree-based GP for multiclass classification problems, in the future we may adopt a different technique for solving each of the sub-problems.

Like what typically happens when training an artificial neural network, GP can also be taught to output a number between 0 and 1 that can be interpreted as representing the likelihood that a given sample belongs to a given class. Joining the solutions of the different sub-problems would then result in a vector containing as many elements (numbers between 0 and 1) as classes involved in the study. This would allow the GP system to perform a fuzzy classification, as opposed to the sharp and clear-cut classification performed by the current solution, which hardly represents the real conditions of most samples collected in the natural environment. This is being considered as future work.

Finally, the fitness function used in this work is still lacking an important feature. It can lead the GP system to the complete separation of classes, but once it gets there it does not promote any further separation, meaning it does not reward the solutions that present a larger distance between the classes. The result is that there is only a thin range of values from where to choose the thresholds of the final solution, something that will probably impair the performance in new data sets. The current fitness function can be extended to promote a larger distance between classes. Currently, it assumes values between 100 and 0, where the null value corresponds to the best cases, with no superposition of classes. In the future, the range of possible fitness values may reach below 0, where lower values represent cases with no superpositions and larger distances between classes, the truly ideal situation. But there are, of course, many other possibilities for improving the fitness function and the general performance of the entire GP system.

6 Conclusions

In this paper we have illustrated the usage of Genetic Programming on the classification of seafloor habitats. Dividing the multiple-class problem into several easier two-class problems, we have proposed a solution that was able to achieve nearly perfect results. From the five classes involved in our classification problem, only two could not be completely separated from each other. The quality of these results is a motivation for performing further validation with new data, and for developing other GP techniques appropriate for solving harder problems of seafloor habitat classification.

Acknowledgements

This work was partially supported by Fundação para a Ciência e a Tecnologia, Portugal (SFRH/BD/14167/2003). The authors acknowledge their PhD supervisors: Ernesto Costa from University of Coimbra, Portugal; Alexander Gavrilov and Alec Duncan from Curtin University of Technology, Australia. Thank you also to Michael Harwerth from Germany, for many fruitful discussions and suggestions regarding this work. Great thanks go to the 3-D still image system leader, Andrew Woods. The data collection was funded by Cooperative Research Centre for Coastal Zone, Estuary & Waterway Management. The construction of the synchronized 3-D still image system with the SIMRAD EQ60 echo sounder was mainly done in the Centre for Marine Science and Technology based at Curtin University of Technology under the Epi-benthic Scattering Project (ESP), which is a subproject of a bigger Coastal Water Habitat Mapping project. Thanks go to all ESP members, especially project leader Rob McCauley and project manager John Penrose.

References

1. Koza, J.R.: Genetic programming - on the programming of computers by means of natural selection. The MIT Press, Cambridge, Massachusetts (1992)
2. Banzhaf, W., Nordin, P., Keller, R.E., Francone, F.D.: Genetic programming - an introduction. Morgan Kaufmann, San Francisco (1998)
3. Smart, W., Zhang, M.: Probability based genetic programming for multiclass object classification. In: Zhang, C., W. Guesgen, H., Yeap, W.-K. (eds.) PRICAI 2004. LNCS (LNAI), vol. 3157, pp. 251–261. Springer, Heidelberg (2004)
4. Zhang, Y., Zhang, M.: A multiple-output program tree structure in genetic programming. In: McKay, R. (ed.) Proceedings of 2004 Asia-Pacific Workshop on Genetic Programming, Cairns, Australia, Springer, Heidelberg (2004)
5. Smart, W., Zhang, M.: Using genetic programming for multiclass classification by simultaneously solving component binary classification problems. In: Keijzer, M., Tettamanzi, A.G.B., Collet, P., van Hemert, J.I., Tomassini, M. (eds.) EuroGP 2005. LNCS, vol. 3447, pp. 227–239. Springer, Heidelberg (2005)
6. Tegowski, J., Gorska, N., Klusek, Z.: Statistical analysis of acoustic echoes from underwater meadows in the eutrophic puck bay (southern baltic sea). Aquatic Living Resources 16, 215–221 (2003)
7. Chakraborty, B., Mahale, V., de Sousa, C., Das, P.: Seafloor classification using echo-waveforms: a method employing hybrid neural network architecture. IEEE Geoscience and Remote Sensing Letters 1(3), 196–200 (2004)
8. Li, D., Azimi-Sadjadi, M.R., Robinson, M.: Comparison of different classification algorithms for underwater target discrimination. IEEE Transactions on Neural Networks 15(1), 189–194 (2004)
9. Siwabessy, P.J.W., Tseng, Y.-T., Gavrilov, A.N.: Seabed habitat mapping in coastal waters using a normal incident acoustic technique. In: ACOUSTICS-2004. Australian Acoustical Society, Gold Coast, Australia (2004)
10. Tseng, Y.-T., Gavrilov, A.N., Duncan, A.J., Harwerth, M., Silva, S.: Implementation of genetic programming toward the improvement of acoustic classification performance for different seafloor habitats. In: Oceans 2005 Europe, Brest, France, IEEE Press, Los Alamitos (2005)
11. Sun, R., Tsung, F., Qu, L.: Combining bootstrap and genetic programming for feature discovery in diesel engine diagnosis. International Journal of Industrial Engineering 11(3), 273–281 (2004)
12. Silva, S., Tseng, Y.-T.: Classification of seafloor habitats using genetic programming. In: GECCO 2005 Late Breaking Papers (2005)
13. Silva, S.: GPLAB - a genetic programming toolbox for MATLAB (2005), `http://gplab.sourceforge.net`
14. Silva, S., Costa, E.: Dynamic Limits for Bloat Control. In: Deb, K., et al. (eds.) GECCO 2004. LNCS, vol. 3103, pp. 666–677. Springer, Heidelberg (2004)
15. Luke, S., Panait, L.: Lexicographic parsimony pressure. In: Langdon, W.B., Cantú-Paz, E., Mathias, K., et al. (eds.) GECCO 2002, pp. 829–836. Morgan Kaufmann, San Francisco (2002)

Selecting Local Region Descriptors with a Genetic Algorithm for Real-World Place Recognition

Leonardo Trujillo[1], Gustavo Olague[1],
Francisco Fernández de Vega[2], and Evelyne Lutton[3]

[1] EvoVisión Project, CICESE Research Center, Ensenada, B.C. México
[2] Grupo de Evolución Artificial, Universidad de Extremadura, Mérida, Spain
[3] APIS Team, INRIA-Futurs, Parc Orsay Université 4, ORSAY Cedex, France
trujillo@cicese.mx, olague@cicese.mx, fcofdez@unex.es,
evelyne.lutton@inria.fr

Abstract. The basic problem for a mobile vision system is determining where it is located within the world. In this paper, a recognition system is presented that is capable of identifying known places such as rooms and corridors. The system relies on a bag of features approach using locally prominent image regions. Real-world locations are modeled using a mixture of Gaussians representation, thus allowing for a multimodal scene characterization. Local regions are represented by a set of 108 statistical descriptors computed from different modes of information. From this set the system needs to determine which subset of descriptors captures regularities between image regions of the same location, and also discriminates between regions of different places. A genetic algorithm is used to solve this selection task, using a fitness measure that promotes: 1) a high classification accuracy; 2) the selection of a minimal subset of descriptors; and 3) a high separation among place models. The approach is tested on two real world examples: a) using a sequence of still images with 4 different locations; and b) a sequence that contains 8 different locations. Results confirm the ability of the system to identify previously seen places in a real-world setting.

1 Introduction

Building an artificial system that is capable of answering the question *"Where am I?"* is one of the central problems studied in computer vision research. This task has only been partially solved in constrained real-world situations. To solve an instance of this problem, and many vision problems in general, three design issues must be accounted for [1] : 1) What information should be extracted from the output of visual sensors?, 2) How is the information extracted?, 3) How should the information be represented?, 4) How will it be used to solve higher-level tasks?

This contribution introduces a system that performs place recognition using only local image information and probabilistic models for each location. The design questions stated above are addressed in the following manner. Questions 1 and 4 are answered using common computer vision techniques that are applicable to different types of problems. The extracted information are local image regions, what corresponds to a *bags of features* approach [2,3,4]. In this way, the system can be robust to occlusions and avoids the need for prior segmentation. The information gathered from the images is used to create

M. Giacobini et al. (Eds.): EvoWorkshops 2008, LNCS 4974, pp. 325–334, 2008.

a probabilistic mixture model for each of the known locations. On the other hand, questions 2 and 3 were answered using design techniques based on evolutionary computation. More precisely, information is extracted using GP-evolved region detectors [5,6,7]. In addition, local image regions are represented by a subset of statistical descriptors that are selected by a genetic algorithm (GA) [4]. The proposal is related to vision based recognition, where a model correspondence is sought. This contrasts with image indexing approaches where specific image instances are retrieved or used for comparison.

This paper is organized as follows. Section 2 reviews related research and gives a working definition for the problem of place recognition. Section 3 presents a detailed description of the proposed method. Section 4 describes the experimental setup, the test sets used and the obtained results. Finally, concluding remarks and future perspectives are outlined in Section 5.

2 Related Work and Problem Definition

This section presents some examples of place recognition systems. Then, a more precise problem statement is given that illustrates the difficulty of the problem.

Related Work. Due to paper size considerations the review of past works is not exhaustive, however it does give a general overview of the type of approaches that are currently being used to solve the problem of place recognition. For instance, Torralba *et al.* [8] present a combined place and object recognition system that uses context dependent information. The system employs global features for place recognition, and object detection is contingent on the location that has been identified. The system also relies on the spatial relations between different locations, basically employing a topological map of the environment represented by a hidden Markov model. On the other hand, the present work is concerned with place recognition that only employs visual information without a Markovian assumption regarding the temporal and spatial relationships between frames of an image sequence. Therefore, the proposed method is closely related with the problem of object class recognition. Furthermore, instead of using a holistic image description, the current work relies on a sparse and local description of each image [2,3,4]. Another relevant example is the work by Wang *et al.* [9], where vision-based localization is performed for a mobile robotic system. In that work, the authors utilize scale invariant features detected using the Harris-Laplace detector and characterize each local region using the popular SIFT descriptor. Their approach is to utilize an image indexing technique, which contrasts with the recognition based approach presented here. Moreover, the use of SIFT limits the type of information that the system can employ for recognition purposes, more notably it excludes any type of color information. Both aforementioned examples show how different types of modelling are possible. Therefore, a more concrete definition for the problem of place recognition is necessary to clearly express the goal behind the current work.

Problem Statement. The place recognition problem can be defined as follows. Given a set of l physical real-world locations L, and a set of n representative images for each location, train a system capable of recognizing which location is being viewed in each frame of a sequence of test images that are different from the images used for training.

Fig. 1. The problem of place recognition. 1) First row: different views of the same location (Research Lab); 2) Second row: Images from four different locations (Students Lab, Computer Lab, Research Lab, and Office).

The only constraint is that the testing sequence only contains images from the l locations learned during training. This constraint could be easily relaxed by adding an *unknown* class to the list of locations.

In order to grasp the difficulty of the problem some visual examples are shown in Figure 1. The first row contains images from different views of the same location. A high degree of variability exists within this location, what represents a single class. The second row, contains images from four different locations; nevertheless, it is possible to observe that many features are shared amongst them. Indeed, finding the set of features that can best discriminate between classes and at the same time capture properties that are special to each class, represents a complex search problem. This work addresses these issues using an evolutionary search and a probabilistic modelling.

3 Outline of the Recognition System

With a clear understanding of the place recognition task, it is now possible to introduce each aspect of the current proposal. In accordance with the introductory discussion, this section starts by giving a description of each of the main design choices. Then, the GA learning loop is described, and finally the proposed recognition criteria are given.

Extracting Local Image Information. As stated above, the system employs sparsely distributed local image regions, also know as a "bags of features" approach. Salient image regions are extracted based on their distinctive properties that make them unique when compared with neighboring regions. Employing this approach has two principal advantages. First, relevant image regions are extracted without the need for prior segmentation thereby eliminating what is considered to be a very difficult task. Second, the extraction of these type of regions is robust to partial occlusions or scene variations.

In order to extract locally prominent regions, a scale adapted interest point detector is employed [7]. Selecting a characteristic scale for local image features is a process in which an operator is embedded into a linear scale-space, and extrema of the operator's response are identified at different scales [10]. The interest operator employed in the current work was synthesized with Genetic Programming and is optimized for high repeatability and global region separability [5,6]. The operator is named K_{IPGP1*} and is given by

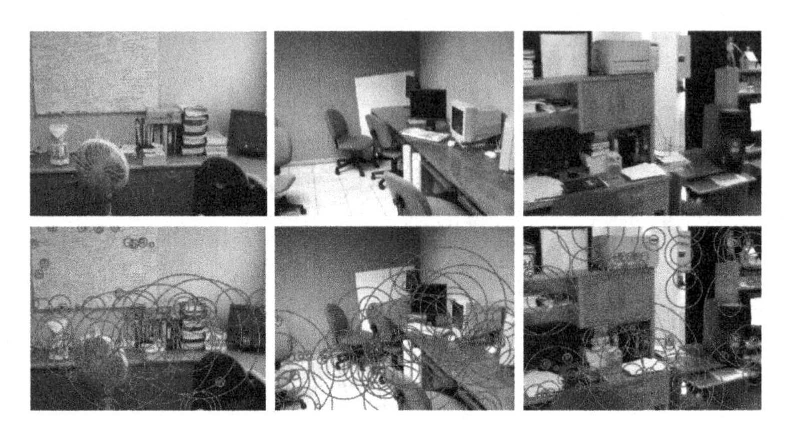

Fig. 2. Scale invariant regions detected on three test images. Top Row: original images; Bottom Row: detected regions. Columns contain a test image from one of the locations in Experiment 2 of this paper, from left to right: Students lab; Computer Lab; Research Lab.

Table 1. The complete feature space Φ

Features	Description
Gradient information	*Gradient*, *Gradient magnitude* and *Gradient Orientation* $(\nabla, \parallel \nabla \parallel, \nabla_\phi)$.
Gabor filter response	The sum of *Gabor filters* with 8 different orientations (gab).
Interest operators †	The response to 3 stable interest operators: *Harris*, $IPGP1$ and $IPGP2$ $(K_{Harris}, K_{IPGP1}, K_{IPGP2})$.
Color information	All the channels of 4 color spaces: *RGB*, *YIQ*, *Cie Lab*, and *rg chromaticity* $(R, G, B, Y, I, Q, L, a, b, r, g)$.

† K_{IPGP1} is proportional to a DoG filter, and K_{IPGP2} is based on the determinant of the Hessian [5,7].

$$K_{IPGP1^*}(\mathbf{x}; t_j) = G_{t_j} * |G_{t_j} * I(\mathbf{x}) - I(\mathbf{x})| \,, \tag{1}$$

with $j = 0, 1, ..., k$, and k is the number of scales to be analyzed, here it is set to $k = 5$. The size of a detected region is proportional to the scale at which it obtained its extrema value. For the sake of uniformity, all regions are scaled to a size of 41×41 pixels using bi-cubic interpolation before region descriptors are computed, as in [11]. Figure 2 shows sample interest regions extracted with the K_{IPGP1^*} operator. It is important to note, however, that the recognition system does not depend on any particular region detector. To exhibit this, the first experimental setup employs the Kadir and Brady detector that relies on a local entropy measure of intensity values [12].

Feature Space. Local image regions can be more discriminantly characterized using different types of numerical descriptors. In the current work, the space of all possible descriptors Φ includes 18 different modes of color and texture related information, see Table 1. To characterize the information contained along different information channels, six statistical descriptors are computed: *mean* μ, *standard deviation* σ, *skewness* γ_1, *kurtosis* γ_2, *entropy* H and *log energy* E. This yields a total of 108 possible descriptor values that can be used to model regions from each location.

Place Models. It is necessary to model how regions are mapped to descriptor space. It is expected that regions extracted from images of the same location, each offering a different view, will create distinctive clusters in Φ. Hence, when a test image is obtained and local regions are extracted, to determine which location those regions correspond with it is only necessary to map those regions to descriptor space and compute their class membership. This is a classification problem, and a Gaussian mixture model (GMM) is chosen to solve this task. GMMs are able to represent multimodal data, a property that can be expected from different image regions taken from a real-world location, see Figure 1. Formally, a GMM pdf is defined as,

$$p(\mathbf{x}; \Theta) = \sum_{c=1}^{C} \alpha_c \mathcal{N}(\mathbf{x}; \mu_\mathbf{c}, \Sigma_c) \,, \tag{2}$$

where $\mathcal{N}(\mathbf{x}; \mu_\mathbf{c}, \Sigma_c)$ is the *cth* multivariate Gaussian component with mean μ_c, covariance matrix Σ_c, and an associated weight α_c. Estimation of the mixture model parameters is done using the EM algorithm when a fixed number of components is assumed. Alternatively, if a variable number of component is desired, with a maximum bound, it is possible to use Figueiredo-Jain (FJ) algorithm [13]. Classification with GMMs can be easily done employing Bayes rule. Additionally, it is possible to estimate the amount of separation between two class models using a closed-form expression.

Fisher's Linear Discriminant. Fisher defined the separation between two distributions $\mathcal{N}_i$ and $\mathcal{N}_j$ as the following ratio

$$S_{i,j} = \frac{(\mathbf{w}(\mu_i - \mu_j))^2}{(\mathbf{w}^T(\Sigma_i + \Sigma_j)(\mathbf{w}))} \,, \tag{3}$$

where $\mathbf{w} = (\Sigma_i + \Sigma_j)^{-1}(\mu_i - \mu_j)$ [14]. Note that S is defined for unimodal pdfs, hence a weighted version $\widehat{S}$ that accounts for the weight α_i and α_j of the associated Gaussian components in a GMM is proposed, such that

$$\widehat{S}_{i,j} = \frac{S_{i,j}}{1 + \alpha_i + \alpha_j} \,. \tag{4}$$

Hence, the separation between components with a small combined weight (they have less influence over their associated models) will *appear* to be larger with respect to the separation between components with larger weights. Therefore, let C_a and C_b represent the number of components of $p_a(\mathbf{x}; \Theta_a)$ and $p_b(\mathbf{x}; \Theta_b)$ respectively, then $\mathcal{S}^{a,b}$ represents the *apparent* separation matrix of size $C_a \times C_b$ that contains the weighted separation $\widehat{S}_{i,j}$ of every component of p_a with respect to every component of p_b. The final *apparent* separation measure S between p_a and p_b is defined as

$$S^{a,b} = inf(\mathcal{S}^{a,b}) \,. \tag{5}$$

3.1 The GA Training Algorithm

The system recognizes a total of l different locations. For each location L, n different images are taken as *representative* views, these images are used for training. For every representative image, scale invariant regions are extracted and for all such regions the complete 108 descriptor values are computed off-line. Then, the GA performs feature selection and learns appropriate GMMs, one for each location, in a single run. The GA employs fitness proportional selection, mask crossover, single bit mutation and an elitist survival strategy. Solution representation and fitness evaluation are described next.

Solution Representation. Each individual in the population is coded as a binary string $B = (b_1, b_2, ...b_{108})$ of 108 bits. Each bit is associated with one of the statistical descriptors in Φ. Therefore, if bit b_i is set to 1 its associated descriptor will be selected, with the opposite being true if $b_i = 0$. The feature vector $\mathbf{x}_\lambda$ for each region λ is given by the concatenation of all the selected descriptors.

Fitness Evaluation. Here is where object models are learned and fitness is assigned. For every physical location L a corresponding GMM $p_L(\mathbf{x}; \Theta_L)$ is trained using an *all vs. all* [1] strategy. Only 70% of the regions extracted from the representative views are used for learning the GMMs, employing the descriptor values selected by B. This generates a set $\mathcal{P} = \{p_L(\mathbf{x}; \Theta_L)\}$ of GMMs, one for each location, with $|\mathcal{P}| = l$. Afterwards, the remaining 30% of image regions are used for validation to compute an accuracy score $\mathcal{A}$ using Bayes rule. Fitness is given for minimization by,

$$f(B) = \begin{cases} \dfrac{B^{\alpha}_{ones} + 1}{\mathcal{A}^2 \cdot inf(\mathcal{S}^{p_i,p_j})} \ \forall\, p_i, p_j \in \mathcal{P}\ , i \neq j\ , \ when\ \mathcal{A} > 0\ , \\[2ex] \dfrac{B_{ones} + 1}{\varepsilon} \qquad otherwise\ . \end{cases} \tag{6}$$

In the above equation, B_{ones} is the number of ones in string B, $\varepsilon = 0.01$, and α is a weight parameter. The first case in Eq. 6 is applied when the GMM training algorithm converges; fitter individuals will minimize the number of selected descriptors and maximize the average validation accuracy $\mathcal{A}$. Furthermore, the term $inf(\mathcal{S}^{p_i,p_j})$ promotes between-class model separation by selecting the infimum of all the apparent separation measures between all models in $\mathcal{P}$. The second case in Eq. 6 is applied when the training algorithm fails to converge.

Pruning the Descriptor Space with a Two-Stage GA. Previous results of a similar algorithm [4], used for object recognition, suggests that the described approach is capable of solving complex recognition problems. However, several limitations were noticeable. For instance, space Φ is quite large, thus all GA runs would converge towards models with only one Gaussian component in a large subspace of Φ, using between 27 and 43 dimensions. This made the use of GMMs completely unnecessary. Moreover, this makes the obtained solutions less desirable because of the large amount of numerical descriptors that they require. Therefore, in order to overcome these limitations a two-stage GA is proposed.

In the first stage, the GA runs using the process described above, with the parameter $\alpha = 2$ in Eq. 6. In this way, the term B_{ones} will have a greater influence on the fitness score. Thus, the GA will favor solutions that use a small subset of Φ. Additionally, due to the large dimensions of Φ, the EM algorithm is used for training with only one

[1] All vs all learning implies that all class models are learned in a single step or EM execution.

Fig. 3. Sample images for each of the four locations used in Experiment 1

Gaussian component in each model. After a fixed number of generations, set to 50 in all experiments, the best solution identifies a subspace of Φ denoted by Φ^*.

The second stage works the same as the first with three modifications. First, the search space employed is defined by Φ^* instead of Φ, what is normally a substantially more compact search space. Second, with the smaller search space the dimensions of the Gaussian components are expected to be smaller thereby encouraging a more multimodal characterization. Hence, the FJ algorithm is used for training the GMMs with maximum of 10 components per GMM. Third, the weight parameter in Eq. 6 is set to $\alpha = 1$ thereby reducing the influence that B_{ones} has over fitness and focusing fitness on the accuracy term $\mathcal{A}$. The output of this two-stage GA is a set $F \subset \Phi^*$ of descriptors that best characterizes the regions from each location L, and a set of trained GMMs $\mathcal{P}^o$ used to classify unseen test images.

Place Recognition. The final place recognition process proceeds as follows. Given an *unseen* image I from one of the known places, interest regions are detected and their corresponding descriptors, specified in F, are computed. The extracted regions are classified using Bayes rule with the models in $\mathcal{P}^o$. Afterwards, if a majority of the regions are classified to a model $p_L \in \mathcal{P}^o$ then it is said that location L is viewed in the imaged scene I.

4 Experimental Setup and Results

This section is divided in two parts, one for each experimental configuration.

Experiment 1. The first test for the proposed recognition system contains four locations: 1) Room, 2) WC, 3) Diner, and 4) Lounge. The training and test images were chosen from the same image sequence of 1 Mb color photos, representative images of each location are shown in Figure 3. The size of the images is larger than what is normal for this type of system, however the larger image allows the region detector to extract more image patches for training and testing. Regions were extracted using the entropy-based Kadir & Brady detector. In order to simplify the learning of GMM parameters, a max number of training regions was set to 3,500, which were randomly selected from all the regions extracted from the training images. From this subset, 30% are used in validation and the rest with the learning algorithm (EM or FJ) for the GMMs. Table 2 gives further details regarding the number of photos per location, the cardinality of each GMM learned, the total of test images,the number of misclassified images. Additionally, the first row of Table 3 presents the characteristics of the best individual found by the two-stage GA, describing: the fitness value, the size of Φ^*, the cardinality of F, the descriptors in F, and validation accuracy $\mathcal{A}$.

Table 2. Description of Experiment 1, setup and results

Location	Training Im.	GMM components	Test Im.	Error
Room	7	3	2	0
WC	5	6	2	0
Diner	9	4	2	0
Lounge	9	8	2	0

Table 3. Performance and selected features; see text for further details

Experiment	Fitness	$\|\Phi^*\|$	$\|F\|$	**Features**	$\mathcal{A}$
1)	0.0061	30	7	$\nabla_{\phi(\sigma)}, K_{IPGP1(\gamma_1)}, R_{(\mu,\sigma)}, Q_{(\gamma_2)}, a_{(\sigma)}, b_{(\mu)}$	70.75%
2)	0.0053	36	14	$gab_{(\mu,\gamma_2)}, K_{IPGP2(H)}, G_{(\sigma)}, B_{(\sigma)}, Y_{(\gamma_1)}, I_{(\mu,\sigma)}, Q_{(\sigma)}, L_{(\sigma)}, a_{(H)}, b_{(\mu,H)}, r_{(\sigma)}$	74.73%

Fig. 4. Sample images for each of the eight locations used for Experiment 2

Experiment 2. The second setup contains eight locations from our research center, these are: 1) Students Lab, 2) Computer Lab, 3) Research Lab, 4) Lockers, 5) 1st Floor, 6) 2nd Floor, 7) Office, and 8) Mail. Two sequences of images were taken from each location, one during the morning and the other in the afternoon thus providing different lighting conditions. Sample images of each location are shown in Figure 4. All images are color jpeg photos of size 320 × 340. In this experimental setup the K_{IPGP1^*} scale invariant detector is used, without any restrictions in the amount of regions that are used for training; Table 4 gives further details. Some locations have more training regions that do others, this is a result of the fact that some regions have many textured objects, such as the Office and all the Labs, while other locations are much simpler, such as the corridors on each floor, the Mail area, and the Locker area. The second row of Table 3 describes the best individual found by the GA. Additionally, Table 5 presents the confusion matrix for this experiment, here it is possible to see that most of the recognition errors occur with the simpler less textured places. This suggests that the

Table 4. Description of Experiment 2, setup and results

Location	Training Im.	Training Regions	GMM components	Test Im.	Error
Students Lab	17	5469	3	17	4
Computer Lab	14	5477	3	14	0
Research Lab	27	4941	2	26	12
Lockers	14	507	3	14	14
1st Floor	13	4534	3	12	11
2nd Floor	20	2263	2	19	12
Office	17	6756	2	16	1
Mail	10	1139	3	10	10

Table 5. Confusion matrix for Experiment 2

	Students	**Computer**	**Research**	**Lockers**	**1st Floor**	**2nd Floor**	**Office**	**Mail**
Students Lab	13	1	0	0	0	0	3	0
Computer Lab	0	14	0	0	0	0	0	0
Research Lab	0	2	14	0	0	0	10	0
Lockers	0	4	0	0	0	0	10	0
1st Floor	1	5	0	0	1	0	5	0
2nd Floor	0	0	2	0	0	7	10	0
Office	0	1	0	0	0	0	15	0
Mail	0	1	2	0	0	0	7	0

the learning algorithm builds more discriminant models for those regions with more training regions, something that can be expected beforehand.

5 Conclusions and Future Work

This paper described a learning approach to place recognition which is an essential task for any mobile vision system, such as those used by autonomous robots. This proposal relies on local scale invariant regions and builds probabilistic models using mixtures of Gaussians. The regions are described using statistical values related to texture and color information. The numerical descriptors used are chosen by a two-stage GA from a maximum of 108 different values. The evolutionary learning process searches for the smallest possible subset of descriptors, while also attempting to maximize classification accuracy and the distinctiveness of the GMMs that represent each physical location. Experimental results confirm the validity of the approach by solving two real-world problems of place recognition, and doing so using only visual information. However, results from Experiment 2 indicate that building a recognition system that only relies on visual cues represents a very difficult problem because of self-similarities between locations within most office buildings. Future extensions of this work will center on integrating the system with an autonomous robot in order to facilitate localization during *kidnapping* events in real time. Moreover, restrictions should be included, such as spatial relationships between different locations thereby making the recognition process more robust. Additionally, it would be of interest to expand the amount and type of descriptors available to the GA search.

Acknowledgements. Research funded by the Ministerio de Educación y Ciencia (project Oplink - TIN2005-08818-C04), the LAFMI project, and the Junta de Extremadura Spain. First author supported by scholarship 174785 from CONACyT México.

References

1. Faugeras, O.: Three-Dimensional Computer Vision (Artificial Intelligence). The MIT Press, Cambridge (1993)
2. Sivic, J., Russell, B.C., Efros, A.A., Zisserman, A., Freeman, W.T.: Discovering objects and their localization in images. In: Proceedings of the 10th IEEE International Conference on Computer Vision (ICCV 2005), Beijing, China, 17-20 October 2005, vol. 1, pp. 370–377. IEEE Computer Society, Los Alamitos (2005)
3. Willamowski, J., Arregui, D., Csurka, G., Dance, C., Fan, L.: Categorizing nine visual classes using local appearance descriptors. In: Proceedings of ICPR 2004, Workshop on Learning for Adaptable Visual Systems, Cambridge, United Kingdom, 23-26 August 2004, IEEE Computer Society, Los Alamitos (2004)
4. Trujillo, L., Olague, G., de Vega, F.F., Lutton, E.: Evolutionary feature selection for probabilistic object recognition, novel object detection and object saliency estimation using gmms. In: BMVC 2003: Proceedings of the 18th British Machine Vision Conference. British Machine Vision Association, vol. 2, pp. 630–639 (2007)
5. Trujillo, L., Olague, G.: Synthesis of interest point detectors through genetic programming. In: Cattolico, M. (ed.) Proceedings of GECCO 2006, vol. 1, pp. 887–894. ACM, New York (2006)
6. Trujillo, L., Olague, G.: Using evolution to learn how to perform interest point detection. In: Proceedings of ICPR 2006, Hong Kong, China, 20-24 August 2006, vol. 1, pp. 211–214. IEEE Computer Society, Los Alamitos (2006)
7. Trujillo, L., Olague, G.: Scale invariance for evolved interest operators. In: Giacobini, M. (ed.) EvoWorkshops 2007. LNCS, vol. 4448, pp. 423–430. Springer, Heidelberg (2007)
8. Torralba, A., Murphy, K.P., Freeman, W.T., Rubin, M.A.: Context-based vision system for place and object recognition. In: ICCV 2003: Proceedings of the Ninth IEEE International Conference on Computer Vision, Washington, DC, USA, p. 273. IEEE Computer Society, Los Alamitos (2003)
9. Wang, J., Zha, H., Cipolla, R.: Coarse-to-fine vision-based localization by indexing scale-invariant features. IEEE Transactions on Systems, Man, and Cybernetics, Part B 36(2), 413–422 (2006)
10. Lindeberg, T.: Feature detection with automatic scale selection. International Journal of Computer Vision 30(2), 79–116 (1998)
11. Mikolajczyk, K., Schmid, C.: A performance evaluation of local descriptors. IEEE Trans. Pattern Anal. Mach. Intell. 27(10), 1615–1630 (2005)
12. Kadir, T., Brady, M.: Saliency, scale and image description. International Journal of Computer Vision 45(2), 83–105 (2001)
13. Figueiredo, M.A.T., Jain, A.K.: Unsupervised learning of finite mixture models. IEEE Trans. Pattern Anal. Mach. Intell. 24(3), 381–396 (2002)
14. Fisher, R.A.: The use of multiple measurements in taxonomic problems. Annals of Eugenics 7, 179–188 (1936)

Object Detection Using Neural Networks and Genetic Programming

Barret Chin and Mengjie Zhang

School of Mathematics, Statistics and Computer Science
Victoria University of Wellington, P.O. Box 600, Wellington, 6140, New Zealand
{chinbarr,mengjie}@mcs.vuw.ac.nz

Abstract. This paper describes a domain independent approach to the use of neural networks (NNs) and genetic programming (GP) for object detection problems. Instead of using high level features for a particular task, this approach uses domain independent pixel statistics for object detection. The paper first compares an NN method and a GP method on four image data sets providing object detection problems of increasing difficulty. The results show that the GP method performs better than the NN method on these problems but still produces a large number of false alarms on the difficult problem and computation cost is still high. To deal with these problems, we develop a new method called *GP-refine* that uses a two stage learning process. The new GP method further improves object detection performance on the difficult detection task.

1 Introduction

This paper is concerned with object detection, the task of finding objects of interest in large images. Object detection has many applications such detecting clones in satellite images, finding tumours in a set of X-ray images, and finding a particular human face from a set of images containing human photographs.

Neural networks (NNs) and genetic programming (GP) are two powerful techniques in computational intelligence [1,2]. Since the 1990s, NNs and GP have been used in many object detection tasks [3,4,5,6,7]. In most existing approaches, high level features are used as inputs to NNs and genetic programs. While such approaches have achieved some success, they often involve a time consuming investigation of important specific features and a hand crafting of feature extraction programs. Another problem of using these two methods for object detection is that they often produce a large number of false alarms.

The goal of this paper is to investigate a domain independent approach to the use of NNs and GP for object detection problems. Instead of using raw pixels as inputs to neural networks and genetic programs, this approach uses domain independent pixel statistics for object detection to avoid too large networks or too large evolved programs. The two methods will be examined on four image data sets providing object detection problems of increasing difficulty. We will investigate whether NNs and GP can achieve good performance on the four problems and how the object detection performance can be improved using a set of good features automatically selected by GP.

M. Giacobini et al. (Eds.): EvoWorkshops 2008, LNCS 4974, pp. 335–340, 2008.

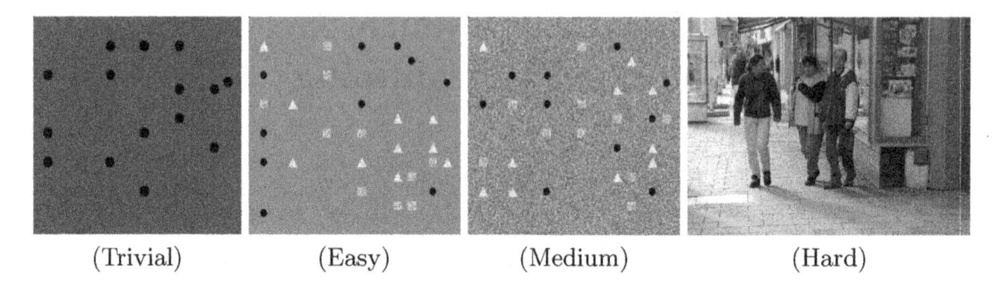

(Trivial) (Easy) (Medium) (Hard)

Fig. 1. Object Detection Problems

2 Image Data Sets

Four different data sets providing object detection problems of increasing difficulty were used in the experiments. Figure 1 shows example images for each data set. The three shapes data sets were generated to simulate a particular obstacle in object detection. Data set 1 is a trivial problem containing only one type of shape against a uniform background. Data set 2 (Easy) introduces three different types of shape objects on a relatively uniform background, but the detection program will need to regard all the three types of objects as a single type. This makes the problem harder but is still easy. In data set 3 (Medium), the objects are similar to those in data set 2, but the background is very noisy and the intensities of the grey squares are very similar to the background. Data set 4 (Hard) consists of a set photographic images taken from the PASCAL OR Database collection [8]. The detection problem here is very hard as there exist many different types of backgrounds as well as faces of varying shapes and sizes.

In the three shape data sets, 300 "objects" were cut out from the images in each data set, 150 for the objects of interest and 150 for background samples. For the hard face data set, 106 objects were cut out from the images where half were viewable faces and half were background examples.

3 The Baseline Approach: NNs vs GP

3.1 Overall Baseline Object Detection Approach

1. Assemble a database of images in which the locations and classes of all the objects of interest are manually determined. These full images are divided into two sets: a training set and a test set.
2. A classification image data set is to be created by cutting out squares of size $n \times n$ from the training set. Each object cutout image either contains a single object or a background example.
3. Use either NNs or GP to learn a classifier that can well separate the object examples from the background examples in the classification data set.
4. The trained classifier (either a trained NN or an evolved genetic program) is then used as a detector, in a moving window template fashion, to locate the objects of interest in the full images in the test set. An object is reported

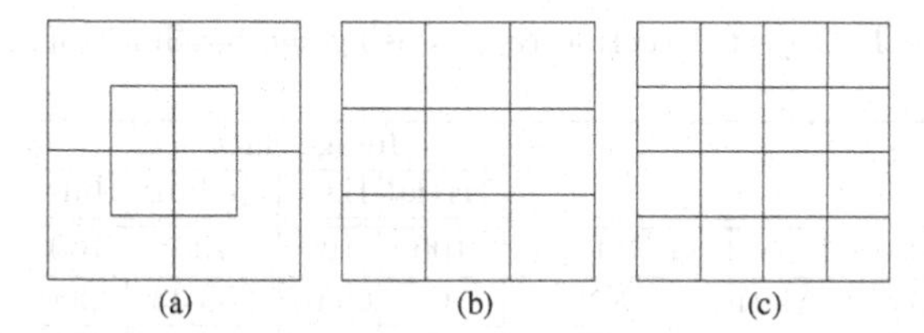

Fig. 2. Three sets of regions. (a) six rectilinear regions; (b) nine square regions; (c) 16 square regions.

based on the network activation values (the NN method) or the output value of the genetic program (GP method).
5. Measure the object detection performance by calculating the detection rate and false alarm rate (number of false alarms per object) in the test set.

3.2 The NN Method

In the NN method, we used the three-layer feed forward network architecture. The number of input nodes is the number of image features and the number of output nodes is 2 (one for object and the other for background). The number of nodes in the single hidden layer was determined based on an empirical search through experiments. The networks were trained by the back propagation algorithm [1] without momentum. The winner-takes-all strategy was applied to the activation values of the output nodes for classification: the class with the larger value is considered the class of the object.

To meet the requirements of domain independence, we used low level pixel statistics as image features. As shown in figure 2 (a), we consider six regions (the whole window, the central square, and the four rectilinear regions) from which the mean and standard deviation pixel statistics are extracted as the features. This gives a total number of 12 features. The network architectures and parameter values can be seen from [9].

3.3 The GP Method

In the GP method, we used tree structures to represent genetic programs[10]. The ramped half-and-half method was used for generating the programs in the initial population. The proportional selection mechanism and the reproduction, crossover and mutation operators were used in the learning process.

The terminal set used in GP consists of the 12 features extracted from the six regions described earlier and a random constant. The function set consisted of the four standard arithmetic operators and a conditional operator: $\{+, -, *, /, if\}$. For the `if` operator, if the first argument is positive, it returns its second argument; otherwise, it returns its third argument.

The GP method used the classification accuracy as the fitness function. A genetic program produces a floating point number where this output value determines whether an input field in the large test image contains an object or not.

Table 1. Object detection results using the baseline approach

		Image data sets			
		Trivial	Easy	Medium	Hard
Best Detection Rate(%)		100	100	100	100
False Alarm Rate (%)	NN	0	66.67	253.33	3300
	GP	0	0	0	250

If the output of a genetic program is positive, then the input field is considered to contain an object; otherwise, it is considered background.

The important parameter values used in this paper can be seen from [9]. The learning/evolutionary process is terminated at a pre-defined maximum number of generations unless a successful solution is found, in which case the evolution is terminated earlier.

3.4 Results and Discussion

For both the NN and GP methods, the experiments were repeated 20 times and the average detection results are shown in table 1. Both methods achieved 100% detection rate on all the four data sets, meaning that all the objects of interest were successfully detected from the large images. Except for the Trivial data set, while the NN method produced a large number of false alarms for all the other three data sets, the GP method did much better particularly for the Easy and the Medium data sets where ideal results were also achieved.

However, the GP method still has two key limitations. The first is the object detection performance for the Hard Faces data set still did not reach an acceptable level. The second is the computational cost for the evolutionary process was still quite high. The next section address these limitations by introducing an initial feature refinement phase to the GP method.

4 GP-Refiner

Further inspection of the evolved programs generated by the baseline GP method reveals that there were some very complex combinations of the different features in the programs. We suspect that one of the reasons behind this is that the six regions used earlier were too abstract and the use of more regions might help. In addition, many evolved genetic programs do not use all of the 12 available features and some features were chosen multiple times. This implies that some features are more important than others.

Based on the two observations, we proposed a new method called *GP-Refiner* to address the limitations of the baseline GP method. The main idea of the new method is to use more local region features as terminals and introduce a "feature refinement" phase into the evolutionary process for training the classifier in the GP method. So the new GP-refine method is a two-phase approach.

Table 2. Object detection results using the GP-Refine method

		Image data sets			
		Trivial	Easy	Medium	Hard
Best Detection Rate(%)		100	100	100	100
False Alarm Rate (%)	9-regions	0	33.33	86.67	1050
	16-regions	0	0	6.67	50

The first phase is called the *feature refinement* phase, where all the local region features are used as the terminal set and the GP evolution is performed over the classification data set just as in the baseline GP method. Two new sets of local regions from which the features were extracted to form two terminal sets, as shown in Figure 2 (b) and (c). In both sets, the mean and standard deviations of each region are used as the features, so 18 and 32 region features construct the two terminals sets respectively. The best programs at all generations are recorded and the statistical usage frequency of the individual features is reserved. Based on this information, top ten features are chosen to form a new terminal set. We expect that these features are more important than other features and they can do a good job for object detection.

The second phase is the same as the evolutionary training phase in the baseline GP method except that only the refined features selected from the first phase are used in the new terminal set. Using a smaller terminal set, together with a small program depth limit allowed to grow during evolution, the evolutionary training process would be more efficient. Since the calculation of the "redundant" features from the input field in the large testing images will be omitted, the object detection process would also be more efficient. The parameter values and termination criteria can be seen from [9] for detail.

4.1 Results and Discussion

Table 2 shows the average detection results of the new GP-refine method with the two sets of region features. Compared with the baseline GP results in table 1, the new GP-refine method achieved worse results with nine region features but achieved better overall results when 16 local regions were used in the feature refinement phase. While the results for the medium data set obtained by the new GP method with 16 regions were slightly worse than the baseline GP method, the difference was really small. In particular, it achieved much better detection results on the Hard face data set. This suggests that with sufficient number of local region features, the new GP-refine method can successfully select the important features for a particular task and achieve better results for object detection. We also observed that the computation costs for evolutionary learning and object detection testing were reduced by about 30–35% compared with the baseline GP method (details are not shown due to page limit). Another observation is that the new GP-refine method with the two local region sets achieved better results than the baseline NN method.

5 Conclusions

The goal of this paper was to develop a domain independent approach to object detection using neural networks and genetic programming. The goal was successfully achieved by using domain independent low-level pixel statistics, investigating a baseline NN method and a baseline GP method, and developing a new GP-refine method tested on four object detection problems of increasing difficulty. The results show that the baseline GP method performed better than the NN method on these problems but still produced a large number of false alarms on the difficult problem and computational cost was still high. The new GP-refine method achieved better object detection performance and the computational cost was also reduced over the baseline NN and GP methods.

There are a number of interesting points derived from this work, including which regions are more important than others for a particular task, and what types of features are more important for specific tasks. Our initial analyses suggest that the trained object classifier favoured the use of standard deviations over the mean features when the detection problem got more difficult, but further investigation is needed in the future.

References

1. Rumelhart, D.E., Hinton, G.E., Williams, R.J.: Learning internal representations by error propagation. In: Rumelhart, D.E., McClelland, J.L. (eds.) the PDP research group: Parallel distributed Processing, Explorations in the Microstructure of Cognition: Foundations, vol. 1, The MIT Press, Cambridge (1986)
2. Koza, J.R.: Genetic programming: on the programming of computers by means of natural selection. MIT Press, Cambridge (1992)
3. Shirvaikar, M.V., Trivedi, M.M.: A network filter to detect small targets in high clutter backgrounds. IEEE Transactions on Neural Networks 6, 252–257 (1995)
4. Casasent, D.P., Neiberg, L.M.: Classifier and shift-invariant automatic target recognition neural networks. Neural Networks 8, 1117–1129 (1995)
5. Roberts, M.E., Claridge, E.: Cooperative coevolution of image feature construction and object detection. Technical report, University of Birmingham, School of Computer Science (2004)
6. Howard, D., Roberts, S.C., Brankin, R.: Target detection in SAR imagery by genetic programming. Advances in Engineering Software 30, 303–311 (1999)
7. Zhang, M., Andreae, P., Pritchard, M.: Pixel statistics and false alarm area in genetic programming for object detection. In: Raidl, G.R., Cagnoni, S., Cardalda, J.J.R., Corne, D.W., Gottlieb, J., Guillot, A., Hart, E., Johnson, C.G., Marchiori, E., Meyer, J.-A., Middendorf, M. (eds.) EvoWorkshops 2003. LNCS, vol. 2611, pp. 455–466. Springer, Heidelberg (2003)
8. The pascal object recognition database collection (2007), `http://www.pascal-network.org/challenges/VOC/databases.html`
9. Chin, B., Zhang, M.: Object detection using neural networks and genetic programming. Technical report, School of MSCS, Victoria University of Wellington (2007), `http://www.mcs.vuw.ac.nz/comp/Publications/CS-TR-07-3.abs.html`
10. Koza, J.R.: Genetic Programming II: Automatic Discovery of Reusable Programs. MIT Press, Cambridge (1994)

Direct 3D Metric Reconstruction from Multiple Views Using Differential Evolution

Luis G. de la Fraga and Israel Vite-Silva

Cinvestav, Computer Science Department.
Av. Instituto Politécnico Nacional 2508. 07360 México, D.F. México
fraga@cs.cinvestav.mx

Abstract. In this work we propose the use of Differential Evolution as a numerical method to estimate directly the orientation and position of several views taken by a same camera. First, from two views, the camera parameters are estimated and also the other parameters (orientation and position) for both views. We can estimate 3D points in this configuration and then, in a second step, new views, and also new 3D points, are added to the initial reconstruction calculated in the first step. We tested our approach with a simulation using four views. The results clearly show the good performance of the proposed approach.

Keywords: Computer Vision, Metric Reconstruction, Parameters Estimation, Direct Metric Reconstruction, Multiple Views, Differential Evolution.

1 Introduction

One of the main problems in Computer Vision, that has been deeply studied, is the three-dimensional metric reconstruction from two images using uncalibrated cameras [1,2,3,4]. Horn in [5] obtained a metric reconstruction from cameras with known intrinsic parameters. It is well known that it is possible to obtain a 3D projective reconstruction using point correspondences from only two images with the aid of the fundamental matrix [1,2,3]. A metric reconstruction can be performed if the intrinsic parameters are calculated. For this reason, the main problem in the metric reconstruction process has been the camera calibration (to estimate the intrinsic parameters) and retrieve the three dimensional points in a Euclidean space.

In this article we present a new approach to obtain a direct metric reconstruction –without using the fundamental matrix– from several views, using the Differential Evolution (DE) as a method to estimate all the parameters involved to obtain a 3D reconstruction.

Hartley in [2] proposed the direct metric reconstruction, however, this is a non linear problem and if a conventional numerical method is used (e.g. Levenberg-Marquardt) an initial solution is necessary, very close to the final solution, to get the convergence when only have two views.

M. Giacobini et al. (Eds.): EvoWorkshops 2008, LNCS 4974, pp. 341–346, 2008.

This is not a problem for DE because it is an evolutionary algorithm, and also DE can deal with the estimation of many parameters directly, as it is the case in the direct metric reconstruction using initially two views and then adding more views.

The DE is a search heuristic technique developed by Rainer Storn and Kenneth Price in 1995 [6] to minimize linear, non linear, and non differentiable functions over continuous spaces.

For DE implementation we used the rand/1/bin because it is robust and provides the best results for different kind of benchmarks and real optimization problems [7].

2 Problem Formulation

Under an image system based in a pinhole camera, exists a perspective projection that transforms 3D coordinates to 2D points over an image. This is represented by $\lambda\mathbf{p} = M\mathbf{P}$, where $\mathbf{p}$ is a homogeneous 2D point, $\mathbf{P}$ is a homogeneous 3D point, λ is a scale factor, and M is a matrix of size 3×4 and range 3, that can be decomposed as:

$$M = K[R|\mathbf{t}] \tag{1}$$

where $[R|\mathbf{t}]$ is the 3×4 matrix determining the relative orientation R and position $-R^T\mathbf{t}$ of the camera with respect to the world coordinates, and K is the 3×3 matrix defining the pinhole camera:

$$K = \begin{bmatrix} f & 0 & u_0 \\ 0 & f & v_0 \\ 0 & 0 & 1 \end{bmatrix} \tag{2}$$

where f stand for the focal length, and u_0 and v_0 are the coordinates of the principal point, or the coordinates of the intersection of the optical axis with the image plane. These parameters contained in K are called the *intrinsic camera parameters*. The parameters of R and $\mathbf{t}$, which relate the camera orientation and position to a world coordinate system are called the *extrinsic* parameters.

The matrix M transforms a 3D point $\mathbf{P} = [x_w, y_w, z_w, 1]^T$ to a point $\mathbf{p} = [x, y, 1]^T$ on a image using $\lambda[x, y, 1]^T = M\mathbf{P}$.

The two optimization problems that we need to solve are:

Problem 1: Given a set of n corresponding points $\mathbf{p} \Leftrightarrow \mathbf{p}'$ over two images, find the parameters f, u_0, v_0 for the camera, α, β, γ, (orientation parameters), t_1, t_2, t_3 (position parameters) for the first view; and α', β', γ', t_1', t_2', t_3' (for the second view), subject to minimize:

$$g = \sum_{i=1}^{n} \left[d(\mathbf{p}_i, \hat{\mathbf{p}}_i) + d(\mathbf{p}_i', \hat{\mathbf{p}}_i') \right], \tag{3}$$

where $\hat{\mathbf{p}} = \hat{M}\hat{\mathbf{P}}$, $\hat{\mathbf{p}}' = \hat{M}'\hat{\mathbf{P}}$, M is expressed as in (1), function $d(*, *)$ denotes the Euclidean distance, $R = R_z(\gamma)R_y(\beta)R_z(\alpha)$, and $\mathbf{t} = [t_1, t_2, t_3]^T$. Also, for every pair of corresponding points it is necessary to estimate a 3D point $\hat{\mathbf{P}}$. Therefore, for n pairs of corresponding points, we need to estimate $3n$ values for n 3D points, 6 parameters for each view, and 3 for the camera, thus it is a total of $3n + 15$ parameters.

Problem 2: For a new view with n $\mathbf{p}''$ points. We suppose that we have $\mathbf{P}$, a set of fixed 3D points (they could be the result of solve problem 1), and a camera with known parameters, represented as K. Then we need find α'', β'', γ'', and t_1'', t_2'', t_3'' (six parameters) subject to minimize:

$$h = \sum_{i=1}^{n} d(\mathbf{p}_i'', \hat{\mathbf{p}}_i''), \tag{4}$$

where $\hat{\mathbf{p}}'' = K[\hat{R}|\hat{\mathbf{t}}\ ]\mathbf{P}$, and $\hat{R}$ is calculated as $\hat{R} = R_z(\gamma'')R_y(\beta'')R_z(\alpha'')$, and $\hat{\mathbf{t}} = [t_1'', t_2'', t_3'']^T$.

3 Algorithms for Direct 3D Metric Reconstruction

To obtain a metric 3D reconstruction from a set of m views, and k points per view, we use Algorithm 1.

Algorithm 1. Direct 3D reconstruction for n views

Require: n views and k_n points selected in each view.

Ensure: The intrinsic camera parameters, the extrinsic parameters corresponding to each view, and the set $\mathbf{P}$ of 3D points.

1: Choose two views and solve Problem 1 with them.
2: Print intrinsic parameters f, u_0, and v_0
3: Print extrinsic parameters for view 1 and 2
4: **for** $3 \leq i \leq n$ **do**
5: For the rest of views, solve Problem 2. The set of k_i points can belong to different 3D points than the obtained before. Therefore new 3D points are added to the set $\mathbf{P}$ of 3D points.
6: Print extrinsic parameters for view i and new 3D points.

The DE algorithm is composed by a population with m individuals. Each individual $\mathbf{I}_{(g,i)}$ in the generation g and $1 \leq i < m$, is associated with a vector $\mathbf{v}$ of the problem variables, and their values are a possible solution. To perform line 1 in Algorithm 1 a procedure to evaluate each individual in the DE algorithm is necessary. This procedure is shown in Algorithm 2.

To calculate line 5 in Algorithm 1 using DE, only a procedure to evaluate one individual is necessary, and this procedure is show in Algorithm 3. Note in line 4 of Algorithm 3 that we add a new scale parameter into individuals in order to calculate the estimated 2D points.

Algorithm 2. One individual evaluation to reconstruct from two views

Require: An individual $\mathbf{I}_{(g,i)}$.
Ensure: The value of the function g, in (3), for the given individual.
1: Set the extrinsic parameters for the first view (α, β, γ) and (t_1, t_2, t_3) from $\mathbf{I}_{(g,i)}$.
2: Set the parameters for the second view $(\alpha', \beta', \gamma')$ and (t'_1, t'_2, t'_3) from $\mathbf{I}_{(g,i)}$.
3: Set the focus, f from $\mathbf{I}_{(g,i)}$, and the coordinates of the principal point (u_0, v_0); these parameters are the same for both views.
4: Generate the two projection matrices $M = K[R|\mathbf{t}]$ and $M' = K[R'|\mathbf{t}']$.
5: **for** $1 \leq j \leq k$ **do**
6: Estimate the 3D points $\mathbf{X}_j$ with the lineal non-homogeneous triangulation method (it uses normal equations), with $\mathbf{x}_j$, $\mathbf{x}'_j$, M, and M'.
7: Calculate the estimated 2D points, $\hat{\mathbf{x}}$ and $\hat{\mathbf{x}}'$.
8: Calculate the value of function g in (3).

Algorithm 3. One individual evaluation to add more views

Require: An individual $\mathbf{I}_{(g,i)}$.
Ensure: The value of the function h, in (4), for the given individual.
1: Set the extrinsic parameters for the new view (α, β, γ) and (t_1, t_2, t_3) from $\mathbf{I}_{(g,i)}$.
2: Set the intrinsic camera parameters, focus f, principal point (u_0, v_0) coordinates, from $\mathbf{I}_{(g,i)}$.
3: Set the scale factor s from $\mathbf{I}_{(g,i)}$.
4: Calculate the projection matrix $M = K[R|s\mathbf{t}]$.
5: Calculate the estimated points $\hat{\mathbf{x}}$ using M and the fixed 3D points.
6: Calculate the reprojection error using (4).

4 Experiments and Results

We performed an experiment with simulated data to test our approach. The values for DE were: population size: 30; recombination constant: 0.9; difference constant: 0.84. For the Problem 1 we used 10,000 generations. The 3D points are not estimated directly by DE. To solve Problem 2 we used 5,000 generations.

Both algorithms used the following intervals for the variables employed in the problem. Extrinsic parameters: α, β, γ (the rotation angles) have the interval $[0, 2\pi]$; t_1, t_2, t_3 (the translation vector) have the interval $[-1, 1]$. Intrinsic parameters: the focal length f has the interval $[50, 1500]$; each coordinate of the principal point, (u_0, v_0) has the interval $[4w/10, 6h/10]$, where w is the image width and h is the image height.

We built a synthetic object that is an arc of 100 degrees and is shown in Fig. 1. This arc has 60 three-dimensional points, but for the experiment only 48 points were reconstructed. We generated four views with different projection matrices. For each view, we take only the bidimensional visible points. The Fig. 2 shows the four views with the visible points. We also add RMS Gaussian noise (1, 2, and 3 pixels) for each 2D point in order to test the robustness of the algorithm. With these conditions, we execute 40 times the proposed approach.

In Fig. 3(a) we show the 3D reconstruction obtained from two views and without noise. The Fig. 3(b) shows the new 3D reconstruction resulting from adding the third view; here 12 points were added to the previous reconstruction. These new 3D points were obtained from corresponding points in the second and third views (and these corresponding points are different than the corresponding points on the first and second views). Finally, Fig. 3(c) shows the re-updated 3D reconstruction after adding the fourth view, and also 12 3D points were added from corresponding points on the third and fourth views.

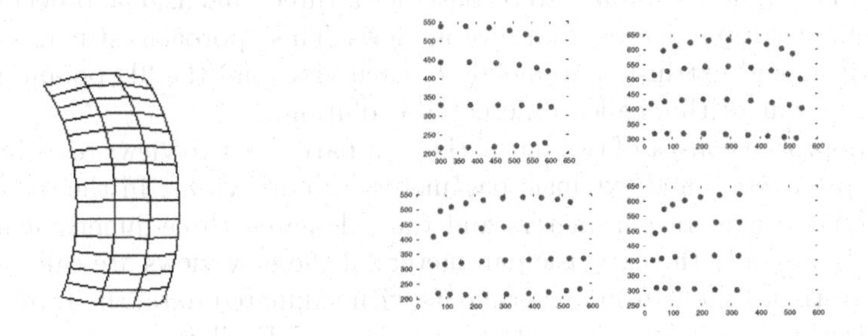

Fig. 1. The 3D synthetic object used in the experiment

Fig. 2. The four views used in the experiment

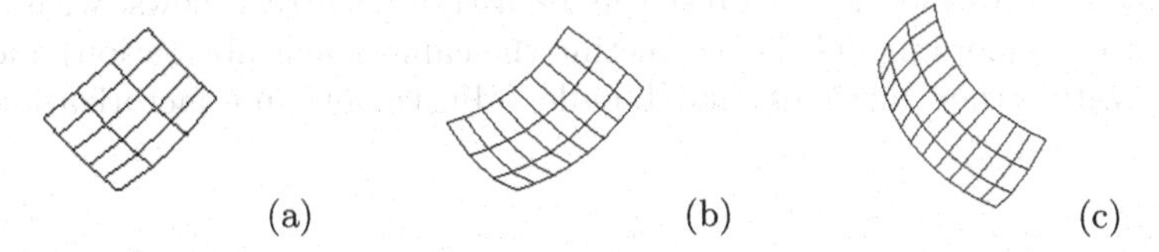

(a) (b) (c)

Fig. 3. 3D reconstruction obtained from 1st and 2nd views in (a). In (b) the reconstruction from three views, and in (c) the final reconstruction obtained from the points without noise.

Table 1. Statistics for the mean of the error resulted in the 40 runs of the proposed algorithm by each part of the algorithm

Algorithm	Points	Noise free	1 pixel of noise	2 pixels of noise	3 pixels of noise
Two Views	24	0.00	0.76	1.39	2.10
Third view	12	0.06	0.92	2.20	3.34
Fourth view	12	0.05	0.82	1.65	2.73

The table 1 shows the statistics of the mean error from the 40 runs. The error is the value of function in (3) divided by the total number of points. Also, table 1 shows the error obtained per each part of the algorithm: the error in the reconstruction obtained with two views, the error only in the third view, and the error in the fourth view. The table 2 shows the mean error for the overall method for each algorithm. We can see that obtained error is almost the same than the used noise, therefore the 3D reconstruction is correct.

Table 2. Reconstruction mean error after adding each view

Algorithm	Points	Noise free	1 pixel of noise	2 pixels of noise	3 pixels of noise
Two views	24	0.00	0.76	1.39	2.10
Three views	36	0.03	1.22	2.49	3.77
Four views	48	0.03	1.08	2.21	3.42

5 Conclusions and Future Work

We presented a new approach to reconstruct a three-dimensional object directly from point correspondences from several views. This approach estimates directly the intrinsic and extrinsic parameters of each view and the 3D points using an evolutionary algorithm called differential evolution.

The approach consist of two parts: the first part uses two views to estimate the camera parameters and extrinsic parameters of both views. In the second part the estimated camera parameters and the calculated three-dimensional points are fixed, and only the extrinsic parameters of the new views are calculated.

The statistics show, from experiments with simulated data, the robustness of the algorithm even with noisy 2D points. Using DE allows us measure the reprojection error as the sum of Euclidean distances, and the problem of minimize this error can not be solved with a conventional numerical method because the square root function has not a continuos derivative.

As future works we need to test the algorithm with real views, with a more complicated camera model (i.e. correcting the camera lens aberration) and compare it with other methods like bundle adjustment and factorization with RANSAC strategy.

References

1. Longuet-Higgins, H.: A computer algorithm for reconstructing a scene from two projections. Nature 293, 133–135 (1981)
2. Hartley, R.I.: Euclidean reconstruction from uncalibrated views. In: Proceedings of the Second Joint European - US Workshop on Applications of Invariance in Computer Vision, pp. 237–256. Springer, London (1994)
3. Sturm, P.: A case against kruppa's equations for camera self-calibration. IEEE Trans. Pattern Anal. Mach. Intell. 22(10), 1199–1204 (2000)
4. Kanatani, K., Nakatsuji, A., Sugaya, Y.: Stabilizing the focal length computation for 3D reconstruction from two uncalibrated views. Int. J. Comput. Vision 66(2), 109–122 (2006)
5. Horn, B.K.P.: Relative orientation revisited. Journal of the Optical Society of America 8(10), 1630–1638 (1991)
6. Price, K.V., Storn, R.M.: Differential evolution. Dr. Dobbs Journal 22(4), 18–24 (1997)
7. Mezura, E., Velázquez, J., Coello, C.A.: A comparative study of differential evolution variants for global optimization. In: GECCO 2006: Proceedings of the 8th annual conference on Genetic and evolutionary computation, pp. 485–492. ACM Press, New York (2006)

Discrete Tomography Reconstruction through a New Memetic Algorithm

Vito Di Gesù[1,2], Giosuè Lo Bosco[1], Filippo Millonzi[1], and Cesare Valenti[1]

[1] Dipartimento di Matematica e Applicazioni
[2] Centro Interdipartimentale Tecnologie della Conoscenza
Università di Palermo, via Archirafi 34, 90123, Italy
{digesu,lobosco,millonzi,cvalenti}@math.unipa.it

Abstract. *Discrete tomography* is a particular case of computerized tomography that deals with the reconstruction of objects made of just one homogeneous material, where it is sometimes possible to reduce the number of projections to no more than four. Most methods for standard computerized tomography cannot be applied in the former case and ad hoc techniques must be developed to handle so few projections.[1]

1 Introduction

Computerized tomography consists of the recovering of a three-dimensional object from its projections. If the object consists of materials with different densities, then 500-1000 projections are necessary, while if the object is made of just one homogeneous material, sometimes it is possible to use just a few projections, so defining the so called *discrete tomography.* This field of research is young and no *general* reconstruction algorithm has been developed so far. The complexity of the reconstruction process is NP-hard for at least three projections along non parallel directions [1]. Moreover, the number of images that satisfy two projections is normally huge [2], while the exact reconstruction needs many projections and a model of the object.

This paper considers the reconstruction of binary images, corresponding to two densities, through a memetic algorithm and four projections (see Sect. 2). A comparison with another evolutionary approach is presented in Sect. 3, with preliminary results that consider noisy projections. Sect. 4 concludes with future progresses and possible applications.

2 A New Memetic Reconstruction Method

Discrete tomography algorithms have been developed to deal with particular binary images such as *hv-convex polyominoes* [3,4,5], which are connected sets with

[1] This work makes use of results produced by the PI2S2 Project managed by the Consorzio COMETA, a project co-funded by the Italian Ministry of University and Research (MIUR) within the Piano Operativo Nazionale "Ricerca Scientifica, Sviluppo Tecnologico, Alta Formazione" (PON 2000-2006). More information is available at http://www.pi2s2.it and http://www.consorzio-cometa.it.

M. Giacobini et al. (Eds.): EvoWorkshops 2008, LNCS 4974, pp. 347–352, 2008.

4-connected rows and columns, and *periodic images* [6], which have repetitions of pixels along some directions. The effects of quantization and instrumental errors have been evaluated in [7]. Genetic algorithms have been applied to reconstruct circular or elliptical objects [8] and images from interferometric fringes [9]. A memetic approach, different from our method has been proposed in [10]. Our fitness function evaluates the *quality* of the individuals and the selection pressure improves each *agent* (i.e. chromosome, in genetic terminology) to direct the whole population toward the solution. It induces a *vertical evolution* (i.e. between consecutive generations) and a *horizontal evolution* (i.e. within the same generation). All operators are applied in the following order, without probability constraints, on each generation only if they improve the agent. The initial population is created by using their corresponding network flows thus to satisfy two of the input projections of the image to be reconstructed.

2.1 Basic Notations

Let us represent a binary image I with $n \times m$ pixels by a matrix $A = \{a_{ij}\}$ with elements equal to 0, if the corresponding pixel in I is black (i.e. it belongs to the background), or equal to 1, if the pixel is white (i.e. it belongs to the foreground). The *projection line* passing through a_{ij} with direction $\mathbf{v} \equiv (r, s)$, where $r, s \in \mathbb{Z}$ and $|r|+|s| \neq 0$, is the subset of A (see Fig. 1):

$$\ell_{\mathbf{v}}(i,j) = \{a_{i'j'} \in A \ : \ i' = i + zs, \ j' = j - zr \text{ with } z \in \mathbb{Z}\}.$$

To simplify the method, we considered only $\mathbf{v}_1 \equiv (1,0)$, $\mathbf{v}_2 \equiv (0,1)$, $\mathbf{v}_3 \equiv (1,1)$, $\mathbf{v}_4 \equiv (1,-1)$. Let $t(\mathbf{v})$ be the number of projections parallel to $\mathbf{v}$ and $\mathcal{L}_k^{\mathbf{v}}$ be one of these projections, with $k = 1, ..., t(\mathbf{v})$. If $p_k^{\mathbf{v}}$ indicates the *k-th projection* of A along $\mathbf{v}$, hence $p_k^{\mathbf{v}}$ is the number of 1's on $\mathcal{L}_k^{\mathbf{v}}$, then the *projection* along $\mathbf{v}$ is:

$$P_{\mathbf{v}} = (p_1^{\mathbf{v}}, p_2^{\mathbf{v}}, ..., p_{t(\mathbf{v})}^{\mathbf{v}}) \quad \text{where} \quad p_k^{\mathbf{v}} = \sum_{a_{ij} \in \mathcal{L}_k^{\mathbf{v}}} a_{ij}.$$

The image I can be represented suitably by the network flow G with one source S, one sink T and two layers of edges between S and T: the first layer of *row-nodes* $\{R_1, R_2, ..., R_m\}$ and the second one of *column-nodes* $\{C_1, C_2, ..., C_n\}$. It is noteworthy that, to construct G, we consider only the horizontal and vertical directions $\mathbf{v}_1$ and $\mathbf{v}_2$. Each edge $\widehat{XY}$ of G has flow $f_{\widehat{XY}}$ and capacity $c_{\widehat{XY}}$:

$$f_{\widehat{SR_i}} = p_i^{\mathbf{v}_1}, \ f_{\widehat{R_iC_j}} = a_{ij}, \ f_{\widehat{C_jT}} = p_j^{\mathbf{v}_2} \quad \text{and} \quad c_{\widehat{SR_i}} = p_i^{\mathbf{v}_1}, \ c_{\widehat{R_iC_j}} = 1, \ c_{\widehat{C_jT}} = p_j^{\mathbf{v}_2}$$

where $i = 1, ..., m$ and $j = 1, ..., n$ (see Fig. 1). The Ford-Fulkerson method can be applied to compute the *maximum flow* through G, which corresponds to an image that satisfies $P_{\mathbf{v}_1}$ and $P_{\mathbf{v}_2}$. We do not carry out a breadth first search to find the augmenting paths, but we randomly select the edges with the biggest residual capacity; this does not guarantee the correct maximum flow, but assures a variability in the initial population and speeds up the whole method. Should the flow not be maximal, then white pixels are randomly added, thus to reach the correct number w of white pixels that are present in the input image.

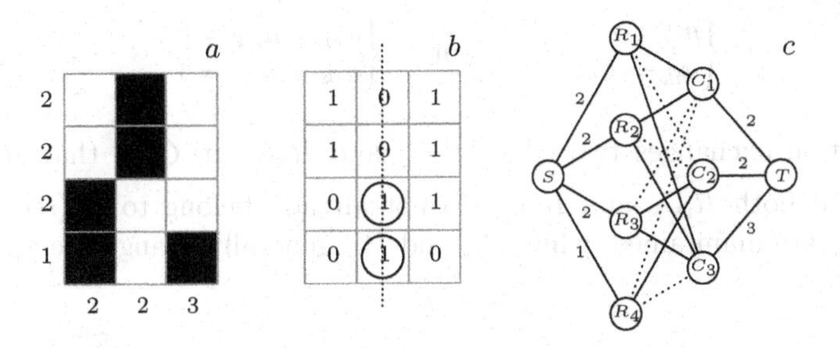

Fig. 1. A binary image with $P_{\mathbf{v}_1}$ and $P_{\mathbf{v}_2}$ (a); its matrix (b) and network flow (c). Projection lines $\ell_{\mathbf{v}_2}(3,2)$ and $\ell_{\mathbf{v}_2}(4,2)$ are equivalent because intercept the same pixels. Dashed edges $\widehat{R_iC_j}$ indicates black pixels; remaining edges correspond to white pixels.

2.2 Fitness Function

The fitness function of an agent, represented by the matrix A, is defined as:

$$\mathcal{F}(A) = \sum_{i=1}^{4} \sum_{k=1}^{t(\mathbf{v}_i)} |p_k^{\mathbf{v}_i} - q_k^{\mathbf{v}_i}|$$

where $P_{\mathbf{v}_i}$ is the actual projection of A and $Q_{\mathbf{v}_i}$ is the input projection, both taken along direction $\mathbf{v}_i$: starting from $Q_{\mathbf{v}}$, the algorithm has to reconstruct a binary image with projections $P_{\mathbf{v}}$ equal to $Q_{\mathbf{v}}$; therefore, it has to minimize $\mathcal{F}$.

2.3 Crossover

We are going to define the a vertical crossover that generates two offsprings; the analogous horizontal version just operates on the transpose of the matrix. The offsprings B_1 and B_2 are obtained by swapping homologous columns of their parents A_1 and A_2. Formally, these columns $A_1^{(j)}$ and $A_2^{(j)}$, with $j = 1, ..., n$, are located by a *crossover mask* $M = (M_1, M_2, ..., M_n)$ of random binary values:

$$B_1^{(j)} = \begin{cases} A_1^{(j)} \text{ if } M_j = 1 \\ A_2^{(j)} \text{ if } M_j = 0 \end{cases} \qquad B_2^{(j)} = \begin{cases} A_2^{(j)} \text{ if } M_j = 1 \\ A_1^{(j)} \text{ if } M_j = 0 \end{cases}$$

2.4 Mutation

Our mutation operator modifies no more than 5% of the pixels; this threshold has been set experimentally to obtain better fitness values. In particular, this operator swaps $\rho = min\{\lfloor \frac{m \times n}{20} \rfloor, m \times n - w, w\}$ white and black pixels.

2.5 Switch

Images that satisfy the same set of two projections can be transformed among themselves by a finite sequence of elementary switches [11]. Our *switch* operator on A swaps a_{ij} with a_{ik} and a_{hj} with a_{hk}, where:

$$\begin{cases} a_{ij} = a_{hk} = 1 \\ a_{ik} = a_{hj} = 0 \end{cases} \quad \text{or} \quad \begin{cases} a_{ij} = a_{hk} = 0 \\ a_{ik} = a_{hj} = 1 \end{cases}$$

This operation exchanges two edges $\widehat{R_iC_j}$ and $\widehat{R_hC_k}$ in G so that $R_i \neq R_h$, $C_j \neq C_k$ and both $\widehat{R_iC_k}$ and $\widehat{R_hC_j}$ do not already belong to G (see Fig. 2). $P_{\mathbf{v}_1}$ and $P_{\mathbf{v}_2}$ are maintained, while $P_{\mathbf{v}_3}$ and $P_{\mathbf{v}_4}$ generally change their values.

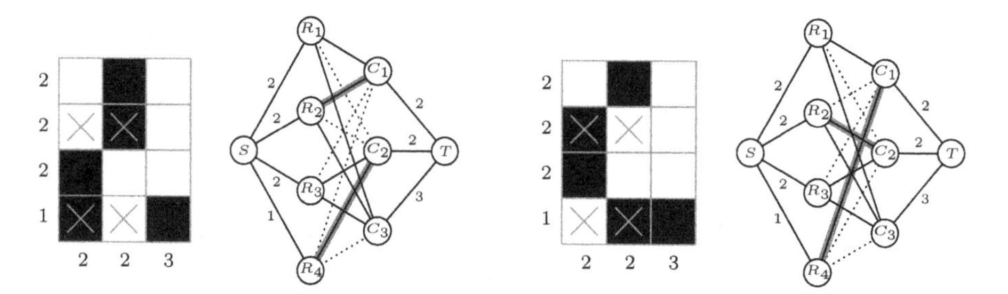

Fig. 2. Image and the corresponding net representation after a switch operation

2.6 Compactness Constraint

Some *isolated* pixels (i.e. surrounded by 8 pixels with opposite color) are usually present and worsen the final image. The compatness operator eliminates as many as possible isolated pixels. This approach does not guarantee a better fitness value, but nonetheless tends to let the image satisfy the input projections.

3 Experimental Study

In real cases only the projections of the input image I will be available. To evaluate the robustness of our algorithm we have computed the Hamming distance $\varepsilon(I, I')$ between I and its reconstruction I', normalized to the size of I. The number ($ng = 500$) of generations and the size ($na = 1000$) of the population have been calculated to achieve a *satisfactory* final error by performing the algorithm 100 times per image. For each session, we have computed the average error $\bar{\varepsilon}$ and the number $N(\bar{\varepsilon})$ of solutions I' with $\varepsilon(I, I') < \bar{\varepsilon}$. To estimate the effect of an instrumental noise (IN), we have incremented or decremented some samples $q_i^{\mathbf{v}}$ of the input projections $Q_{\mathbf{v}_i}$. The probability (from 1% to 5%) of finding an error has been considered as a percentage of the samples (i.e. it is maximal in case of bigger samples). This means all samples of $Q_{\mathbf{v}_i}$ can be modified, if this remains compatible with the given percentage of IN.

According to our knowledge, no standard benchmark of images exists to compare different discrete tomography reconstruction algorithms; we used the same hv-convex polyominoes described in [7], but real medical tests show objects with holes and concavities. We have proved elsewhere that our memetic method can reconstruct a variety of different objects, but here we have compared it with the

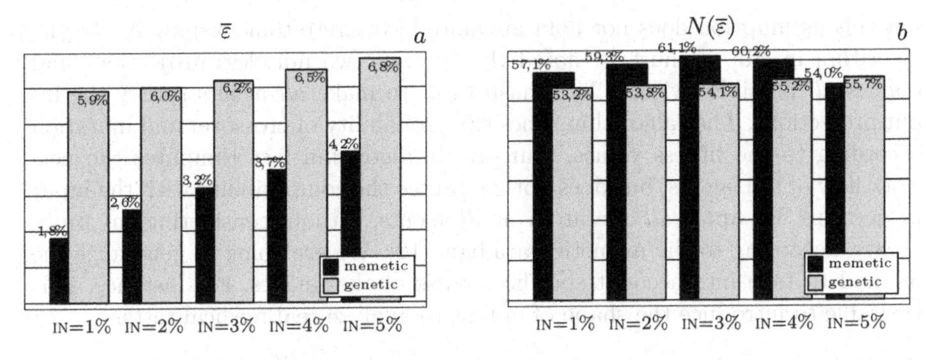

Fig. 3. Genetic vs memetic algorithm, with different amounts of instrumental noise IN

genetic algorithm presented in [7], to study the reconstruction stability in case of noisy projections. We just sketch the main operators described in [7]:

• A uniform crossover produces one child from pairs of random parents: their genes are selected to enhance the resulting chromosome. Two of these three individuals are chosen, according to their fitness, to keep the size of the population.
• A mutation operator tries to speed up the convergence of the genetic algorithm, by moving pixels from dense to sparse zones.
• A black and a white pixel, chosen at random, are swapped by a simpler mutation which ensures a variability of the genomes, though it may worsen the fitness of the new chromosome.
• As for our algorithm, another (morphological) mutation applies the convexity constraint to swap a black with a white isolated pixel.
• The elitist selection guarantees the presence of the best individual.

Fig. 3 summarizes the results obtained by both the genetic and memetic algorithms on 100 images with 100×100 pixels. The genetic implementation is stable and produces good results, but our new memetic method returns better results, especially for low degrees of instrumental noise IN. Fig. 3(a) reports the average reconstruction error $\overline{\varepsilon}$; therefore we wish for shorter bars. Vice versa, Fig. 3(b) shows the percentage $N(\overline{\varepsilon})$; therefore we are interested in higher bars.

4 Conclusions and Further Works

At present, there is no real discrete tomography scanner, but we hope that it will become a powerful tool for non-invasive medical imaging. Other applications include industrial quality control, crystallography and preliminary investigations for the restoration of works of art.

We have introduced here a memetic method, tested on hv-convex polyominoes. Different evolutionary operators have been verified, but we present the results gained by a set of particular crossover, mutation, compactness and switch. In the case of computerized tomography, instrumental noise is usually distributed across the whole reconstructed image, due to the huge number of projections, but this assumption does not hold anymore in discrete tomography. As for the algorithm in [10], it must be noted that it uses two noiseless projections and applies *all* possible switches by a hash table to make *all* agents satisfy the input projections. That algorithm tunes the probability of crossover and mutation according to the fitness values. Our greedy algorithm fast computes the network flow of the agents, but does not guarantee the compatibility with the input projections. We apply *all* operators on *all* agents, without considering any probability, according to the memetic paradigm [12]. We are going to generalize the method to take into account specific models of the images. For instance, it is desirable to introduce the shape of organs to analyze real medical tests.

References

1. Gardner, R.J., Gritzmann, P., Prangenberg, D.: On the computational complexity of reconstructing lattice sets from their X-rays. Discrete Mathematics 202, 45–71 (1999)
2. Wang, B., Zhang, F.: On the precise number of (0,1)-matrices in A(R,S). Discrete Mathematics 187, 211–220 (1998)
3. Kuba, A.: The reconstruction of two-directional connected binary patterns from their two orthogonal projections. Computer Vision, Graphics, and Image Processing 27, 249–265 (1984)
4. Barcucci, E., Del Lungo, A., Nivat, M., Pinzani, R.: Medians of polyominoes: A property for the reconstruction. Int. J. Imaging Syst. Technol. 9, 69–77 (1998)
5. Chrobak, M., Dürr, C.: Reconstructing hv-convex polyominoes from orthogonal projections. Information Processing Letters 69, 283–291 (1999)
6. Frosini, A., Nivat, M., Vuillon, L.: An introductive analysis of periodical discrete sets from a tomographical point of view. Theoretical Computer Science 347(1–2), 370–392 (2005)
7. Valenti, C.: A Genetic Algorithm for Discrete Tomography Reconstruction. Genetic Programming and Evolvable Machines. Springer, Heidelberg (in print)
8. Venere, M., Liao, H., Clausse, A.: A Genetic Algorithm for Adaptive Tomography of Elliptical Objects. IEEE Signal Processing Letters 7(7) (2000)
9. Kihm, K.D., Lyons, D.P.: Optical tomography using a genetic algorithm, Optics Letters. 21(17), 1327–1329 (1996)
10. Batenburg, J.K.: An evolutionary algorithm for discrete tomography. Discrete Applied Mathematics 151, 36–54 (2005)
11. Ryser, H.J.: Combinatorial mathematics. The carus mathematical monographs. In: MAA, vol. 14, ch. 6 (1963)
12. Moscato, P.: Memethic algorirhms: A short introduction. New ideas in optimization, pp. 219–234. McGraw-Hill, New York (1999)

A Fuzzy Hybrid Method for Image Decomposition Problem

Ferdinando Di Martino[1,2], Vincenzo Loia[1], and Salvatore Sessa[2]

[1] Università degli Studi di Salerno, Dipartimento di Matematica e Informatica, Via Ponte Don Melillo, 84081 Fisciano (Salerno), Italy
loia@unisa.it

[2] Università degli Studi di Napoli Federico II, Dipartimento di Costruzioni e Metodi Matematici in Architettura, Via Monteoliveto 3, 80134 Napoli, Italy
{fdimarti,sessa}@unina.it

Abstract. We use an hybrid approach based on a genetic algorithm and on the gradient descent method in order to decompose an image. In the pre-processing phase the genetic algorithm is used for finding two suitable initial families of fuzzy sets that decompose R in accordance to the well known concept of Schein rank. These fuzzy sets are successively used in the descent gradient algorithm which determines the final fuzzy sets, useful for the reconstruction of the image. The experiments are executed on some images extracted from the the SIDBA standard image database.

Keywords: Image decomposition, Fuzzy relation, Genetic algorithm, Descent gradient algorithm.

1 Introduction

Let $R: X \times Y \rightarrow [0,1]$ be a fuzzy relation, being X and Y two referential finite sets with cardX=m and cardY=n. Any image can be considered as a fuzzy relation R if each its pixel is normalized with respect to the length of the grey scale used. Then many techniques of image processing based, for instance, on fuzzy relation equations were developed (cfr., e.g., [1] ÷[7]). In this paper we decompose a fuzzy relation R in the smallest possible number "c" of pairs of fuzzy sets $\{A_i,B_i\}$, $A_i:X \rightarrow [0,1]$, $B_i : Y \rightarrow [0,1]$, $i = 1,\ldots,c$, namely we approximate R with another fuzzy relation $\tilde{R}$: $X \times Y \rightarrow [0,1]$, such that

$$R(x,y) \approx \tilde{R}(x,y) = \vee_{i=1}^{c}\left[A_i(x) \wedge B_i(y)\right] \tag{1}$$

for all $(x,y) \in X \times Y$. It is well known that the parameter c ($\leq \min\{m,n\}$) is defined as the Schein rank of $\tilde{R}$ and this problem was already formalized as an optimization task in [8], [9], [10]. In [8], [10] the determination of $\tilde{R}$ is based on the minimization of the cost function:

M. Giacobini et al. (Eds.): EvoWorkshops 2008, LNCS 4974, pp. 353–358, 2008.

$$Q(R,\tilde{R}) = \sum_{(x,y)\in XxY} \left(R(x,y) - \tilde{R}(x,y)\right)^2 \tag{2}$$

by fixing a priori the value of "c" and the initial values $A_i^0(x)$ and $B_i^0(y)$ of the fuzzy sets. Further a slight extension of (2) is introduced in [9] in order to reduce the pepper noise in the reconstructed image. Here we have used the same cost function of [9] by adopting an hybrid approach, that is we firstly consider a genetic algorithm in the pre-processing phase in order to determine the best initial values $A_i^0(x)$ and $B_i^0(y)$. The population of candidate chromosomes consists of arrays of dimension c·(m+n), where the first c·m (resp. latter c·n) genes represent the fuzzy sets $A_1^0,\ldots,A_c^0$ (resp. $B_1^0,\ldots,B_c^0$). The fitness function is defined via the inverse of the cost function given in [9]. The fitness, selection, crossover and mutation operations are iterated until the number of iteration "Iter" is equal a to a predefined value "Imax". The values $A_i^0(x)$ and $B_i^0(y)$ are successively used in the gradient descent algorithm for determining the final fuzzy sets (A_i, B_i) and the correspondent reconstructed image $\tilde{R}$. Our hybrid approach gives better quality of the reconstructed images with respect to the method presented in [9]. Fig. 1 illustrates the genetic algorithm used in the pre-processing phase.

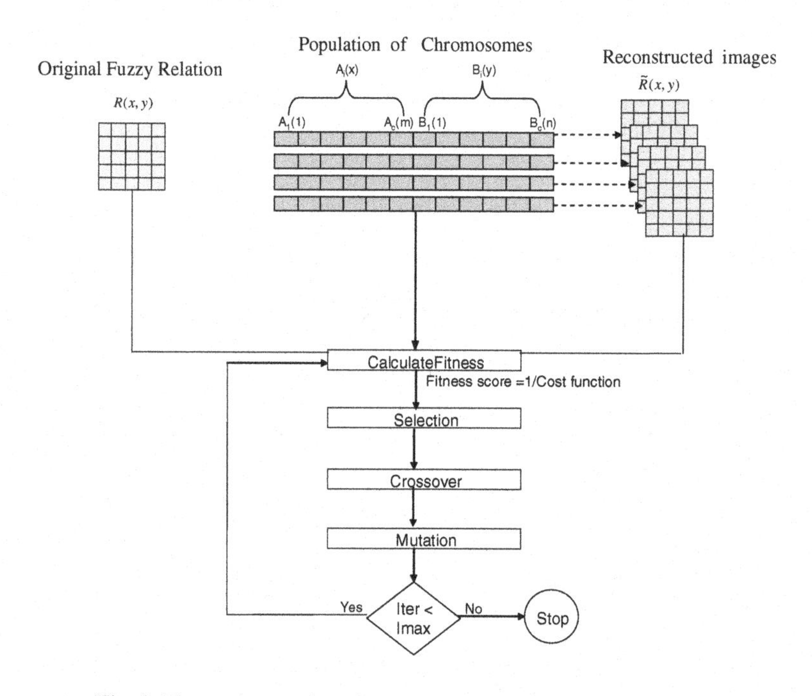

Fig. 1. The pre-processing phase concerning the genetic algorithm

2 The Pre-processing Phase

In the genetic algorithm we use genes whose allele is an integer value of the set {0,1,2,…, 255}, but we normalize these values in the calculus of the fitness. A crucial problem is the choice of the fitness derived from the cost function (2). If $\tilde{R}_k$ is the approximated fuzzy relation determined for the k-th chromosome and the related cost function is Q(R, $\tilde{R}_k$), we define as fitness score of the k-th chromosome the real number FS($\tilde{R}_k$)=F(R, $\tilde{R}_k$)/$\sum_j F(R,\tilde{R}_j)$, where F(R, $\tilde{R}_k$)=1/Q(R, $\tilde{R}_k$) and the sum is extended to all the j-th chromosomes $\tilde{R}_j$. We have used the roulette-wheel method for the selection process and after the selection operator, the crossover operator is applied to the population of the mating pool for the generation of a child from two parent chromosomes. We utilize a fixed-length array Single Point Crossover with probability p_{cross} and successively a mutation flip operator is applied for transforming each chromosome with probability p_{mut} (for further details, see, e.g., [11], [12]). After an assigned number of selection, crossover and mutation processes is performed, the initial pairs (A_i^0, B_i^0), i=1,…,c, are determined and the descent gradient algorithm given in [9] starts with these pairs of fuzzy sets. For brevity, this last algorithm is not presented here.

3 Experimental Results

We have used in our experiments some test images of sizes 256 ×256 extracted from SIDBA image data set [13] by using the C++ library class GALIB [14]. For the pre-processing phase we have used a population number N = 100, a crossover probability p_{cros} = 0.6 and a mutation probability p_{mut} = 0.01. We tested the image decomposition process by using different values of the iteration number Imax until to 200 iterations. We have calculated the classical Peak Signal to Noise Ratio (PSNR) by setting

$$PSNR(R,\tilde{R}) = 20\log_{10}\frac{255}{\sqrt{\dfrac{\sum_{(x,y)\in X\times Y}\left[R(x,y)-\tilde{R}(x,y)\right]^2}{255\times 255}}} \tag{3}$$

The square root at second member of (15) is the SQE (SQuare Error). We present the results obtained with the image Lena. Fig. 2 (resp. Fig. 3) shows the trend of the SQE parameter with respect to the iteration number Iter under c=200 by using the descent gradient method (for short, DG) (resp. our hybrid method (for short, GA+DG)). Indeed, for c = 200, by using Imax=5×10^4 iterations in GA and approximately Iter=200 iterations for DG, we obtain a SQE less than 200 for the reconstructed image (cfr. Fig. 3); instead the usage of the only DG with

Table 1. SQE and PSNR values for some reconstructed images of Lena

Method	Iter = 60		Iter = 100		Iter = 200	
	SQE	PSNR	SQE	PSNR	SQE	PSNR
DG	2726.23	13.78	998.21	18.14	784.39	19.19
GA+DG	1079.11	17.80	299.31	23.97	113.94	27.56

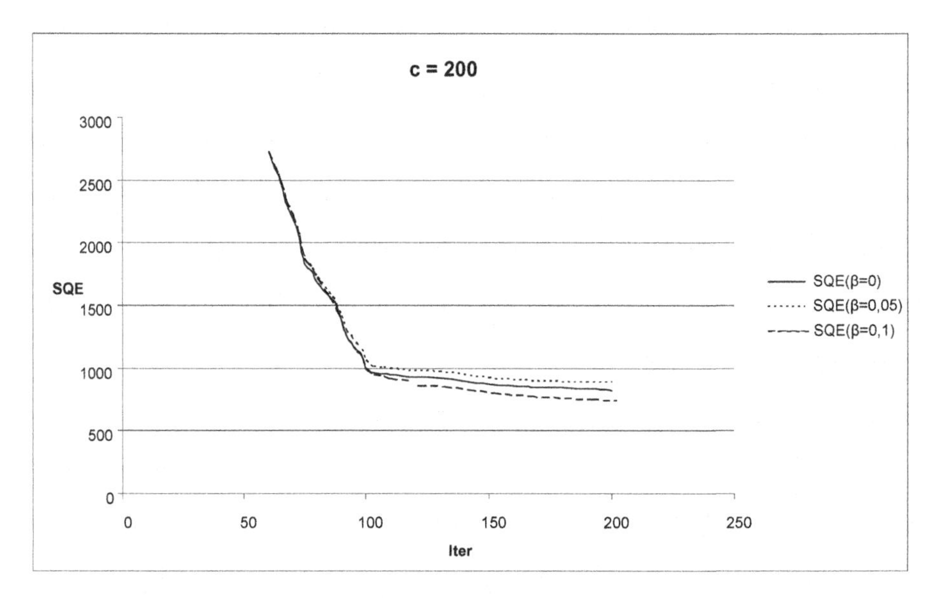

Fig. 2. SQE with respect to iteration number Iter for c= 200 by using only DG

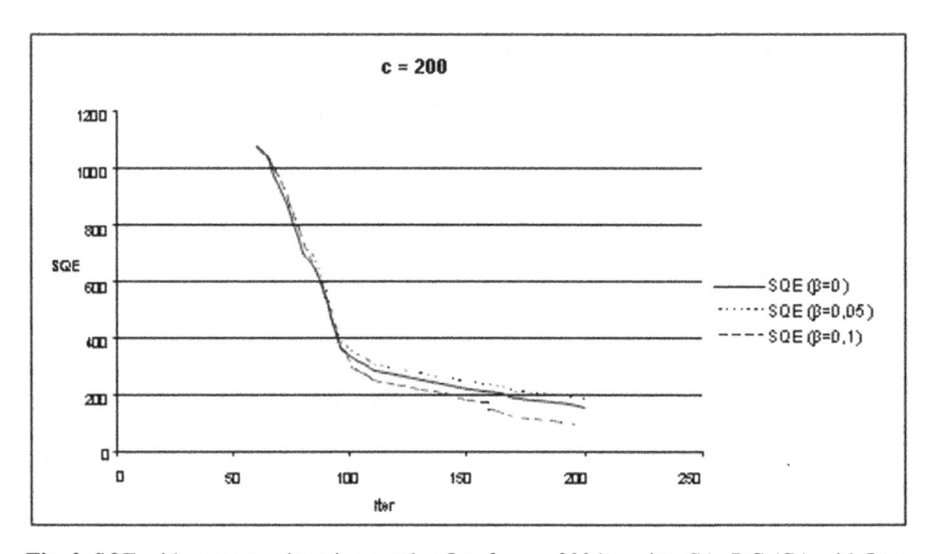

Fig. 3. SQE with respect to iteration number Iter for c = 200 by using GA+DG (GA with Imax = 5×10^4 iterations)

approximately Iter=200 iterations gives a SQE greater than 750 (cfr. Fig. 2). β is a learning rate introduced in the calculation of Q(R, $\tilde{R}$) (cfr. [9] for further details). In Table 1 we show the SQE and PSNR values obtained for the gray image Lena with DG and GA+DG by using c = 200, N = 40, Imax=5×10^4, respectively for 60, 100 and 200 iterations in DG. Therefore the additional usage of GA improves the quality of the reconstructed image with respect to the only usage of DG.

Fig. 4 is the original image Lena and Figures 5a, 5b, 5c show the reconstructed images obtained with only DG method of [9] for c=200 and Iter=60, 100, 200, respectively. Figures 5a, 5b, 5c show the reconstructed images obtained with our GA+DG method by using an initial population of chromosomes N = 40, Imax=5×10^4, c=200 and Iter=60, 100, 200, respectively.

Fig. 4. Original image Lena

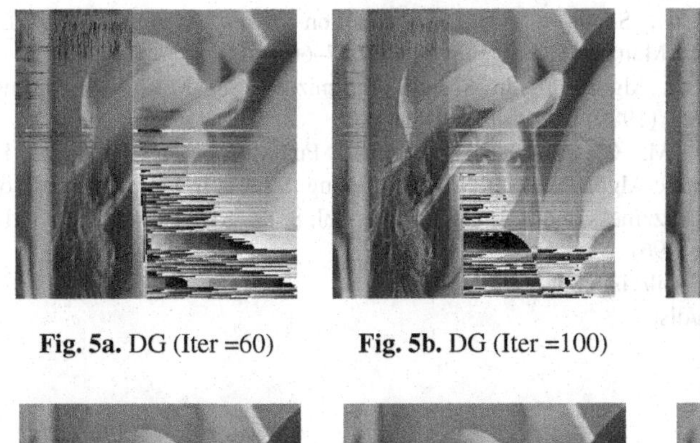

Fig. 5a. DG (Iter =60) **Fig. 5b.** DG (Iter =100) **Fig. 5c.** DG (Iter =200)

Fig. 5a. GA+DG (Iter =60) **Fig. 5b.** GA+DG (Iter =100) **Fig. 5c.** GA+DG (Iter =200)

References

1. Di Martino, F., Sessa, S.: A Genetic Algorithm Based on Eigen Fuzzy Sets for Image Reconstruction. In: Masulli, F., Mitra, S., Pasi, G. (eds.) WILF 2007. LNCS (LNAI), vol. 4578, pp. 342–348. Springer, Heidelberg (2007)
2. Nobuhara, H., Hirota, K., Pedrycz, W.: Fast Solving Method of Fuzzy Relational Equations and its Application to Lossy Image Compression. IEEE Transactions of Fuzzy Systems 8(3), 325–334 (2000)
3. Nobuhara, H., Takama, Y., Hirota, K.: Fast Iterative Solving Method of Various Types of Fuzzy Relational Equations and its Application to Image Reconstruction. Internat. J. Adv. Comput. Intell. 5(2), 90–98 (2001)
4. Nobuhara, H., Bede, B., Hirota, K.: On Various Eigen Fuzzy Sets and their Application to Image Reconstruction. Information Sciences 176, 2988–3010 (2006)
5. Di Nola, A., Pedrycz, W., Sessa, S.: When is a Fuzzy Relation Decomposable in Two Fuzzy Sets? Fuzzy Sets Syst. 16, 87–90 (1985)
6. Di Nola, A., Pedrycz, W., Sessa, S.: Decomposition Problem of Fuzzy Relations. Internat. J. Gen. Syst. 10, 113–123 (1985)
7. Pedrycz, W.: Optimization Schemes for Decomposition of Fuzzy Relations. Fuzzy Sets Syst 100, 301–325 (1998)
8. Nobuhara, H., Hirota, K., Pedrycz, W., Sessa, S.: Two Iterative Methods of Decomposition of a Fuzzy Relation for Image Compression/Decomposition Processing. Soft. Comput. 8(10), 698–704 (2004)
9. Nobuhara, H., Hirota, K., Pedrycz, W., Sessa, S.: Efficient Decomposition Method of Fuzzy Relation and Their Application to Image Decomposition. Applied Soft Computing 5, 399–408 (2005)
10. Pedrycz, W., Hirota, K., Sessa, S.: A Decomposition of Fuzzy Relations. IEEE Transactions on Systems, Man and Cybernetics 31(4), 657–663 (2001)
11. Goldberg, D.E.: Genetic Algorithms in Search, Optimization & Machine Learning. Addison-Wesley, Reading (1989)
12. Pedrycz, W., Reformat, M.: Genetic Optimization with Fuzzy Coding. In: Herrera, F., Verdegay, J. (eds.) Genetic Algorithms and Soft Computing. Studies in Fuzziness and Soft Computing. Studies in Fuzziness and Soft Computing, vol. 8, pp. 51–67. Springer, Berlin Heidelberg New York (1996)
13. http://www.cs.cmu.edu/~cil/vision.html
14. http://lancet.mit.edu/ga/dist

Triangulation Using Differential Evolution

Ricardo Landa-Becerra and Luis G. de la Fraga

Cinvestav, Computer Science Department
Av. I.P.N. 2508, 07360, Mexico City, Mexico
rlanda@computacion.cs.cinvestav.mx, fraga@cs.cinvestav.mx

Abstract. Triangulation is one step in Computer Vision where the 3D points are calculated from 2D point correspondences over 2D images. When these 2D points are free of noise, the triangulation is the intersection point of two lines, but in the presence of noise this intersection does not occur and then the best solution must be estimated. We propose in this article a fast algorithm that uses Differential Evolution to calculate the optimal triangulation.

1 Introduction

One of the most important problems in 3D Computer Vision is to calculate the three-dimensional reconstruction of the visible surface of a 3D object from a pair of images [1,2]. The problem is reduced to calculate the 3D point $\mathbf{P}$ that best adjustes to two points $(\mathbf{p}, \mathbf{p}')$ over the images; this problem is known as *triangulation*. If the projection matrices (M, M'), from which each image was obtained, are known, it is possible to project two lines from the 2D point of each image. These lines must intersect in a reconstructed 3D point $\mathbf{P}'$, which is the same that the real 3D point $\mathbf{P}$, in the absence of noise in the 2D points and the projection matrices. However, those conditions are not found in real images, thus it is necessary to apply other methods to estimate the best 3D point.

The triangulation methods described in this paper are applied to three types of reconstructions: the projective reconstruction, in which no kind of metric nor parallelism exists; the affine reconstruction, where the concept of parallelism exists, but there is no specific metric on each coordinate axis; and the metric reconstruction, in which there is a specific metric on the three axes and also exists the concept of parallelism [2, Ch. 2].

In Sec. 2 we are going to see a brief review about solving the triangulaction problem with evolutionaty algorithms. In Sec. 3 we describe a new triangulation algorithm that uses DE. And in 4 we show the results of some experiments with the proposed algorithm. Finally some conclusions are drawn.

2 Previous Work on Evolutionary Algorithms for the Triangulation Problem

In a previous work [3], a genetic algorithm, a particle swarm approach and a multiobjective NSGA-II-based approach were used to solve the triangulation

M. Giacobini et al. (Eds.): EvoWorkshops 2008, LNCS 4974, pp. 359–364, 2008.

problem. The authors noted that single-objective approaches require a very high number of generations in order to obtain partially good solutions. From this result, they hypothesized that the search space is very rugged with an objective function of sum of errors, so they experimented with a recent proposal to deal with rugged landscapes: the so called multiobjectivization [4], which consists of decomposition of the original objective function, or the addition of "helper" objectives, in order to diversify the search, and alleviate in a certain degree the search difficulties. This is possible, because the search can progress either in the direction of one objective or the other.

Using such a method, the objective function of sum of errors was split, considering the error from each image as an objective, and optimizing them simultaneously. The authors of [3] adopted the NSGA-II [5] as multiobjective optimizer. The results show that this problem is a good candidate for the multiobjectivization approach, reducing the number of objective functions evaluations needed to obtain good results. Moreover, the results obtained are compared with the classical linear least squares (LLS), and the Poly-Abs methods [1]. The NSGA-II clearly outperforms the Poly-Abs method in all the cases, and presents a good performance in several cases where the LLS method exhibits some difficulties (mainly, when the levels of noise are high).

So, the results are very good, and the trade-off points may help at a certain stage of the search. But one may think that some of the computational effort spent in generating multiple points as output is not necessary, since we only need a single point as the result. Another possible issue is that a given user or designer may want to integrate all the solutions for its use with other implementations, not necessarily in C language, but previously adopted multiobjective evolutionary algorithms, like the NSGA-II, are somewhat difficult to implement.

Differential evolution (DE) is a recently developed evolutionary algorithm proposed by Price and Storn [6], whose main design emphasis is real parameter optimization. The DE is very easy to implement; easier than the NSGA-II and other multiobjective evolutionary approaches. It does not require excessive effort to be implemented in a variety of languages, or even at hardware level.

In this paper we are concerned about the two previous issues, and propose a modified multi-objective algorithm that works with only one objective at a time, which performs the search in different directions at different phases of the execution, with the aim of obtaining the ability to escape from the accidents of the search space (as the multiobjectivization-based approaches), but only producing a single point as output. Additionally, this approach is based on the differential evolution algorithm, which is a technique very easy to implement by anyone who prefers to use his/her own code.

3 Proposed Approach

In a previous work [3], taking as input a set of pairs $(\mathbf{p}, \mathbf{p}')$ (where $\mathbf{p}$ represents a point from the first image, and $\mathbf{p}'$ is the corresponding point from the second image) the multiobjectivization has been done splitting the original objective:

$$f = \sum (d(\mathbf{p}, \hat{\mathbf{p}}) + d(\mathbf{p}', \hat{\mathbf{p}}')), \quad (1)$$

in two objectives:

$$f_1 = \sum d(\mathbf{p}, \hat{\mathbf{p}}), \quad \text{and} \quad f_2 = \sum d(\mathbf{p}', \hat{\mathbf{p}}'), \quad (2)$$

where $\hat{\mathbf{p}}$ and $\hat{\mathbf{p}}'$ are the reconstructed points obtained as $\hat{\mathbf{p}} = M\mathbf{P}$ and $\hat{\mathbf{p}}' = M'\mathbf{P}$; $d(\cdot, \cdot)$ represents the Euclidian distance; M and M' are 3×4 matrices, and $\mathbf{P}$ is a 3D point. The sums are perfomed on all elements of the set of pairs $(\mathbf{p}, \mathbf{p}')$, and the aim is to obtain a 3D point $\mathbf{P}$ for each pair of 2D points.

In this previous approach, some computational cost is spent in generating the whole Pareto front, even when we are only interested in the point at the knee of the curve. In this paper we attempt to reduce the computational cost needed, avoiding the generation of the whole Pareto front, but trying to keep some of the advantages of multiobjectivization. This is done by optimizing one of the objectives at a time, while maintaining the other below a certain threshold (such objective is considered as a constraint); this way, when difficulties are found in one of the objectives (accidents in the search space, for example a local optimum), the technique switches the objective to move the search in the other direction, expecting to overcome such difficulty in the first objective.

The point obtained by the optimization of one objective, while maintaining the others as constraints, will lead us to a point over the Pareto front, as is demonstrated in [7] (this problem is equivalent to the ε-constraint problem). But how to obtain the point on the knee of the curve? As each objective is measured in the same space and with the same dimensions, the magnitudes of both objectives is expected to be similar. Thus, the point with the minimum sum of errors is very likely to be near the point on the Pareto front with $f_1 = f_2$ in (2), and we can search for that point with a proper setting of the constraints.

Given such considerations, the proposed approach proceeds as follows: arbitrarily choose one objective, and perform an unconstrained optimization considering only that objective, during k generations. Then, switch the objectives for first time, optimizing the other objective, and constraining the value of the first. The value of the first objective should not be larger than $0.75(f_1 + f_2)$.

That is, let f_1 be the first objective to be optimized without constraints. After such first step, let t_1 be the best value for f_1 in the population, and t_2 be the best value for f_2; is expected that $t_1 < t_2$. The problem to tackle then is:

$$\begin{array}{l} \text{minimize } f_2 \\ \text{subject to } f_1 \leq 0.75(t_1 + t_2) \end{array}$$

This process is performed until the value of f_2 improves the value $0.75t_1$, or no progress in the search can be obtained Then, the search continues with the other objective function, and the t values are updated. The process will lead us to a point very near the knee of the curve of the Pareto front, because the distances of the newer thresholds will be lower at each step.

The procedure can be resumed as that presented in Algorithm 1. The procedure opt(f_i, P) performs the optimization using f_i as objective, with the other

Algorithm 1. Approach for Objective Switching

```
initialize(P = {x_1, x_2, ..., x_μ})
g = 0; t_1 = t_2 = ∞; P = opt(f_1, P)
t_1 = min_j(f_1(x_j)), t_2 = min_j(f_2(x_j))
while g < G_max do
  if t_1 < t_2 then
    P = opt(f_2, P)
  else
    P = opt(f_1, P)
  Update t values
```

as constraint with the current t values. At the end of the algorithm, the result is the point in the final population with the lowest value of $f_1 + f_2$.

4 Experimental Results

Data for the experiments were obtained from synthetic images of a regular polyhedron, where each point was perturbed with Gaussian noise. The reconstructions were performed using 8, 12, 16 and 20 pairs of points, randomly chosen from the images. In the case of a projective reconstruction, the fundamental matrix was obtained executing the 8-point algorithm [2], and then obtaining the projection matrices. In the case of the affine reconstruction, the same steps were performed, plus the generation of vanishing points, in order to calculate the homography to transform the space of the points, into an affine space.

The parameters of the algorithm are: population size $\mu = 50$, number of generations $G_{max} = 300$, for the differential evolution: $R = 1$ and $F = U(0.5, 1)$, and the maximum generations for objective switching is $k = 10$. The algorithm performed $15,000$ objective function evaluations. These parameter values were obtained after several experiments. The algorithm was executed 30 times.

4.1 Comparison of Results

The results of the modified differential evolution are compared with the results obtained by the NSGA-II. Data for the NSGA-II were extracted from [3], where the NSGA-II required $30,000$ evaluations per execution (the proposed approach here required $15,000$ evaluations). A graphical comparison of results is presented in Figures 1 and 2, where the obtained average error versus the standard deviation of the Gaussian noise applied are plotted. In Figure 1 are plotted the results for the projective reconstructions with 8, 12, 16 and 20 points. As we expected, the error is lower as the number of points used for the calculation of the projection matrix increases. Also, the error is lower when the noise has a lower standard deviation.

In Figure 2 are shown the results for the affine reconstruction. In general, a slightly larger error was obtained compared with the projective reconstruction.

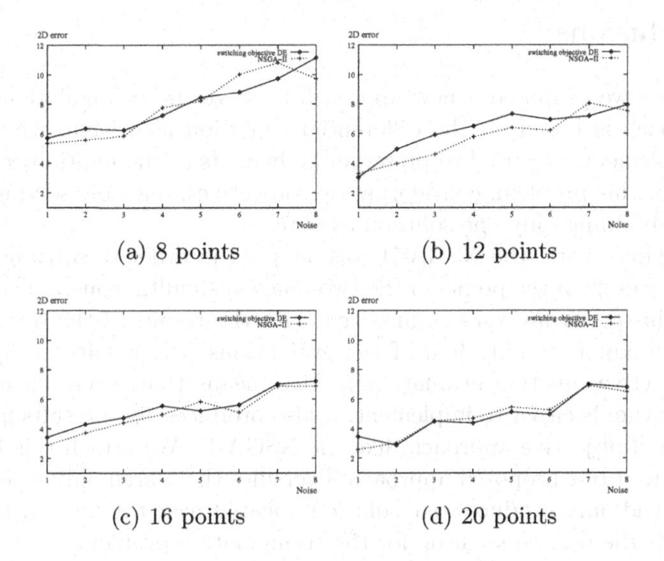

Fig. 1. Results for the projective reconstructions

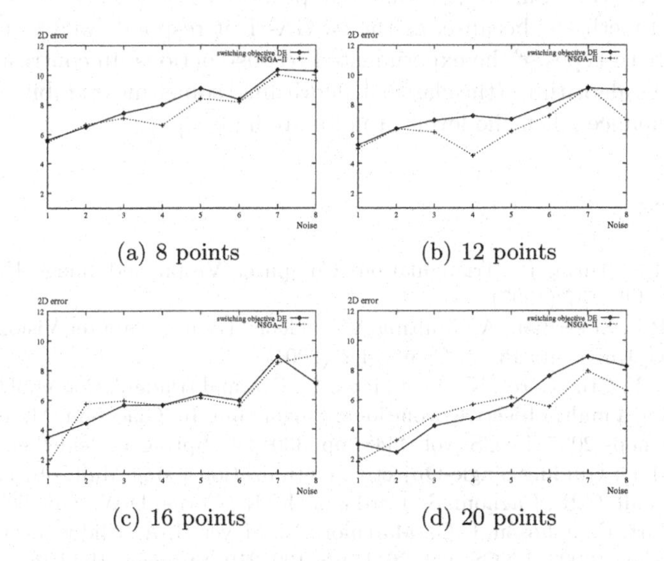

Fig. 2. Results for the affine reconstructions

The results are good in all the cases, obtaining an error lower than 10. Moreover, the algorithm is very robust, obtaining good results in a wide variety of conditions, including projection matrices calculated with few points, a strong noise added, and different types of reconstruction (projective and affine). The results are competitive with those obtained by the NSGA-II, but requiring fewer objective function evaluations, and also with an approach easier to implement.

5 Conclusions

In this paper we proposed a new approach to solve the triangulation problem. This approach is based on the differential evolution algorithm, which is very easy to implement. We tried to preserve the benefits of the multiobjectivization approach for this problem, considering two objectives, but only solving one at a time, and obtaining only one solution as outcome.

In a previous work, the NSGA-II was adopted in order to solve both objectives (distances from the points of the two images) simultaneously. The reported results in this paper are very competitive with the results of the NSGA-II, but require approximately only half of the evaluations (the results in [3] required $30,000$ objective function evaluations). This means that, even when the proposed algorithm is easier to implement, it also produces good results faster than a totally multiobjective approach, like the NSGA-II. We attach this behaviour to the fact that our proposed approach diversifies the search only when needed, but at the end only produces one solution, located near the knee of the Pareto front (this is the desired solution for the triangulation problem).

By comparing the results presented here with those of the NSGA-II, and based on the comparisons on [3] of the NSGA-II with the LLS and Poly-Abs algorithms, we can roughly say that our proposed approach also outperforms the classical methods, because, as the NSGA-II, it responds well to high levels of noise in both types of the experimented reconstructions. In contrast, the LLS and Poly-Abs algorithms (the classical algorithms) sometimes exhibit a relatively poor performance when the levels of noise are high [3].

References

1. Hartley, R., Sturm, P.: Triangulation. Computer Vision and Image Understanding 68(2), 146–157 (1997)
2. Hartley, R.I., Zisserman, A.: Multiple View Geometry in Computer Vision, 2nd edn. Cambridge University Press, Cambridge (2004)
3. Vite Silva, I., Cruz Corés, N., de la Fraga, L.: Optimal triangulation in 3D computer vision using a multi-objective evolutionary algorithm. In: Giacobini, M., et al. (eds.) EvoWorkshops 2007. LNCS, vol. 4448, pp. 330–339. Springer, Heidelberg (2007)
4. Jensen, M.T.: Guiding Single-Objective Optimization Using Multi-objective Methods. In: Raidl, G.R., Cagnoni, S., Cardalda, J.J.R., Corne, D.W., Gottlieb, J., Guillot, A., Hart, E., Johnson, C.G., Marchiori, E., Meyer, J.-A., Middendorf, M. (eds.) EvoWorkshops 2003. LNCS, vol. 2611, pp. 199–210. Springer, Heidelberg (2003)
5. Deb, K., Pratap, A., Agarwal, S., Meyarivan, T.: A Fast and Elitist Multiobjective Genetic Algorithm: NSGA–II. IEEE Transactions on Evolutionary Computation 6(2), 182–197 (2002)
6. Price, K.V.: An introduction to differential evolution. In: Corne, D., Dorigo, M., Glover, F. (eds.) New Ideas in Optimization, pp. 79–108. McGraw-Hill, London (1999)
7. Miettinen, K.M.: Nonlinear Multiobjective Optimization. Kluwer Academic Publishers, Boston (1999)

Fast Multi-template Matching Using a Particle Swarm Optimization Algorithm for PCB Inspection

Da-Zhi Wang[1,2], Chun-Ho Wu[1], Andrew Ip[1], Ching-Yuen Chan[1], and Ding-Wei Wang[2]

[1] Department of Industrial and Systems Engineering (ISE)
The Hong Kong Polytechnic University, Hung Hom, Kln, Hong Kong
[2] College of Information Science and Engineering
Northeastern University, Shenyang 110004, China
{mfwangdz,mfwhip,mfcychan}@inet.polyu.edu.hk,
jack.wu@polyu.edu.hk, dwwang@mail.neu.edu.cn

Abstract. Template matching is one of the image comparison techniques which is widely applied to determine the existence and location of a component within a captured image in the printed circuit board (PCB) industry. In this research, an efficient auto-detection method using a multi-template matching technique for PCB components detection is described. In many cases, the run time of template matching applications is dominated by repeating the similarity calculation, locating multi-templates, and exploring of the optimum result. A new approach using accelerated species based particle swarm optimization (SPSO) for multi-template matching (MTM) is proposed. To test its performance, our proposed SPSO-MTM algorithm is compared with other approaches by using the real captured PCB image. The SPSO-MTM method is proven to be superior to the others in both efficiency and effectiveness.

Keywords: PSO, Multi-temple Matching, PCB Inspection.

1 Introduction

Referential methods execute a point-to-point comparison whereby the reference data from the image of an "error-free" sample is stored in a database. Template matching, one of the referential methods, consists of calculation for every position of the image by examining a function that measures the similarity between the template and a portion of the captured image [1]. Normalized cross correlation (NCC) is widely used as the similarity function in template matching [2, 3]. It can greatly reduce the data storage and also reduces the sensitivity to acquired images when compared with traditional image subtraction, and hence enhances the robustness of the system [4]. As the circuit pattern density becomes higher, the traditional manual checking method can not meet the needs of large scale production of PCB. In this research, an efficient approach is presented for detection of the PCB components. The method proposed is based on template matching and an accelerated species based particle swarm

M. Giacobini et al. (Eds.): EvoWorkshops 2008, LNCS 4974, pp. 365–370, 2008.

optimization (SPSO) search. This embedded species approach can greatly improve the PSO performance in locating multiple components.

2 Multi-template Matching for PCB Inspection

Template matching can be used to find out how well a sub-image (template image) matches with the window of a captured image [5]. The degree of matching is often determined by evaluating the normalized cross correlation (NCC) value. The NCC has long been an effective similarity measurement method in feature matching.

Correlation based template matching has been commonly used in different applications [6]. Its sliding operation, however, is time consuming, especially as the captured image is large and multiple portions of the image need to be matched. The research has been motivated by the need to improve the efficiency of the correlation based template matching, hence, the idea of multi-template matching (MTM) to solve such a problem was mooted. In multi-template matching, at each point (x, y), the MTM-NCC value is defined as the maximum of a set of calculated NCC values which is obtained using corresponding templates. $\overline{f}$ is the average grey-level intensity of the captured image region coincident with the kth template image. $\overline{w}_k$ is the average grey-level intensity of the kth template image. The value γ is scaled from -1 to 1 and is independent of scales changes in the captured and template images. A "perfect" match between f and w_k will result in a maximum value. For example, at point(x, y), NCC values were obtained, NCC_1 was computed by using template 1, and NCC_2, calculated by using template 2 respectively. The MTM- NCC value at (x, y) can be obtained through the formula (2):

$$NCC_k(x,y) = \frac{\sum_{i=0}^{m-1}\sum_{j=0}^{n-1}[f(x+i,y+j)-\overline{f}]\cdot[w_k(i,j)-\overline{w}_k]}{\{\sum_{i=0}^{m-1}\sum_{j=0}^{n-1}[f(x+i,y+j)-\overline{f}]^2\cdot\sum_{i=0}^{m-1}\sum_{j=0}^{n-1}[w_k(i,j)-\overline{w}_k]^2\}^{1/2}} \quad (1)$$

$$MTM-NCC(x,y) = \max\{NCC_1(x,y), NCC_2(x,y), \ldots\ldots, NCC_k(x,y)\} \quad (2)$$

When the multi-template method is applied, it is clear that all objects can be matched with the most likely template simultaneously, thus, the components can be located. Applying MTM, one can create a NCC value space of a real PCB image. The captured PCB image, with 712 pixels x 612 pixels, and the six template images of the resistors are shown in Figure 1 (left). On the right side, the NCC value searching space is presented in a 3D plot. Every NCC value is calculated exhaustively at each pixel by the MTM approach. There are six global optima and many local optima. Every global optimum represents a preferred matching of the template and the target object. The problem of locating multiple resistors can be solved when all the global peaks can be detected.

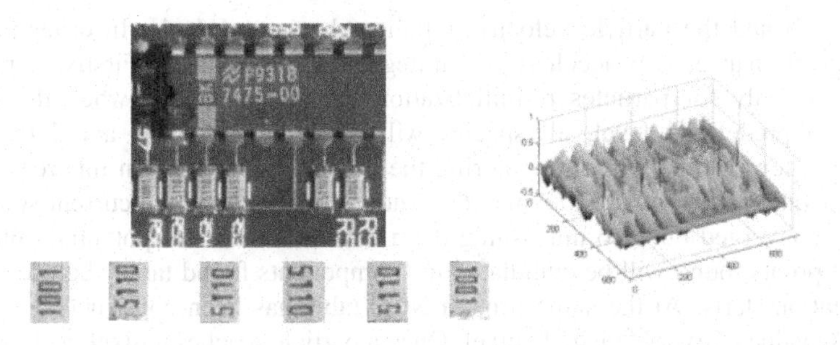

Fig. 1. Image of PCB-A with 6 templates of the resistors & 3D plot of NCC searching space

3 Development of a Species Based Particle Swarm Optimization

A PSO algorithm maintains a population of particles (swarm), where each particle represents a location in a multidimensional search space. The particles start at a random initial position and search for the minimum or maximum of a given objective function by moving through the search space. The manipulation of the swarm can be written as Equation (3) and (4). Equation (3) updates the velocity of each particle, whereas Equation (4) updates each particle's position in the search space, and c_1, c_2 are cognitive coefficients and r_1, r_2 are random numbers between [0, 1]; p_{id} is the personal best position and $lbest_i$ is the neighborhood best of particle i.

$$v_{id} = w \cdot v_{id} + c_1 \cdot r_1 \cdot (p_{id} - x_{id}) + c_2 \cdot r_2 \cdot (lbest_i - x_{id}) \tag{3}$$

$$v_{id} = w \cdot v_{id} + c_1 \cdot r_1 \cdot (p_{id} - x_{id}) + c_2 \cdot r_2 \cdot (lbest_i - x_{id}) \tag{4}$$

3.1 Species Based PSO (SPSO)

In recent years, the PSO has shown to be an effective technique for solving complex and difficult optimization problems in different areas. However, the original PSO fails to locate multi optima since the principle of PSO uses the "best known positions" to guide the whole swarm to converge to a single optimum in the search space. Li (2004) proposed a species based PSO which allows the particles to search for multiple optima (either local or global) simultaneously to overcome the deficiency of the original PSO in multimodal optimization problems [7]. The notion species in this study can be defined as a group of particles that share the same similarities, and are measured by a radius parameter r_s. The centre of a species, called species seed is the fittest particle in the species and all the particles in the species are confined within a circle with radius r_s.

4 Experimental Results

This section shows the test results of locating all the resistors in Figure 1 by the proposed method. The SPSO-MTM algorithm parameters in the setup are w= 0.5,

$c_1 = c_2 = 1.5$ and the particle velocity initializes between [-4, 4]; In order to have a better performance, two acceleration strategies are introduced. Firstly, a restarting strategy is only for particles re-initialization which means that when the iteration reaches a predefined level, all species will be re-initialized to avoid them from tracking a few optima without exploring the potential optimum. In this research, the re-initialization interval is 500 which does not only guarantee the current species can be fully converged but also unleashing them to search for more potential optima. All the best points found will be cumulated in "components found table" before every re-initialization starts. At the same time, a NCC table has been constructed for storing the NCC value of every "visited" pixel. Once a particle reaches a pixel, a chosen NCC value will be obtained and stored in the table during the entire searching process. So, when other particles "visit" the same pixel, the NCC value can be loaded directly from the table without any calculation. It leads to a dramatic improvement in the efficiency of the matching process, as duplicated calculation for a single pixel can be reached at 7000 times, without using the NCC table.

Table 1. Experimental results of six components detection based on SPSO-MTM

Particle number	r_s	Min. Run Time (s)	Max. Run Time (s)	Successful Rate*	Time (Mean & Std. err.)	Checked Pixels (Mean & Std. err.)
30	80	3.120	15.196	100%	7.06 ± 2.72	43008 ± 18394
	100	1.498	16.350	100%	6.40 ± 3.10	42586 ± 17307
	120	**2.150**	**13.260**	**100%**	**5.46 ± 2.48**	**39346 ± 14463**
50	80	3.818	17.666	100%	7.26 ± 2.84	39679 ± 13900
	100	2.762	16.158	100%	7.24 ± 2.98	38906 ± 14929
	120	2.258	13.562	100%	6.28 ± 2.86	39510 ± 18031

* *Successful rate means how many times that all 6 resistors can be successfully detected out of 100 times.*

The algorithm is coded in FORTRAN and executed on an Intel Pentium M 1.7GHz CPU with a 1G RAM notebook computer. In order to get an optimal parameter set for the test, a statistical analysis has been performed with different particle numbers and species radii. The mean execution time is derived from 100 independent runs, and Table 1 shows the experimental results. The species based PSO uses r_s to determine species and species seed; therefore it plays an important role in the performance of SPSO. From the experiments, one can see that the whole process becomes more robust in that all the components have been located within 23 seconds, when the r_s changes from 80 to 120. A larger r_s can improve the convergence speed. It is evident that 30 particles can fully assure the SPSO-MTM method successfully locates all components and gets a satisfactory result efficiently (39346 out of 435744 pixels are visited only). In summary, the best average processing duration for detecting all six resistors is 5.46 seconds, which is better than the values obtained by the other approaches (Table 2). The corresponding parameter setting, population size 30 with radius 120, is said to be the optimal setting. Figures 2 shows the snapshots of the simulated searching process with a particle number of 50 and r_s of 100.

Although each component has fixed size, the captured images of components do have significant different grey-level appearances between images and a single template for each component, such as non-uniform illumination and minor differences in the printed labels etc. A generalized template, using a set of template images for each component, is suggested to solve this problem. It is combined linearly by a series template images to average the statistical variation. Generalized template matching (GTM) combined with our proposed searching method has been tested for detecting three resistors in another PCB image (Figure 3). It has been found empirically that an NCC value of 0.94 is a good termination criteria for stopping the search with a match found.

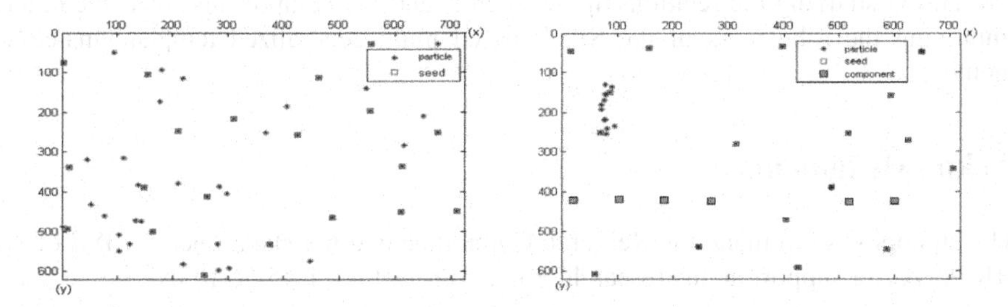

Fig. 2. The searching process from left to right: at the 1st & 8500th iteration (6 resistors detected)

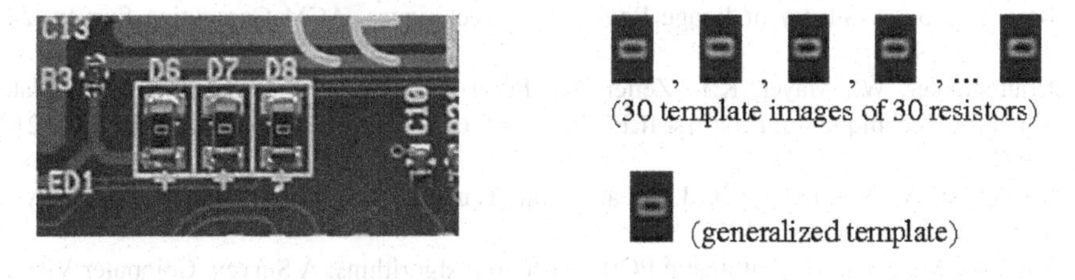

Fig. 3. Image of PCB-B & the generalized templates of the resistors to be located

Table 2. Results of different approaches

Method	PCB-A	Method	PCB-B
NCC-MTM exhaustive search	69.25 sec.	NCC-GTM exhaustive search	9.48 sec.
NCC GA-MTM[a]	37.72[b] sec.	NCC GA-GTM[d]	8.78[e] sec.
NCC SPSO-MTM	5.46[c] sec.	NCC SPSO-GTM	1.20[f] sec.

[a] The parameter setting of GA (with NCC table): Population-160; Cross rate-0.75; Mutation-0.35.
[b] Mean of 100 runs with std. err. of +/- 8.28 [c] Mean of 100 runs with std. err. of +/- 2.48.
[d] The parameter setting of GA (with NCC table): Population-120; Cross rate-0.75; Mutation-0.35.
[e] Mean of 100 runs with std. err. of +/- 2.64 [f] Mean of 100 runs with std. err. of +/- 0.32.

5 Conclusion

In this research, a species based PSO (SPSO) for multi-template matching (MTM) to solve a multiple components PCB detection problem is proposed. Experimental results showed that the proposed algorithm successfully located all the components in all runs, in a shorter time than in other approaches. SPSO-MTM, for locating multiple objects, is more efficient than GA-MTM and the exhaustive search in the given searching space, because there are less than 10% of the pixels of the whole space (PCB-A) which need to be checked. The results indicate that the SPSO is effective in template matching. However, the captured images of components do differ significantly between images and the single template for each component. Generalized template is proven as a simple but effective method to encode the variation. The method developed in this research is a start and has many other potential applications, such as object tracking [8] and robotic vision. Future research will aim at studying the relationship between r_s and the components' placement, and improving the robustness of the SPSO based multi-generalized template matching method.

Acknowledgments

The authors wish to thank the Research Committee and the Department of ISE of the HK PolyU for support in this research project (Project no. 1-45-56-RGNY).

References

1. Brown, L.G.: A Survey of Image Registration Techniques. ACM Computing Surveys 24, 325–376 (1992)
2. Krattenthaler, W., Mayer, K.J., Zeiler, M.: Point Correlation: A Reduced-cost Template Matching Technique. In: Proc. 1st IEEE Int. Conf. on Image Processing, vol. I, pp. 208–212 (1994)
3. Rosenfeld, A., Vanderburg, G.J.: Coarse-Fine Template Matching. IEEE Trans. on Sys., ManandCyb. 7, 104–197 (1977)
4. Moganti, M., Ercal, F.: Automatic PCB Inspection Algorithms: A Survey. Computer Vision and Image Understanding 63(2), 287–313 (1996)
5. Seul, M., O'Gorman, L., Sammon, M.J.: Practical Algorithms References for Image Analysis: Description, Examples and Code, pp. 106–110. Cambridge University Press, Cambridge (2000)
6. Stefano, L.D., Mattoccia, S., Tombari, F.: An Algorithm for Efficient and Exhaustive Template Matching. In: Proc. International Conference on Image Analysis and Recognition, pp. 408–415. Springer, Heidelberg (2004)
7. Li, X.D.: Adaptively Choosing Neighbourhood Bests Using Species in a Particle Swarm Optimizer for Multimodal Function Optimization. In: Proc. Genetic and Evolutionary Computation, pp. 105–116. Springer, Seattle (2004)
8. Cagnoni, S., Mordonini, M., Sartori, J.: Particle Swarm Optimization for Object Detection and Segmentation. In: Giacobini, M. (ed.) EvoWorkshops 2007. LNCS, vol. 4448, pp. 241–250. Springer, Heidelberg (2007)

A Generative Representation for the Evolution of Jazz Solos

Kjell Bäckman and Palle Dahlstedt

IT University, Gothenburg, Sweden

Abstract. This paper describes a system developed to create computer based jazz improvisation solos. The generation of the improvisation material uses interactive evolution, based on a dual genetic representation: a basic melody line representation, with energy constraints ("rubber band") and a hierarchic structure of operators that processes the various parts of this basic melody. To be able to listen to and evaluate the result in a fair way, the computer generated solos have been imported into a musical environment to form a complete jazz composition. The focus of this paper is on the data representations developed for this specific type of music. This is the first published part of an ongoing research project in generative jazz, based on probabilistic and evolutionary strategies.

1 Introduction

The most important feature of a good jazz musician is to be able to keep an entire solo together as an entity, i.e. to build up the solo phrase by phrase in collaboration with the other musicians, where each phrase is a natural continuation of the previous one and leads up to a climax of intensity. After the climax the solo should be rounded off. A longer solo might contain several climaxes, but they should then be organized in a musically meaningful way. A good improviser is not expected to drop the focus and give way to meaningless cascades of notes or producing routine phrases for lack of artistic ideas. The challenge is to be able to plan the structure of the solo already from start, and then stick to the plan during the entire solo. There are some excellent examples in the history of jazz with this ability, such as John Coltrane, Miles Davies, Keith Jarrett, Bill Evans, Claes Crona and possibly some more.

This project aims at making the computer build up a solo based on these principles. This is done using evolutionary principles on a genome structure consisting of a raw melody line split up into small melody fragments (delta phrases) and a structure of operators applied hierarchically on the delta phrases. Initially, the raw melody line is built up according to a "rubber band" principle, where each pitch interval is constructed using energy constraints much like the tension of a stretched rubber band. After application of the operators, the delta phrases will be somewhat different, however hopefully preserving some kind of musical idea. The aim is to ensure logical development of a consistent material during the solo and thus reflect the feature of a well-planned solo.

There are others working with similar concepts. Al Biles [1] has developed a system, GenJam, which uses phrases played by the "master" soloist as basis for the

M. Giacobini et al. (Eds.): EvoWorkshops 2008, LNCS 4974, pp. 371–380, 2008.

computer played solo. GenJam has three types of improvisation: whole chorus, chase improvising and collective improvising. The two last types are typical for older jazz forms like New Orleans jazz and the Swing era, while the first type is more relevant to modern jazz forms from the second half of the 20'th century. In GenJam the function of listening to motives from fellow musicians has been solved by means of an Analog-To-MIDI converter device. However, GenJam does not have the grand format principle of building up a solo from a low intensity level to a climax and rounding it off at the end, which is the long-term aim of our research.

Francois Pachet [9] has developed a system where the user plays a melodic material, from which the system builds its solo according to Markov chain probability calculation. The user can at any time introduce new melodic material to which the system responds. The sounding result is remarkably good and does not suffer from any technical instrumental constraints. Pachet's system sounds like a well-trained musician.

Dahlstedt [3][4] uses small melody fragments and combines them using operators in a recursively generative tree structure. Thywissen [13] uses generative music grammars in his GeNotator project. Dahlstedt and Thywissen apply their theories to classical music composition. In our experiment we try to apply similar theories to improvised jazz music. Also Manning's [8] exploration of MIDI technologies and Dean's [5] work on hyperimprovisation have been valuable.

Robert Rowe has in his two volumes, Interactive Music Systems [10] and Machine Musicianship [11] some interesting features such as scales connected to certain chord types, which have been valuable for the development of this system.

2 The Algorithmic Process

A jazz solo in this project consists of the melodic raw material and a hierarchical structure of operators. The melodic raw material can be auditioned separately. It is split into small portions, delta phrases, to allow processing at a lower level. Each operator of the operator structure processes a single delta phrase. For instance, one operator can add a note to the delta phrase, another operator can transpose the delta phrase a stipulated interval, and still another operator can invert a delta phrase. After processing of the delta phrases by the operators the result will be a modified melody possible to play back. The operators are applied hierarchically, i.e., one operator is applied to the whole melodic material, then one to each half, etc.

Fig. 1 shows how the raw material is split into delta phrases and then processed by the operators of the operator tree.

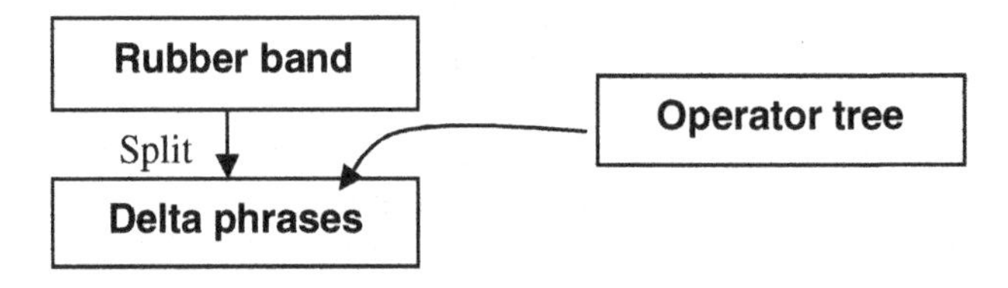

Fig. 1. Delta phrases are created by splitting the rubber band. The operator tree then processes the delta phrases.

The creation of the raw material and the application of the operators to the delta phrases are described in the subsequent sections.

2.1 Rubber Band Principle

The rubber band principle utilizes the contrapuntal [12] aspect of consuming a certain amount of energy to make an interval jump. The larger interval, the more energy is required. This is also to some extent depending on the length and volume of the destination note.

A maximum amount of energy for the whole melody line is allowed to be consumed. Having spent much energy, the melody can collect new energy by making less energy-consuming movements for a while, which is accumulated to the available energy reservoir. This will make the melody go up and down in intensity.

The creation of the raw material uses a similar technique as used in the mid-point displacement algorithm for landscape generation. It is created by originating from start and end pitches, then dividing the interval recursively. The middle pitch is stored for each interval division. It is represented as a deviation from the mean between the start and end pitch. So the representation is a binary tree of deviation figures from the mean line.

The raw material is created as follows. First we calculate a start pitch and an end pitch (fig. 2).

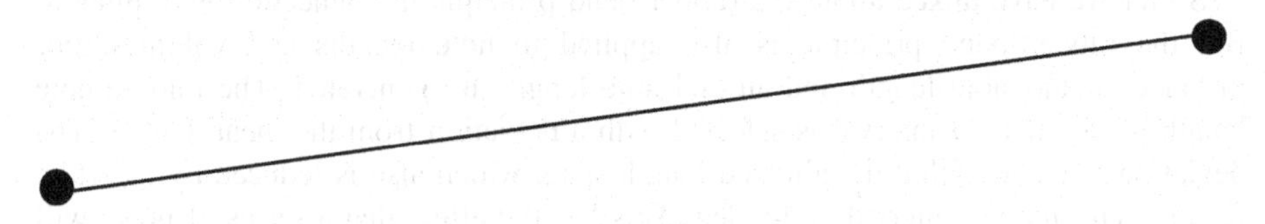

Fig. 2. The start and end points of the rubber band

Then we generate a pitch in the middle of the time span, which is allowed to deviate from the mean pitch line by a maximum pitch span (fig. 3) achieved by experimentation.

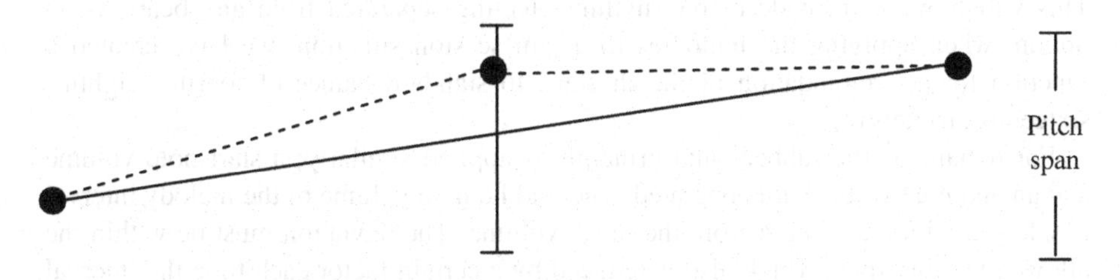

Fig. 3. The middle point is created within the allowed span

Then we split the time span into two equal time spans and repeat the process of generating a new note in the middle of each time interval (fig. 4).

Fig. 4. The complete rubber band

For each recursive subdivision, the maximum allowed pitch span is reduced by a certain factor, also achieved by experimentation.

For each new note the required energy is calculated, which is dependent on the deviation from the mean pitch line, the length of the new note, and the volume of the new note. Notes with a high energy are delayed some microseconds to reflect the situation when a real musician prepares himself prior to making the big pitch jump. A big pitch jump is also prepared, according to contrapuntal theory [12], by playing some ornamentation around the source note before making the big jump. The bigger the pitch jump, the more ornamentation is played. Also, according to contrapuntal theory, the gap between the source and destination pitches is filled with further notes after the big jump in order not to leave an empty hole in the melody.

The pitches can be accommodated to a given chord progression, which is created by a separate evolutionary process, to be described elsewhere.

So far we have talked about the rubber band principle in connection with pitches. But the rubber band principle is also applied to note lengths and volumes. For instance, a start note length and an end note length are generated. The middle note length of the melody interval is selected with a deviation from the mean length. The deviation must be within the allowed length span, which also is reduced by a certain factor each time the interval is divided. This has the effect that a series of notes will have about the same length, however by modifying the length span factor, this can be adjusted to achieve sudden burst outs of short notes.

The calculation of note lengths is not quantized to the standard rhythmical values whole notes, half notes, fourths, eighths, triplets etc. The lengths can have any MIDI ticks value. The reason for this is to not being tied up to traditional musical thinking concerning rhythm, but to concentrate on melody shapes and intensity fluctuations. This will however provide a free-rhythmic feeling separated from any beat. As an option, when applying the melodies to a jam session situation, we have created a function for accommodation of the rhythms to standard values of fourths, eighths, sixteenths, triplets etc.

For dynamics, the rubber band principle is applied similarly; a start note volume and an end note volume are generated. The middle note volume of the melody interval is selected with a deviation from the mean volume. The deviation must be within the allowed volume span, which also is reduced by a certain factor each time the interval is divided. By modifying the volume span factor you can achieve more or less smooth volume shapes.

By combining the rubber band principles for pitch, length and volume we achieve pitch shapes, length shapes and volume shapes operating independent of each other.

The technical representation of the raw material is MIDI pitch, length and volume for the start note and end note. The contour is represented as a tree structure of relative values of pitch, duration and volume (relative to the mean of the end points of the current time subdivision), which when applied recursively will recreate the exact contour.

2.2 Delta Phrases

When the raw material has been created it is split into delta phrases. A delta phrase is a series of notes with pitch, length and volume. The number of notes per delta phrase is given by the number of notes in the rubber band divided by the number of delta phrases, which is a global parameter. Suppose we get n delta phrases, ∂Ph_0 - ∂Ph_{n-1} (fig. 5).

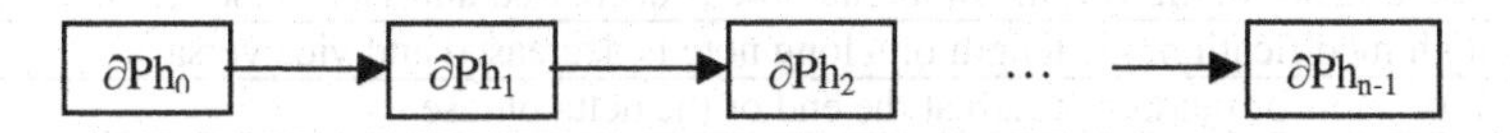

Fig. 5. Division of the rubber band in delta phrases

2.3 Operator Tree

The purpose of organizing operators hierarchically into an operator tree is to allow each delta phrase to be processed hierarchically by a series of operators. Fig. 6 shows the structure of an operator tree.

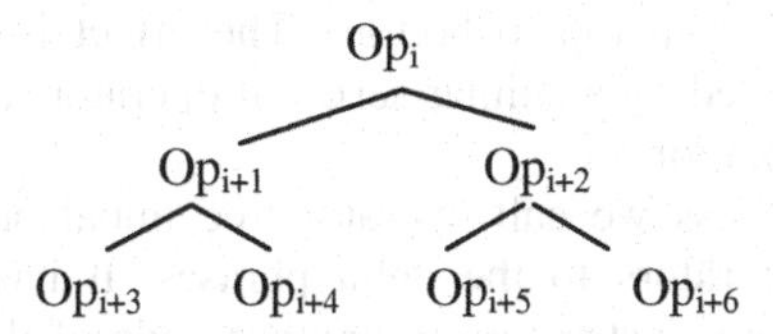

Fig. 6. Structure of an operator tree

The operator at the top level is applied to all delta phrases. The operators at level 2 are applied to half of the delta phrases each. The 4 operators at level 3 are applied to ¼ of the delta phrases. The division by 2 for each level is continued until there is one single operator to each delta phrase. The effect of this is that each delta phrase is processed by a series of operators from top to bottom of the operator tree, one operator per level. Since each operator performs the same operation to each delta phrase, this process introduces conformity over the whole solo, and as the recursive process branches out, variation is introduced between sections.

Each operator modifies a delta phrase in one particular way. The options are given in table 1.

The composition of operators in the operator tree is randomly created based on probability percents per each type of operator in table 1.

Table 1. Operator options

Copy, leave the delta phrase unmodified.
Transposition of the entire delta phrase by a random number of halftones
Transposition of a random single note by a random number of halftones
Addition of a note at a random position of the delta phrase within the pitch interval given by the delta phrase
Removal of a randomly selected note in the middle of the delta phrase
Augmentation of each interval in the delta phrase by a random number of halftones
Diminuation of each interval in the delta phrase by a random number of halftones
Retrograde, the delta phrase is reversed
Inversion, the pitches of the delta phrase are mirrored around its average pitch
Rhythm modification of a randomly selected tone. The amount of time to add/delete is randomly selected
Volume modification, the volume of a loud note is decreased and vice versa
Note length modification, the length of a long note is decreased and vice versa
Insertion of a rest of random length at the end of the delta phrase
Repetition of part of the delta phrase. Delta phrase is divided into three segments, and one of them is repeated
Polyphony, the highest pitch of the delta phrase is calculated, then some notes later an extra note is added, a random number of halftones higher than that note.
Pitch bend, if a slope of five ascending intervals is followed by three descending intervals, the top note will be subject to pitch bend, which is performed by starting a halftone below and sliding up to the top pitch

After all operator levels have been processed, each delta phrase has been modified by a series of operators from top to bottom. The effect is that two adjacent delta phrases have been processed by a similar series of operators and consequently should have some features in common.

The purpose of a process we call operator tree imbalance is to acquire a more varied application of operators to the delta phrases. It has been implemented by moving a part of the operator structure to another node of the tree, thus achieving a deeper operator level in some parts of the operator tree (fig. 7). The amount of imbalance (how many pair of operators to be moved) is controlled by a global parameter.

Fig. 7. Operator tree imbalance

With the aim of acquiring melody fragments according to some kind of ABA form, the experiment has been equipped with "trinary" functionality besides the binary structure of the operator tree, which means grouping of operators three by three, where the third operator is set equal to the first. This means that the first operator in a pair of operators is set equal to the first operator in the previous pair. This is made at the bottom level of the operator tree only. The amount of "trinarity" (the number of times to do this) is controlled by a global parameter.

To incur more operator processing to a delta phrase than accomplished by one series of operators from top to bottom of the operator tree, a delta phrase is allowed with a certain probability to "jump back" in the operator tree and follow another branch of operators (fig. 8). The frequency of doing this is controlled by a global parameter.

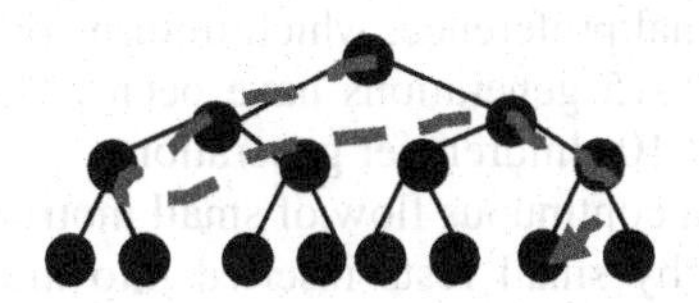

Fig. 8. The operator tree jump-back mechanism

2.4 Evolution of Solos

The generative representation described above has been used in an interactive evolutionary application, where a population of solos (typically about ten) are auditioned, selected and reproduced. The selection is a manual process using our personal preference, which stems from our background as jazz musicians. Solos can be saved at any time as MIDI files. Stored solos can be brought back into the process for further breeding. The genetic operators are described below.

A genome consists of the raw melody material and the operator tree. For each new generation two genomes (raw material melodies and the corresponding parent operator trees) are combined by crossover and mutated to generate a set of child genomes.

The two parent raw material melodies are combined by selecting start and end notes from the two parents interchangeably. Child no 1 will then get the deviation figures for pitch, length and volume for the first of two halves from parent 1 and for the last of two halves from parent 2. Child no 2 will get the figures for the first of two halves from parent 2 and for the last of two halves from parent 1. Child no 3 and 4 will get their figures from parent 1 and 2 interchangeably per each of 4 parts. Child no 5 and 6 will get their figures per each of 8 parts and child no 7 and 8 will get their figures per each of 16 parts.

A mutation is performed to the child raw material melody by increasing the deviation figure for pitch by 3 halftones up or down depending on whether the deviation is positive or negative, the deviation figure for MIDI note length by 12 ticks and the MIDI volume by 28 units, depending on whether the deviations are positive or negative. This is done for six notes in a sequence starting on a random position of the rubber band.

The new operator trees are combined by extracting branches from the two parent trees. This is done by originating from the parent 1 operator tree and copying random branches from parent 2 operator tree into parent 1 operator tree until 50% of the operators have been copied. When an operator in the operator tree of parent 2 is selected, all operators of the sub-tree under that operator will also be copied.

Mutations are applied to the child operator trees by modifying a random operator by the "jump back" series of operators, and applying slight random changes to the parameters of the operators (transposition, rest length, etc.), with a certain probability.

This gives a specific operator tree and a specific raw material melody per child.

3 Results

Some sounding examples are provided, where the title of each sound file gives an indication of the basic parameter setting. The manual selection of children has been carried out based on personal preference, which in turn springs from our background as jazz musicians. About 5-15 generations have been processed for each parameter setting with a population of 10 children per generation.

The sounding output is a continuous flow of small motives hooked onto each other, now and then interrupted by small rests inserted into the flow. You can trace the rubber band-like shapes of pitch, note length and volume individually, which in combination cause the intensity fluctuations. Thus the melody has some kind of intensity curves rolling up and down, trying to imitate melody curves of good music in general and jazz improvisation solos in particular.

The sounding output provides an interesting sequence of thematic material according to a slowly developing process, giving a feeling of recognition since each delta phrase has some kind of resemblance with adjacent delta phrases.

There is also an interesting polyrhythmic feature sometimes caused by a repetitive sequence of an odd number of notes not matching the natural beat (3 against 4, 5 against 4 etc.), and sometimes also caused by accentual volume effects to individual notes.

The "trinary" functionality did not provide the expected result to acquire some sort of ABA form, so the functionality could probably be dismissed without any loss of musical quality. The tree imbalance functionality also has a doubtful impact on the quality and could probably be excluded without any recognizable negative impact. The "jump back" functionality, on the other hand, provides a richer variation of the sounding output and deserves to be kept and further developed.

A set of unaccompanied sound examples can be heard at this link; in General MIDI format: http://oden.ei.hv.se/genmel.

A link to an example with computer generated drums and bass accompaniment is also available: http://oden.ei.hv.se/genmel/midi_evolv1.mid

4 Conclusions

Does this system provide any valuable artistic material? Yes, at least some sounding examples are of interest, maybe not of high professional musician class, but provide interesting and unexpected artistic output.

An improviser often uses standard phrases and motives trained during a long time of practicing and concerting. He relies on routines built up through repeated usage of similar muscular movements that are well accommodated to the physical design of the instrument. Some musicians are very strongly tied up to this behaviour, which makes them sound somewhat cliché-like and limited. There are some examples of this kind of musician in jazz history. They use the same kind of motives whatever style they play in, and may sound technically very brilliant and swinging, but if you transcribe their solos and try to play them with a good fingering, you will realize that they are astonishingly easy to learn to play.

The main purpose of using computers to produce jazz improvisation is that it opens your mind to new thinking and frees you from old habitual paces of playing. Hopefully it can enrich your improvisation style with new kinds of musical material.

The process of listening to many children in each generation may be boring after some time and it is easy to loose your concentration of separating between musically meaningful melodies and not so meaningful. Another problem is that the musical properties you are looking for might not be exactly the same in the beginning of a session as it is later in the session. Therefore, an intelligent setting of basic parameters and consistent evaluation conditions are required. The basic idea with evolution is that it works over many generations, but a concentrated listening to 10 children of more than 10 generations is probably not possible. So an automated fitness procedure would be valuable. The work of developing a computer based automatic fitness procedure has been started, but is not included in this paper.

Since the rubber band principle is applied to note lengths with no restrictions as concerns the traditional note values of whole notes, half notes, fourths, eighths etc., you cannot trace any particular beat or tempo, which is critical to all jazz music performed in jam session groups. This is easily remedied by quantization of the note lengths to the nearest traditional note value according to some logic, and thereby maybe also taking bar lengths in account. But by omitting this regulation, it is easier to concentrate on the real melodic value and not be distracted by side effects like swing or rhythmic effects.

5 Future Work

A mentioned, an automated fitness function would enable us to utilize the full strength of the evolutionary process, including a large population and many generations. Development has started, with promising results, which will be published elsewhere in the near future.

Since the a final goal of this project is to use the generated melodies as improvised solos in jam sessions with both virtual instruments and acoustic instruments, it will be necessary to accommodate the rhythm to the beat of the tune being played, and also to the periodicity and chorus lengths.

Evolutionary algorithms can also be used to produce new harmonies, drum rhythms, walking bass figures, piano and guitar accompaniment chord arrangement, and basic tune themes. This work has already been initiated, and will be published in the future.

Communication between musicians is very important in live jazz music. It would be possible to implement this in a computer generated jam session by letting the soloist, the drummer and the accompanying pianist "listen" to each other and reuse motives and rhythmic accents. Hopefully this will render not only a communicative feature but also a feeling of collective improvisation, where no particular soloist is leading the others but instead a situation where all musicians have the same value and contribute to the musical result on an equality basis.

References

1. Biles, J.A.: GenJam: a genetic algorithm for generating jazz solos. In: Proceedings of the 1994 International Computer Music Conference, ICMA, San Fransisco, pp. 131–137 (1994)
2. (anonymized), Evolutionary Jazz Improvisation. Masters thesis (2006)
3. Dahlstedt, P.: Sounds Unheard of – Evolutionary algorithms as creative tools for the contemporary composer. PhD thesis, Chalmers University of Technology, Gothenburg (2004)
4. Dahlstedt, P.: Autonomous Evolution of Complete Piano Pieces and Performances. In: Almeida e Costa, F., Rocha, L.M., Costa, E., Harvey, I., Coutinho, A. (eds.) ECAL 2007. LNCS (LNAI), vol. 4648, Springer, Heidelberg (2007)
5. Dean, T.: Hyperimprovisation: Computer-Interactive Sound Improvisation. A-R Editions Inc.,Middleton, Wisconsin (2003)
6. Levine, M.: The Jazz Piano Book. SHER MUSIC CO. Petaluma, CA, USA (1989)
7. Levine, M.: The Jazz Theory Book. SHER MUSIC CO. Petaluma, CA, USA (1995)
8. Manning, P.: Electronic and Computer Music. Oxford University Press, New York (2004)
9. Pachet, F.: Interacting with a Musical Learning System: The Continuator. SONY-CSL, Paris, France (Accessed 2 March 2006) (2002) , http://www.csl.sony.fr/pachet
10. Rowe, R.: Interactive Music Systems. The MIT Press, Cambridge (1993)
11. Rowe, R.: Machine Musicianship. The MIT Press, Cambridge (2001)
12. Söderholm, V.: Arbetsbok i kontrapunkt. Eriks. Cop., Stockholm (1980)
13. Thywissen, K.: GeNotator: An environment for investigating the application of generic algorithms in computer assisted composition. In: Proceedings of International Computer Music Conference 1996 (ICMC 1996), Hong Kong, pp. 274–277 (1996)

Automatic Invention of Fitness Functions with Application to Scene Generation

Simon Colton

Department of Computing, Imperial College, London
180 Queens Gate, London, SW7 2RH, UK
sgc@doc.ic.ac.uk

Abstract. We investigate the automatic construction of visual scenes via a hybrid evolutionary/hill-climbing approach using a correlation-based fitness function. This forms part of The Painting Fool system, an automated artist which is able to render the scenes using simulated art materials. We further describe a novel method for inventing fitness functions using the HR descriptive machine learning system, and we combine this with The Painting Fool to generate and artistically render novel scenes. We demonstrate the potential of this approach with applications to cityscape and flower arrangement scene generation.

1 Introduction

The Painting Fool (www.thepaintingfool.com) is an automated artist which has been designed primarily to produce interesting and aesthetically pleasing artworks, but also to test certain high level theories about how people perceive software as creative or not, within the computational creativity paradigm of Artificial Intelligence. The Painting Fool has so far been employed to take digital images and ground truths about those images, segment the images into paint regions and render them stroke-by-stroke by simulating natural media such as pencils, pastels and acrylic paints in a diverse range of styles.

We describe here an addition to The Painting Fool's capabilities. In particular, we describe a novel application where the input to The Painting Fool is not images, but rather a high-level description of a complex scene such as a cityscape. It then uses a hybrid hill-climbing/evolutionary approach to generate a segmentation according to the scene specification, and proceeds to render this in an artistic style. As described in section 2, generic scenes are specified as a set of elements and a set of desired correlations between aspects of the elements. The hybrid evolutionary/hill-climbing search methods we tested for constructing such scenes are described in section 3. Inspired by [2], we took this application to the meta-level by using the HR machine learning system [4] to invent novel fitness functions for scene elements, as described in section 4, with some illustrative results given in section 5. To conclude, we argue that with this project, and others, The Painting Fool has exhibited behaviour which is skillful, appreciative and imaginative, and as such, could be described as creative.

M. Giacobini et al. (Eds.): EvoWorkshops 2008, LNCS 4974, pp. 381–391, 2008.

2 Problem Description

We are interested in the problem of automatically generating scenes which potentially contain hundreds of similar scene elements. We use a simple representation of scenes as an ordered list of *scene elements.* To describe a generic scene to the system, the user/developer must provide the following:

- high level details of the scene, namely the pixel height and width of the scene and how many scene elements it should contain.
- a specification of a *scene element* in terms of a set of numerical attributes and a range of values for each attribute.
- an (optional) set of attributes to be calculated from the specifying attributes of a scene element, with suitable calculation methods.
- a set of desired correlations which the numerical attributes must conform to, and a weighting indicating the importance of each correlation.
- a method for using the attributes of each element to transform it into a set of colour segments as used by The Painting Fool.

With respect to the desired correlations that the user specifies, we calculate the Pearson product-moment correlation of two numerical variables v_1 and v_2 to approximate how they are related. This produces a value of: 1 if the variables are positively correlated, i.e., if v_1 increases, then so does v_2; -1 if the variables are negatively correlated, i.e., if one increases then the other decreases and vice-versa; and 0 if they are not correlated. In addition to specifying which two attributes of scene elements should be assessed by correlation, the user also supplies the correlation value they require. For instance, if the user requires two attributes to be completely positively correlated, they specify that the correlation should score 1, but if they want the two attributes to have some, but not total positive correlation, they might specify 0.5 as the required correlation score.

Using the terminology of Ritchie [9], our *inspiring example* has been scenes approximating Manhattan-style skylines. To achieve such scenes, we described them as being 1000×200 pixels in dimension with 300 elongated rectangular scene elements. Each element is described by the following attributes: (i) x-coordinate, of range 0-1000 (ii) y-coordinate, of range 60-100 (iii) width, of range 5-20 (iv) height, of range 20-100 (v) hue, of range 0-360 (vi) saturation, of range 0-1 and (vii) brightness, of range 0-1. We also provided methods to calculate (a) the *edge distance*, i.e., the minimum distance of the scene element from the left or right hand side of the scene and (b) the *depth* of the scene element, which is the number of scene elements that are both earlier than the element in the ordered list and overlap it. In addition, we specified 7 correlations with equal weights that the scene element attributes should adhere to:

- the edge distance of a scene element should have a correlation of 1 with the element's (i) width, (ii) height, (iii) y-coordinate and (iv) saturation.
- the depth of a scene element should have a correlation of: (v) 1 with the element's saturation (vi) -1 with the element's height and (vii) -0.5 with the element's brightness.

These correlations were chosen – after some experimentation – so that buildings in the middle of the scene appear bigger than those at the edges, and that elements at the back of the scene are taller – so that they can be seen – and less saturated, which is also true of the scene elements at the edge of the scene. This gives the impression that the buildings at the edges and at the back are further away than those in the centre of the scene (which is similar to the Manhattan skyline as viewed from the Staten Island ferry).

3 Search Methods for Scene Generation

Recall that the user supplies a set of n correlation specifications, which can be thought of as quadruples $\langle u, v, r, w\rangle$, where r is the required value that the Pearson product-moment calculation $p(u, v)$ should return for ordered lists of real-numbered values u and v, and w is a weight such that $\sum_1^n w_i = 1$, with the weights indicating the relative importance of the correlation. These quadruples can be expressed as a fitness function for scenes S which returns values in the range 0 to 1 by calculating:

$$f(S) = \sum_i^n w_i \left(1 - \frac{|r_i - p(u_i, v_i)|}{max\{|1 - r_i|, |-1 - r_i|\}}\right)$$

This scores 1 if all specified pairs of scene element attributes in a scene are perfectly correlated as per the specification, and 0 if their correlation score is as far away as it can be from the specified value.

We first looked at an evolutionary approach to generating scenes which maximise the fitness function. As in a standard evolutionary approach, the user specifies the population size, number of generations, mutation rate and crossover method (1-point or 2-point). An initial population of scenes each containing the requisite number of scene elements is then generated randomly. This is done by choosing values for each scene element attribute in each scene randomly from those allowed by the user-specified range. The fitness $f(S)$ for each scene S in the population is calculated, and S is given an overall score $e(S)$ by dividing $f(S)$ by the average of f over the entire population. $\lfloor e(S) \rfloor$ copies of S are placed into an intermediate population, with an extra one added with a probability of $e(S) - \lfloor e(S) \rfloor$. Pairs are chosen randomly from the intermediate population and an offspring is produced from them via crossover. Doing so is a relatively straightforward matter, as each scene is an ordered list of scene elements of a fixed size, hence sequential blocks of scene elements are simply swapped from the parents into the offspring. Offspring are mutated and added to the new population until there are the requisite number of individuals in the population. Each scene element in an offspring is chosen for mutation with a probability proportionate to the mutation rate. To perform a mutation, the chosen scene element is removed, then a completely new one is generated and inserted into the ordered list at a random position.

We also experimented with a hill-climbing approach. Here, a random scene is generated as above and each scene element is altered in turn in a single pass.

This is done by choosing a random value for each numerical attribute (and position in the ordered list) and checking whether this improves the fitness of the entire scene. If so, the new value is kept, but if not, the original value is re-instated. The user specifies a repetition factor for hill-climbing. This dictates the number of times a random value is chosen for each attribute of each scene element, and is usually in the range 1 to 100. With initial experimentation using simple cityscape scene descriptions, we found that both approaches were able to produce scenes with fitness over 0.9 in search sessions lasting only a few minutes. However, with our inspiring example of Manhattan cityscapes, which uses 7 correlations, we found that we could only achieve fitnesses between 0.8 and 0.9, using population sizes and generation numbers that resulted in run-times over 10 minutes. Moreover, on inspection of the scenes produced with fitness less than 0.9, we found that they were not of sufficiently high quality for our purposes. Hence, we also experimented with a hybrid approach, whereby an evolutionary approach is first attempted, and the most fit individual ever recorded becomes the basis of a hill-climbing search. We found that this performed better than both stand-alone methods.

To determine the optimal search strategy in terms of the highest achievable fitness in a reasonable time, we searched for solutions to the Manhattan example with 56 different search setups. For the evolution-only approach, we varied population size (100, 500 and 1000), number of generations (100, 500), mutation rate (0, 0.1, 0.01 and 0.001) and crossover method (1 and 2 point). For hill-climbing, we tested 1, 5, 10 and 100 repetition steps. We then examined the results and chose four hybrid setups to test. The best 20 setups are presented in figure 1 along with the fitness achieved and the time taken (all experiments were performed on a 2.4Ghz machine running Mac OS X).

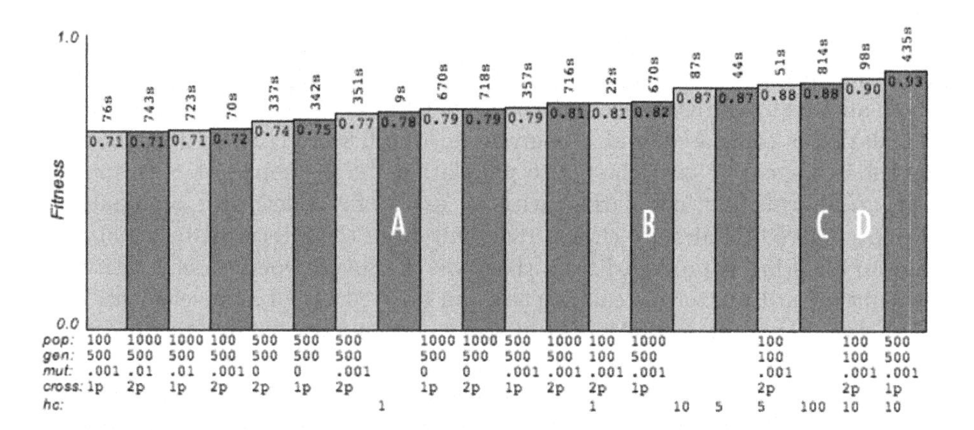

Fig. 1. Best 20 search setups for generating Manhattan-style scenes. Time taken in seconds and fitness are indicated at the top of each bar, with the search setup parameters given below each bar: pop=population size, gen=number of generations, mut=mutation rate, cross=crossover method [(1p)oint and (2p)oint], hc=hill-climbing repetition.

There were some interesting findings in these results. Firstly, we noted that anything but a 0.001 mutation rate resulted in premature convergence, which – as suggested by a reviewer – is probably due to the roughness of the fitness landscape. We found little discernable difference between 1 point and 2 point crossover, and as expected, with evolution only, the fitness achieved was largely proportional to the population size and number of generations. The quickest setup (labelled setup A in figure 1) achieved a fitness of 0.78 in only 9 seconds, using hill-climbing with repetition factor 1. We found that hill-climbing with a repetition rate of 100 achieved a fitness of 0.88, but took more than 10 minutes to achieve (setup C). In comparison, the best evolved (non-hill-climbing) scene had a fitness of 0.82 and took 670 seconds to produce (setup B). In contrast, an evolutionary approach with population size 100 for 100 generations followed by a hill-climbing session with repetition factor 10 achieved fitness 0.9 in only 98 seconds (setup D). It seems likely that we could achieve better results with different settings, and perhaps by using an iterative hill-climbing approach. However, as our main focus is on automatically generating fitness functions, and setup D performs adequately for that task, we have not experimented further yet.

In figure 2, we show the scene generated using search setup D, which achieved a fitness of 0.9. We see that, while there are a few outliers amongst the scene elements, the desired properties of the scene are there. That is, the buildings at the left and right of the scene are smaller in both width and height, less saturated and higher, which gives them the appearance of distance. Also, the buildings at the back of the scene are taller, less saturated and slightly brighter, again giving the impression of distance. In figure 2, we also present two rendered versions of the cityscape. The first is rendered using simulated coloured pencil outlining (with a reduced palette of urban colours) of the buildings over a simulated pastel base on art paper. The second is rendered using simulated acrylic paints over a pastel base, on primed canvas, giving a slightly three dimensional effect.

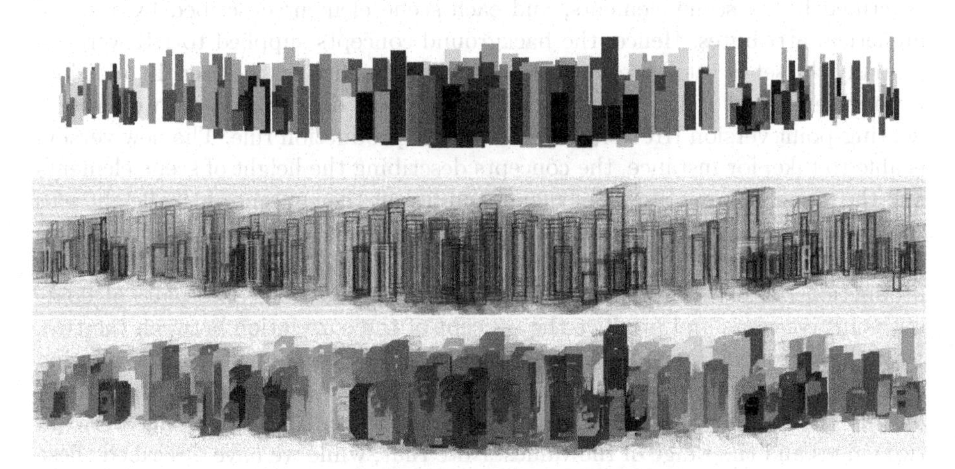

Fig. 2. Evolved setup D cityscape scene, rendered with: block shapes; simulated pastels and pencils; and simulated acrylic paints

4 Automatic Invention of Fitness Functions

One of the defining aspects of Cohen's AARON program [8] is that it invents scenes. This has been a motivation for development of the scene generation abilities of The Painting Fool. However, with the approach described above, it is difficult to state that The Painting Fool fully invents the scenes it paints, because the user must specify both the scene elements and some required properties of the scene as a whole, i.e., by specifying a fitness function in the form of a set of weighted correlations. We have also been motivated by Buchanan [2] and others in their opinion that meta-level reasoning is essential for creative behaviour. For this reason, we decided to use the HR system to invent fitness functions so that the user is able to specify only the types of elements a scene should contain.

The HR system is a descriptive machine learning program which starts with minimal information about a domain and generates examples, invents concepts which categorise the examples, makes hypotheses which relate the concepts, and proves these hypotheses using third party reasoning systems. HR has been described extensively elsewhere, e.g., [4,6]. For our purposes here, we need to know only that it is able to take background concepts which describe objects of interest and invent new concepts via a set of generic production rules. For instance, in number theory [3,5], HR is given just the ability to multiply two numbers together. It then invents the concept of divisors, using its *exist* production rule, then the concept of the number of divisors of an integer, using its *size* production rule, then the concept of integers with exactly two divisors (prime numbers), using its *split* production rule. In this way, HR invented novel integer sequences which, in the terminology of Boden [1], were H-creative. HR has been used in a number of other mathematical domains, in particular finite algebras [10].

To enable HR to invent fitness functions for scene generation, we made the objects of interest in theory formation sessions the scenes, with each object described by its scene elements, and each scene element described by a set of numerical attributes. Hence, the background concepts supplied to HR were essentially the set of attributes of scene elements specified by the user. In addition, we have extended HR to work with floating point data. To do so, we wrote a floating-point version HR's existing *arithmetic* production rule. The new version is able to take, for instance, the concepts describing the height of scene elements and the x-coordinate of scene elements and invent the concept of the height *plus* the x-coordinate of scene elements. In a similar manner, HR can subtract, multiply and divide pairs of floating point concepts. We also implemented two new production rules. Firstly, the *correlation* rule is able to take two pairs of floating point concepts and produce the concept of the correlation between the two. Secondly, the *float-summary* production rule is able to take in a single floating point concept and produce a concept with summary details about the values, namely the minimum/maximum/average value, the smallest difference between two values and the range of the values. Note that, while we have described these new rules in terms of scenes and scene elements, they are generic and would work in any other domain with floating point background concepts.

Input
S: scene overview specifications (number of scene elements, scene width & height)
E: scene element attributes and ranges for the attributes
R: rendering specifications
Algorithm
1. Using E, R and S, TPF generates five scenes randomly with 10 elements
2. TPF translates the scenes into a HR input file
3. TPF invokes HR to produce a theory using 1000 steps
4. HR translates its theory into a Java class, J, and compiles it
5. TPF constructs a random population, P of 100 scenes according to S
6. while ($maxfitness(P, F) < 0.4$ or $maxfitness(P, F) > 0.8$) TPF builds a fitness function F by choosing 3 to 6 correlation concepts from J and giving them equal weights
7. TPF evolves a best scene, B, according to F using setup D (see section 3)
8. TPF hill-climbs to improve B using repetition factor 10 (setup D again)
9. TPF translates B into a segmentation, G
10. TPF renders G according to R

Fig. 3. Overview of the interaction between HR and The Painting Fool (TPF)

In figure 3, we describe in overview how a combination of The Painting Fool and HR were used to automatically construct a fitness function, use this to search for a scene which maximises the fitness, and then render the scene. Note that, to keep HR's run-time to around two minutes, it is only supplied with 5 scenes of 10 elements, which is enough for it to form a theory. Note also that, in step 6, The Painting Fool builds a fitness function using a Java class that HR has generated. The Java class is able to take a scene and calculate a fitness for it. The Painting Fool constructs the fitness function as an equally-weighted sum of between 3 and 6 correlation concepts randomly chosen from HR's theory. It checks that the best fitness in a randomly generated population of 100 scenes is above 0.4 (so that there is a chance of evolving a scene to above 0.9 fitness), and below 0.8 (which ensures that a random scene will not be output). If this is not true of the fitness function it has constructed, it tries again until it succeeds. As an example, the first cityscape scene generated in a session described in the next section was generated using a fitness function with the five equally weighted correlations below. The scene generated using this is given in figure 4.

cor(brightness, x-coord) = 1	
cor(y-coord - x-coord, brightness) = -1	cor(brightness, height) = 1
cor(y-coord * x-coord, height) = 1	cor(saturation, height) = -1

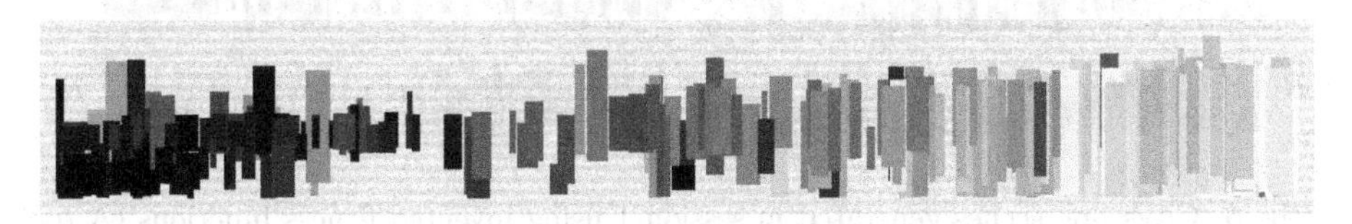

Fig. 4. Scene generated with an automatically invented fitness function

5 Illustrative Results

We present here two scene generation sessions with The Painting Fool/HR. Firstly, we used the cityscape scene specification as above, and the algorithm in figure 3 to generate a fitness function. We repeated this 10 times, with the results from the session given in figure 5, alongside two randomly generated scenes for comparison. In each of the 10 generated scenes, there are at least two noticeable patterns, e.g., in scene E, the buildings are more colourful at the edges and wider on the left than on the right. In scenes B and G, two distinct clusters of buildings were generated, which we did not expect, and in scene C, the fitness function forced some buildings to be placed vertically on top of each other, which was also not expected. The fitness of the scenes were in the range 0.88 to 0.94.

In the second session, we tested the flexibility and ease by which the system can be given a new task, using arrangements of flowers, as this is quite different to cityscapes. We chose 17 closely cropped digital images of flowers, and used The Painting Fool to generate the segmentations of these. We then described the generic scene in terms of scene elements with the following properties: size (30 to 150), x and y coordinates (both 0 to 400), and flower number (1 to 17). We wrote code so that when The Painting Fool translates the scenes into segmentations, it simply retrieves the segmentation for the appropriate flower number from file. We also wrote code able to calculate (i) the distance of the scene element from the centre of the scene and (ii) the position in the ordered list that the scene element appears at. To test the scene specification, we used a fitness function with three correlations: cor(centre-distance, flower-number) = 1; cor(centre-distance, size) = -1; cor(centre-distance, pos) = -1. This meant that the scenes had flowers around the edge which were smaller (with the smaller ones being painted later than the larger ones), and that the flowers portrayed changed from the centre to the outside of the scene.

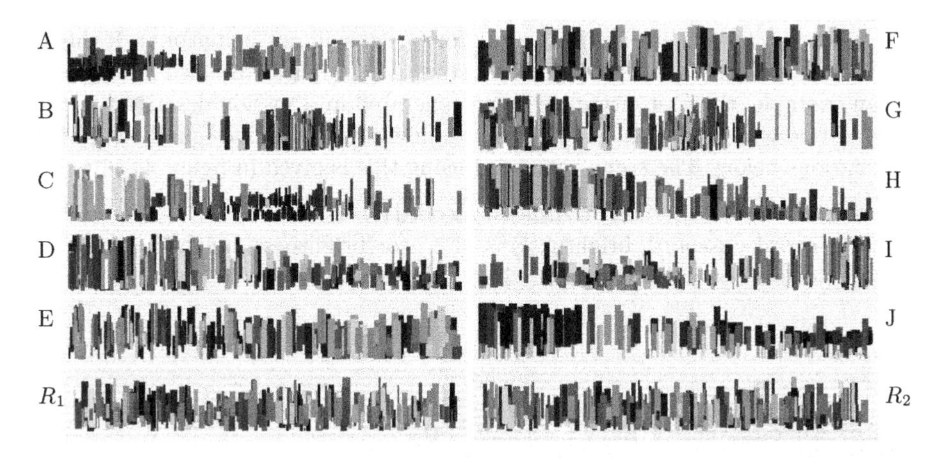

Fig. 5. Cityscape scenes generated in session 1 using invented fitness functions (A to J), compared to random cityscapes (R_1 and R_2)

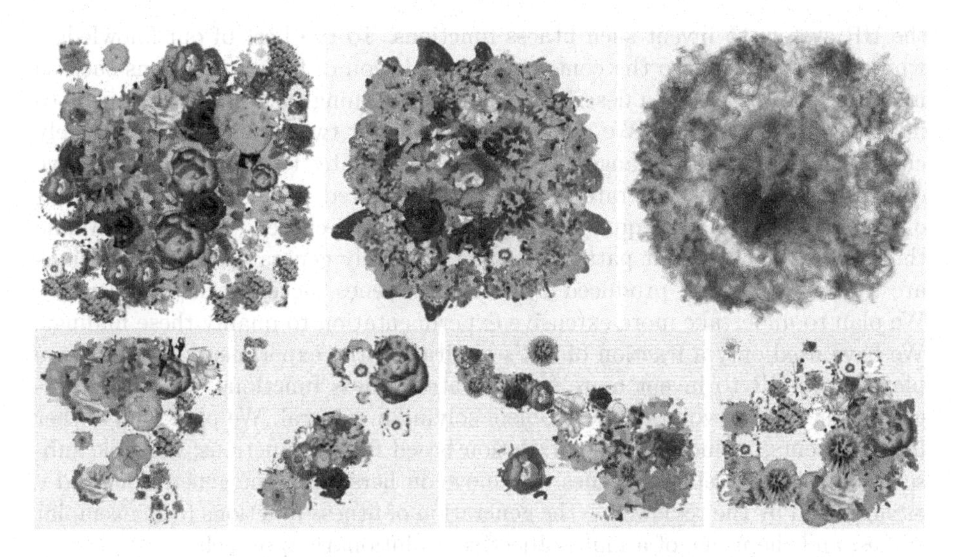

Fig. 6. Cartesian flower arrangement scene; polar coordinate scene 1, rendered with simulated acrylic paints onto a painting of leaves; polar coordinate scene 2, rendered with simulated pastels and pencils; four invented flower arrangement scenes

We ran a search using setup D to generate a scene with 150 flowers in it, with the resulting scene given in figure 6. We also repeated this experiment twice using polar instead of Cartesian coordinates for the scene elements, to give a circular arrangement of the flowers. In figure 6, we present artistic renderings of the two polar coordinate scenes. Using the Cartesian setup, we then ran the automatic invention of fitness function routine 10 times, but to produce scenes with only 50 elements, and we chose the four most interesting to show in figure 6 (an entirely subjective choice by the author). As with the invented cityscapes of figure 5, each of the four scenes clearly exhibits a pattern. However, of the six other scenes from the session (which are not shown), with four of them, we could discern no obvious scene structure. These scenes scored less than 0.8 for fitness, and on inspection, this was because the fitness functions needed to achieve contradictory correlations. With fewer attributes to seek correlations between in this application than in the previous one, the likelihood of generating such contradictory fitness functions was higher. We aim to get The Painting Fool to avoid such cases in future. Including the time taken to find and segment the flower photographs and for us to write the necessary code for attribute calculation and segmentation translation, the entire session took around 4 hours.

6 Conclusions and Further Work

We have described a hybrid evolutionary/hill-climbing approach to the construction of scenes, via a correlation based fitness function, and a method which uses the HR system to invent such fitness functions. To the best of our knowledge, while this work fits into the context of co-evolution of fitness functions such as in [7], it is the first time a descriptive machine learning system has been used to invent a fitness function. We demonstrated this for two types of scene, namely cityscapes and flower arrangements. As suggested by two reviewers, we have compared the scenes generated using HR's invented fitness functions against ones generated using an equal weighting of randomly chosen correlations. While these scenes showed clear patterns, as the randomly generated fitness functions are a subset of those produced by HR, the scenes lacked variety somewhat. We plan to undertake more extensive experimentation to qualify these findings. We have used only a fraction of HR's abilities in the experiments here, and we plan to use HR to invent more sophisticated fitness functions, and to investigate using HR in evolutionary problem solving in general. We plan many other improvements, including: non-correlation based fitness functions; multiple sub-scenes; post-processing of scenes to remove outliers; 3d scene generation; and – as suggested by the reviewers – the generation of fitness functions from exemplar scenes, and the usage of a multi-objective evolutionary approach.

As described in [?], we believe that software exhibiting skill, appreciation and imagination should be considered creative, and we are building The Painting Fool along these lines. In two other projects (the Amelie's Progress gallery and the Emotionally Aware application described on `www.thepaintingfool.com`), The Painting Fool has exhibited behaviour which could be considered appreciative. In enabling it to invent fitness functions and generate scenes using them, we have implemented some more imaginative behaviour, and we hope to have added a little to the case that the software is creative in its own right.

Acknowledgements

We wish to thank the anonymous reviewers for their excellent comments, which improved this paper and suggested many interesting avenues for further work.

References

1. Boden, M.: The Creative Mind. Weidenfeld and Nicolson (1990)
2. Buchanan, B.: Creativity at the meta-level: Presidential address at AAAI-2000. AI Magazine 22(3), 13–28 (2001)
3. Colton, S.: Refactorable numbers - a machine invention. Journal of Integer Sequences 2 (1999)
4. Colton, S.: Automated Theory Formation in Pure Mathematics. Springer, Heidelberg (2002)
5. Colton, S., Bundy, A., Walsh, T.: Automatic invention of integer sequences. In: Proceedings of the Seventeenth National Conference on Artificial Intelligence (2000)
6. Colton, S., Muggleton, S.: Mathematical applications of Inductive Logic Programming. Machine Learning 64, 25–64 (2006)
7. Maher, M.L., Poon, J.: Co-evolution of the fitness function and design solution for design exploration. In: IEEE International Conference on Evolutionary Computation (1995)

8. McCorduck, P.: AARON's Code: Meta-Art, Artificial Intelligence, and the Work of Harold Cohen. W.H. Freeman and Company, New York (1991)
9. Ritchie, G.: Assessing creativity. In: Wiggins, G. (ed.) Proceedings of the AISB 2001 Symposium on AI and Creativity in Arts and Science, pp. 3–11 (2001)
10. Sorge, V., Meier, A., McCasland, R., Colton, S.: Automatic construction and verification of isotopy invariants. In: Journal of Automated Reasoning (2007) (forthcoming)

Manipulating Artificial Ecosystems

Alice Eldridge, Alan Dorin, and Jon McCormack

Centre for Electronic Media Art, Monash University, Australia, 3800
{alice.eldridge,alan.dorin,jon.mccormack}@infotech.monash.edu.au

Abstract. Artificial ecosystems extend traditional evolutionary approaches in generative art in several unique and attractive ways. However some of these traits also make them difficult to work with in a creative context. This paper addresses the issue by adapting predictive modelling tools from theoretical ecology. Inspired by the ecological concept of *specialism*, we construct a parameterised *fitness curve* that controls the relative efficacy of generalist and specialist strategies. We use this to influence the population's trajectory through phenotype space. We also demonstrate the influence of environmental structure in biasing evolutionary outcomes. These ideas are applied in a creative ecosystem, *ColourCycling* which generates abstract images.

1 Introduction

The concept of the *creative ecosystem* is gaining in significance as a conceptual approach to evolutionary music and art [1], [2], [3], [4]. Creative ecosystems implement metaphors from ecology and evolutionary theory, the idea being that the complex interactions and evolutionary dynamics exhibited by real ecosystems offer rich practical and conceptual possibilities for generative art.

Ecosystem models offer a number of advantages over existing Evolutionary Computing (EC) techniques [3]. These include the demonstration of complex dynamics over fine and coarse timescales, the support of non-linear interactions and implicit environmental fitness evaluation. The emphasis shifts from explicit fitness evaluation of phenotypes to the creative design of complex environments. These advantages and possibilities also bring new conceptual challenges. For the generative artist adopting standard EC approaches, the extensive body of research in engineering applications provides guidance on general design issues such as selecting mutation rates or crossover schemes. For the artist wishing to explore ecosystemic models, there is no such equivalent.

The dynamic complexity of the ecosystem is highly desirable in a creative context, but its potential may be difficult to realise unless these complex dynamics can be harnessed for the purposes of *creative discovery*. We must strike a balance between the ecosystem's generative autonomy and our artistic control over the aesthetic relevance of the search trajectory. To date, a predominantly "build-it-and-see" approach has been adopted by artists working with ecosystemic metaphors. This paper addresses the lack of a "user manual" for creative ecosystems and describes a strategy to control the diversity of an artificial ecosystem's population over evolutionary time-scales.

M. Giacobini et al. (Eds.): EvoWorkshops 2008, LNCS 4974, pp. 392–401, 2008.

1.1 General Strategies for Ecosystem Control

Examining existing ecosystemic artworks we can identify two distinct design approaches: *agent-centric* and *enviro-centric*. The agent-centric approach gives primacy to the agent itself. Agents are parameterised by a complex set of physical and behavioural characteristics which are commonly user-specified [5], [1]. The system's behaviour emerges from the interactions of the agents and is governed by the ability of the creator to design and control them.

An alternative enviro-centric approach takes a global perspective, varying the environmental conditions or physics of the world. This "broad brush" strategy does not require micro-management of the ecosystem but allows it an appealing degree of autonomy. Variation of environmental conditions modifies the system dynamics without requiring specific, low-level tinkering [6], [7]. We favour this more elegant approach to working with creative ecosystems and take cues from theoretical ecology, adapting predictive modelling techniques to create methods of manipulating the evolution of an abstract artificial ecosystem.

An overview of some relevant work in ecology is given in Section 2. Section 3 describes an abstract model in which agents are specified according to the niche they occupy. We show that by adapting inter-agent competitiveness and the structure of the environment we can influence the evolutionary outcomes. Section 4 provides a graphical application of this method.

2 Insights from Theoretical Ecology

Within ecology, competition is recognised as a major structuring force of ecosystems [8], it is therefore a natural starting point for our own exploration of creative ecosystem control algorithms. Competition can be measured using the concept of the *niche*: the role of an organism in its community [9]. The structure of individual niches and the relationships between them can be used to predict the species composition of a community. The width of a niche describes the degree of specialisation of an individual or species: a *specialist* has a narrow niche, it occupies a limited and particular portion of habitat; a *generalist* occupies a broad niche and can make use of a wider expanse of habitat. Niche overlap provides a measure of competition between species.

Two relevant insights arise from this conceptualisation. Firstly, when there is no cost to habitat selection, generalists are more competitive than specialists [10], [8]. Secondly, coexisting species cannot occupy identical niches. It follows that communities of specialists can support a greater diversity of species. Within theoretical ecology specialism provides a conceptual tool for building models of the origins and maintenance of species diversity [11]. This is a key issue for artists who work with evolutionary techniques.

2.1 Factors Promoting Specialism or Generalism

A review of models of specialist-generalist competition identified several factors that affect the viability of each strategy. These include the shape of the fitness set, negative density dependance, the cost of activity selection and the temporal variability of the environment [12]. The impact of environmental structure has also been examined in an explicitly spatial model that suggests specialism becomes more competitive in a heterogenous environment [8]. The factors we consider in our study are the shape of the fitness set and the impact of environmental structure.

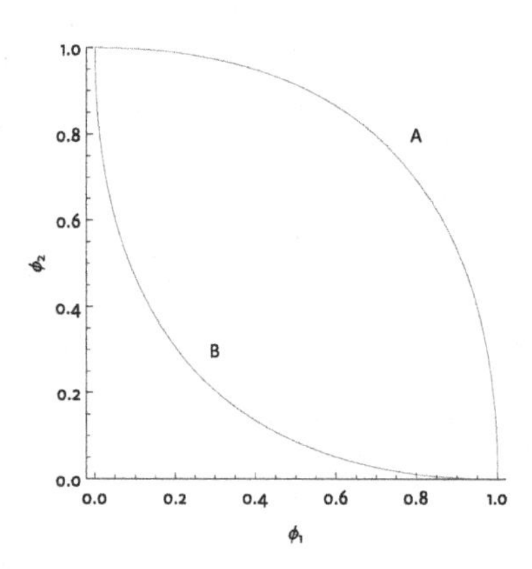

Fig. 1. A fitness set in two dimensions. The horizontal and vertical axes represent an individual's fitness in each of the two niches (ϕ_1, ϕ_2). Individuals are represented as points along the curves.

The Fitness Set. Levins [13] introduced the *fitness set* to represent the fitnesses of individuals in multi-dimensional niche space. As shown in Figure 1 each axis of the fitness set corresponds to the fitness of individuals in particular niches (ϕ_1 and ϕ_2). Specialists lie at the extremes of the axes, generalists in the middle. The negative slope of both lines reflects the assumption that any increase in fitness in one niche demands a decrease in fitness associated with the other: a Jack-of-all-trades can be master of none. In the convex case (A) the generalist is only slightly inferior to a specialist in either niche. For the concave curve (B) the disparity is increased. This fitness set has been used in a variety of different models which conclude that it must be concave for the advantages of specialisation to outweigh the costs [13], [14], [15].

The fitness set is used in ecological studies to represent and predict population structure: for our purposes it suggests a means of manipulating them. We represent the set as a variable curve used to determine ϕ_2 as a function of ϕ_1, where ϕ_1 is genetically specified. Changing the convexity of this curve lets us manipulate the relative competitiveness of generalists and specialists and influences the course of evolution. We call this the *fitness curve* and demonstrate its application in Section 3.

Environmental Structure. The fitness set describes the fitness of an individual within a niche but does not address the relationship between niches. In addition to the effects of competition from the rest of the population (which is modelled by default in evolutionary agent-based modelling) the relationship *between* niches is determined by environmental structure.

Spatial models can behave differently to non-spatial models of the same phenomenon [16], [17]. Those studies of specialist-generalist competition that have considered environmental variability have modelled either temporal or spatial variation in resource abundance. In both cases the scale over which resources vary is key in favouring either specialism or generalism. Heterogenous environments favour specialism and vice versa[1].

3 Experimental Manipulation of Population Structure

To explore the impact of the fitness curve and environmental structure on a population, a series of evolutionary agent-based models was developed. The environment contains two resource types (0 and 1) and an agent population. Agents have a single, real-valued gene determining their resource preference and degree of specialisation (see below). The effects of our experimental manipulations were measured by observing the genotypic distribution of the evolved population once it had stabilised.

All models are initialised with agents whose gene and initial energy store values are drawn from a uniform random distribution. At each timestep all agents are updated in random order and a fixed metabolic cost is deducted from their energy value. The probability of an individual acquiring an available resource is determined by its gene value and the fitness curve as described in Section 3.1. Resource availability differs between models as described in the relevant sections below.

When a resource is acquired, one unit of energy is transferred from the environment to the agent. Asexual reproduction occurs when an agent increases its initial energy endowment by 50%. If the current population exceeds a specified maximum, either the newborn or a randomly selected agent is killed with a fixed probability. Agents donate half of their energy to their offspring. The child's gene is mutated using creep mutation with probability 0.1 and intervals drawn from a Normal distribution. An agent dies if it reaches a globally defined maximum age or its energy level falls to zero.

3.1 Varying the Fitness Curve

The gene value of each individual represents both its resource preference and its degree of specialism – in the current model this means its capacity for acquiring

[1] If a preferred resource is in constant supply throughout an organism's reproductive lifetime specialists need never engage in sub-optimal behaviour. This is the case when resource density satisfies an agent's energetic needs from birth to reproduction or if the period of temporal variation exceeds this. In conditions where resource abundance varies within an organism's lifetime, generalists are at an advantage because they can switch between habitats.

its preferred resource, ϕ_1. Values above 0.5 denote a preference for resource type 1, values below 0.5 for resource type 0. (E.g. a gene of 0.8 denotes a preference for resource type 1 and an 80% chance of acquiring it.) Capacity is symmetric about 0.5, i.e. an individual with gene value 0.2 has an 80% chance of acquiring resource type 0. We modelled the fitness set described by Levins [13] using Equation 1, (Figure 1). This equation's control parameter, α, varies the curve from concave ($\alpha < 0.0$) to convex ($\alpha > 0.0$). Equation 1 determines an individual's capacity for acquiring resources of the *non*-preferred type, ϕ_2. The overall fitness of an individual (independent of environmental or population effects) is given by the sum of its gene value (ϕ_1) and the value determined by the fitness curve (ϕ_2). Adjusting α alters the relative competitiveness of generalism or specialism and so influences the evolutionary pressure on the population.

$$\phi_2 = \begin{cases} 1 - \phi_1 & \text{if } \alpha = 0, \\ \frac{\sqrt{1+4\alpha(1+\alpha-2\phi_1)}+2\alpha\phi_1-1}{2\alpha} & \text{otherwise.} \end{cases} \quad (1)$$

The effect of the fitness curve was first examined in the absence of any spatial component or direct competition between agents using the basic model described in Section 3. Resource availability was set at 80% of that required by the population. At each run individuals were randomly selected and presented with one of the two resource types (with equal probability) until resources were depleted.

Figure 2(a) shows the distribution of genes at generation 2000 for concave, flat and convex curves. At $\alpha = 0.0$ the curve becomes a flat line such that the overall fitness of generalists and specialists are equal. This provides a control case for comparison between models.

As expected, the convex curve favours generalism (gene values converge on the centre), concave curves promote the evolution of specialism and a flat line (the control case) produces an even distribution across genotype space. The graphical results are supported by examination of the interquartile ranges (IQRs) given under each plot which decrease as the curve moves from concave to convex. These results suggest the fitness curve can be used to manipulate the viability of different levels of specialism and influence the genetic structure of the population.

3.2 Adding Competition

Ecological theory suggests that in a resource-limited, structureless environment generalists will out-compete specialists. To ensure that our model is in line with these predictions, we altered the resource acquisition algorithm to include direct competition. Tournament selection was carried out between two individuals selected at random from the population, until all resources were depleted. The two resource types were selected with equal probability and resource levels were 80% of that required by the population. In this scenario, the individual with the higher capacity for the offered resource is more likely to win any particular tournament.

As evident in Figure 2(b), the generalist strategy is more competitive in this situation. For all three fitness curves, generalist genotypes dominate the

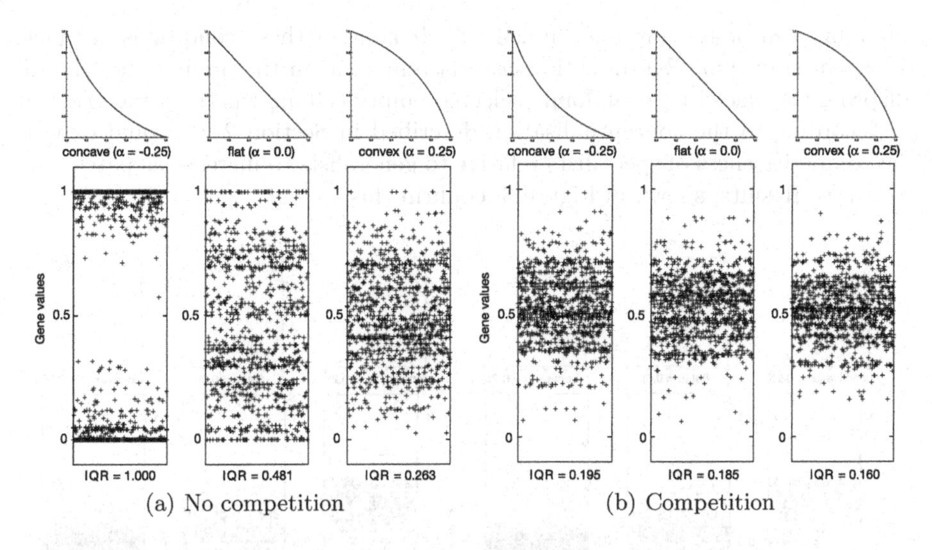

(a) No competition (b) Competition

Fig. 2. Scatter plots of gene values in the population at generation 2000 for non-spatial models for concave (left), flat (centre) and convex (right) fitness curves. Results are shown in the (a) absence and (b) presence of competition.

population. The slight variation in IQRs suggests that the curve has a limited effect (the population spreads out as the curve becomes concave), but generalists dominate even for convex curves[2].

3.3 Varying Environmental Structure

As noted in Section 2, resource distribution can alter the cost of habitat selection. This can be modelled by a heterogenous distribution of resources in finite, variable sized *patches*. The competitiveness of specialism increases with patch size [8]. In order to test whether these findings hold in an agent-based evolutionary model we spatialised our model and manipulated resource distribution as described in [17] where the environment is segregated into variable-size patches and gaps. In this case a torroidal plane was covered in a patchwork of equally sized, equally spaced squares of alternate resource types. Patch resource type is fixed at the start of each run.

Agents are initialised randomly to unique cells on the grid. At each timestep resource levels are uniformly incremented and agents updated in random order. Each agent examines all of its eight neighbouring cells in random order and preferentially selects: (a) an unoccupied cell containing sufficient resources of its preferred type; (b) an unoccupied cell containing *either* resource, rather than

[2] The reason for this becomes clear if we consider the discrete case where specialists are able to acquire energy from one resource type exclusively. Whilst they are more likely to win the resource when it *is* available they can *never* obtain energy from their non-preferred type. Generalists, however, have *some* chance of winning either resource and can therefore accrue energy, and so reproduce, more rapidly.

an empty cell or (c) any unoccupied cell. If none of these conditions are met the agent stays put. Reproduction criteria remain as in the original model and offspring are placed in a randomly selected empty cell anywhere on the grid.

According to the conceptualisation described in Section 2 we would expect the competitiveness of specialists relative to generalists to increase as patch size increases. Results, shown in Figure 3, confirm this.

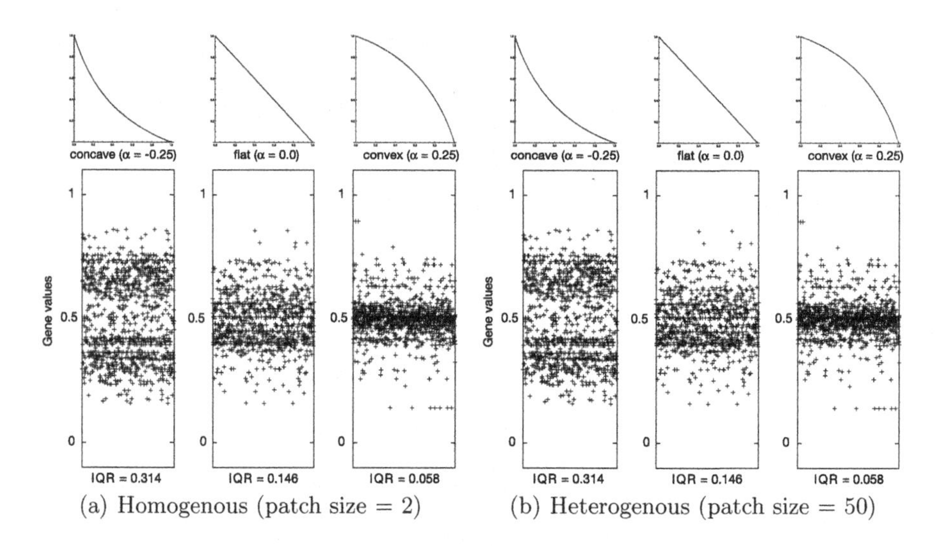

(a) Homogenous (patch size = 2) (b) Heterogenous (patch size = 50)

Fig. 3. Scatter plots of gene values in the population at generation 2000 for spatial models for concave (left), flat (centre) and convex (right) fitness curves. Results are shown for (a) homogenous and (b) heterogenous environments.

A **homogenous environment** was created by setting patch size to two and gap size to one. Resource renewal rates were adjusted such that agents were forced to move between patches to survive. Given the alternating pattern of resources, we would expect this situation to favour generalists as they can acquire energy from both resource types. As shown in Figure 3(a) the population converges to be predominantly generalist. Increasing the concavity of the fitness curve increases the genetic spread of the population slightly but few extreme specialists evolve.

A **heterogenous environment** was created with a patch size of fifty and gap size one. Resource renewal rates and population density were kept constant. Figure 3(b) shows the genotypic distributions. As expected this situation favours the evolution of specialism as individuals can spend an entire lifetime within their preferred patch. Although there is no direct competition between agents, specialists accrue energy more rapidly than generalists and therefore reproduce more prolifically[3].

[3] Initial investigations into temporal variation of resource abundance suggest it has a similar effect on the relative fitnesses of specialists and generalists.

4 ColourCycling, a Creative Application

ColourCycling is a creative ecosystem that demonstrates the enviro-centric control provided by the fitness curve. Images are created by agents moving in a 2D space and metabolising pixel hue, whilst altering brightness and saturation values. The piece is visually and methodologically similar to artworks such as *E-Volver* (Verstappen & Driessens 2006) and ant pheremone paintings (e.g. [18]).

Agent initialisation and distribution, movement rules and the evolutionary scheme are as described in the spatial model of Section 3.3, except that offspring are preferentially placed in a pixel neighbouring their parent. The hue value of pixels are the resources: as before, each agent's binary resource preference, and its chance of acquiring this is genetically determined by ϕ_1 (Section 3.1). ϕ_1 values are mapped across the 360° of the colour spectrum. Rather than simply depleting resources, agents redeposit 'waste', altering pixels from which they acquire energy. Red specialists process red pixels most efficiently and shift them towards blue; orange specialists are optimised for orange pixels and shift them towards green. Generalists may have a blue or green bias but can acquire energy from across the spectrum. The amount by which a pixel is altered is directly proportional to agent type: extreme specialists change a pixel by 36°, true generalists change it by 18°. Waste produced by one agent cannot be consumed until a fixed degradation time has elapsed.

Agents also deposit a coloured trail of brightness and saturation, the values of which are determined by their type. In contrast to the cumulative effect of agent's metabolic activity, these trails set pixel values absolutely.

Figure 4 shows the outcome of manipulating the fitness curve in this context. Even this simple system demonstrates how agent-environment interactions are steered by the fitness curve through different dimensions of phenotypic space. Most directly, the population structure varies – reflected here in differences in brightness and saturation. Indirectly the population structure affects the dynamics of movement – seen in the emergent 2D patterns – and the net effect of the populations' metabolic activity – seen here as the variation in hue.

The concave curve produces large bands or patches of (bright and dark) specialists (a). The two populations are mutually supportive - the waste of one is an ideal resource for the other. The result is a continuous flipping of the hue either side of the central cyan band. The flat curve generates a more even distribution of strategies which increases the level of brightness detail. The increased number of generalists also creates a broader colour range since they change hue values in smaller increments (b). This allows pixels to be shifted multiple times in the same (hue) direction. These effects are even more pronounced for a convex curve (c). More detail in mid-range brightness is evident, and the predominance of generalists creates localised pixel-forms across the entire hue range.

ColourCycling illustrates an enviro-centric method of influencing the evolution of abstract images, providing an alternative and complimentary approach to other aesthetic fitness-based methods such as used in *E-volver* or [18].

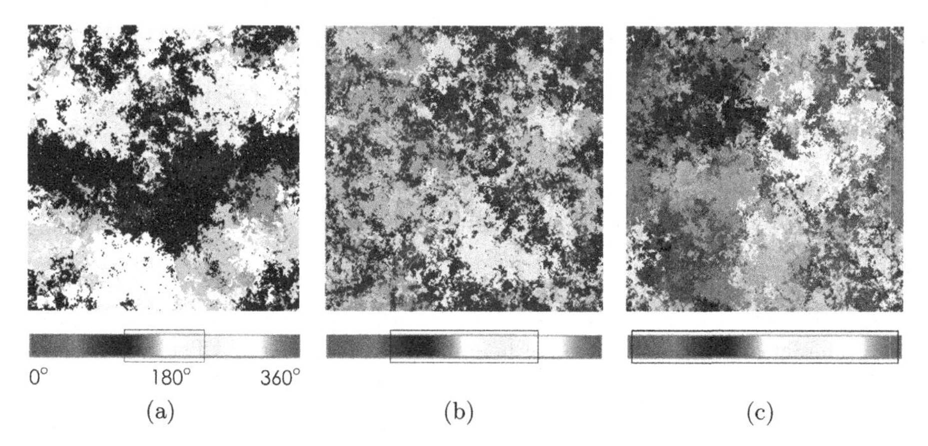

Fig. 4. Images generated by *ColourCycling* with (a) concave, (b) flat and (c) convex fitness curves. An Increased level of detail in brightness and larger variation in hue is observed as the curve becomes more convex. The range of hue in each image is illustrated by the width of the window on the spectrum below each image. The spectrum runs from red (0°) through green, cyan (90°), blue and back to red (360°).

5 Conclusion

The appropriation of predictive modelling techniques as control mechanisms is a promising route to develop our understanding and increase the sophistication of our creative ecosystems. This paper demonstrates novel enviro-centric methods for steering the evolution of an artificial ecosystem. Inspired by conceptual tools of theoretical ecology we developed a method for influencing the distribution of the population in genotype space. By implementing this in a simple generative graphics system, we showed how the rich interactions of the ecosystem model amplify the effect of this control mechanism, creating distinctive differences in image quality in a number of dimensions. This takes us a step closer toward the development of creative ecosystemic tools for generative art practice and illustrates much sort-after alternatives to interactive or explicit fitness evaluation in evolutionary art.

Acknowledgements

This research is supported by Australian Research Council Discovery Project grant DP0772667.

References

1. Sommerer, C., Mignonneau, L.: "A-Volve" an evolutionary artificial life environment. In: Proceedings of A-Life V, Nara, Japan, pp. 167–175 (1996)
2. di Scipio, A.: Sound is the interface: From interactive to ecosystemic signal processing. Organised Sound 8(3), 269–277 (2003)

3. Dorin, A.: The virtual ecosystem as generative electronic art. In: Raidl, G.R., Cagnoni, S., Branke, J., et al. (eds.) Proceedings of 2nd European Workshop on Evolutionary Music and Art, Applications of Evolutionary Computing: EvoWorkshops, pp. 467–476 (2004)
4. McCormack, J.: Artificial ecosystems for creative discovery. In: Thierens, D., et al. (eds.) Genetic and Evolutionary Computation Conference, pp. 301–307 (2007)
5. Prophet, J.: Technosphere: "Real" time "artificial" life. Leonardo: The Journal of the International Society for The Arts, Sciences and Technology 34(4), 309–312 (2001)
6. McCormack, J.: Eden: An evolutionary sonic ecosystem. In: Kelemen, J., Sosík, P. (eds.) ECAL 2001. LNCS (LNAI), vol. 2159, pp. 133–142. Springer, Heidelberg (2001)
7. Dorin, A.: Artificial life, death and epidemics in evolutionary, generative electronic art. In: Rothlauf, et al. (eds.) Proceedings of 3rd European Workshop on Evolutionary Music and Art, Applications of Evolutionary Computing, pp. 448–457 (2005)
8. Lanchier, N., Neuhauser, C.: A spatially explicit model for competition among specialists and generalists in a heterogeneous environment. Annals of Applied Probability 16(3), 1385–1410 (2006)
9. Elton, C.S.: Animal ecology. In: Sidgwick and Jackson, London (1927)
10. Brown, J.S.: Habitat selection as an evolutionary game. Evolution 44(3), 732–746 (1990)
11. Rosenzweig, M.L.: Habitat selection as a source of biological diversity. Evolutionary Ecology 1, 315–330 (1987)
12. Wilson, D.S., Yoshimura, J.: On the coexistence of specialists and generalists. The American Naturalist 144(4), 692–707 (1994)
13. Levins, R.: Theory of fitness in a heterogeneous environment I. The fitness set and adaptive function. American Naturalist 96, 361–373 (1962)
14. Lawlor, L., Maynard Smith, J.: The coevolution and stability of competing species. Journal of the American Naturalist 110(79–99) (1976)
15. Wilson, D.S., Turello, M.: Stable underdominace and the evolutionary invasion of empty niches. American Naturalist 127, 835–850 (1986)
16. Neuhauser, C., Pacala, S.: An explicitly spatial version of the Lotka-Volterra model with interspecific competition. Annals of Applied Probability 9, 1226–1259 (1999)
17. Pepper, J.: An agent-based model of group selection. In: Group Selection Workshop, ALIFEVII Workshop Proceedings (2000)
18. Monmarchè, N., Mahnich, I., Slimane, M.: Artificial art made by artificial ants. In: The Art of Artificial Evolution: A Handbook on Evolutionary Art and Music, pp. 227–247. Springer, Heidelberg, Berlin (2008)

Evolved Diffusion Limited Aggregation Compositions

Gary Greenfield

University of Richmond, Richmond VA 23173, USA
ggreenfi@richmond.edu
http://www.mathcs.richmond.edu/~ggreenfi/

Abstract. Diffusion limited aggregation (DLA) is a simulation technique for modeling dendritic growth. It has seen limited use for artistic purposes. We consider an evolutionary scheme for evolving DLA compositions with multiple seed particles. As a consequence we are led to consider robustness and stability issues related to the use of evolutionary computation whose phenotypes invoke inherently random processes.

1 Introduction

Diffusion limited aggregation (DLA) describes a rule-based process first introduced by Witten and Sander in 1981 [15] for simulating dendritic growth. It has seen widespread application in the physical, biological, and social sciences. It has also been used to implement special effects in computer graphics and has been investigated as a nonphotorealistic rendering technique. In [6] Greenfield explored the use of image compositing of DLA structures to develop DLA compositions. In this paper we introduce evolutionary techniques to evolve DLA compositions with multiple seed particles. Fitness assignment is based on canvas coverage. Since simulating dendritic growth is inherently governed by random processes this raises interesting questions about the interpretation of maximal fitness of a phenotype and — since the image phenotypes are intended as aesthetic artifacts — image stability.

This paper is organized as follows. In this section we provide background and motivation. In Section 2 we provide details of our DLA simulation. In Section 3 we present our evolutionary framework. In Section 4 we give our results. In Section 5 we give our conclusions and suggest future work.

1.1 DLA Applications

Diffusion limited aggregation has been used to model electrodeposition [8], urban cluster growth [1], root system growth [2], and even aspects of string theory [7]. Mathematically, much of the interest in DLA structure formation has centered on its fractal-like nature [14], with considerable effort having been devoted to measuring the fractal dimension of DLA formations [8]. DLA simulations have

M. Giacobini et al. (Eds.): EvoWorkshops 2008, LNCS 4974, pp. 402–411, 2008.

also been used in computer graphics for imaging fractal-like phenomena such as ice formation and root formation [2] or lightning [11], and in nonphotorealistic rendering as a stylistic digital halftoning effect [11].

1.2 A Simple DLA Model

Conceptually, DLA structures are formed when individual particles belonging to systems of particles moving under the influence of Brownian motion collide with and subsequently adhere to existing structures. Existing structures are spawned by *seed particles*. To simplify modeling a collection of particles moving, colliding, and adhering in the plane, in two dimensions Kobayashi et al. [8] propose simulating the DLA process by placing particles one at a time on an integer lattice and letting them undergo a random walk until either they meet an existing structure and adhere, they wander too far away, or they exceed a specified time limit. Although it is not necessary to do so, Kobayashi et al. suggest using the Manhattan metric i.e., taxicab geometry for measuring distance. Recall that this implies the distance between points (x_1, y_1) and (x_2, y_2) on the integer lattice is defined to be $d = |x_1 - x_2| + |y_1 - y_2|$. Assuming a single seed particle is placed at the origin, and the existing structure has radius R, the pseudocode for the random walk of a particle P in this model would be as follows.

```
place P at initial distance d = 2R from the origin
while time not expired
    take random step
    if d > 3R
        kill P
        exit
    if P adjacent to existing structure
        adhere P
        update R
        exit
    increment time
```

An example of a dendritic structure produced using this algorithm is given in Figure 1.

1.3 DLA in Music and Art

Due in part to variations in DLA models, the exact nature of the role DLA has previously played in fractal art and fractal music in unclear [14]. Digital prints of DLA simulations by Lomas from his "Aggregation" series [9] have been widely exhibited. Casselman [4] informs us that the lattice-free simulation phase of one of Lomas' images took 182 hours and rendering at $8,192 \times 8,192$ pixels took an additional 28 hours. In fact, this DLA simulation is three dimensional with particles raining down from a virtual hemisphere sitting over the canvas [10]. In [6] we presented the results of a more modest effort that coupled the

Fig. 1. The basic dendritic structure from the single seed particle diffusion limited aggregation model. In this model subsequent particles adhere as a result of Brownian motion simulated using random walks.

Kobayashi et al. model with digital image compositing so that layers of dendritic structures could be developed into a DLA composition. Figure 2 shows an example. Note how use of the Manhattan metric, where the "ball" of radius R centered at the origin is the "diamond" in the Euclidean plane determined by the vertices $(0, \pm R)$ and $(\pm R, 0)$, causes particles to build-up on diagonals. Image compositing using multiple seeds informs the work described here.

2 DLA with Multiple Seed Particles

If particles are released using the simple model when there are *multiple* dendritic structures present then these existing structures can function as both attractors and repellers. On one hand, it is difficult for a particle to reach a given structure if there is a second structure blocking its path for the simple reason that it is difficult to *randomly* walk around the intervening structure. On the other hand, if a particle approaches two existing structures that are roughly the same distance away, since the chances of being captured by either are roughly the same, the two structures grow at approximately the same rate, whence for the reasons cited above "dead zones" are established between the structures. The question that motivated the work described here was: How should one place multiple seed particles on the canvas in such a way that interesting DLA compositions arise?

Shortcomings of the Kobayashi et al. model for deciding this question include the nature of the unit ball in the Manhattan metric and the dead zones that build up around central dendritic structures. Since the distance metric used dictates what the particle build-up will look like at the boundaries, the first issue is easy to address. If we use the supremum norm where the distance between points (x_1, y_1) and (x_2, y_2) on the (integer) lattice is defined to be $d = \max(|x_1 - x_2|, |y_1 - y_2|)$ instead of the taxicab metric, then the ball of radius R centered at the origin will be the square with vertices $(\pm R, \pm R)$. The second issue motivates why we conducted our experiments using the total number of particles aggregated during the DLA simulation as our fitness assignment. But providing evolutionary computation with mechanisms that would allow fitness to improve was more problematic.

The dead zones caused by neighboring structures that form around the central seed particle at the origin cause the critical value of R, the inner most radius

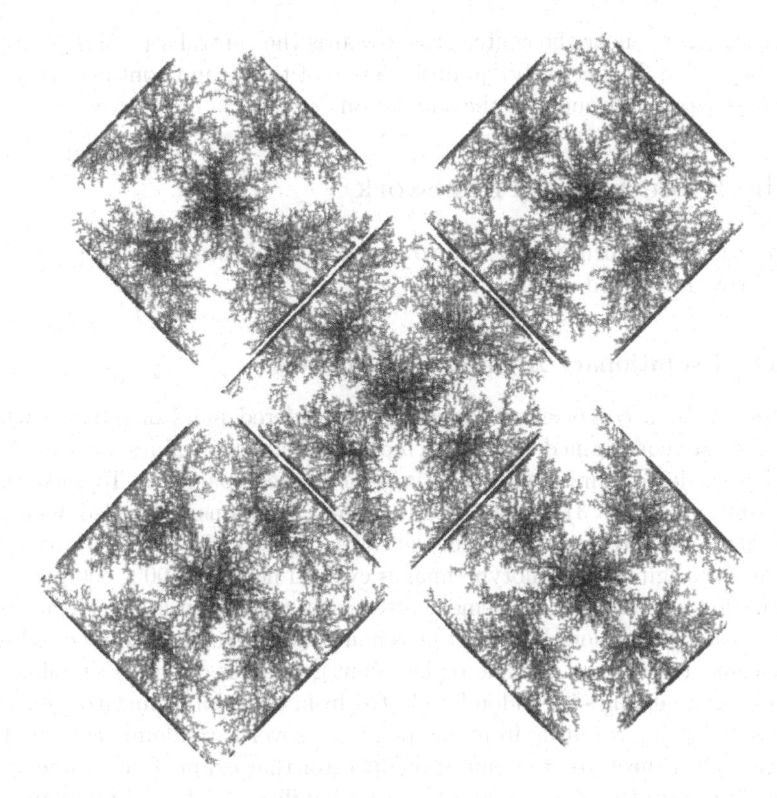

Fig. 2. A composited diffusion limited aggregation painting after Greenfield [6]

that is used to determine the intermediate radius R_I of the ball where particles are released from and the outer most radius R_M of the ball where particles escape from, to become frozen. When this happens the released particles eventually build up at the constrained *canvas* distance and this blocks the flow of subsequent particles to the canvas. Our first attempt to remedy this problem was to decrease the particle *release* distance R_I, initialized to $2R$ after each R update, whenever R became stagnant. Regardless of whether or not the maximum allowable particle distance R_M, initialized to $3R$ after each R update, was decreased concurrently this only caused thick borders to build up at the canvas boundary, with the thickness depending on the length of time the simulation was allowed to run for. The solution we finally settled upon was to restore the particle release distance R_I to $2R$, and the maximum particle distance R_M to $3R$, each time R increased while making the decrease of R_I and R_M sensitive to a new parameter E that tracked the number of particles released since R's last increase. For the results described here, R_I and R_M were decremented by one each time a particle was released once E exceeded the threshold $4R$. With this new model it is easy to lose sight of the fact that the current value of what is being viewed as the inner most radius R will very quickly no longer be the bounding radius of the central dendritic structure. Using multiple seed particles, as other structures near the center grow towards the central structure, early on in the simulation R effectively "jumps" to one of them, and continues to jump to other structures throughout the simulation.

3 The Evolutionary Framework

The evolutionary algorithm we used to evolve multiple seed DLA compositions is fairly straightforward.

3.1 The Evolutionary Algorithm Parameters

The genome for a composition consists of K ordered pairs of integers whose coordinates are constrained so that each ordered pair determines a canvas location. These ordered pairs are the locations for the seed particles. To make them more visible, small (3×3) neighborhoods at these locations are seeded. One seed particle is also placed at the center of the canvas. The center of the canvas serves as the *lattice* origin. All phenotype images evolved here are 600×600 pixels.

Population size is 24. Replacement after each generation is elitest, meaning that after each generation the 8 most fit genomes are preserved and used to breed 16 replacement genomes. A single replacement genome is obtained as follows: A distinct pair of genomes is randomly selected from the pool of survivors and one of the two offspring resulting from one-point crossover is randomly chosen. This offspring is then subjected to a mutation operator that examines each one of the $2K$ coordinates in the genome, and has a probability of 0.1 of altering any one of them. When ccordinate mutation is invoked, with probability 0.9 this (point) mutation operator will slightly perturb the existing coordinate; with probability 0.1 it will replace it. We did not tune these mutation parameters. They are *ad hoc.* A heuristic argument suggests our results should not be sensitive to their values. By running the evolutionary algorithm for 30 generations, the evolutionary algorithm evaluates $24 + (30)(16) = 504$ DLA compositions. At the end of each run the most fit genome is preserved.

3.2 The DLA Simulation Parameters

Inherently, DLA simulation depends on random processes. Particles are released from random locations with each particle allowed to take up to 100,000 random steps before its time period expires. Therefore, for fairness reasons, the termination condition for our DLA simulation is not the number of particles released but the total number of steps alloted (i.e. virtual time) for all random walks. Here this value is 6×10^8. This value was chosen because during testing, it was found to be sufficient to let the boundary of the composition fill in. During an evolutionary run, when a genotype is evaluated, the random number generator is always (re)set to the "master" seed, so that any bias in the random number sequence will be shared by all genomes. Stated in a more positive way, this means that if simulated evolution (in the early stages?) can benefit from knowledge about the random number sequence, it has the capacity to do so. We will return to this point again later. As a subtlety, we note that we were careful to continually (re)seed the random generator in such a way that this random number generation stability criterion imposed during DLA simulation execution did not, as an unintended side effect, influence the genome replacement algorithm.

3.3 The Fitness Calculation

As mentioned previously, our fitness function was simply the number of particles that adhered during execution of the DLA simulation algorithm. The heuristic was to achieve canvas "coverage", and the challenge for the evolutionary algorithm was to find particle seed placements to maximize coverage. This was a naive first choice — hoping to find interesting aesthetics — during the course of studying stochastic evolutionary processes.

4 Results

We made evolutionary runs using genome sizes $K = 30, 50, 70, 90$. Table 1 gives the maximal fitness values we obtained. Figure 3 shows the corresponding evolved DLA compositions for these maximally fit genomes. Although reproduced here in black and white, the compositions are in color so that the seed particles are visible against the adhered particles. Given the nature of the DLA simulation it is not surprising that it was difficult to evolve seed particle placements such that the resulting dendritic structures themselves aggregated. More importantly, it is not surprising that the evolved seed particle placements are approximately uniformly spaced. One could appeal to an entropy argument to support this observation.

Table 1. Table of maximal fitness obtained from a genome of size K. The genome determines seed particle placement for K dendritic structures separate from the central dendrite spawned from a seed particle at the lattice origin.

Number of Genes	Maximal Fitness
30	70127
50	72271
70	74462
90	75250

4.1 Analysis

We selected one of our maximally fit genomes and conducted further experiments to learn more about the effect of random walk processes on the visual and numerical stability of the evolved images. Figure 4 shows four DLA compositions that were generated using the $K = 30$ genome of Figure 3 (top left)

Fig. 3. Clockwise from top left, the most fit evolved images for genomes with $K = 30$, 50, 70, and 90 seed particle location genes

by (re)seeding the random number generator each time with a new seed. Space prohibits showing more examples. The uniformity of the results is reassuring.

This visual stability test is supplemented by the numerical data we obtained in Table 2. The table shows average, minimum, maximum, and 95% confidence intervals for the fitness based on a sample of 100 DLA compositions obtained from random number generator reseeding using each of our maximally fit genomes. From this table we see that even though the fitness of our evolved genomes was exceeded in all of the samples except $K = 70$, their fitness was always in the top decile. More impressively, the table indicates that the fitness value obtained using evolved seed placement was robust.

4.2 Artistic Effects

Figure 5 illustrates the use of two techniques we used to enhance the aesthetics of our DLA compositions. For this example we used the maximally fit genome of size $K = 50$ from Figure 3 (top right). First, we colored each particle as

Fig. 4. Four images using the genome of the top left image in Figure 3 imaged using different random number generator seedings

Table 2. Table of fitness statistics for the best genomes (see Figure 4) based on samples of n = 100 fitness calculations for each genome of sizer K using different random number generator seedings

Genes	Fitness			
K	Mean	95% CI	Min	Max
30	68019.12	[67794.07, 68244.17]	64089	70795
50	70119.24	[69932.33, 70306.15]	67750	72321
70	73300.60	[73179.33, 73421.87]	70456	74300
90	74435.26	[74343.92, 74526.60]	73039	75594

it adhered according to its distance d from the canvas center. Figure 5 (left) weights the assigned color by a factor of d, while Figure 5 (right) weights it by a factor of $\sqrt{1-d}$. Second, we diffused color throughout the 3×3 neighborhood of the pixel where the particle adhered. At the resolution here, the latter effect is somewhat subtle. It is more noticeable in Figure 2 where it was also used.

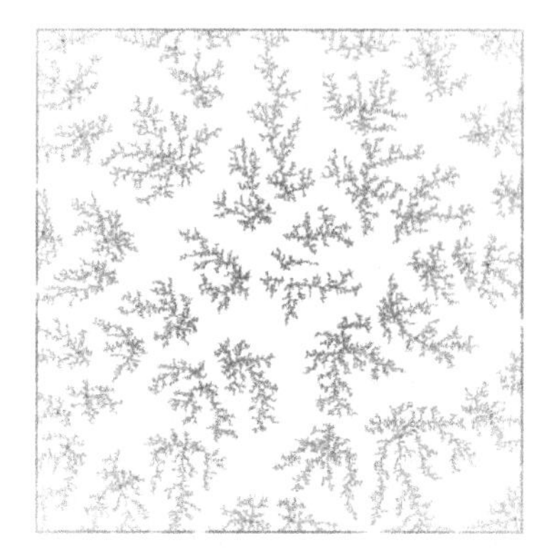

Fig. 5. Using the most fit genome with $K = 50$ seed location genes, the results of experiments imposing a color grading on the DLA compositions according to the distance metric and also incorporating color diffusion for artistic effect

5 Conclusions and Future Work

We have shown how evolutionary computation based on the genetic algorithm can be instantiated for a DLA simulation with multiple seed particles in order to evolve DLA compositions. Since the genotype to phenotype mapping invokes inherently random processes we have also examined the visual and numerical robustness of our results. Even though compelling artistic results (e.g. Lomas [9]) are costly to compute and difficult to control, the simplified Kobayashi et al. [8] model with appropriate modifications for handling multiple seeds and arbitrary distance metrics provides an efficient computational framework for future exploratory work.

DLA simulation is a global process. To improve its effectiveness as an artistic technique, it must be made sensitive to local constraints. For example, seed particle placements that will result in local structures aggregating to form desired visual icons, features, patterns, or shapes, Long [11] uses path planning rather than particle methods for this purpose. Our future work will consider modifications of out algorithms so that they are sensitive to topological constraints on particles instead. Due to the interest in digital special effects where image enlargements have pixels with visual content, a theme popularized by Silvers' photomosaics [13], we also feel this is an area where DLA simulation using multiple seeds could yield interesting results.

References

1. Batty, M.: Cities and Complexity. MIT Press, Cambridge (2005)
2. Bourke, P.: Constrained limited diffusion aggregation in 3 dimensions. Computers and Graphics 30(4), 646–649 (2006)
3. Bourke, P.: Diffusion Limited Aggregtion (accessed March 2007) (2007), `http://local.wasp.uwa.edu.au/~pbourke/fractals/dla/`

4. Casselman, B.: About the cover Aggregation 22. Notices of the American Mathematical Society 54(6), 800 (2007)
5. Gaylord, R., Tyndall, W.: Diffusion limited aggregation. Mathematica in Education 1(3), 6–10 (1992) (accessed October 2007), http://library.wolfram.com/infocenter/Articles/2866/
6. Greenfield, G.: Composite diffusion limited aggregation paintings. In: Sarhangi, R., Barrallo, J. (eds.) BRIDGES 2007 Conference Proceedings, pp. 15–20 (2007)
7. Halsey, T.: Diffusion-limited aggregation: a model for pattern formation. Physics Today 53(11), 36–47 (2000) (accessed October 2007), http://www.physicstoday.org/pt/vol-53/iss-11/p36.html
8. Kobayashi, Y., Niitsu, T., Takahashi, K., Shimoida, S.: Mathematical modeling of metal leaves. Mathematics Magazine 76(4), 295–298 (2003)
9. Lomas, A.: 2006 Bridges Exhibit of Mathematical Art, London (2006) (accessed October, 2007), http://www.bridgesmathart.org/art-exhibits/bridges06/lomas.html
10. Lomas, A.: Private communication (2006)
11. Long, J.: Modeling dendritic structures for artistic effects. MSc. Thesis University of Saskatchewan (2007) (accessed October 2007), http://www.cs.usask.ca/grads/jsl847/
12. Ramachandran, V., Hirstein, W.: The science of art: a neurological theory of aesthetic experience. Journal of Consciousness Studies 6(1–2), 15–52 (1999)
13. Silvers, R.: Photomosiacs. Henry Holt and Company, New York (1997)
14. Voss, R.: Fractals in nature: From characterization to simulation. In: Peitgen, H., Saupe, D. (eds.) The Science of Fractal Images, pp. 36–38. Springer, New York (1988)
15. Witten, T., Sander, L.: Diffusion-limited aggregation, a kinematic critical phenomenon. Physical Review Letters 47, 1400–1403 (1981)
16. Zeki, S.: Inner Vision, An Exploration of Art and the Brain. Oxford University Press, New York (1999)

Scaffolding for Interactively Evolving Novel Drum Tracks for Existing Songs

Amy K. Hoover, Michael P. Rosario, and Kenneth O. Stanley

Evolutionary Complexity Research Group
School of Electrical Engineering and Computer Science
University of Central Florida
Orlando, FL 32836
ahoover@cs.ucf.edu, michael.rosario@yahoo.com, kstanley@cs.ucf.edu
http://eplex.cs.ucf.edu/neatdrummer

Abstract. A major challenge in computer-generated music is to produce music that sounds natural. This paper introduces NEAT Drummer, which takes steps toward natural creativity. NEAT Drummer evolves a kind of artificial neural network called a Compositional Pattern Producing Network (CPPN) with the NeuroEvolution of Augmenting Topologies (NEAT) method to produce drum patterns. An important motivation for this work is that instrument tracks can be generated as a *function* of other song parts, which, if written by humans, thereby provide a *scaffold* for the remaining auto-generated parts. Thus, NEAT Drummer is initialized with inputs from an existing MIDI song and through interactive evolution allows the user to evolve increasingly appealing rhythms *for that song.* This paper explains how NEAT Drummer processes MIDI inputs and outputs drum patterns. The net effect is that a compelling drum track can be automatically generated and evolved for any song.

Keywords: compositional pattern producing networks, CPPNs, computer-generated music, interactive evolutionary computation, IEC, NeuroEvolution of Augmenting Topologies, NEAT.

1 Introduction

Because music is at heart patterns and motifs that vary and repeat over time, it should be possible to generate such patterns automatically. Yet in practice, the most significant problem with computer-generated music is that it sounds *artificial* [1]. Subjectively, this problem is plain to the human ear. Yet its cause is subtle and difficult to establish objectively, making the solution elusive.

This paper takes a first step toward confronting the problem by showing that natural-sounding musical parts *can* be generated when they are constrained by a *scaffold*, i.e. an existing support structure, that is itself produced by competent humans. In particular, in experiments in this paper, the instrumental parts of

M. Giacobini et al. (Eds.): EvoWorkshops 2008, LNCS 4974, pp. 412–422, 2008.

an existing song are supplied as the scaffold upon which drum tracks are automatically generated, that is, the drum tracks are a *function* of the instrumental parts. The result is original drum sequences for existing songs that are often indistinguishable from human art.

All music has rhythm [2] and because rhythm is simpler than melody or harmony, rhythm generation is an appropriate place to begin to investigate the problem of producing natural patterns. The novel method for generating rhythms in this paper is implemented in a program called NEAT Drummer. It accepts existing human compositions as input to a type of artificial neural network (ANN) called a Compositional Pattern Producing Network (CPPN) and outputs drum patterns to accompany the instruments. The inputs to NEAT Drummer are specific parts of a Musical Instrument Digital Interface (MIDI) file (e.g. the lead guitar, bass guitar, and vocals) and the outputs are drum tracks that are played along with the original MIDI file. That way, outputs are a function of the original MIDI file inputs, forcing synchronization with the MIDI.

To take into account the user's own inclinations, NEAT Drummer allows the user to interactively evolve rhythms from an initial population of drum tracks with the NeuroEvolution of Augmenting Topologies (NEAT) algorithm, which can evolve increasingly complex CPPN-encoded patterns.

The results are drum patterns that sound like genuine human compositions when played with their associated songs. The ability to automatically generate musical parts that sound natural demonstrates an important lesson: As long as *part* of a song sounds natural and other instrument tracks are represented as functions of that part, they can *inherit* the plausibility inherent in the original design, suggesting a new path for computer-generated music.

2 Background

This sections explains how interactive evolutionary computation (IEC) relates to computer generated music and introduces the NEAT method.

2.1 Interactive Evolutionary Computation

NEAT Drummer refines its original drum patterns through a process called interactive evolutionary computation (IEC), which means a human, rather than a predefined fitness function, selects the parents of the next generation [3]. Richard Dawkins first popularized IEC with Biomorphs, a visual representation of designs based on an L-System encoding [4,5]. Like most IEC programs, to start a Biomorph run, a randomized population of biomorphs are generated. The user then selects which individual should be allowed to reproduce to evolve increasingly complex creatures. This idea naturally led to other visual IEC like the L-system-encoded Mutator [6], Karl Sims' genetic art [7], and eventually the first musical IEC application, Sonomorphs [8].

2.2 Computer Generated Music

Although music generated by computers can be interesting, typically it also sounds artificial, often because it lacks a global structure that progresses from the beginning to the end of the song [1,9].

It is common for music generators, such as Sonomorphs, the first Biomorph-inspired music evolver, to focus on short phrases rather than on the entire song [8,10]. These short phrases, which are selected and evolved by the user, may be extended through looping or manual juxtaposition.

Early connectionist approaches also evolved short phrases and represent change over time through recurrence [11]. Todd and Loy [11] first applied recurrent ANNs to music generation by training them to reproduce patterns with Real Time Recurrent Learning (RTRL). Chen and Miikkulainen [12] later combined this recurrent learning approach with evolution based on the idea that a simple recurrent network (SRN) can capture a general style of music and then vary it through evolution. This approach succeeded in producing melodies in the style of Bela Bartok on a local level; however, even with recurrence it is difficult to capture global structure.

NEAT Drummer avoids this problem by generating its rhythms from already-existing instrumental parts that span entire songs, thereby precluding the need to represent patterns over time through recurrence. The next section describes the NEAT method that implements evolution in NEAT Drummer.

2.3 NeuroEvolution of Augmenting Topologies (NEAT)

NEAT Drummer evolves a neural-based encoding of drum patterns. The NEAT method was originally developed to solve difficult control and sequential decision tasks. The ANNs evolved with NEAT control agents that select actions based on their sensory inputs. While previous methods that evolved ANNs, i.e. neuroevolution methods, evolved either fixed topology networks [13,14], or arbitrary random-topology networks [15], NEAT is the first to begin evolution with a population of small, simple networks and complexify the network topology over generations into diverse species, leading to increasingly sophisticated behavior. This section briefly reviews the NEAT method; Stanley and Miikkulainen [16,17] provide complete introductions.

NEAT is based on three key principles. First, in order to allow ANN structures to increase in complexity over generations, a method is needed to keep track of which gene is which; otherwise, it is not clear in later generations which individual is compatible with which, or how their genes should be combined to produce offspring. NEAT solves this problem by assigning a unique historical marking to every new piece of network structure that appears through a structural mutation. The historical marking is a number assigned to each gene corresponding to its order of appearance over the course of evolution. The numbers are inherited during crossover unchanged, and allow NEAT to perform crossover without the need for expensive topological analysis.

Second, NEAT traditionally speciates the population so that individuals compete primarily within their own niches instead of with the population at large, which protects topological innovations. However, because the *user* performs selection in interactive evolution instead of the evolutionary algorithm itself, speciation is not applicable in NEAT Drummer and therefore not utilized.

Third, unlike other systems that evolve network topologies and weights [15], NEAT begins with a population of simple networks with no hidden nodes. New structure is introduced incrementally as structural mutations occur, and only those structures survive that are found to be useful through fitness evaluations. This way, NEAT searches through a minimal number of weight dimensions and finds the appropriate complexity level for the problem. NEAT Drummer lets the user evolve patterns of increasing complexity through this approach.

Finally, in NEAT Drummer, NEAT evolves a kind of ANN called a Compositional Pattern Producing Network (CPPN), which is designed to compactly represent patterns with regularities, such as pictures and songs [18]. What distinguishes CPPNs from ANNs is that in addition to traditional sigmoid functions, CPPN hidden nodes can include several classes of functions, including periodic functions (like sine) for repetition and symmetric functions (like Gaussian) for symmetry. An individual network can contain a heterogeneous set of functions in its nodes, which are evolved along with the weights. In this way, drum tracks with regular patterns are easily discovered quickly.

3 Approach

NEAT Drummer produces drum tracks for songs through two primary functions. First, it generates an initial set of candidate songs from a set of user-defined inputs. Second, it allows the user to further evolve the initial candidates.

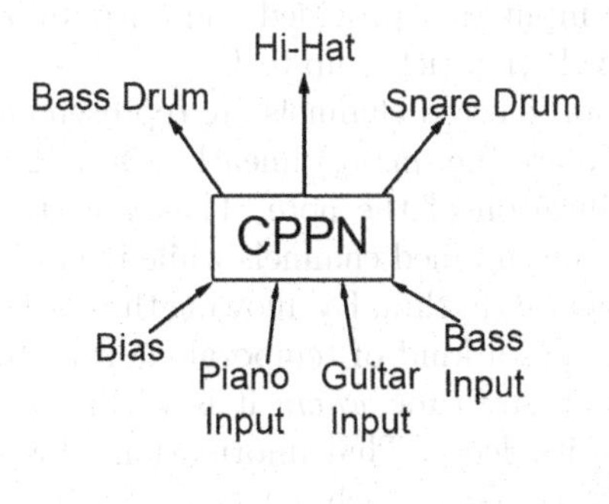

(a) Inputs and Outputs

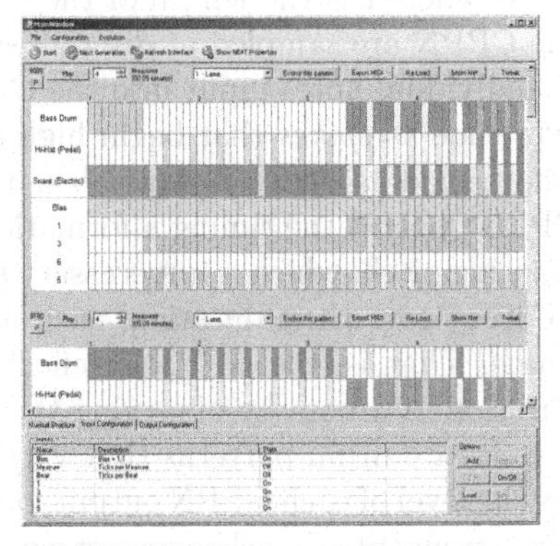

(b) NEAT Drummer Screenshot

Fig. 1. **Using NEAT Drummer.** (a) The user selects both a set of inputs from among the channels in the MIDI file and a set of outputs corresponding to specific drums. (b) NEAT drummer presents an interactive evolution interface where visual representations of drum patterns help the user to decide whether to listen to each candidate and then select their favorites.

3.1 CPPN Rhythm Generation

To generate the initial drum tracks, the user first specifies the inputs and outputs of the CPPN (figure 1a) through a GUI provided by the program (figure 1b).

Because NEAT Drummer is based on the philosophy that a natural sound can be produced from a scaffold that is human-produced, the CPPN inputs must provide this scaffolding. From these inputs the CPPN derives its original patterns, which are therefore *functions* of the scaffolding.

For drum tracks in particular, the most natural scaffolding is the music itself (e.g. melody and harmony), from which the drum pattern can be derived. Thus, the user selects any combination of *channels* from a MIDI file to be input into the CPPN. These channels represent individual instrumental parts from the song. The main idea is that NEAT Drummer generates a rhythm that is a function of these channels, such that it is constrained by, though not identical to, their intrinsic patterns. Thus, it is important to choose instruments that play salient motifs in the song so that the drum pattern can be derived from the richest structures available. Further texture can be achieved by inputting more than one channel, e.g. bass and guitar, so that the rhythm is a function of both.

The user also chooses the percussion instruments that will play the rhythm. Each such instrument is represented by a single output on the CPPN. For example, one output may be a bass drum, one a snare, and the final a hi-hat. Any number of drums, and hence any number of outputs, are permissible.

To produce the initial patterns, a set of random initial CPPNs with a minimal initial topology (following the NEAT approach) and the chosen inputs and outputs are generated. NEAT Drummer then inputs the selected channels into the CPPN over the course of the song in sequential order and records the consequent outputs, which represent drums being struck. Specifically, from time $t = 0$ to $t = l$, where l is the length of the song, the inputs are provided and outputs of the CPPN are sampled at discrete subintervals (i.e. ticks) up to l.

Individual notes input into the CPPN from selected channels are represented over time as *spikes* that begin high and decrease (i.e. *decay*) linearly (figure 2). The period of decay is equivalent to the duration of the note. That way, the CPPN "hears" the timing information from the supplied channels while in effect ignoring pitch, which is unnecessary to appreciate rhythm. By allowing the spikes to decay over their duration, each note becomes a kind of temporal coordinate frame. That is, the CPPN in effect knows at any time *where* it is within the duration of a note by observing the stage of its decay. That information allows it to create drum patterns that vary over the course of each note.

The level of each CPPN output, on the other hand, is interpreted as the *volume* (i.e. strength) of each drum strike. That way, NEAT Drummer can produce highly nuanced effects through varying softness. Two consecutive drum strikes one tick after another are interpreted as two separate drum strikes (as opposed to one long strike). To produce a pause between strikes, the CPPN must output an inaudible value for some number of intervening ticks. Because the CPPN has one output for each drum, the end result of activating the network over t ticks is a drum sequence for each drum in the ensemble.

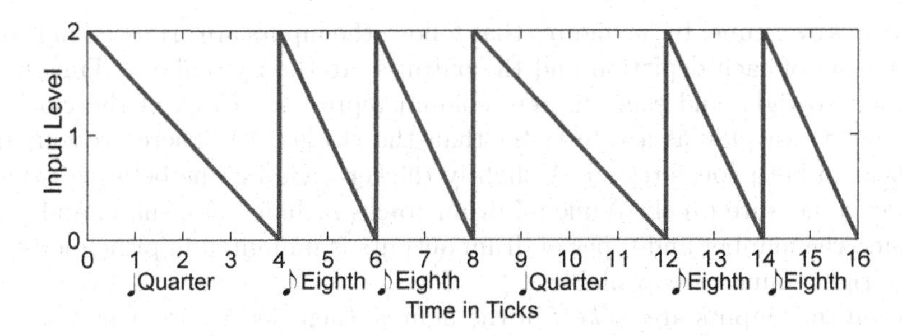

Fig. 2. Channel Input Encoding. Regardless of the instrument, each note in a sequence in any channel input to the CPPN is encoded as a spike that decays over the duration of note. The pattern depicted in this figure shows how eighth notes decay faster than quarter notes, thereby conveying timing information to the CPPN, which samples this pattern at discrete ticks.

3.2 Drum Pattern Interactive Evolution

As shown in figure 1b, NEAT Drummer displays the set of candidate patterns visually after they are generated. It is important to note that unlike in many evolutionary experiments, patterns in the initial generation *already* sound appropriate because they are functions of other parts of the song. This initial high quality underscores the contribution of the existing tracks to the structure of the generated patterns. The aim of evolution is thus to elaborate on such patterns. The user can choose to listen to any displayed pattern. When listening, the user can listen to the drum track alone or the drum track with its associated song. The visual presentation allows the user to quickly identify unappealing patterns without wasting time listening to them (e.g. wherein the bass is hit over and over again without pause).

Then either the user rates the individual patterns from which NEAT Drummer chooses parents or the user selects a single parent of the next generation. Further rounds of selecting and breeding continue until the user is satisfied. In this way, drum tracks evolve interactively. Because of complexification in NEAT, they can become increasingly elaborate as evolution progresses.

To encourage rapid elaboration over generations, the probability of adding a connection or node in NEAT was 90%. While these high probabilities would be deleterious in typical NEAT experiments [16], because drum tracks are constrained to follow the music no matter what, this domain supports adding structure quickly. The mutation power, i.e. the maximum magnitude of weight change, was 0.1 and the probability of mutating an individual connection was 90%.

4 Experimental Results

Results in this section are reported through figures that are designed to demonstrate the relationship between the CPPN inputs and outputs as the song progresses over time. In the figures that follow, the inputs are arranged in rows at the bottom of each depiction and the outputs are the rows above. Time moves from left to right and each discrete column represents a tick of the clock. No instrument can play at a rate faster than the clock ticks. There are four ticks per beat in both songs tested. A slightly thicker dividing line between columns denotes a measure break. While all drum tracks include bass, snare, and hi-hat outputs, the number and types of drum outputs is unlimited in principle as long as the right sounds are available.

Recall that inputs are *spikes*; in the figures, their decays are depicted as decreasing darkness. In contrast, outputs represent *volumes*, wherein darker shading indicates higher volume. The main difference between inputs and outputs is that a single note in the input may straddle several columns during its decay. Outputs on the other hand are played as *separate* notes for every solid column. For an output drum to last for more than a single tick before the next drum attack, it must be followed by white (empty) columns.

To appreciate the results in this section it is important to judge the subjective natural quality of the generated rhythms. Thus MIDI files for every experiment in this section are available at `http://eplex.cs.ucf.edu/neatdrummer/`.

Figure 3 shows individuals from generations one and 11 generated for the folk song *Johnny Cope*. The relationship between the the bass, hi-hat, and snare and the three input channels is complex because each drum is related to all three inputs. Note however that the *instrumental* patterns in measures three and four are highly related though not identical. Slight differences exist between the piano pattern in measure three and measure four; this difference is reflected in the snare in both generations one and 11, which both slightly differ between the early parts of measures three and four. Thus, the drum pattern's subtle variation is *correlated* to the music because of their coupling, which evokes a strong subjective sense of appropriate stylistic sophistication and creativity.

At measure 23, the song changes sharply by eliminating the piano part. Consequently, the CPPN outputs also diverge from their previous consistent motifs. This strongly coupled divergence that is carried both in the tune and in the drums creates a sense of purposeful coordination, again yielding a natural, sophisticated feel. In this way, the functional relationship represented by the CPPN causes the drums to follow the contours of the music seamlessly.

Generation 11, which evolved 12 additional connections and six additional nodes, reacts particularly strongly to the elimination of the piano by significantly altering its overall pattern. In generation 1, the shift is less dramatic, showing how the user interactively pushed evolution towards a sharper transition over those ten generations. Generation 11 also elaborates on the snare, making it harder-hitting in the later measures than in earlier ones.

Results from generation 25 of *Oh! Susanna* are shown in figure 4. NEAT Drummer produces similarly natural and style-appropriate rhythms for this song as well, suggesting its generality. Because style is inherent in the scaffold, it transfers seamlessly to the drum track without any need for explicit stylistic

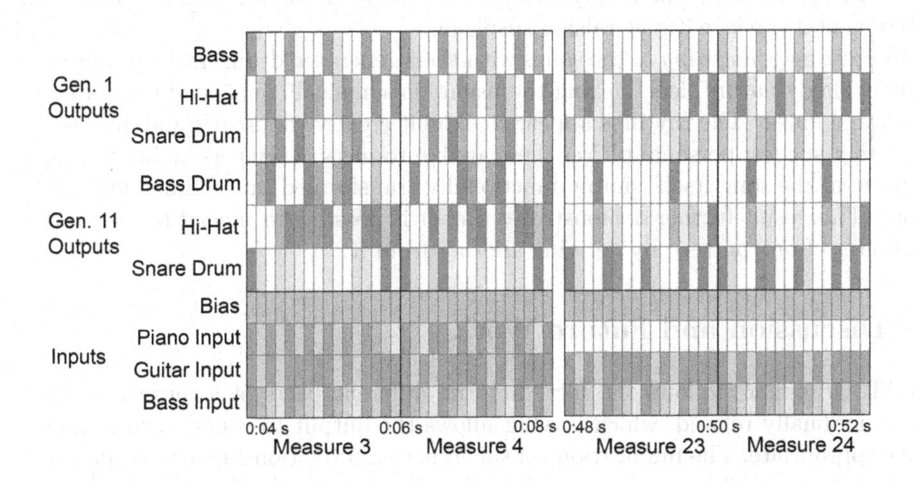

Fig. 3. Johnny Cope results. Results are depicted from two different generations at two different parts of the song. The inputs from the original song are shown at bottom. The motif in measures three and four is typical of the first part of the song until measure 23, when it switches to a different motif in both generations. The main conclusion is that the output is a function of the input that inherits its underlying style and character. (These tracks are available at `http://eplex.cs.ucf.edu/neatdrummer/`).

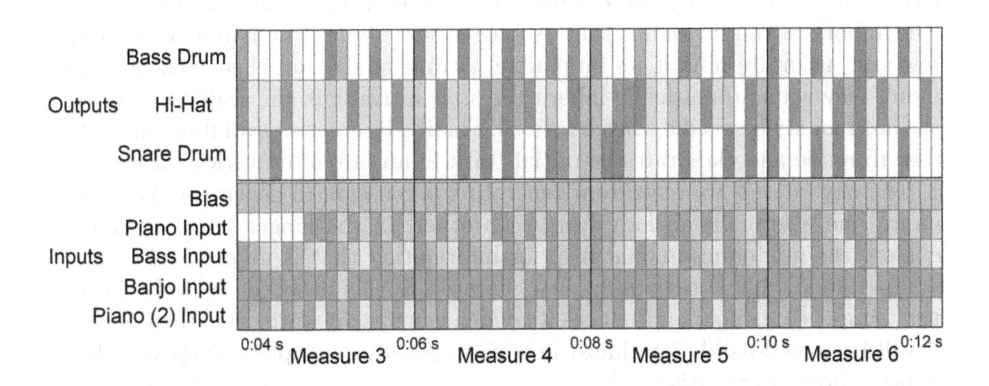

Fig. 4. Oh! Susanna outputs. This pattern from measures three through six of *Oh! Susanna* is from generation 25 of evolution. The network evolved 15 hidden nodes and 41 connections. Near the end of measure four is a particularly improvisational riff in the snare that transitions to measure five. This riff is caused by variation in the piano and other inputs at the same time. As with Johnny Cope, the drum pattern sounds natural and styled correctly for this upbeat song. (This track is available at `http://eplex.cs.ucf.edu/neatdrummer/`).

rules. The result is an entertaining sound that could be attached to the original instrumental tracks without raising suspicion.

Interestingly, *listening* to the songs with their generated drum tracks produces a surprisingly natural feel, lacking the usual "mechanical" quality of computer-generated music. The rhythms appear to exhibit creative flourishes and interjections that suggest personality. Yet these are byproducts of the personality that is implicit in the song itself, simply functionally transformed into a different local motif. This result further demonstrates that it is possible to *inherit* the natural character of one pattern by deriving another from it.

5 Discussion and Future Work

NEAT Drummer exploits the fact that the different parts of a musical piece are functionally related, which is what allows the output to sound natural and song-appropriate. The implications of this functional relationship are significant because it means that many parts of a song are less complex than they seem to be when put in the context of the rest of the instruments.

Formally, any pattern over time can be described as a function of time, $f(t)$. However, a good pattern for a particular drum alone may be highly complex with respect to t, making its discovery prohibitive. Yet given another part $p(t)$ that varies over time, because the parts of a song are related to each other, it may be easier to discover $g(p(t))$ than $f(t)$ even though they produce the same output. In effect, p provides a *scaffold* upon which other parts can be attached.

In this way, the drums are a function of the melody and the harmony, and, interestingly, the melody and harmony are also an inverse function of the drums. The important implication is that as long as *something* is human-generated, the rest will naturally inherit its intangible natural quality on their own. Furthermore, they will automatically inherit the *style* intrinsic in the scaffold, removing the need for style-specific considerations. Finally, interactive evolution allows the user to refine and elaborate the initial track in an unlimited variety of ways.

The next step is to move from generating percussion to generating bass, harmony, and eventually melody from each other. In the longer term, we aim to explore the minimum amount of prior information that is necessary to form a scaffold upon which other parts are generated. If this basis can be reduced to a very simple core encoding that even laymen can specify with ease, then perhaps it will become possible for almost anyone to generate complete songs with little human effort or expertise.

6 Conclusion

This paper introduced NEAT Drummer, a program that can generate novel drum tracks for songs sequenced in MIDI. The main idea is to input instrumental tracks directly into a kind of neural network called a Compositional Pattern Producing Network (CPPN), which produces a pattern that is a function of its input. In this way, the drum sequences output by the CPPN are related to the song inputs, which thereby supply a scaffold for the drums. Human users can then evolve and thereby elaborate the track to satisfy their specific tastes. Drum tracks generated for two popular folk songs exhibit a natural style appropriate to each song that lacks the familiar computerized feel of computer-generated music. In the future, the scaffolding approach may apply to generating instrumental tracks as well, and eventually may enable nearly-autonomous music generation.

Acknowledgments

Special thanks to Barry Taylor for granting special permission to utilize his own MIDI productions of folk music in this work. Barry Taylor originally sequenced both Johnny Cope and Oh! Susanna (both without percussion), which are the two songs for which drum tracks were generated in Section 4. This research was supported in part by NSF grants IIS-REU: 0647120 and IIS-REU 0647018.

References

1. McCormack, J.: Open problems in evolutionary music and art. In: Rothlauf, F., Branke, J., Cagnoni, S., Corne, D.W., Drechsler, R., Jin, Y., Machado, P., Marchiori, E., Romero, J., Smith, G.D., Squillero, G. (eds.) EvoWorkshops 2005. LNCS, vol. 3449, pp. 428–436. Springer, Heidelberg (2005)
2. London, J.: Rhythm (November 2007), http://www.grovemusic.com
3. Takagi, H.: Interactive evolutionary computation: Fusion of the capacities of EC optimization and human evaluation. Proceedings of the IEEE 89(9), 1275–1296 (2001)
4. Dawkins, R.: The Blind Watchmaker. Longman, Essex, U.K (1986)
5. Lindenmayer, A.: Mathematical models for cellular interaction in developement parts I and II. Journal of Theoretical Biology 18, 280–299 & 300–315 (1968)
6. Todd, S., Latham, W.: Evolutionary Art and Computers. Academic Press, London (1992)
7. Sims, K.: Artificial evolution for computer graphics. In: Proceedings of the 18th Annual Conference on Computer Graphics and Interactive Techniques (SIGGRAPH) 1991, pp. 319–328. ACM Press, New York (1991)
8. Nelson, G.L.: Sonomorphs: An application of genetic algorithms to growth and development of musical organisms. In: 4th Biennial Art and Technology Symp., pp. 155–169 (March 1993)
9. Husbands, P., Copely, P., Eldridge, A., Mandelis, J.: 1. Evolutionary Computer Music. Springer, London (2007)
10. Biles, J.A.: 2. Evolutionary Computer Music Springer. Springer, London (2007)
11. Todd, P.M., Loy, D.G.: Music and Connectionism. MIT Press, Cambridge (1991)
12. Chen, C.C.J., Miikkulainen, R.: Creating melodies with evolving recurrent neural networks. In: Proceedings of the 2001 International Joint Conference on Neural Networks (IJCNN-2001, Washington, DC), pp. 2241–2246. IEEE Press, Washington (2001)
13. Gomez, F., Miikkulainen, R.: Solving non-Markovian control tasks with neuroevolution. In: IJCAI-1999, KAUF-ADDR, KAUF, pp. 1356–1361 (1999)
14. Saravanan, N., Fogel, D.B.: Evolving neural control systems. IEEE Expert, 23–27 (June 1995)

15. Yao, X.: Evolving artificial neural networks. Proceedings of the IEEE 87(9), 1423–1447 (1999)
16. Stanley, K.O., Miikkulainen, R.: Evolving neural networks through augmenting topologies. Evolutionary Computation 10, 99–127 (2002)
17. Stanley, K.O., Miikkulainen, R.: Competitive coevolution through evolutionary complexification. JAIR 21, 63–100 (2004)
18. Stanley, K.O.: Compositional pattern producing networks: A novel abstraction of development. Genetic Programming and Evolvable Machines Special Issue on Developmental Systems 8(2), 131–162 (2007)

AtomSwarm: A Framework for Swarm Improvisation

Daniel Jones

Lansdown Centre for Electronic Arts, Middlesex University, London EN4 8HT, UK
dan@erase.net

Abstract. This paper introduces AtomSwarm, a framework for sound-based performance using swarm dynamics. The classical ruleset for flocking simulations is augmented with genetically-encoded behaviours, hormonal flows, and viral 'memes', creating a complex sonic ecosystem that is capable of temporal adaptation and self-regulation. The architecture and sound design methodologies are summarised here, with critical reference to its biomimetic design process, sonic spatialisation and self-organising capabilities. It is finally suggested that the system's life-likeness is a product of its relational complexity, creating empathic engagement purely through abstract formal structures.

1 Introduction

It is well-established that there are compelling parallels between the complex behaviours demonstrated by swarms of social animals and those found within the realm of free musical improvisation [1,2]: simple, local interactions spontaneously give rise to complex global forms, with no a priori organisational principles, creating aesthetic structures that cannot easily be reduced to the sum of their parts. In support of these ideas, the author's research at the Lansdown Centre for Electronic Arts has revolved around situating swarms of simulated agents in the context of live musical performance, as active and reactive software agents.

The outcome of this research is AtomSwarm [3], a framework for sound-based composition and performance. AtomSwarm expresses the dynamics of a swarm both visually and as a spatialised sonic ecology, with each agent corresponding to a genetically-determined synthesis graph, whose parameters are modified based on its movements and interactions. It can be used as an independent generative system, developing and evolving through its internal resource regulation and temporal cycles, or as a complex interactive instrument, 'played' in an improvisatory fashion by modulating the rules that govern the system's interactions.

From the ground up, AtomSwarm's architecture and design has been conducted with biomimetics in mind; both the swarming engine and sound design draw extensively on biology, biochemistry and evolutionary theory. The drive behind the system's development, however, is essentially aesthetic, and certain contributions and abstractions have been made in order to develop the anthromorphic qualities of a performance. A consequence of this is that the system

M. Giacobini et al. (Eds.): EvoWorkshops 2008, LNCS 4974, pp. 423–432, 2008.

architecture features a degree of complexity that is at odds with classical alife objectives[1], seen here as a contingent sacrifice for expressive richness and behavioural familiarity.

In this paper, we first review the system's architecture and design methodologies (Sections 2 and 3). We then position and assess AtomSwarm along three axes: as a spatiotemporal environment (Section 4); as a generative, interactive instrument (Section 5); and as a self-organising ecosystem (Section 6). We conclude with an anecdotal account of the ability of a symbolic system to induce an empathic response in an audience.

2 Swarming Engine

The core of AtomSwarm's infrastructure is its object-orientated swarming framework, developed entirely using Processing[2] [5]. Working from the basis of Craig Reynolds' Boid algorithm [6], the engine is extended in complexity by introducing a number of artificial hormones ($h_x \in [-1..1]$) to each agent. Each virtual hormone's behaviours are modelled on the qualities of one or more actual chemical hormones or neurotransmitters, and is modulated by certain interactions and cycles analogous to its counterpart. In turn, it then affects the ruleset for local interactions that each agent follows as it traverses the swarm space, by modifying the strength and threshold values of each rule.

Table 1. Types of artificial hormone

Name	*Modelled on*	*Function*
h_t	Testosterone	Increases with age and crowdedness; decreases upon giving birth. Causes an increase in the likelihood of reproduction.
h_a	Adrenaline	Increases with overcrowding; decreases as a result of internal regulation over time. Causes a greater rate and variance of movement.
h_s	Serotonin	Increases during 'day' cycles; decreases during 'night' cycles, and as a result of hunger. Causes a greater social attraction towards other agents.
h_m	Melatonin	Increases during 'night' cycles; decreases during 'day' cycles. Decreases rate of movement.
h_l	Leptin	Increases upon eating 'food' deposits; decreases steadily at all other times. Signifies how well-fed an agent is, and causes downwards regulation of h_s when depleted, and greater attraction to food deposits.

In addition, each agent has a static *genome*, comprised of a series of floating-point values $g_x \in [0..1]$. These encode various qualities of the agent's behaviour, and remain invariant over time, though individual genes can be overwritten by viral infection (described in Section 6). The full set of genes is summarised in Table 2.

[1] See Christopher Langton's edict in the inaugural workshop on Artificial Life that its systems are expressed using "*simple* rather than *complex* specifications" (original emphasis) [4].

[2] Processing is a graphically-orientated extension of Java, incorporating a programming language and standalone IDE.

Table 2. Types of gene

Name	*Name*	*Function*
g_{col}	Colour	The hue of colour used to depict the agent.
g_{age}	Age	The rate at which the agent 'ages'.
g_{int}	Introspection	The degree to which the agent is attracted to social groups, or otherwise.
g_{perc}	Perception	The range at which the agent can perceive and respond to its peers.
g_{son_x}	Sonic parameters	The sonic behaviours exhibited by the agent, described in the following section.
g_{cyc_x}	Hormone cycles	The strength or speed of each hormonal cycle; for example, g_{cyc_s} determines the amount that the agent's h_s level increases during a 'day' cycle.
g_{up_x}	Hormone uptakes	The quantitative increase in hormone level experienced when uptake occurs; for example, g_{up_a} determines the degree of h_a increase following a collision with another agent.

The g_{cyc_x} and g_{up_x} gene sets, determining hormonal cycles and sensitivities, are critical to the ecosystem's diversity and evolution. For example, as a consequence of their effects, a particular agent may be prone to sudden increases in h_a levels, resulting in it 'fleeing' an overcrowded area and locating new food deposits. Given the presence of of agents containing suitable genetic traits, the swarm is thereby able to respond appropriately to a wide range of states.

Thus, in distinction to the time-invariant behaviour of the original Boid algorithm, there are now a number of feedback mechanisms in place: the genotype of an agent determines its hormonal fluctuations; the genotype and hormone levels co-determine its response to each of the set of physical rules (cohesion, separation, etc); and events caused by following these rules (eating, colliding, becoming overcrowded) feed back to modulate hormone levels. The interactions between these three planes of codification quickly become very complex, and result in diverse and shifting collective behaviours over time. Moreover, they create the ability for the ecosystem to adapt and self-regulate its population, as outlined in Section 6.

The size of the population is also subject to continuous fluctuations due to the processes of reproduction and death. The agents are asexual, and give birth to a single offspring after their h_t level reaches a fixed threshold. The genome of the offspring is a duplicate of its parent's, subject to minor variance (wherein each gene g_x is altered by up to ± 0.1) and a degree of genetic mutation (with a probability of 0.05, a gene may be replaced with a uniformly random value $X \in [0..1]$). The offspring's behaviour is therefore usually similar to that of its parent, but may occasionally exhibit radical alterations, opening up the possibility of advantageous anomalies.

Deaths can be caused by hormone imbalances, representing starvation, depression and metabolic overload, or simply by old age, determined relative to the g_{age} gene.

The system additionally incorporates the concept of 'food', an arbitrary resource placed in scattered collections at uniformly random intervals, whose presence introduces a constraint on the swarm's collective resource levels; and 'day' and 'night' periods, which cycle over the course of a few minutes, reflecting the length of a typical performance. These constructs serve to modulate the swarm's hormonal levels over time, in a fashion analogous to natural metabolic systems, and thus structure the sonic form into subtly-defined movements of tempo and intensity.

3 Sound Synthesis and Mappings

The sound generation components of AtomSwarm are handled by SuperCollider's powerful synthesis engine, communicating with the swarming process via Open Sound Control. As the system's design requires fine-grained control over the musical output from the ground up, sonic behaviours are also codified extensively in the swarming engine itself.

At the sound design stage, a palette of 'generator' and 'processor' synthesis modules was defined, each of which is comprised of a number of SuperCollider's primitive DSP units and accepts up to two arbitrary parameters ($x_a, x_b \in [0..1]$). Each agent is then assigned one generator and one processor unit, determined by its genetic content, which are instantiated on the server and combined in serial. This compound graph provides it with an identifiable sonic signature – or to use the language adopted by Dennis Smalley, its "physiognomy" [7].

Alongside its synthesis graph, each agent's genome encodes a 'trigger' mode and threshold, which together determine the point at which the synth is instructed to generate output. This may be, for example, upon collision with another agent, or when the agent reaches a velocity of X pixels per second, where X is genetically specified.

For continuous sonic modulation, the genome also determines a fixed mapping from the agent's movements to a given property of its synthesis graph; for example, its velocity (normalised to $[0..1]$) may be mapped to the generator's x_a parameter, which, for some generator units, is then assigned to its amplitude. In a more complex case, the relative-shift parameter, reflecting the rate at which the agent is moving towards or away from a peer, could be assigned to the frequency parameter of an oscillator. This creates an approximation of the Doppler effect, exemplified by the familiar drop in pitch of a passing ambulance siren. In testing, this was found to also give a convincing approximation of the pitch oscillations of a swarm of buzzing bees.

AtomSwarm is orientated towards the composition of textural and quasi-rhythmic forms, often making use of repetitive cycles of short, non-tuned sound objects. The single pitched synthesis class is a pure sine wave, with an amplitude envelope for gradual onset and release. Combined with the potential combinatorical complexity of a genetically-selected processor unit and motion mapping, even this can give rise to a diverse range of output.

3.1 Biomimetics in Sound Design

The sound design for the synthesis components was a broadly biomimetic process. An earlier incarnation of the framework used sound recordings of a number of natural phenomena: human bodily functions, cicada calls, and metallic, drip-like impulses. The transition to pure synthesis was made by observing the spectral qualities of these classes of sound, and translating them into functions of basic oscillators and signal processing units. Each generator component thus has a distinct morphological identity, texturally modified by its processor, but retaining sufficient qualities to be identifiable as being from the same source.

Why was this approach taken? Rather than simply using the ordered structures of motion for composition within an existing framework, like the melody-orientated generative composition of Blackwell's early research [8], a conscious decision was made to evoke the qualities of an ecosystem "as it could be" [4] – supporting the existing visual and conceptual narratives, and suggesting immersion within a possible, quasi-biological world. The intention was that, even without the visual depiction of the ecosystem, the sonic design alone would suggest that the source of the sound is organic in nature. With the addition of heavily synthetic processor units, this reference is warped and distended to suggest a bio-technological hybrid.

4 Spatiotemporality

AtomSwarm's agents are located within a continuous 2-dimensional space, bounded by the limits of the visual display. Early prototypes used n-dimensional vectors, following Blackwell's swarm composition research [9] which frequently uses up to 7 spatial dimensions to compelling effect.

However, in the case of AtomSwarm's sonic expression, we are less interested in positional data, instead focusing on the overall dynamics of the swarm's motion and its status as a continually generating ecosystem. Furthermore, as we are not working with the traditional axes of pitch/amplitude/duration, we have no need to capture this number of positional values in parallel; we have a sufficiently wide combinatorical space of timbral qualities to be content with one motion mapping per agent, which quickly results in complex sonic interactions even with a relatively small swarm. Even this one mapping may not be positional, instead taking values from velocity or relative movements. In this way, we hope to express a greater range of the dynamics of the swarm. A crowded, fast-moving group may be expressed by heavy layers of high-frequency ticks, a series of sine waves undergoing rapidly fluctuating Doppler shifts, or by frequent percussive pulses given off by collisions between agents.

The most prominent use of the swarm space, however, is in its identification with the space surrounding the viewer. The single agent present from the very start of a performance is known as the 'Listener', visually identifiable by its red outer ring. Effectively, the viewer hears the swarm's motions from the perspective of the Listener, using vector-amplitude panning [10] for simulated sound source positioning on an arbitrary number of output speakers. On a 2-dimensional multichannel speaker set, sound events to the left of the Listener are heard to the left of the viewer; events displayed above the Listener are heard straight ahead. As the Listener moves around the world, therefore, the viewer's soundscape shifts accordingly.

This is further supported by a global reverberation unit, whose reflection parameters are adjusted according to the Listener's mean distance from its peers to simulate the reverberant qualities of distant sounds within a large space. Though this technique lacks precision, it serves to support the notions of distance and proximity, both important criteria for the faithful sonification of a distributed population.

These methods encourage the viewer to identify with an agent inside the space, shifting them from a position outside of the system to one immersed within it. Indeed, 'immersivity' is intended to be key to the experience of a performance, reinforcing empathic engagement with the spectator. These ideas are a continuation of the drive to realize a possible space "as it could be". Yet, though this space is rendered perceptible around the audience, the system's digital manifestation indelibly marks the experience with the grain of non-reality.

5 Performance Interface and Human Intervention

In the context of a live performance, AtomSwarm is projected onto a screen visible to the audience, with audio distributed via a multi-channel speaker system. Control over the environment is limited to a basic MIDI interface, through which the human 'conductor' is able to create and destroy agents, add food deposits, and manipulate the weightings of the physical rules governing the swarm's movements. Thus, the only control mechanism is wholly indirect, with no scope for determining its sonic behaviours, nor even manipulating the individual agents themselves[3]. Three layers of interactions serve to mediate the conductor's influence over the soundscape: between the rule weightings and the swarm's hormone levels; between the relative positions of each of the agents; and between each agent's motion dynamics and the sonic mappings described by its genome. Through these layers of mediation, it is often the case that that attempts at influencing the system go unheeded; increasing the 'Cohesion' rule, for example, may be ignored entirely by a swarm made up of highly introverted agents.

As far as modulating the current behaviour is concerned, therefore, the conductor's role is limited by constraints imposed within the system. A constant tension emerges between order and chaos, with the human input in continual threat of being outweighed by the balance of internal forces. This is the same "dynamic network of relations" as described by Lev Manovich [11], in which current trends are vulnerable to being swept away by amplified oscillations towards a new structural equilibrium. The resultant experience is almost game-like, in that the aesthetic 'fitness' of the collective sonic output may be at odds with the fitness criteria of its constitutive agents. For example, a clustered group may be generating a rich, compelling timbre, but this cannot be sustained if its collision rate is too high (wherein h_t overload will kill many of the agents), or hunger

[3] It is for these reasons that the term 'conductor' has been adopted: as in an orchestra or choir, the conductor maintains real-time control over the unified ensemble, gesturally influencing its flow and dynamics en masse.

levels rise to the point at which the agents ignore the 'cohesion' rule and depart to seek food.

An alternative approach to performance is to allow the ecosystem to develop and regulate itself independently, and engage in total autopoiesis. In the absence of human intervention to supervise its growth, the swarm will still engage in self-regulating behaviour as a consequence of its hormonal requirements, limited resource supplies and aging processes. Evolutionary narratives unfold according to the interconnected rulesets that determine the genome-hormone-ruleset interactions; spectators can select whether to engage on a macroscopic scale, with the synchronised movement and sonification of the swarm as a whole, or on a microscopic scale, in the interactions of individual agents.

6 Self-organization and Emergence in AtomSwarm

The flows of resources within AtomSwarm are carefully balanced to ensure that interactions occur at appropriate levels and rates, so that, for example, a population will not normally die of starvation within a few seconds. As a consequence of this equilibrium, it is now demonstrably capable of exhibiting a range of non-trivial self-organising behaviours, many of which were not anticipated when these interactions were first implemented.

On one level of resource flow, each agent attempts to maintain an internal homeostasis: as a hormone quantity is amassed or depleted, its rule-following behaviours will be slowly weighted towards those actions that will assist its regulation (eating, reproducing, seeking isolation). Above this, on the macro-scale of the swarm as a whole, a "homeorhesis" [12] occurs, or the *regulation of flow* of resources between agents, with only a limited quantity of food deposits available. If the population grows too large, insufficient food supplies result in downwards regulation due to deaths from starvation. If it shrinks, food is abundant and the population is free to increase. Yet, this is no guarantee of survival: agents which are excessively sociable risk death from the h_t overload caused by excessive collisions; a nomadic tendency may be useful for finding isolated deposits of food, but can result in h_s depletion and a lack of h_t, and thus the inability to reproduce.

Another emergent surprise, and one which genuinely instilled the rewarding, unexpected sensation of the "something extra" emphasised by Whitelaw [13] in his cross-section of a-life art, is the swarm's ability to effectively discover food deposits. Each deposit comprises of up to 10 food particles, each of which is sufficient to satiate an agent's hunger for a short period. In one case, a fairly tight-knit swarm was located far away from any food resources. One nomadically-disposed agent was moving separately from this cluster, with sufficient random motion to quickly encounter a food deposit. After consuming a particle, it lingered near the deposit. The remainder, following the rule of cohesion to the swarm's centre of mass, gradually moved across the space to join the nomad, and in doing so discovered and consumed the food deposit.

This food-finding ability through nomadic exploration is clearly not something that was programmed into the individual actions of the agents. It is purely the result of a circular feedback loop between the ecosystem's internal states, via positive feedback through the regulation processes of the individual agents. Yet it is manifest as an adaptive and peculiarly intelligent-seeming behaviour. Despite our awareness that this is simply the result of a set of interactions taking place within a wholly deterministic machine, it is difficult to avoid anthropomorphizing this on-screen behaviour and taking pleasure in its lifelike form.

6.1 Sonic Self-organisation

Because each agent's sonic behaviours are encoded in its genome, which is passed down to child agents and selected through generations of fitness-driven evolution, a significant degree of sonic ordering can be perceived through focusing on the auditory representation of the environment. The sonic spatialisation, described in detail shortly, gives a richly accurate sense of movement and change from within the swarm's frame of reference. Given an agent with a distinctive sound signature, we can hear the result of its reproduction through the sudden introduction of a similar-sounding signature. Population growth is accompanied by an increase in the density and spectral depth of the output.

This is supplemented by the presence of viral *memes*[4], a recent evolution to the AtomSwarm framework itself. An agent will very occasionally develop a temporary meme corresponding to one of its sonic synthesis chromosomes, which can then infect nearby agents through collisions, with statistical probability based on the meme's arbitrary 'strength' rating. An infected agent will adopt this same single genetic value, permanently overwriting the value from its existing genome, and so its sound signature will immediately be transformed to resemble that of the infector. If the population's density is sufficiently high, a strong meme can spread between the agents extremely rapidly, and so the sonic landscape may suddenly switch to a chorus of unified chirping.

Is this sonic self-organisation? Insofar as the sonic terrain frequently orders itself into spectral unison, from a chaotic starting point, then it could certainly be classed as such. Moreover, consider the fact that the population of the swarm is bounded by the limited availability of resources. As the production of sound objects is directly proportional to the swarm's population size, this same bound is placed upon the sonic density; a period of high activity (expressed by high amplitude levels across the spectrum) cannot be sustained.

However, one of the critical principles for non-trivial[5] self-organisation is that of positive feedback: a circular interaction between components that proceeds to amplify a change [15]. In the example of the ant colony, this is manifest in the increase in pheromone trails after locating food. As further ants proceed to follow the pheromone gradient and arrive at the food deposit, the trail is strengthened, amplifying the feedback loop.

---

[4] Taken from the terminology of Richard Dawkins [14].

[5] Francis Heylighen draws a continuum between simple and complex instances of self-organisation; certain traits "will only be exhibited by the more complex systems, distinguishing for example an ecosystem from a mere process of crystallization" [15].

Through the interaction of metabolic systems within AtomSwarm, this class of feedback occurs at several points, such as in the example described in the previous section wherein the swarm can discover food deposits based on its shifting centre of mass. Though these interactions do have a direct result on the sonic output, this cannot be classed as *sonic* self-organisation for the fundamental reason that the this feedback *does not occur in the same frame of reference* as the relations that constitute the plane of sound generation. For true sonic self-organisation to occur, changes in sound synthesis must be reinforced and amplified based on properties of the sound itself. The distribution of sonic artefacts via memes can be modestly viewed as an organisational process, but not one that is linked to an evaluatory procedure based on auditory criteria.

In fact, no richly meaningful form of sonic self-organisation can take place without an internal concept of 'fitness' in the same frame of reference. This immediately poses the old problem of creating an objective assessment of essentially aesthetic criteria. Given that whether something 'sounds good' is an inherently subjective judgement, how can a symbolic system provide positive or negative feedback on its current auditory state?

It is out of the scope of this paper to review methods of evaluating sound-based fitness. In a similar context, however, Blackwell and Young provide an elegant solution by placing 'attractors' in the swarm space, whose locations are determined by the attributes of a musical source that is analysed in real-time (such as pitch, amplitude, and duration). As agents swarm towards these attractors, their output – which is parametrised along the same axes – tends towards being relationally similar to the input. Assuming the musical source is a human musician, this swarming can then be positively reinforced by playing more notes in a similar vein, or negatively reinforced by modulating playing style – say, by switching to a different pitch register.

A similar procedure could here be adopted based on timbral analysis of a sound source. However, due to the heterogeneity of each agent's sonic behaviours, no universal parametrisation of timbral qualities is possible. It is one of the future research directions of this project to consider how the output of audio analysis might result in environmental modifications of other types.

7 Conclusion: Swarming and Complexity

AtomSwarm, like any complex dynamical system, is fundamentally a staging ground for a continuous parallel flux of interactions, between forces, agents and resources. Convoluted feedback loops arise between the multiple planes of interaction (human input, rules, hormones and genomes), with sufficient complexity to evoke the organic (in)stability of a natural ecosystem.

It is only in virtue of this complexity that the system becomes open to the types of anthromorphism that are frequently demonstrated when an audience encounters AtomSwarm's digital population. Though this is clearly a symbolic ecology – which its sparse, geometrical rendering does little to dispel – whose behaviours are generated by a set of deterministic algorithms, it consistently induces a significant empathic response in audiences. One public performance was concluded with two agents seemingly engaged in a form of dance, pursuing each other in swooping curves. This was left to continue until, eventually, one reached its natural lifespan and died, disappearing from the display. In recognition of the situation, an audible sigh of loss emanated from the auditorium.

This willingness to emotionally engage with a symbolic community, whose resemblance to a living system is limited to its relational structures, is an indicator that concepts are being applied beyond those intrinsic to an abstract generative system. Using the terms of Mitchell Whitelaw [16], a "system story" is being imposed through the the viewer's imaginative faculties, perceiving the underlying biological origins of the system from its formal isomorphism. The spectator takes joy in this familiar-yet-strange image of "life as it could be" [4].

References

1. Blackwell, T., Young, M.: Swarm Granulator. In: Raidl, G.R., Cagnoni, S., Branke, J., Corne, D.W., Drechsler, R., Jin, Y., Johnson, C.G., Machado, P., Marchiori, E., Rothlauf, F., Smith, G.D., Squillero, G. (eds.) EvoWorkshops 2004. LNCS, vol. 3005, pp. 399–408. Springer, Heidelberg (2004)
2. Borgo, D.: Sync or Swarm: Musical improvisation and the complex dynamics of group creativity. In: Futatsugi, K., Jouannaud, J.P., Meseguer, J. (eds.) Algebra, Meaning and Computation: Essays dedicated to Joseph A. Goguen on the Occasion of His 65th Birthday, pp. 1–24. Springer, New York (2006)
3. Jones, D.: AtomSwarm (2007), `http://www.erase.net/projects/atomswarm`
4. Langton, C.: Artificial life. In: Langton, C. (ed.) Artificial Life: The proceedings of an interdisciplinary workshop of the synthesis and simulation. Santa Fe Institute Studies in the Sciences of Complexity, vol. 6, pp. 1–44. Addison-Wesley, Reading (1989)
5. Reas, C., Fry, B.: Processing (2007), `http://www.processing.org/`
6. Reynolds, C.: Flocks, herds, and schools: A distributed behavioral model. ACM SIGGRAPH Computer Graphics 21(4), 25–34 (1987)
7. Smalley, D.: Defining timbre – refining timbre. Contemporary Music Review 10(2), 35–48 (1994)
8. Blackwell, T.: Swarm music: Improvised music with multi-swarms. In: Proceedings of the 2003 AISB Symposium on Artificial Intelligence and Creativity in Arts and Science, pp. 41–49 (2003)
9. Blackwell, T., Young, M.: Self-organised Music. Organised Sound 9(2), 137–150 (2004)
10. Pulkki, V.: Virtual sound source positioning using vector base amplitude panning. Journal of the Audio Engineering Society 45(6), 456–466 (1997)
11. Manovich, L.: Abstraction and Complexity (2004), `http://www.manovich.net/DOCS/abstraction_complexity.doc`
12. Hasdell, P.: Artificial ecologies: Second nature emergent phenomena in constructed digital - natural assemblages. Leonardo Electronic Almanac 14(7–8) (2006)
13. Whitelaw, M.: Metacreation. MIT Press, Cambridge (2004)
14. Dawkins, R.: The Selfish Gene. Oxford University Press, Oxford (1990)
15. Heylighen, F.: The science of self-organization and adaptivity. The Encyclopedia of Life Support Systems (2003), `http://pespmcl.vub.ac.be/papers/EOLSS-Self-Organiz.pdf`
16. Whitelaw, M.: System stories and model worlds: A critical approach to generative art. In: Goriunova, O. (ed.) Readme 100: Temporary Software Art Factory, pp. 135–154. Books on Demand GMBH, Norderstedt (2005)

Using DNA to Generate 3D Organic Art Forms

William Latham[1], Miki Shaw[1], Stephen Todd[1], Frederic Fol Leymarie[1],
Benjamin Jefferys[2], and Lawrence Kelley[2]

[1] Computing, Goldsmiths College, University of London, UK
{w.latham,ffl}@gold.ac.uk
[2] Bioinformatics, Imperial College, London, UK

Abstract. A novel biological software approach to define and evolve 3D computer art forms is described based on a re-implementation of the *FormGrow* system produced by Latham and Todd at IBM in the early 1990's. This original work is extended by using DNA sequences as the input to generate complex organic-like forms. The translation of the DNA data to 3D graphic form is performed by two contrasting processes, one intuitive and one informed by the biochemistry. The former involves the development of novel, but simple, look-up tables to generate a code list of functions such as the twisting, bending, stacking, and scaling and their associated parametric values such as angle and scale. The latter involves an analysis of the biochemical properties of the proteins encoded by genes in DNA, which are used to control the parameters of a fixed *FormGrow* structure. The resulting 3D data sets are then rendered using conventional techniques to create visually appealing art forms. The system maps DNA data into an alternative multi-dimensional space with strong graphic visual features such as intricate branching structures and complex folding. The potential use in scientific visualisation is illustrated by two examples. Forms representing the sickle cell anaemia mutation demonstrate how a point mutation can have a dramatic effect. An animation illustrating the divergent evolution of two proteins with a common ancestor provides a compelling view of an evolutionary process lost in millions of years of natural history.

1 Introduction

We present a novel biological approach to define and evolve 3D art forms. The work combines a re-implementation of the *FormGrow* system of Todd and Latham [1] with an external source to define the shapes: DNA sequences. *FormGrow* is a virtual machine producing 3D computer art forms or designs. It embodies the particular organic aesthetics favored by Latham together with a shape grammar made of primitives (horn-like structures), transforms and assembly rules, and a number of parameters encoding color, scale, texture. We have re-visited the *FormGrow* system adapting it to a modern implementation taking advantage of standard graphics libraries and portable coding, and putting the emphasis on bringing this system closer to the realm of biology.

Two methods for using real DNA data, in the form of nucleotide sequences, were devised. In the first, sequences are directly transformed via a series of empirically designed tables to become readable by *FormGrow*. These tables process nucleotides as codon triplets of data as would ribosomes in a live cell. Notions of "start," "stop," and

M. Giacobini et al. (Eds.): EvoWorkshops 2008, LNCS 4974, pp. 433–442, 2008.

"junk" DNA code are also embedded in our system. We explore the application of our method to generate 3D organic art forms in the visualisation of particular genetic defects, and present as a case study the well-known sickle cell anaemia mutation. The second method for interpreting DNA sequences is to look at the biochemical properties of the amino acids which are encoded by the codons in a gene. Simple counts of the amino acids properties types are turned into parameters for a fixed *FormGrow* structure. We illustrate the use of such an interpretation with an animation of the evolution of a pair of related proteins over millions of years. The precise sequence and structure of the ancestral proteins is unknown, but sophisticated tools and an artistic interpretation of the data gives a glimpse of a process which is lost in eons of evolutionary history.

2 Background

In the 1980's, while at the Royal College of Art in London, Latham devised a rule-based hand-drawn evolution system called *FormSynth* [2]. He then joined forces with Stephen Todd at IBM to develop the *FormGrow* and *Mutator* systems from 1987 to 1993 [3,4,1,5,6]. Our work started from an open-ended aim to revisit this project which had been untouched for about twelve years.

FormGrow is a kind of building blocks kit for creating organic-style 3D computer generated forms. It uses a hierarchical system, building up complex forms from primitive shapes. The central *FormGrow* construct is a horn made of n ribs: repeated primitive shapes. Variants of the basic horn are created by applying elementary transforms: stack, bend, twist, and grow. *Mutator* allows forms to be grown using life-like techniques such as cross-fertilisation (marriage) and mutation. A form — as obtained via the *FormGrow* system — is expressed as a sequential set of instructions, which constitute its encoding. *Mutator* reads *FormGrow* instructions to be combined (if coming from various parents) or modified (simulating mutations). In the original work, the survival of a form was governed by human selection — typically embodied by the artist Latham seen as a kind of gardener of art-forms — or by closeness to some pre-defined measure. Latham and Todd's work during the period of 1987–93 coincided in particular with the works of Sims on 2D and 3D forms [7,8], of Prusinkiewicz and Lindenmayer on plants [9], and of Leyton on process grammars [10,11]. Discussion of the properties, differences and similarities of such systems are covered, *e.g.*, in [6,12,13].

A mathematical and computational formalism which unites these shape generative systems is that of *shape grammars* whereby objects of various complexities can be generated by an iteration of a finite number of simple outline transformation instructions, such as *FormGrow*'s bends, twists and stacks. Various shape grammars have been developed in the literature. For example, the generation of self–similar fractal objects is possible with very simple grammars [14, §8.1]. Selection rules, which forbid the addition of a sub-unit under certain conditions, have been used by Ulam to generate less regular patterns [14, §8.2]. Trees and river systems, crystals, tessellations and space filling organisations are other examples of domains of applications of shape grammars as object generators [14].

An important example of early work is to be found in *L-systems*, also called Lindenmayer systems or *parallel string-rewrite systems*, which are made from productions rules used to define a tracing of piecewise linear segments with joints parameterised by rotation angles [15,9]. These rules also are a compact way to iteratively repeat constructive sequences in the description of fractals, often used to model groups of plants, flowers, leaves, and so on [16].

One can generalise shape grammars within the context of cellular automata where some randomisation is introduced in the manifestation of the rules leading to *dynamic* shapes; for example see the works of Wolfram *et al.* [17] and more recent studies in biological pattern genesis [18]. The combination of dynamical L-systems with cellular automata has been considered, in particular in the works of Jon McCormack with application to art form genesis [19].

Another possible generalisation is in the context of genetic programming where mutations and the natural mixing of a pool of genes (possibly representing shape parts or features) is used to obtain evolving *natural* or organic-looking shapes; *e.g.*, see the early works of Dawkins on biomorphs [20], and, again, of Latham and Todd on genetic art [1], and more recent works in art [12], design [6] and biological sciences [21].

Our motivation for re-visiting Latham and Todd's work is that it is a powerful system which offers the possibility of generating organic-like shapes and which from its origins was meant as a metaphor to nature's way of evolving forms. In re-visiting this work, on the one hand we bring up-to-date the technology developed in [1] in the context of recent advances in graphics and computational geometry, and on the other hand we bring it much closer to biology via the recent advances made in understanding the working of nature in the fields of genomics and proteomics, the focus of this paper.

3 Use of DNA in *FormGrow*

DNA can be thought of as a shape-specification language residing in the cells of every living organism, encoding proteins which constitute the body's key builders and building blocks. The DNA molecule is essentially a very long string of much smaller molecules, the nucleotides, which come in four varieties (A, C, T, G).

How does this apparently simple string of nucleotides encode the complex form of a protein? A protein is also a string of simpler molecules: the amino acids. As there are 20 types of amino acids and only 4 types of nucleotides, the DNA translation mechanism looks at nucleotides in groups of three, triplets called *codons*; every codon translates to a single amino acid [21]. Working down the chain of DNA generates the corresponding chain of amino acids, yielding a protein. The codon-amino acid equivalences can be represented in a translation table (Tbl.1).

Following this model, we created an analogous translation system to convert DNA sequences into *FormGrow* code. At a coarse level, *FormGrow* code can be viewed as a series of function calls, with each function requiring a small number of arguments (this number varies from 0 to 3 depending on the particular function). Thus, we created 2 translation tables: the *transform table*, which translates from codons to transformational functions (Tbl.2); and the *number table* (Tbl.3), which translates from codons to numerical arguments (integers in the range 0 to 63). Given our input sequence, we translate the first codon into a function using the transform table, and then generate numerical arguments for that function by translating the following codons into numbers,

Table 1. Translating codons to amino acids

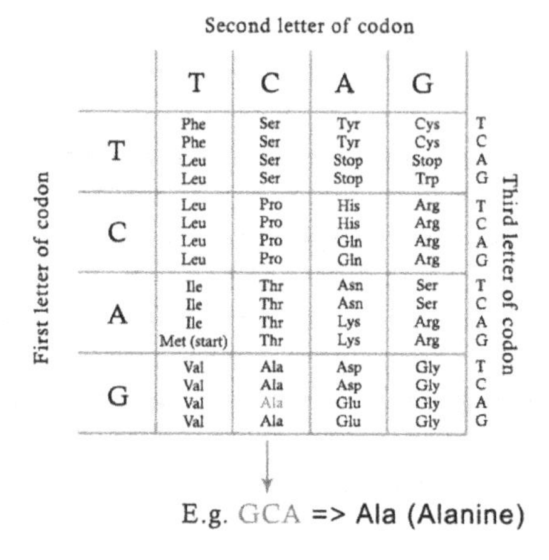

First letter of codon	Second letter: T	C	A	G	Third letter of codon
T	Phe	Ser	Tyr	Cys	T
	Phe	Ser	Tyr	Cys	C
	Leu	Ser	Stop	Stop	A
	Leu	Ser	Stop	Trp	G
C	Leu	Pro	His	Arg	T
	Leu	Pro	His	Arg	C
	Leu	Pro	Gln	Arg	A
	Leu	Pro	Gln	Arg	G
A	Ile	Thr	Asn	Ser	T
	Ile	Thr	Asn	Ser	C
	Ile	Thr	Lys	Arg	A
	Met (start)	Thr	Lys	Arg	G
G	Val	Ala	Asp	Gly	T
	Val	Ala	Asp	Gly	C
	Val	Ala	Glu	Gly	A
	Val	Ala	Glu	Gly	G

E.g. GCA => Ala (Alanine)

Table 2. Codons to *FormGrow* transforms

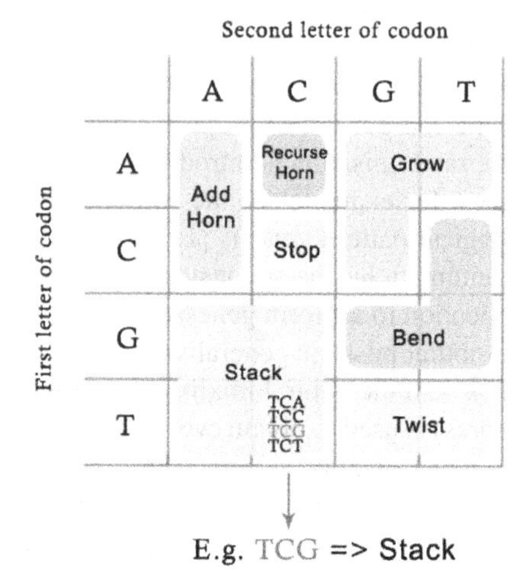

Table 3. Translating codons to numbers

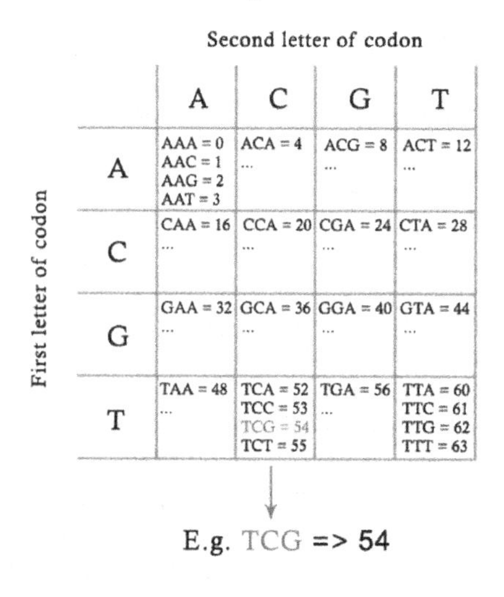

First letter of codon	Second letter: A	C	G	T
A	AAA = 0 AAC = 1 AAG = 2 AAT = 3	ACA = 4 ...	ACG = 8 ...	ACT = 12 ...
C	CAA = 16 ...	CCA = 20 ...	CGA = 24 ...	CTA = 28 ...
G	GAA = 32 ...	GCA = 36 ...	GGA = 40 ...	GTA = 44 ...
T	TAA = 48 ...	TCA = 52 TCC = 53 TCG = 54 TCT = 55	TGA = 56 ...	TTA = 60 TTC = 61 TTG = 62 TTT = 63

E.g. TCG => 54

Table 4. Translating a codon sequence to *FormGrow* code

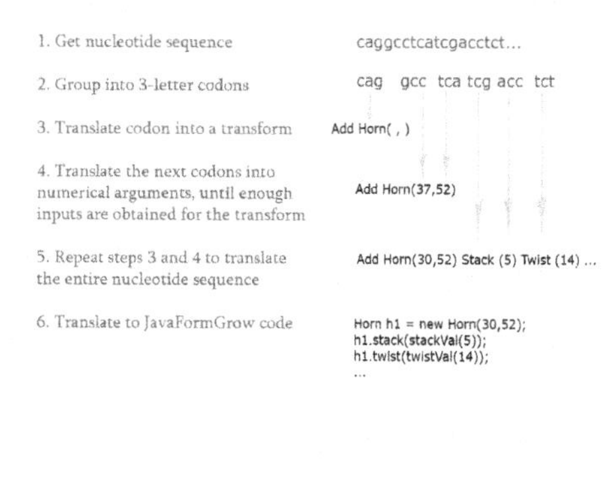

1. Get nucleotide sequence	caggcctcatcgacctct...
2. Group into 3-letter codons	cag gcc tca tcg acc tct
3. Translate codon into a transform	Add Horn(,)
4. Translate the next codons into numerical arguments, until enough inputs are obtained for the transform	Add Horn(37,52)
5. Repeat steps 3 and 4 to translate the entire nucleotide sequence	Add Horn(30,52) Stack (5) Twist (14) ...

6. Translate to JavaFormGrow code	Horn h1 = new Horn(30,52);
	h1.stack(stackVal(5));
	h1.twist(twistVal(14));
	...

using the number table. Once we have sufficient arguments, we return to the transform table to generate our next function, and so the cycle continues (Tbl.4). Finally we render the generated *FormGrow* code to produce a 3D shape. Figures 1 and 2 show some images generated from genuine DNA sequences.

It is interesting to note some similarities between nature's translation method and ours, these being features which we needed in our system and realised that the biological precedents were well worth adopting. In the original translation table there is a "start" codon (ATG) which signals that a new protein is being specified. Likewise, in our transform table, the "add horn" transform flags the beginning of a new shape. The "stop" codon is also mirrored in our system. This instructs termination of the current

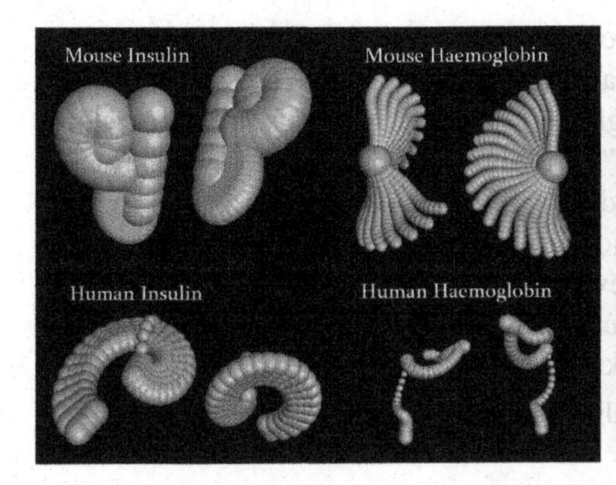

Fig. 1. Images generated from real DNA

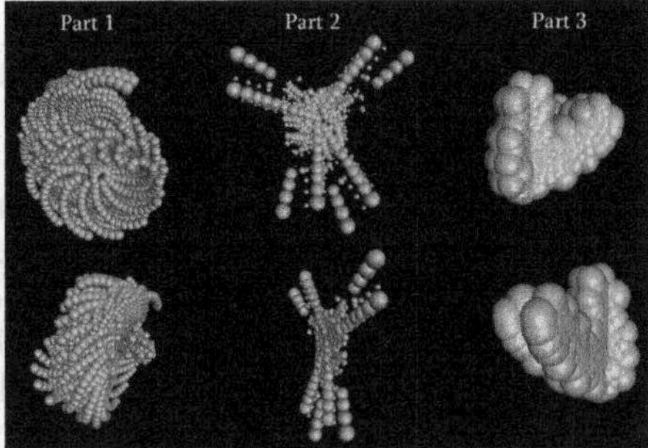

Fig. 2. Images generated from sections of the mouse keratin DNA

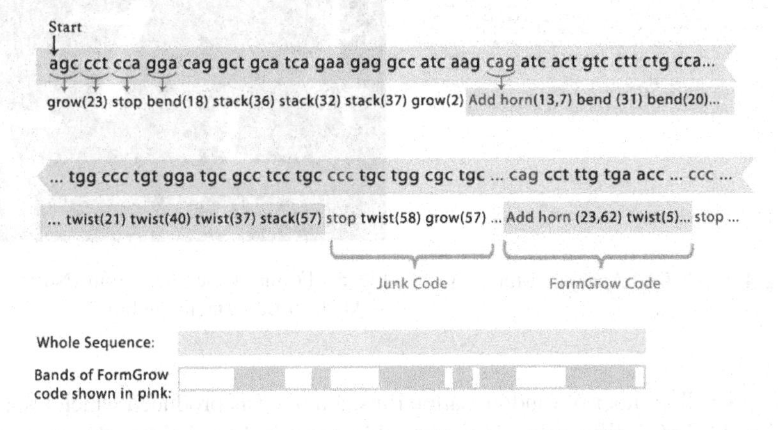

Fig. 3. Translating an example DNA sequence (Human Insulin)

protein or shape. A side effect of adopting the "start" and "stop" mechanism is that we end up with large sections of "junk code," *i.e.*, code which generates no proteins or shapes because it lies in a non-coding section of the sequence (Fig.3). By changing the layout of the transform table we can affect the proportion of junk code produced. We experimented with producing a few different iterations of the transform table in order to get a balance of functions that would produce a visually interesting variety of shapes.

3.1 Case Study: Sickle Cell Anaemia

Using this novel method of converting DNA into 3D shapes, we wondered if we could compare different DNA sequences. We selected as our case study the gene for sickle cell anaemia. This inherited disease affects millions of people worldwide. It damages the red blood cells which deliver oxygen to vital organs, resulting in anaemia and further complications. It is particularly common in malarial regions, because it offers some protection against malaria. All this is caused by one faulty gene. The problem appears as a single point mutation in the beta haemoglobin gene: a single "A" nucleotide is changed to a "T" (Fig.4). The reason that this single nucleotide substitution is so influential is because "GTG" encodes a different amino acid to "GAG." And this amino acid switch changes the physical behaviour of haemoglobin in the body. In our initial transform table, there is no difference between "GAG" and "GTG," so both the normal and sickle cell forms of the DNA sequence generated identical shapes. It is not unusual for a single mutation to go unregistered. Both the amino acid table and the transform table exhibit some redundancy — in fact, there is an evolutionary advantage in this redundancy, as it makes DNA more resistant to minor changes. However the transform table only produces 7 different output functions (unlike the amino acid table which has 20), thus more repetition is inevitable. Rather than adjusting the table further by hand, we applied a procedure to randomise it.

Human Haemoglobin Beta DNA Sequence:

att tgc ttc tga cac aac tgt gtt cac tag caa cct caa aca gac acc atg gtg cat ctg
act cct gag gag aag tct gcc gtt act gcc ctg tgg ggc aag gtg aac gtg gat gaa gtt
ggt ggt gag gcc ctg ggc agg ctg ctg gtg gtc tac cct tgg acc cag agg ttc ttt gag
tcc ttt ggg gat ctg tcc act cct gat gct gtt atg ggc aac cct aag gtg aag gct cat
ggc aag aaa gtg ctc ggt gcc ttt agt gat ggc ctg gct cac ctg gac aac ctc aag ggc
acc ttt gcc aca ctg agt gag ctg cac tgt gac aag ctg cac gtg gat cct gag aac ttc
agg ctc ctg ggc aac gtg ctg gtc tgt gtg ctg gcc cat cac ttt ggc aaa gaa ttc acc
cca cca gtg cag gct gcc tat cag aaa gtg gtg gct ggt gtg gct aat gcc ctg gcc cac
aag tat cac taa gct cgc ttt ctt gct gtc caa ttt cta tta aag gtt cct ttg ttc cct aag
tcc aac tac taa act ggg gga tat tat gaa ggg cct tga gca tct gga ttc tgc cta ata
aaa aac att tat ttt cat tgc

Normal: ... act cct gag gag aag ...

Mutant (Sickle Cell Anaemia): ... act cct gtg gag aag ...

Fig. 4. Sickle Cell Anaemia Mutation

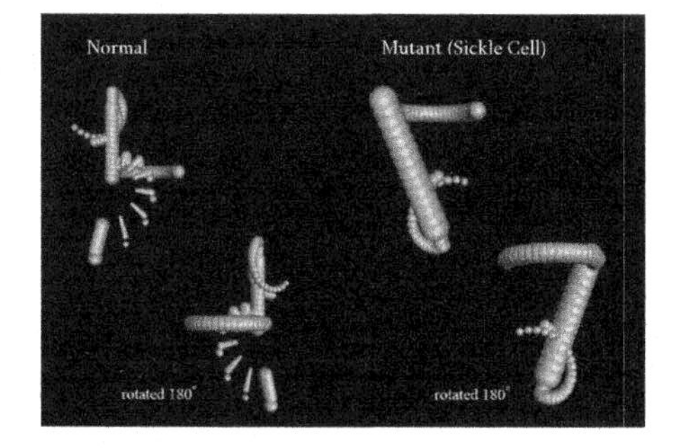

Fig. 5. Forms generated from Normal and Sickle Cell Beta Haemoglobin

After a small number of randomisation runs, a table was produced which translated "GTG" and "GAG" differently. This new table generated the images shown in Fig.5. The two forms are easily distinguished, though many similarities can be seen in their component parts. Effectively we had produced an alternative way of visualising this genetic mutation. It is an artistic impression of how a point mutation can have such a dramatic effect on phenotype. Ideally we would like to optimise the table such that the visualisation reflects the sequence in some expected or sensible manner for a large set of proteins and mutations. This optimisation may take place using any standard algorithm, for example an evolutionary method might be appropriate, making stepwise changes in the table and asking a human (artist or biologist) if the resulting visualisations are an improvement on those produced by the parent transform table.

4 Use of Amino Acid Biochemistry in *FormGrow*

The approach of directly using DNA sequences (interpreted as codons) to generate *FormGrow* shapes ignores the biochemical characteristics of the amino acids which the codons represent. A protein, in the form of a chains of amino acids, folds into a specific shape, governed by these properties of the amino acids. The relationship between the amino acid sequence of a protein and the way the protein folds is complex and is probably the most fundamental unsolved problem in biology. Sometimes a point mutation, such as that in sickle cell anaemia, has a fundamental and devastating effect on the protein structure. But sometimes it has no effect, or a very small effect. It is often impossible to predict which will happen, and therefore adjust the *FormGrow* output to reflect the importance of a mutation. But we can say that the general nature of the folds depends upon the general make-up of the amino acid chain. Therefore, if we summarise the amino acids content of a protein into a set of numbers, this provides a reasonable overview of the nature of the protein, and how it is related to other proteins. The 20 amino acids used in proteins can be grouped many ways according to their biochemical characteristics. These groupings are illustrated in Fig.6 (after [22]). This approach leads to a two-step process for creating *FormGrow* structures from genes. First, the DNA sequence for the gene is converted into a *histogram* denoting the relative frequencies of each amino acid type and grouping. This produces a histogram *profile* of the protein's amino acid content. Secondly, these values are used as input into a fixed *FormGrow* structure. The process is illustrated in Fig.7.

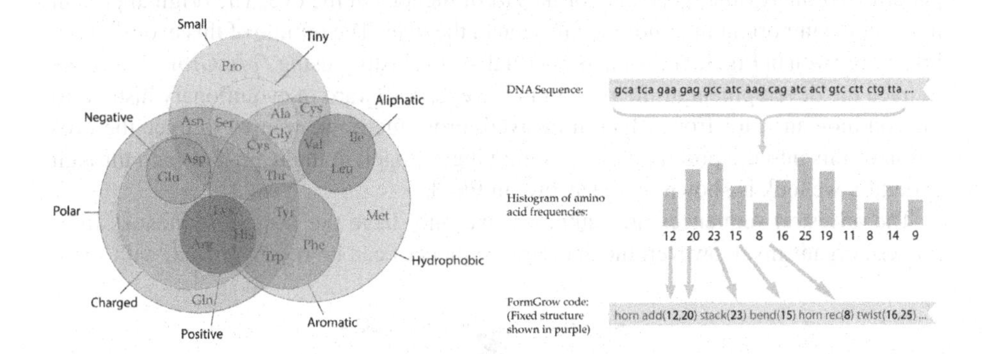

Fig. 6. Venn diagram: amino acids grouped according to their unique biochemical properties (after [22])

Fig. 7. Histogram of biochemical properties derived from the DNA sequence for the gene is used to control parameters of a fixed *FormGrow* structure

The advantages of this approach are twofold. Firstly, the most reliable significant information from the protein is captured in the histogram which captures features of the entire sequence. Any further interpretation of the sequence would rely on predictive techniques which carry some uncertainty. This technique is used in bioinformatics and is related to the "spectral decomposition" approach [23]. It is alike a Fourier transform in that the amino acid frequencies are similar to first order terms of a Fourier transform, the dipeptide frequencies are akin to second order terms, etc. Furthermore, we do not need to worry about frame-shifts of the sequence using this technique, which was one problem with the original simpler codon model. Secondly, using a fixed *FormGrow* structure means two shapes can be directly compared visually, and can be morphed in-between to create a smooth animation. Therefore, the differences in amino acid composition of two proteins can be shown in a compelling new way.

4.1 Case Study: Evolution of Related Proteins

Proteins evolve within organisms, with random mutations in DNA either causing the death of the organism, or increasing its chances of survival, or (usually) having no particular effect. A protein with one function can evolve into one with a different function through accumulation of mutations. Sometimes a whole gene is duplicated, meaning that one copy can continue to perform the original function, whilst the other can evolve to do something else. We looked at one such case as an interesting test of our histogram-based translation. Argininosuccinate lyase is a protein involved in producing arginine, one of the 20 amino acids from which proteins are made. In producing arginine, it consumes nitrogen, a product of many activities in the cell which can be toxic to an organism if it turns into ammonia. This protein exists in most organisms, from single-celled bacteria to humans and other apes. Approximately 450 million years ago, this protein was duplicated in a common ancestor of all animals, possibly some mobile multi-cellular organism. The duplicate accumulated mutations which eventually turned the protein into a structure, Delta crystallin, which, when fitted together in a specific pattern with many other proteins, forms part of the lens of the eye. The original protein has become important in removing nitrogen in the liver. The outline of this evolutionary history is given in Fig.8. We wanted to visualise this history using *FormGrow*. We chose to trace the development of the protein in the eye, backward in evolutionary history, to the common ancestor from which it evolved. From there, we wanted to trace the evolution of this into the protein used for removing nitrogen in the liver. The path through evolution we took is shown by the arrows in Fig.8.

The ancestral sequences are unknown: we only have the protein sequences from modern organisms. However, the ancestral sequences can be reconstructed, with some

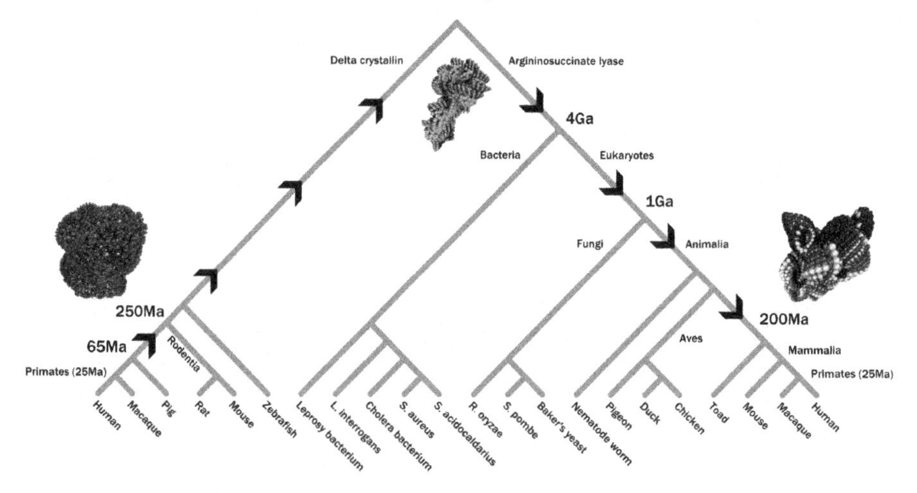

Fig. 8. Divergent evolution of two proteins. This is not a conventional evolutionary tree. It shows how *ancescon* [24] thinks the proteins are related, based upon the sequence of proteins found in modern organisms. This does not necessarily correspond to the true evolutionary tree, since the signal of evolution in protein sequence is sometimes lost in the "noise" of mutations over millions of years. The arrows show the route of the History of the Species film in visualising the devolution of one protein then the evolution of another.

uncertainty, from the sequences in modern day organisms. For example, the protein sequence of 25 million years ago can be reconstructed by combining the sequence from humans with the sequence from the crab-eating macaque. The result is the sequence as it may have existed in a common ancestor of the primates. The sequence from 65 million years ago can be reconstructed from that sequence by combining it with the sequence from the pig — we can repeat this process to produce other ancestral sequences. We used *ancescon* [24] to construct the ancestral sequences. Initial sequences came from a number of sources, including Pfam [25], and UniProt [26].

We can then morph through forms created for all the ancestral sequences from the human eye, back to the earliest ancestral sequence, and then forward to the human liver. Stills from the resulting film are shown in Fig.9. The film is available to view at [http://hos.mrg-gold.com] and covers up to 50 million years per second. The soundscape for the film was generated from the forms themselves, therefore both the audio and visual elements of the film are inspired by biology.

Fig. 9. Stills from the animation "History of the Species"

5 Discussion/Conclusion

At the core of this paper is a simple idea of feeding DNA data sequences into a rich 3D form generator called *FormGrow*, to generate organic-looking 3D growth structure, creating an equivalence of the DNA mapped into an alternative multi-dimensional space. How useful this mapped equivalence may be will become clearer as we work closer with biologists and engage in further cross-fertilization of ideas. Could this methodology have more direct and short-term scientific applications as well? While our shapes bear no resemblance to the proteins that the genes encode, they are still being driven by the same initial DNA sequences, and, therefore, it is possible that we could use our system as a visualisation tool. Conceivably, our tool could enable users to identify whether two given sequences are similar or identical. The advantage of this tool being that it is faster and easier for the human eye to compare shapes than repetitive string sequences. The primary method used by researchers in bioinformatics is to look at a multiple sequence alignment, *i.e.*, all the sequences are simply lined up one beneath the other according to an empirical scoring function. This alignment is often turned into a statistical profile or a hidden Markov model which are useful for sequence matching and structure prediction, but there is no attempt at visualisation.

Additionally, our system is deterministic. Thus, given a sequence and transform, the same shape will result every time. But, in the case of direct transformation of a DNA sequence using a table, redundancy means that some small changes may go undetected.

To rectify this we could produce more transform variants and copy the layout of the amino acid table, so that the redundancy locations are the same.

FormGrow produces shapes which are nothing like the proteins which are actually encoded by genes. Proteins inhabit a completely different "shape space" to the multi-cellular organisms which *FormGrow* is inspired by. Could shapes inspired by proteins rather than entire organisms be as rich a tool for artistry and visualisation as *FormGrow*? We intend to answer this question in future work.

References

1. Todd, S., Latham, W.: Evolutionary Art and Computers. Academic Press, London (1992)
2. Latham, W.: Form Synth. In: Computers in Art, Design and Animation, Springer, Heidelberg (1989)
3. Latham, W., Todd, S.: Computer sculpture. IBM Systems Journal 28(4), 682–688 (1989)
4. Burridge, J.M., et al.: The WINSOM solid modeller. IBM Systems Journal 28(4) (1989)
5. Todd, S., Latham, W.: Artificial life or surreal art? In: Varela, F.J., Bourgine, P. (eds.) Toward a Practice of Autonomous Systems (A Bradford Book), pp. 504–513. MIT Press, Cambridge (1992)
6. Bentley, P.J. (ed.): Evolutionary Design by Computers. Morgan Kaufmann, San Francisco (1999)
7. Sims, K.: Artificial evolution for computer graphics. Computer Graphics 25(4) (1991)
8. Sims, K.: Evolving 3D morphology and behavior. In: Proc. of Artificial Life IV (1994)
9. Prusinkiewicz, P., Lindenmayer, A.: The Algorithmics Beauty of Plants. Springer, Heidelberg (1990)
10. Leyton, M.: A process grammar for shape. A.I. Journal 34(2), 213–247 (1988)
11. Leyton, M.: A Generative Theory of Shape. LNCS, vol. 2145. Springer, Heidelberg (2001)
12. Whitelaw, M.: Metacreation — Art and Artificial Life. MIT Press, Cambridge (2004)
13. Leymarie, F.F.: Aesthetic computing and shape. In: Fishwick, P. (ed.) Aesthetic Computing. Leonardo Books, pp. 259–288. MIT Press, Cambridge (2006)
14. Lord, E.A., Wilson, C.B.: Math. Description of Shape and Form. Halsted Press (1984)
15. Lindenmayer, A.: Mathematical models for cellular interactions in development: Parts I and II. Journal of Theoretical Biology 18, 280–315 (1968)
16. Ferraro, P., et al.: Toward a quantification of self-similarity in plants. Fractals 13(2) (2005)
17. Wolfram, S.: Cellular Automata and Complexity: Collected Papers. Addison-Wesley, Reading (1994)
18. Deutsch, A., Dormann, S.: Cellular Automaton Modeling of Biological Pattern Formation. In: Modeling and Simulation in Science, Engineering and Technology, Birkhäuser (2005)
19. McCormack, J.: Aesthetic evolution of L-systems revisited. In: Raidl, G.R., et al. (eds.) EvoWorkshops 2004. LNCS, vol. 3005, pp. 477–488. Springer, Heidelberg (2004)
20. Dawkins, R.: The Blind Watchmaker. Penguin Books (1986)
21. Kumar, S., Bentley, P.J. (eds.): On Growth, Form and Computers. Elsevier, Amsterdam (2003)
22. Taylor, W.R.: The classification of amino acid conservation. J. Theor. Biology 119 (1986)
23. Shamim, M.T.A., et al.: Support vector machine-based classification of protein folds. Bioinformatics 23(24), 3320–3327 (2007)
24. Cai, W., Pei, J., Grishin, N.V.: Reconstruction of ancestral protein sequences and its applications. BMC Evolutionary Biology 4(33) (2004)
25. Finn, R.F., et al.: Pfam: clans, web tools & services. Nucleic Acids Res. 34, D247–51 (2006)
26. Wu, C.H., et al.: The universal protein resource. Nucleic Acids Res. 34, D187–91 (2006)

Towards Music Fitness Evaluation with the Hierarchical SOM

Edwin Hui Hean Law and Somnuk Phon-Amnuaisuk

Music Informatics Research Group,
Faculty of Information Technology, Multimedia University,
Jln Multimedia, 63100 Cyberjaya, Selangor Darul Ehsan, Malaysia
edwin@mmu.edu.my, somnuk.amnuaisuk@mmu.edu.my

Abstract. In any evolutionary search system, the fitness raters are most crucial in determining successful evolution. In this paper, we propose a Hierarchical Self Organizing Map based sequence predictor as a fitness evaluator for a music evolution system. The hierarchical organization of information in the HSOM allows prediction to be performed with multiple levels of contextual information. Here, we detail the design and implementation of such a HSOM system. From the experimental setup, we show that the HSOM's prediction performance exceeds that of a Markov prediction system when using randomly generated and musical phrases.

Keywords: Fitness Evaluation for Music Generation, Self Organizing Map, Hierarchical SOM.

1 Introduction

The generation of music is a very interesting and particularly challenging problem. Researchers over the years have sought a solution to this problem via the application of various techniques. In this paper, we will propose an extension to the technique introduced in [13]. Particular attention will be given to the selection and development of the appropriate fitness functions in order to evaluate pieces of music. We will propose a technique that utilizes a hierarchically organized SOM to capture the various levels of contextual information within a musical piece. This Hierarchical SOM (HSOM) can then be used to form a model for the evaluation, prediction and generation of new musical sequences.

The proceeding sections provides some background to the work that has been performed in [13] thus far. Section 3 explains the motivation for a HSOM system with a short brief on the Memory Prediction Framework proposed in [7], [6]. Section 4 provides a detail explanation of the HSOM, including its design and implementation. Next, an experiment and its results is shown in Section 5. We end with discussion and some proposed future directions for this work.

2 Background

In any evolutionary based system, the fitness function plays the most important role in determining the direction of the evolution and thus the resulting output

M. Giacobini et al. (Eds.): EvoWorkshops 2008, LNCS 4974, pp. 443–452, 2008.

produced. The selection of an appropriate fitness function is one of the most important aspects of evolutionary computation. In a domain such as music, the problem of selecting an appropriate fitness function becomes very problematic because we do not have a complete theory that allows us to objectively evaluate a particular piece of music. What constitutes a desirable piece of music? How do we quantify this desirability? How can we assign a fitness score to an individual in a population of musical phrases?

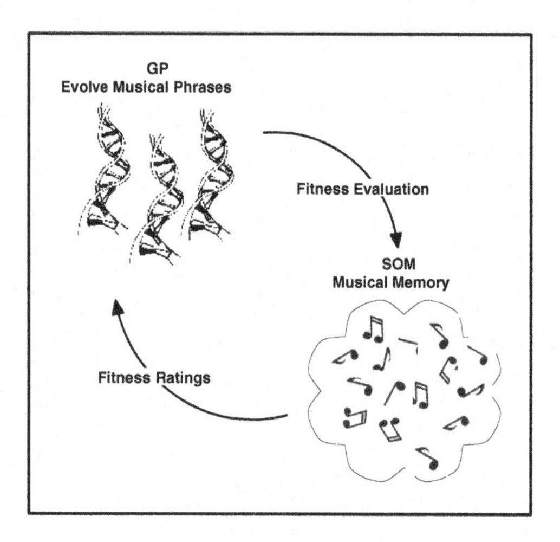

Fig. 1. GP-SOM music generation system

In [13], we proposed a solution to the fitness problem by utilizing a Self Organizing Map(SOM) [8]. In this system, a SOM was trained with musical phrases chosen from existing pieces of music. Once trained, the SOM will capture the qualities of these phrases and use them as the criteria in which to evaluate the outputs produced by a GP based music generator. Fig. 1 illustrates this system. The fitness evaluation was performed on the basis of the degree of similarity between the output produced via evolution and the clusters contained in the SOM. Higher fitness is assigned to individuals with high degrees of similarity.

After further experimentation with the GP-SOM, we believe that further improvements could be achieved with the system as the current SOM system was not able to adequately capture the large amounts of information contained within a musical phrase. A more sophisticated system was required in order to improve on the performance of the system.

3 Memory, Prediction and the Neocortex

In [7], a high level conceptual model of the human brain dubbed the *Memory Prediction Framework* is proposed. In this model, it is suggested that human intelligence is predicated upon our ability to make predictions based on the information contained within the extremely large memory store that is our brain. This memory, which is a result of our accumulated experiences with the world at large, shapes our perceptions and actions towards the world.

In this paper, we propose a technique for music sequence prediction based on the underlying principles of the Memory Prediction Framework - a *hierarchical* and *auto-associative* memory, *invariance*, and *bi-directional traversal* of the hierarchy via *feedback*. This system will be built based on the Hierarchical SOM which we will discuss in section 4. It is our belief that such a sequence prediction system will be a good way to model the perception of musical sequences in humans.

3.1 Musical Sequence Prediction

A human listener's perception and reactions towards a piece of music is shaped by the sum total of a lifetime of musical experiences. This set of musical experiences produces a set of expectations towards music. This influences what is perceived by the listener as acceptable or unacceptable music. In other words, a musical piece is acceptable if it fits within our general expectations, while adverse reactions are elicited if it does not.

A memory based sequence predictor allows us to model the musical *expectations* of a listener. Different musical phrases will produce different levels of activation within the trained memory depending on the inputs. *These varying levels of activation can then be used as fitness values in terms of how much the test input fit into the system's expectation.* Musical prediction systems have also been presented using statistical analysis [4], and neural networks [11]. The multiple viewpoint system presented in [4] in particular shares a number of similarities with our proposed method but is focused towards the statistical aspects of music.

In any system that seeks to make predictions about the future, or about things that are unknown, the quality of this prediction is intimately linked to the amount of information that is available to the system. In the most commonly applied Markov Model, the statistical transitional probabilities of the prior information is used to help determine the predicted outcome. Such systems however can only *see* the information within a predefined level of abstraction. Because of this, the quality of the predictions will invariably suffer.

In the hierarchically organized system like the HSOM, we are able to refer to a wider scope of contextual information for our predictions. The hierarchical levels allow us to look at each prediction in varying levels of abstraction. Each prediction is thus made based on information that can be seen in multiple contexts. In terms of music, we can think of this in terms of looking at the note to note, chord to chord, phrase to phrase transitions and up to the style of the music, before making a prediction on a missing note.

The hierarchical musical analysis approach has also been discussed extensively in the earliest form in terms of Schenkerian Analysis [5]. Subsequently, these methods were expanded into the Generative Theory of Tonal Music [10], the Implication and Realization Model [12] and more recently in [1]. The work presented in [2] also falls within this general the scope of such analysis.

4 The Hierarchical SOM (HSOM)

The Hierarchical SOM system proposed in this paper is an extension of the traditional SOM. Our implementation shares some similarities to previous hierarchical SOM implementations [9]. However, whilst most previous implementations were focused on the efficiencies of a hierarchical system, this system's main focus is on the facilitating predictions using a hierarchical architecture.

4.1 Basic Self Organizing Maps

The Self-Organizing Map [8] is an unsupervised neural network maps high dimensional data into a lower dimensional space. Each neuron (grid location) in the SOM is associated with a weight vector of length n, $\mathbf{w}_i = [w_{i1}, w_{i2}, ..., w_{in}]^T \in R^n$. During training, an input vector of the same length $\mathbf{x}_i = [x_1, x_2, ..., x_n]^T \in R^n$ is projected onto the 2D SOM grid.

The algorithm that leads to this self organization is based on a *winner takes all* strategy which can be summarized in 2 steps. Firstly, a winning neuron (best matching unit) is found. This bmu is defined as:

$$\|x(t) - w_{bmu}(t)\| \leq \|x(t) - w_i(t)\| \text{ for all } i.$$

After a bmu is chosen, it and its neighboring nodes are updated using the following rule:

$$w_i(t+1) = w_i(t) + \eta(t)h_{c,i}(t)(x(t) - m_i(t)),$$

where η is the learning rate and $h_{c,i}(t)$ is the neighborhood function. These two steps are performed repeatedly with a decaying learning rate and neighborhood function until the map converges, or for a pre-defined number of iterations.

4.2 HSOM Architecture

The Hierarchical SOM proposed here is a system of SOMs which is organized in a hierarchy of multiple levels, with each subsequent level encompassing a higher level of abstraction over the one below it. The inputs to the bottom level of each hierarchical SOM will consist of the original input vectors. At the higher levels of the hierarchy, inputs will be captured based on the outputs of SOM from the level immediately before it. The number of levels in the hierarchy are not fixed and can be easily changed depending on the needs of the input data.

Each input vector that is fed into the bottom level SOM will produce one set of corresponding outputs. These outputs consist of the position (x and y coordinates on a 2D SOM) of the node which was activated when each output was shown to it (the BMU for the input vector). A series of two or more of these outputs are then concatenated and sent to the SOM immediately above it. The relationship of each SOM with the SOM immediately above and below it is not strictly one to one, multiple SOMs can exists at any level and connections can be made across them from a higher level. This is shown in Fig. 2.

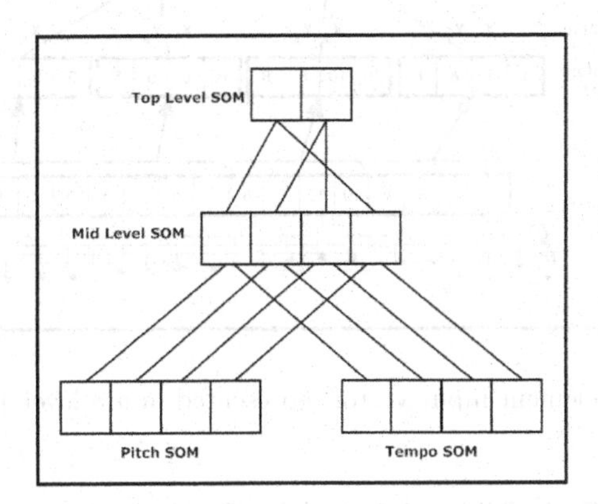

Fig. 2. Hierarchical SOM connection accross multiple lower level SOMs

This flexibility of connection across multiple lower level SOMs allows us to connect together different types of data that require significantly different representations. The different data can be trained into different SOMs and this reduces the time needed to design complex methods to put the data together.

Data Representation. The data representation scheme in the HSOM is not significantly different from that of the basic SOM. As in the SOM, the input is a fixed length vector that contains a series of weights representing the training data. However, this input vector is not fed directly into the SOM but is instead *sliced* into smaller parts for training onto the HSOM. This slicing process is entirely dependent on the chosen dimensions of the HSOM.

For example, Fig. 3 shows how a 16 element vector will be presented to a three level HSOM with four weights at each level. Each node at the bottom level captures four elements of the input, eight at the mid level (4 elements x 2 bottom nodes) and sixteen at the top level (4 elements x 2 bottom 2 mid nodes).

In terms of music representation, a time windowing method is used to obtain the pitch and duration values at each point of time (measured in terms of a semiquaver, quaver, crotchet etc.). Pitch values will be represented in terms of their midi equivalents and each element of the input vector will represent a fixed duration within the score. Unlike the representation used in [13] the pitch and tempo representations no longer need to be tied together. They are trained into separate SOMs with both maintaining a common window duration length for each vector element.

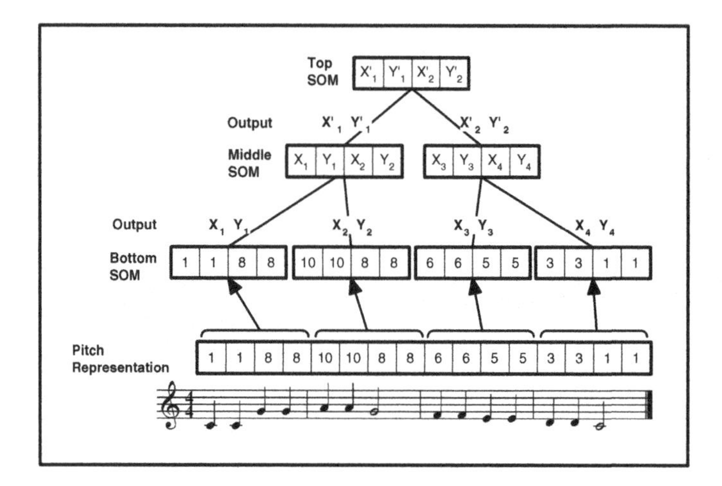

Fig. 3. A 16 element input vector,represented in a 3 level 4x4x4 HSOM

Training the HSOM. The training of the HSOM is similar to the training of a normal SOM. The training of each level of the HSOM can be done independently (by saving each level's outputs) or concurrently as each input vector is provided. Within each level, the training of the SOM is based on the basic SOM algorithm. The bottom level SOM is trained with the input vectors, while the higher levels are trained with the outputs from the level immediately before it.

The concurrent training algorithm for an N level SOM

```
Train(SOM N){
    If(N==1){ //1 = Bottom Level
        Inputs = Parse(Data_File);
        Start_Basic_SOM_Training(Inputs);
    }
    Else{
        Train(SOM N-1);
        Inputs = Get_Outputs(SOM N-1));
        Start_Basic_SOM_Training(Inputs);
        Update_Connection_Weights();
    }
}
```

In the case of multiple lower level SOMs, each SOM at the same level is trained successively(there is no data dependency within the same level) and the outputs are interleaved before being presented as inputs. With this training, the lowest level SOMs will produce clusters of the lowest level units of data. Whilst each subsequently higher level SOM will capture the interactions between multiple units of data as represented at the SOM one level below it.

4.3 Sequence Prediction with the HSOM

When making predictions of an oncoming note or missing note, these single element predictions are made at the bottom level SOMs. The predictions are made using the information available at the current level together with the information that is provided by the levels that are hierarchically above it.

A binary input vector with 1 missing element is given as

$$10x1010110110101$$

with x being the missing element. This can used on a 4x4x4 SOM as shown in Fig. 3. A prediction on the missing element x will be performed on 3 separate levels based on the 3 level hierarchy. At the first level, the prediction is done on a local basis where only 4 elements $10x1$ are used to predict x.

This prediction is done by obtaining a partial best matching unit from the bottom level SOM. The partial bmu is found using the bmu formula shown in Section. 4.1, the only difference being that only the known elements are considered when measuring the bmu. x', w' denote the input and weight vectors where the elements at the missing positions have been omitted.

$$\|x'(t) - w'_{bmu}(t)\| \leq \|x'(t) - w'_i(t)\| \text{ for all } i.$$

If multiple possibilities are available, this is resolved at the Mid Level by looking at 8 elements $10x10101$, this Mid level prediction provides a greater context to the prediction being performed at the lowest level. If more ambiguity is observed, we can then proceed to propagate this prediction up to the Top Level where the full 16 element vector will be referred to during prediction. The same partial bmu technique is applied to finding the matches at these higher levels.

From Sequence Prediction to Fitness Evaluation. The sequence prediction algorithm can be adapted to be used as a fitness evaluator without making any major modifications to the system architecture. In prediction, the aim is to be able to produce correct estimations of the element that will fit into a particular position in the sequence. This ability to fill in the gaps shows that the sequence predictor is able to *evaluate* what an acceptable sequence should *look like* given the inputs that it has been trained with.

With this information at hand, we can then make adjustments to the sequence predictor so that it produces a score that evaluates if a particular sequence shown to it fits within its expectations. This can be achieved by making measurements at each level of the HSOM. A score can be assigned to the sequence where the expectations of the HSOM are measured against the sequence presented. The scores at each level can then be totaled to produce the fitness for that sequence.

5 Experiments on Sequence Prediction

In our experimental setup, we will be using a 3 level hierarchical SOM. In this experiment, two input data sets with 20 vectors of length 16 is used. The first set is generated randomly and each element of this vector will contain an integer value ranging from 0 to 2. The second set consists of the first 20 bars from Bach's Fugue in G Minor BWV 578 with values ranging from 0-23. The aim of the experiment is to show that by using a HSOM model we are able to predict the full vector sequence with reasonable accuracy when incomplete inputs are fed into the system. The HSOM model is compared with a first order Markov model based system.

Fig. 4. The first 4 bars of J.S. Bach, Fugue in G Minor, BWV 578

This experiment is divided into 2 parts. In the first part, the HSOM is trained with the input data for 1000 iterations, this produces an organized mapping of the data. Similarly, the table containing the transition probabilities of the input data is produced by the Markov system. In the second part, the actual prediction experiment is performed. Here, a vector is selected amongst the 20 inputs and n number of elements will be randomly removed from this vector. Our predictor systems are then required to predict these missing elements based on the elements that it is provided in the vector.

5.1 Experimental Results

Fig. 5 shows the results of the prediction experiment. The accuracy of the prediction is measured in terms of the full sequence. In order for the prediction to score a hit, it must provide an accurate prediction of 100%. The prediction is performed with 0,1,2 4 and 8 missing elements and each run was repeated for a total of 1000 times.

From the results, it is encouraging to see that the HSOM is able to achieve significantly better results than first order Markov prediction. This improvement is more pronounced if we look at sequences with larger numbers of missing elements. The ability to reference the different levels of contextual information allows the HSOM to make more accurate predictions especially when there are large numbers of missing or corrupted data within the lower level structures.

When comparing the performance with random data against the data extracted from the Fugue, the HSOM models shows better performance whilst the almost no correct prediction was able to be made using the Markov model. It is clear that the first order markov model is not able to handle the increased complexity of the fugue data. The HSOM on the other hand showed increased performance because of 2 factors.

Firstly, the HSOM stores and organizes *unique* instances of information (measured in terms of the the structure at each level of the HSOM). New information which is similar to data already stored is organized within the same grouping.

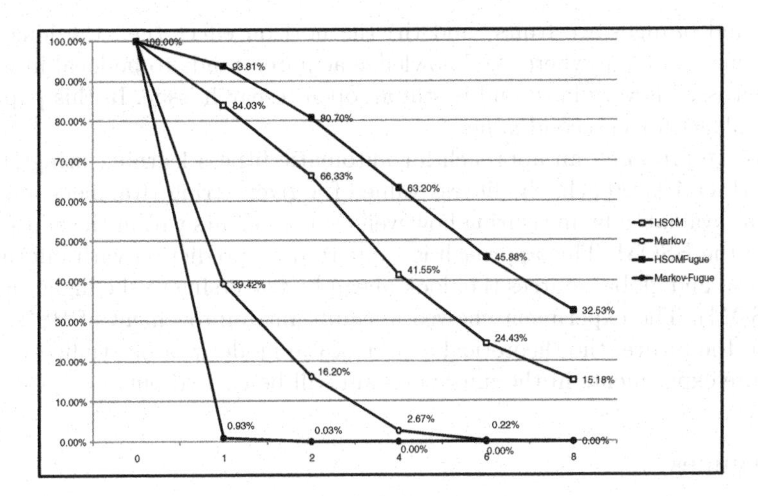

Fig. 5. The prediction results for HSOM and Markov Prediction with 0,1,2,4,6 and 8 missing elements

Thus, the performance of the HSOM should only be affected by the increase in the number of unique data found in the data set. Secondly, real world data like music and natural languages are repetitive in nature. This repetition occurs at multiple levels and the HSOM exploits that tendency. Because frequencies of unique structures decrease in repetitive data, the HSOM's performance increases.

We believe that with further fine-tuning of the prediction algorithm, better results will be achieved with the HSOM. With that, it will be possible to adapt the HSOM to be used as a fitness evaluator for individuals evolved from a evolutionary based melody generator where instead of producing predictions, the HSOM can be adapted to produce evaluations by measuring how far these individuals fall from the expectations of the HSOM at the different levels.

6 Discussions and Future Work

Applying machine learning techniques in tasks of knowledge elicitation has advantages in terms of flexibility but at the same time suffers from the lack of specificity in the learned knowledge. Alternatively, in the case of traditional knowledge engineering approaches, a successful implementation of the system requires the knowledge engineer to be a domain expert. This approach is expensive and usually not flexible enough to handle changes in the domain knowledge.

In our previous work, we employed SOM to learn music input and used the SOM map as a fitness evaluation function for new input generated by the GP. The two major issues we faced were (i) lack of semantic continuity in the SOM map (e.g., the word *apples* may be interpreted as closer to *boomer* (i.e., in terms of distance according to alphabetical orders) or closer to *oranges* (i.e., both apples and oranges are fruits) and (ii) the horizon effect (i.e., the long range dependency problem)where the knowledge acquired tends to be local in nature. The first issue is very hard and is still an open research issue. In this paper, we have dealt with the second issue.

Here, we proposed an approach for automatic fitness learning using HSOM. In our HSOM approach, the fitness value of a given string (fragment of music) might be evaluated by measuring how well that string fits within the expectation store in the HSOM. This approach is supported by prediction with information from local and global sources (i.e. look ahead by traversing to the higher layer of the HSOM). The experiment showed a significant improvement of HSOM over MM. In the future, the theoretical aspect of the model will be studied in detail and more experiments in the music domain will be carried out.

References

1. Cambouropoulos, E.: Towards a General Computational Theory of Musical Structure. PhD Thesis, University of Edinburgh (1998)
2. Cope, D.: Virtual Music: Computer Synthesis on Musical Style. MIT Press, Cambridge (2004)
3. Conkin, D.: Representation and Discovery of Vertical Patterns in Music. In: Anagnostopoulou, C., Ferrand, M., Smaill, A. (eds.) ICMAI 2002. LNCS (LNAI), vol. 2445, pp. 33–42. Springer, Heidelberg (2002)
4. Conklin, D., Witten, I.: Multiple viewpoint systems for music prediction. Journal of New Music Research 24, 51–73 (1995)
5. Forte, A., Gilbert, S.E.: Introduction to Schenkerian Analysis. W. W. Norton, New York (1982)
6. George, D., Hawkins, J.: A bayesian model of invariant pattern recognition in the visual cortex. In: Proceedings of the International Conference on Neural Networks, Montreal, Canada (2006)
7. Hawkins, J., Blakeslee, S.: On intelligence. Henry Holt, New York (2004)
8. Kohonen, T.: Self-organising Maps, 2nd edn. Springer, Heidelberg (1997)
9. Lampinen, J., Oja, E.: Clustering Properties of Hierarchical Self-Organizing Maps. Journal of Mathematical Imaging and Vision 2, 261–272 (1992)
10. Lerdahl, F., Jackendoff, R.: A Generative Theory of Tonal Music. MIT Press, Cambridge (1983)
11. Mozer, M.C.: Neural Network Music Composition by Prediction: Exploring the Benefits of Psychoacoustic Contraints and Multi-scale Processing. In: Griffith, N., Todd, P.M. (eds.) Musical Networks, pp. 227–260. MIT Press, Cambridge (1999)
12. Narmour, E.: The Analysis and Cognition of Basic Melodic Structures: The Implication-Realization Model. University of Chicago Press, Chicago (1990)
13. Phon-Amnuaisuk, S., Law, E.H.H., Ho, C.K.: Evolving Music Generation with SOM-Fitness Genetic Programming. In: Giacobini, M. (ed.) EvoWorkshops 2007. LNCS, vol. 4448, pp. 557–566. Springer, Heidelberg (2007)

Evolutionary Pointillist Modules: Evolving Assemblages of 3D Objects

Penousal Machado and Fernando Graca

CISUC, Department of Informatics Engineering, University of Coimbra
Polo II of the University of Coimbra, 3030 Coimbra, Portugal
machado@dei.uc.pt, fejg@student.dei.uc.pt
http://eden.dei.uc.pt/~machado

Abstract. A novel evolutionary system for the creation of assemblages of three dimensional (3D) objects is presented. The proposed approach allows the evolution of the type, size, rotation and position of 3D objects that are placed on a virtual canvas, constructing a non-photorealistic transformation of a source image. The approach is thoroughly described, giving particular emphasis to genotype–phenotype mapping, and to the alternative object placement strategies. The experimental results presented highlight the differences between placement strategies, and show the potential of the approach for the production of large-scale artworks.

Keywords: Evolutionary Art, Evolutionary Image Filters, Non-Photorealistic Rendering, Assemblage.

1 Introduction

The main goal of the research presented in this paper is the creation of large-scale artworks through the assemblage of 3D virtual objects. The main artistic sources of inspiration for this work are: pointillism, mixed media assemblage of objects, and ornamentation techniques (e.g. similar to the ones found in Gustav Klimt works). From a scientific point of view, areas such as evolutionary non-photorealistic rendering and artistic filter evolution, are of particular relevance.

Our approach can be seen as the evolution of a filter that transforms an input source image. More exactly, the evolved individuals take as input (through a terminal node) an image, and produce as output the type, scale, rotation and placement of the objects, which will be placed on the virtual canvas. The color of each object is determined by the color of the corresponding pixel of the source image. In this way, an assemblage of 3D objects, which constitutes a non-photorealistic portrayal of the source image is obtained. Finally, the 3D assemblage is rendered using a raytracer.

We begin with a short survey of related work. Next, in the third section, we make an overview of the different modules of the system. In the fourth section, we describe the evolutionary module, giving particular emphasis to genotype–phenotype mapping, and to the description of the object placement strategies. The experimental results are presented and analyzed in the fifth section. Finally, we draw some conclusions and discuss aspects to be addressed in future work.

M. Giacobini et al. (Eds.): EvoWorkshops 2008, LNCS 4974, pp. 453–462, 2008.

2 Related Work

The use of evolutionary algorithms to create image filters and non-photorealistic renderings of source images has been explored by several researchers. Focusing on the works where there was an aesthetic goal, we can mention the research of: Neufeld and Ross [1,2], where Genetic Programming (GP) [3], multi-objective optimization techniques, and an empirical model of aesthetics are used to automatically evolve image filters; Lewis [4], which evolved live-video processing filters through interactive evolution; Machado et al. [5], where GP is used to evolve image coloring filters from a set of examples; Yip [6], which employs Genetic Algorithms (GAs) to evolve filters that produce images that match certain features of a target image; Collomosse [7,8], which uses image salience metrics to determine the level detail for portions of the image, and GAs to search for painterly renderings that match the desired salience maps. Several other examples exist, however a thorough survey is beyond the scope of this article.

3 Overview of the System

Figure 1 presents the architecture of the system, which is composed of two main components: an *Evolutionary* module and a *Previewing and Rendering* module.

The evolutionary module is an expression–based GP [9] interactive breeding tool. It comprises a *Function Visualizer* that depicts a grayscale visualization of the individuals' expression trees. As is usually the case in expression based GP, the grayscale value of a pixel at the (x,y) coordinates (in our case, $x, y \in [-1, 1]$) is determined by the output value of the individuals' expression trees for (x, y). Each individual is an assemblage of 3D objects. Therefore, usually, this visualization mode does not provide enough information to allow educated choices by the user. As such, the system also provides a 2D and 3D previewer.

The 2D previewer runs on the master computer. It evaluates the genotypes and places objects accordingly. However, as the name indicates, it doesn't take into consideration the 3D nature of the objects, lighting effects, shadows, etc. The 3D previewer resorts to a Condor–based [10] render farm. The master creates and submits several Condor jobs for each individual of the population. Each job is responsible for: converting the genotype in a Persistence of Vision (POV) 3D scene file[1]; rendering a slice of the resulting 3D scene using POV-Ray; transferring the rendered image slice to the master. The master gathers and merges the rendered image slices, displaying the images as they become available. By changing the settings of the POV-Ray initialization file, the user can adjust the quality and size of the renderings, thus also adjusting the speed.

As mentioned above, we use an interactive breeding approach, i.e., instead of assigning fitness, the user selects two parents, which generate offsprings through crossover and mutation. We also provide a *chromosome replication* operator, which allows the user to select a specific chromosome and transfer mutated versions of it to all the individuals of the population.

[1] `http://www.povray.org/`

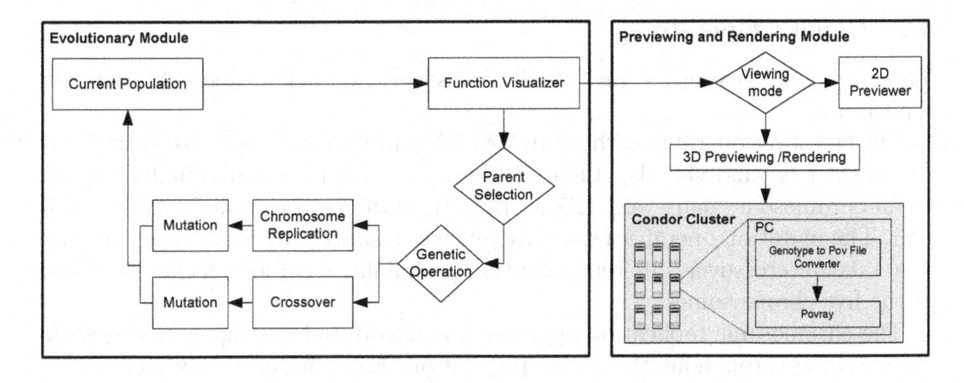

Fig. 1. The main modules of the system

4 Evolutionary Module

In this section, we describe the evolutionary module, focusing on aspects such as: representation, genetic operators and genotype–phenotype mapping.

4.1 Representation

The genotype of each individual has five chromosomes: <type, rotation, size, x-position, y-position>. Each chromosome is an expression tree, encoding a particular aspect of the 3D assemblage of objects, as follows:

<type> – The output value of the *type* expression tree determines what object, from a pool of available ones, will be placed;
<rotation> – The rotation that will be applied to the object;
<size> – The scaling applied to the object.
<x-position> and <y-position> – The x and y coordinates where the object will be placed;

The output of the trees is calculated for each of the pixels of the source image. In the experiments described in this paper, the function set is:

$$\{sin, cos, max, min, abs, +, -, \times, \%, diff\},$$

where *sin* and *cos* are the usual trigonometric operations; *max* and *min* take two arguments returning, respectively, the maximum and minimum value; *abs* returns the absolute value; $\{+, -, \times\}$ are the standard arithmetic operations; % the protected division operator [3]; $diff$, a function that takes two arguments $(\Delta x, \Delta y)$ and returns the difference between the value of the (x, y) pixel of the source image and the pixel at $(x + \Delta x, y + \Delta y)$. The terminal set used is:

$$\{x, y, image, random_{constants}\},$$

where x and y are variables; *image* is a zero–arity operator that returns the value of the x, y pixel of the source image; $random_{constants}$ are floating point random values between -1 and 1.

4.2 Genetic Operators

Three genetic operators are used: crossover, mutation and chromosome replication.

The crossover operator is the standard GP sub-tree exchange crossover [3]. If we consider two individuals, A and B, this operator will be individually applied to all chromosome pairs (e.g. (A_{type}, B_{type})), with a given probability for each pair. The mutation operator randomly replaces a subtree by a randomly created one. Like for crossover, different mutation probabilities can be specified for each of the five chromosomes.

The chromosome replication operator was introduced to propagate a specific chromosome throughout the entire population. E.g., the user may feel particularly pleased with the rotations applied to the objects in one individual, and wish to use the same rotation expression in all individuals. Alternatively, the user may wish to test small variations of a specific chromosome without changing the remaining ones. To address these needs, the replication operation copies mutated versions of a chromosome, selected by the user, to all individuals of the population.

4.3 Genotype–Phenotype Mapping

In this section, we describe the genotype–phenotype mapping procedure. To illustrate our explanation we resort to: the genotype presented in Fig. 2, and the source image presented in Fig. 3(a). For the time being, we will assume that the objects are placed following a regular 32×32 grid, and that three types of objects are available: squares, circles and triangles.

The first chromosome, type, determines which type of object is placed. In this case, values in $]0, 0.33]$ correspond to cubes, in $[0.33, 0.66]$ to spheres, and in $[0.66, 1[$ to triangular shapes. The application of this chromosome, alone, would produce the 3D scene depicted in Fig. 4(a). Likewise, the rotation determines the rotation that will be applied to each object, and size determines the scaling that will be applied to the object. Figures 4(b) and (c) depict the results of independently applying these chromosomes, using, respectively, triangular shapes and

Type	Rotation	Size	X-position	Y-position
max(1.79,+(image,x))	min(x,-(1.8, sin(max(y,1.9))))	min(y,-(min (x,x),sin(max (x,1.9))))	abs(x)	-(sin(y),x)

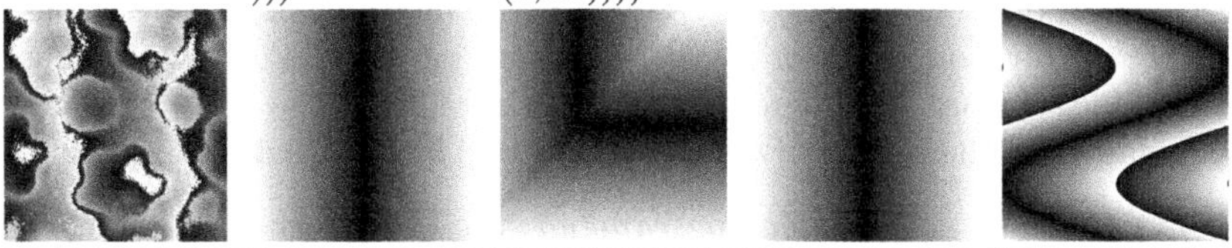

Fig. 2. Chromosomes of a sample genotype and the visualization of the corresponding functions over $[-1, 1]$, considering the source image presented in Fig. 3(a)

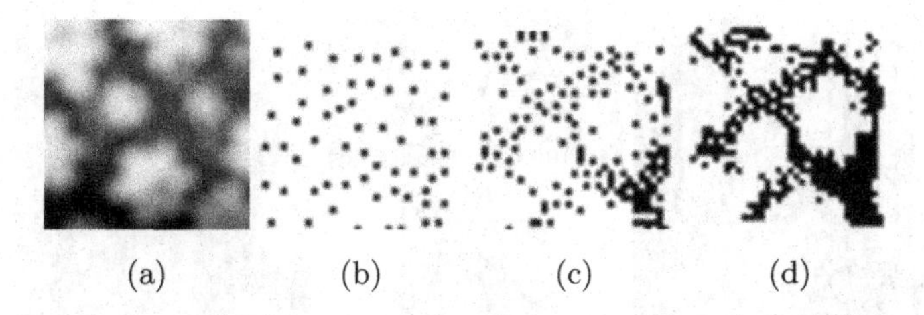

Fig. 3. (a) Source Image; (b) to (d) Dither masks for stages 1 to 3

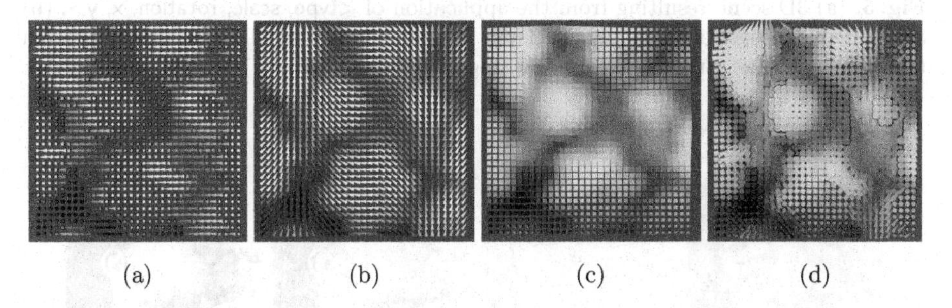

Fig. 4. Images resulting from the application of type (a), rotation (b), size (c), and <type, rotation, size> (d) to the source image of Figure 3(a), assuming a regular grid placement of the objects

cubes for easier viewing. The joint application of type, rotation and size would produce the 3D scene presented in Fig. 4(d).

Object Placement. So far we considered that the objects are placed on a regular grid. This type of placement has characteristics that we wish to avoid, namely: i) the regularity of the grid can become a visual distraction; ii) it only allows a homogeneous distribution of the objects, making it impossible to ignore regions of the image, and, to clutter objects on certain regions. To overcome this limitation, we introduced the x- and y-position chromosomes, which determine the coordinates where the objects are placed (see Fig. 5(a)).

The number of objects placed is also relevant. To address this issue, we resort to masks. A modified version of a space-filling curve dither algorithm [11,12] is applied to the source image. By establishing different parameter settings, three dither masks are created (see Fig. 3). The phenotype is rendered in three stages, each using a different dither mask. In each stage, the positions of the objects are calculated using the x- and y-position chromosomes, but an objects is only placed if the mask allows it. In Fig. 5(b), we present the 3D scene corresponding to each rendering stage, and in Fig. 5(c), the one resulting from the combination of the three stages. The color of each object is determined by the color of the area of the source image where the object is placed, which tends to avoid excessive distortions of the original image.

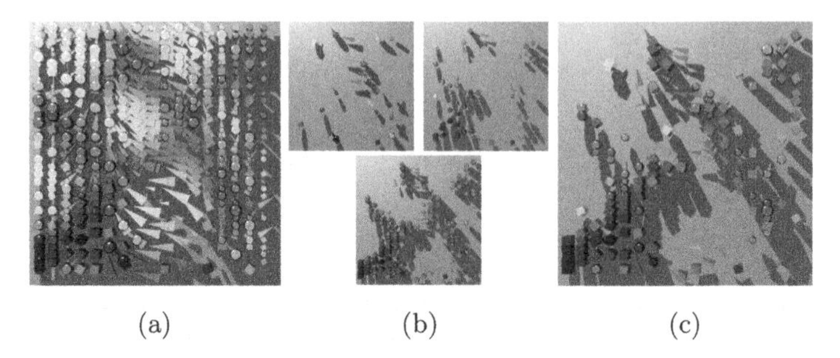

Fig. 5. (a) 3D scene resulting from the application of <type, scale, rotation, x, y>; (b) 3D scene resulting from the stage 1, 2 and 3 dither masks; (c) 3D scene resulting from the combination of the three rendering stages

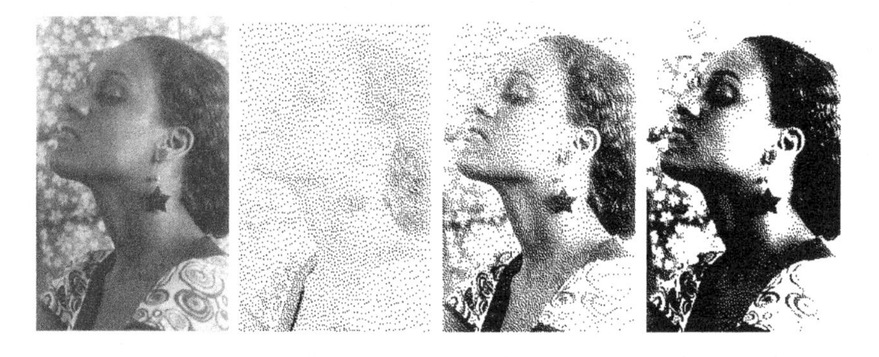

Fig. 6. Source image and corresponding dither masks

5 Experimental Results

The analysis of the experimental results attained by evolutionary art systems, specially user driven ones, entails a high degree of subjectivity. In our case, there is an additional difficulty — the approach is thought for large-scale formats. Therefore, is close to impossible to adequately convey the real look of the evolved images in the space and format available for their presentation. Considering these difficulties, we focus on the comparison between different object placement strategies, since an analysis of the effects of type, rotation and size would require a higher level of detail. Due to space restrictions and to the visual nature of the results, we chose to focus on a single evolutionary run.

5.1 Experimental Setup

We used the following experimental settings: Function–set = $\{sin, cos, max, min, abs, +, -, \times, \%, diff\}$; Terminal–set = $\{x, y, image, random_{constants}\}$ (see Sect. 4.1); Population size = 20; Number of Generations = 40; Crossover probability = 0.6; random–subtree mutation probability = 0.2; node–change

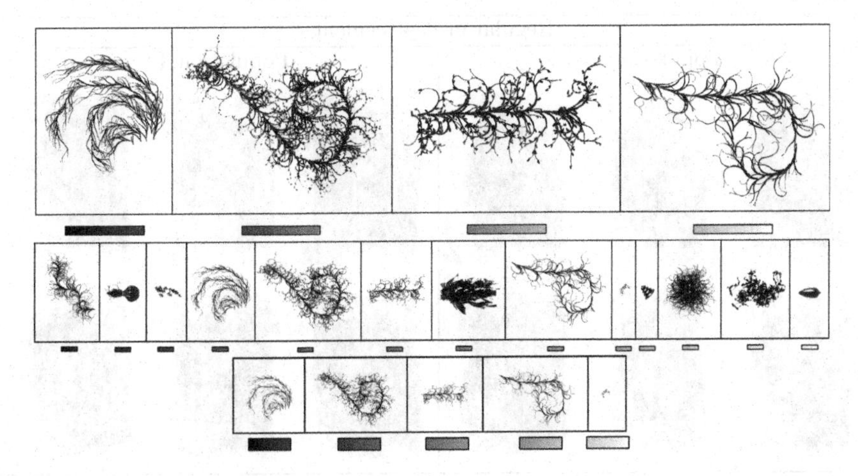

Fig. 7. Pool of objects used in stages 1–3. When no dither masks are used, the object pool is equal to the one of stage 2.

mutation probability = 0.02 per node; Population initialization method = Ramped half–and–half.

The experiments were performed on an Intel Core2Duo, 2.8GHz, Windows master computer. During the course of evolution, an heterogeneous Condor cluster – with 40-70 available machines – was used for the 3D previewing of the populations. To produce large-scale renderings, we used a dedicated Condor cluster with networked file system, composed of 24 Intel Core2Duo, 2.8GHz, running Ubuntu. For previewing, we used a resolution of 800×600 pixels, and the images were rendered without anti-aliasing; the resolution of the large-scale renderings ranged from 3200×2400 to 16000×12000. Typically, previewing took 15 to 30 minutes, and large-scale rendering took 4 to 80 hours (depending on the resolution and number of objects).

In Fig. 6, we present the source image used in the course of these experiments, and the corresponding dither masks for three stage rendering. Figure 7 depicts the pool of objects used in each of the rendering stages, and the associated grayscale gradients. When we follow a object placement strategy that doesn't resort to masks (regular grid or non–masked x- and y-position chromosomes), the object pool is equal to the one of stage 2.

5.2 Results and Analysis

Figure 8, displays the 3D preview of a subset (individuals 1 to 6) of the 1*st* and the 40*th* populations, rendered using: regular grid placement; evolved x- and y-position chromosomes; three stage rendering with evolved x- and y-position chromosomes and dither masks.

In general, all object placement strategies produced interesting images. During the evolutionary run, the selections of the user were based on the dithered previews. As such, a comparison of object placement approaches, based on the

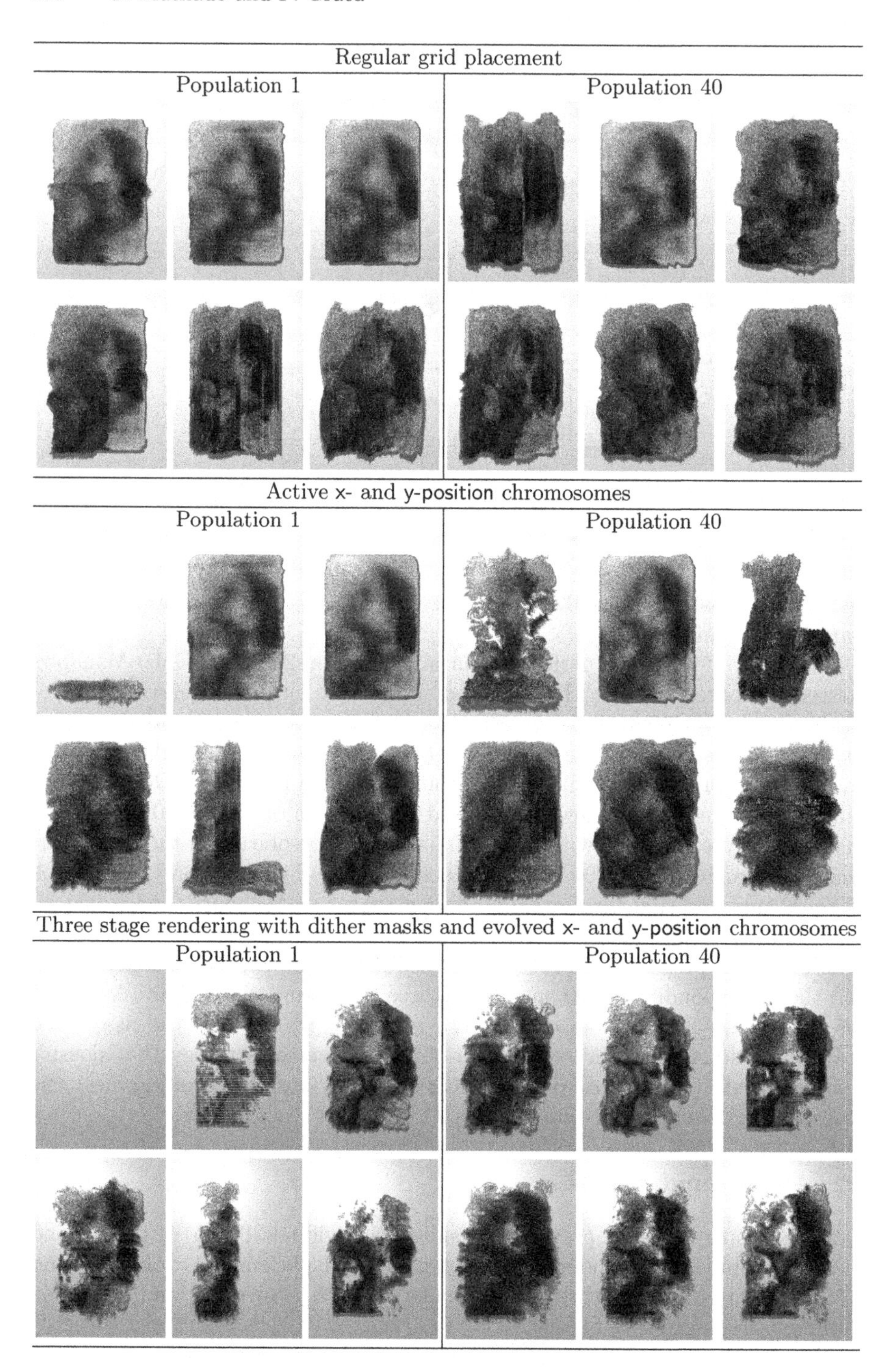

Fig. 8. The first 6 individuals of populations 1 and 40 rendered using different object placement strategies

individuals of 40*th* population, would be biased. Therefore, the differences between placement strategies are better perceived by observing the images of the 1*st* population.

The limitations of regular grid placement are not visible in this small-scale renderings. The artificial vertical and horizontal artifacts induced by regular placement only become a distraction at larger resolutions. The variations in size and rotation of the objects also help overcoming this limitation, producing, for instance, irregular image boundaries.

As it can be observed, evolving the coordinates of the objects can lead to peculiar images, where significant portions of the image are missing. This expected outcome, is particularly frequent in the beginning of the evolutionary run.

The images produced by three stage rendering with dither masks were the ones that better matched our expectations. The introduction of these masks allowed the "abstraction" of regions of the image, providing the heterogenous level of detail that we wanted.

The comparison of the dithered renderings of the 1*st* and 40*th* population shows that there is less variability in the 40*th* population, indicating some degree of convergence. Unfortunately, the size of the images presented doesn't allow the reader to observe the subtler differences among images, in what concerns object type, rotation, size and placement. Nevertheless, it is safe to say that, according to the opinion of the users guiding the experiment, the images of the 40*th* population are significantly more expressive than the ones from the 1*st*. In other words, the evolutionary algorithm, guided by the user, was able to find regions of the search space that were considered more promising.

Larger versions of some of the individuals, the results of applying a given individual to different source images, and several other images can be found online at: `http://evolving-assemblages.dei.uc.pt`

6 Conclusions and Future Work

We presented a novel evolutionary approach for the generation of assemblages of 3D objects, and compared the results attained with three different object placement strategies. The experimental results show the potential of the approach, indicating that the proposed three stage rendering, with evolved x- and y-position chromosomes and dither masks, achieves better results.

One of the main limitations of our approach is the computational effort required to preview and render the individuals. To overcome it, we used a Condor-based cluster to distribute the rendering tasks. Nevertheless, this process is still time consuming, which becomes a limitation (specially for previewing, since final renderings can be made offline). In the experiments presented, we used fairly complex objects. Using simplified versions of the same objects would greatly reduce previewing time. This is one one of the aspects that will be addressed in the future.

Although the object placement strategies we explored were able to produce interesting results, evolving the masks, would allow greater flexibility. Additionally, image salience analysis can play an important role. The identification of salient detail could be used to guide object placement, in order to promote the placement of objects in areas with salient detail.

Finally, a larger set of experiments – with different types of source images and object pools – is also necessary to better assess the strengths and weakness of the approach.

Acknowledgments. We would like to express our gratitude towards Jennifer Santos, which posed for the photographs employed in this paper. We also acknowledge the contribution of Paulo Marques provided valuable support in the setup of the clusters.

References

1. Ross, B.J., Ralph, W., Hai, Z.: Evolutionary image synthesis using a model of aesthetics. In: Yen, G.G., Lucas, S.M., Fogel, G., Kendall, G., Salomon, R., Zhang, B.T., Coello, C.A.C., Runarsson, T.P. (eds.) Proceedings of the 2006 IEEE Congress on Evolutionary Computation, Vancouver, BC, Canada, pp. 1087–1094. IEEE Press, Los Alamitos (2006)
2. Neufeld, C., Ross, B., Ralph, W.: The evolution of artistic filters. In: Romero, J., Machado, P. (eds.) The Art of Artificial Evolution, Springer, Heidelberg (2007) (in Press)
3. Koza, J.R.: Genetic Programming: On the Programming of Computers by Natural Selection. MIT Press, Cambridge (1992)
4. Lewis, M.: Aesthetic video filter evolution in an interactive real-time framework. In: Raidl, G.R., Cagnoni, S., Branke, J., Corne, D.W., Drechsler, R., Jin, Y., Johnson, C.G., Machado, P., Marchiori, E., Rothlauf, F., Smith, G.D., Squillero, G. (eds.) EvoWorkshops 2004. LNCS, vol. 3005, pp. 409–418. Springer, Heidelberg (2004)
5. Machado, P., Dias, A., Cardoso, A.: Learning to colour greyscale images. The Interdisciplinary Journal of Artificial Intelligence and the Simulation of Behaviour – AISB Journal 1, 209–219 (2002)
6. Yip, C.: Evolving Image Filters. Master's thesis, Imperial College of Science, Technology, and Medicine (2004)
7. Collomosse, J.P., Hall, P.M.: Genetic paint: A search for salient paintings. In: Applications of Evolutionary Computing, EvoWorkshops 2005, Lausanne, Switzerland, pp. 437–447 (2005)
8. Collomosse, J.P.: Supervised genetic search for parameter selection in painterly rendering. In: Applications of Evolutionary Computing, EvoWorkshops 2006, Budapest, Hungary, pp. 599–610 (2006)
9. Sims, K.: Artificial evolution for computer graphics. ACM Computer Graphics 25, 319–328 (1991)
10. Tannenbaum, T., Wright, D., Miller, K., Livny, M.: Condor – a distributed job scheduler. In: Sterling, T. (ed.) Beowulf Cluster Computing with Linux, MIT Press, Cambridge (2001)
11. Velho, L., de Miranda Gomes, J.: Digital halftoning with space filling curves. SIGGRAPH Comput. Graph. 25, 81–90 (1991)
12. Shiraishi, M., Yamaguchi, Y.: An algorithm for automatic painterly rendering based on local source image approximation. In: NPAR 2000: Proceedings of the 1st international symposium on Non-photorealistic animation and rendering, pp. 53–58. ACM, New York (2000)

An Artificial-Chemistry Approach to Generating Polyphonic Musical Phrases

Kazuto Tominaga and Masafumi Setomoto

School of Computer Science, Tokyo University of Technology
1404-1 Katakura, Hachioji, Tokyo 192-0982 Japan
tomi@acm.org

Abstract. Many techniques in the evolutionary-computing and artificial-life fields have been applied to generating musical phrases. Recently, artificial chemistries, which are virtual models of chemical systems, have come into use for the purpose. In this paper, we attempt to produce polyphonic phrases using an artificial chemistry. The molecules and chemical equations are so designed as to implement a basic set of rules in counterpoint, a technique to compose polyphonic music. We perform a preliminary experiment using a simulator for the artificial chemistry; in the experiment, some good polyphonic phrases are generated, which confirms that the present approach is promising.

Keywords: artificial chemistry, music composition, polyphony, counterpoint, trial and error.

1 Introduction

Generating music by computers is always of great interest. Many techniques in various areas have been applied to achieving the goal, including those of evolutionary computation. Specifically, genetic algorithms (GAs) are extensively used and have achieved great successes in various applications [1]. Modelling techniques in the field of artificial life, such as cellular automata and swarm, have also been used to generate musical phrases [2,3]. Recently, *artificial chemistries* have come into use for the purpose [4]. An artificial chemistry is a model of a (virtual or real) system comprising a large number of molecules that interact with each other according to a given set of rules similar to chemical equations [5]. We have our artificial chemistry [6], and studied on generating homophonic phrases with it [7]. In this paper, we address a more challenging problem — generating *polyphonic* musical phrases — in order to investigate the applicable range of artificial chemistries in music generation.

2 Artificial Chemistry

This section briefly explains our artificial chemistry [6][1]. It has *v-atoms* and *v-molecules*, corresponding to natural atoms and molecules, respectively.

[1] This paper uses different terminology and notation from the ones in [6].

M. Giacobini et al. (Eds.): EvoWorkshops 2008, LNCS 4974, pp. 463–472, 2008.

V-molecules react with each other according to *recombination rules*, which correspond to chemical equations but are expressed in terms of patterns. The artificial chemistry employs the "well-stirred tank reactor" model: the reaction pool has no spatial structure.

Atoms and Molecules. A v-atom is denoted by a capital letter followed by a sequence of lower-case letters and/or numbers. For example, A, B, C2, Ix and Term can be v-atoms. A v-molecule is a sequence of v-atoms, or a stack of such sequences. Example v-molecules are shown in Fig. 1, with their string representations. In the notation, lines are delimited by slashes (/). Each line is preceded by a number called *displacement* (the number before the hash (#)), which is the offset of the line relative to the top line.

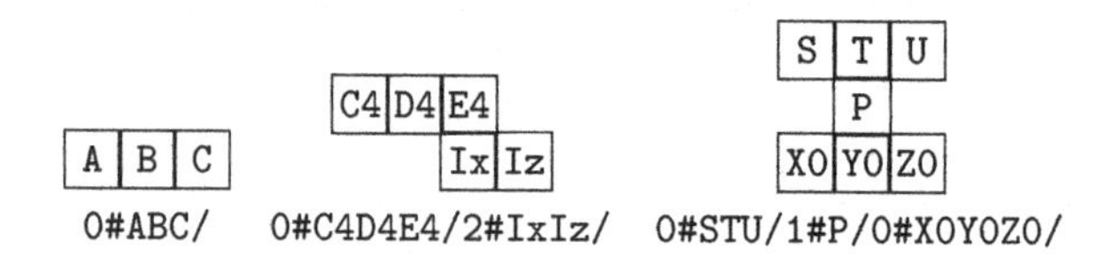

Fig. 1. Example v-molecules

Patterns and Recombination Rules. A *pattern* matches (or does not match) a v-molecule. A pattern can include *wildcards*. There are two types of wildcards: an *atomic wildcard*, denoted by <*n*> (*n* is a number), which matches any v-atom; and a *sequence wildcard*, denoted by <**n*> or <*n**>, which matches a sequence of zero or more v-atoms of any kinds (the star (*) indicates the direction that the wildcard can extend). *Recombination rules* are defined in terms of patterns. For example, the following recombination rule

0#<*1>AB/1#CD/ + 0#AB<2*> → 0#<*1>ABAB<2*>/1#CD/ (1)

recombines v-molecules, 0#ABAB/3#CD/ and 0#ABABAB/, to produce a v-molecule, 0#ABABABABAB/3#CD/. In this case, the wildcard <*1> matches the sequence of v-atoms AB, and <2*> matches ABAB.

Dynamics. A *system* is defined in the artificial chemistry, which consists of the set of v-atoms, the set of recombination rules, and the multiset of initial v-molecules (or *initial pool*)[2]. The system operates as follows.

1. The working multiset (called *pool*) is initialised to be the initial pool.
2. Apply a recombination rule to a collection of v-molecules in the pool.
3. Go to Step 2.

Recombination rules and v-molecules are selected stochastically on Step 2 when the system is run on a simulator.

[2] The artificial chemistry has other components called *sources* and *drains*, but we omit them since they are not used in this study.

3 Composition of Polyphonic Music

Polyphonic music consists of two or more individual melodies (called *parts*) played simultaneously; it is typified by works in the Baroque period, such as ones by J. S. Bach. Homophonic music, on the other hand, has one main melody, accompanied by chords; modern popular music is mostly homophonic.

Automatic generation of polyphonic phrases is expected to be more difficult than generating homophonic ones. In homophonic music, the melody is principal and the chords support it, so the chords can be added to the melody after the melody is complete. But in polyphonic music, each part should be a good melody, and at the same time, they should sound good when they are played simultaneously.

This study attempt to generate two-part polyphonic phrases by implementing rules in counterpoint in the artificial chemistry. Counterpoint is a technique for composing polyphonic music. Roughly speaking, counterpoint is a set of rules that decide what phrase is good and what is not. For example, a basic rule prohibits a progression of part by any augmented interval. We choose two sets of basic rules for two-part counterpoint for this study, which are shown below.

Rules for the Progression of One Part. Each part of the two parts should obey the following rules. Let *CPRuleset 1* refer to this set of rules ("CP" stands for counterpoint).

1. An augmented interval is not good.
2. A diminished one is good.
3. A disjunct motion should be 6th or less.
4. The 7th note of the scale, the leading tone, should be followed by the tonic.
5. But it can be used transitionally in the sequence of tonic→7th→6th.
6. It can also be used in an arpeggio.

Rules for Two-Parts. The two parts must conform to the following rules about intervals between them. Let *CPRuleset 2* refer to this set of rules.

1. Vertical intervals between the two parts should be mainly 3rd (10th) or 6th.
2. A musical piece must start with one of the three chords I, V and IV; the interval between the parts should be unison, octave, 5th, 3rd (major or minor) or 6th (major or minor).
3. The end of the piece must be an authentic cadence; the lower part should end with the tonic, and the upper should end with unison, octave, 3rd or 10th to the lower.
4. The two parts must not progress in more than three notes in row with keeping the interval of 3rd (10th); same for 6th.

From now, we will use the notation "CPR *m*-*n*" to refer to the *n*-th rule of CPRuleset *m*.

4 Implementing Rules in the Artificial Chemistry

Let us have a v-atom represent a note, e.g., having a v-atom A3 represent a note with the 440Hz pitch; a melody can be represented by a sequence of such v-atoms. In this study, we use only the C-major scale: the pitches are represented by C*n*, D*n*, E*n*, F*n*, G*n*, A*n* and H*n*, where *n* is a number. The pitch C4 is higher than C3 by the interval of an octave. A3 is next to G3 (not G2), and C4 is next to H3 (not H4) (Fig. 2(a)). A two-line v-molecule can then represent a two-part phrase. For example, the v-molecule 0#E3F3G3A3/0#C3D3E3F3/ represents the phrase shown in Fig. 2(b).

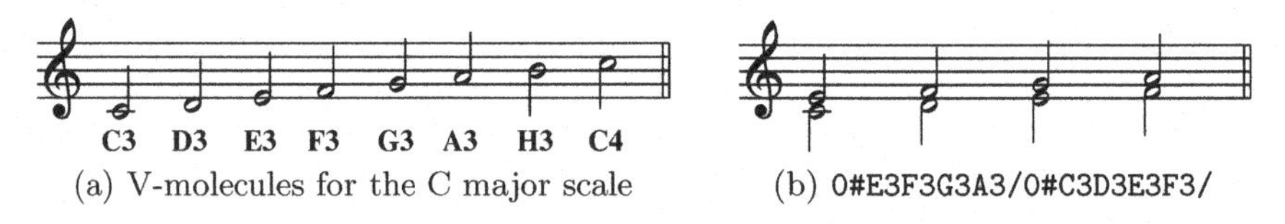

(a) V-molecules for the C major scale (b) 0#E3F3G3A3/0#C3D3E3F3/

Fig. 2. Representing notes as v-atoms

The set of recombination rules are so designed that the simulator generates phrases by trial and error. The overview of the designed procedure is as follows.

- Seeds of phrases are given as initial v-molecules (according to CPR 2-2).
- A note or a sequence of notes is added to the shorter part, according to the rules for horizontal intervals between notes (CPRuleset 1).
- If the vertical interval between notes is a prohibited one (as per CPR 2-1 and 2-4), one of the two notes, along with its succeeding notes, is separated from the phrase under production and is returned to the pool.
- If the phrase can connect to one of the given cadences, let it do so to finish the phrase (CPR 2-3). We use four cadences for major keys (Sec. 4.4).

The subsequent sections describe how each process is implemented.

4.1 Seeds of Phrases

CPR 2-1 requires a vertical interval between two notes of the two parts to be mainly 3rd, 6th or 10th. We use the range of pitches from G2 to C5 (approximately two and a half octaves), expecting many combinations of notes may appear in a phrase, and also in order to limit the problem size. CPR 2-2 specifies the beginning of a phrase; we choose C4 for the upper note and C3 for the lower. Let the notation P/Q denote a vertical pair of notes P and Q; the beginning of the phrase shown in Fig. 2(b) is denoted by E3/C3, for example.

The v-molecule that represents the first notes is 0#IpIp/0#Y0C4/0#Y0C3/. The second line (Y0C4) is the upper part, and the third line (Y0C3) is the lower. The v-atom Y0 represents the beginning of a phrase, which prevents the phrase from becoming a trailing part of another phrase. The first line is allotted for v-atoms that express vertical intervals, such as I1 (unison), I2 (2nd), etc. (Sec. 4.3).

4.2 Adding Notes to a Part

In order to achieve the requirements stated in CPRuleset 1, we employ a simple method: providing v-molecules that represent the good progressions in the initial pool. For example, the motion from C4 to D4 is good, so v-molecules of the form `0#C4D4/` are given to the pool. V-molecules `0#H3A3/` are not provided because this motion is prohibited (CPR 1-4), but v-molecules of the form `0#C4H3A3/` are given (CPR 1-5) and so are those of `0#H3G3D3/` (CPR 1-6).

These v-molecules are added to a phrase by such recombination rules as the one shown below. This rule attaches a one-line v-molecule (a sequence of notes) that starts with `C4` to the upper part of the phrase that ends with `C4`.

```
0#<*1><2*>/0#<*3>C4/1#<*5><6*>/ + 0#C4<4*>/
                  → 0#<*1><2*>/0#<*3>C4<4*>/1#<*5><6*>/   (2)
```

This rule can add, for example, a note sequence `0#C4H3C4E4/` to the upper part of a phrase `0#IpIpI6I6/0#YOC4D4C4/0#YOC3F3E3D3/`; this results in a longer phrase `0#IpIpI6I6/0#YOC4D4C4H3C4E4/0#YOC3F3E3D3/`. The rule that adds notes to the lower part is described similarly.[3]

4.3 Handling Vertical Intervals

In order to deal with vertical intervals easily, we add an extra line to the two-line representation of a phrase, as we have already seen. The line consists of the following v-atoms.

- V-atoms that represent the interval from unison to 10th: `I1`, `I2`, `I3`, `I4`, `I5`, `I6`, `I7`, `I8`, `I9` and `Ix`.
- V-atom to represent other (bad) intervals: `Iz`.
- The dummy interval to protect the notes: `Ip`.
- V-atom to indicate the interval has not yet been given: `S`.

An example phrase under production is

```
0#IpIpI6IzS/0#YOC4E4G4E4C4/0#YOC3G3D3C3H2/ .   (3)
```

`Ip` v-atoms protect the beginning of phrase from the separation explained below. The interval between G3 and E4 is 6th, which is represented by `I6`. The interval between D3 and G4, 11th, is not used in the current study, so the v-atom for "bad" interval, `Iz`, is given. The v-atom `S` indicates that the both parts have notes so the interval must be calculated; another `S` will come next to this v-atom.

Adding `S` to a phrase is performed by the following rule.

```
0#<*1>/0#<*2><3><4*>/0#<*5><6><7*>/
                  → 0#<*1>S/0#<*2><3><4*>/0#<*5><6><7*>/   (4)
```

[3] We relaxed the restriction that each recombination must conserve v-atoms [6], since it is not essential in this study; modifying the rules to conform to it is easy.

Then each `S` is replaced with the interval of notes. The rule below replaces `S` with a v-atom that represents the interval between C3 and A3, 6th (`I6`).

```
0#<*1>S<2*>/0#<*3>A3<4*>/0#<*5>C3<6*>/
                    → 0#<*1>I6<2*>/0#<*3>A3<4*>/0#<*5>C3<6*>/   (5)
```

For each combination of notes, this type of rule is given.

If a vertical pair of notes has a "bad" interval (in line with CPR 2-1), the sequence after (and including) the bad note is separated from the phrase. The following rules carry out this separation for a bad interval, 4th.

```
0#<*1><2>I4<3*>/0#<*4><5><6><7*>/0#<*8><9><10><11*>/
 → 0#<*1><2>/0#<*4><5><6><7*>/0#<*8><9>/ + 0#<9><10><11*>/   (6)
0#<*1><2>I4<3*>/0#<*4><5><6><7*>/0#<*8><9><10><11*>/
 → 0#<*1><2>/0#<*4><5>/0#<*8><9><10><11*>/ + 0#<5><6><7*>/   (7)
```

Rule (6) cuts the lower part, and (7) cuts the upper. The sequence separated from the phrase is returned to the pool and will be reused to construct another phrase. The interval information after `I4` is thrown away, since only one part has notes in the portion.

These types of rules are given to each of the "bad" interval v-atoms including `Iz`. No such rule is given to the "good" intervals, to leave them intact.

CPR 2-4 is also implemented using the interval information. A long sequence of 3rd (i.e., the two parts move in parallel, keeping the 3rd interval) is cut by the following rules.

```
0#<*1>I3I3I3<2*>/0#<*3><4..6><7*>/0#<*8><9..11><12*>/ + 0#Dec/
        → 0#<*1>I3I3/0#<*3><4..6><7*>/0#<*8><9><10>/
                                   +0#<10><11><12*>/ + 0#Dec/   (8)
0#<*1>I3I3I3<2*>/0#<*3><4..6><7*>/0#<*8><9..11><12*>/ + 0#Dec/
        → 0#<*1>I3I3/0#<*3><4><5>/0#<*8><9..11><12*>/
                                      +0#<5><6><7*>/ + 0#Dec/   (9)
```

`<4..6>` is an abbreviation of `<4><5><6>`. When a phrase that has more than three consecutive pairs of notes with the 3rd interval collides with `0#Dec/`, that portion of the phrase is broken and the trailing portion is released. We give similar rules for 6th and 10th.

4.4 Finishing a Phrase with a Cadence

In Sec. 4.2, the rules to add sequence of notes to a phrase are given. In a similar way, rules to join two phrases (each of which has the upper and lower parts) are given. The following rule joins a phrase that ends with A4/F4 and another phrase that starts with A4/F4.

```
0#<*1><2><3>/0#<*4><5>A4/0#<*6><7>F4/+
0#<8><9><10*>/0#A4<11><12*>/0#F4<13><14*>/ →
0#<*1><2><3><9><10*>/0#<*4><5>A4<11><12*>/0#<*6><7>F4<13><14*>/
                                                          (10)
```

This type of rule is given for each possible combination of vertical pair.

Then a cadence such as 0#IpIpIpIpIp/0#A4G4F4E4Z0/0#F4H3D4C4Z0/ can connect to a phrase when the first pair of the cadence is the same as the last pair of the phrase; thus arbitrary cadences can be given in the pool. The dummy v-atom Z0 prevents another phrase from joining to the finished phrase. We give the following four types of cadence molecules to the pool.

- 0#IpIpIpIpIp/0#A4G4F4E4Z0/0#F4H3D4C4Z0/
- 0#IpIpIpIpIp/0#A3G3F3E3Z0/0#F3H2D3C3Z0/
- 0#IpIpIpIpIp/0#F4E4D4C4Z0/0#A3G3H2C3Z0/
- 0#IpIpIpIpIp/0#D4C4H3C4Z0/0#F3A3G3C3Z0/

We give an additional rule (shown below) to finish the prolongation of a phrase when it becomes long enough.

```
0#Ip<1..17><18*>/0#Y0<19..35><36*>/0#Y0<37..53><54*>/ →
        0#Term/0#Ip<1..12>/0#Y0<19..30>/0#Y0<37..48>/
                  + 0#<30..35><36*>/ + 0#<48..53><54*>/   (11)
```

5 System Description

The designed system to generate polyphonic phrases is as follows. The initial pool comprises the following v-molecules. The number in braces ({ }) is the number of v-molecules of the specified form.

- Phrase seeds: 0#IpIp/0#Y0C4/0#Y0C3/ {20}
- Authentic cadences: {20 for each}
  - 0#IpIpIpIpIp/0#A4G4F4E4Z0/0#F4H3D4C4Z0/
  - 0#IpIpIpIpIp/0#A3G3F3E3Z0/0#F3H2D3C3Z0/
  - 0#IpIpIpIpIp/0#F4E4D4C4Z0/0#A3G3H2C3Z0/
  - 0#IpIpIpIpIp/0#D4C4H3C4Z0/0#F3A3G3C3Z0/
- Sequences for conjunct motion (2nd-interval motion): 0#C3D3/, 0#C3H2/, etc.; 20 v-molecules for every motion within the range from G2 to C5, except for the motion prohibited by CPR 1-4.
- Sequences for disjunct motion (non-2nd-interval motion): 0#C3E3/, 0#C3F3/, etc.; 4 v-molecules for every motion allowed by CPR 1-2 and 1-3 within the range from G2 to C5, except for the motion prohibited by CPR 1-1 and 1-4.
- Motion by CPR 1-5: 0#C5H4A4/, 0#C4H3A3/ and 0#C3H2A2/ {4 for each}
- Motion by CPR 1-6: 0#H4G4D4/ and 0#H3G3D3/ {4 for each}
- V-molecules to cut a long parallel motion: 0#Dec/ {50}

The following recombination rules are given to the system.

- Rules to add notes: for each note in the range from G2 to C5, a recombination rule similar to Rule (2) is given.
- Rules to join two phrases: for each possible vertical pair of notes, a recombination rules similar to Rule (10) is given.
- Rule (4): the rule to add `S` in the line for interval information.
- Rules to provide interval information: rules to replace `S` with an interval v-atom (`I1`, `I2`, etc.), such as Rule (5).
- Rules to destroy bad vertical intervals: for each undesirable vertical interval, rules similar to Rules (6) and (7) are given.
- Rules to cut a long parallel motion of the two parts: in addition to Rules (8) and (9), similar rules are given for 6th and 10th.
- Rule (11): the rule to terminate a long generation.

6 Preliminary Experiment

The description given in the previous section is executed on a simulator of the artificial chemistry. The total number of v-molecules in the initial pool is 1178, and that of recombination rules is 708. The numbers may seem big, but most part of the description is mechanically generated by a script program. V-molecule species for the same purpose, such as conjunct motion, have the same form; they are different only in their pitches. Same applies to the recombination rules.

The simulator was implemented with the Cocoa Framework and Objective-C, and was run in Mac OS X 10.4 on Intel Core 2 Duo 2 GHz with 2 GB memory. During the run, we noticed that v-molecules of the types `0#H2C3/`, `0#H3C4/` and `0#H4C5/` were lacking (because the conjunct motion C→H and the disjunct motion from the leading tone are prohibited), so we supplied 40 v-molecules of each type while the simulator was running. Except this, there was no human intervention. The run took about nine hours, and produced ten phrases. Fig. 3 shows obtained good phrases (chosen by human ears). The phrase (a) has a cadence at the end, while (b) and (c) do not (i.e., prolongation terminated). Neither of them violates the CPRulesets. Some of the phrases not shown here violate CPR 1-4; they seem to be fragments from v-molecules given for CPR 1-5 and 1-6, produced by cutting. The description should be improved to fix this flaw.

7 Discussion

In this approach, notes are encoded as v-atoms, and then phrases are encoded as v-molecules. These mappings are straightforward. The computing process directly deals with the v-molecules. This approach does not need a complex mapping scheme, such as interpreting a geometric position as sound, which is used in some music-generating systems [3,4].

Fig. 3. Generated two-part polyphonic phrases

In the traditional programming, a procedure is usually so designed as to take correct steps of computation, avoiding errors. In contrast, this approach allows temporary errors, or utilises it to search for a better solution. Some rules try to make as good phrases as possible; some others modify (or destroy partly) a phrase if it is not good. Trial and error are useful when a deterministic algorithm is not available, such as the case of composing music.

Genetic algorithms usually use randomness for selection, crossover and mutation, and have been used in studies for generating musical phrases [1]. A crucial part of GA is fitness evaluation: the fitness function should be designed carefully to have the system work in the desired way. Once the fitness of an individual is evaluated, it is selected as a parent for the next generation (or simply discarded), and crossover and/or mutation are performed. But designing crossover and mutation for a specific application is not an easy task. In contrast, our approach enables programming in modifying a v-molecule to make a better phrase. For example, if a phrase has a bad vertical interval, the system can remove only the problematic portion, or even can replace one of the notes with a good one in a programmed way. We think this facilitates more flexible control in computation.

The simulator employs stochastic collision as the reactor algorithm, and the number of v-molecules affects the probability of reaction in which the v-molecule species is involved. Therefore, increasing or decreasing the number of v-molecules controls the behaviour of the system. For example, if a phrase consisting mainly of conjunct motion is desired, giving many v-molecules that represent conjunct motion will increase the probability of generating such phrases.

In our approach, good components can be given in the initial pool, such as arpeggios and cadences. Furthermore, as we did in our experiment, it is easy to add v-molecules to the pool during the execution. This is not a typical way to control a computation, but it is similar to adding insufficient reagent during a chemical experiment. Moreover, removing v-molecules and adding/removing recombination rules are easy, since it is an "artificial" chemistry.

Nevertheless, the current approach has some difficulties. One is that it is hard to numerically evaluate how good a phrase is: although phrases that comply with the implemented Rulesets are generated, it is difficult to tell which of them is better than others. One way to achieve such evaluation is to combine this approach with GAs: the initial population generated by the artificial chemistry are fed to GA to evolve. Another difficulty is the low performance of the current implementation. Since we put the understandability of description before the performance, the description of the present system has not been refined for efficiency. Refining recombination rules, as well as adjusting the numbers of v-molecules, would improve the performance.

8 Concluding Remarks

This study implemented a basic set of rules in counterpoint in the artificial chemistry, and generated polyphonic phrases using the simulator. Each rule in counterpoint is realised as steps of recombination, which is similar to programming in procedural language. At the same time, the artificial chemistry deals with multiple v-molecules (i.e., phrases) and decomposes or modifies bad phrases; this realises a kind of generate-and-test method like GAs. Since a process of composing music should include trial and error, we think this dual nature of artificial chemistry will be useful in implementing music-generating systems.

Some generated phrases (including those generated in our previous works) can be listened to in our web site: `http://www.tomilab.net/alife/music/`.

References

1. Burton, A., Vladimirova, T.: Generation of musical sequences with genetic techniques. Computer Music Journal 23(4), 59–73 (1999)
2. Blackwell, T., Bentley, P.: Improvised music with swarms. In: Proceedings of the 2002 Congress on Evolutionary Computation, Piscataway, NJ, vol. 2, pp. 1462–1467. IEEE Press, Los Alamitos (2002)
3. Miranda, E.: On the music of emergent behavior: What can evolutionary computation bring to the musician? Leonardo 36(1), 55–59 (2003)
4. Beyls, P.: A molecular collision model of musical interaction. In: Soddu, C. (ed.) Proceedings of the 8th International Conference on Generative Art (GA 2005), Milan, AleaDesign, pp. 375–386 (2005)
5. Dittrich, P., Ziegler, J., Banzhaf, W.: Artificial chemistries — a review. Artificial Life 7(3), 225–275 (2001)
6. Tominaga, K., Watanabe, T., Kobayashi, K., Nakamura, M., Kishi, K., Kazuno, M.: Modeling molecular computing systems by an artificial chemistry — its expressive power and application. Artificial Life 13(3), 223–247 (2007)
7. Miura, T., Tominaga, K.: An approach to algorithmic music composition with an artificial chemistry. In: Explorations in the Complexity of Possible Life (Proceedings of the 7th German Workshop on Artificial Life (GWAL-2007), Akademische Verlagsgesellschaft Aka GmbH, Berlin, Germany, pp. 21–30 (2006)

Implicit Fitness Functions for Evolving a Drawing Robot

Jon Bird, Phil Husbands, Martin Perris, Bill Bigge, and Paul Brown

Centre for Computational Neuroscience and Robotics
University of Sussex, Brighton, UK

Abstract. We describe an approach to artificially evolving a drawing robot using *implicit* fitness functions, which are designed to minimise any direct reference to the line patterns made by the robot. We employ this approach to reduce the constraints we place on the robot's autonomy and increase its utility as a test bed for synthetically investigating creativity. We demonstrate the critical role of neural network architecture in the line patterns generated by the robot.

1 Introduction

The Drawbots project is a multidisciplinary investigation into creativity involving philosophers, adaptive systems researchers and an artist. A theoretical goal is to investigate the question: what is the simplest mechanism that can be described as creative? To this end we artificially evolve wheeled robots that move around an arena making pen marks on the floor. These 'embodied thought experiments' help clarify some of the necessary conditions for 'minimal creativity' (autonomy, novelty and evaluation) and how they can be embodied in a robot (see [1] for a detailed consideration of these issues). An artistic goal of the project is to generate aesthetically interesting line drawings that are suitable for exhibition. This is distinct from, and potentially at odds with, the theoretical goal. For example, by incorporating artistic knowledge into fitness functions we might enhance the aesthetic appeal of the resulting line markings, but at the expense of compromising the autonomy (and therefore 'minimal creativity') of the robots. This paper describes our use of implicit fitness functions to evolve a drawing robot where we minimise any direct reference to the line patterns and our focus is on investigating minimal creativity.

2 Evolutionary Robotics

The main synthetic, bottom-up methods used in the project are those of evolutionary robotics (ER). ER is a biologically inspired approach to the automatic design of autonomous robots. The field encompasses a wide range of work where one or more (sometimes all) of the following aspects of robot design are in the hands of an evolutionary search algorithm: the control system; the overall body morphology; and sensor and actuator properties. The evolutionary process uses

M. Giacobini et al. (Eds.): EvoWorkshops 2008, LNCS 4974, pp. 473–478, 2008.

a fitness measure based on how good a robot's behaviour is according to some evaluation criteria: a key distinction here is between *implicit* and *explicit* fitness functions [2]. An explicit fitness function rewards specific behavioural elements - such as travelling in a straight line or maximum velocity achieved - and hence shapes the overall behaviour from a set of predefined primitives. Implicit fitness functions operate at a more indirect, abstract level - reward is given for completing some task but the robot is free to achieve it in any possible way. The number of variables and constraints defined in a fitness function determine where it falls on the implicit-explicit dimension: implicit fitness functions have no or very few such components. Fitness is tested either in simulation, in the real world or using a combination of the two. Typically some form of artificial neural network acts as the nervous system of the robot; properties of the network will invariably be evolved even if other aspects of the robot design are not. By artificially evolving control architectures from suitably low level primitives, the final controller "need not be tightly restricted by human designers' prejudices" [3, p.83]: ER can therefore potentially generate novel models of creativity and art-making machines that are not necessarily constrained by the artist's (systems designer's) stylistic 'signature'.

3 Implicit Fitness Function Experiments

In this section we describe two sets of ER experiments that aimed to minimise our influence on the resulting robot behaviour by using *implicit* fitness functions that did not specify the types of marks that a robot should make. The first 'sensory-motor correlation' fitness function was tested in simulation; the second 'ecological' fitness function was initially tested in simulation but some of the resulting controllers have also been successfully transferred and tested on the Drawbot (Figure 3).

3.1 Sensory-Motor Correlation Fitness Function

Initial experiments were carried out in simulation using an accurate model of a Khepera robot, a standard ER platform, augmented with a drawing pen placed between its drive wheels. In the simulation, each robot controller was a neural

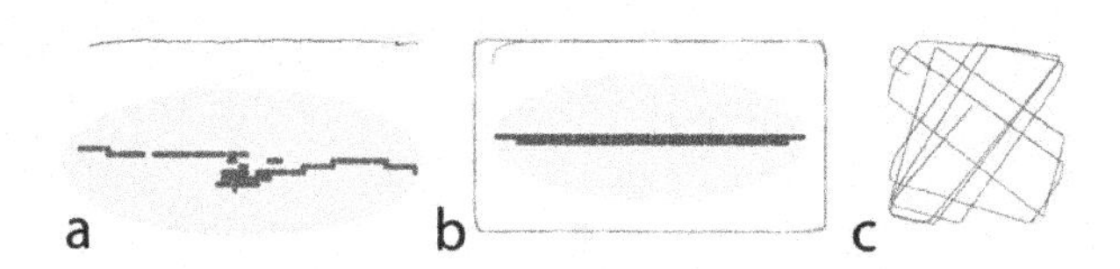

Fig. 1. Results from the implicit fitness function that rewarded correlated activity between the pen movement (up/down) and line detector (on/off). a) mid-fitness individual; b) high fitness individual; c) the patterns that result from adding a selection pressure to mark the entire arena.

network consisting of six motor neurons: two for each of the left wheel, right wheel and pen position (up or down) motors. At each time step in the simulation, the most strongly activated neuron of each motor pair controlled its associated actuator. The robot has seven sensors (six frontal IR sensors and one line detector positioned under the robot that can detect marks made by the pen). Each of the seven sensors was connected to each of the six motor neurons. A genetic algorithm was used to determine the strength of each of these connections and the bias of each of the motor neurons. The fitness function rewarded controllers that correlated the changes in state of their line detector and pen position. For example, if a line was detected and the robot's pen was then raised or lowered within a short time window, the robot accumulated fitness. This fitness function resulted in robots that used the arena walls (a constant feature of the environment) to guide their drawing behaviour. Mid-fitness individuals follow a wall to a corner and then gain fitness by repetitively moving forwards and backwards over a mark and appropriately co-ordinating the movement of their pen (Figure 1a). High fitness individuals initially follow the arena walls for one circuit making a continuous line and on their second circuit raise and lower their pen making marks adjacent to the initial line (Figure 1b). Different behaviours evolve when the fitness function also rewards robots for the extent to which they mark the whole area of the arena: the robots turn away from the walls at angles and mark the central parts of the arena as well (Figure 1c). In all these experiments crashing into walls is penalised by stopping the evaluation and thereby giving the robots less time to accumulate fitness.

3.2 Ecological Fitness Function

The controllers evolved in the experiments briefly described in this section were Continuous Time Recurrent Neural Networks (CTRNNs), a rather more complex network than those used in the earlier experiments described above.

The networks consisted of either 40 or 20 fully connected nodes. The connection weights, time constants, biases and gains were encoded as a string of real numbers in the range [0,1] and linearly scaled to values in the range [-5,5], [0.04,4], [-10, 10] and [0.01, 10.01] respectively. The state of each neuron was

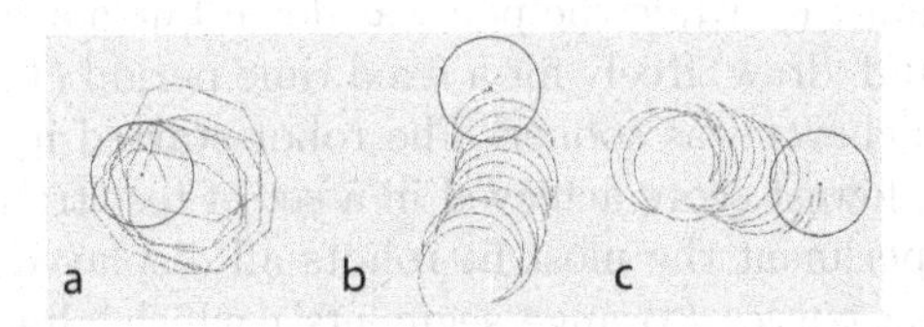

Fig. 2. Line patterns generated in simulation using an ecological fitness function and a simple motor model. a) is the typical pattern generated by a 20 neuron network; b) is an 'orange segment' pattern occasionally (approximately 30%) generated by a 20 neuron network; c) is the typical pattern generated by a 40 neuron network (which after further evolution looks like Figure 3). The robot is the circle with the dot at its centre.

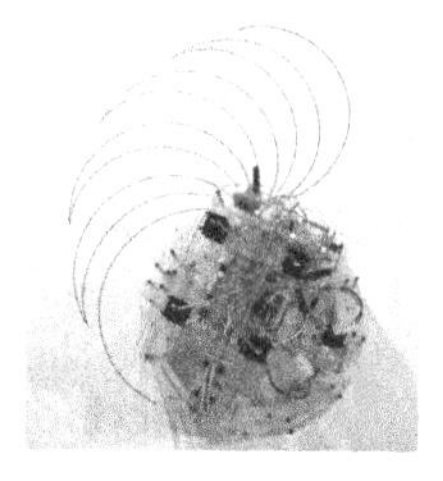

Fig. 3. Top-down view of an 'orange segment' line pattern generated by a Drawbot in the real world which was evolved in simulation using the implicit ecological fitness function, a 40 neuron network and a simple motor model.

initially set to 0 plus a small random value. 6 of the neurons had external inputs from the sensors and 3 neurons acted as motor outputs: one for each wheel and one to lower and raise the pen. For full details see [4].

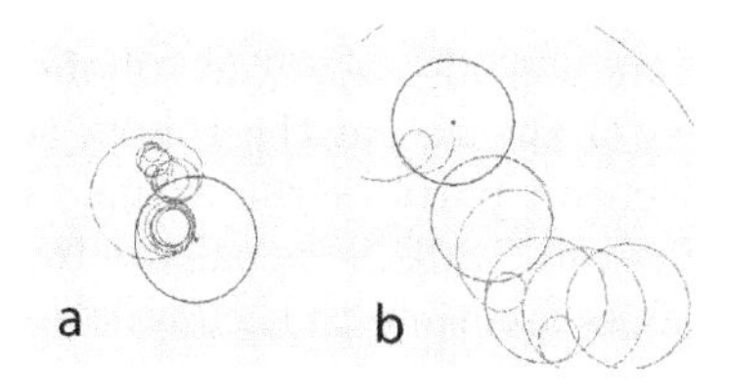

Fig. 4. Circle patterns generated in simulation using a 40 neuron network and a more sophisticated motor model with inertia and momentum. The robot is the circle with the dot at its centre.

Robot controllers were initially evolved in simulation using an 'ecological' fitness function. Small circular pieces of 'food' were randomly scattered in a target area of the arena (either a central rectangle or a semi-circle adjacent to a wall). Fitness was gained when a line drawn by the pen intersected one of the food particles. Each robot started with a fixed amount of energy which was used up at a constant rate while the pen was down but not while it was up; the robot could move and 'draw' freely for a fixed time period (1 minute) or until its energy ran out, whichever was sooner. The robot started in a random position and fitness was the lowest score achieved in a set of test trials.

In the initial experiment the most fit robots all displayed similar behaviour: they made sweeping curves ('orange segments') which alternated in direction and fanned out over a reasonable area of the target area (Figures 2c and 3). In the patterns generated by the fittest individuals, the separation of the segments is just larger than the diameter of a food particle. This is a good strategy for systematic coverage of an area without crossing a food particle more than once (an individual can only score one point per food particle, regardless of the number of times its lines intersect it). The image produced by the real robot

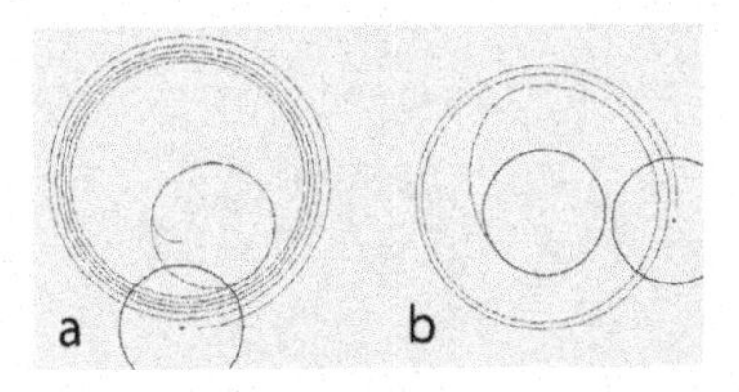

Fig. 5. Line patterns generated in simulation with a 20 neuron network and a more sophisticated motor model with inertia and momentum. The robot is the circle with the dot at its centre.

(Figure 3) is qualitatively very similar to those found in the simulation but the semi-circular curves are closer together and the robot tends to draw a full circle at the start. We halved the number of neurons in the CTRNNs from 40 to 20. When driven by the simple motor model the 20 neuron controllers tended to produce looping patterns (Figure 2a) and occasionally overlapping 'orange segments' (2b) - the pattern always generated by a 40 neuron network (2c - with further evolution the segments stop overlapping and look like 3). Although the 40 neuron controllers transferred well, the simulation did not model inertia or momentum and the robots were restricted to high speeds. In order to overcome these limitations a further series of experiments were carried out with a more sophisticated motor model. The change in motor model resulted in 40 neuron controllers generating circular patterns of varying diameter (Figure 4) and 20 neuron controllers generating spirals (Figure 5) - an effective solution for covering an area and minimising multiple intersections of the same food particle if the gap between the spirals is larger than the food particle diameter.

In all the above experiments the target area was located in the centre of the arena and although the controllers use the light as an energy source (they stop working if the light is switched off) they did not use if for directing their movement. We therefore conducted an experiment where the location of the target area varied in each trial and was always placed adjacent to a wall so that robots had to actively use their IR sensors to avoid crashing. A light was placed above the wall to indicate the centre of the semi-circular target region. The fitness of a robot was determined by its ability to perform phototaxis as well as the number of food particles it drew over. Crashing was again penalised. We found that a more distributed architecture facilitated the evolution of successful controllers in this task. The pen neuron was only connected to the light sensors and two other neurons and the threshold above which the pen was lowered was also evolved. Successful individuals make the majority of their marks in the target region, regardless of their starting position and orientation in the arena. The looping line patterns are less structured than the circles, spirals and orange segments prodcued in previous experiments (Figure 6), again illustrating that the robot's embodiment (change in network architecture), as well as the environment, influence the line patterns generated.

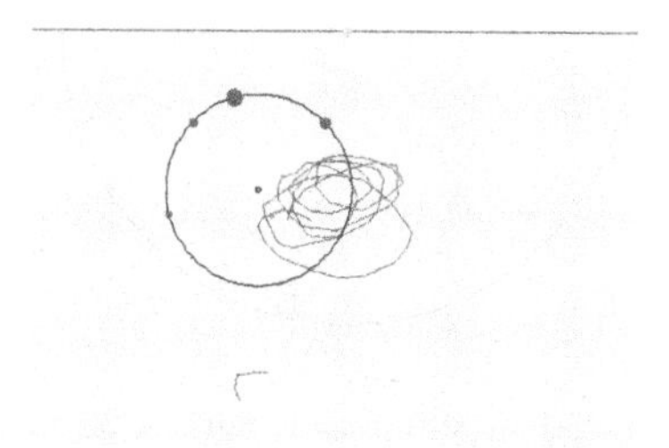

Fig. 6. Line patterns generated by a robot that had to perform obstacle avoidance and phototaxis in order to find the target regions where its lines would gain the maximum fitness. The top of the image is the arena wall, the dot on this wall represents the light and the semicircular area is the target region. Note that the robot marks a small curved line on the way to the light. The robot is the circle with the dot at its centre.

4 Conclusion

When investigating minimal creativity, our working hypothesis is that is advantageous to use an implicit fitness function in order to maximise the robot's autonomy. If we want to exhibit work produced by the robots, then a more explicit fitness function that embodies artistic knowledge about 'aesthetically pleasing' line patterns seems worth exploring and this is the focus of current research. However, even the patterns generated by implicit fitness functions can have an artistic impact, especially if the drawing process underlying the drawings is made evident, for example, by exhibiting the robots behaving in an arena rather than displaying the resulting drawings on a wall.

This paper forms part of the research supported by AHRC grant number B/RG/AN8285/APN19307.

References

1. Bird, J., Stokes, D.: Minimal Creativity, Evaluation and Pattern Discrimination. In: Cardoso, A., Wiggins, G. (eds.) Proceedings of the 4th International Joint Conference on Computational Creativity, pp. 121–128 (2007)
2. Nolfi, S., Floreano, D.: Evolutionary Robotics: The Biology, Intelligence, and Technology of Self-Organizing Machines. MIT Press/Bradford Books, Cambridge (2000)
3. Cliff, D., Harvey, I., Husbands, P.: Explorations in Evolutionary Robotics. Adaptive Behavior 2, 73–110 (1993)
4. Perris, M.: Evolving Ecologically Inspired Drawing Behaviours. MSc dissertation, Department of Informatics, University of Sussex (2007)

Free Flight in Parameter Space: A Dynamic Mapping Strategy for Expressive Free Impro

Palle Dahlstedt[1,*] and Per Anders Nilsson[2]

[1] Dep. of Applied IT, Göteborg University, 41296 Göteborg, Sweden
[2] Acad. of Music and Drama, Göteborg University, Box 210, 40530 Göteborg, Sweden
palle@ituniv.se, pan@hsm.gu.se

Abstract. The well-known difficulty of controlling many synthesis parameters in performance, for exploration and expression, is addressed. Inspired by interactive evolution, random vectors in parameter space are assigned to an array of pressure sensitive pads. Vectors are scaled with pressure and added to define the current point in parameter space. Vectors can be scaled globally, allowing exploration of the whole space or minute timberal expression. The vector origin can be shifted at any time, allowing exploration of subspaces. In essence, this amounts to mutation-based interactive evolution with continuous interpolation between population members. With a suitable sound engine, the system forms a surprisingly expressive performance instrument, used by the electronic free impro duo pantoMorf in concerts and recording sessions over the last year.

1 Introduction

Sound synthesis typically involves a large number of parameters. With conventional mappings, due to limitations in motor control and cognitive capacity, we can consciously control only a small number. This applies both to sound design, i.e., exploration of unknown sound spaces, and to performance expression.

Vast unknown soundspaces can be explored by interactive evolution (e.g., [1,2]). This has proven efficient for sound design and composition. But this approach is unsuitable as a live instrument. While it has been used live by the first author, the stepwise nature of the process excludes gestural expression. The user may like a specific variation of a mother sound, but he cannot access the continuum between the parent and the offspring. Also, the evaluation process may not be suitable for the audience's ears. On a traditional instrument mistakes can be covered or exploited by quick adjustment or by developing them into new themes. The stepwise control of conventional interactive evolution does not allow such flexibility. In the presented system, performer-controlled interpolation between population members (vector end points) allows continuous exploration of the subspace around the current parent sound, and selection is carried out by shifting the vector mechanism to a new chosen point.

* This research is part of a project funded by the Swedish Research Council. The paper was written during a visit to CEMA, Monash Univ., Melbourne, Australia.

M. Giacobini et al. (Eds.): EvoWorkshops 2008, LNCS 4974, pp. 479–484, 2008.

This system is also an attempt to reintroduce musicianship in electronic music performance. In our system, the performer has continuous control over up to 30 parameters, allowing great musical expressivity, thanks to direct connection (in time and direction) between gesture and sound, and between physical effort and sound (no hands – no sound). The physical connection between the instrument and the body is important, as are the microvariations of human activities and the sense of strain, which creates a natural phrasing between effort and rest in all kinds of acoustic musical performance.

These thoughts are not new. The importance of effort has, e.g., been stressed by Ryan [3], while Wessel et al emphasize intimacy between gestural controller and sound generation [4]. Important collections on the subject of gestural control in general [5] and mappings in particular [6] have established the terminology and provide a foundation for discussion.

The musical and performance-related motivations for this project will be elaborated in other publications, while this paper concentrates on one of the developed mapping strategies, and its evolutionary aspects.

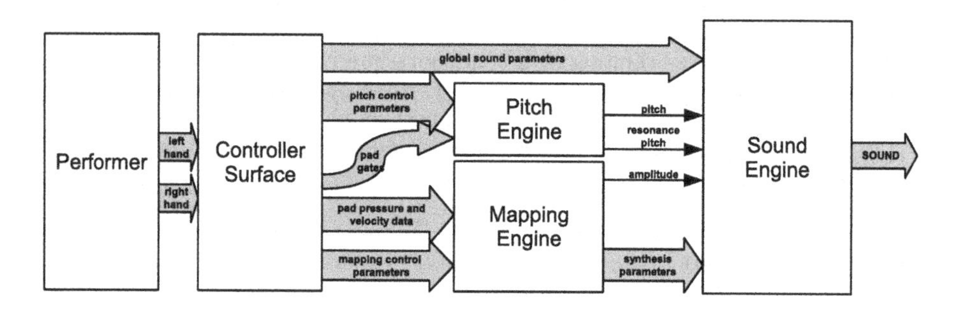

Fig. 1. A schematic of the performance system

2 The Instrument

The general idea behind this mapping strategy is to assign a random vector in synthesis parameter space to each controller. Each vector is scaled by its controller and by a global scaling control, and the vectors are added to arrive at a single point in parameter space, effectively deciding the synthesis parameters of the monophonic sound. In this way, the parameter space around the current origin (which can be shifted) can be explored in all directions. These ideas are applicable to a wide range of controllers and sound engines. We have used the M-Audio Trigger Finger, a 4x4 array of pressure sensors (each transmitting on/off gate, initial velocity and continuous pressure), allowing precise control of vector scaling by finger pressure. This particular implementation has been used, tested and refined over more than a year. The sound engines are interchangeable, and hence only briefly discussed.

The system consists of the controller surface, a mapping engine, a pitch engine and a sound engine (see Fig. 1). A few knobs and sliders regulate the behavior of the mapping and pitch engines, and some are routed directly to the sound engine for control of global synthesis parameters.

2.1 The Mapping Engine - Timbral Control

A pad i is assigned a vector $\mathbf{r}_i$ in $\mathbb{R}^n$, where n is the number of synthesis parameters. The vector components $r_{i,1}, r_{i,2}, ..., r_{i,n}$ are initialized to random floating point values in the range $[-1, 1]$, with equal distribution. Each vector is scaled by the current pressure p_i of the pad and a global pressure scaling knob p_{amt}. It is also affected by an attack-decay envelope function $\beta(t) \in [0, 1]$ scaled by the initial velocity v_i of the pad and a global envelope scaling knob v_{amt}. Short attack and decay times are preferred, since a slow envelope would reduce the performer's direct control over the sound. The main purpose is to add a distinct transient to the sound, enabling more rhythmical playing. At the outset, the departure point in synthesis parameter space $\mathbf{o} = (0, 0, ..., 0)$. The resulting point $\mathbf{s}$ in $\mathbb{R}^n$ is calculated by the following formula:

$$\mathbf{s} = \mathbf{o} + \sum_{i=1}^{m} (p_{amt} p_i + v_{amt} v_i \beta(t_i)) \mathbf{r}_i \,, \quad p_{amt}, v_{amt}, v_i, p_i \in [0, 1]$$

where m is the number of pads and t_i is the time since pad i was last pressed. The components $s_1, s_2, ..., s_n$ of $\mathbf{s}$ are the synthesis parameters sent to the sound engine. Adjusting p_{amt} allows exploration of the whole sound space in search of interesting regions, or minute expressions around a particular point, while different values of v_{amt} allow different playing techniques, with more or less percussive attacks. The total pressure $\sum p_{1...m}$ is mapped to the overall amplitude of the sound.

At any time during performance, e.g., when an interesting region in sound space has been found, $\mathbf{o}$ can be shifted to the current sounding point $\mathbf{s}$ with a momentary switch/pedal, effectively recentering the whole vector exploration mechanism to the subspace around the current $\mathbf{s}$. This is often followed by a reduction of p_{amt} to enable finer movement around the new center point. With another switch, $\mathbf{o}$ can be reset to $(0, 0, ..., 0)$, if one is lost in sound space. With each shifting of $\mathbf{o}$, the vectors can if desired be reinitialized to new random values. It is a choice between fresh exploration of the space and retained consistency of mapping directions.

This mapping can be used to control timbre in many different ways. Pads can be played one at a time, or many together. Pads can be held while other pads are varied. Rhythmic pulses can be played on one pad while other pads are varied smoothly. By alternating and overlapping two or more pads, complex trajectories in sound space can be realized. The navigational controls allow zooming in to interesting subspaces, or instant escape from them if stuck. Since vector components can be negative, pads can have the effect of damping or muting, which can be exploited musically by the player.

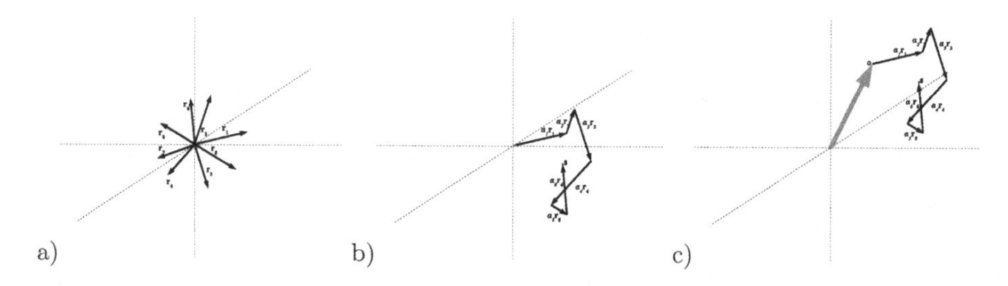

Fig. 2. a) A set of vectors in synthesis parameter space, each corresponding to a pressure pad. **b)** Vectors are scaled and added to arrive at the sounding point, **s**. **c)** The origin **o** can be shifted, allowing exploration of different subspaces.

2.2 Pitch Control and Sound Engines

Pitch is controlled by an additive pitch mechanism. Each pad is assigned a chromatic interval, increasing to the right and upwards. Intervals of pressed pads are added, similar to the valves of a trumpet. The sum is scaled by a slider (0 to 1), and added to a base pitch controlled by another slider. This allows for a continuum from no pitch control via microtonality to chromatic pitch, over the whole audible range. It is relatively easy to press a pad without any significant pressure, so pitch can to a certain extent be controlled separately from timbre.

We have used a number of different sound engines, of 7-30 synthesis parameters. Because of the random exploratory nature of the system, it is suitable for wild, noisy sound engines, but equally capable of smooth expressive playing. Designing a sound engine for this system is similar to designing one for interactive evolution – one has to think in potential, and all combinations of parameter values have to be considered, which of course is impossible.

3 Discussion

This mapping strategy was inspired by years of experience developing and using systems for interactive evolution of synthesis parameter sets, and there are some close resemblances. Obviously, this is not an optimization process, but rather an exploration, which is often true for interactive evolution. The current origin is the parent sound, and the end point of each random vector corresponds to an offspring by mutation – the vector actually *is* the mutation, i.e., the difference between parent and child (at full pad pressure). The global vector scaling has the effect of a mutation rate. By pressing one pad, you hear one offspring, but you can also hear combinations of different offspring (by interpolation, not recombination) by pressing several pads. When a good sound is found, it is reproduced by shifting the origo and re-randomizing the vectors, making the current point the parent sound. This process can be iterated. So, selection is applied by one hard choice per generation, similar to the systems described in [1,2]. In essence, this amounts to mutation-based interactive evolution of synthesis parameter sets with continuous expressive interpolation between population members. One notable difference is that each offspring actually represents a direction of mutation, not a specific step.

This system is not better or easier to use than existing systems for interactive evolution of synthesis parameters. Instead, it is a way of applying the idea of interactive evolutionary exploration to a performance instrument, with continuous interpolation providing a means for musical nuances and expression. Essentially, it is a merge of coupled gestural mappings and exploratory evolution.

Disregarding the possibility to shift the vector origin, this is a linear many-to-many mapping, which is not so common but has been discussed [7]. Experiments have shown that coupled mappings are easier to learn and more engaging to play [8], and this is here taken to its extreme. When playing, you cannot think in parameters. You have to use your ear, all of the time.

Choi *et al* emphasize the feedback process through which we learn how an instrument works, and define the word intuitive to mean "the experience of participating in feedback systems where the participant can draw compatible conclusions from a small number of trials..." [9]. This is related to issues of consistency (same gesture, same result), continuity (small gesture, small effect) and coherence (works like the physical world) [10]. Our system has short term repetitivity and is mostly continuous (depending on sound engine). There is nothing random in the realtime process, but the mapping is dynamic. After an origin reset or a change of vector scaling, gestures give different results. When global sound parameters are changed, the space is "warped" and all sounds are different. Still, based on months of playing, it is clear that one can learn and internalize the way the system works on a higher level. In a conventional instrument the aural feedback teach you the correspondence between gestures and effects. Here, the learning is on another conceptual level – you learn how to extrapolate from the current behavior of the system into meaningful musical gestures. The immediate feedback between fingers and ear is essential. The vector mapping is random and complex, but it feels easy to play, with a natural alternation between exploratory (long vectors) and expressive playing (short vectors). The sound can be controlled with great precision, in a way unlike any other electronic instrument the authors have experienced. The real life evaluation has consisted of weekly rehearsals, a number of concerts and the recording of an album.

The system is very suitable for improvisation, where the random aspects of the instrument is an advantage. They provide an unpredictable component, in the same way as an extra performer, and one can react literally instantly, *in* the sound, not after, to cover up or develop it into something useful.

The numbers of control and synthesis parameters have been roughly equal. With more synthesis parameters, the vectors would not span the whole space. To maximize the volume of the search space, the random vectors could be orthogonalized. But since they are re-randomized at origin shift, all dimensions will eventually be explored. An analogy is the concept of mutations – one does not require every generation to have a mutation in every locus, but trusts chance to supply, sooner or later, variation in all directions.

This mapping is neither explicit nor generative, but *dynamic*. Since the performer has direct control over essential mapping parameters (vector scaling, origin shift), the mapping can be changed at will, and these changes become integrated into the way of playing. Meta-control becomes part of control, and exploration and expression are merged into one process.

3.1 Conclusions

We have described a kind of mapping from a control surface to synthesis parameters that merges exploration of a sound space with expressive playing, overcoming the expressive limitations of conventional interactive evolutionary approaches. It is essentially different from previous mappings, in that it is based on random correlations and that it is reconfigurable during performance, continuously and stepwise. Still, it satisfies criteria such as low learning threshold, short term repeatability and possibility of developing skill over time. Thanks to the direct and continuous connection between physical gesture and sonic output, it feels, behaves, and appears like a traditional instrument, and is very suitable for free improvisation with intimate control over timbre, pitch and dynamics. It is not your average musical instrument, but extensive playing has proven that, while complex, it is a fascinating and expressive partner on stage, much like acoustic instruments, and it deserves to be further developed.

References

1. Dahlstedt, P.: Creating and exploring huge parameter spaces: Interactive evolution as a tool for sound generation. In: Proceedings of the 2001 International Computer Music Conference, Habana, Cuba, pp. 235–242 (2001)
2. Dahlstedt, P.: Evolution in creative sound design. In: Miranda, E.R., Biles, J.A. (eds.) Evolutionary Computer Music, pp. 79–99. Springer, Heidelberg (2007)
3. Ryan, J.: Effort and expression. In: Proceedings of the 1992 International Computer Music Conference, San Jose, California, USA, pp. 414–416 (1992)
4. Wessel, D., Wright, M.: Problems and prospects for intimate musical control of computers. In: Workshop in New Interfaces for Musical Expression, Seattle (2001)
5. Wanderley, M., Battier, M. (eds.): Trends in gestural control of music (CDROM), Ircam/Centre Pompidou, Paris (2000)
6. Wanderley, M. (ed.): Organised Sound, 7(2) (2002)
7. Hunt, A., Wanderley, M.M., Kirk, R.: Towards a model for instrumental mapping in expert musical interaction. In: Proceedings of the 2000 International Computer Music Conference, pp. 209–212 (2000)
8. Hunt, A., Kirk, R.: Mapping strategies for musical performance. In: Wanderley, M., Battier, M. (eds.) Trends in Gestural Control of Music (CDROM), Ircam/Centre Pompidou, Paris (2000)
9. Choi, I., Bargar, R., Goudeseune, C.: A manifold interface for a high dimensional control space. In: Proceedings of the 1995 International Computer Music Conference, Banff, pp. 89–92 (1995)
10. Garnett, G.E., Goudeseune, C.: Performance factors in control of high-dimensional spaces. In: Proceedings of the 1999 International Computer Music Conference, Beijing, China (1999)

Modelling Video Games' Landscapes by Means of Genetic Terrain Programming - A New Approach for Improving Users' Experience

Miguel Frade[1], F. Fernandez de Vega[2], and Carlos Cotta[3]

[1] School of Technology and Management, Polytechnic Institute of Leiria, Portugal
mfrade@estg.ipleiria.pt
[2] Centro Universitario de Merida, Universidad de Extremadura, Spain
fcofdez@unex.es
[3] ETSI Informática, Campus de Teatinos, Universidad de Málaga, Spain
ccottap@lcc.uma.es

Abstract. Terrain generation algorithms can provide a realistic scenario for video game experience and can help keep users interested in playing by providing new landscapes each time they play. Nowadays there are a wide range of techniques for terrain generation, but all of them are focused on providing realistic terrains. This paper proposes a new technique, Genetic Terrain Programming, based on evolutionary design with GP to allow game designers to evolve terrains according to their aesthetic feelings or desired features. The developed application produces Terrains Programs that will always generate different terrains, but consistently with the same features (e.g. valleys, lakes).

Keywords: terrain generation, video games, evolutionary art, genetic programming.

1 Introduction

To achieve richer human-machine interaction video games must be dynamic, they need to present game players with novel plots, intelligent artificial opponents, different goals and even scenario changes. Terrain generation algorithms have an important role in the video games' dynamics. They help keep users interested in playing by providing new landscapes each time they play.

Nowadays there are a wide range of techniques for terrain generation (see Sect. 2), but all of them have constrains. More elaborated methods depend highly upon designer's skills, time and effort to obtain acceptable results. The simpler methods generate only a narrow variety of terrain types and do not allow the control over terrain features. A common feature with current techniques is that they all try to generate realistic terrains. Although this is important, imagine the possibility of a game designer to evolve their own terrain accordingly to their aesthetic feelings. This can lead to the generation of more exotic terrains at the cost of realism, but might also give players a feeling of awe and increase their interest in playing. This paper presents a new technique, that we designated as

M. Giacobini et al. (Eds.): EvoWorkshops 2008, LNCS 4974, pp. 485–490, 2008.

Genetic Terrain Programming, that allows the generation of terrains based on aesthetic evolutionary design with Genetic Programming (GP).

Sect. 2 presents some background on the current terrain generation techniques and an overview of evolutionary systems applied to terrain generation and similar domains. The details of the Genetic Terrain Programming technique are described in Sect. 3 and some results are presented in Sect. 4. Finally, Sect. 5 presents the conclusions.

2 Background

A terrain can be represented in many ways, but one of the most common is the height map, which is equivalent to a grayscale image. Height maps have the limitation that they cannot represent structures with multiple heights for the same coordinates (such as caves), but are sufficient for many uses and can be highly optimised for rendering and object collision detection [1].

Current terrain generation techniques can be divided in three main categories: measuring, modeling and procedural. In the measuring techniques elevation data is derived from real-world measurements to produce Digital Elevation Models [1]. Measuring has the advantage of producing highly realistic terrains, but at the expense of designer control. The designer might have to search extensively to find real-world data that meets his goals.

Modeling is by far the most flexible technique for terrain generation. A human artist manually models the terrain morphology using a 3D modeling program (e.g. 3D Studio[2], Blender[3]). With this approach the designer has unlimited control over the terrain design and features, but this might be also a disadvantage. This technique imposes high requirements on the designer in terms of time, effort and skills.

The procedural category can be divided into physical, spectral synthesis and fractal techniques. Physically-based techniques simulate the real phenomena of terrain evolution trough effects of physical processes such as erosion [2], or plate tectonics. These techniques generate highly realistic terrains, but require an in-depth knowledge of the physical laws to implement and use them effectively. The spectral synthesis approach calculates random frequency components and then the inverse Fast Fourier Transform (FFT) is computed to convert the frequency components into altitudes [2]. The fractal techniques exploit the self-similar characteristic that terrains present to a limited extent [3]. These algorithms are the favourite ones by game's designers, mainly due to their speed and simplicity of implementation. There are several tools available, such as Terragen[4] and GenSurf[5]. However, terrains generated by fractal techniques can be recognised

[1] http://rockyweb.cr.usgs.gov/nmpstds/demstds.html
[2] http://www.autodesk.com/fo-products
[3] http://www.blender.org
[4] http://www.planetside.co.uk/terragen
[5] http://tarot.telefragged.com/gensurf

because of the self-similarity characteristic. The spectral synthesis and fractal techniques do not allow the designer to control the terrain outcome features.

Teong Ong et al. proposed an evolutionary design optimisation technique to generate terrains [4]. Their approach breaks down the terrain generation process into two stages: the terrain silhouette generation phase and the terrain height map generation phase. A database of height map samples, representative of the different terrain types, is used to search an optimal arrangement of elevation data that approximates the map generated in the first phase.

Like the techniques described previously, the solution proposed in [4] is focused only on generating realistic terrains. It cannot generate terrains accordingly to the designer's aesthetic appeal, or with an alien look. On the opposite side there are artists. They have used evolutionary art systems, for many years, to generate aesthetically pleasing forms rather than realistic ones. Because of the equivalence between height maps and grayscale images, it is possible to apply the same principle of evolutionary art to terrain generation. However, the phenotype will be a terrain surface instead of an image.

GP has been the most fruitful evolutionary algorithm applied to evolve images interactively. Karl Sims used GP to create and evolve computer graphics by mathematical equations [5]. Since then several researchers extended his work. The *SBART* (Simulated Breeding ART) allowed the creations of collages [6] and *NEvAr* (Neuro Evolutionary Art) [7] focused on automatic seeding procedures.

We believe that the use of evolutionary art systems with GP will allow the creation of both aesthetic and real terrains. Additionally some control over localised terrain features will be possible trough the use of several Terrain Programmes to compose the full landscape. It will also require less effort and time than modeling techniques to create complex terrains and the result will not be solely dependent on the designer's skills.

3 Genetic Terrain Programming

Our main goals are to create TPs (Terrain Programmes) capable of generating different terrains but consistently with the same features and to obtain TPs capable of generating realistic or aesthetic terrains. We propose the use of aesthetic evolutionary design with GP to achieve them. This approach consists of Interactive Evolution (IE) of terrains accordingly to a desired feature or aesthetic appeal. We have coined the term Genetic Terrain Programming to denote this technique. Although the concept is similar to the one used in evolutionary art systems, to the best of our knowledge, the application to terrain generation has never been applied.

Our application, *GenTP* (Generator of Terrain Programs) has been developed with GPLAB[6], an open source GP toolbox for Matlab. The initial population is created randomly with trees depth size limited to 20 and a fixed population size of 12. The number of generations is decided by the designer, who can stop the application at any time. The designer can select one or two individuals to create

[6] http://gplab.sourceforge.net/

the next population and the genetic operators used depend upon the number of selected individuals. If one individual is selected only the mutation operator will be used. In case the designer chooses to select two individuals both the standard crossover and mutation operators [8] will be applied. Like in others IE systems, the fitness function relies exclusively on designers' decision, either based on his aesthetic appeal or on desired features.

$$F = \{plus_2; minus_2; multiply_2; myDivide_2; log_1; sin_1; cos_1; tan_1; atan_1; myMod_2; power_2; mySqrt_1; smooth_1; incline_1; FFT_1\} \ . \quad (1)$$

$$T = \{squares; rectangles; circles; triangles; planes; rand; fftGen\} \ . \quad (2)$$

Each GP individual is a tree composed by mathematical functions from Eq. (1), which are applied as image processing functions instead of strictly per-pixel functions, and terminals, from Eq. (2), which are height maps rather than scalar values. While in [5,6,7] the mathematical equations are used to calculate both the pixel value and its coordinates, in *GenTP* only the height will be calculated. The (x, y) coordinates will be dictated by the matrix position occupied by the height value.

Some terminals depend upon a Random Ephemeral Constant (REC) to define some characteristics, such as inclinations of *planes* and spectrum values of *fftGen*. All terminals depend upon a random number generator, which means that consecutive calls of one TP will always generate different terrains. This is a desired characteristic because we want to be able create different terrains with each TP, but that will share the same features.

4 Experimental Results

Two kind of experiments were conducted, the first one consisted on obtaining aesthetic appealing terrains (regardless of their realism) and the second one to achieve a realistic terrain with a specific feature in mind. On the first kind of experiments we were able to get aesthetic appealing terrains after about 30 to 70 generations. On those experiments we were able to obtain very different kinds of terrains types. For example, the TP represented in Eq. (3) creates terrains with a bank of knolls with two ridges that give them an alien look. Fig. 1 has examples of terrains generated from three different TPs. Each row has pictures of terrains generated by 3 consecutive executions of the same TP.

$$H = log(incline(sin(mySqrt(smooth(fftGen(1.25)))))) \ . \quad (3)$$

On the second kind of experiments we tried to obtain TPs to generate terrains with specific features, such as mountains, cliffs or corals (see Fig. 2). In this case the number of necessary generations varies widely until we are able to get acceptable results. These number is highly dependent on the initial population and could vary between 10 to more than 100 generations. When running the

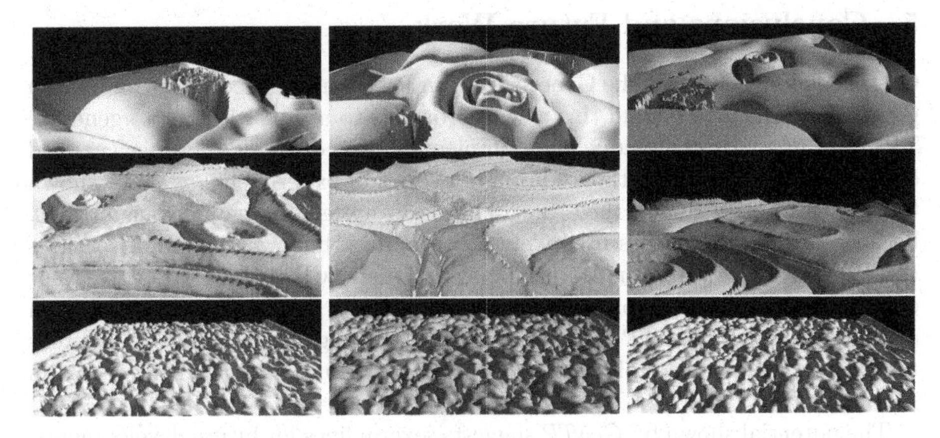

Fig. 1. Exotic terrains generated by 3 different TPs (rendered with 3DS Max). The pictures of the third row were generated by Eq. (3).

experiments, if after a number of generations an interesting result is not obtained, we have preferred to cancel the experiment and begin again, avoiding this way a long run. We also verified that, for realistic landscapes, the range of terrains types were narrower than in the first experiment. Eq. (4) has an example of a TP that was evolved having the aim of generating a coral looking terrain.

$$H = log(minus(fftGen(2.75), log(minus(smooth(fftGen(1.5)), fftGen(2.5))))) \ . \qquad (4)$$

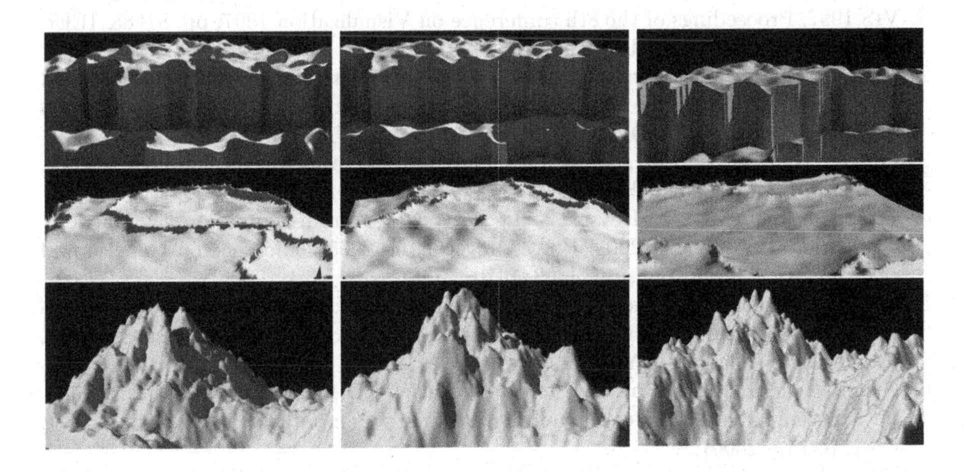

Fig. 2. TPs evolved with specific features in mind (rendered with 3DS Max). From top to bottom row: cliffs, corals (Eq. (4)) and mountains.

5 Conclusions and Future Work

On this paper we present the Genetic Terrain Programming technique for terrain generation. The idea behind this new approach is to use IE with GP to generate TPs. To employ this technique a first implementation of *GenTP* has been carried out with GPLAB and Matlab. Through a series of experiments we have shown that multiple execution of the same TP will allways generate differrent terrains, because of the randomness present in its terminals, but with the same features (Fig. 1 and 2). With a single technique designers will be able to evolve very different kinds of TPs, from real looking terrains to more exotic ones with an alien semblance. Those TPs can be inserted in video games to generate different terrains each time a user plays and consequently help to keep users interested in playing.

The potential shown by *GenTP* suggests several lines for future developments. One of them is to augment the GP terminal set in order to try to obtain a wider range of realistic terrain types with fewer generations. More features could be added to our technique so that whole scenarios, including vegetation and buildings, can be generated.

Acknowledgements. This work was partially funded by National Nohnes project TIN2007-68083-C02-01 Spanish MCE. We also would like to deeply thank Nuno Monteiro for his helpful assistance with 3D Studio Max.

References

1. Duchaineau, M., et al.: ROAMing terrain: Real-time optimally adapting meshes. In: VIS 1997: Proceedings of the 8th conference on Visualization 1997, pp. 81–88. IEEE Computer Society Press, Los Alamitos (1997)
2. Olsen, J.: Realtime procedural terrain generation - realtime synthesis of eroded fractal terrain for use in computer games. In: Department of Mathematics And Computer Science (IMADA), University of Southern Denmark (2004)
3. Peitgen, H.O., Jürgens, H., Saupe, D.: Chaos and Fractals - New Frontiers of Science, 2nd edn. Springer, Heidelberg (2004)
4. Ong, T.J., Saunders, R., Keyser, J., Leggett, J.J.: Terrain generation using genetic algorithms. In: GECCO 2005: Proceedings of the 2005 conference on Genetic and evolutionary computation, pp. 1463–1470. ACM, New York (2005)
5. Sims, K.: Artificial evolution for computer graphics. In: SIGGRAPH 1991: Proceedings of the 18th annual conference on Computer graphics and interactive techniques, pp. 319–328. ACM, New York (1991)
6. Unemi, T.: SBART 2.4: an IEC tool for creating 2D images, movies, and collage. In: Proceedings of 2000 Genetic and Evolutionary Computational Conference, NV, USA, p. 153 (2000)
7. Machado, P., Cardoso, A.: NEvAr - the assessment of an evolutionary art tool. In: Wiggins, G. (ed.) Proceedings of the AISB'00 Symposium on Creative & Cultural Aspects and Applications of AI & Cognitive Science 2000, Birmingham, UK (2000)
8. Koza, J.R.: Genetic programming. on the programming of computers by means of natural selection. The MIT Press, Cambridge (1992)

Virtual Constructive Swarm Compositions and Inspirations

Sebastian von Mammen[1], Joyce Wong[2], and Christian Jacob[1]

University of Calgary
[1] Department of Computer Science
[2] Faculty of Fine Arts
2500 University Drive NW
T2N 1N4 Calgary, Alberta, Canada
{s.vonmammen,wongjky,cjacob}@ucalgary.ca

Abstract. This work is an example of the interplay between computer-generated art work by a computer scientist and traditional paintings on canvas by an artist. We show that computer-generated swarm constructions can obtain great expressiveness and exhibit liveliness, rhythm, movement, tension, contrasts, organic looks, and rigid forms. These characteristics lend themselves to complement traditional paintings when swarm constructions are integrated into the according works. The interplay between computationally generated and traditional art is even furthered when artistic conceptualizations are governed by swarm constructions.

1 Introduction

When swarms were first modeled in the virtual realm in the form of flocking birds (boids) [12], spectators awed at animations of the interaction dynamics of flocking agents. Soon the cooperative swarm paradigm and the intrinsic dynamics of swarm models were transferred to other fields, most prominently those of artificial intelligence, where they became versatile optimizers [2,9].

A typical example of swarm intelligence in nature are the stunning construction capabilities of insect societies. Ants, wasps, and termites intrigue with magnificent compositions made from leaves, mud, and sand. Models of these distributed, decentralized construction processes are able to reproduce those natural structures in virtual, computer-generated scenarios [13,11,16]. In the same context, *swarm grammars* have been shown to elegantly capture constructive swarm methods. Here the individuals flock like traditional boids [12,8], but they also reproduce in accordance with a grammatical production system [14]. Swarm agents, as they split and specialize during the construction process, give rise to innovative and artistic structures [6,15].

We have extended the swarm grammar concept by an event-driven rule application, thus creating a more versatile constructive swarm system. In Section 3 we show how constructions of this extended swarm model become part of traditionally crafted art works, thereby offering a complementary perspective.

M. Giacobini et al. (Eds.): EvoWorkshops 2008, LNCS 4974, pp. 491–496, 2008.

The interplay between computer-generated and human art is deepened when impressions of swarm constructions give rise to modern artworks in Section 4.

2 Related Work

In 1987 Craig Reynolds presented a computational model of swarms that exhibit natural flocking behavior [12]. An individual, or *boid*, accelerates according to its *flocking urges—separation:* heading away from the neighbors, *alignment:* adjusting one's velocity towards the neighbors, and *cohesion:* being attracted towards one's perceived neighbors. In later work, additional urges and specific configurations were introduced that result in diverse flocking formations such as 'V'-shaped formations [4] or line, circular and figure-eight patterns [8].

Emergent choreographic flocking of bio-inspired swarms have influenced many art works. While spontaneous creativity of swarms is reflected in many paintings [10], their potential to coordinate and to show surprising vividness is, for instance, applied in automatic and assisted music generation [1]. The same features render them ideal as interacting units of interactive swarm art installations [5] that exhilarate large audiences.

But a swarm's abilities can surpass choreographic flocking. For example, a very successful model suggests that termites, ants and wasps construct their nests in a step by step fashion according to stimuli in their nearby environment [2,11]. Importantly, the construction is not driven by a blueprint or by explicit communication between the individuals. Instead, the configuration of the environment (*template*) determines the next steps of constructional swarm activity: A *stigmergic* chain-reaction is triggered where environmental changes lead to subsequent constructional efforts. Wasp nests emerge through *qualitative* or *discrete* stigmergy [7] depicting discrete stimuli that trigger activity. Termite mounds rise through *quantitative* stigmergy where the amount or intensity of a stimulus is reflected in the construction behavior [3].

In our constructive swarm model we embedded a generalized rule system into swarm grammars [6] to allow for stigmergic communication. Now, neighborhood stimuli, random events and timers determine the reproduction and construction activity of a swarm individual while flocking parameters determine the flight behavior in accordance with the boids model [12]. The swarm agent must know about the configuration of its offspring and about the attributes of its construction elements and templates. Hence, the genotype of an individual comprises the configuration parameters and reproduction rules of possible offspring and configurations of the construction elements and templates it can build.

3 Compositions

In this section, we present two compositions of traditional paintings and printed digital structures that were constructed by swarm grammars [6]. Figure 1(a) shows: [Generative], where desolate ambiance is reduced to its simplest form (20" x 26" Pencil, pencil crayon on Mayfair). Tension is built by swift vibrant strokes and radiant aggregations. The swarm construction placed at the bottom-right of the painting *bridges back and foreground* and strengthens the impression of *movement*. Figure 1(b) shows [A Splinter of Blue in a Sea of Red], a nonsensical piece based on rhythms (20" x 28" Gouache, watercolor, charcoal, transparent paper). Here, rhythmic themes are complemented by *wavelike* (center-right) and *circular* structures (bottom) that arose through dancing swarms. In both cases the embedded swarm constructions add *dynamics* to the paintings. The constructions leave the impression of vibrant or rhythmical movements because the lively building processes of the swarms are solidified in the respective sculptures.

The swarms utilized in the presented compositions were bred by interactive evolution: Upon the display of several phenotypes the breeder drove the evolutionary process by manual fitness assignment. The genotypical information comprised the flocking parameters, the attributes of the construction elements, and the grammatical reproduction rules of the swarm agents.

Fig. 1. Compositions of printed computer-generated swarm constructions and brush work: (a) [Generative], (b) [A Splinter of Blue in a Sea of Red]

4 Inspirations

In the previous section swarm constructions solely complemented traditional art work. Now, digital swarm designs become the basis to inspire artistic studies. Thereby, the relationship between computer-generated and traditional art is emphasized.

Virtual constructive swarms were configured to serve loose artistic conceptions that resulted in swarm constructions (Fig. 2). These computer-generated constructions paved the way for further art work fleshed out by knives, pens and brushes (Fig. 3). In the following paragraph, we successively describe the pairs of swarm constructions depicted in Fig. 2 and inspired art works displayed in Fig. 3.

A steady trace of green spheres with regular branches creates the impression of an *organic structure* (Fig. 2(a)). The artistic realization extrapolates from organic matter towards human life and problems in our relationship with the environment: [Aftermath] (Fig. 3) is a study on the condition of human consumption (8" x 8" acrylic on matte board).

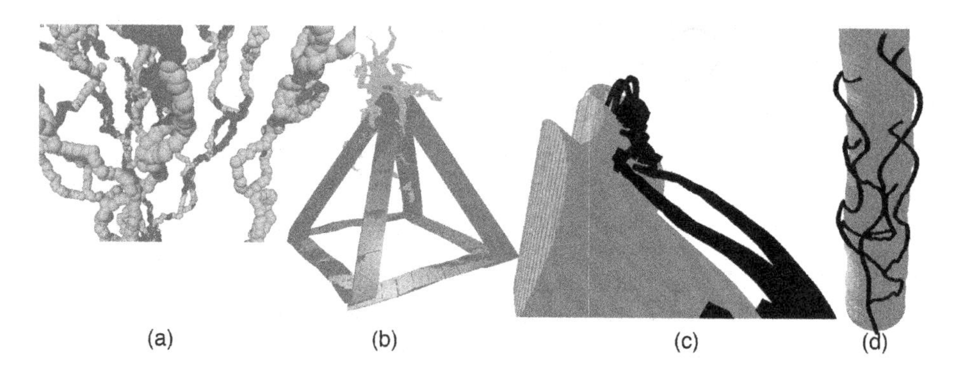

Fig. 2. Constructions of swarms programmed according to loose artistic conceptions. From left to right: (a) Branching constructions of spherical shapes, (b) a pyramidal construction with swarming agents in green, (c) and (d) a wave-like and a stem-like construction are outlined by a thin, black, branching thread of swarm agents.

The base of the pyramidal swarm construction (Fig. 2(b)) is *rigid* and shines in *cold, steel-blue* colors. In contrast, tentacle-forming swarms *wind* from its peak. The *explosive polarity* of the pyramid inspired the piece [Manifest] (Fig. 3) that places the swarm pyramid into a new context, in which unrequited thoughts seek ways to escape (12" x 24" black gesso, acrylic on Masonite). Painted layers were scratched away to reveal the raw Masonite surface. Soft swipes led to a semi-transparent reflection of the rigid pyramid foundation. Energetic cuts at the pyramid's top lend the painting real structure.

The intrinsic swarm dynamics and the *contrast* between *hard, structural manifestations* and *delicate outlines* (Fig. 2 (c) and (d)) are reflected in the remaining four pieces in Figure. 3. In [Skip Cross] a swarmette[1] makes a playful leap when facing a crossroad (8" x 16" acrylic on Masonite). The swarmette's flight is retraced with swift motions and underlined with an emphasis on positive and negative space. [Unravel] is the title of a piece where a Chinese calligraphic character reveals itself beneath a surface (8" x 8" ink on rice paper). The color themes and associated forms are directly inherited from the underlying swarm constructions. [Net Sky] inverts the predominance of the original swarm constructions: A web is woven in a *coordinated* fashion. The black threads *agglutinate* in front of dissolving purple and yellow clouds (8" x 10" acrylic on Masonite). In the first panel of the diptych [Outlining Blues] a distortion of the swarmettes formulates a Whale-like specimen as it swims in a peaceful surrounding. The second panel leaps backward in time and depicts first organisms coming into existence (both 12" x 24" oil paint and rusting agents on metal).

The presented works isolate and abstract from a variety of phenomena that materialize in swarm constructions, examples are: Organic looks (Aftermath, Outlining Blues), indications of lively movements (Manifest, Skip Cross), polarities and contrasts of shapes and looks (Manifest, Unravel, Net Sky), as well

[1] A swarmette is a swarm agent within our swarm grammar system.

Fig. 3. Swarm-inspired art — descriptions of the pieces are found in the text. Top-left to bottom-right: *Aftermath, Manifest, Skip Cross, Unravel, Net Sky,* Diptych: *Outlining Blues.*

as structural coalescence (Net Sky, Outlining Blues). Additionally, many of the presented pieces put an emphasis on spatial dimensionality — with an interplay of positive and negative space (Skip Cross), the elaboration on real structural depth (Aftermath), or via the creation of crinkly textures (Unravel).

5 Summary and Future Work

In many cases swarm constructions conserve the dynamical processes of their creation. Thus, swarm constructions show a remarkable variety of features that are interesting for artistic works. We have shown that rule-based virtual constructive swarms are capable to create structures that become part of compositions and inspire traditionally created art works. The characteristics of swarm constructions displayed by the presented art are hard to evaluate computationally in the sense of a 'fitness function'. In order to foster these qualities (such as vividness or coalescence) in constructive swarms, interactive evolution is an ideal approach. Therefore it would be interesting to render them the objectives of interactive evolutionary breeding — to maximize, for instance, vividness or coalescence. However, swarms could also develop diverse and innovative designs by themselves. As we have also demonstrated, instead of interactive supervision loose pre-defined constraints could drive purely non-human art.

References

1. Blackwell, T.: Swarming and music. In: Miranda, E.R., Biles, J.A. (eds.) Evolutionary Computer Music, pp. 194–217. Springer, London (2007)
2. Bonabeau, E., Dorigo, M., Theraulaz, G.: Swarm Intelligence: From Natural to Artificial Systems. Oxford University Press, New York (1999)
3. Camazine, S., et al.: Self-Organization in Biological Systems. Princeton University Press, Princeton (2003)
4. Flake, G.W.: The Computational Beauty of Nature: Computer Explorations of Fractals, Chaos, Complex Systems, and Adaption. A Bradford Book, MIT Press, Cambridge (1999)
5. Jacob, C., Hushlak, G., Boyd, J.E., Sayles, M., Pilat, M.: Swarmart: Interactive art from swarm intelligence. LEONARDO 40(3), 248–254 (2007)
6. Jacob, C., von Mammen, S.: Swarm grammars: Growing dynamic structures in 3d agent spaces. Digital Creativity: Special issue on Computational Models of Creativity in the Arts 18 (2007)
7. Karsai, I., Penzes, Z.: Comb building in social wasps: Self-organization and stigmergic script. Journal of Theoretical Biology 161(4), 505–525 (1993)
8. Kwong, H., Jacob, C.: Evolutionary exploration of dynamic swarm behaviour. In: Congress on Evolutionary Computation, Canberra, Australia, IEEE Press, Los Alamitos (2003)
9. Liu, Y., Passino, K.M.: Swarm intelligence: Literature overview, `http://www.ece.osu.edu/~passino/swarms.pdf`
10. Moura, L.: Website of leonel moura (December 2007), `http://www.lxxl.pt`
11. Pilat, M.: Wasp-inspired construction algortihms. Technical report, University of Calgary (2004)
12. Reynolds, C.W.: Flocks, herds, and schools: A distributed behavioral model. In: SIGGRAPH 1987 Conference Proceedings, vol. 4, pp. 25–34 (1987)
13. Theraulaz, G., Bonabeau, E.: Modelling the collective building of complex architectures in social insects with lattice swarms. Journal of Theoretical Biology 177(4), 381–400 (1995)
14. von Mammen, S.: Swarm grammars - a new approach to dynamic growth. Technical report, University of Calgary (May 2006)
15. von Mammen, S., Jacob, C.: Genetic swarm grammar programming: Ecological breeding like a gardener. In: Srinivasan, D., Wang, L. (eds.) 2007 IEEE Congress on Evolutionary Computation. IEEE Computational Intelligence Society, pp. 851–858. IEEE Press, Singapore (2007)
16. von Mammen, S., Jacob, C., Kokai, G.: Evolving swarms that build 3d structures. 2005 IEEE Congress on Evolutionary Computation 2, 1434–1441 (2005)

New-Generation Methods in an Interpolating EC Synthesizer Interface

James McDermott[1], Niall J.L. Griffith[1], and Michael O'Neill[2]

[1] CCMCM, Dept. CSIS, University of Limerick
[2] NCRA, School of CSI, University College Dublin
jamesmichaelmcdermott@gmail.com, niall.griffith@ul.ie, m.oneill@ucd.ie

Abstract. This paper describes work on a graphical user interface (GUI) for sound synthesizers based on interactive evolutionary computation. The GUI features user-controlled interpolation for fast auditioning and evaluation of relatively large populations. Interpolation behaviour is considered with reference to usability and the psychoacoustical "just noticeable difference". Different methods of generating new populations are tested in a formal usability experiment: a method giving a type of consistency of behaviour and increased population diversity is shown to give improved performance over two methods lacking these characteristics.

1 Introduction

The problem of setting sound synthesizer control parameters to achieve a desired sound is often difficult and unintuitive. Interactive evolutionary computation (IEC) can be applied to make the problem easier by removing from the user some of the burden of understanding and manipulating the parameters [1,2,3].

This paper gives the results of further study of an IEC synthesizer graphical user interface (GUI) described previously [4]: this GUI uses a simple interpolation scheme to allow the user to create and audition a large number of individuals quickly, using direct (interactive) selection and dispensing with an explicit fitness function. Its performance has been shown in previous work to be competitive with existing IEC and non-EC GUIs, and the aim here is to study some variations experimentally. The behaviour of the interpolations is investigated with regard to the psychoacoustical concept of the *just noticeable difference*. A formal usability experiment is run to test the performance in *target-matching* tasks of four versions of the GUI. The main results are that a version emphasising a type of consistency of behaviour and increased population diversity gives improved performance over two versions lacking these characteristics.

2 Previous Work

IEC is often used in aesthetic systems as a way to avoid the problem of defining explicit fitness functions for hard-to-define goals, and several authors have used this approach to build IEC systems for synthesis control. A good example is

M. Giacobini et al. (Eds.): EvoWorkshops 2008, LNCS 4974, pp. 497–502, 2008.

given by Johnson [2]: the GUI consists of buttons (to hear the sounds in the current population) and sliders (to assign them fitness values). After evaluation of a population, the user clicks an "evolve" button: the algorithm is iterated and a new generation produced using standard EC methods. Other examples include *Genophone* [3] and *MutaSynth* [1].

Despite the sophistication and success of these and other systems, some problems seem to remain. It is very difficult to measure the success of IEC systems, and this is reflected in a relative lack of experimental results in the literature. IEC applications suffer from a *fitness evaluation bottleneck* [5] — the fact that usually a human will be orders of magnitude slower than a computer in evaluating an individual's fitness. Users may become fatigued or bored even with relatively small populations and numbers of generations.

2.1 Previous Work in This Project

The project of which this paper is a part uses several methods in an attempt to measurably improve IEC performance. The "Sweep" GUI proposed in previous work and studied further here uses simple interpolation between individuals to allow the fast auditioning of many sounds with user control (*sweeping*); direct selection to avoid requiring numerical fitness evaluations of the user; and a sequenced triggering of the sound to avoid mouse-gestural overhead.

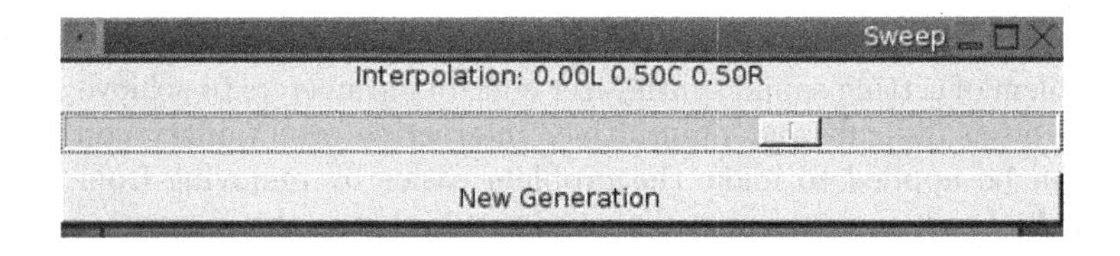

Fig. 1. Sweeping IGA GUI

The GUI is depicted in Fig. 1. The nominal population consists of three points L, C and R at the left, centre and right of the single slider, each corresponding to an array of synthesis parameter values (i.e. a *patch*). This array is the genetic representation. By moving the slider to intermediate positions the user can access (or create) a much larger effective population, which varies between close similarity to the centre point and complete dissimilarity. The output sound corresponds to an output patch X, which is formed by interpolation of individual parameters: when the slider is at a proportion x of the distance from L to C, $X = L + x(C - L)$ (with the analogous relation for points between C and R). The details on issues such as logarithmically-mapped parameters, and varying parameter ranges, have been dealt with in previous work. When the user is satisfied that the current generation has been sufficiently explored, he or she sets the slider at a preferred point X_0 and clicks the "New Generation" button. X_0 becomes the new C, and new points L and R are created. The possible methods of creating them are the main subject of this paper.

3 Open Questions

Previous work [4] has demonstrated that a GUI of the type discussed above can be competitive both with more typical interactive EC GUIs and with non-EC GUIs. New questions are raised on issues such as the creation of the end-points. The *just noticeable difference* (the smallest distance between sounds which is yet noticeable) is central to these questions and is considered first.

3.1 Just Noticeable Differences

Intuitively, there are some sound-changes which are noticeable, and some which are too small to notice, and therefore there is some sort of threshold between them. The *just noticeable difference* (JND) is a psychoacoustical term referring to the smallest change in a sound which is still capable of being perceived by a listener. The JNDs for changes in many timbral and non-timbral aspects of sound have been measured experimentally [6].

A listening test to determine JNDs for many synthesis parameters individually is a very large undertaking; however, for the purposes of studying the "Sweep" GUI it is not required. Rather, a JND is defined *over interpolations* as follows: it is the smallest distance from a point L along a parameter-space interpolation projected through a point C such that a listener can perceive that the sound corresponding to the point X at that distance is different from L. Thus JND is a distance between two points (L and X) in the synthesizer parameter space, and can be measured as a distance in parameter space. The experiment described in Sect. 4 both investigates the interpolation-JND experimentally and makes use of the concept in usability testing.

3.2 Methods of Generating End-Points

A natural question concerning the "Sweep" GUI is whether methods, other than random generation, of creating new end-points can improve performance. Previous work investigated the idea of *background evolution*, where good results from an ongoing non-interactive search process were copied to the end-points. Another possibility is random generation with *distance constraints*: here, the distances between the elite centre point C and the end-points L and R, the distances between the end-points and the individuals in previous generations, and the distance between the end-points themselves can be constrained to be either less than, equal to, or greater than a given distance (distance being measured in the parameter space). Motivations for this type of scheme include increasing population diversity and preventing new individuals from being too similar to bad individuals rejected in previous generations. The idea of the JND impacts on this scheme in that the JND provides an (approximate) distance which must be enforced between points to ensure that they are distinct. Four algorithms (given in Sect. 4.1) were chosen for testing in a formal experiment, described next.

4 Experimental Protocol

The experiment was composed of 10 JND tests and 8 target-matching tasks, described in detail below. There were 20 subjects, 12 female and 8 male, mixed in synthesizer ability. The synthesizer was a slightly modified version of Xsynth [7], with 29 parameters. Subjects performed the experiment one at a time using a laptop, headphones and a mouse. They were given written and informal spoken instructions, and allowed to ask questions during a trial run, before the experiment proper. They were paid a nominal fee of €5.

The aim of the JND tests was to gather experimental data on the interpolation-JND as defined above. Each subject performed 10 JND tests using a GUI derived from the "Sweep" GUI itself. The subject heard two synthesized sounds alternating, one of which was controlled by the slider, and was required to move the slider to the point at which the two sounds became just noticeably different. The parameter-space distance between the two was then recorded.

4.1 Target-Matching Tasks

The aim of the target-matching tasks was to compare directly the usability and performance of four "Sweep" GUIs. They differed only in the underlying endpoint-generation algorithms, as follows:

GUI 0: Random Generation. The end-points L and R for each new generation were generated randomly, as in previous work.

GUI 1: Construction of Nearby Individuals. L and R constructed randomly such that the parameter distance between each of them and C was equal to 0.06, the largest non-outlier value found in JND pilot experiments.

GUI 2: Construction of Distant Individuals. L and R constructed randomly such that the parameter distance between each of them and C was exactly equal to 0.45, which is close to the largest value (0.5) constructible in general.

GUI 3: Multiple Distance Constraints. L and R randomly generated such that the parameter distance between them was greater than 0.06, and the distance between each of them and each individual in preceding populations was greater than 0.06.

Two target sounds were chosen from among the synthesizer's built-in presets. This ensured that the target sounds were achievable using the given synthesizer, and were "desirable" sounds. They were a typical "synth strings" sound and a percussive xylophone-like sound.

There were eight target-matching tasks consisting of the four GUIs crossed with two target sounds, the presentation order of the GUIs being varied among the subjects. During the task, the subject could freely switch the synthesizer to play either the target sound or the current candidate sound (that under the subject's control). At the end of each task, subjects were required to give a User Rating expressing the perceived similarity of the final result to the target on a scale of 1-7. The Elapsed Time was also recorded.

5 Results

JND results are plotted in Fig. 2: the mean JND was about 0.02, with non-outlying data between (effectively) 0.0 and 0.06 (in parameter distance). Fig. 2 also shows the distribution of parameter distance between adjacent pairs of points in the "Sweep" GUI. Comparing the two distributions shows that in the vast majority of cases, the smallest possible movement in the "Sweep" GUI (between adjacent points) is smaller than the interpolation-JND. The conclusion is drawn that the control afforded by the "Sweep" GUI is sufficiently fine.

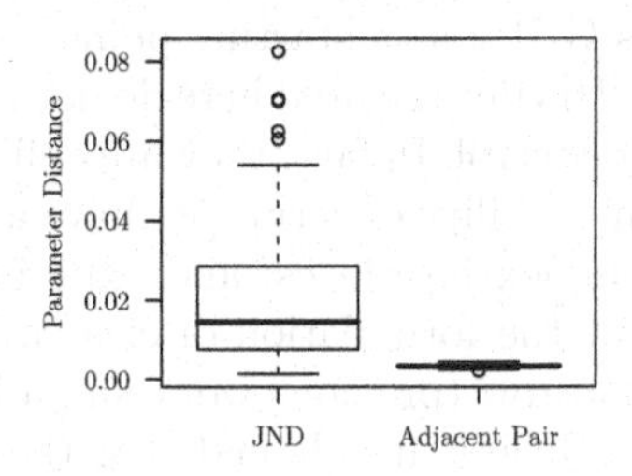

Fig. 2. Distribution of JNDs and distances between adjacent pairs of points in the "Sweep" GUIs, using the parameter distance measures

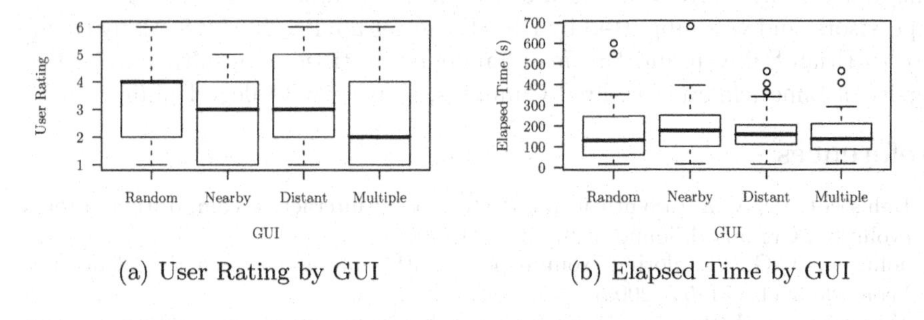

(a) User Rating by GUI (b) Elapsed Time by GUI

Fig. 3. Results analysed by GUI

Figure. 3 shows the User Rating and Elapsed Time results of the target-matching tasks, analysed by GUI. A two-way repeated measures ANOVA was run to analyse the results. There were significant differences ($p < 0.01; F = 4.3$) in User Rating by GUI. Post-test pairwise t-tests showed that GUI 2 performed better than GUIs 1 and 3 at the $p < 0.05$ confidence level, after a Bonferroni correction of a factor of 20 to allow for multiple statistical tests on the same data. The differences between GUI 0 and all of the others were *not* found to be significant. The synth-strings sound was easier to achieve than the xylophone, as judged by User Rating ($p < 0.001; F = 123.0$). Neither the GUI nor the target sound affected the Elapsed Time results.

6 Conclusions and Future Work

The fact that statistical significance is found in some results (GUI 2 beating GUIs 1 and 3) is evidence that performance improvements can be detected where they exist, and gives strong support for the methodology adopted in this experiment. It also demonstrates that high diversity is essential to good performance.

Although performance for GUI 2 was statistically equivalent to that for GUI 0, the mean User Rating was slightly higher; and it may give the benefit of more *consistent* behaviour of interpolations. The parameter distance represented by a given mouse movement is consistent between generations and between trials. A paper in preparation on two-dimensional and higher-population extensions of the method therefore takes GUI 2 as a starting point.

It will be clear that the GUIs investigated here do not represent the state of the art either in EC synthesizer control. In fact, they are deliberately very simplified, for two reasons. Subjects in usability experiments have a limited amount of time and capacity to absorb complex interfaces; and experimental set-ups differing only in a single aspect allow the formulation of clear hypotheses. The scope of these experiments is quite narrow (previous work and a follow-up paper expand the scope somewhat). Nevertheless, it is hoped that the experimental approach and results may be applied to other IEC applications.

Acknowledgements

The first co-author gratefully acknowledges the guidance of his co-authors and supervisors, and was supported by IRCSET grant no. RS/2003/68. Many thanks also to Brian Sullivan and Dr. Jean Saunders, Statistics Consulting Unit, University of Limerick, and Dr. Fred Cummins, University College Dublin.

References

1. Dahlstedt, P.: A MutaSynth in parameter space: interactive composition through evolution. Organised Sound 6(2), 121–124 (2001)
2. Johnson, C.G.: Exploring sound-space with interactive genetic algorithms. Leonardo 36(1), 51–54 (2003)
3. Mandelis, J., Husbands, P.: Genophone: Evolving sounds and integral performance parameter mappings. In: Raidl, G.R., Cagnoni, S., Cardalda, J.J.R., Corne, D.W., Gottlieb, J., Guillot, A., Hart, E., Johnson, C.G., Marchiori, E., Meyer, J.-A., Middendorf, M. (eds.) EvoIASP 2003, EvoWorkshops 2003, EvoSTIM 2003, EvoROB/EvoRobot 2003, EvoCOP 2003, EvoBIO 2003, and EvoMUSART 2003. LNCS, vol. 2611, pp. 535–546. Springer, Heidelberg (2003)
4. McDermott, J., Griffith, N.J., O'Neill, M.: Evolutionary GUIs for sound synthesis. In: Giacobini, M. (ed.) EvoWorkshops 2007. LNCS, vol. 4448, pp. 547–556. Springer, Heidelberg (2007)
5. Takagi, H.: Interactive evolutionary computation: Fusion of the capabilities of EC optimization and human evaluation. Proc. of the IEEE 89(9), 1275–1296 (2001)
6. Zwicker, E., Fastl, H.: Psycho-acoustics: Facts and Models. Springer, Heidelberg (1990)
7. Bolton, S.: XSynth-DSSI (2005) (accessed 2 March 2006), `http://dssi.sourceforge.net/`

Composing Music with Neural Networks and Probabilistic Finite-State Machines

Tomasz Oliwa and Markus Wagner

Artificial Intelligence Center, University of Georgia
Institute for Computer Science, University of Koblenz-Landau
oliwa@uga.edu, wagnermar@uni-koblenz.de

Abstract. In this paper, biological (human) music composition systems based on Time Delay Neural Networks and Ward Nets and on a probabilistic Finite-State Machine will be presented. The systems acquire musical knowledge by inductive learning and are able to produce complete musical scores for multiple instruments and actual music in the MIDI format. The quality of our approaches is analyzed in objective and subjective manner with existing techniques.

Keywords: Biological Inspired Music, Music Composition, Representation Techniques, Comparative Analysis, Time Delay Neural Networks, Finite State Machines, Inductive Learning.

1 Introduction

Artificial music composition systems have been created in the past using various paradigms. Approaches using Recurrent Neural Networks [7] and Long-Short Term Memory (LSTM) Neural Networks [3] architectures to learn from a dataset of music and to create new instances based on the learned information have been taken as well as approaches with genetic algorithms. The latter ones focus on semi-objective [9], i.e. combined computational and human evaluation of the songs, or fully objective fitness functions [8] to generate new songs. Associative Memories have also been tried [5] using a context-sensitive grammar.

Classical Algorithm-based automatic music composition systems, which aim at following predefined rules to construct music, stand or fall by human implementation of the underlying algorithms, which leaves the cumbersome task to derive sets of musical creation rules completely to the human designer. Another approach is to modify existing melodies by applying specific noise function to create new melodies [4], thus possibly reducing the dominance of the human factor. Heuristic Search Algorithms like Genetic Algorithms, on the other side, suffer from the fitness bottleneck [2][6], a gigantic, and in terms of music mostly unusable, search space.

Our machine learning systems extract important features/key elements from a dataset of music (created by humans) and are able to produce new song material which inherits these ideas. They compose music based on the extracted information gained by inductive learning. In both of our following approaches, we

M. Giacobini et al. (Eds.): EvoWorkshops 2008, LNCS 4974, pp. 503–508, 2008.

use machine learning techniques for the feature selection of musical information from the music database.

1.1 Music Background

Western Music can be defined as the chronology of notes, a note itself by its pitch and length (which we consider as our atomic unit, and with a "note", we always mean the combined information of note length and note pitch).

In classical music theory, a piece of music is written in a specific musical scale, which defines a subset from the set of all possible notes. Our machine learning systems use an unambiguous mapping of notes to our internal representation, which means that every note can be learned, regardless of its pitch or length.

The music which we considered as our music database for the melody where 19 classical and folk songs[1]. As an addition, we included 77 drum patterns in a drum training dataset from Metallica's rock song "Creeping Death" for the finite state machine (FSM) approach to come up with a multi-instrument song.

2 The Finite-State Machine Approach

2.1 Stochastic Uniform Sampling with Accumulated Probabilities

Our system parsed all songs from the song database and constructed multidimensional hashmaps (our knowledge base), which contain the probability of all possible note sequences and their successors. Figure 1 shows the underlying algorithm. The use of the accumulated probabilities simplifies the search for the successor(s), based on the idea from the stochastic uniform sampling method. The learned subsequent structures are similar to the "grouping structures" [1].

3 Music Representation

The 19 songs in our database were written in *abc* language [10]. Conversion from and to the *abc* language format from the MIDI format can be done using the abcMIDI package[2]. MIDI (Musical Instrument Digital Interface)[3] defines a communications protocol for instruments and computers to transfer data.

3.1 Feature Extraction

An illustrative example of assigning distinct integer values to notes, pauses, triplets etc. of the beginning of the song "Claret and Oysters" by Andy Anderson is shown in Figure 3.1 in the corresponding *abc* language and our integer representation. The richness of this representation stands in contrast to other approaches with a more restricted search space (like [9], which has no pauses or triplets). We found more diversified music because of our richer representation and thus bigger search space.

[1] http://abc.sourceforge.net/abcMIDI/original/MIDI.zip
[2] http://abc.sourceforge.net/abcMIDI/
[3] http://www.midi.org/

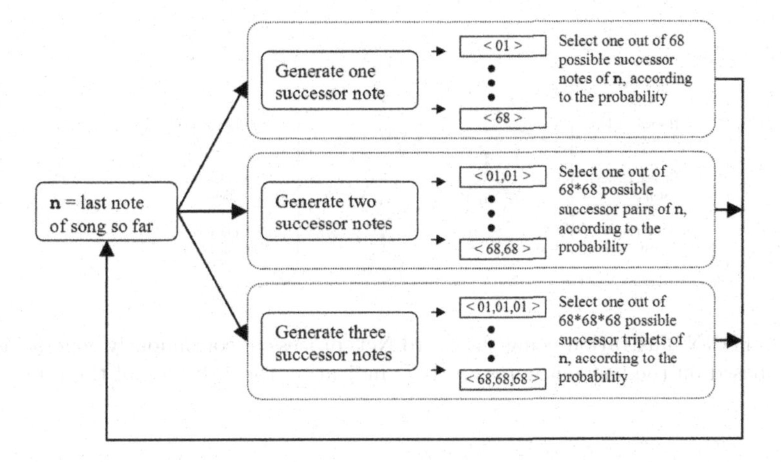

Fig. 1. Example of our algorithm, a Probabilistic FSM for a possible successor (sequence). In practice up to 4 base nodes and 10 successor notes are possible.

```
E2 G2 G2 G F         10, 18, 18, 20, 16,
G2 A2 B2 c2          18, 22, 26, 30,
d2 e2 g e d B        34, 38, 48, 40, 36, 28,
G2 G F G2 A2         18, 20, 16, 18, 22
```

Fig. 2. Beginning of "Claret and Oysters" in the *abc* language (left) and our integer representation (right)

4 The Neural Network Approach

An artificial neural network (NN) is a layered network of artificial neurons and is able to learn real-valued functions from examples (in this case, the subsequent nodes). The temporal characteristics of music are exploited with a Time Delay Neural Network (TDNN) architecture, where the input patterns are successively delayed in time. The best result we had was using a Ward Net architecture, as implemented in the NeuroShell 2[4] package and modified it into a TDNN-Ward Net, which is shown in Figure 3. The entire song database was used as the training set.

5 Experimental Results

The reference song "Claret and Oysters" and one song made by the FSM are visualized in Figure 4 with our integer representation, with the integer numbers on the x-axis and the time (sequential numbers) on the y-axis. As can be seen, there exist repeating patterns in the graph, the "landscape" of the song shares similar "hills", for example the notes 50-57 and 80-87.

[4] http://www.wardsystems.com/products.asp?p=neuroshell2

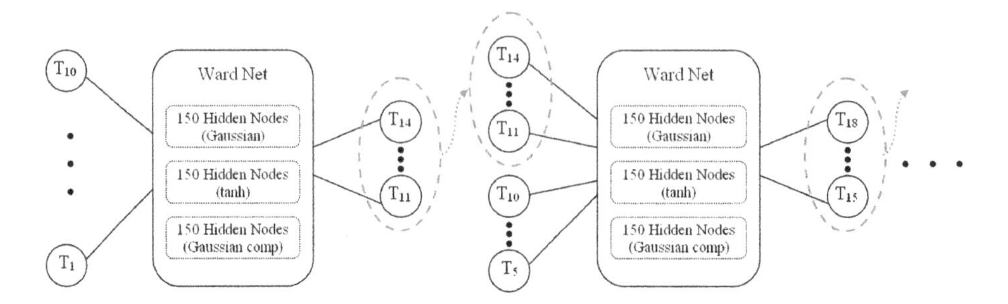

Fig. 3. TDNN illustration using the Ward Net, process of continuously generating four notes, based on the last ten notes, with T_i indicating the i-th note of the song

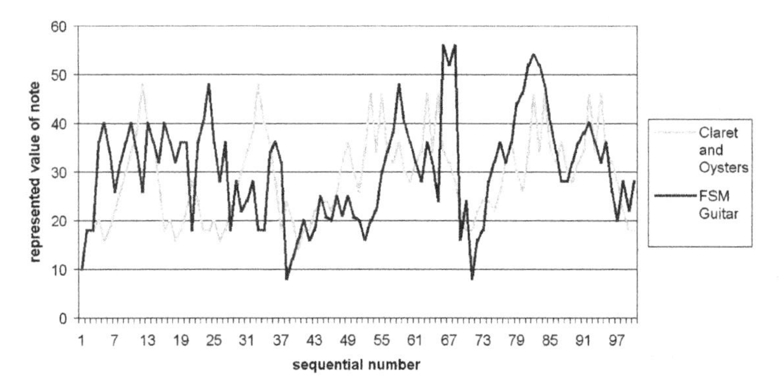

Fig. 4. Visualization of the internal representation

5.1 Results with the FSM

Several smaller and longer learned patterns can be recognized by ear and eye, not only from "Claret and Oysters" in Figure 4, but from different songs of the entire database as well. It is noteworthy that the overall quality of the song does not change over time (in contrast to the NN songs, described in the next section). The beginning of this song is shown in Figure 5 as musical score.

5.2 Results with the NN

In Figure 6, the result from a TDNN-Ward Net, which was trained over 72000 epochs with an average error of 0.0006714 on the training set (the song "Claret and Oysters"), is shown.

In the second half of the song, after a given starting seed, the NN is oscillating more often between extreme notes than in the beginning of the song, it can not predict the song any more. Including multiple songs to the knowledge database did not significantly change the result. That means that even with a large number of epochs the network architecture was not able to memorize the whole song.

Fig. 5. Sample song generated by the FSM in its musical score

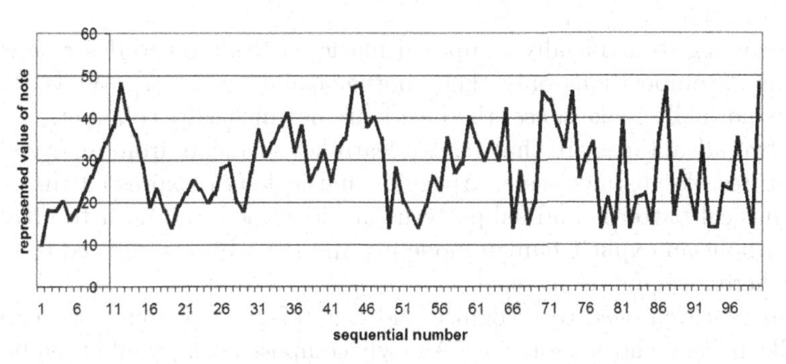

Fig. 6. Sample song generated by the NN, trained on one song (internal representation)

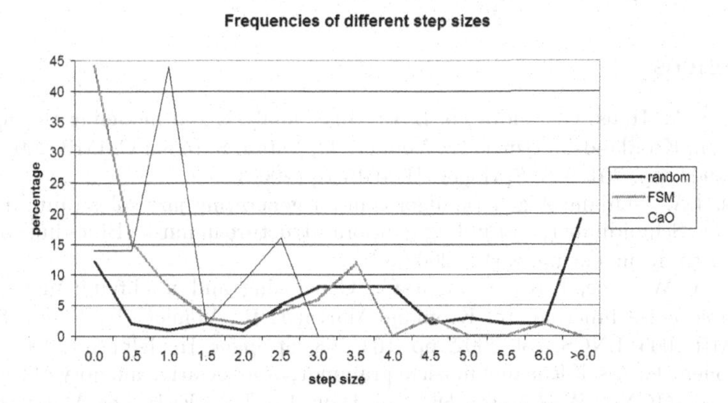

Fig. 7. Frequencies of the intervals of a random generated song, a FSM song and "Claret and Oysters"

6 Comparison with Other Approaches

In total, three songs (which feature drums violas, electric guitars played by FSM and TDNN Ward Nets) created by our systems have been made public[5]

[5] http://www.uni-koblenz.de/~wagnermar/evomusart

and we encourage the reader to form his/her own opinion about the music's quality.

6.1 Musical Intervalls - Consonances and Dissonances

Statistics of music intervals, the difference in pitch between two notes, can be provided. In Figure 7, it can be seen that a randomly generated song does not have any preference for any interval. Although higher intervals were used by the FSM occasionally, preferences for medium and low size intervals were observed. "Claret and Oysters" has a preference for lower intervals as well.

7 Conclusion and Future Work

When listening to artificially composed music, critics tend to describe the creations as "..compositions only their mother could love.." [7]. While the local contours normally make sense, the pieces are not musically coherent.

Our two advantages are that (a) we learn key elements from analyzed songs and features like distinct scale, arpeggios, musical style expressed through frequent musical patterns and subpatterns are identified and used to create new songs without an explicit human modeling and (b) a bias is induced to produce short coherent sequences at once.

Future research need to recognize and use different coherent structures of a song (like refrain, chorus, solo etc.) A newly composed song would then be structured by the learned musical structures and inherit features which are present only in specific parts of songs, such as the refrain.

References

1. Baratè, A., Haus, G., Ludovico, L.A.: Music analysis and modeling through petri nets. In: Kronland-Martinet, R., Voinier, T., Ystad, S. (eds.) CMMR 2005. LNCS, vol. 3902, pp. 201–218. Springer, Heidelberg (2006)
2. Biles, J.A.: Genjam: A genetic algorithm for generating jazz solos (June 15, 1994)
3. Eck, D., Schmidhuber, J.: Finding temporal structure in music: Blues improvisation with lstm recurrent networks (2002)
4. Jeon, Y.-W., Lee, I.-K., Yoon, J.-C.: Generating and modifying melody using editable noise function. In: Kronland-Martinet, R., Voinier, T., Ystad, S. (eds.) CMMR 2005. LNCS, vol. 3902, pp. 164–168. Springer, Heidelberg (2006)
5. Kohonen, T.: A self-learning musical grammar, or associative memory of the second kind. In: IJCNN, Washington DC, vol. I, pp. I–1–I–6, IEEE, Los Alamitos (1989)
6. Miranda, E.R., Biles, J.A. (eds.): Evolutionary Computer Music. Springer, Heidelberg (2007)
7. Mozer, M.: Neural network music composition by prediction: Exploring the benefits of psychoacoustic constraints and multi-scale processing (1994)
8. Schoenberger, J.: Genetic algorithms for musical composition with coherency through genotype. Spring (2002)
9. Unehara, M., Onisawa, T.: Construction of music composition system with interactive genetic algorithm (October 2003)
10. Walshaw, C.: The abc notation system (1993), http://abcnotation.org.uk

Trans<->Former #13: Exploration and Adaptation of Evolution Expressed in a Dynamic Sculpture

Gunnar Tufte[1] and Espen Gangvik[2]

[1] The Norwegian University of Science and Technology
Department of Computer and Information Science
Sem Selandsvei 7-9, 7491 Trondheim, Norway
gunnart@idi.ntnu.no

[2] Trondheim Electronic Arts Centre
Fjordgata 20, 7010 Trondheim, Norway
espen@gangvik.no

Abstract. *Trans<->Former #13* is a dynamic sculpture with a set of equal sub-structures with a behaviour sought that continuously redraws its own positions in space. To achieve the artistic vision the sculpture is considered as a non-uniform 2D cellular automata. By including the adaptive and explorative properties of a Cellular Genetic Algorithm in the expression of the moving sculpture *Trans<->Former #13* is capable of generating a continuous emerging behaviour. The global behaviour is a result of a local EA and control system in each sub-structure. The principles of the work are demonstrated in two experimental approaches using 3D simulation.

1 Introduction

The project herein originates in an idea of a sculpture, *Trans<->Former #13*, capable of expressing different visual expressions by applying a set of constrained rules for how to assemble a given set of equal shaped sub-structures (see Figure 1). In addition to assemble static sculptures the rules can be applied to a sculpture in order to have a dynamic structure. The goal of this approach is expressed in the artist Espen Gangvik's vision:

> "*Trans<->Former #13* is a sculpture that continuously redraws its own positions in space. New shapes are being created in an eternal process, where the movements itself takes part as the sculpting force." [1]

If the idea of a reshaping sculpture built of a set of equal sub-structures is pursued into the domain of computer science each sub-structure may not only be a physical structure but also an autonomous unit running the assembly rules and the EA locally.

The results and methods described herein are an attempt towards a physical sculpture consisting of autonomous elements capable of movements going on

M. Giacobini et al. (Eds.): EvoWorkshops 2008, LNCS 4974, pp. 509–514, 2008.

forever, thus restricted by the geometrical constraints given by the artist. To fulfil this requirement the power of adaptivity and exploration of evolution itself are part of the dynamic of the sculpture. The never ending process of evolution is the responsible factor ensuring diversity in the geometrical expression of the sculpture.

To obtain autonomous sub-structure each element should be capable of running a local copy of the EA. The algorithm chosen is Sipper's Cellular Genetic Algorithm (CGA) [2] an algorithm that can be autonomous to each element and only need local communication between the elements. This approach is not population based. Instead each element have a part of the genetic pool and the evolutionary search is based on exchange of genetic material between the local EAs over time.

Unlike an evolved static artefact [3] where each evolved individual can be evaluated by how well a single structure performs its task evaluation of a constantly moving structure where the path of movements is as important as a given goal behaviour require a different evaluation strategy. Life time evaluation [4], a strategy used in artificial development to discover interesting phases of behaviour in an emerging growing organism is chosen. The evaluation is a continuous process of evaluating the behaviour of the sculpture as the genetic material change as a result of the local exchange and manipulation.

The project was realized in BREVE [5], a 3D simulation environment with a defined programming language. As stated *Trans<->Former #13* is a project of an artist. This implies that the artist's vision for the sculpture is the most important priority. The chosen tool gives the artist a possibility to ensure that the visual expressions of the sculpture are within scope of the artist.

A sculpture was made as a functional 3D model with a local control system [6] and CGA in each sub-structure. Two experimental approaches were carried out to explore the power of the CGA and investigate if the control system chosen could steer the sculpture with realism acceptable to the artist.

The article is laid out as follows: The sculpture *Trans<->Former #13* is presented in Section 2. Section 3 deals with the cellular genetic algorithm and the control system of the sculpture. Experiments of simulated behaviour and evolved results are presented in Section 4. Finally Section 5 concludes the work.

2 Trans<->Former #13

Trans<->Former #13 can be a static solid sculpture, as shown in the photo of an exhibited early version in Figure 1(a) or a sculpture that continuously redraws its own positions in space. The later is illustrated by two different shapes in Figure 1(b) showing how the sculpture alter by reshaping.

The sculpture is made by assembling equal sub-structures. Each sub-structure consists of two box elements connected by a knee joint. One end of the two elements is capable of rotation. In Figure 2 enhanced details of sub-structure $n+1$ illustrate the mechanics of *Trans<->Former #13*. Two possible movements are included in each sub-structure: a shaft capable of rotating and a knee-joint

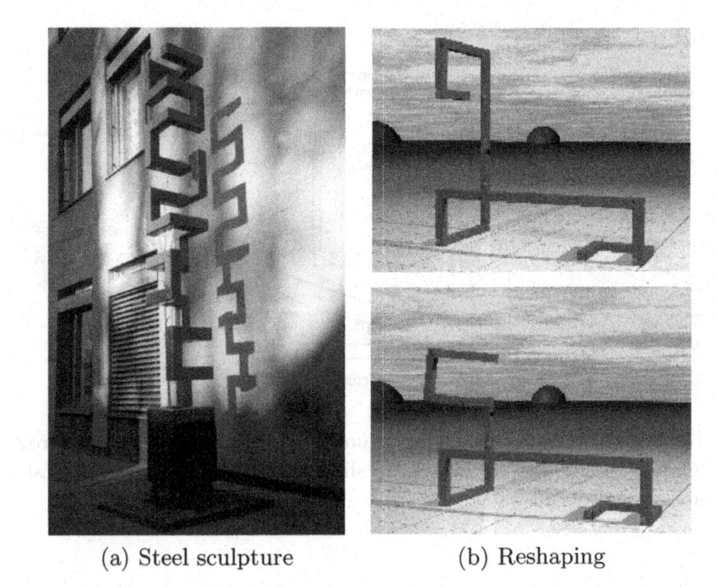

(a) Steel sculpture (b) Reshaping

Fig. 1. *Trans<->Former #13* shown as an exhibited static sculpture and as a dynamic reshaping structure

making bending possible. The two movements are termed rotation-angle and pitch-angle. As such, each sub-structure can be moved in two degrees of freedom.

3 Evolutionary Control System

A population based EA is not a favourable option for this work. This is due to the fact that individuals in a population can code for very diverging behaviour. As such, the evaluation of different individuals makes the property of a sculpture that "continuously redraws its own position" hard to meet due to possible large jumps in the search space.

The sculpture is considered as a 2D non-uniform Cellular Automata (CA) [2]. Each sub-structure is a cell. Figure 2 shows the local control system implemented as a 2D non-uniform CA and the local EAs. The CA control system consists of CA rules giving the output as change of pitch and rotation angle. Possible output values are: -90, 0, 90 for pitch and -180, -90, 0, 90, 180 for rotation. The inputs to a cell are the current state of itself and its neighbouring elements. Each sub-structure runs a copy of the EA. The CPA communication enables exchange of genetic material and fitness values between the neighbours. The fitness value is calculated locally and reflects the sub-structure's performance.

A run is performed by first initializing all sub-structures with random generated rules. The CA runs for a given amount of iterations. A CA iteration is a parallel update of all sub-structures outputs, i.e. pitch and rotation angle, based on the input from the previous CA iteration. After the apportioned number of iterations a local fitness value is calculated in all sub-structures. The fitness is

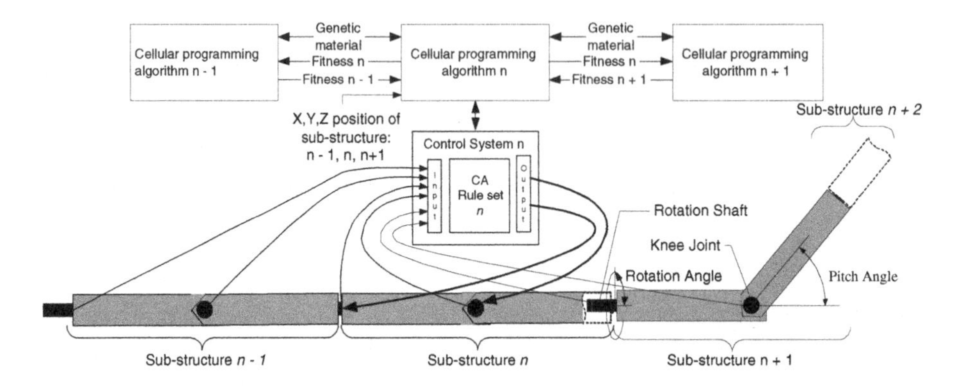

Fig. 2. Three sub-structures shown with control system and cellular programming algorithm. Only a single control system is shown. Sub-structure $n + 1$ is shown with enhanced mechanical details.

used to decide what genetic operators to apply to each rule set according to the CPA. This sequence is thought to be repeated forever in an actual sculpture making an eternal process of new shapes.

4 Experiments

Two experiments with different goals are presented. The fitness function used by the CPA is the fitness function used in the selection process. The evaluation value is based on the performance of each sub-structure in relation to its neighbours. This local evaluation criterion does not reflect the global behaviour of the sculpture over time. Therefore, a second function was used to monitor the global emerging behaviour.

In both experiments the iterations was read out in seconds. This is due to the tool BREVE. Each sub-structure was given the possibility to move every tenth BREVE iteration. The fitness was calculated for every tenth possible move. As such, crossover and mutation was carried out every hundredth BREVE iteration. The results presented are examples of typical results.

4.1 Go Far

The Go far experiment target was to move as far as possible from the initial position. The distance was given by the sub-structures placement in the x, z plane relative to the starting position. As such the distance and the fitness value is given by: $\sqrt{x^2 + z^2}$ were x and z is the current distance along the axes from the initial starting point. The measurement monitored was the average distance of all sub-structures from the starting point. The result of a run is shown in Figure 3(a). The distance measure used is in the coordinates used in the internal world of BREVE.

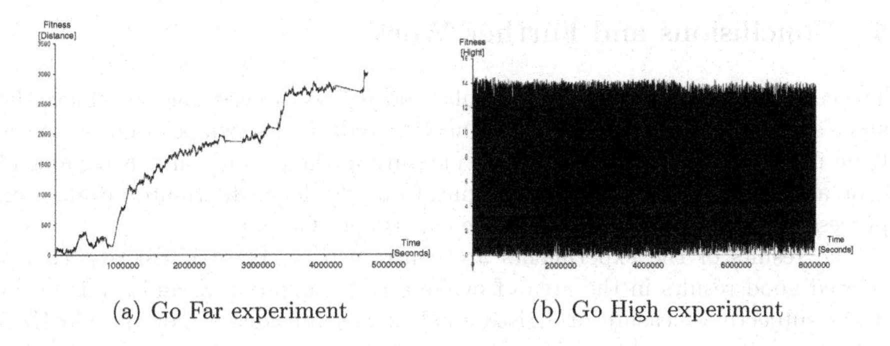

(a) Go Far experiment (b) Go High experiment

Fig. 3. Results for the Go Far and Go High experiments

4.2 Go High

In this experiment the sculpture's target behaviour was to stretch toward the sky. The fitness was given by height above ground, i.e. along the Y-axis, for each sub-structure compeered to its neighbours height. In Figure 3(b) a result of a run is shown. The height measure in the plot is for the most elevated sub-structure. The height is in BREVE coordinates along the Y-axis. It is hard to get any meaningful information out of the plot. However, the behaviour responsible for the fluctuations was found.

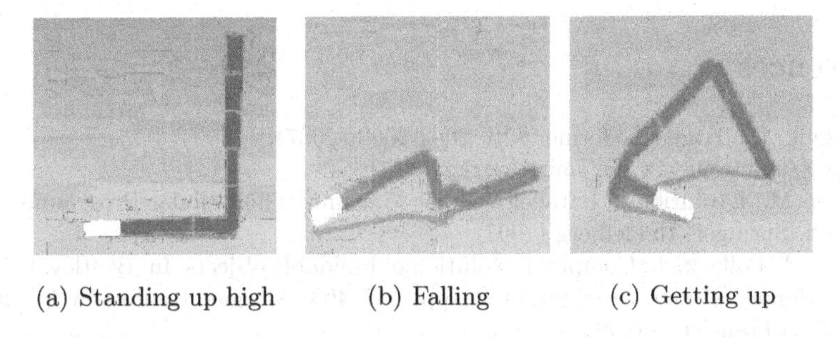

(a) Standing up high (b) Falling (c) Getting up

Fig. 4. Screen shoots from the "Go high" experiments showing an example of behaviour causing the large fluctuation in the fitness plot

In Figur 4 three screen shots of the sculpture are shown. In the first, Figure 4(a) the sculpture balances and get a high fitness score. However, the sculpture is not able to keep the balance and falls over. Resulting in the screen shot shown in Figure 4(b). Here the sculpture is close to the ground resulting in a low fitness score. In the last screen shoot shown in Figure 4(c) the sculpture is reshaping and the fitness increase as the sculpture try to get up standing. As such, the fluctuations are due to losing balance when the sculpture reaches high.

5 Conclusions and Further Work

The combination of a CPA and a cellular control system was able to achieve the artist's vision of a sculpture that continuously redraws its own positions in space. Even if the EA uses no global fitness measure in the evaluation it is capable of generating a global behaviour emerging from the local distributed evaluation processes. This is clearly shown in the experiment Go Far.

The results of the experiments are contradictory. The Go Far experiment showed good results in the area of evolutionary computation and a fair result if the subjective measure of artistic quality was considered. For the Go High experiment it was contrary. The result viewed in an evolutionary computation setting is hardly a result; it indicates that there was no working evolutionary search. However, the Go High experiment showed behaviour very appealing in an artistic setting. As such, a bad result in the world of computer science may be a nice result if the artistic quality is considered.

Further work involves building the active sub-structures and implementation of the systems described to be able to build a working prototype. Currently working scale models are built to get a better impression of the artistic quality of the approach described herein. In addition the mechanical system for a full size sculpture is in an early design phase. This work can be followed at *www.gangvik.no*.

Acknowledgment. Thanks to Karl Hatteland for BREVE implementation and simulations.

References

1. Gangvik, E.: Trans<->former#13, 2002-2008 (2007), `http://www.gangvik.no/works/works.html`
2. Sipper, M.: Evolution of Parallel Cellular Machines The Cellular Programming Approach. Springer, Heidelberg (1997)
3. Funes, P., Pollack, J.: Computer evolution of buildable objects. In: Bentley, P.J. (ed.) Evolutionary Design by Computers, pp. 387–403. Morgan Kaufmann Publishers, Inc., San Francisco (1999)
4. Tufte, G., Haddow, P.C.: Identification of functionality during development on a virtual sblock fpga. In: Congress on Evolutionary Computation(CEC 2003), pp. 731–738. IEEE, Los Alamitos (2003)
5. Klein, J.: Breve: A 3d simulation enviroment for multi-agent simulations and artificial life. Technical report, Complex Systems Group at Department of Physical Resource Theory, Goteborg Univeristy Sweeden and Cognitive Science, Hampshire College Amherst (2005), `http://www.spiderland.org/breve/`
6. Sims, K.: Evolving 3d morphology and behavioar by competition. In: Artificial Life IV, 1994, pp. 28–39. MIT Press, Cambridge (1994)

Multiobjective Tuning of Robust PID Controllers Using Evolutionary Algorithms

J.M. Herrero, X. Blasco, M. Martínez, and J. Sanchis

Department of Systems Engineering and Control,
Polytechnic University of Valencia, Spain
Tel.: +34-96-3879571; Fax: +34-96-3879579
{juaherdu,xblasco,mmiranzo,jsanchis}@isa.upv.es

Abstract. In this article a new procedure to tune robust PID controllers is presented. To tune the controller parameters a multiobjective optimization problem is formulated so the designer can consider conflicting objectives simultaneously without establishing any prior preference. Moreover model uncertainty, represented by a set of possible models, is considered. The multiobjective problem is solved with a specific evolutionary algorithm ($\epsilon\nearrow$−MOGA). Finally, an application to a non-linear thermal process is presented to illustrate the technique.

1 Introduction

PIDs are the most common controllers in industry since they have simple control structures and are relatively easy to maintain and tune. In fact, 95% of the control loops use PID and most of them are PIs [2]. Furthermore even if more sophisticated advanced control strategies are implemented, low level control loops are still implemented with PIDs. By this reason, tuning PID controllers efficiently is up to this time an interesting research.

A lot of tuning methods have been presented in the literature, some of them based on empirical methods [2], analytical [12] or optimal ones [10,11,1]. Nevertheless most of these methods have a lack of performance because they are based on a linear model, which is usually adjusted around an operating point. When the process operates outside the validity zone of the model - where differences between model and process behavior increase - poor control performance is obtained since the tuning is suboptimal and the closed loop could be unstable. To avoid this, a robust PID controller is needed.

When model uncertainties are taken into account in the controller to cover non-modeled dynamics - such as non linearities, high frequency dynamics, etc - and measurement noise the PID is called *robust* [13]. The simpler the process model is, the bigger the uncertainties are, therefore robust PID tuning methods offer controllers too conservative. This fact produces bad control performances.

In this work, a procedure to minimize the stability-performance trade off of the control loop is stated. The PID tuning methodology proposed has the following features:

M. Giacobini et al. (Eds.): EvoWorkshops 2008, LNCS 4974, pp. 515–524, 2008.

- It uses non-linear parametric models with uncertainty. The uncertainty is considered by means of a set of models, the Feasible Parameter Set (FPS^*). Although the real process is not known, it is assumed that it lies within the FPS^*.
- It solves the PID controller tuning as a Multiobjective Optimization (MO) problem . It permits to take into account different and possibly conflicting specifications, such as good tracking of input signals, low energy of the control signals, etc. that cannot be cast into a single norm [8].

Optimal tuning considers not only a nominal model but the FPS^*, adjusting the controller parameters for the worst case (the most unfavorable model). Moreover, because the tuning method has to consider conflicting objectives, a MO problem is stated where each objective minimizes the maximum cost function for all the models belonging to the FPS^*.

MO techniques presents advantages as compared with single objective optimization techniques due to the possibility of giving a set of solutions with different trade-offs among different individual problem objectives so that the Decision Maker (DM) can select the best final solution. The presence of multimodal MO functions and non-convex constrained spaces have a big influence on the optimization method selected. For instance, the use non-linear programming based on Gauss-Newton methods could present problems to obtain optimal solutions. An effective way to solve any kind of MO problems is the use of stochastic optimizers such as Evoluationary Algorithms (EAs) that can work well with multi-modal and non-convex problems. The good results obtained with Multiobjective Evolutionary Algorithms (MOEAs) and their capacity to handle a wide variety of problems explain why they are currently one of the areas where a lot of progress is being made within the EAs field [4,3]. Although they involve a significant computational cost, the PID tuning presented in this work will be performed off-line and therefore the time will not be important.

This paper is organized as follows. Section 2 presents the PID tuning procedure proposed which is *robust* since it uses the FPS^* to tune the PID parameters and *multiobjective* because several objectives will be taking into account simultaneously. Section 3 describes briefly the ϵ-MOGA algorithm which is used to obtain the Pareto front. Section 4 illustrates the PID tuning procedure with an example of PID tuning to control a non linear thermal process. Finally, some concluding remarks are reported in section 5.

2 Multiobjective Tuning of Robust PID

Let's assume the following model structure:

$$\dot{\mathbf{x}}(t) = f(\mathbf{x}(t), u(t), \theta), \quad \hat{y}(t, \theta) = g(\mathbf{x}(t), u(t), \theta) \tag{1}$$

where $f(.)$ and $g(.)$ are the non-linear functions of the model; $\theta \in R^L$ is the vector of unknown model parameters; $\mathbf{x}(t) \in R^n$ is the vector of model states; $u(t)$ is the process input and $\hat{y}(t, \theta)$ the process output.

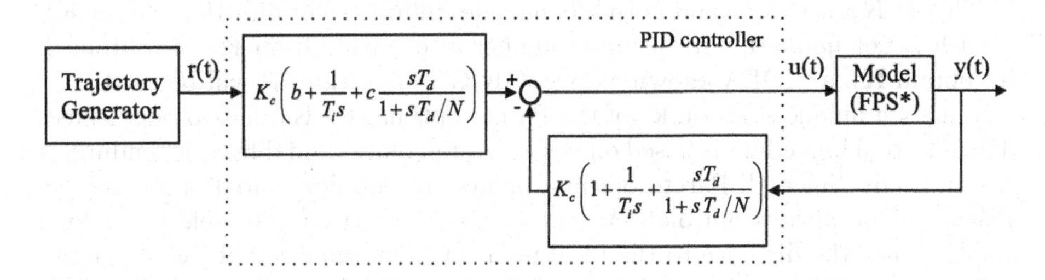

Fig. 1. Control structure for PID tuning where $r(t)$ is the setpoint signal, $y(t)$ is the process output and s represents the Laplace Transform

Assume that model uncertainty is represented by means of a set of models FPS^*

$$FPS^* := \{\theta_1, \ldots, \theta_p\} \tag{2}$$

The control structure is a two degrees of freedom PID in the ISA standard form (Fig. 1). The controller parameters to tune are $\mathbf{k} = \{K_c, T_i, T_d, b, c, N\}$ where K_c is the proportional constant, T_i and T_d are the integral and the derivative times respectively; b and c are the weighting of proportional and derivative actions respectively and N is the filter parameter of the derivative action.

To obtain the controller parameters $\mathbf{k}$, the following MO problem is stated:

$$\min_{\mathbf{k} \in D \in R^6} \mathbf{J}(\mathbf{k}) = \min_{\mathbf{k} \in D \in R^6} [J_1(\mathbf{k}), J_2(\mathbf{k}), \ldots, J_s(\mathbf{k})] \tag{3}$$

where $J_i(\mathbf{k})$, $i \in B := [1 \ldots s]$, are the objectives to minimize and $\mathbf{k}$ is a solution inside the 6-dimensional solution space D.

Since each objective to minimize has to take into account the model uncertainty FPS^* then

$$J_i(\mathbf{k}) = \max_{\theta \in FPS^*} \phi_i, \quad i \in B := [1 \ldots s] \tag{4}$$

where the function ϕ_i is the design objective to minimize for the worst model case belonging to FPS^*. Some typical criteria are related to:

- A norm of the control actions: $\phi_i = ||\mathbf{u}||$.
- A norm of the rate of change of the control actions: $\phi_i = ||\dot{\mathbf{u}}||$.
- A norm of the output errors: $\phi_i = ||\mathbf{r} - \mathbf{u}||$.

Anyway, to solve the MO problem the Pareto optimal set $\mathbf{K}_P$ (where no solution dominates another one) must be found. Pareto dominance is defined as:

A solution $\mathbf{k}_1$ dominates another solution $\mathbf{k}_2$, denoted by $\mathbf{k}_1 \prec \mathbf{k}_2$, iff

$$\forall i \in B, J_i(\mathbf{k}_1) \leq J_i(\mathbf{k}_2) \wedge \exists n \in B : J_n(\mathbf{k}_1) < J_n(\mathbf{k}_2) .$$

Therefore the Pareto optimal set $\mathbf{K}_P$, is given by

$$\mathbf{K}_P = \{\mathbf{k} \in D \mid \nexists\ \tilde{\mathbf{k}} \in D : \tilde{\mathbf{k}} \prec \mathbf{k}\} . \tag{5}$$

The set $\mathbf{K}_P$ is unique and normally includes infinite solutions. Hence a set $\mathbf{K}_P^*$ (which is not unique), with a finite number of elements from $\mathbf{K}_P$, is obtained. To obtain $\mathbf{K}_P^*$ a MOEA known as the ϵ-MOGA algorithm [7] will be used.

Finally a unique solution $\mathbf{k}^*$ of the Pareto optimal set $\mathbf{K}_P^*$ has to be selected. The selection procedure is based on designer preferences and differs depending on design needs. Since all Pareto optimal points are non-dominated, any selection made will be always optimal. A very common alternative to select the final solution uses the distance to the ideal point in the normalized objective space.

Defining $J_i^{min} = \min_{\mathbf{k}\in D} J_i(\mathbf{k})$ and $J_i^{max} = \max_{\mathbf{k}\in D} J_i(\mathbf{k})$, the normalized objective space is obtained by means of:

$$\bar{J}_i(\mathbf{k}) = \frac{J_i(\mathbf{k}) - J_i^{min}}{J_i^{max} - J_i^{min}}, i \in B := [1 \dots s] \tag{6}$$

and the selected compromise solution $\mathbf{k}^*$ can be calculated as:

$$\mathbf{k}^* = \arg \min_{\mathbf{k}\in\mathbf{K}_P^*} ||\bar{\mathbf{J}}(\mathbf{k})|| \tag{7}$$

where $\bar{\mathbf{J}}(\mathbf{k}) = [\bar{J}_1(\mathbf{k}), \bar{J}_2(\mathbf{k}), \dots, \bar{J}_s(\mathbf{k})]$.

3 ϵ-MOGA Algorithm

ϵ-MOGA is an elitist multiobjective evolutionary algorithm based on the concept of ϵ-dominance [9]. It obtains an ϵ-Pareto set, $\mathbf{K}_P^*$, which converges towards the Pareto optimal set $\mathbf{K}_P$ in a distributed manner around Pareto front $\mathbf{J}(\mathbf{K}_P)$ - with limited memory resources. Next a brief description of the $\epsilon-MOGA$ algorithm [7] is presented.

$\epsilon-MOGA$ adjusts the limits of the Pareto front $\mathbf{J}(\mathbf{K}_P^*)$ dynamically and prevents the solutions belonging to the ends of the front from being lost. For this reason, the objective space is split up into a fixed number of boxes n_box_i, for each dimension i. This grid preserves the diversity of $\mathbf{J}(\mathbf{K}_P^*)$ since one box can be occupied by only one solution. This fact prevents that the algorithm converges towards just one point or area inside the objective space (see Fig. 2).

The algorithm is composed of three populations: The main population, $P(t)$, which explores the search space D during the algorithm iterations (t) - its population size is $Nind_P$. The auxiliary population $G(t)$ -its size is $Nind_G$, which must be an even number. And the archive, $A(t)$, which stores the solution $\mathbf{K}_P^*$ - its size $Nind_A$ can vary but it will never be higher than (8)

$$Nind\_max\_A = \frac{\prod_{i=1}^{s}(n\_box_i + 1)}{n\_box_{max} + 1} \tag{8}$$

where $n\_box_{max} = \max[n\_box_1, \dots, n\_box_s]$.

The pseudocode of the $\epsilon-MOEA$ algorithm is given by

```
1. t:=0
2. A(t):=∅
```

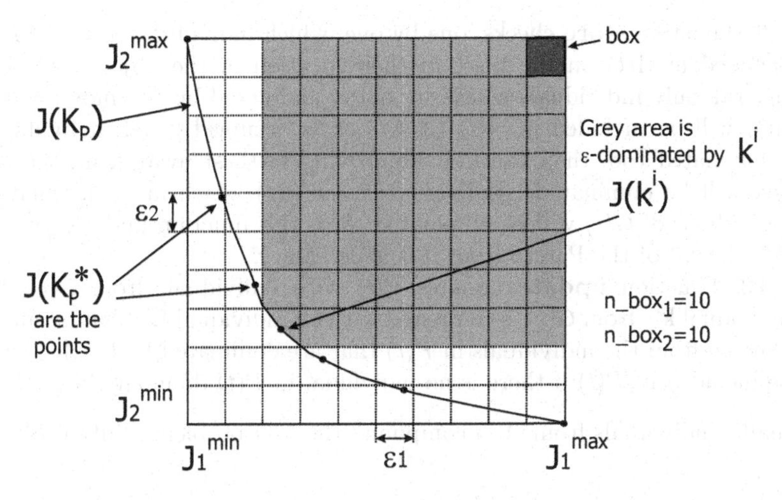

Fig. 2. The concept of ϵ-dominance. ϵ-Pareto Front $\mathbf{J}(\mathbf{K}_P^*)$ in a two-dimensional problem. J_1^{min}, J_2^{min}, J_1^{max}, J_2^{max}, Pareto front limits; ϵ_1, ϵ_2 box widths; and n_box_1, n_box_2, number of boxes for each dimension.

```
3. P(t):=ini_random(D)
4. eval(P(t))
5. A(t):=store_ini(P(t),A(t))
6. while t<t_max do
7.      G(t):=create(P(t),A(t))
8.      eval(G(t))
9.      A(t+1):=store(G(t),A(t))
10.     P(t+1):=update(G(t),P(t))
11.     t:=t+1
12. end while
```

The main steps of the algorithm are detailed as follows:

Step 3. $P(0)$ is initialized with $Nind_P$ individuals (solutions) that have been randomly selected from the search space D.

Steps 4 and 8. Function **eval** computes function value $\mathbf{J}(\mathbf{k})$ for each individual in $P(t)$ (step 4) and $G(t)$ (step 8).

Step 5. Function **store**$_{ini}$ checks individuals of $P(t)$ that might be included in the archive $A(t)$ as follows:

1. Non-dominated $P(t)$ individuals are detected, $\mathbf{K}_{ND}$.
2. Pareto front limits J_i^{max} and J_i^{min} are calculated from $\mathbf{J}(k), \forall \mathbf{k} \in \mathbf{K}_{ND}$.
3. Individuals in $\mathbf{K}_{ND}$ are analyzed, one by one, and those that are not ϵ-dominated by individuals in $A(t)$, will be included in $A(t)$.

Step 7. With each iteration, the function **create** creates individual of $G(t)$ by using extended linear recombination technique and random mutation with Gaussian distribution [7].

Step 9. Function **store** checks, one by one, which individuals in $G(t)$ must be included in $A(t)$ on the basis of their location in the objective space. In general, only individuals which are not ϵ-dominated by any individual from $A(t)$ will be included in $A(t)$ (if its box is occupied by an individual not ϵ-dominated too, then the individual lying farthest away from the center box will be eliminated). Individuals from $A(t)$ which are ϵ-dominated by individuals of $G(t)$ will be eliminated. Also this function updates the limits J_i^{max}, J_i^{min} of the Pareto front if it is necessary.

Step 10. Function **update** updates $P(t)$ with individuals from $G(t)$. Every individual $\mathbf{k}^G$ from $G(t)$ is compared with an individual $\mathbf{k}^P$ that is randomly selected from the individuals in $P(t)$ that are dominated by $\mathbf{k}^G$. $\mathbf{k}^G$ will not be included in $P(t)$ if there is no individual in $P(t)$ dominated by $\mathbf{k}^G$.

Finally, individuals from $A(t)$ compound the MO problem solution $\mathbf{K}_P^*$.

4 Robust PID Tuning for a Thermal Process Control

This section describes the application of the robust PID tuning procedure to a thermal process control. A scale furnace with a resistance placed inside is considered. A fan continuously introduces air from outside (air circulation) while energy is supplied by an actuator controlled by voltage.

The dynamics of the resistance temperature can be modeled by

$$\dot{x}(t) = \frac{\left(\theta_1 u(t)^2 - \theta_2\left(x(t) - T_a(t)\right) - \frac{\theta_3(273+x(t))^4}{100^4}\right)}{1000}, \quad \hat{y}(t) = x(t), \qquad (9)$$

where $\dot{x}(t)$ is the model state; $u(t)$ is the input voltage with rank 0 - 100 (%); $\hat{y}(t)$ is the resistance temperature (^oC) (model output); $T_a(t)$ is the air temperature (^oC) and $\theta = [\theta_1, \theta_2, \theta_3]^T$ are the model parameters.

To obtain the set of models which forms the Feasible Parameter Set (FPS^*) - the model uncertainty representation - the robust identification method presented in [5,6] is applied. The FPS^* is the discrete set of models which keeps the model prediction errors bounded for certain norms and bounds. In this example both ∞-norm and absolute norm are simultaneously used to determine the FPS^*. In [5], the bounds are selected to ensure that the FPS^* model predictions errors are not greater than 2^oC and their average values are not greater than 0.8^oC. These bounds produce a FPS^* containing 304 models (Fig. 3).

When the same input is applied to each model belonging to the FPS^* a different output is obtained. The Fig. 4 shows the envelope generated by the models belonging to the FPS^*. Notice how the real response of the process is inside the limits produced by the FPS^*. It means that the process dynamics is correctly captured.

Once the model uncertainty is represented, the controller parameters have to be selected. For the industrial PID Omron E5CK which implements the controller of the figure 1 with $b = 1$, $c = 0$, $N = 3$ and proportional band $BP = 20/K_c$, the tuning parameters are $\mathbf{k} = \{BP, T_i, 0, 1, 0, 3\}$ (because of

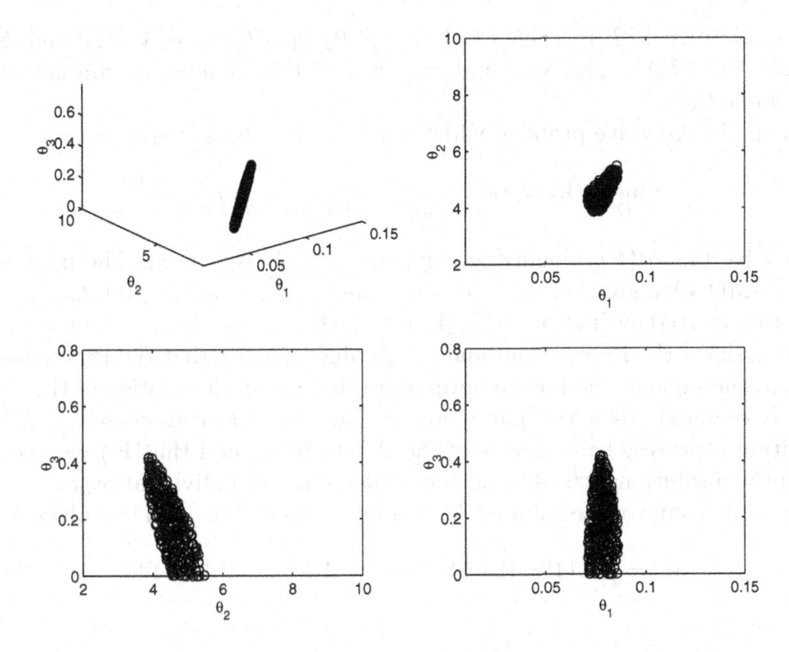

Fig. 3. Top-left: the FPS^* models marked with $\circ$. Rest of figures: the FPS^* model projections.

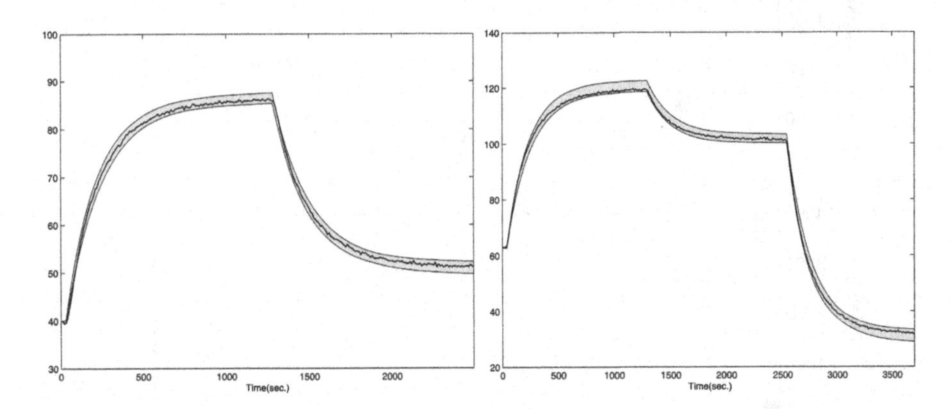

Fig. 4. Two model validation experiments. Real output of the process (black). Simulated outputs for each model belonging to the FPS^* (grey).

measurement noise the derivative action is disconnected). The search space D is a subspace of R^2 and it is defined by $BP \in [0.1 \dots 999.99]\%$ and $T_i \in [1 \dots 3999]$ seconds (range of BP and T_i in the Omron E5CK controller).

To complete the MO problem, we define the functions ϕ_1 and ϕ_2 as follows:

$$\phi_1 = \frac{||\mathbf{r} - \mathbf{y}||_1}{N}, \phi_2 = ||\mathbf{\Delta u}||_1$$

with $\mathbf{r} = [r(0), r(1 \cdot T_s) \ldots r(N \cdot T_s)]$, $\mathbf{y} = [y(0), y(1 \cdot T_s) \ldots y(N \cdot T_s)]$ and $\mathbf{\Delta u} = [\Delta u(0), \Delta u(1 \cdot T_s) \ldots \Delta u(N \cdot T_s)]$, being $N = 250$ the number of samples and T_s the sample time.

The multi-objective problem which needs to be solved is given as:

$$\min_{\mathbf{k} \in D} [J_1(\mathbf{k}), J_2(\mathbf{k})] = \min_{\mathbf{k} \in D} [\max_{\theta \in FPS^*} \phi_1, \max_{\theta \in FPS^*} \phi_2]$$

To solve this MO problem the algorithm ϵ-MOGA is used. The parameters of the ϵ-MOGA algorithm were set to: $Nind_G = 4$; $Nind_P = 100$; $t_{max} = 1000$ (resulting in 4100 evaluations of $J_1(\mathbf{k})$ and $J_2(\mathbf{k})$) and $n\_box_1 = n\_box_2 = 200$.

Fig. 5 shows the Pareto front and set obtained. Notice that the Pareto front is disjoint, the same as the Pareto optimal set. If a better description of the Pareto front is required, then the parameter n_box_i has to be increased. ϵ-MOGA algorithm captures the extremes of the Pareto front, and thus $\mathbf{K}_P^*$ will contain the optimal solutions $\mathbf{k}^{J_i}$ of each J_i considered on an individual basis.

Then the compromise solution $\mathbf{k}^*$ can be calculated as in (7) resulting in

$$\mathbf{k}^* = [12.11\%, 128.51\ sec.] \Rightarrow \mathbf{J}(\mathbf{k}^*) = [3.137, 149]$$

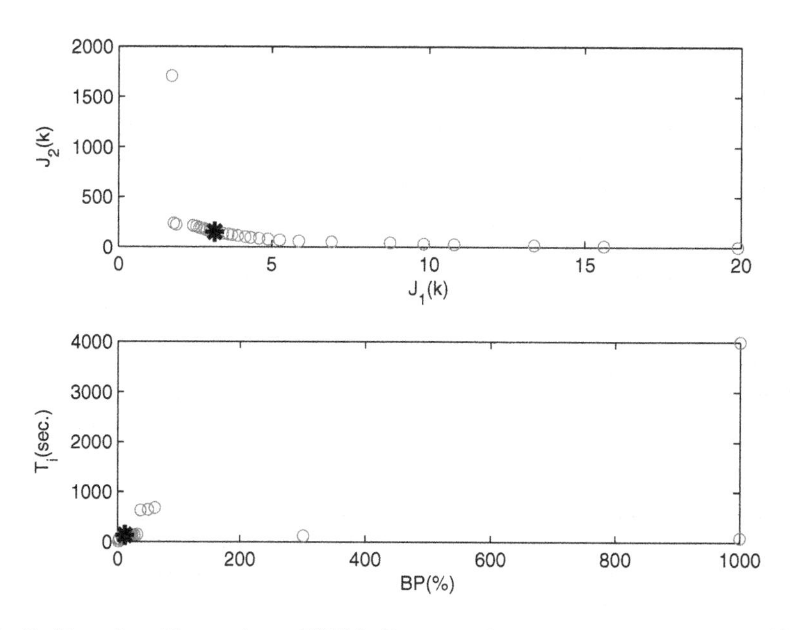

Fig. 5. Top: the ϵ-Pareto front $\mathbf{J}(\mathbf{K}_P^*)$. Bottom: the Pareto optimal set $\mathbf{K}_P^*$. (*) Compromise solution obtained, $\mathbf{k}^*$ and $\mathbf{J}(\mathbf{k}^*)$.

Fig. 6 depicts the closed loop simulation results obtained when each process model of the FPS^* is controlled with the calculated compromise controller $\mathbf{k}^*$. Notice how the controller $\mathbf{k}^*$ designed is able to control all the models belonging to the FPS^* acquiring the qualifier of *robust*.

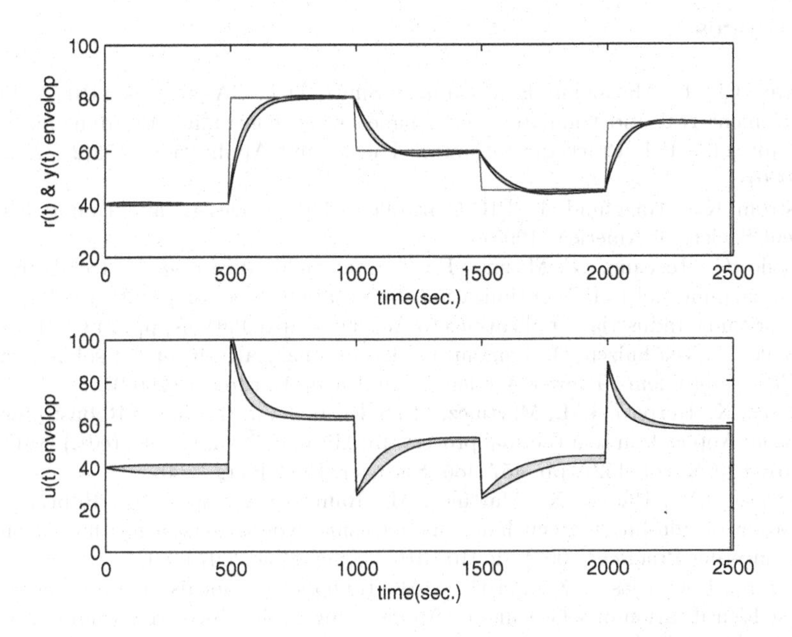

Fig. 6. Top: Desired set point $r(t)$ and the envelope of the outputs $y(t)$ when the controller $\mathbf{k}^*$ is applied to the FPS^* models. Bottom: control actions.

5 Conclusions

A new methodology, based on Evolutionary Algorithms, has been developed to tuning robust PIDs from an MO point of view. The methodology presents the following features:

- Assuming parametric uncertainty, all kind of processes can be considered.
- Since model uncertainty has been calculated using a robust identification method less conservative controllers are obtained than those designed with other uncertainty approaches.
- Any kind of design objectives can be used simultaneously to tune the PID controller resulting in a MO Problem. Thanks to the ϵ-MOGA algorithm would be possible to characterize all kind of Pareto fronts in a well-distributed manner with bounded memory resources.
- The computational burden can be high but it is not very important since the tuning methodology is made off line.
- Any kind of parametric controller can be use instead of PID, for instance, a Predictive controller, Fuzzy and so on.

Acknowledgments

Partially supported by MEC (Spanish government) and FEDER funds: projects DPI2005-07835, DPI2004-8383-C03-02 and GVA-026.

References

1. Alfaro-Cid, E., McGookin, E.W., Murray-Smith, D.J.: GA-optimised PID and pole placement real and simulated performance when controlling the dynamics of a supply ship. IEE Proceedings-Control Theory and Applications 153(2), 228–236 (2006)
2. Aström, K.J., Hägglund, T.: PID Controllers: Theory, Design, and Tuning. Instrument Society of America (1995)
3. Coello, C., Toscano, G., Mezura, E.: Current and future research trends in evolutionary multiobjective optimization, Information Processing with Evolutionary Algorithms: Industrial Applications to Academic Speculations, pp. 213–231 (2005)
4. Coello, C., Veldhuizen, D., Lamont, G.: Evolutionary algorithms for solving multiobjective problems. Kluwer Academic Publishers, Dordrecht (2002)
5. Blasco, X., Herrero, J.M., Martínez, M., Salcedo, J.V.: Non-linear Robust Identification: Application to a thermal process. In: Mira, J., Álvarez, J.R. (eds.) IWINAC 2007. LNCS, vol. 4527, pp. 457–466. Springer, Heidelberg (2007)
6. Herrero, J.M., Blasco, X., Martínez, M., Ramos, C., Sanchis, J.: Robust identication of non-linear green house model using evolutionary algorithms. Control Engineering Practice (2007), doi:10.1016/j.conengprac.2007.06.001
7. Herrero, J.M., Blasco, X., Martínez, M., Ramos, C., Sanchis, J.: Non-linear Robust Identification of a Greenhouse Model using Multi-objective Evolutionary Algorithms. Biosystems Engineering 98(3), 335–346 (2007)
8. Herreros, A., Baeyens, E., Perán, J.R.: Design of PID-type controllers using multiobjective genetic algorithms. ISA Transactions 41, 457–472 (2002)
9. Laumanns, M., Thiele, L., Deb, K., Zitzler, E.: Combining convergence and diversity in evolutionary multi-objective optimization. Evolutionary computation 10(3), 263–282 (2002)
10. Marlin, T.E.: Process Control. Designing Processes and Control Systems for Dynamic Performance. McGraw-Hill, New York (1995)
11. Mitsukura, Y., Yamamoto, T., Kaneda, M.: A Design of PID Controllers Using a Genetic Algorithm. Transactions of the Society of Instrument and Control Engineers 36(1), 75–81 (2000)
12. Ogata, K.: Modern Control Engineering, 4th edn. Prentice-Hall, Englewood Cliffs (2002)
13. Reinelt, W., Garulli, A., Ljung, L.: Comparing different approaches to model error modelling in robust identification. Automatica 38(5), 787–803 (2002)

Truncation Selection and Gaussian EDA: Bounds for Sustainable Progress in High-Dimensional Spaces

Petr Pošík

Czech Technical University in Prague
Faculty of Electrical Engineering, Department of Cybernetics
Technická 2, 166 27 Prague 6, Czech Republic
posik@labe.felk.cvut.cz

Abstract. In real-valued estimation-of-distribution algorithms, the Gaussian distribution is often used along with maximum likelihood (ML) estimation of its parameters. Such a process is highly prone to premature convergence. The simplest method for preventing premature convergence of Gaussian distribution is enlarging the maximum likelihood estimate of σ by a constant factor k each generation. Such a factor should be large enough to prevent convergence on slopes of the fitness function, but should not be too large to allow the algorithm converge in the neighborhood of the optimum. Previous work showed that for truncation selection such admissible k exists in 1D case. In this article it is shown experimentaly, that for the Gaussian EDA with truncation selection in high-dimensional spaces no admissible k exists!

1 Introduction

Estimation of Distribution Algorithms (EDAs) [1] are a class of Evolutionary Algorithms (EAs) that do not use the crossover and mutation operators to create the offspring population. Instead, they build a probabilistic model describing the distribution of selected individuals and create offspring by sampling from the model.

In EDAs working in real domain (real-valued EDAs), the Gaussian distribution is often employed as the model of promising individuals ([2], [3], [4], [5]). The distribution is often learned by maximum likelihood (ML) estimation. It was recognized by many authors (see e.g. [3], [6], [7]) that such a learning scheme makes the algorithm very prone to premature convergence: in [8], it was shown also theoretically that the distance that can be traversed by a simple Gaussian EDA with truncation selection is bounded, and [9] showed similar results for tournament selection.

Many techniques that fight the premature convergence were developed, usually by means of artificially enlarging the ML estimate of variance of the learned distribution. In [6], the variance is kept on values greater than 1, while [7] used self-adaptation of the variance. Adaptive variance scaling (AVS), i.e. enlarging

M. Giacobini et al. (Eds.): EvoWorkshops 2008, LNCS 4974, pp. 525–534, 2008.

the variance when better solutions were found and shrinking the variance in case of no improvement, was used along with various techniques to trigger the AVS only on the slope of the fitness function in [10] and [11].

However, in the context of the simple Gaussian EDAs there exists a much simpler method for preventing premature convergence (compared to the above mentioned techniques): enlarging the ML estimate of standard deviation each generation by a factor k that is held constant during the whole evolution. The resulting EDA is depicted in Fig. 1. Although the parameter learning schemes suggested in the works mentioned above are more sophisticated, the approach studied in this article is appealing mainly from the *simplicity* point of view. To the best of the author's knowledge, nobody has shown yet that this approch does not work.

1. Set the initial values of parameters $\mu^0 = (\mu_1^0, \ldots, \mu_D^0)$ and $\sigma^0 = (\sigma_1^0, \ldots, \sigma_D^0)$, D is the dimensionality of the search space. Set the generation counter $t = 0$.
2. Sample N new individuals from the distribution $\mathcal{N}(\mu^t, \sigma^t I_D)$, I_D is the identity matrix.
3. Evaluate the individuals.
4. Select the τN best solutions.
5. Estimate new values of parameters μ^{t+1} and σ^{t+1} based on the selected individuals independently for each dimension.
6. Enlarge the σ^{t+1} by a constant factor k.
7. Advance generation counter: $t = t + 1$.
8. If termination condition is not met, go to step 2.

Fig. 1. Simple Gaussian EDA analysed in this article

The algorithm in Fig. 1 was studied in [12] where the minimal values of the 'amplification coefficient' were determined by search in 1D case. In [13], the theoretical model of the algorithm behavior in 1D was used to derive the minimal and maximal admissible values for k. In this paper, the behaviour of the algorithm in multidimensional case is studied with the aim of developing the bounds for values of k if they exist.

The rest of the paper is organized as follows: in Sec. 2, the relevant results from [13] are surveyed. Section 3 presents the experimental methodology used to determine the bounds for k in more-dimensional spaces, the results of the experiments and their discussion. Finally, Sec. 4 summarizes and concludes the paper, and presents some pointers to future work.

2 Fundamental Requirements on EDA

When analysing the behaviour of evolutionary algortithms, the fitness landscape is often modelled as consisting of slopes and valleys (see e.g. [10], [14], [12]).

There are two fundamental requirements on the development of the population variance that ensure reasonable behavior of the algorithm as a whole:

1. The variance *must not shrink on the slope.* This ensures that the population position is not bounded and that it eventually finds at least a local optimum.
2. The variance *must shrink in the valley.* In the neighborhood of the optimum, the algorithm must be allowed to converge to find the optimum precisely.

These two conditions constitute the bounds for the variance enlarging constant k which must be large enough to traverse the slopes, but must not be too large to focus to the optimum. In this article, the slopes are modelled with a linear function (Eq. 1) that takes into account only the first coordinate of the individual, and the valleys are modelled as a quadratic bowl (Eq. 2):

$$f_{\text{lin}}(\mathbf{x}) = x_1 \tag{1}$$

$$f_{\text{quad}}(\mathbf{x}) = \sum_{d=1}^{D} x_d^2 \tag{2}$$

Since the fitness function influences the algorithm only by means of the truncation selection, the actual values of the fitness function do not matter, only their order. In 1D space, the results derived for the linear function thus hold for all monotonous functions, in more-dimensional spaces for all functions with parallel contour lines and monotonous function values along the gradient. Similarly, the results derived for the quadratic function also hold for any unimodal function isotropic around its optimum with function values monotonously increasing with the distance from the optimum.

2.1 Bounds for k

In this paper, it is assumed that the development of variance can be modelled with the following recurrent equation:

$$(\sigma^{t+1})^2 = (\sigma^t)^2 \cdot c, \tag{3}$$

where c is the ratio of the population variances in two consecutive generations, t and $t+1$.

As already said in the introduction, the simplest method of preventing premature convergence is to enlarge the estimated standard deviation σ by a constant factor k. Thus

$$\sigma^{t+1} = k \cdot \sigma^t \cdot \sqrt{c} \tag{4}$$

In order to prevent the premature convergence on the slope, the ratio of the consecutive standard deviations should be at least 1, i.e.

$$\frac{\sigma^{t+1}}{\sigma^t} = k \cdot \sqrt{c_{\text{slope}}} \geq 1 \tag{5}$$

$$k \geq \frac{1}{\sqrt{c_{\text{slope}}}} \stackrel{\text{def}}{=} k_{\min} \tag{6}$$

On the other hand, to be able to focus to the optimum, the model must be allowed to converge in the valley. The ratio of the two consecutive standard deviations should be lower than 1, i.e.

$$\frac{\sigma^{t+1}}{\sigma^t} = k \cdot \sqrt{c_{\text{valley}}} < 1 \tag{7}$$

$$k < \frac{1}{\sqrt{c_{\text{valley}}}} \stackrel{\text{def}}{=} k_{\max} \tag{8}$$

Joining these two conditions together gives us the bounds for the constant k:

$$k_{\min} = \frac{1}{\sqrt{c_{\text{slope}}}} \leq k < \frac{1}{\sqrt{c_{\text{valley}}}} = k_{\max} \tag{9}$$

In this paper, the value of k is called *admissible* if it satisfies condition 9.

2.2 Model of EDA Behaviour in 1D

For 1D case, the bounds of Eq. 9 are known and can be computed theoretically (see [13]). The behavior of the simple Gaussian EDA with truncation selection in 1D space can be modelled by using statistics for the truncated normal distribution ([13], [10]). In one iteration of the EDA, the variance of the population changes in the following way:

$$(\sigma^{t+1})^2 = (\sigma^t)^2 \cdot c(z_1, z_2) \tag{10}$$

where

$$c(z_1, z_2) = 1 - \frac{z_2 \cdot \phi(z_2) - z_1 \cdot \phi(z_1)}{\Phi(z_2) - \Phi(z_1)} - \left(\frac{\phi(z_2) - \phi(z_1)}{\Phi(z_2) - \Phi(z_1)}\right)^2, \tag{11}$$

$\phi(z)$ and $\Phi(z)$ are the probability density function (PDF) and cumulative distribution function (CDF) of the standard normal distribution $\mathcal{N}(0, 1)$, respectively, and z_i are the truncation points (measured in the standard deviations) which are constant during the EDA run for both f_{lin} and f_{quad}.

The effect of the truncation selection on the population distribution is different depending on the following situations:

- The population is on the slope of the fitness function.
- The population is in the valley of the fitness function.

First, suppose *the population is on the slope* of the fitness function (which is modelled by linear function). The best $\tau \cdot N$ individuals are at the left-hand side of the distribution, i.e.

$$z_1 \to -\infty, \text{ and } z_2 = \Phi^{-1}(\tau), \tag{12}$$

where $\Phi^{-1}(\tau)$ is the inverse cumulative distribution function of the standard normal distribution. We can thus define

$$c_{\text{slope}}(\tau) = c\left(-\infty, \Phi^{-1}(\tau)\right). \tag{13}$$

In the second case, when *the population is centered around the optimum*, the selected $\tau \cdot N$ individuals are centered around the mean of the distribution, thus

$$z_1 = \Phi^{-1}\left(\frac{1-\tau}{2}\right), \text{ and } z_2 = \Phi^{-1}\left(\frac{1+\tau}{2}\right). \tag{14}$$

and we can again define

$$c_{\text{valley}}(\tau) = c\left(\Phi^{-1}\left(\frac{1-\tau}{2}\right), \Phi^{-1}\left(\frac{1+\tau}{2}\right)\right). \tag{15}$$

The above equations are taken from [13], but were already presented (in a bit different form) in previous works (Eq. 13 e.g. in [8] and Eq. 15 in [14]).

The bounds for k in 1D case computed using equations 9, 13, and 15 are depicted on Fig. 2, the same values are shown in tabular form in Table 1.

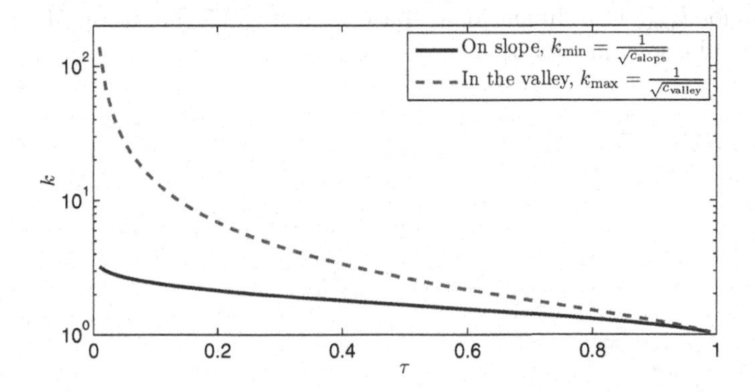

Fig. 2. Minimum and maximum values for setting the enlarging parameter k for various values of τ

Table 1. The bounds for the standard deviation enlargment constant k for various values of τ in 1D case

τ	0.01	0.10	0.20	0.30	0.40	0.50	0.60	0.70	0.80	0.90	0.99
$k_{\min}$	3.213	2.432	2.139	1.944	1.791	1.659	1.539	1.424	1.310	1.185	1.033
$k_{\max}$	138.195	13.798	6.866	4.540	3.364	2.648	2.159	1.797	1.511	1.267	1.040

These results show that in 1D case for each value of the selection proportion τ, there is an interval of admissible values of k ensuring that the algorithm will not converge prematurely on the slope of the fitness function, and will be able to focus (i.e. decrease the variance) in the neighborhood of the optimum. It is thus highly appealing to study these features in more-dimensional case.

3 Bounds for k in Multidimensional Space

To derive bounds for k in multidimensional case theoretically (as in case of 1D space) is much more complicated. In this article, an experimental approach is used. The lower bound is found by using the linear fitness function. During the experiments, the value of standard deviation of coordinate x_1 is tracked and it is checked if it increases or decreases (on average). The bisection method is used to determine the value of k for which the variance stays the same (with certain tolerance). Similarly, the upper bound is found by experiments with quadratic function.

3.1 Experimental Methodology

The population size 1,000 was used in all experiments. To determine each particular $k_{\min}$ (and $k_{\max}$), 10 independent runs of 100 generations were carried out. During each run, the standard deviation of x_1 was tracked; this gives 10 values of st.d. for each of 100 generations. Examples of the development of σ when searching for $k_{\max}$ in 5-dimensional space with $\tau = 0.5$ (i.e. using 5D quadratic fitness) can be seen in Fig. 3.

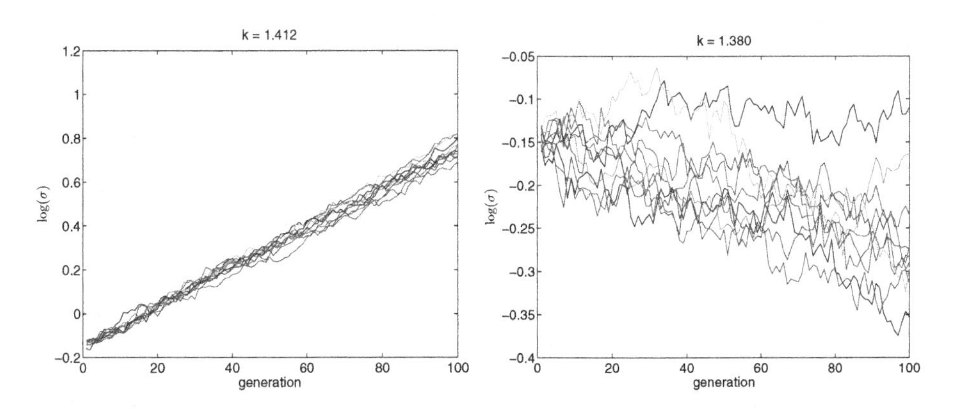

Fig. 3. Examples of the development of standard deviation during the search for $k_{\max}$ in 5-dimensional space with $\tau = 0.5$. On the left: k is by far higher than $k_{\max}$. On the right: k is slightly lower than $k_{\max}$.

To this data, a linear function of the form $E(\log(\text{st.d.})) = a \cdot \text{gen} + b$ was fitted ('gen' is the generation counter) using simple linear regression which should be adequate type of model. The sign of the learned parameter a was used to decide, if the variances increase or decrease during the run. To find the critical value of k (for which the parameter a changes its sign), the bisection method was used with the precision given by 24 iterations.

3.2 Results and Discussion

The minimal and maximal admissible values of the variance enlarging factor k are displayed in Tab. 2 in tabular form, and in Fig. 4 in graphical form.

Table 2. Minimum and maximum admissible values for the enlarging factor k for various values of τ and dimensions 1, ..., 6

Dim		τ										
		0.01	0.1	0.2	0.3	0.4	0.5	0.6	0.7	0.8	0.9	0.99
1	$k_{\min}$	3.477	2.443	2.140	1.947	1.794	1.660	1.539	1.423	1.309	1.184	1.033
	$k_{\max}$	130.911	13.755	6.862	4.540	3.363	2.647	2.157	1.796	1.510	1.266	1.040
2	$k_{\min}$	3.501	2.443	2.139	1.947	1.791	1.660	1.540	1.424	1.310	1.186	1.033
	$k_{\max}$	14.239	4.396	3.049	2.441	2.068	1.804	1.603	1.438	1.293	1.159	1.024
3	$k_{\min}$	3.520	2.439	2.141	1.946	1.791	1.659	1.540	1.424	1.309	1.185	1.033
	$k_{\max}$	6.667	2.976	2.300	1.958	1.734	1.568	1.434	1.320	1.218	1.120	1.018
4	$k_{\min}$	3.411	2.438	2.142	1.945	1.790	1.660	1.539	1.425	1.310	1.185	1.034
	$k_{\max}$	4.607	2.432	1.981	1.740	1.576	1.452	1.349	1.260	1.179	1.099	1.015
5	$k_{\min}$	3.509	2.442	2.147	1.946	1.794	1.660	1.540	1.424	1.310	1.185	1.033
	$k_{\max}$	3.672	2.145	1.803	1.614	1.485	1.382	1.297	1.223	1.154	1.085	1.013
6	$k_{\min}$	3.481	2.449	2.141	1.948	1.788	1.659	1.539	1.424	1.310	1.184	1.033
	$k_{\max}$	3.140	1.966	1.690	1.533	1.422	1.336	1.262	1.198	1.137	1.076	1.011

Comparing the theoretical values of $k_{\min}$ and $k_{\max}$ from Tab. 1 and experimental values from Tab. 2 for 1D case, we can see that they do not differ substantially which constitutes at least partial confirmation that the model is in accordance with experiments.

Another (expectable) observation is the fact that the minimal as well as maximal bound for k decreases with decreasing selection pressure (with increasing selection proportion τ). It is worth to note, that the lower bound of k does not change with dimensionality (the solid lines in Fig. 4 are almost the same) since only the first coordinate is taken into account in the selection process.

The upper bound of k, however, changes substantially with the dimensionality of the search space. In 1D case, the upper bounds of k for all selection proportions τ lie above the lower bounds, and it is thus possible to find admissible k for every τ. With increasing dimensionality, the upper bound falls below the lower bound for increasingly larger interval of τ (which means that not for every τ we are able to find admissible k). For dimensionality 5 and above, admissible k actually does not exist regardless of τ!

The fact that $k_{\max}$ drops with increasing dimensionality can be explained as follows. Consider the quadratic bowl f_{quad} as the fitness function and the situation when the population is centered around origin. In that case, the individuals selected by truncation selection are bounded by a D-dimensional hypersphere. Its radius (the maximum distance from origin) can be computed as $r = \sqrt{CDF^{-1}_{\chi^2}(\tau, D)}$, where $CDF^{-1}_{\chi^2}(\tau, D)$ is the inverse cumulative distribution function of χ^2 distribution with D degrees of freedom and τ is the selection proportion. As can be seen in Fig. 5 for $\tau = 0.5$, the hypersphere radius grows with dimensionality. It means that the data points after selection are more spread

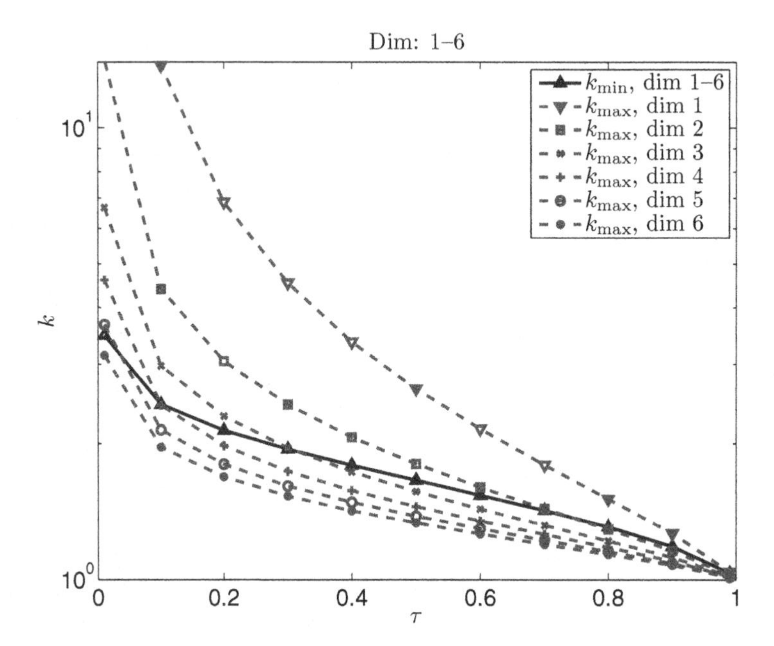

Fig. 4. Minimum and maximum admissible values for the enlarging factor k for various values of τ and dimensions 1, ..., 6, graphically

along the axes and the ML estimate of the standard deviation σ is higher. It is thus necessary to multiply the ML estimate of σ with a lower value of k only, not to overshoot the value 1.

4 Summary and Conclusions

This paper analysed some of the convergence properties of a simple EDA based on Gaussian distribution and truncation selection. Specifically, the simplest way of preventing premature convergence was studied: each generation, the ML estimate of the population standard deviation σ was enlarged by a factor k held constant during the whole evolution. Previous theoretical results for 1D case suggested there should exist an interval of admissible values of k that are

- sufficiently large to allow the algorithm to traverse the slope of the fitness function with at least nondecreasing velocity, and
- sufficiently low to allow the algorithm to exploit the neighborhood of the optimum once it is found.

However, as the experimental results in more dimensional spaces suggest, *the interval of admissible values of* k *gets smaller with increasing dimensionality and eventually vanishes*! In other words, in more-dimensional spaces there is no single value of the variance enlarging factor k that would allow the simple Gaussian

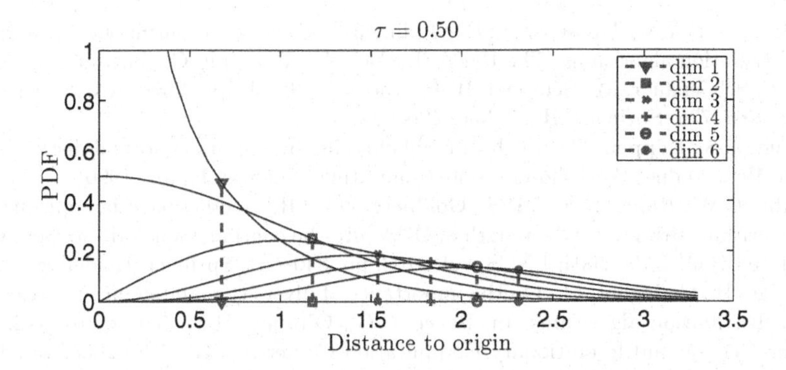

Fig. 5. The radius of hypersphere containing τN best individuals increases with dimensionality. Blue lines: probability density functions (PDF) of the distribution of distances of points sampled from D-dimensional standard normal distribution. Red dashed lines: the hypersphere radiuses for dimensions $1, \ldots, 6$.

EDA with truncation selection to behave reasonably on slopes and in the valleys of the fitness function in the same time!

This result is important on its own, but also serves as another supporting evidence to favor dynamic, adaptive, or self-adaptive control strategies for the standard deviation of the sampling distribution (e.g. those mentioned in Sec. 1), when truncation selection is used. Nevertheless, if the constant enlarging factor should be used, then I suggest at least 2-stage search procedure: in the first stage, set the factor $k > k_{\min}$, so that the optimum is localized; after a few iterations in which the progress is slow, switch to the second phase and set the factor $k < k_{\max}$ (or use directly the ML estimate of σ) to fine tune the solution.

Similar study of other selection mechanisms remains as the future work. If the interval of admissible values exists for them, having the bounds for the factor k is very profitable from the experimenter's and EA designer's point of view. These bounds could be also used in various adaptive variance scaling schemes as safeguards.

Acknowledgments. The author is supported by the Grant Agency of the Czech Republic with the grant no. 102/08/P094 entitled "Machine learning methods for solution construction in evolutionary algorithms".

References

1. Larrañaga, P., Lozano, J.A. (eds.): Estimation of Distribution Algorithms. GENA. Kluwer Academic Publishers, Dordrecht (2002)
2. Larrañaga, P., Lozano, J.A., Bengoetxea, E.: Estimation of distribution algorithms based on multivariate normal distributions and gaussian networks. Technical Report KZZA-IK-1-01, Dept. of Computer Science and Artificial Intelligence, University of Basque Country (2001)

3. Bosman, P.A.N., Thierens, D.: Expanding from discrete to continuous estimation of distribution algorithms: The IDEA. In: Deb, K., Rudolph, G., Lutton, E., Merelo, J.J., Schoenauer, M., Schwefel, H.-P., Yao, X. (eds.) PPSN 2000. LNCS, vol. 1917, pp. 767–776. Springer, Heidelberg (2000)
4. Rudlof, S., Köppen, M.: Stochastic hill climbing by vectors of normal distributions. In: First Online Workshop on Soft Computing, Nagoya, Japan (1996)
5. Ahn, C.W., Ramakrishna, R.S., Goldberg, D.E.: Real-coded bayesian optimization algorithm: Bringing the strength of BOA into the continuous world. In: Deb, K., et al. (eds.) GECCO 2004. LNCS, vol. 3103, pp. 840–851. Springer, Heidelberg (2004)
6. Yuan, B., Gallagher, M.: On the importance of diversity maintenance in estimation of distribution algorithms. In: Beyer, H.G., O'Reilly, U.M. (eds.) Proceedings of the Genetic and Evolutionary Computation Conference GECCO-2005, vol. 1, pp. 719–726. ACM Press, New York (2005)
7. Koumoutsakos, P., Očenášek, J., Hansen, N., Kern, S.: A mixed bayesian optimization algorithm with variance adaptation. In: Yao, X., Burke, E.K., Lozano, J.A., Smith, J., Merelo-Guervós, J.J., Bullinaria, J.A., Rowe, J.E., Tiňo, P., Kabán, A., Schwefel, H.-P. (eds.) PPSN 2004. LNCS, vol. 3242, pp. 352–361. Springer, Heidelberg (2004)
8. Grahl, J., Minner, S., Rothlauf, F.: Behaviour of UMDAc with truncation selection on monotonous functions. In: IEEE Congress on Evolutionary Computation, CEC 2005, vol. 3, pp. 2553–2559 (2005)
9. Gonzales, C., Lozano, J., Larranaga, P.: Mathematical modelling of UMDAc algorithm with tournament selection. International Journal of Approximate Reasoning 31(3), 313–340 (2002)
10. Grahl, J., Bosman, P.A.N., Rothlauf, F.: The correlation-triggered adaptive variance scaling IDEA. In: Proceedings of the 8th annual conference on Genetic and Evolutionary Computation Conference - GECCO 2006, pp. 397–404. ACM Press, New York (2006)
11. Bosman, P.A.N., Grahl, J., Rothlauf, F.: SDR: A better trigger for adaptive variance scaling in normal EDAs. In: GECCO 2007: Proceedings of the 9th annual conference on Genetic and Evolutionary Computation, pp. 492–499. ACM Press, New York (2007)
12. Yuan, B., Gallagher, M.: A mathematical modelling technique for the analysis of the dynamics of a simple continuous EDA. In: IEEE Congress on Evolutionary Computation, CEC 2006, Vancouver, Canada, pp. 1585–1591. IEEE Press, Los Alamitos (2006)
13. Pošík, P.: Gaussian EDA and truncation selection: Setting limits for sustainable progress. In: IEEE SMC International Conference on Distributed Human-Machine Systems, DHMS 2008, Athens, Greece, IEEE, Los Alamitos (2008)
14. Grahl, J., Bosman, P.A.N., Minner, S.: Convergence phases, variance trajectories, and runtime analysis of continuous EDAs. In: GECCO 2007: Proceedings of the 9th annual conference on Genetic and evolutionary computation, pp. 516–522. ACM Press, New York (2007)

Scalable Continuous Multiobjective Optimization with a Neural Network–Based Estimation of Distribution Algorithm

Luis Martí, Jesús García, Antonio Berlanga, and José M. Molina

Universidad Carlos III de Madrid, Group of Applied Artificial Intelligence
Av. de la Universidad Carlos III, 22 — Colmenarejo, Madrid 28270, Spain
{lmarti,jgherrer}@inf.uc3m.es, {aberlan,molina}@ia.uc3m.es
http://www.giaa.inf.uc3m.es/

Abstract. To achieve a substantial improvement of MOEDAs regarding MOEAs it is necessary to adapt their model building algorithm to suit this particular task. Most current model building schemes used so far off–the–shelf machine learning methods. However, the model building problem has specific requirements that those methods do not meet and even avoid.

In this we work propose a novel approach to model building in MOEDAs using an algorithm custom–made for the task. We base our proposal on the growing neural gas (GNG) network. The resulting model–building GNG (MB–GNG) is capable of yielding good results when confronted to high–dimensional problems.

1 Introduction

Although multiobjective optimization evolutionary algorithms (MOEAs) [1] have successfully solved many complex synthetic and real–life problems, the majority of works have been concentrated on low dimensional problems [2]. When advancing to higher dimensions MOEAs suffer heavily from the curse of dimensionality [3], requiring an exponential increase of the resources made available to them (see [1] pp. 414–419).

A possible way of alleviating this scaling issue is to resort to more efficient evolutionary approaches. Estimation of distribution algorithms (EDAs) [4] can be used for that objective. EDAs have been hailed as landmark in the progress of evolutionary algorithms. They replace the application of evolutionary operators with the creation of a statistical model of the fittest elements of the population. This model is then sampled to produce new elements.

The extension of EDAs to the multiobjective domain has lead to what can be denominated multiobjective EDAs (MOEDAs). However most MOEDAs have limited themselves to port single objective EDAs to the multiobjective domain by incorporating features taken from MOEAs. Although MOEDAs have proved themselves as a valid approach to the MOP, this later fact hinders the achievement of a significant improvement regarding “standard” MOEAs in high–dimensional problems.

M. Giacobini et al. (Eds.): EvoWorkshops 2008, LNCS 4974, pp. 535–544, 2008.

Adapting the model building algorithm is one of the ways of achieving a substantial advance. Most model building schemes used so far by EDAs use off–the–shelf machine learning methods. However, the model building problem has specific requirements that those methods do not meet and even avoid. In particular, in those algorithms, outliers are treated as invalid data, where in the model building problem they represent newly discovered regions of the search space. Similarly, an excess of resources is spent in finding the optimal size of the model.

In this work we propose a novel approach to model building in MOEDAs using an algorithm custom–made for the task. We modify the growing neural gas (GNG) network [5] to make it particularly suitable for model building. The resulting model–building GNG (MB–GNG) addresses the above described issues with success. Therefore it is capable of yielding better results when confronted to high–dimensional problems.

The rest of this work first deals with the theoretical background that supports our proposal. We then proceed to describe MB–GNG and the EDA framework in which it is embedded. After that, a series of well–known test problems are solved with a set of other state–of–the–art algorithms and our approach. Each problem is scaled regarding the number of optimization functions in order to assess the behavior of the algorithms when subjected to extreme situations. As a conclusion some final remarks, comments and lines of future development are presented.

2 Theoretical Background

The concept of multiobjective optimization refers to the process of finding one or more feasible solutions of a problem that corresponds to the extreme values (either maximum or minimum) of two or more functions subject to a set of restrictions. More formally, a multiobjective optimization problem (MOP) can be defined as

Definition 1 (Multiobjective Optimization Problem).

$$\left.\begin{array}{r}\text{minimize } \boldsymbol{F}(\boldsymbol{x}) = \langle f_1(\boldsymbol{x}), \ldots, f_M(\boldsymbol{x})\rangle\,, \\ \text{subject to } c_1(\boldsymbol{x}), \ldots, c_R(\boldsymbol{x}) \leq 0\,, \\ \text{with } \boldsymbol{x} \in \mathcal{D}\,,\end{array}\right\} \tag{1}$$

where $\mathcal{D}$ is known as the *decision space*. The functions $f_1(\boldsymbol{x}), \ldots, f_M(\boldsymbol{x})$ are the objective functions. Their corresponding image set, $\mathcal{O}$, of $\mathcal{D}$ is named *objective space* ($\boldsymbol{F} : \mathcal{D} \rightarrow \mathcal{O}$). Finally, inequalities $c_1(\boldsymbol{x}), \ldots, c_R(\boldsymbol{x}) \leq 0$ express the restrictions imposed to the values of $\boldsymbol{x}$.

In general terms, this class of problems does not have a unique optimal solution. Instead an algorithm solving (1) should produce a set containing equally good trade–off optimal solutions. The optimality of a set of solutions can be defined relying on the so–called *Pareto dominance relation* [6]:

Definition 2 (Dominance Relation). *Under optimization problem (1) and having* $\boldsymbol{x}_1, \boldsymbol{x}_2 \in \mathcal{D}$, $\boldsymbol{x}_1$ *is said to dominate* $\boldsymbol{x}_2$ *(expressed as* $\boldsymbol{x}_1 \prec \boldsymbol{x}_2$*) iff* $\forall f_j$, $f_j(\boldsymbol{x}_1) \leq f_j(\boldsymbol{x}_2)$ *and* $\exists f_i$ *such that* $f_i(\boldsymbol{x}_1) < f_i(\boldsymbol{x}_2)$.

The solution of (1) is a subset of $\mathcal{D}$ that contains elements are not dominated by other elements of $\mathcal{D}$, or, in more formal terms,

Definition 3 (Pareto–optimal Set). *The solution of problem (1) is the set* $\mathcal{D}^*$ *such that* $\mathcal{D}^* \subseteq \mathcal{D}$ *and* $\forall \boldsymbol{x}_1 \in \mathcal{D}^* \nexists \boldsymbol{x}_2$ *that* $\boldsymbol{x}_2 \prec \boldsymbol{x}_1$.

Here, $\mathcal{D}^*$ is known as the *Pareto–optimal set* and its image in objective space is called *Pareto–optimal front*, $\mathcal{O}^*$.

2.1 Evolutionary Approaches to Multiobjective Optimization

MOPs have been addressed with a variety of methods [7]. Among them, evolutionary algorithms (EAs) [8] have proved themselves as a valid and competent approach from theoretical and practical points of view. This has led to what has been called multiobjective optimization evolutionary algorithms (MOEAs) [1]. Their success is due to the fact that EAs do not make any assumptions about the underlying fitness landscape. Therefore, it is believed they perform consistently well across all types of problems, although it has been shown that they share theoretical limits imposed by the no–free–lunch theorem [9]. Another important benefit arises from the parallel search. Thanks to that these algorithms can produce a set of equally optimal solutions instead of one, as many other algorithms do.

2.2 Estimation of Distribution Algorithms

Estimation of distribution algorithms (EDAs) have been claimed as a paradigm shift in the field of evolutionary computation. Like EAs, EDAs are population based optimization algorithms. However in EDAs the step where the evolutionary operators are applied to the population is substituted by construction of a statistical model of the most promising subset of the population. This model is then sampled to produce new individuals that are merged with the original population following a given substitution policy. Because of this model building feature EDAs have also been called probabilistic model building genetic algorithms (PMBGAs) [10]. A framework similar to EDAs is proposed by the iterated density estimation evolutionary algorithms (IDEAs) [11].

The introduction of machine learning techniques implies that these new algorithms lose the biological plausibility of its predecessors. In spite of this, they gain the capacity of scalably solve many challenging problems, significantly outperforming standard EAs and other optimization techniques.

First EDAs were intended for combinatorial optimization but they have been extended to continuous domain (see [4] for a review). It was then only a matter of time to have multiobjective extensions. This has led to the formulation of multiobjective optimization EDAs (MOEDAs).

3 Scalability: A Salient Issue

One topic that remains not properly studied inside the MOEA/MOEDA scope is the scalability of the algorithms [12]. The most critical issue is the dimension of the objective space. It has been experimentally shown to have an exponential relation with the optimal size of the population (see [1] pp. 414–419). This fact implies that, with the increase of the number of objective functions an optimization algorithm needs an exponential amount of resources made available to it.

In order to achieve a substantial improvement in the scalability of these algorithms it is therefore essential to arm them with a custom–made model building algorithm that pushes the population towards newly found zones in objective space. So far, MOEDA approaches have mostly used off–the–shelf machine learning methods. However, those methods are not meant specifically for the problem we are dealing with here. This fact leads to some undesirable behaviors that, although justified in the original field of application of the algorithms, might hinder the performance of the process, both in the accuracy and in the resource consumption senses. Among these behaviors we can find the creation of an excessively accurate model and the loosing of outliers.

For the model building problem there is no need for having the most accurate model of the data. However, most of the current approaches dedicate a sizable effort in finding the optimal model complexity. For instance, for cluster–based models its not required to find optimal amount of clusters just to find a "fair" amount such that the data set to be modeled is correctly covered.

In the other case, outliers are essential in model building. They represent unexplored areas of the local optimal front. Therefore they not only should be preserved but perhaps even reinforced. A model building algorithms that primes outliers might actually accelerate the search process and alleviate the rate of the exponential dimension–population size dependency.

In the next section we introduce a MOEDA that uses a custom–made model building algorithm to overcome the problems here described.

4 Model Building Growing Neural Gas

Clustering algorithms [13] have been used as part of the model building algorithms of EDAs and MOEDAs. However, as we discussed in the previous section a custom–made algorithm might be one of the ways of achieving a significative improvement in this field.

For this task we have chosen the growing neural gas (GNG) network [5] as a starting point. GNG networks are intrinsic self–organizing neural networks based on the neural gas [14] model. This model relies in a competitive Hebbian learning rule [15]. It creates an ordered topology of inputs classes and associates a cumulative error to each. The topology and the cumulative errors are conjointly used to determine how new classes should be inserted. Using these heuristics the model can fit the network dimension to the complexity of the problem being solved.

Our model building GNG (MB–GNG) is an extension of the original (unsupervised) GNG. MB–GNG is a one layer network that defines each class as a local Gaussian density and adapts them using a local learning rule. The layer contains a set of nodes $\mathcal{C} = \{c_1, \ldots, c_{N^*}\}$, with $N_0 \leq N^* \leq N_{\max}$. Here N_0 and $N_{\max}$ represent initial and maximal amount of nodes in the network.

A node c_i consists of a center, $\boldsymbol{\mu}_i$, deviations , $\boldsymbol{\sigma}_i$, an accumulated error, ξ_i, and a set of edges that define the set of topological neighbors of c_i, $\mathcal{V}_i$. Each edge has an associated age, $\nu_{i,j}$.

MB–GNG creates a quantization of the inputs space using a modified version of the GNG algorithm and then computes the deviations associated to each node. The dynamics of a GNG network consists of three concurrent processes: network adaptation, node insertion and node deletion. The combined use of these three processes renders GNG training Hebbian in spirit [15].

The network is initialized with N_0 nodes with their centers set to randomly chosen inputs. A training iteration starts after an input $\boldsymbol{x}$ is randomly selected from the training data set. Then two nodes are selected for being the closest ones to $\boldsymbol{x}$. The *best–matching node*, c_b, is the closest node to $\boldsymbol{x}$. Consequently, the *second best–matching node*, $c_{b'}$, is determined as the second closest node to $\boldsymbol{x}$.

If $c_{b'}$ is not a neighbor of c_b then a new edge is established between them $\mathcal{V}_b = \mathcal{V}_b \cup \{c_{b'}\}$ with zero age, $\nu_{b,b'} = 0$. If, on the other case, $c_{b'} \in \mathcal{V}_b$ the age of the corresponding edge is reset $\nu_{b,b'} = 0$.

At this point, the age of all edges is incremented in one. If an edge is older than the maximum age, $\nu_{i,j} > \nu_{\max}$, then the edge is removed. If a node becomes isolated from the rest it is also deleted.

A clustering error is then added to the best–matching node error accumulator, ξ_b.

After that, learning takes place in the best–matching node and its neighbors with rates ϵ_{best} and ϵ_{vic}, respectively. For c_b adaptation follows the rule originally used by GNG,

$$\Delta\boldsymbol{\mu}_b = \epsilon_{\text{best}} \left(\boldsymbol{x} - \boldsymbol{\mu}_b\right). \tag{2}$$

However for c_b's neighbors a cluster repulsion term [16] is added to the original formulation, that is, $\forall c_v \in \mathcal{V}_b$,

$$\Delta\boldsymbol{\mu}_v = \epsilon_{\text{vic}} \left(\boldsymbol{x} - \boldsymbol{\mu}_v\right) + \beta \exp\left(-\frac{\|\boldsymbol{\mu}_v - \boldsymbol{\mu}_b\|}{\zeta}\right) \frac{\sum_{c_u \in \mathcal{V}_b} \|\boldsymbol{\mu}_u - \boldsymbol{\mu}_b\|}{|\mathcal{V}_b|} \frac{(\boldsymbol{\mu}_v - \boldsymbol{\mu}_b)}{\|\boldsymbol{\mu}_v - \boldsymbol{\mu}_b\|}, \tag{3}$$

Here β is an integral multiplier that defines the amplitude of the repulsive force while ζ controls the weakening rate of the repulsive force with respect to the distance between the nodes' centers. We have set them to $\beta = 2$ and $\zeta = 0.1$ as suggested in [17].

After a given amount of iterations a new node is inserted to the network. First, the node with largest error, c_e, is selected the node. Then the worst node among its neighbors, $c_{e'}$, is located. N^* is incremented and the new node, c_{N^*}, is inserted equally distant from the two nodes. The edge between c_e and $c_{e'}$ is removed and two new edges connecting c_{N^*} with c_e and $c_{e'}$ are created. Finally, the errors of all nodes are decreased by a factor δ_G,

$$\xi_i = \delta_\mathrm{G}\xi_i,\ i = 1, .., N^*. \tag{4}$$

After training has ended the deviations, $\boldsymbol{\sigma}_i$, of the nodes must be computed. For this task we employ the unbiased estimator of the deviations.

5 Embedding MB–GNG in an EDA Framework

To test MB–GNG is it essential to insert it in an EDA framework. This framework should be simple enough to be easily understandable but should also have a sufficient problem solving capacity. It should be scalable and preserve the diversity of the population.

Our EDA employs the fitness assignment used by the NSGA–II algorithm [18] and constructs the population model by applying MB–GNG. The NSGA–II fitness assignment was chosen because of its proven effectivity and its relative low computational cost.

The algorithm's workflow is similar to other EDAs. It maintains a population of individuals, P_t, where t is the current iteration. It starts from a random initial population P_0 of z individuals. It then proceeds to sort the individuals using the NSGA–II fitness assignment function. A set $\hat{P}_t$ containing the best $\lfloor \alpha \left| P_t \right| \rfloor$ elements is extracted from the sorted version of P_t ,

$$\left| \hat{P}_t \right| = \alpha \left| P_t \right|. \tag{5}$$

A MB–GNG network is then trained using $\hat{P}_t$ as training data set. In order to have a controlled relation between size of $\hat{P}_t$ and the maximum size of the network, N_max, these two sizes are bound,

$$N_\mathrm{max} = \left\lfloor \gamma \left| \hat{P}_t \right| \right\rfloor. \tag{6}$$

The resulting Gaussian kernels are used to sample new individuals. An amount of $\lfloor \omega \left| P_t \right| \rfloor$ new individuals is synthesized. Each one of these individuals substitute a randomly selected ones from the section of the population not used for model building $P_t \setminus \hat{P}_t$. The set obtained is then united with best elements, $\hat{P}_t$, to form the population of the next iteration P_t.

Iterations are repeated until a given stopping criterion is met. The output of the algorithm is the set of non–dominated individuals of P_t.

6 Experiments

We now focus on making a preliminary study that validates our proposal from an experimental point of view. For comparison purposes a set of known and

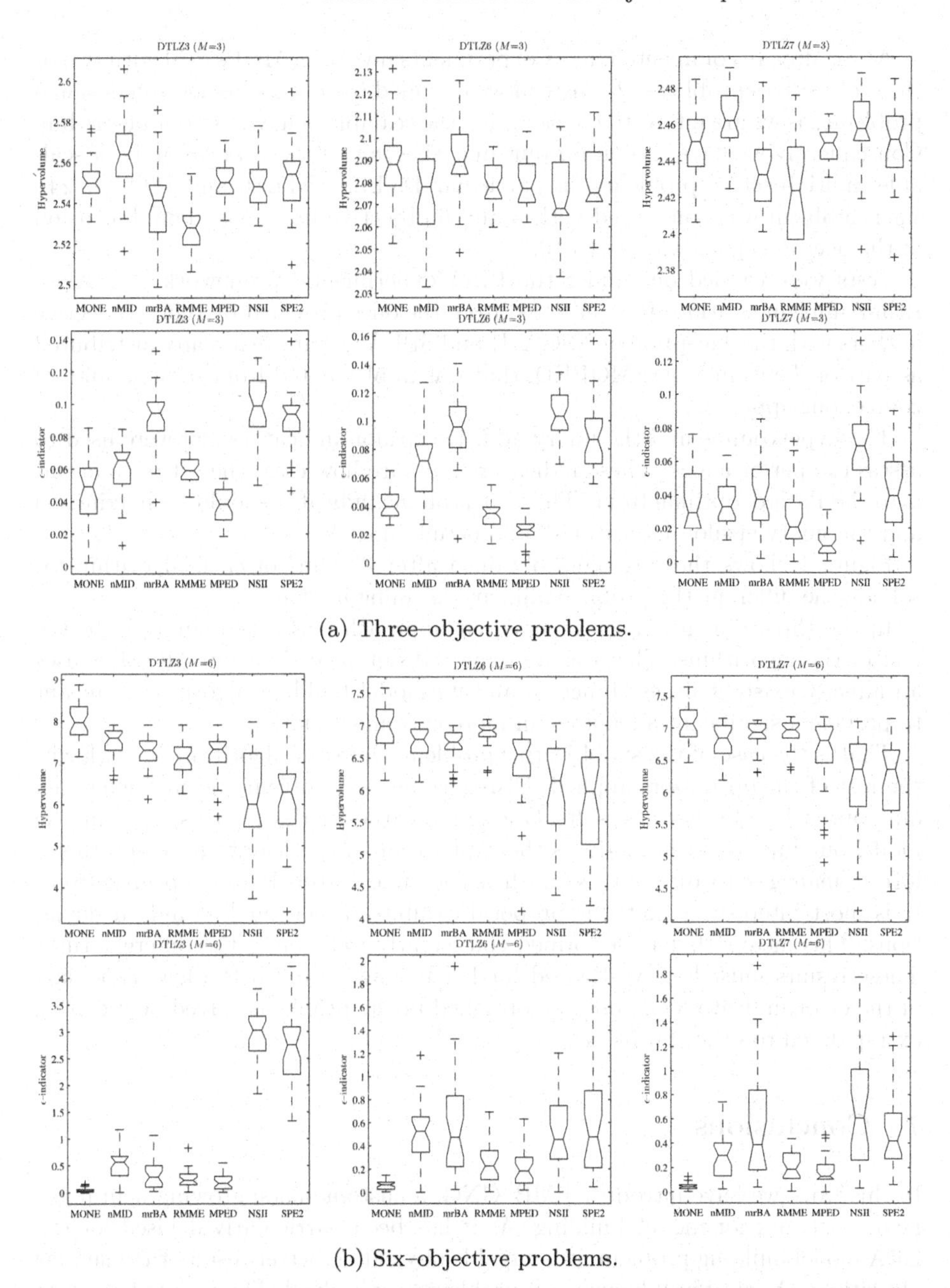

(a) Three–objective problems.

(b) Six–objective problems.

Fig. 1. DTLZ3, DTLZ6 and DTLZ7 test problems with naïve MIDEA (nMID), mrBOA (mrBA), RM–MEDA (RMME), MOPED (MPED), NSGA–II (NSII), SPEA2 (SPE2) and our approach (MONE)

competent evolutionary multiobjective optimizers are also applied. In particular, naïve MIDEA [19], mrBOA [20], RM–MEDA [21], MOPED [22], NSGA–II, SPEA2 [23].

As we already commented, most experiments involving MOPs deal with two or three objectives problems [2]. Instead we intended to address higher–dimensional problems, as we are interested in assessing the scalable behavior of our algorithm. Consequently we have chosen for our analysis some of the DTLZ family of scalable multiobjective problems, in particular DTLZ3, DTLZ6 and DTLZ7 [24]. Each problem was configured with 3 and 6 objective functions. The dimension of the decision space was set to 10.

Tests were carried out under the PISA experimental framework [25]. Algorithm' implementations were adapted from the ones provided by their respective authors with the exception of NSGA–II and SPEA2, that are already distributed as part of the framework; MOPED, that was implemented from scratch, and, of course, our approach.

The hypervolume and the unary additive epsilon indicators [26] were used to assess the performance. These indicators measure how close the set of solutions is to the Pareto–optimal front. The hypervolume indicator should be maximized and the unary epsilon indicator should be minimized.

Figure 1 shows the box plots obtained after 30 runs of each algorithm for solving the different the problem/dimension configuration.

In the three dimensional problems our approach performed similarly to the rest of the algorithms. This was an expected outcome since our MOEDA uses an already existent fitness function and its model building algorithm is meant to provide a significant advantage in more extreme situations.

That is the case of the six objective problem. Here our algorithm outperforms the rest of the optimizers applied. Disregarding the problem and metric chosen our approach yields better results. One can hypothesize that, in this problem, the model building algorithm induces the exploration of the search space and therefore it manages to discover as much as possible of the Pareto–optimal front. It is most interesting that our proposal exhibits rather small standard deviations. This means that it performed consistently well across the different runs. These results must be investigated further to understand if the low dispersion of the error indicators can only be obtained in the problems solved or if can be extrapolated to other problems.

7 Conclusions

In this work we have introduced MB–GNG, a custom–made growing neural gas neural network for model–building. As it has been particularly devised for the EDA model building problem, it is capable of yielding better results than similar algorithms. A set of well known test problems were solved. The results helped to assert MB–GNG advantages but further tests are obviously necessary. The study here presented must be extended to higher dimensions and other test problems must be solved. However, because of length restrictions we have had to limit ourselves to providing a brief outline of our current research.

These tests will cast light upon the actual processes take place during the model building phase. They should also lead to further improvements on the model building algorithms applied. Regarding this, some other machine learning methods must be extrapolated to this domain in order to compare and stablish the validity of our idea. It is expected that these new approaches will have the search capabilities of MB–GNG while minimizing its computational footprint.

Acknowledgements

This work was supported in part by projects MADRINET, TEC2005-07186-C03-02, SINPROB, TSI2005-07344-C02- 02 and CAM CCG06-UC3M/TIC-0781. The authors wish to thank the anonymous reviewers for their insightful comments.

References

1. Deb, K.: Multi-Objective Optimization using Evolutionary Algorithms. John Wiley & Sons, Chichester (2001)
2. Deb, K., Saxena, D.K.: On finding Pareto–optimal solutions through dimensionality reduction for certain large–dimensional multi–objective optimization problems. Technical Report 2005011, KanGAL (December 2005)
3. Bellman, R.E.: Adaptive Control Processes. Princeton University Press, Princeton (1961)
4. Larrañaga, P., Lozano, J.A. (eds.): Estimation of Distribution Algorithms. A new tool for Evolutionary Computation. Genetic Algorithms and Evolutionary Computation. Kluwer Academic Publishers, Boston/Dordrecht/London (2002)
5. Fritzke, B.: A growing neural gas network learns topologies. In: Tesauro, G., Touretzky, D.S., Leen, T.K. (eds.) Advances in Neural Information Processing Systems, vol. 7, pp. 625–632. MIT Press, Cambridge (1995)
6. Pareto, V.: Cours D'Economie Politique. F. Rouge, Lausanne (1896)
7. Miettinen, K.: Nonlinear Multiobjective Optimization. International Series in Operations Research & Management Science, vol. 12. Kluwer, Norwell (1999)
8. Bäck, T.: Evolutionary algorithms in theory and practice: Evolution Strategies, Evolutionary Programming, Genetic Algorithms. Oxford University Press, New York (1996)
9. Corne, D.W., Knowles, J.D.: No Free Lunch and Free Leftovers Theorems for Multiobjective Optimisation Problems. In: Fonseca, C.M., Fleming, P.J., Zitzler, E., Deb, K., Thiele, L. (eds.) EMO 2003. LNCS, vol. 2632, pp. 327–341. Springer, Heidelberg (2003)
10. Pelikan, M., Goldberg, D.E., Lobo, F.: A survey of optimization by building and using probabilistic models. IlliGAL Report No. 99018, University of Illinois at Urbana-Champaign, Illinois Genetic Algorithms Laboratory, Urbana, IL (1999)
11. Bosman, P.A.: Design and Application of Iterated Density-Estimation Evolutionary Algorithms. PhD thesis, Universiteit Utrecht, Utrecht, The Netherlands (2003)
12. Coello Coello, C.A.: Evolutionary multiobjective optimization: A historical view of the field. IEEE Computational Intelligence Magazine 1(1), 28–36 (2006)
13. Xu, R., Wunsch II, D.: Survey of clustering algorithms. IEEE Transactions on Neural Networks 16(3), 645–678 (2005)
14. Martinetz, T.M., Berkovich, S.G., Shulten, K.J.: Neural–gas network for vector quantization and its application to time–series prediction. IEEE Transactions on Neural Networks 4, 558–560 (1993)

15. Martinetz, T.M.: Competitive Hebbian learning rule forms perfectly topology preserving maps. In: International Conference on Artificial Neural Networks (ICANN 1993), Amsterdam, pp. 427–434. Springer, Heidelberg (1993)
16. Timm, H., Borgelt, C., Doring, C., Kruse, R.: An extension to possibilistic fuzzy cluster analysis. Fuzzy Sets and Systems 147(1), 3–16 (2004)
17. Qin, A.K., Suganthan, P.N.: Robust growing neural gas algorithm with application in cluster analysis. Neural Networks 17(8–9), 1135–1148 (2004)
18. Deb, K., Pratap, A., Agarwal, S., Meyarivan, T.: A Fast and Elitist Multiobjective Genetic Algorithm: NSGA–II. IEEE Transactions on Evolutionary Computation 6(2), 182–197 (2002)
19. Bosman, P.A., Thierens, D.: The Naïve MIDEA: A baseline multi–objective EA. In: Coello Coello, C.A., Hernández Aguirre, A., Zitzler, E. (eds.) EMO 2005. LNCS, vol. 3410, pp. 428–442. Springer, Heidelberg (2005)
20. Ahn, C.W.: Advances in Evolutionary Algorithms. Theory, Design and Practice. Springer, Heidelberg (2006)
21. Zhang, Q., Zhou, A., Jin, Y.: A regularity model based multiobjective estimation of distribution algorithm. IEEE Transactions on Evolutionary Computation (in press 2007)
22. Costa, M., Minisci, E., Pasero, E.: An hybrid neural/genetic approach to continuous multi–objective optimization problems. In: Apolloni, B., Marinaro, M., Tagliaferri, R. (eds.) WIRN 2003. LNCS, vol. 2859, pp. 61–69. Springer, Heidelberg (2003)
23. Zitzler, E., Laumanns, M., Thiele, L.: SPEA2: Improving the Strength Pareto Evolutionary Algorithm. In: Giannakoglou, K., Tsahalis, D., Periaux, J., Papailou, P., Fogarty, T. (eds.) EUROGEN 2001. Evolutionary Methods for Design, Optimization and Control with Applications to Industrial Problems, Athens, Greece, pp. 95–100 (2002)
24. Deb, K., Thiele, L., Laumanns, M., Zitzler, E.: Scalable Test Problems for Evolutionary Multiobjective Optimization. Advanced Information and Knowledge Processing. In: Evolutionary Multiobjective Optimization: Theoretical Advances and Applications, pp. 105–145. Springer, Heidelberg (2004)
25. Bleuler, S., Laumanns, M., Thiele, L., Zitzler, E.: PISA—A Platform and Programming Language Independent Interface for Search Algorithms. In: Fonseca, C.M., Fleming, P.J., Zitzler, E., Deb, K., Thiele, L. (eds.) EMO 2003. LNCS, vol. 2632, pp. 494–508. Springer, Heidelberg (2003)
26. Knowles, J., Thiele, L., Zitzler, E.: A tutorial on the performance assessment of stochastic multiobjective optimizers. TIK Report 214, Computer Engineering and Networks Laboratory (TIK), ETH Zurich (2006)

Cumulative Step Length Adaptation for Evolution Strategies Using Negative Recombination Weights

Dirk V. Arnold and D.C. Scott Van Wart

Faculty of Computer Science, Dalhousie University
Halifax, Nova Scotia, Canada B3H 1W5
{dirk,wart}@cs.dal.ca

Abstract. Cumulative step length adaptation is a mutation strength control mechanism commonly employed with evolution strategies. When using weighted recombination with negative weights it can be observed to be prone to failure, often leading to divergent behaviour in low-dimensional search spaces. This paper traces the reasons for this breakdown of step length control. It then proposes a novel variant of the algorithm that reliably results in convergent behaviour for the test functions considered. The influence of the dimensionality as well as of the degree of ill-conditioning on optimisation performance are evaluated in computer experiments. Implications for the use of weighted recombination with negative weights are discussed.

1 Introduction

Evolution strategies [6, 13] are nature inspired, iterative algorithms most commonly used for solving numerical optimisation problems. Such problems typically require that the mutation strength, which controls the step length of the strategies, be adapted in the course of the search. Several adaptation mechanisms have been proposed. Among them are the 1/5th success rule [11], mutative self-adaptation [11, 13], and cumulative step length adaptation [10]. The latter algorithm is particularly significant as it is the step length adaptation mechanism used in covariance matrix adaptation evolution strategies (CMA-ES) [7, 9].

In the basic form of evolution strategies, all of the selected candidate solutions are weighted equally. Weighted recombination, i.e., having candidate solutions enter recombination with different weights that are based on their ranks in the set of all offspring, has been known for some time to be an effective means for speeding up local convergence. Rudolph [12] discovered that assigning negative weights to especially unfavourable candidate solutions can result in a further speed-up. Underlying the use of negative weights is the assumption that the opposite of an unfavourable step is likely to be favourable. More recently, it has been seen that for the idealised environments of the quadratic sphere and the parabolic ridge, if the search space dimensionality is high, weighted recombination with optimally chosen weights (half of which are negative) can speed up convergence by a factor of up to 2.5 compared to unweighted recombination [2, 3, 4]. Moreover, optimal weights agree for both of those cases.

However, cumulative step length adaptation is not without problems when used in connection with negative weights. It can be observed that if the dimensionality of the

M. Giacobini et al. (Eds.): EvoWorkshops 2008, LNCS 4974, pp. 545–554, 2008.

optimisation problem at hand is low, cumulative step length adaptation may fail to generate useful step lengths and result in divergent behaviour for objective functions as simple as the sphere model. Moreover, for some problems the use of negative weights in combination with cumulative step length adaptation results in significantly reduced performance even if the search space dimensionality is high. Presumably, these problems are the reason that CMA-ES refrain from using negative weights altogether [7, 9].

In this paper, we propose a modification to the cumulative step length adaptation mechanism that enables evolution strategies employing negative weights to converge reliably in low-dimensional search spaces. Its remainder is organised as follows. Section 2 describes multirecombination evolution strategies with cumulative step length adaptation. In Section 3, the reason for the breakdown of cumulative step length control when using negative recombination weights in low-dimensional search spaces is discussed, and a modification to the mechanism is proposed. In Section 4, we evaluate the performance of the novel algorithm using several test functions. Section 5 concludes with a brief discussion.

2 Strategy

For an N-dimensional optimisation problem with objective function $f : \mathbb{R}^N \to \mathbb{R}$, the state of a weighted multirecombination evolution strategy with cumulative step length adaptation as proposed in [10] is described by search point $\mathbf{x} \in \mathbb{R}^N$, search path $\mathbf{s} \in \mathbb{R}^N$ (initialised to the zero vector), and mutation strength $\sigma \in \mathbb{R}$. The strategy repeatedly updates those quantities using the following five steps (where $\leftarrow$ denotes the assignment operator):

1. Generate λ offspring candidate solutions $\mathbf{y}^{(i)} = \mathbf{x} + \sigma\mathbf{z}^{(i)}$, $i = 1, \ldots, \lambda$. The $\mathbf{z}^{(i)}$ are vectors consisting of N independent, standard normally distributed components and are referred to as mutation vectors. The nonnegative mutation strength σ determines the step length of the strategy.
2. Determine the objective function values $f(\mathbf{y}^{(i)})$ of the offspring candidate solutions and order the $\mathbf{y}^{(i)}$ according to those values.
3. Compute the progress vector

$$\mathbf{z}^{(\mathrm{avg})} = \sum_{k=1}^{\lambda} w_{k,\lambda}\mathbf{z}^{(k;\lambda)} \tag{1}$$

   where index $k;\lambda$ refers to the kth best of the λ offspring. The $w_{k,\lambda}$ are used to weight the mutation vectors and depend on the rank of the corresponding candidate solution in the set of all offspring.
4. Update the search point and search path according to

$$\mathbf{x} \leftarrow \mathbf{x} + \sigma\mathbf{z}^{(\mathrm{avg})} \tag{2}$$

$$\mathbf{s} \leftarrow (1-c)\mathbf{s} + \sqrt{\mu_{\mathrm{eff}}\, c(2-c)}\,\mathbf{z}^{(\mathrm{avg})} \tag{3}$$

   where $c \in (0, 1)$ is a cumulation parameter and where $\mu_{\mathrm{eff}} = 1/\sum_{k=1}^{\lambda} w_{k,\lambda}^2$ denotes the "variance effective selection mass" [7].

5. Update the mutation strength according to

$$\sigma \leftarrow \sigma \exp\left(c\,\frac{\|\mathbf{s}\| - \chi_N}{\chi_N d}\right) \tag{4}$$

where $d > 0$ is a damping constant and where $\chi_N = \sqrt{2}\,\Gamma((N+1)/2)/\Gamma(N/2)$ denotes the expected value of a χ_N-distributed random variable.

Using mutative self-adaptation as a starting point, Hansen and Ostermeier [9] describe the development of cumulative step length adaptation as a sequence of steps aimed at derandomising the former mechanism. While mutative self-adaptation adapts the mutation strength based solely on differences in the length of mutation steps, cumulative step length adaptation attempts to measure correlations between the directions of successive progress vectors. The search path $\mathbf{s}$ as updated in Eq. (3) implements an exponentially fading record of steps taken by the strategy. The cumulation parameter c determines how quickly the memory of the strategy expires, with larger values resulting in a more rapid decay of the information present in $\mathbf{s}$. Higher-dimensional problems require a longer memory, and c is often chosen to be asymptotically inversely proportional to N. The coefficient in Eq. (3) which the progress vector is multiplied with is chosen such that under random selection (i.e., on a flat fitness landscape), after initialisation effects have faded the search path has independent, normally distributed components with zero mean and unit variance.

Cumulative step length adaptation relies on the assertion that ideally, consecutive steps of the strategy should be uncorrelated. Uncorrelated random steps lead to the search path having expected length χ_N. According to Eq. (4), a search path of that length results in the mutation strength remaining unchanged. If the most recently taken steps are positively correlated, they tend to point in similar directions and efficiency could be gained by making fewer but longer steps. In that case, the length of the search path exceeds χ_N and Eq. (4) acts to increase the mutation strength. Similarly, negative correlations between successive steps (i.e., the strategy stepping back and forth) lead to the mutation strength being reduced. Larger values of the damping constant d in Eq. (4) act to moderate the magnitude of the changes in mutation strength. Lower-dimensional problems in particular may require significant damping. On the other hand, too large a damping constant can negatively impact the performance of the strategy as it prevents rapid adaptation of the mutation strength and therefore rapid convergence for "simple" problems. Typically, d is chosen to be asymptotically independent of N.

Several settings for the weights $w_{k,\lambda}$ occurring in Eq. (1) can be found in the literature:

– The most common choice is to use unweighted recombination in connection with truncation selection. The corresponding weights are

$$w_{k,\lambda} = \begin{cases} 1/\mu & \text{if } k \le \mu \\ 0 & \text{otherwise} \end{cases} \tag{5}$$

for some $\mu \in \{1, \dots, \lambda - 1\}$. Typically, $\mu \approx \lambda/4$. The resulting strategy is referred to as $(\mu/\mu, \lambda)$-ES [6].

- Hansen and Kern [7] set $\mu = \lfloor \lambda/2 \rfloor$ and use weighted recombination with weights

$$w_{k,\lambda} = \begin{cases} \frac{\ln(\mu+1)-\ln(k)}{\mu \ln(\mu+1)-\sum_{i=1}^{\mu} \ln(i)} & \text{if } k \leq \mu \\ 0 & \text{otherwise} \end{cases}. \tag{6}$$

  They note that strategies employing this choice of weights "only slightly outperform [$(\mu/\mu, \lambda)$-style recombination with] $\mu \approx \lambda/4$" when used in the context of the CMA-ES.
- In [2], setting

$$w_{k,\lambda} = \frac{E_{k,\lambda}}{\kappa} \tag{7}$$

  for some $\kappa > 0$ is proposed, where

$$E_{k,\lambda} = \frac{1}{\sqrt{2\pi}} \frac{\lambda!}{(\lambda-k)!(k-1)!} \int_{-\infty}^{\infty} x \mathrm{e}^{-\frac{1}{2}x^2} [\Phi(x)]^{\lambda-k} [1-\Phi(x)]^{k-1} \mathrm{d}x$$

  denotes the expected value of the $(\lambda + 1 - k)$th order statistic of a sample of λ independent, standard normally distributed random variables and Φ denotes the cumulative distribution function of the standard normal distribution. In [2, 3, 4], this setting has been seen to be the optimal choice of weights for both the quadratic sphere and parabolic ridge models in the limit of very high search space dimensionality. The resulting strategy is referred to as $(\lambda)_{\text{opt}}$-ES and in both of those environments outperforms the $(\mu/\mu, \lambda)$-ES with optimally chosen μ by a factor of up to 2.5 in the limit $N \to \infty$. The scalar quantity κ in Eq. (7) is referred to as the rescaling factor and can be used to control the length of the progress vectors relative to that of the mutation vectors.

3 Direction Based Cumulative Step Length Adaptation

While for sufficiently high search space dimensionality the $(\lambda)_{\text{opt}}$-ES often converges faster than evolution strategies that use nonnegative recombination weights, it may fail to converge altogether even for simple test functions if N is small. Figure 1 contrasts the performance of the $(\lambda)_{\text{opt}}$-ES with $\lambda = 10$ with that of a positively weighted strategy when optimising the quadratic sphere model (see Table 1 for a definition). For that model, evolution strategies converge linearly provided that the mutation strength is adapted successfully. The normalised quality gain Δ^* as defined for example in [1, Chapter 6] is a measure for the speed of convergence[1]. The positively weighted strategy employs weights $w_{k,\lambda} = \max(0, E_{k,\lambda}/\kappa)$ (i.e., the weights are as prescribed by Eq. (7) for ranks $k \leq \lambda/2$, and they are zero for the remaining candidate solutions). The rescaling factor κ is set to 3.68 as for $\lambda = 10$, that setting leads to approximately the same value of μ_{eff} as Eq. (6) does. The resulting choice of weights is nearly identical to that proposed by Hansen and Kern [9], and the strategy is referred to as $(\lambda)_{\text{pos}}$-ES. The $(\lambda)_{\text{opt}}$-ES differs from the $(\lambda)_{\text{pos}}$-ES only in that it uses negative weights as prescribed by Eq. (7) for candidate solutions with ranks $k > \lambda/2$. The solid lines in Fig. 1 show

[1] For the sphere model, this definition agrees with the definition of the log-progress rate in [5].

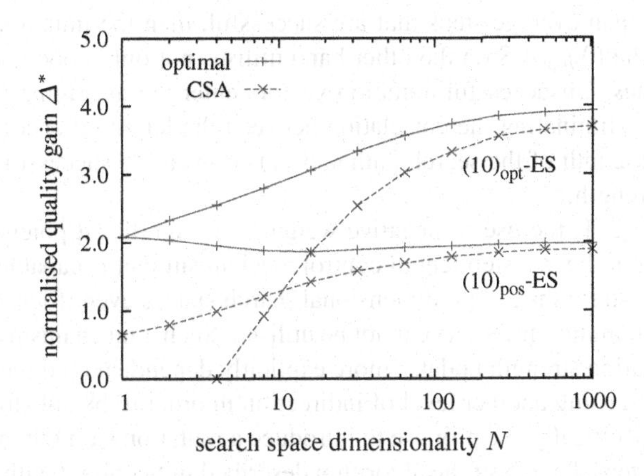

Fig. 1. Normalised quality gain Δ^* for the sphere model of $(10)_{\text{pos}}$-ES and $(10)_{\text{opt}}$-ES plotted against search space dimensionality N. Shown are both the quality gain achieved with optimal step lengths and the quality gain achieved with cumulative step length adaptation.

the maximal quality gain that can be achieved if at every step the mutation strength is set optimally. The data points have been obtained by numerically optimising the normalised mutation strength of the strategy (which, of course, does not constitute a viable step length control algorithm in general as it assumes knowledge of the distance of the search point from the location of the optimum). It can be seen that the $(\lambda)_{\text{opt}}$-ES is capable of nearly twice the normalised quality gain of the $(\lambda)_{\text{pos}}$-ES if the search space dimensionality is high. For smaller values of N the potential advantage of the negatively weighted strategy is smaller, but it is present for N as small as two. The dashed lines in Fig. 1 show the normalised quality gain of the strategies when using cumulative step length adaptation. In accordance with [9], the cumulation parameter c that appears in Eq. (3) is set to $4/(N+4)$. The damping constant d has been optimised numerically. It can be seen that for large N the performance of the adaptive strategies appears to approach optimal behaviour for both $(\lambda)_{\text{pos}}$-ES and $(\lambda)_{\text{opt}}$-ES. However, while for low search space dimensionalities cumulative step length adaptation results in convergence for the $(\lambda)_{\text{pos}}$-ES, the $(\lambda)_{\text{opt}}$-ES fails to generate positive quality gain.

It is not immediately clear why cumulative step length adaptation enables the $(\lambda)_{\text{pos}}$-ES to converge for the sphere model while the $(\lambda)_{\text{opt}}$-ES does not when N is small. It is certainly true that the opposite of a bad mutation is not always a good mutation, and the use of negative weights may contribute to bad steps being made. However, Fig. 1 clearly illustrates that the negatively weighted strategy is capable of convergence even for small N, and that the observed failure to converge is due to imperfect step length adaptation rather than to the use of negative weights per se. Interestingly, not shown, the $(\lambda)_{\text{pos}}$-ES employing cumulative step length adaptation is capable of convergence on the one-dimensional sphere even if the cumulation parameter c is set to 1. As for $c = 1$ no cumulation takes place, it cannot be correlations between consecutive steps that enable the $(\lambda)_{\text{pos}}$-ES to converge. Instead, the strategy is able to use information with regard to the length of the selected mutation vectors. If successful mutation vectors are short, then so is the search path and Eq. (4) acts to reduce the mutation strength. If it is the longer than average steps that are successful, then the mutation strength will be increased. The $(\lambda)_{\text{opt}}$-ES on the other hand utilises not only good mutation vectors but also bad ones. Unsuccessful mutation vectors enter the averaging in Eq. (1) with a negative sign, eliminating the correlation between the length of successful mutation vectors and the length of the search path and thus preventing successful adaptation of the mutation strength.

As seen in Fig. 1, the use of negative weights has significant potential benefits. It is thus desirable to have a step length control mechanism that is capable of generating useful mutation strengths in low-dimensional search spaces even if length information about successful mutation vectors cannot be utilised. Such a mechanism can be devised by making mutation strength updates more explicitly dependent on directions of recent progress vectors, using another level of indirection in order to be able to cope with the noisy signal. Specifically, we propose to introduce a mutation strength modifier $h \in \mathbb{R}$, and to replace steps 4 and 5 of the algorithm described in Section 2 with:

4'. Update the search point, mutation strength modifier, and search path according to

$$\begin{aligned} \mathbf{x} &\leftarrow \mathbf{x} + \sigma \mathbf{z}^{(\text{avg})} \\ h &\leftarrow (1 - c_h)h + c_h \mathbf{s} \cdot \mathbf{z}^{(\text{avg})} \\ \mathbf{s} &\leftarrow (1 - c)\mathbf{s} + \sqrt{\mu_{\text{eff}}\, c(2 - c)}\mathbf{z}^{(\text{avg})} \end{aligned} \tag{8}$$

where $c \in (0, 1)$ and $c_h \in (0, 1)$ are cumulation parameters.

5'. Update the mutation strength according to

$$\sigma \leftarrow \sigma \exp\left(\frac{h}{Nd}\right) \tag{9}$$

where $d > 0$ is a damping constant.

The inner product $\mathbf{s} \cdot \mathbf{z}^{(\text{avg})}$ in Eq. (8) is greater than zero if the progress vector points in the predominant direction of the previous steps accumulated in the search path. It is negative if the direction of $\mathbf{z}^{(\text{avg})}$ is opposite to that of $\mathbf{s}$. Rather than immediately using the inner product to update the mutation strength, $\mathbf{s} \cdot \mathbf{z}^{(\text{avg})}$ contributes to the mutation strength modifier h. This additional level of indirection is reminiscent of the use of a momentum term and acts as a low pass filter. It is useful as it reduces fluctuations of the mutation strength in low-dimensional search spaces, and it is without relevance if N is large. We have found setting $c_h = 0.1$ useful in all of our experiments.

4 Experimental Evaluation

In order to evaluate the usefulness of the step length control mechanism proposed in Section 3, we conduct computer experiments involving a subset of the test functions employed in [9]. The functions used are listed in Table 1. The first four are convex quadratic and have a unique optimum at $(0, 0, \ldots, 0)$. For each of them, the search point of the evolution strategies is initialised to $\mathbf{x} = (1, 1, \ldots, 1)$. While the eigenvectors of the Hessians of f_{cigar}, f_{discus}, and $f_{\text{ellipsoid}}$ are aligned with the coordinate

Table 1. Test functions

sphere	$f_{\text{sphere}}(\mathbf{x}) = \sum_{i=1}^{N} x_i^2$
cigar	$f_{\text{cigar}}(\mathbf{x}) = x_1^2 + \sum_{i=2}^{N} (a x_i)^2$
discus	$f_{\text{discus}}(\mathbf{x}) = (a x_1)^2 + \sum_{i=2}^{N} x_i^2$
ellipsoid	$f_{\text{ellipsoid}}(\mathbf{x}) = \sum_{i=1}^{N} \left(a^{\frac{i-1}{N-1}} x_i\right)^2$
Rosenbrock	$f_{\text{Rosen}}(\mathbf{x}) = \sum_{i=1}^{N-1} \left(100\left(x_i^2 - x_{i+1}\right)^2 + (x_i - 1)^2\right)$

axes, none of the strategies considered here make use of the separability of those functions, and the same results would be observed if an arbitrary rotation were applied to the coordinate system. The parameter a controls the degree of ill-conditioning of the functions. For $a = 1$ all of them are identical to the sphere function. Rosenbrock's function is characterised by a long, bent valley that needs to be followed in order to arrive at the global optimum at location $(1, 1, \ldots, 1)$. For f_{Rosen}, the initial search point is $\mathbf{x} = (0, 0, \ldots, 0)$. For all of the functions, the optimal objective function value is zero. The initial mutation strength is $\sigma = 1$ for all functions but Rosenbrock's, for which it is $\sigma = 0.1$ as this setting effectively prevents convergence to the local optimum the existence of which is noted in [9]. Optimisation proceeds until an objective function value $f(\mathbf{x}) \leq 10^{-10}$ is reached.

Throughout this section, the strategies with step length adaptation as described in Section 2 are referred to as $(\lambda)_{\text{opt}}$-CSA-ES and $(\lambda)_{\text{pos}}$-CSA-ES, depending on whether negative weights are used or not. The negatively weighted strategy with direction based cumulative step length adaptation as described in Section 3 is referred to as $(\lambda)_{\text{opt}}$-dCSA-ES. In all runs, $\lambda = 10$ and $\kappa = 3.68$. Setting the cumulation parameter to $c = 4/(N + 4)$ has proven useful for all strategy variants. The damping constant is set to $d = 1 + c$ for the $(\lambda)_{\text{pos}}$-CSA-ES according to a recommendation in [9]. For the $(\lambda)_{\text{opt}}$-CSA-ES and the $(\lambda)_{\text{opt}}$-dCSA-ES, parameter settings of $d = 0.25 + 4/N$ and $d = 1 + 7/N$, respectively, have been employed after some numerical experimentation.

We first conduct a series of experiments with the goal of comparing the performance of the step length adaptation mechanisms for mildly ill-conditioned problems. This is motivated by the CMA-ES striving to learn covariance matrices that locally transform arbitrary objective functions into the sphere model [9]. If covariance matrix adaptation is successful, it can be hoped that objective functions can locally be approximated by convex quadratic functions with low condition numbers of their Hessians. Figure 2 compares the optimisation performance of the three strategy variants for the four convex quadratic test functions, where $a = 4$ for the cigar, discus, and ellipsoid functions. It can be seen that the strategies employing negative weights are significantly superior to the $(\lambda)_{\text{pos}}$-CSA-ES if the search space dimensionality is high, outperforming the latter by a factor between 1.9 and 2.8. Moreover, while the $(\lambda)_{\text{opt}}$-CSA-ES fails to converge for small values of N for three of the four test functions, direction based cumulative step length adaptation is successful in that the $(\lambda)_{\text{opt}}$-dCSA-ES consistently either matches or outperforms the $(\lambda)_{\text{pos}}$-CSA-ES even in low-dimensional search spaces.

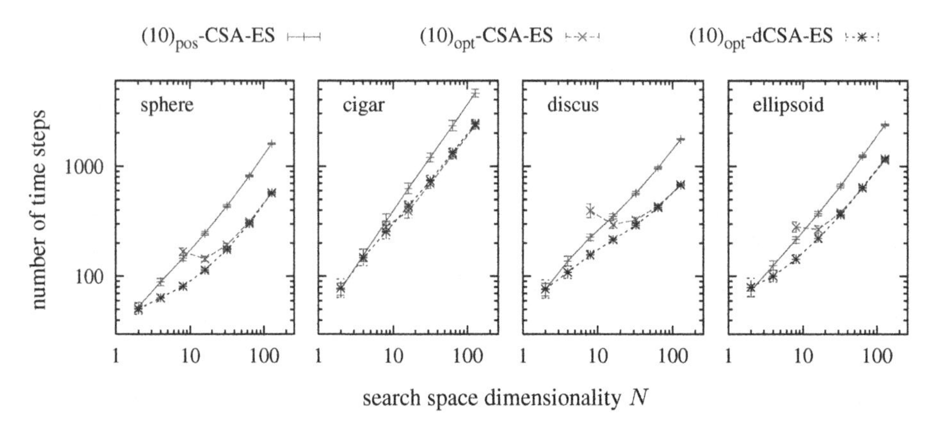

Fig. 2. Number of time steps required to reach an objective function value of $f(\mathbf{x}) = 10^{-10}$ plotted against the dimensionality N of the search space for $a = 4$. Results are averaged over 100 independent runs, with error bars indicating the standard deviation of the measurements.

Next, we investigate the effect of the degree of ill-conditioning on optimisation performance. Figure 3 contrasts the performance of $(\lambda)_{\text{pos}}$-CSA-ES and $(\lambda)_{\text{opt}}$-dCSA-ES for the anisotropic convex quadratic objective functions. It can be seen that the advantage afforded by the use of negative weights that is present for small values of a generally decreases or even turns into a disadvantage as the degree of ill-conditioning increases. While the use of negative weights generally seems to do no harm for the cigar function, refraining from weighting mutation vectors negatively results in significantly better performance for the discus and ellipsoid functions unless a is sufficiently small or N is sufficiently large.

Finally, we are interested in the potential of the use of negative recombination weights for "less artificial" objective functions. While the degree of ill-conditioning of

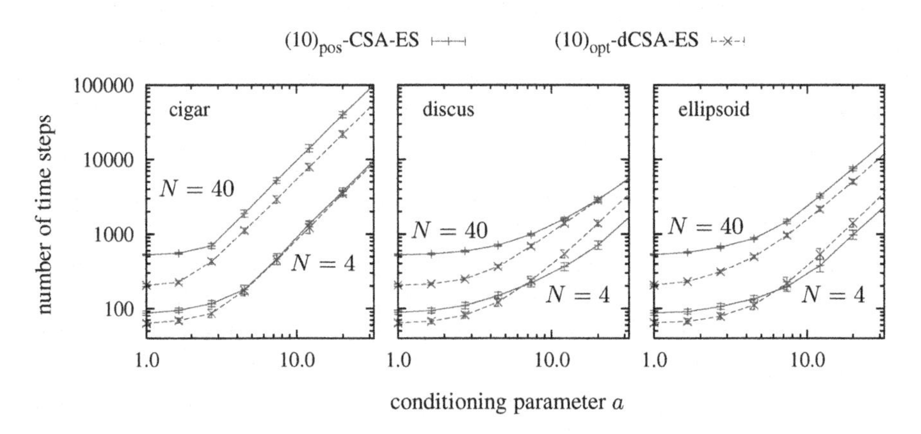

Fig. 3. Number of time steps required to reach an objective function value of $f(\mathbf{x}) = 10^{-10}$ plotted against the conditioning parameter a for $N = 4$ and $N = 40$. Results are averaged over 100 independent runs, with error bars indicating the standard deviation of the measurements.

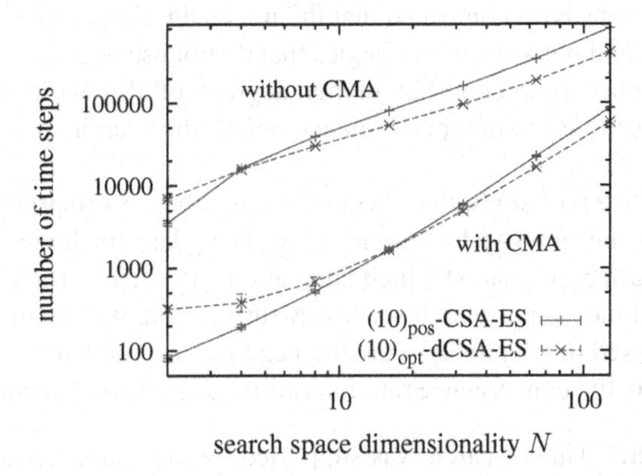

Fig. 4. Number of time steps required to reach an objective function value of $f(\mathbf{x}) = 10^{-10}$ plotted against the search space dimensionality N for Rosenbrock's function. Results are averaged over 100 independent runs, with error bars indicating the standard deviation of the measurements.

f_{cigar}, f_{discus}, and $f_{\text{ellipsoid}}$ can be increased indefinitely, it is unclear how relevant such large condition numbers are for real-world optimisation. As noted above, a covariance matrix adaptation algorithm may lead to the strategies "seeing" only moderately ill-conditioned problems, which the experiments above suggest negative weights may be beneficial for. We have thus equipped the evolution strategies with the covariance matrix adaptation algorithm described in [9]. The settings of the cumulation parameters of the $(\lambda)_{\text{pos}}$-ES are as described in that reference. For the $(\lambda)_{\text{opt}}$-ES, the parameter c_{cov} that governs the rate at which the covariance matrix is updated is reduced from $2/(N+\sqrt{2})^2$ to $2/(N+8)^2$ in order to achieve reliable convergence for small values of N.

Figure 4 contrasts the performance of the $(\lambda)_{\text{opt}}$-dCSA-ES with that of the $(\lambda)_{\text{pos}}$-CSA-ES, both with and without covariance matrix adaptation, for Rosenbrock's function. It can be seen that the use of negative weights when optimising f_{Rosen} is generally beneficial if the search space dimensionality is high. Without covariance matrix adaptation, the $(\lambda)_{\text{opt}}$-dCSA-ES outperforms the $(\lambda)_{\text{pos}}$-CSA-ES for $N > 4$. With covariance matrix adaptation, the slower adaptation of that matrix in the $(\lambda)_{\text{opt}}$-dCSA-ES that results from the different setting of c_{cov} leads to the $(\lambda)_{\text{pos}}$-CSA-ES enjoying a performance advantage up to $N = 16$.

5 Conclusions

To summarise, in low-dimensional search spaces the existing approach to cumulative step length adaptation implicitly combines information with regard to the length of selected mutation vectors with directional information about consecutive steps. For evolution strategies using negative recombination weights, the progress vector does not contain useful length information and cumulative step length adaptation is prone to failure. This paper has introduced a novel variant of cumulative step length adaptation that more explicitly relies on directional information, and that introduces an additional level of indirection in order to improve robustness in low-dimensional search spaces.

Computer experiments have confirmed that the novel algorithm reliably results in convergent behaviour and outperforms strategies that do not use negative weights for moderately ill-conditioned problems. For higher degrees of ill-conditioning, the use of negative weights can deteriorate performance unless the search space dimensionality is high.

In future work we will study the effect of the number of offspring on the findings made here. Using the improved covariance matrix update for large populations proposed in [8] may affect the rate at which adaptation can occur for negatively weighted strategies in low-dimensional search spaces. A further goal will be to devise a mechanism that can be used to adaptively scale the negative recombination weights with the goal of maximising the convergence rate for various degrees of ill-conditioning.

Acknowledgement. This research was supported by the Natural Sciences and Engineering Research Council of Canada (NSERC).

References

[1] Arnold, D.V.: Noisy Optimization with Evolution Strategies. Kluwer Academic Publishers, Dordrecht (2002)
[2] Arnold, D.V.: Optimal weighted recombination. In: Wright, A.H., et al. (eds.) Foundations of Genetic Algorithms 8, pp. 215–237. Springer, Heidelberg (2005)
[3] Arnold, D.V.: Weighted multirecombination evolution strategies. Theoretical Computer Science 361(1), 291–308 (2006)
[4] Arnold, D.V., MacDonald, D.: Weighted recombination evolution strategies on the parabolic ridge. In: Proc. of the 2006 IEEE Congress on Evolutionary Computation, pp. 411–418. IEEE Press, Los Alamitos (2006)
[5] Auger, A., Hansen, N.: Reconsidering the progress rate theory for evolution strategies in finite dimensions. In: Proc. of the 8th Annual Conference on Genetic and Evolutionary Computation, pp. 445–452. ACM Press, New York (2006)
[6] Beyer, H.-G., Schwefel, H.-P.: Evolution strategies — A comprehensive introduction. Natural Computing 1(1), 3–52 (2002)
[7] Hansen, N., Kern, S.: Evaluating the CMA evolution strategy on multimodal test functions. In: Yao, X., Burke, E.K., Lozano, J.A., Smith, J., Merelo-Guervós, J.J., Bullinaria, J.A., Rowe, J.E., Tiňo, P., Kabán, A., Schwefel, H.-P. (eds.) PPSN 2004. LNCS, vol. 3242, pp. 282–291. Springer, Heidelberg (2004)
[8] Hansen, N., Müller, S.D., Koumoutsakos, P.: Reducing the time complexity of the derandomized evolution strategy with covariance matrix adaptation (CMA-ES). Evolutionary Computation 11(1), 1–18 (2003)
[9] Hansen, N., Ostermeier, A.: Completely derandomized self-adaptation in evolution strategies. Evolutionary Computation 9(2), 159–195 (2001)
[10] Ostermeier, A., Gawelczyk, A., Hansen, N.: Step-size adaptation based on non-local use of selection information. In: Davidor, Y., Männer, R., Schwefel, H.-P. (eds.) PPSN 1994. LNCS, vol. 866, pp. 189–198. Springer, Heidelberg (1994)
[11] Rechenberg, I.: Evolutionsstrategie — Optimierung technischer Systeme nach Prinzipien der biologischen Evolution. Friedrich Frommann Verlag (1973)
[12] Rudolph, G.: Convergence Properties of Evolutionary Algorithms. Verlag Dr. Kovač (1997)
[13] Schwefel, H.-P.: Evolution and Optimum Seeking. John Wiley & Sons, Chichester (1995)

Computing Surrogate Constraints for Multidimensional Knapsack Problems Using Evolution Strategies

José Luis Montaña[1,*], César Luis Alonso[2,**], Stefano Cagnoni[3], and Mar Callau[1]

[1] Universidad de Cantabria, Spain
{montanjl,callaum}@unican.es
[2] Universidad de Oviedo, Spain
calonso@aic.uniovi.es
[3] Università di Parma, Italy
cagnoni@ce.unipr.it

Abstract. It is an important task to obtain optimal solutions for multidimensional linear integer problems with multiple constraints. The surrogate constraint method translates a multidimensional problem into an one dimensional problem using a suitable set of surrogate multipliers. In general, there exists a gap between the optimal solution of the surrogate problem and the original multidimensional problem. Moreover, computing suitable surrogate constraints is a computationally difficult task. In this paper we propose a method for computing surrogate constraints of linear problems that evolves sets of surrogate multipliers coded in floating point and uses as fitness function the value of the ϵ-approximate solution of the corresponding surrogate problem. This method allows the user to adjust the quality of the obtained multipliers by means of parameter ϵ. Solving $0-1$ multidimensional knapsack problems we test the effectiveness of our methodology. Experimental results show that our method for computing surrogate constraints for linear $0-1$ integer problems is at least as effective as other strategies based on Linear Programming as that proposed by Chu and Beasley in [6].

1 Introduction

This paper deals with the task of computing suitable surrogate constraints for $0-1$ integer linear problems. A surrogate constraint is an inequality implied by the constraints of an integer program, and designed to capture information that cannot be extracted from the original constraints individually but as a consequence of their conjunction. The use of such constraints is proposed to solve the $0-1$ integer programming problem whose most representative instance is the $0-1$ Multidimensional Knapsack Problem (MKP) that can be stated as follows. We are given a set of n objects where each object yields p_j units of profit and requires a_{ij} units of resource consumption in the i-th knapsack. The goal is to find a subset of the objects such that the overall profit is maximized without exceeding the resource capacities of the knapsacks. More precisely, given $A = (a_{ij})$, $\overline{p} = (p_1, \ldots, p_n)$, and $\overline{b} = (b_1, \ldots, b_m)$, an instance of MKP can be described as:

* Partially supported by the Spanish grant TIN2007-67466-C02-02.
** Partially supported by the Spanish grant MTM2004-01176.

M. Giacobini et al. (Eds.): EvoWorkshops 2008, LNCS 4974, pp. 555–564, 2008.

$$\text{maximize } f(x_1, \ldots, x_n) = \sum_{j=1}^{n} p_j x_j \quad (1)$$

$$\text{subject to constraint } C_i : \quad \sum_{j=1}^{n} a_{ij} x_j \leq b_i \quad i = 1, \ldots, m \quad (2)$$

where variables $(x_1, \ldots, x_n)$ take values in $\{0, 1\}^n$.

A surrogate constraint is defined to be any inequality $\sum a_j x_j \leq b_0$ implied by the MKP constraints of Equation 2. Several exact algorithms based on the computation of surrogate constraints for MKP are known (see [10]). But these methods become not applicable for large values of m and n and other strategies must be introduced. Many heuristic approaches for the MKP have appeared during the last decades following different ideas: greedy-like assignment ([8], [16]); Linear Programming based search ([3]); surrogate duality information ([16]) are some examples. The genetic approach has shown to be well suited for solving large MKP instances. Chu and Beasley (see [6]) have developed a genetic algorithm that searches only into the feasible search space. They use a repair operator based on the surrogate multipliers of some suitable surrogate problem calculated using linear programming. Following the same idea as Chu and Beasley, in [1] a genetic algorithm is proposed to compute a good set of surrogate multipliers.

In this paper we propose an evolution strategy for computing surrogate multipliers that extends previous work ([1]). The main novelties of our new method are the following: a) the computation of the multipliers is done by means of an evolution strategy that evolves sets of multipliers coded in floating point (this method is better adapted to this problem reducing time complexity); b) we use as fitness function the ϵ- approximate solution of the corresponding one-dimensional surrogate problem instead of the commonly used Linear Programming (LP) relaxation problem. When the value ϵ is set to 0.5 we recover the fitness function proposed in [1]. We test the effectiveness of our new method solving MKP instances. Our experimental results reported in Section 5 support the idea that our method for computing surrogate multipliers is competitive with methodologies based on LP ([6]) for values of ϵ in the interval $[0.1, 0.5]$. In almost all tested examples taken from the OR-Library proposed in [4], for $\epsilon \in [0.1, 0.5)$, the average number of evaluations required to find the best solution found by our algorithm is smaller than the average number of evaluations required when using the algorithm by Chu and Beasley described in [6].

2 Evolutionary Algorithms to Compute Surrogate Multipliers (EASM)

First we briefly describe the mathematical background related with surrogate problems. A more detailed explanation can be found in [1]. According to equations (1) and (2) we represent an instance of the MKP as a 5-tuple $K = (n, m, \overline{p}, A, \overline{b})$ where n and m are both natural numbers representing (respectively) the number of objects and the number of constraints; $\overline{p}$ is a vector of n positive real numbers representing the profits; A is a

$m \times n$-matrix of non-negative real numbers corresponding to the resource consumptions and $\overline{b}$ is a vector of m positive real numbers representing the knapsack capacities. As usual, a bit-vector $\alpha_K^{opt} \in \{0,1\}^n$ is a solution of instance K if it is a feasible solution, i.e., it satisfies all the constraints, and for all feasible solution $\alpha \in \{0,1\}^n$ of instance K, $f(\alpha) \leq f(\alpha_K^{opt})$ holds, where $\alpha[j]$ stands for the value of variable x_j.

If $K = (n, m, \overline{p}, A, \overline{b})$ is an instance of MKP and $\overline{\omega}$ is a vector of m positive real numbers, the surrogate constraint for K associated to $\overline{\omega}$, is defined as:

$$\sum_{j=1}^{n}(\sum_{i=1}^{m} \omega_i a_{ij})x_j \leq \sum_{i=1}^{m} \omega_i b_i \tag{3}$$

The vector $\overline{\omega} = (\omega_1, \ldots, \omega_m)$ is called the vector of surrogate multipliers. Next, the surrogate problem for K is defined as the following 0–1 one-dimensional knapsack problem where $\overline{b}'$ denotes the transposition of $\overline{b}$

$$SR(K, \overline{\omega}) = (n, 1, \overline{p}, \overline{\omega} \cdot A, \overline{\omega} \cdot \overline{b}') \tag{4}$$

We shall denote by $\alpha_{SR(K,\overline{\omega})}^{opt}$ a solution of the surrogate instance $SR(K, \overline{\omega})$.

The best possible upper bound for $f(\alpha_K^{opt})$ using surrogate constraints is computed by finding the minimum value $min_{\overline{\omega}}\{f(\alpha_{SR(K,\overline{\omega})}^{opt})\}$. We shall call optimal any vector of surrogate multipliers that provides the above minimum. Following Gavish and Pirkul [10] we point out that, from this optimal set of surrogate multipliers, the generated upper bound on $f(\alpha_K^{opt})$ is less than or equal to the bounds generated by LP-relaxation. Unfortunately , in terms of time complexity, computing the optimal surrogate multipliers is an NP-hard problem [10].

Next, we present an evolution strategy to obtain approximate values for $\overline{\omega}$.

2.1 Individual Representations and Fitness Functions

One possible representation of an individual $\overline{\omega} = (\omega_1, \ldots, \omega_m)$ is by means of a 0–1 binary string as appears in [1] and [2]. Since the surrogate constraint in inequality (3) represents a homogenous relation, one can assume without loss of generality that $\overline{\omega} \in (0,1]^n$. In this case each ω_i is represented as a q-bit binary substring, where q is a parameter that determines the desired precision for $\omega_i \in (0,1]$.

A representation which fits the problem better and is suggested by the continuous nature of the surrogate problem is to deal with $\overline{\omega}$ as a vector of floating point numbers in $(0,1]^n$.

One way of defining the fitness of a candidate vector $\overline{\omega}$ of surrogate multipliers, fitness$_1$ of $\overline{\omega}$, is to compute the value of the optimal solution of the relaxed version of $SR(K, \overline{\omega})$. In the relaxed version, variables can take values in the whole interval $[0,1]^n$. A minimizer of this fitness can be computed by Linear Programming (as in [6]) or by a genetic algorithm (as in [1] and [2]).

We propose here a fitness function based on the notion of relative error and approximate solution.

Definition 1. *Let* $K = (n, m, \overline{p}, A, \overline{b})$ *be an instance of the MKP. For every feasible solution* $\alpha \in \{0,1\}^n$, *the relative error of* α *is defined by:*

$$\mathcal{E}_\alpha = \frac{f(\alpha_K^{opt}) - f(\alpha)}{f(\alpha_K^{opt})} \tag{5}$$

Given $\epsilon \in (0,1)$, *an* ϵ*–approximate solution for* K *is any feasible solution* α_K^ϵ *satisfying* $\mathcal{E}_{\alpha_K^\epsilon} \leq \epsilon$.

Definition 2. *Given an instance of MKP,* $K = (n, m, \overline{p}, A, \overline{b})$, $\epsilon \in (0,1)$, *and a representation of a vector of surrogate multipliers* $\overline{\omega} = (\omega_1, \ldots, \omega_m)$, *we define the fitness of* $\overline{\omega}$ *as the value of the* ϵ*-approximate solution of the surrogate problem associated to* $\overline{\omega}$. *More precisely:*

$$fitness(\overline{\omega}) = f(\alpha^\epsilon_{SR(K,\overline{\omega})}) = \sum_{j=1}^{n} p_j \alpha^\epsilon_{SR(K,\overline{\omega})}[j] \tag{6}$$

The computation of the fitness associated to an individual is done by calculating an ϵ–approximate solution $\alpha^\epsilon_{SR(K,\overline{\omega})}$ of the surrogate problem $SR(K, \overline{\omega})$ implementing the fully polynomial time approximation scheme for the one-dimensional knapsack problem (see [5]). The computational complexity of this algorithm is $O(\frac{n^2(1+\epsilon)}{\epsilon})$, where n is the number of objects.

2.2 Selection and Variation Operators

Selection. We use tournament with size 2 as selection procedure. The initial population is randomly generated.

Mutation. We define a non-uniform mutation operator adapted to our search space $(0,1]^m$ which is convex ([15]). This avoids the need for a repair operator as could happen when dealing with the more popular Gaussian mutation. Our mutation operator is defined as follows, taking for p the value 0.5.

$$\omega_k^{t+1} = \omega_k^t + \Delta(t, 1 - \omega_k^t), \text{ with probability } p \tag{7}$$

and

$$\omega_k^{t+1} = \omega_k^t - \Delta(t, \omega_k^t), \text{ with probability } 1 - p \tag{8}$$

for $k = 1, \ldots, m$ and t being the generation number. The function Δ is defined as $\Delta(t, y) := y \cdot r \cdot (1 - \frac{t}{T})$ where r is a random number in [0,1] and T represents the maximum number of generations. Note that function $\Delta(t, y)$ returns a value in $[0, y]$ such that the probability of $\Delta(t, y)$ being close to 0 increases as t increases. This property lets this operator search the space uniformly initially (when t is small), and very locally at later stages. In experiments reported by Michalewicz et al. ([15]), the following function was used $\Delta(t, y) := y \cdot r \cdot (1 - \frac{t}{T})^b$ where b is a parameter determining the degree of non-uniformity. In both cases mutation is applied according to a given probability p_m.

Crossover. We use arithmetic crossover (c. f.[19]). Here two vectors $\overline{\omega}_1$ and $\overline{\omega}_2$ produce two offsprings, $\overline{p}$ and $\overline{q}$, which are linear combinations of their parents, i.e., $\overline{p} = \lambda \cdot \overline{\omega}_1 + (1-\lambda) \cdot \overline{\omega}_2$, $\overline{q} = \lambda \cdot \overline{\omega}_2 + (1-\lambda) \cdot \overline{\omega}_1$. In our implementation we set λ to $\frac{1}{2}$ (average crossover); so in this case we only obtain one offspring.

3 A Genetic Algorithm for the MKP (EAMKP)

Given an instance of MKP, $K = (n, m, \overline{p}, A, \overline{b})$, and a set of surrogate multipliers, $\overline{\omega}$, we will briefly describe a steady state evolutionary algorithm for solving K with a repair operator for the infeasible individuals. The repair operator uses as heuristic the set of surrogate multipliers. The management of infeasible individuals and their improvement are done following ([6]). This algorithm structure is also used in [1] and [2].

3.1 Individual Representation and Fitness Function

We choose the standard 0–1 binary representation since it represents the underlying 0–1 integer values. A chromosome α representing a candidate solution for $K = (n, m, \overline{p}, A, \overline{b})$ is an n-bit binary string. A value $\alpha[j] = 0$ or 1 in the jth bit means that variable $x_j = 0$ or 1 in the represented solution. The fitness of chromosome $\alpha \in \{0,1\}^n$ is defined by

$$f(\alpha) = \sum_{j=1}^{n} p_j \alpha[j] \tag{9}$$

3.2 Repair Operator and Local Improvement Procedure

Let $K = (n, m, \overline{p}, A, \overline{b})$ be an instance of MKP and let $\overline{\omega} \in (0,1]^m$ be a vector of surrogate multipliers. The utility ratio for the variable x_j is defined by $u_j = \frac{p_j}{\sum_{i=1}^{m} \omega_i a_{ij}}$. Given an infeasible individual $\alpha \in \{0,1\}^n$ we apply the following procedure to repair it.

Procedure DROP(see[6])
input: $K = (n, m, \overline{p}, A, \overline{b})$; $\overline{\omega} \in (0,1]^m$ and a chromosome $\alpha \in \{0,1\}^n$

```
begin for j=1 to n compute u_j
  P:=permutation of (1,...,n) with
        u_P[j] <= u_P[j+1]
  for j=1 to n do
   if (alpha[P[j]]=1 and infeasible(alpha))
         then   alpha[P[j]]:=0
end
```

Once we have transformed α into a feasible individual α', a second phase is applied in order to improve α'. This second phase is denominated the ADD phase.

Procedure ADD(see [6])
input: a chromosome $\alpha \in \{0,1\}^n$

```
begin
  P:=permutation of (1,...,n) with
   u_P[j] >= u_P[j+1]
   for j=1 to n do
     if alpha[P[j]]=0 then alpha[P[j]]:=1
     if infeasible(alpha) then alpha[P[j]]:=0
end
```

3.3 Genetic Operators and Initial Population

We use the roulette wheel rule as selection procedure, the uniform crossover as replacement operator and bitwise mutation with a given probability p. So when mutation must be applied to an individual α, a bit j from α is randomly selected and flipped from its value $\alpha[j] \in \{0,1\}$ to $1-\alpha[j]$. We construct the individuals of the initial population generating random permutations from $(1,\ldots,n)$ and applying the DROP procedure to the chromosome $\alpha = (1,\ldots,1)$, following the permutation order, to get feasibility. Then we apply the ADD procedure to the resulting chromosomes to improve their fitness.

4 Strategies of Combination of the Evolutionary Algorithms

The natural method to combine the computation of the surrogate multipliers and to solve an MKP instance, is to compute first a "good" set of surrogate multipliers with the corresponding algorithm, and then to run, with that set as input, the evolutionary algorithm for the MKP. But this sequential execution of both evolutionary algorithms could waste some computational effort in finding the surrogate multipliers. This happens when a "bad" set of surrogate multipliers is enough to obtain "good" solutions for the given MKP instance. Following the approach appearing in [7] we propose here a cooperative strategy between both evolutionary algorithms as described below.

Let a *turn* be the isolated and uninterrupted evolution of one population for a fixed number of generations or individual evaluations. The cooperative strategy consists of executing turns of EAMKP until it does not yield fitness improvements for a turn. In this case a execution of a turn of EASM will be done. We display below the algorithm describing this strategy:

Procedure co-DGA

```
begin
  initialize EAMS population
  initialize EAMKP population with the best
  set of surrogate multiplier from EAMS population
  repeat
    repeat
```

```
      execute one turn of EAMKP
    until the best fitness has not improved or
    termination condition
    if not termination condition then
     execute one turn of EASM and
     assign the new set of surrogate multipliers to EAMKP
  until termination condition
end
```

Using this strategy EASM evolves until it finds a set of surrogate multipliers from which EAMKP can obtain a candidate solution. Both strategies, sequential and cooperative, with their variants, have been tested and the experimental results are reported in the next section.

5 Experimental Results

The experiments were performed to demonstrate the effectiveness of the EASM algorithm proposed in Section 2. To this aim we have designed two kinds of experiments. The first one consists in a sequential execution of EASM algorithm (with different values of ϵ) followed by the execution of EAMKP. The second one consists in a cooperation between EASM and EAMP according to the procedure co-DGA described in Section 4.

We have executed our evolutionary algorithms to solve the problems included in the OR-Library proposed in [6][1]. This library contains randomly generated instances of MKP with number of constraints $m \in \{5, 10, 30\}$, number of variables $n \in \{100, 250, 500\}$ and tightness ratios $r \in \{0.25, 0.5, 0.75\}$. The tightness ratio r fixes the capacity of i-th knapsack to $r \sum_{j=1}^{n} a_{ij}$, $1 \leq i \leq m$. There are 10 instances for each combination of m, n and r giving a total of 270 test problems. We have run our genetic algorithms setting the parameters to the following values. For the sequential strategy EASM uses population size 25, probability of mutation 0.1 and a maximum number of 8000 generations to finish (this parameters are suggested by a previous experimentation which reveals that 8000 generations and 25 individuals are enough for the range of problems considered in this paper); and EAMKP uses population size 100, probability of mutation equal to 0.1 and 10^6 non-duplicate evaluations to finish (this selection of the parameters is done to reproduce the conditions of the algorithm given in [6]). For the cooperative algorithm we have fixed a turn in 600 generations for EASM and 60000 non-duplicate evaluations for EAMKP, maintaining the remaining parameters as above. In this case the termination condition is reached when 10^6 non-duplicate individuals have been evaluated by EAMKP. We have experimented with values of $\epsilon \in \{0.5, 0.25, 0.1, 0.01\}$. For values of $\epsilon < 0.01$ the execution time is too large and the quality of the obtained solution does not improve significantly.

Since the optimal solution values for most of these problems are unknown the quality of a solution α is measured by the percentage gap of its fitness value with respect

[1] Publicly available on line at http://www.brunel.ac.uk/depts/ma/research/jeb/info.html

Table 1. Computational results for $DGA\epsilon$, $co - DGA\epsilon$ and CHUGA comparing average %gap

Problem	%gap					
m–n	$DGA_{\epsilon=0.5}$	$DGA_{\epsilon=0.25}$	$DGA_{\epsilon=0.1}$	$DGA_{\epsilon=0.01}$	$co - DGA_{\epsilon=0.1}$	CHUGA
5–100	0.60	0.60	0.59	0.58	0.53	0.59
5–250	0.14	0.14	0.14	0.14	0.14	0.16
5–500	0.06	0.05	0.05	0.05	0.04	0.05
10–100	1.04	1.03	0.98	0.96	0.94	0.94
10–250	0.32	0.32	0.31	0.30	0.32	0.35
10–500	0.16	0.14	0.14	0.14	0.16	0.14
30–100	1.71	1.70	1.68	1.66	1.68	1.74
30–250	0.76	0.74	0.67	0.66	0.74	0.73
30–500	0.49	0.45	0.39	0.39	0.40	0.40

to the fitness value of the optimal solution of the LP-relaxed problem $f(\alpha^{opt}_{KLP})$ defined by: $\%gap = 100\frac{f(\alpha^{opt}_{KLP})-f(\alpha)}{f(\alpha^{opt}_{KLP})}$. We estimate the computational complexity of our algorithms by means of the number of evaluations required to find the best obtained solution. This measure is independent of the execution platform.

The results of our experiments are displayed in Tables 1 and 2 based on five independent executions for each instance. In both tables, our algorithms are compared with that of Chu et. al (CHUGA) ([6]). To our knowledge this is the best known genetic algorithm for MKP based on surrogate constraints. The first column of each table identifies the set of 30 instances that correspond to each combination of m, n. In the first table the remaining columns contain the average %gap of our algorithms: DGA_ϵ holds for the sequential execution of EASM and EAMKP and $co - DGA\epsilon$ denotes the cooperative execution of EASM and EAMKP. The parameter ϵ indicates the degree of approximation chosen for EASM. The last column indicates the corresponding average %gap of the CHUGA algorithm. The last three columns of Table 2 show, for the

Table 2. Average number of Evaluations to find the Best encountered Solution (A.E.B.S.) for MKP

Problem	A.E.B.S.		
m–n	$DGA_{\epsilon=0.1}$	$co - DGA_{\epsilon=0.1}$	CHUGA
5–100	38320	149716	24136
5–250	157312	298364	218304
5–500	265558	537664	491573
10–100	123762	248179	318764
10–250	283500	381543	475643
10–500	224221	503130	645250
30–100	103502	157593	197855
30–250	297657	703943	369894
30–500	301469	611735	587472

algorithms that are being compared, the average number of evaluations required before the best individual was encountered (A.E.B.S).

From our experimentation we deduce that our algorithms DGA and co-DGA perform as well as the CHUGA algorithm in terms of the average %gaps for $\epsilon < 0.25$: DGA uses less amount of computational effort when $\epsilon \geq 0.1$ because the best solution is reached within a considerably smaller number of individual evaluations in almost all tested cases as shown in Table 2; however, co-DGA performs better on small instances in terms of the %gap but needs more evaluations. Algorithm EASM has a computational asymptotic cost $O(\frac{n^2(1+\epsilon)}{\epsilon}E)$, where E is the number of evaluated individuals. As a consequence of this observation, taking into account the three factors, %gap, A.E.B.S and cost of computing surrogate multipliers , we conclude that 0.1 is the best value for the parameter ϵ both in the sequential and the cooperative algorithms.

6 Conclusive Remarks and Future Research

The main novelty in this work is the use of an evolutionary strategy to compute a suitable set of surrogate multipliers based on the exact computation of ϵ-approximate solutions of the surrogate one-dimensional problem (that can be computed in polynomial time). For the large-sized test data used, our family of genetic algorithms $DGA(\epsilon)$ converged most of the time much faster to high quality solutions (for values of $\epsilon \in [0.1, 0.5]$) than the comparable GA from Chu et al. and this feature can only be explained by the better quality of the surrogate multipliers obtained using our heuristic method. In particular this supports the conjecture that evolutionary computation is also well suited when tackling optimization problems of parameters with variables in continuous domains, as the case of computing surrogate multipliers.

Another novel feature of this work is the use of a cooperative strategy between algorithms EASM and EAMKP. In this case the family of algorithms $co-DGA(\epsilon)$ performs very similarly to CHUGA algorithm. This shows in particular that EASM is also well suited to be used in a cooperative strategy for solving integer programming problems. We remark that our cooperative methods converge more slowly than DGA, although they were designed to converge faster. This can be due to the rigidity of our prototype. In this framework we are experimenting a cooperative algorithm with dynamical size turns that seems to be promising for solving MKP instances. Even if our family of DGAs gives better results than the CHUGA algorithm in terms of number of evaluations much more experimentation is needed to show that our algorithm improves on several other evolutionary approaches. This, for instance, implies direct comparison with other evolutionary algorithms which are not based on surrogate constraint techniques such as [21] and [12].

References

1. Alonso, C.L., Caro, F., Montana, J.L.: An Evolutionary Strategy for the Multidimensional 0–1 Knapsack Problem based on Genetic Computation of Surrogate Multipliers. In: Mira, J., Álvarez, J.R. (eds.) IWINAC 2005. LNCS, vol. 3562, pp. 63–73. Springer, Heidelberg (2005)

2. Alonso, C.L., Caro, F., Montana, J.L.: A Flipping Local Search Genetic Algorithm for the Multidimensional 0–1 Knapsack Problem. In: Marín, R., Onaindía, E., Bugarín, A., Santos, J. (eds.) CAEPIA 2005. LNCS (LNAI), vol. 4177, pp. 21–30. Springer, Heidelberg (2006)
3. Balas, E., Martin, C.H.: Pivot and Complement–A Heuristic for 0–1 Programming. Management Science 26(1), 86–96 (1980)
4. Beasley, J.E.: Obtaining Test Problems via Internet. Journal of Global Optimization 8, 429–433 (1996)
5. Brassard, G., Bratley, P.: Fundamentals of Algorithms. Prentice-Hall, Englewood Cliffs (1997)
6. Chu, P.C., Beasley, J.E.: A Genetic Algorithm for the Multidimensional Knapsack Problem. Journal of Heuristics 4, 63–86 (1998)
7. Vanneschi, L., Mauri, G., Valsecchi, A., Cagnoni, S.: Heterogeneous Cooperative Coevolution: Strategies of Integration between GP and GA. In: Genetic and Evolutionary Computation Conference (GECCO 2006), pp. 361–368 (2006)
8. Freville, A., Plateau, G.: Heuristics and Reduction Methods for Multiple Constraints 0–1 Linear Programming Problems. European Journal of Operationa Research 24, 206–215 (1986)
9. Gavish, B., Pirkul, H.: Allocation of Databases and Processors in a Distributed Computing System. In: Akoka, J. (ed.) Management od Distributed Data Processing, pp. 215–231. North-Holland, Amsterdam (1982)
10. Gavish, B., Pirkul, H.: Efficient Algorithms for Solving Multiconstraint Zero–One Knapsack Problems to Optimality. Mathematical Programming 31, 78–105 (1985)
11. Goldberg, D.E.: Genetic Algorithms in Search, Optimization and Machine Learning. Addison-Wesley, Reading (1989)
12. Gottlieb, J.: Permutation-based evolutionary algorithms for multidimensional knapsack problems. In: Proc. of ACM Symposium on Applied Computin (SAC 2000), pp. 408–414. ACM Press, New York (2000)
13. Khuri, S., Bäck, T., Heitkötter, J.: The Zero/One Multiple Knapsack Problem and Genetic Algorithms. In: Proceedings of the 1994 ACM Symposium on Applied Computing (SAC 1994), pp. 188–193. ACM Press, New York (1994)
14. Martello, S., Toth, P.: Knapsack Problems: Algorithms and Computer Implementations. John Wiley & Sons, Chichester (1990)
15. Michalewicz, Z., Logan, T., Swaminathan, S.: Evolutionary operators for continuous convex parameter spaces. In: Proceedings of the 3rd Annual Conference on Evolutionary Programming, pp. 84–97. World Scientific, Singapore (1994)
16. Pirkul, H.: A Heuristic Solution Procedure for the Multiconstraint Zero–One Knapsack Problem. Naval Research Logistics 34, 161–172 (1987)
17. Raidl, G.R.: An Improved Genetic Algorithm for the Multiconstraint Knapsack Problem. In: Proceedings of the 5th IEEE International Conference on Evolutionary Computation, pp. 207–211 (1998)
18. Rinnooy Kan, A.H.G., Stougie, L., Vercellis, C.: A Class of Generalized Greedy Algorithms for the Multi-knapsack Problem. Discrete Applied Mathematics 42, 279–290 (1993)
19. Schwefel, H.-P.: Numerical Optimization of Computer Models (1995), 2nd edn. John Wiley and Sons, New-York (1981)
20. Thiel, J., Voss, S.: Some Experiences on Solving Multiconstraint Zero–One Knapsack Problems with Genetic Algorithms. INFOR 32, 226–242 (1994)
21. Vasquez, M., Hao, J.K.: A hybrid approach for the 0-1 multidimensional knapsack problem. In: Proc. of the 13 Intl. Joint Conference on Artificial Intelligence, pp. 328–333 (2001)

A Critical Assessment of Some Variants of Particle Swarm Optimization

Stefano Cagnoni[1], Leonardo Vanneschi[2],
Antonia Azzini[3], and Andrea G.B. Tettamanzi[3]

[1] Dipartimento di Ingegneria dell'Informazione, Università di Parma
[2] Dipartimento di Informatica Sistemistica e Comunicazione, Università di Milano "Bicocca"
[3] Dipartimento di Tecnologie dell'Informazione, Università di Milano Crema
cagnoni@ce.unipr.it, vanneschi@disco.unimib.it,
{azzini,tettamanzi}@dti.unimi.it

Abstract. Among the variants of the basic Particle Swarm Optimization (PSO) algorithm as first proposed in 1995, EPSO (Evolutionary PSO), proposed by Miranda and Fonseca, seems to produce significant improvements. We analyze the effects of two modifications introduced in that work (adaptive parameter setting and selection based on an evolution strategies-like approach) separately, reporting results obtained on a set of multimodal benchmark functions, which show that they may have opposite and complementary effects. In particular, using only parameter adaptation when optimizing 'harder' functions yields better results than when both modifications are applied. We also propose a justification for this, based on recent analyses in which particle swarm optimizers are studied as dynamical systems.

1 Introduction

Particle Swarm Optimization (PSO) [1,2] has gained popularity thanks to its effectiveness and extremely easy implementation. As reported in [3], searching the IEEExplore (http://ieeexplore.ieee.org) technical publication database by the keyword "PSO" returns a list of much more than 1,000 titles, about one third of which deal with theoretical aspects. Amazingly, about two thirds of them have been published in the last two years.

Despite this intensive activity, much is still to be learned about PSO from a theoretical point of view [4]. Most authors have dealt with modifications of the basic algorithm. The essentiality of the basic equations suggests that further improvements may be introduced modifying the dynamics of the swarm. This can be achieved by adding new terms to the update equations, as well as swarm-related terms, possibly dependent on the context, to the fitness function. Making the algorithm parameters adaptive seems to be beneficial for both convergence speed and robustness with respect to local optima. Nevertheless, much is still to be discovered about the actual impact of parameters on PSO effectiveness and efficiency. Studies about PSO as a dynamical system, performed in stationary conditions, have found "regions of stability" for PSO, within which convergence is guaranteed [5,6].

M. Giacobini et al. (Eds.): EvoWorkshops 2008, LNCS 4974, pp. 565–574, 2008.

Miranda and Fonseca [7] have proposed a modified version of PSO, called EPSO (Evolutionary Particle Swarm Optimization), in which concepts derived from evolutionary computation, and in particular from Evolution Strategies (ES) [8], have been introduced into the PSO algorithm. In particular, they have introduced parameter mutation and particle selection, showing improvements in performance. However, a sufficiently deep analysis of the reasons why EPSO improves PSO performances was not made by the authors. We report results of an experimental study we have made to assess the specific effects of the two main modifications introduced in [7].

The paper is structured as follows: in Section 2 we introduce the details of the modifications to the PSO algorithm we have studied. Section 3 describes the tests we have made along with their results. Finally, Section 4 presents a final discussion and hints for future research.

2 Modified PSO Algorithms

In this section we describe EPSO as presented in [7]. In particular, we focus on the two main modifications to the PSO algorithm, whose effects have been analyzed in our experiments. In basic PSO, velocity and position-update equations for a particle are given as follows:

$$\begin{aligned}\mathbf{V}(t) = & \; w \cdot \mathbf{V}(t-1) \\ & + C_1 \cdot rand() \cdot [\mathbf{X}_{best}(t-1) - \mathbf{X}(t-1)] \\ & + C_2 \cdot rand() \cdot [\mathbf{X}_{gbest}(t-1) - \mathbf{X}(t-1)] \end{aligned} \quad (1)$$

$$\mathbf{X}(t) = \mathbf{X}(t-1) + \mathbf{V}(t) \quad (2)$$

where $\mathbf{V}$ is the velocity of the particle, C_1, C_2 are two positive constants, w is the inertia weight (constriction factor), $\mathbf{X}(t)$ is the position of the particle at time t, $\mathbf{X}_{best}(t-1)$ is the best-fitness point visited by the particle up to time $t-1$, $\mathbf{X}_{gbest}(t-1)$ is the best-fitness point ever found by the whole swarm. In [7], three main changes to equations (1) and (2) were proposed: (i) w, C_1 and C_2 were not considered as constants, but were mutated according to rules inspired by Evolution Strategies; (ii) a selection rule, inspired, as well, by ES, was used to set $\mathbf{X}(t)$, and (iii) Gaussian noise was added to $\mathbf{X}_{gbest}(t-1)$ in equation (1). The combined effects of the three changes were considered, achieving better performances than basic PSO on a set of benchmark functions.

In some preliminary experiments, however, we observed that: (i) such improvements tended to disappear as the problem became harder for PSO, up to some cases in which results were even worse than those obtained by basic PSO; (ii) significant improvements seemed to be obtained by applying only the first change. This behaviour could be justified by observing that, while letting the parameters of the algorithm vary during a run could affect positively the search behaviour of PSO, introducing selection might lead to a "greedy" behaviour, which makes the algorithm more likely to be trapped in local minima of the fitness function.

The experiments described in this paper aim at analyzing such effects more in details, by performing experiments in which: (i) the algorithm parameters are mutated at runtime with rules derived from ES, and (ii) parameter mutation is applied along with an ES-inspired selection scheme. We did not add Gaussian noise to $\mathbf{X}_{gbest}(t-1)$ in equation (1), since we do not aim at comparing our results to the ones obtained by EPSO, but at analyzing more in depth the effects of the main ideas underlying it. This is why, even if the algorithms we tested are very similar to EPSO as is described in [7], we did not exactly replicate it.

Parameter mutation. We used the following update rule for the parameters of each particle P:

- $\sigma_{C_i,P}(t) = \sigma_{C_i,P}(t-1) \cdot e^{2*N(0,1)} \quad i = 1, 2$
- $\sigma_{w,P}(t) = \sigma_{w,P}(t-1) \cdot e^{2*N(0,1)}$
- $C_{i,P}(t) = C_{i,P}(t-1) + N(0, \sigma_{C_i,P}(t)) \quad i = 1, 2$
- $w_P(t) = (C_{1,P}(t) + C_{2,P}(t))/2 - 1 + N(0, \sigma_{w,P}(t))$

N(μ,σ) being a random number generated by a Gaussian distribution having mean μ and standard deviation σ. While the third equation represents a Gaussian mutation of $C_{i,P}$, the fourth one sets w_P at the lower end of the stability interval for such a parameter, defined by [9], before mutating it.

Parameter mutation and selection. In the second PSO variant, we added to parameter mutation a selection scheme proposed in [7]. Selection works as follows:

```
begin
  for each particle i
        - generate R offspring by applying the position
          and velocity update R times to X_i(t-1);
        - set the new position X_i(t) as the offspring
          with the best fitness;
  endfor
end
```

In the next section we report on the results obtained by applying the two modifications described above to a set of functions commonly used as benchmarks and to a hard real-world optimization problem.

Notice that the first PSO variant we tested is just a particular case of the second one, where R is set to 1. In that case, by applying only parameter mutation, with no selection, we generate a single offspring. From now on, even if partially improperly, we are going to term the different versions of the algorithm we tested EPSOx, where x stands for the value assigned to R. EPSO1 therefore means "PSO + parameter mutation" while EPSOx with $x > 1$ means "PSO + parameter mutation + selection among R offspring".

3 Experimental Results

We compared our EPSO algorithms [7] with different numbers of offspring with standard PSO. Here we present our results on three well known benchmark functions, while in Section 3.1 we perform the same comparison on a real-world application. Given a generic particle $P = (p_1, p_2, ..., p_n)$, the benchmark functions used here are:

- Sphere: $f_1(P) = \sum_{i=1}^{n} p_i^2$
- Rosenbrock: $f_2(P) = \sum_{i=1}^{n-1}[100(p_{i+1} - p_i)^2 + (p_i - 1)^2]$
- Rastrigin: $f_3(P) = \sum_{i=1}^{n}[p_i^2 - 10\cos(2\pi p_i) + 10]$

All the experiments presented in this section have been performed running 50 independent experiments for each PSO variant and each function, using the following set of parameters: swarm size of 100 particles; particle coordinates initialized randomly with uniform distribution in the range $[-50, 50]$; velocities initialized randomly with uniform distribution in the range $[-0.5, 0.5]$; the termination criterion for one run was: 10^6 fitness evaluations performed or a global best fitness value lower or equal to 10^{-10}. A PSO run in which at least one particle has reached a fitness lower or equal to 10^{-10} after less than 10^6 fitness evaluations will be called a *successful* run from now on. Basic PSO has been run setting $C_1 = C_2 = 1.49618$ and $w = 0.729844$ as proposed in [5].

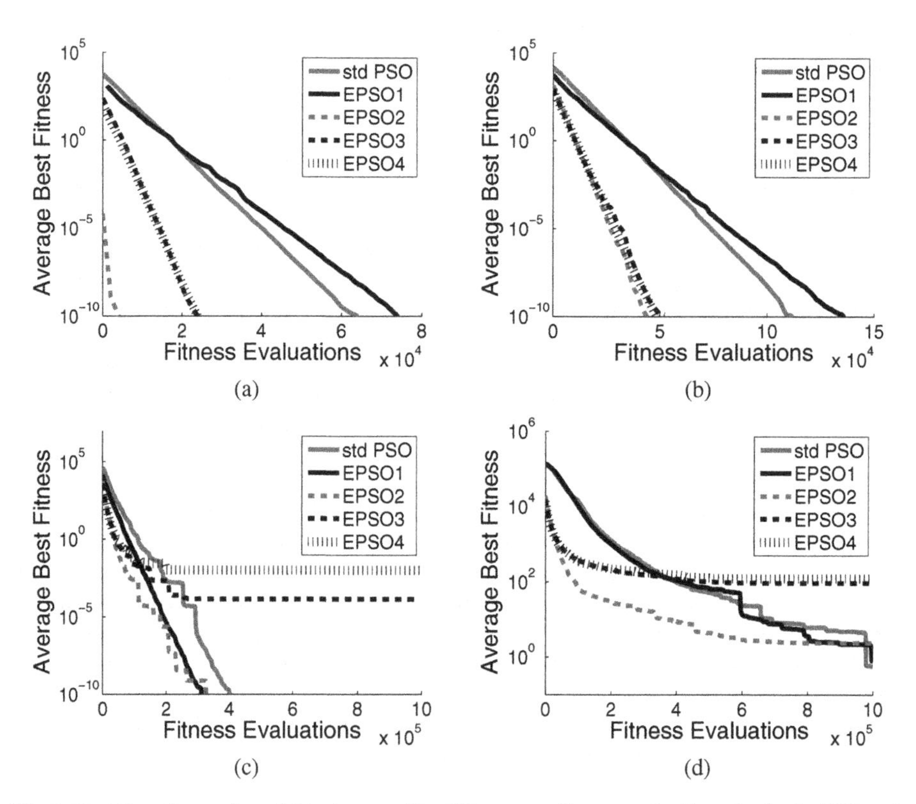

Fig. 1. Test Function: sphere (f_1). Average Best Fitness vs. fitness evaluations for basic PSO and EPSO with offspring size equal to 1, 2, 3, and 4. (a): problem dimension = 30; (b): dimension = 50; (c): dimension = 100; (d): dimension = 250.

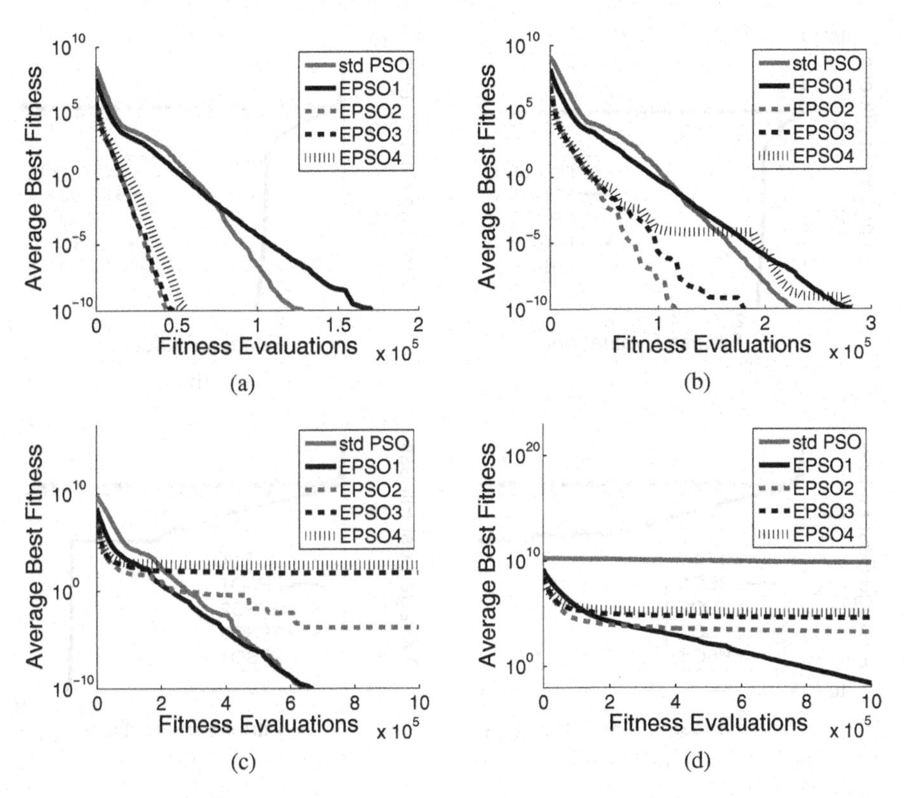

Fig. 2. Test Function: Rosenbrock (f_2). Average Best Fitness vs. fitness evaluations for basic PSO and EPSO with offspring size equal to 1, 2, 3, 4. (a): dimension = 30; (b): dimension = 50; (c): dimension = 100; (d): dimension = 250.

EPSO has been run using clamping for C_1, C_2, $\sigma_{C_1,P}$, $\sigma_{C_2,P}$ and $\sigma_{w,P}$. In particular, the values of C_1 and C_2 have been limited into the range $[0.5, 2.3]$ and values of $\sigma_{C_1,P}$ and $\sigma_{C_2,P}$ have been limited into the range $[0.1, 2]$. No clamping has been applied to particle velocity and position. We compare the performances of the PSO variants by plotting the graphics of Average Best Fitness (ABF) (i.e. the average over all runs of the best fitness values) against fitness evaluations and reporting the success rate (SR) (i.e. the ratio of successful runs with respect to the total number of runs) statistics. Pros and cons of these performance measure are discussed, for instance, in [10].

Figure 1 shows the ABF of the PSO variants over 50 independent runs vs. fitness evaluations for the Sphere benchmark. EPSO2 outperforms PSO and EPSO1 for this test function, even though the improvement appears to be marginal for problem dimensions of 50 and 100. For a problem dimension of 250, no successful runs are recorded for any algorithm but, while EPSO2 finally converges to values similar to those reached by PSO and EPSO1 and these models seem to be still improving their results after 10^6 fitness evaluations, EPSO3 and EPSO4 seem to have already got stuck into local minima in most runs, as happens to them already with a problem dimension of 100. Figure 2 shows the results for the Rosenbrock test function. In this case, when problem

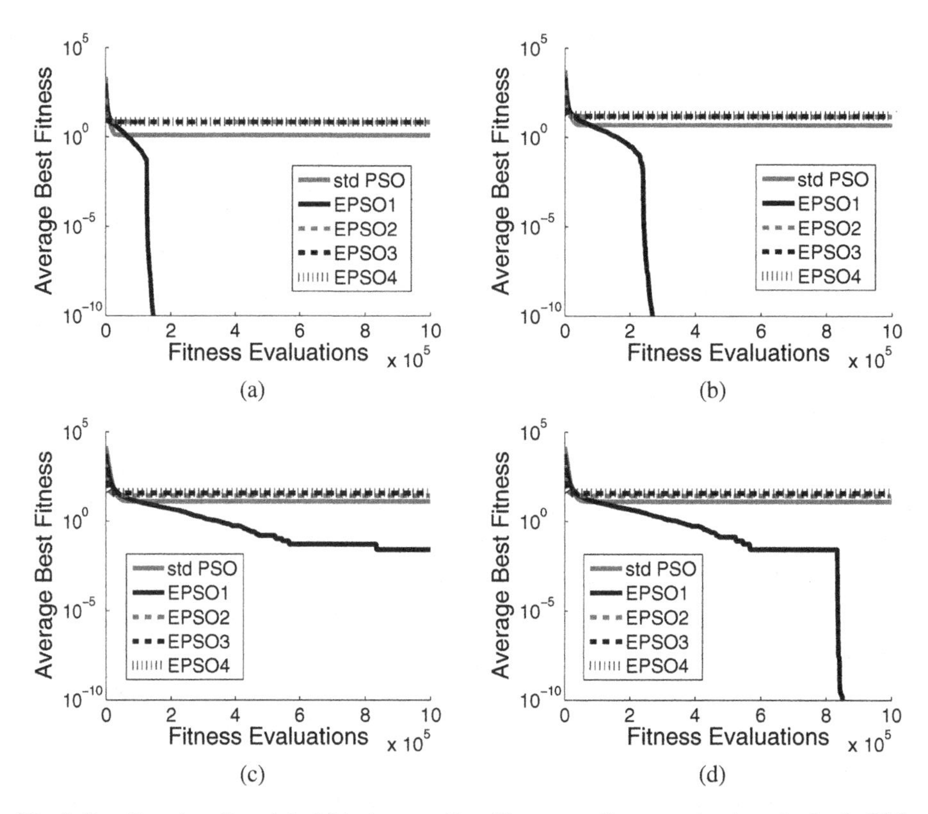

Fig. 3. Test Function: Rastrigin (f_3). Average Best Fitness vs. fitness evaluations for basic PSO and EPSO with offspring size equal to 1, 2, 3, 4. (a): dimension = 20; (b): dimension = 30; (c): dimension = 50; Frame (d) shows the same results as (c), but only the 49 successful EPSO1 runs (out of 50) have been used to compute the black curve, to highlight the actual behaviour in successful runs.

dimension is equal to 30 and 50 EPSO2, EPSO3 and EPSO4 outperform the other models but, for a dimension of 100 and 250, EPSO1 outperforms the other models. In particular, when dimension is equal to 250, EPSO1 is the only PSO variant which is successful in all 50 runs. This seems to indicate that selection has negative effects on convergence when the problem becomes more difficult or of larger dimension. Figure 3 reports the results for the Rastrigin test function. In this case the advantage of EPSO1, compared to the other PSO variants, is remarkable for all problem dimensions. Notice how all other algorithms always get stuck in local minima already on the 20-dimensional problem, while EPSO1 always reaches the pre-set error threshold except for one run of the 50-dimensional one. This seems to confirm that, as the problem becomes "PSO-hard" parameter mutation offers better results, while introducing selection is detrimental for convergence. Finally, Table 1 reports the values of the SR (calculated over the same 50 runs as the results presented in Figures 1, 2 and 3) for the three test functions and for all problem dimensions taken into account. On the Sphere function EPSO3 and EPSO4 have a lower SR than the other PSO variants for problem

Table 1. Success Rate of standard PSO, EPSO1, EPSO2, EPSO3 and EPSO4 on 50 independent runs for each benchmark and problem dimension

Benchmark Problem	Problem Dimension	PSO	EPSO1	EPSO2	EPSO3	EPSO4
Sphere	30	1	1	1	1	1
	50	1	1	1	1	1
	100	1	1	1	0.6	0.34
	250	0	0	0	0	0
Rosenbrock	30	1	1	1	1	1
	50	1	1	1	1	1
	100	1	1	0	0	0
	250	0	0	0	0	0
Rastrigin	20	0	1	0	0	0
	30	0	1	0	0	0
	50	0	0.98	0	0	0

dimension equal to 100. On the Rosenbrock function, SR is equal to 1 for all variants for dimensions of 30 and 50. For dimension equal to 250, the SR of all the models is equal to 0, even if with different trends, as described above. In the 100-dimensional problem, the SR of PSO and EPSO1 is equal to 1, while the SR of EPSO2, EPSO3 and EPSO4 is equal to 0. In this version of the problem, which is clearly PSO-hard, EPSO1 is the method that has achieved the best results (Figure 2(d)). On the Rastrigin function EPSO1 has a SR equal to 1 for the 20- and 30-dimensional problems, and equal to 0.98 for the 50-dimensional one, while for the other models the SR is always equal to 0.

These results show that, at least for the problems studied here, when the problem to solve is "simple" enough and/or the problem dimension is "small" enough, using selection (EPSO with $R \geq 2$) improves PSO performances. However, as the problem gets more complex and/or the problem dimension increases, EPSO1 clearly outperforms standard PSO and the other EPSO variants for both the average quality of the solutions and the percentage of successful runs.

3.1 Human Oral Bioavailability Assessment by Linear Regression

High dimensionality and multimodality are typical features which characterize most real-world problems. Given the results obtained on the test functions, we wanted to verify if the same hints which we could derive from them were confirmed on a real-world problem. Therefore, we performed another set of experiments on a hard real-world problem: assessing human oral bioavailability of drugs.

Human oral bioavailability [11] is a pharmacokinetic parameter that measures the percentage of initial drug dose that reaches the systemic blood circulation after passing through the liver. This parameter is particularly relevant because it is a representative measure of the quantity of active principle that can actually achieve its therapeutic effect. Here, following [11], we have obtained a set of molecular structures and the corresponding bioavailability values using a public database of drugs and drug-like

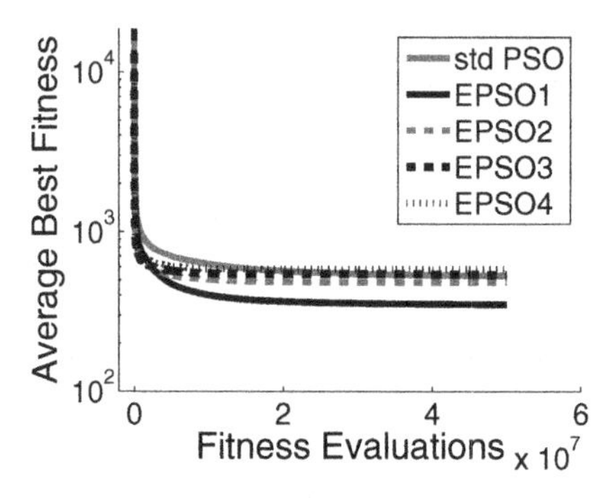

Fig. 4. Results of the linear regression for predicting bioavailability values of a dataset of molecules

compounds [12]. Chemical structures are all expressed as SMILES code (Simplified Molecular Input Line Entry Specification), i.e. strings encoding the two-dimensional molecular structure. The resulting libraries of molecules contain 359 molecules with measured bioavailability values. SMILES strings have been used to compute 241 bi-dimensional molecular descriptors using ADMET Predictor, a software package produced by Simulation Plus, Inc. (http://www.simulationplus.com). We now have a dataset of molecules $H = [H_{(i,j)}]$ of 359 rows and 241 columns and a column vector $b = [b_1, b_2, ..., b_{359}]$ of known bioavailability values for these molecules. We want to perform a linear regression on these data, i.e. we look for the vector of coefficients $c = [c_1, c_2, ..., c_{242}]$ such that for each $i = 1, 2, ..., 359$, $\sum_{j=1}^{241} c_j H_{(i,j)} + c_{242}$ approximates b_i.

We try to solve this problem with the PSO models presented above. In particular, particles are vectors of 242 floating point values representing the coefficients $c = [c_1, c_2, ..., c_{242}]$ of the linear regression. The fitness of each particle has been calculated as the root mean squared error (RMSE) between outputs and targets, i.e.

$$\mathrm{RMSE}(c) = \sqrt{\frac{1}{359} \sum_{i=1}^{359} \left(\sum_{j=1}^{241} c_j H_{(i,j)} + c_{242} - b_i \right)^2}$$

The other parameters we have used are: swarm size equal to 100; each particle coordinate initialized randomly with uniform distribution in the range $[-50, 50]$; velocities initialized randomly with uniform distribution in the range $[-10, 10]$; maximum number of fitness evaluations equal to $5 \cdot 10^7$. The minimum RMSE that can be obtained by a linear regression on this dataset, computed using the least squares regression method, is equal to 338.61. Since the problem is computationally quite expensive, we only performed 10 independent runs for each PSO variant in this experiment.

The ABF vs. fitness evaluations is reported in Figure 4. EPSO1 outperforms EPSO2, EPSO3, EPSO4 and standard PSO. We do not report Success Rate statistics, since any threshold we could set would be a totally arbitrary choice, in this case. Instead, we report, in Table 2, the ABFs with their standard deviations at termination (i.e. when

Table 2. Best Fitness, Worst Fitness, Median and Average Best Fitness values with standard deviation at termination (i.e. when $5 \cdot 10^7$ fitness evaluations have been performed), calculated on the same 10 independent runs as in Figure 4.

Algorithm	BF	WF	MEDIAN	ABF (ST. DEV)
PSO	427.99	772.23	518.53	532.59 (100.76)
EPSO1	349.68	361.51	350.89	352.62 (3.91)
EPSO2	399.81	574.83	484.02	407.20 (61.79)
EPSO3	418.16	647.06	547.60	540.26 (71.57)
EPSO4	454.29	815.23	584.57	584.84 (124.68)

$5 \cdot 10^7$ fitness evaluations have been performed), as well as the best and worst result obtained in the 10 runs for each PSO variant.

Besides finding solutions with lower errors, EPSO1 has also a remarkably lower standard deviation than the other PSO variants and the difference between the results obtained by EPSO1 and by the other ones are statistically significant.

4 Conclusions and Future Work

We have analyzed the modifications to PSO introduced in the PSO variant called Evolutionary Particle Swarm Optimization [7]. Selection pressure is tuned by changing the number of offspring in updating each particle. We have empirically compared results of versions of EPSO characterized by different selection pressures and of the standard PSO algorithm. The hypothesis of this work was that while parameter mutation tends to improve robustness with respect to local optima for both PSO-easy and PSO-hard problems, selection pressure can increase convergence speed, for PSO-easy problems, but may be detrimental for PSO-hard problems. We plan to extend the analysis of the results by testing the algorithms on the notoriously hard benchmark developed for the competition held at CEC2005 [13] and compare the results with those obtained in that occasion.

By now, results seem to confirm the hypothesis. A possible explanation is that assigning the parameters variable values which let the system switch from stable to unstable states during a run keeps the swarm in a high-energy state in which particles have high mobility, enhancing exploration. Results also suggest that "greedy" behaviour induced by selection causes the algorithm to be trapped into local optima more easily and increasingly with the number of offspring considered by the selection scheme.

The PSO variants we considered make PSO mostly independent of parameter choice. We observed that the range within which we clamped the values of C_1 and C_2, which slightly extends the usual range chosen for standard PSO, seems to be effective on all problems on which we have performed our tests. Further investigations will be required to draw final conclusions on topics related with parameter choice. For example, the value to which we set w depending on the values of C_1 and C_2 ($(C_1 - C_2)/2 - 1$) before mutating it, which is one of the main differences between our algorithms and EPSO as proposed in [7], has been chosen based on considerations of studies on PSO stability [9]. Even if some preliminary experiments suggest that changing such a value leads to worse results, a more accurate study is needed to confirm the hypothesis.

The results obtained with a mostly random parameter adaptation scheme as in EPSO1 also suggest that it would be worth trying to make parameter adaptation more deterministic by making inferences about the difficulty of the optimization task which PSO is performing. That would allow the effects of the two modifications to be weighted accordingly. Our future activity will include trying to apply some of the hardness measures presented, for instance, in [14], to accomplish this task or testing a set of PSO variants on a range of theoretically hand-tailored, difficultly tunable fitness landscapes, possibly built using evolutionary algorithms such as Genetic Programming, as suggested in [15].

Acknowledgements

This work was partially supported by the Spanish grant TIN2007-67466-C02-02.

References

1. Kennedy, J., Eberhart, R.: Particle swarm optimization. In: Proc. IEEE Int. conf. on Neural Networks, vol. IV, pp. 1942–1948 (1995)
2. Shi, Y., Eberhart, R.: A modified particle swarm optimizer. In: Proc. IEEE Int. Conference on Evolutionary Computation, pp. 69–73 (1998)
3. Poli, R.: Analysis of the publications on the applications of particle swarm optimisation. Journal of Artificial Evolution and Applications (in press)
4. Kennedy, J., Poli, R., Blackwell, T.: Particle swarm optimisation: An overview. Swarm Intelligence 1(1), 33–57 (2007)
5. Clerc, M. (ed.): Particle Swarm Optimization. ISTE (2006)
6. Poli, R.: The sampling distribution of particle swarm optimizers and their stability. Technical Report CSM-465, Department of Computer Science, University of Essex (2007)
7. Miranda, V., Fonseca, N.: EPSO: Best-of-two-worlds meta-heuristic applied to power system problems. In: Proc. IEEE Congress on Evolutionary Computation 2002, pp. 1080–1085. IEEE Computer Society Press, Los Alamitos (2002)
8. Schwefel, H.P.: Collective phenomena in evolutionary systems. In: 31st Annual Meeting of the International Society for General System Research, Budapest (Preprint 1987)
9. Van den Bergh, F.: An Analysis of Particle Swarm Optimizers. PhD thesis, Dept. of Computer Science, University of Pretoria, South Africa (2003)
10. Eiben, A., Jelasity, M.: A critical note on experimental research methodology in EC. In: Proceedings of the 2002 Congress on Evolutionary Computation (CEC 2002), pp. 582–587. IEEE Press, Piscataway (2002)
11. Archetti, F., Lanzeni, S., Messina, E., Vanneschi, L.: Genetic programming for human oral bioavailability of drugs. In: Cattolico, M. (ed.) Proceedings of the 8th annual conference on Genetic and Evolutionary Computation, Seattle, Washington, USA, pp. 255–262 (July 2006)
12. Wishart, S.D., Knox, C., Guo, A.C., Shrivastava, S., Hassanali, M., Stothard, P., Chang, Z., Woolsey, J.: DrugBank: a comprehensive resource for in silico drug discovery and exploration. Nucleic Acids Research 34 (2006)
13. Suganthan, P., Hansen, N., Liang, J., Deb, K., Auger, Y.P.C.A., Tiwari, S.: Problem definitions and evaluation criteria for the CEC 2005 special session on real-parameter optimization. Technical report, Nanyang Technological University (May 2005)
14. Vanneschi, L.: Theory and Practice for Efficient Genetic Programming. Ph.D. thesis, Faculty of Sciences, University of Lausanne, Switzerland (2004)
15. Langdon, W.B., Poli, R.: Evolving problems to learn about particle swarm optimizers and other search algorithms. IEEE Trans. on Evolutionary Computation 11(5), 561–578 (2007)

An Evolutionary Game-Theoretical Approach to Particle Swarm Optimisation

Cecilia Di Chio[1], Paolo Di Chio[2], and Mario Giacobini[3]

[1] Department of Computer Science, University of Essex, UK
cdichi@essex.ac.uk
[2] Dipartimento di Sistemi e Istituzioni per l'Economia, University of L'Aquila, Italy
pdc@ec.univaq.it
[3] Department of Animal Production, Epidemiology and Ecology
and Molecular Biotechnology Center, University of Torino, Italy
mario.giacobini@unito.it

Abstract. This work merges ideas from two very different areas: Particle Swarm Optimisation and Evolutionary Game Theory. In particular, we are looking to integrate strategies from the Prisoner Dilemma, namely **cooperate** and **defect**, into the Particle Swarm Optimisation algorithm. These strategies represent different methods to evaluate each particle's next position. At each iteration, a particle chooses to use one or the other strategy according to the outcome at the previous iteration (variation in its fitness). We compare some variations of the newly introduced algorithm with the standard Particle Swarm Optimiser on five benchmark problems.

1 Introduction

Particle Swarm Optimisation (PSO) is a method for optimisation of continuous nonlinear functions, discovered through the simulation of a simplified social model (particle swarm) [1,2]. PSO has its roots in two main methodologies: artificial life (in particular flocking, schooling and swarming theory), and evolutionary computation. PSOs use a population of interacting particles (i.e., candidate solutions) that "fly" over the landscape searching for the optimum. Each particle is placed in the parameter space of some problem and evaluates its fitness at the current location. It then determines its movement through the space by combining some aspect of the history of its own fitness values with those of one or more other members of the swarm. Finally, the particle moves through the parameter space with a velocity determined by the locations and processed fitness values of those other members, possibly together with some random perturbations.

More formally, the update equations for force, velocity and position of a particle i are as follows:

$$f_i = \underbrace{\phi_1 R_1 (x_{s_i} - x_i)}_{\text{social component}} + \underbrace{\phi_2 R_2 (x_{p_i} - x_i)}_{\text{individual component}} \tag{1}$$

M. Giacobini et al. (Eds.): EvoWorkshops 2008, LNCS 4974, pp. 575–584, 2008.

$$v_i(t) = v_i(t-1) + f_i \qquad\qquad x_i(t) = x_i(t-1) + v_i$$

where x_i is the current particle's position, v_i is the particle's velocity, x_{s_i} is the swarm's best position, x_{p_i} is the particle's best position, and with $\phi_1 = \phi_2 = 2.05$. The next iteration takes place after all particles have been moved. Eventually the swarm as a whole is likely to move close to the best location.

As with many other search heuristics, PSO has also been used in the context of game theory. Amongst others, Franken and Engelbrecht [3] employed co-evolutionary particle swarm optimisation for the generation of feature evaluation scores for the Seega game. Closer to Evolutionary Game Thoery (EGT), Abdelba and co-authors [4] investigated the application of co-evolutionary training techniques based on PSO to evolve playing strategies for the Iterated Prisoner's Dilemma (IPD), and Pavlidis *et al.* [5] compared the effectiveness of PSO versus covariance matrix adaptation evolution strategies and differential evolution to compute Nash equilibria of finite strategic games.

An interesting combination of PSO and Evolutionary Game Theory (EGT) was attempted by Cui and co-authors [6], who refined Predicted-Velocity PSO, a variant of PSO which uses both velocity and position vectors in the search. Particles are regarded as players in an artificial evolutionary game, and apply different strategies (position and velocity vectors) to generate the historical best position at an arbitrary time. Similarly to a 2-strategy game's probabilistic player, each particle is associated a probability to play one of the two strategies (in their case choosing the position or the velocity vector), and the distribution of the probabilities of all particles is adaptively adjusted.

Our approach shares with [6] this idea that a strategy for a game is given to each particle in a PSO. However, unlike [6], we utilise the Prisoner's Dilemma (PD) framework, associating to each particle one of the two well known cooperative and defective PD strategies, that will be interpreted as the behaviour of particles. Different behaviors will result in different search strategies of the particles and, for this first investigation, the payoff values of the game do not have a direct influence on the PSO dynamics. In a certain sense, our intention is to bring the PSO back to natural swarm theory (their inspirational metaphor), letting the artificial swarm dynamics be influenced not only by attractive forces on the particles, but also by their different social characters. Moreover, these particle characteristics evolve during the swarm's exploration of the fitness landscape, allowing a combination of strategies to emerge that should be beneficial for each problem. This paper presents a preliminary investigation in the direction of integrating PSO and EGT. As will be shown, the comparison of these new variants with a standard PSO suggests that this approach has merit.

The article is organized as follows. In the next session, we describe the underlying idea: i.e., how we intend to integrate EGT concepts into a PSO framework. In section 3, the proposed algorithms are described in detail, stressing their differences. The following section is dedicated to the experimental comparison of the proposed variants to a standard PSO on five classical test functions. Finally, in section 5, we draw our conclusions on this first integration of PSO and EGT, together with some ideas on future research directions.

2 PSO and Evolutionary Game Theory

The goal of Game Theory is to model and analyse conflicting situations such as those that arise in economy, biology, and society in general [7]. In this context, the well-known Prisoner's Dilemma (PD) game has played an extremely important role and has received a lot of attention [8]. The PD has fascinated researchers because it is an interaction where the individual's rational pursuit of self-interest, one of the pillars of game theory, produces a collective result that is self-defeating. Evolutionary Game Theory [9], more generally, considers a large population of players that meet randomly in pairs, and has been shown to be a better framework for analysing behavioural emergence in social and biological societies.

The underlying idea of the present work is to modify the PSO algorithm so that to each particle in the swarm is associated a behaviour (or "strategy") taken from the PD framework - i.e., to cooperate (being cooperative, or C) or to defect (being defective, or D) - that is used to compute its position in the next time step. We have employed two different interpretations of the two strategies C and D.

Social vs. Individual

In the first interpretation, to cooperate means for a particle to have a stronger social component, while to defect means to have a stronger individual component.

Conformist vs. Deviant

In the second interpretation, to cooperate means for a particle to behave as a standard PSO particle, while to defect means to deviate from the standard PSO trajectory in one of the following ways:

- towards the particle's best position (this is the same as PSO being 100% individual);
- in opposite direction with respect to the best of the group;
- in opposite direction with respect to the whole force;
- towards a random point (either retaining the individual part, the social part, or neither).

3 Description of the Algorithms

The two different interpretations of a particle's strategy presented in the previous section reflect different implementations of a PSO integrated with EGT (PSO-EG). We present the formal implementation in the next subsection. Before that, we need to introduce another required aspect of the strategies: the mechanism by which particles change from one strategy to another (i.e., their "strategy dynamics"). There are two variants here, as for the interpretations of C vs. D.

Flip-Strategy

At each time step[1], each particle **A** compares its present fitness with its fitness at the previous time step. If the fitness improves, then **A** keeps its strategy (C or D), while, if the fitness decreases, then **A** changes to the opposite strategy (D or C).

Always C or D (No_flip)

As before, at each time step, each particle **A** compares its present fitness and its fitness value at the previous time step. If the fitness improves, then **A** cooperates (adopts a C strategy), while if its fitness decreases, then **A** defects (adopts a D strategy).

Because the particles are initially positioned in the landscape randomly, they may require a few iterations to "organise" themselves into a proper swarm. We have therefore allowed a certain numbers of iterations during which, even if the particle's fitness decreases, the particle keeps behaving as if the fitness had increased (i.e., according to which strategy dynamics is being used, the particle keeps the strategy it was using or cooperates). We call this parameter the *forgiving delay*.

3.1 Formal Description of the PSO-EG Algorithms

Let us recall from the previous section the two interpretations for the cooperate and the defect strategies (1 = social vs. individual; 2 = conformant vs. deviant):

1. C: the particle is completely social
   D: the particle is completely individual
   (in the results, these will be identified as **single**)
2. C: the particle uses the standard PSO algorithm
   D: the particle deviates from the standard PSO direction according to:
   (a) opposite to the swarm's best x_{s_i} (**oppSB** in the results)
   (b) opposite to the whole force (**oppW** in the results)
   (c) towards a random point (retaining x_{p_i}) (**socR** in the results)
   (d) towards a random point (retaining x_{s_i}) (**indR** in the results)
   (e) towards a random point (100% random) (**rand** in the results)

These two interpretations reflect in various way how the particle's strategy can influence its movements in the search space. Referring to equation (1) in section 1, we obtain:

1.C (100% social particle) the force acting on the particle is only influenced by the social component

$$f_i = \phi_1 R_1 (x_{s_i} - x_i)$$

[1] This is not entirely true. Refer to the next section for a detailed explanation.

1.D (100% individual particle) the force acting on the particle is only influenced by the individual component

$$f_i = \phi_2 R_2(x_{p_i} - x_i)$$

2.C (standard PSO particle) the force acting on the particles is the standard one (equation (1))

2.D.a (particle's direction deviates opposite swarm's best x_{s_i}) the social component is substituted with $(-x_{s_i} + x_i)$, which drives the particle in the opposite direction from the swarm best position

$$f_i = \phi_1 R_1(-x_{s_i} + x_i) + \phi_2 R_2(x_{p_i} - x_i)$$

2.D.b (particle's direction deviates opposite the whole force) the social component is substituted with $(-x_{s_i} + x_i)$ and the individual component is substituted with $(-x_{ps_i} + x_i)$, which drives the particle in the opposite direction from the whole force

$$f_i = \phi_1 R_1(-x_{s_i} + x_i) + \phi_2 R_2(-x_{p_i} + x_i)$$

2.D.c (particle's direction deviates towards a random point but retains x_{p_i}) the social component is substituted with a random factor

$$f_i = R + \phi_2 R_2(x_{p_i} - x_i)$$

2.D.d (particle's direction deviates towards a random point but retains x_{s_i}) the individual component is substituted with a random factor

$$f_i = \phi_1 R_1(x_{s_i} - x_i) + R$$

2.D.e (particle's direction deviates towards a random point) both the social and the individual components are substituted with a random factor (100% random PSO)

$$f_i = R$$

4 Experimental Analysis

We have performed an extensive set of experiments to compare the performance of the various versions of the PSO-EG with the standard PSO (with global topology). The details of the experiments are as in Table 1.

4.1 Results

For each benchmark problem, the standard PSO has been compared with 72 different PSO-EG variants[2] (with the same initial conditions). Because of this large number of experiments, and the limited space available, it is impossible to

[2] 6 forgiving delays × 6 C/D strategies × 2 flip/no_flip dynamics.

Table 1. Experimental settings. ⋆ For those functions whose centre is at 0.0, a shift to the global optimum has been applied.

Test functions (as minimisation problems)	Ackely (world size = 64.0, centre = 0.0 ⋆) Griewank (world size = 1200.0, centre = 0.0 ⋆) Rastrigin (world size = 10.24, centre = 0.0 ⋆) Rosenbrock (world size = 100.0, centre = 1.0) Sphere (world size = 200.0, centre = 0.0 ⋆)
Dimensions	2, 10, 30
Particles	10
Iterations	500
Runs	100
Forgiving delay	$0 \rightarrow 5$

present all the results obtained. We therefore summarise the best results obtained in the following tables.

Table 2 presents the data for the best mean square error (mse) over 100 runs found for each problem for both the best function value (i.e., the value of the function for the swarm best) and the normalised error (i.e., the distance of the swarm best from the global optimum, divided by the number of dimensions).

Table 2. Mean square error (mse) of the value of the function and the normalised error found by each strategy for each function. The three columns refer to the dimensions considered for each test problem.

		2	10	30
Ackley	*fun*	1.972E-31 in PSO	2.816E-01 in socR flip 2	1.721E+01 in socR flip 1
	err	1.294E-32 in oppSB no_flip 3	1.340E-03 in socR flip 2	5.420E-02 in socR flip 1
Griewank	*fun*	3.364E-05 in oppW no_flip 2	2.311E-02 in rand flip 3	3.027E-01 in indR flip 3
	err	2.996E+00 in socR flip 0	3.837E+00 in rand flip 4	8.205E-02 in socR flip 1
Rastrigin	*fun*	9.083E-17 in indR flip 2	3.574E+02 in socR flip 3	1.754E+04 in rand flip 3
	err	1.831E-12 in indR flip 2	1.457E-01 in oppSB no_flip 2	1.139E-01 in rand flip 3
Rosenbrock	*fun*	3.767E-03 in single flip 0	5.155E+05 in socR no_flip 3	4.799E+06 in socR flip 2
	err	1.050E+00 in single flip 0	6.037E+00 in socR no_flip 2	1.077E+00 in indR flip 1
Sphere	*fun*	0.000E+00 in PSO	1.287E-24 in socR no_flip 5	2.353E-01 in indR flip 2
	err	0.000E+00 in PSO	2.594E-15 in socR no_flip 5	3.926E-04 in socR flip 2

Table 3. Average (avg), standard deviation (std) and mean square error (mse) of both the value of the function and the error found for Rastrigin

Dim.		
2	*Best function values*	best avg = 1.470E-09 in indR flip 2
		best std = 9.416E-09 in indR flip 2
		best mse = 9.083E-17 in indR flip 2
	Best errors	best avg = 3.060E-07 in indR flip 2
		best std = 1.318E-06 in indR flip 2
		best mse = 1.831E-12 in indR flip 2
10	*Best function values*	best avg = 1.680E+01 in socR flip 3
		best std = 7.751E+00 in socR no_flip 0
		best mse = 3.574E+02 in socR flip 3
	Best errors	best avg = 2.910E-01 in socR flip 5
		best std = 2.245E-01 in socR no_flip 2
		best mse = 1.457E-01 in oppSB no_flip 2
30	*Best function values*	best avg = 1.277E+02 in rand flip 3
		best std = 2.272E+01 in oppSB no_flip 0
		best mse = 1.754E+04 in rand flip 3
	Best errors	best avg = 2.669E-01 in rand no_flip 5
		best std = 1.973E-01 in rand flip 3
		best mse = 1.139E-01 in rand flip 3

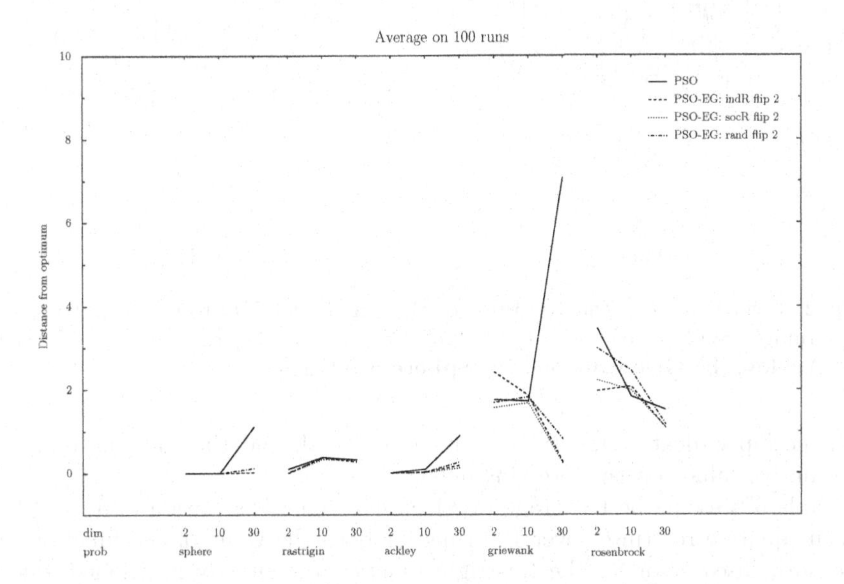

Fig. 1. Mean square error over 100 runs of three of the best PSO-EGs compared with the standard PSO for the normalised error. On the x axis the five benchmark problems grouped by dimension.

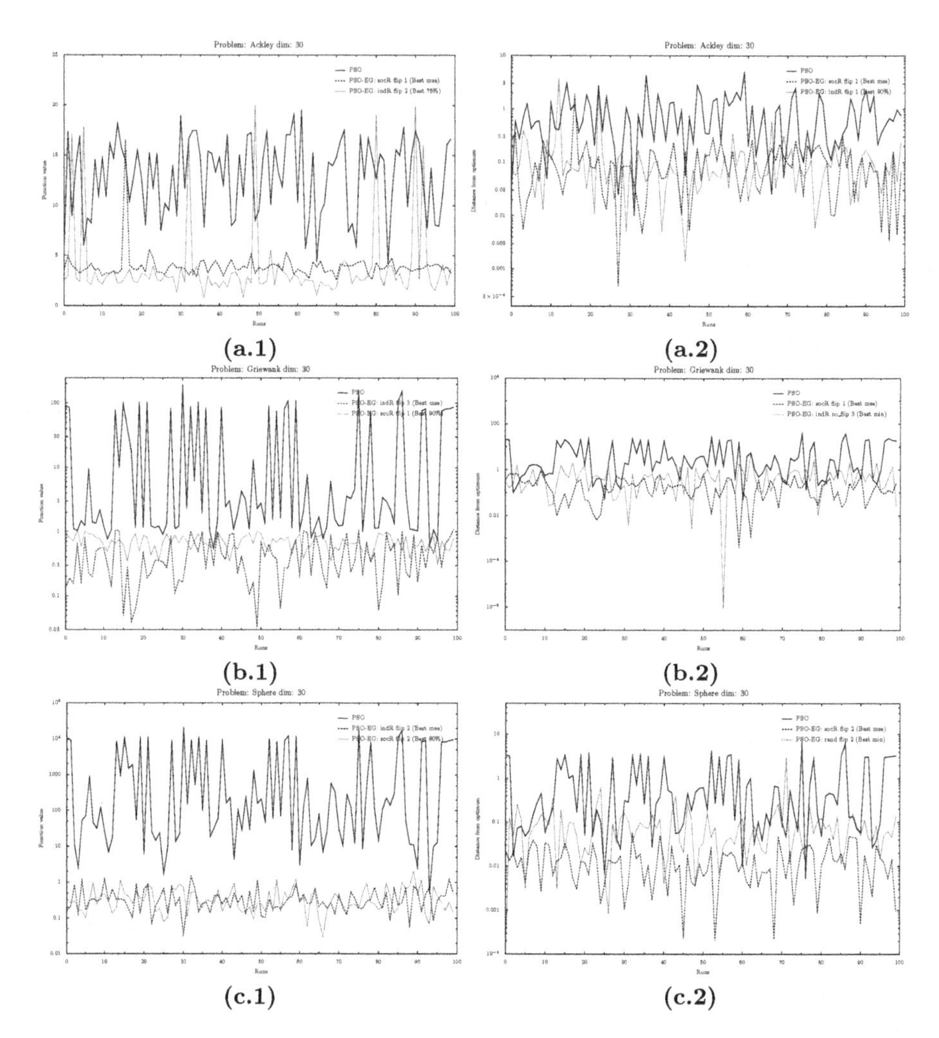

Fig. 2. Behaviour, over the 100 runs, of the standard PSO algorithm and the best performing PSO-EGs for both **(1)** the function value and **(2)** the normalised error for **(a) Ackley**, **(b) Griewank** and **(c) Sphere** in 30 dimensions

Flip/no_flip indicates the strategy dynamics used, and the columns represent alternative values for the forgiving delay.

Table 3 presents best average (avg), best standard deviation (std) and best mean square error (mse) over 100 runs for both the value of the function and the normalised error for the Rastrigin function (as already mentioned, due to space limitation, it is impossible to present the results for all the 5 benchmark problems: this is just an illustrative example).

It is evident from the results that changing behaviour is beneficial for the swarm, especially when the number of dimensions increases, and for the multimodal problems. Rosenbrock is the only case in which, for the smallest dimension, the particles use the **single** strategy (100% social in case of cooperation, and 100% individual in case of defection). Also, the strategy dynamics preferred in this case is the flip one (change strategy if the fitness decreases, and keep the strategy if the fitness improves), and the best performance are obtained when no forgiving factor is used. The reason for this might lie in the shape of the landscape. Further analysis is required to investigate this behaviour.

In general, we can observe that the preferred (best performing) strategies are either **socR** or **indR**. This is interesting, as in both these cases the equation of the force of the PSO is reduced to a single component (either the individual or the social respectively) with the addition of a further random term. This suggests that, for certain types of landscapes, the complete equation of the PSO algorithm might be superfluous and a much simpler formulation might be enough to obtain good (in fact, better) results.

Figure 1 presents, for each benchmark function and for each dimension, the mean square error over 100 runs of three of the best PSO-EGs compared with the standard PSO for the normalised error. It shows how the performance of the PSO and selected PSO-EGs varies when the dimension of the problem increases. For example, for the Griewank function, the PSO algorithm performs quite well in dimensions 2 and 10 ($\tilde{1}.5$ units away from the optimum) and are similar to those of some PSO-EGs. In 30 dimensions however, the PSO's performances worsen drastically while those of the PSO-EGs improve.

Figure 2 (**(a)** – **(c)**) compares, for three of the five benchmark functions in 30 dimensions, the behaviour of the standard PSO algorithm and the two best performing PSO-EGs for both **(1)** the function value and **(2)** the normalised error. This is therefore not an average over the 100 runs, but the actual value of the function and distance from the optimum found by the best particle in the swarm in each of the 100 runs.

5 Conclusions and Future Work

We have looked at how to integrate Prisoner Dilemma style cooperate/defect game theoretic strategies into the Particle Swarm Optimisation algorithm, interpreting them as behaviours of the particles. The two strategies are construed in two ways: firstly, to cooperate means that a particle has a stronger social component, while to defect means a stronger individual component; secondly, to cooperate means that a particle behaves as a standard PSO particle, while to defect means that it deviates from the standard PSO direction. At each iteration, a particle chooses to use one or the other strategy according to the fitness from the previous iteration, where two different mechanisms for this choice were investigated ("strategy dynamics"). The combination of these different interpretations leads to different implementations of the PSO-EG algorithm.

It is evident from the experimental results that changing behaviour is beneficial for the swarm, especially when the number of dimensions of the problem increases and for multimodal problems. In general, we can observe that the best performing strategies are those where the equation of the force of the PSO is reduced to a single component with the addition of a further random term. This suggests that, for certain types of landscapes, the complete equation of the PSO algorithm may be superfluous.

As we made clear at the beginning of this paper, this is still only a preliminary investigation in the direction of integrating PSO and EGT. For example, the payoff values of the game do not have a direct influence on the PSO dynamics: further work will be done in this direction. Although a more thorough analysis is required, the results obtained from the comparison between the standard PSO algorithm and a large number of PSO-EG variants are very promising, motivating further pursuit of this line of research. In particular, we intend to study the evolution of the cooperation among the particles of the swarm. Besides the clear interest in the emergence of cooperative behaviour, this study could result in a way to measure the difficulty of a given problem, relating the cooperative vs. defective strategies to the hardness of the problem.

References

1. Kennedy, J., Eberhart, R.C.: Swarm Intelligence. Morgan Kaufmann Publishers, San Francisco (2001)
2. Kennedy, J., Eberhart, R.C.: Particle Swarm Optimization. In: Proceedings of IEEE International Conference on Neural Networks, pp. 1942–1948 (1995)
3. Franken, N., Engelbrecht, A.P.: Co-Evolutionary Particle Swarm Optimization Applied to the 7×7 Seega Game. IEEE Transactions on Evolutionary Computation 9(6), 562–579 (2005)
4. Abdelbar, A.M., Ragab, S., Mitri, S.: Particle Swarm Optimization Approaches to Coevolve Strategies for the Iterated PrisonerÕs Dilemma. In: Proceedings 2004 IEEE International Joint Conference on Neural Networks, vol. 1, pp. 243–248 (2004)
5. Pavlidis, N.G., Parsopoulos, K.E., Vrahatis, M.N.: Computing Nash Equilibria through Computational Intelligence Methods. Journal of Computational and Applied Mathematics 175, 113–136 (2005)
6. Cui, Z., Cai, X., Zeng, J., Sun, G.: Predicted-Velocity Particle Swarm Optimization Using Game-Theoretic Approach. In: Huang, D.-S., Li, K., Irwin, G.W. (eds.) ICIC 2006. LNCS (LNBI), vol. 4115, pp. 145–154. Springer, Heidelberg (2006)
7. Myerson, R.B.: GameTheory: Analysis of Conflict. Harvard University Press, Cambridge (1991)
8. Axelrod, R.: The Evolution of Cooperation. Basic Books, Inc., New York (1984)
9. Weibull, J.W.: Evolutionary Game Theory. MIT Press, Boston (1995)

A Hybrid Particle Swarm Optimization Algorithm for Function Optimization

Zulal Sevkli[1] and F. Erdoğan Sevilgen[2]

[1] Fatih University, Dept. of Computer Eng.,Büyükçekmece, Istanbul, Turkey
zsevkli@fatih.edu.tr
[2] Gebze Institue of Technology, Dept. of Computer Eng., Gebze, Kocaeli, Turkey
sevilgen@bilmuh.gyte.edu.tr

Abstract. In this paper, a new variation of Particle Swarm Optimization (PSO) based on hybridization with Reduced Variable Neighborhood Search (RVNS) is proposed. In our method, general flow of PSO is preserved. However, to rectify premature convergence problem of PSO and to improve its exploration capability, the best particle in the swarm is randomly re-initiated. To enhance exploitation mechanism, RVNS is employed as a local search method for these particles. Experimental results on standard benchmark problems show sign of considerable improvement over the standard PSO algorithm.

Keywords: Particle Swarm Optimization, Reduced Variable Neighborhood Search, Function Optimization.

1 Introduction

Nature always has been an inspiration for researchers in designing solutions to many problems. Especially, researchers who study on optimization problems take advantage of nature to develop metaheuristic algorithms like genetic algorithm, ant colony optimization and simulated annealing. Particle Swarm Optimization is one of such metaheuristic algorithms. It is inspired of the behaviors of social models like bird flocking or fish schooling. It is introduced by Kennedy and Eberhart [7] as an optimization method for continuous nonlinear functions. Later, it has been applied to wide range of problems due to its conceptual and implementation simplicity.

Previous experiences indicate that PSO may have premature convergence especially for multimodal functions. To overcome this problem and to improve search efficiency, researches proposed several modifications to the standard PSO algorithm. For instance, in the Guaranteed Convergence PSO (GCPSO), a separate velocity update formula is used for the best particle in the swarm [11]. Comprehensive Learning PSO (CLPSO) [13] uses all particles's previous experiences as a potential during calculation of a particle's new velocity. In [4], to avoid getting trapped in local optimum, a mutation operator is embedded in PSO (MPSO). To improve the convergence behavior of PSO, several neighborhood topologies including gbest (*GCPSO-g,MPSO-g*) (fully connected network structure), lbest (*GCPSO-l,MPSO-l*) (*k* nearest neighbors connected network structure), and the Von Neumann topology (*GCPSO-v*) (four

M. Giacobini et al. (Eds.): EvoWorkshops 2008, LNCS 4974, pp. 585–595, 2008.

neighbors connected in a grid network structure) are considered in [10],[4]. Furthermore, various hybrid approaches such as Variable Neighbor PSO (VNPSO) [6] are proposed.

In this paper, we propose a new variation of PSO. To prevent premature convergence, we a separate velocity update mechanism to update the best particles' velocity similar to GCPSO. However, the new velocity of the best particle is determined in a simpler manner, randomly. We call the PSO algorithm using this technique as RePSO. Experimental results show that RePSO performs better than GCPSO for multimodal functions. While RePSO presents good explorative behavior, results for unimodal functions reveals that its exploitation capacity is not satisfactory. To improve exploitation and search efficiency for both unimodal and multimodal functions, a hybrid method of PSO with Reduce Variable Neighborhood Search (RVNS) is proposed. In this hybrid algorithm RVNS is used to perform local search around the best particle. Our hybridization is different from VNPSO; In VNPSO, VNS is used for diversification but we use RVNS for intensification. By this type of use our hybridization is more typical and similar to the Simulated Annealing and PSO hybridization in [1].

In our previous study [12], we applied our hybrid PSO algorithm successfully to an NP-hard combinatorial optimization problem and achieved best known solutions in literature. The purpose of this paper is to present the performance our hybrid PSO algorithm for the continuous problems. We present experimental results for standard benchmark problems which have been widely studied by PSO researchers [2],[4],[9],[10]. Our approach improves the standard PSO and achieves competitive results with other similar PSO variants.

2 Particle Swarm Optimization

Particle Swarm Optimization (PSO) is a stochastic optimization technique based on individual improvement, social cooperation and competition in a population. Since its introduction, PSO becomes an important tool for the optimization problems. It is famous for its fast convergence to a solution close to the optimal, by balancing the local search and global search i.e., exploitation and exploration, respectively.

PSO is a population-based method in which a swarm includes several individuals called particles. Each particle has a position and a velocity vector. Particles can initially be positioned randomly or to predetermined locations. The position of a particle is a candidate solution and it is updated at each iteration by using the particle's current velocity. The velocity of a particle is re-evaluated by using the particle's inertia as well as the social interaction (swarm's experience) and personal experience of the particle. The experience of each particle is usually captured by its local best position (*pbest*). The experience of the swarm is captured by the global best position (*gbest*). In the course of several iterations, particles make use of this experience and are supposed to move towards the optimum position.

Optimization is achieved in the course of several iterations of update and evaluation steps. The stopping condition of the iteration is usually the attainment of a maximum number of iterations or the maximum number of iterations between two improvements. During the update step, for each particle, the velocity and the position vector of the particle at iteration t+1 are calculated by using the Equations (1) and (2). In these equations, $v_{i,j}^{t}$ and $p_{i,j}^{t}$ are the velocity and the position of the jth dimension of the ith particle at iteration t, respectively. The parameters *c1* and *c2* are coefficients of learning factors, which are the weights of contributions of personal experience and social interaction. The stochastic behavior of PSO is achieved by *r1* and *r2* which are random numbers, generally in (0, 1). The parameter *w* is the inertia weight which is the balancing factor between exploration and exploitation. Small inertia weight facilitates more exploitation while large inertia weight enables more exploration. For faster convergence, inertia weight is usually selected to be high at the beginning (which lets PSO to explore) and decreased in course of optimization.

$$v_{i,j}^{t+1} = wv_{i,j}^{t} + c1r1(pbest_{i,j}^{t} - p_{i,j}^{t}) + c2r2(gbest_{j}^{t} - p_{i,j}^{t}) \tag{1}$$

$$p_{i,j}^{t+1} = p_{i,j}^{t} + v_{i,j}^{t+1} \tag{2}$$

After the update step, the fitness function value is calculated for each particle based on its position (candidate solution represented by the particle.) The local best position *pbest* of each particle and the global best position *gbest* are updated using these fitness values.

3 Modified PSO Algorithms

Implementation of above mentioned inertia weighted version of the PSO algorithm is applied to many problems. These studies indicate that PSO can have premature convergence depending on the form of the objective function. Particles may converge (velocities become so small to continue search journey) to local optimums or even other suboptimal values even after few iterations. Consequently, PSO is not able to explore the search space thoroughly. We force converged particles to continue exploration using a very simple method; we use uniform distribution and randomly refresh the velocity of the particle which achieves the best position in the swarm. In the same iteration, the position of the particle is recalculated with its new velocity. In this way, instead of possible convergence, the particle jumps to a complete different solution and continue its search with its past experience. We call this version of PSO as RePSO.

Although, RePSO has better performance than the standard PSO (as it is shown in experimental results section), inherited noble balance between intensification and diversification in PSO is disturbed. Particles tend to perform more exploration than exploitation since they are not allowed to stay around good positions. If a particle finds a good position it is forced to move away and continue to search other places.

Our second modification aims to rectify aforementioned problem in RePSO. This modification is based on the general strategy for many meta-heuristic methods: Repeating exploitation after exploration several times to go over the search space systematically. To achieve similar effect, we embed a local search into the PSO algorithm. During PSO iterations, the local search is conducted for particles which achieve or enhance the global best value. At the end of the local search, the position and *pbest* of the particle and *gbest* of the swarm are updated. The position of the particle is later recalculated with its new velocity. In this way, the particle which achieves or enhances *gbest* is forced to "fly" to different location while both personal experience of the particle and swarm's experience get affected from the local search. Since Reduce Variable Neighborhood Search (RVNS) is used as local search method, this version is called RePSORVNS. Pseudo-code of the RePSORVNS algorithm is illustrated in Fig.1.

```
Procedure RePSORVNS
   Initialize Parameters
   Initialize Population
   For each particle
     Evaluate
     Update local best
   Update global best
   Do
     For each particle
       If (fitness equal to fitness of gbest)
            position←RVNS_LocalSearch(position)
            Update local best
            Create new random velocity
       Else
            Update velocity
       Update position
       Evaluate
       Update local best
    Update global best
   While (Not Terminated)
End Procedure
```

Fig. 1. The RePSORVNS algorithm

In RePSORVNS, basic PSO part performs exploration (identification of promising neighborhoods) and exploitation (in depth search of specific neighborhoods) is realized by RVNS. RVNS is a variation of the Variable Neighborhood Search (VNS) meta-heuristic which is first introduced as an optimization method for combinatorial optimization problems [5]. Basic idea of the VNS algorithm is to search solution space with a systematic change of neighborhood in two main functions (Shake and LocalSearch), systematically. When LocalSearch is removed from VNS, RVNS is obtained. RVNS is usually preferred for problems where local search is expensive.

The pseudo-code of RVNS based local search algorithm is given in Fig. 2, where N_k represents k^{th} neighborhood structure (k=1 or 2), $N_k(s)$ represents the set of solutions in the kth neighborhood of the solution s. Both the choice and the order of neighborhood structures are critical for the performance of the RVNS algorithm. Generally, neighborhoods are ordered from the smallest to the largest. In the course of two nested loops, starting from the initial solution, s, the algorithm selects a random solutions s′ from first neighborhood. If s′ is better than s, it replaces s and the algorithm starts all over again with the same neighborhood. Otherwise, the algorithm continues with the second neighborhood structure until a stopping condition is met. Possible stopping conditions include the maximum CPU time allowed, the maximum number of iterations or the maximum number of iterations between two improvements.

In this study, d-dimensional balls (d is the dimensionality of the solution space) are used as neighborhood structures. Neighborhoods are determined dynamically by using a method similar to local search area determination in [3]. The distance between the position of the particle initiating RVNS and the nearest particle in the swarm is used in the calculation of the neighborhood radius. For the first neighborhood structure, the distance is multiplied with a low percentage ratio in order to find neighbors close to initial solution. The radius of the second neighborhood is calculated by multiplying the distance with a high percentage ratio so, a larger neighborhood is examined. The percentage ratios are two additional parameters of our method which are determined through experimental evaluation.

```
Procedure RVNS_LocalSearch(position p)
  Define neighborhood structures N1 and N2
  Use position of p as initial solution s
  Calculate distance of the nearest particle in the swarm
  while stopping condition is not met do
     k ← 1
     while k ≤ 2 do
        if (k equal 1)
             radius ← low-percentage*distance
             s'← FindNeighbour(s, radius), s' in N1(s)
        if (k equal 2)
             radius ← high-percentage*distance
             s'← FindNeighbour(s, radius), s' in N2(s)
        if (Fitness(s')<Fitness(s))
             s ← s'
             k ← 1
        else
             k ← k+1
     end-while
  end-while
End-Procedure
```

Fig. 2. The RVNS local search algorithm

4 Experimental Setup

Six benchmark problems which have been used in various PSO studies [2], [4], [9], [10] are experimented to test the algorithms. These problems are presented in Table 1 in order of complexity of finding global optimum. The first two problems, Spherical and Rosenbrock, are unimodal while others are multimodal. The target of all problems is to find minimum value which is at 0.

All three versions of PSO mentioned in the previous section (Standard PSO, RePSO, and RePSORVNS) are evaluated experimentally. The gbest PSO model is used in which all particles are considered to find global best value of the swarm. Initial positions are generated randomly, using uniform distribution in the search domain given

Table 1. Initial velocities are also generated randomly, using uniform distribution in the range [-r, r] where r is 4 for all problems except Rastrigin. Since search domain is small in Rastrigin, velocity range is smaller (r=0.5). During the search process, neither position nor velocity vector are allowed to get value from outside their range. If the calculated value is out of the range, a random value is used.

PSO parameters and RVNS parameters are listed in Table 2. At each iteration, inertia-weight (w) is decreased by two percent until the value of 4. The standard value of 2.0 for learning factors (c1 and c2) is used for most of the experiments. However, for some problems, the value of 1.1 results in better performance.

Table 1. Optimization test functions

Name	Formula	Dim	Search Domain
Sphere	$f(x)=\sum_{i=1}^{n} x_i^2$	30	$[-100,100]^n$
Rosenbrock	$f(x)=\sum_{i=1}^{n-1}(100(x_{i+1}-x_i^2)^2+(x_i-1)^2)$	30	$[-30,30]^n$
Griewank	$f(x)=\frac{1}{4000}\sum_{i=1}^{n} x_i^2-\prod_{i=1}^{n}\cos(\frac{x_i}{\sqrt{i}})+1$	30	$[-600,600]^n$
Rastrigin	$f(x)=\sum_{i=1}^{n}(x_i^2-10\cos(2\pi x_i)+10)$	30	$[-5.12,5.12]^n$
Schwefel	$f(x)=418.9829n+\sum_{i=1}^{n} x_i\sin(\sqrt{\lvert x_i\rvert})$	30	$[-500,500]^n$
Ackley	$f(x)=-20\exp(-0.2\sqrt{\frac{1}{n}\sum_{i=1}^{n} x_i^2})-\exp(\frac{1}{n}\sum_{i=1}^{n}\cos(2\pi x_i))+20+e$	30	$[-30,30]^n$

Table 2. Parameter setting for three PSO algorithms

PSO

Swarm Size=40; w=0.9

Stopping Condition=200.000 function evaluation

c1=c2	Sph.	Rbr.	Grie.	Ack.	Sch	Rast.
PSO	2.0	2.0	2.0	2.0	2.0	2.0
RePSO and RePSORVNS	1.1	1.1	1.1	2.0	2.0	2.0

RVNS

# of Neighborhood :2; Low-percentage: 0.2; High-percentage:0.5

Stopping Condition:(problem dimension*5) or 50 times back-to-back no improvements

Experiments are performed on an Intel P4 2.8 GHz PC with 1 GB memory. Each experiment is executed 30 times and the results are compared based on the fitness of the best solution found (Fitness), the number of fitness function evaluations (Eval.) and the time spend (CPU) until the best solution is found for the repetition yielding the best solution (Best), for the repetition yielding the worst solution (Worst), for the repetition yielding the median solution (Median) and averaged over all repetitions (Mean).

5 Experimental Results

The results of three PSO versions for unimodal problems (Spherical and Rosenbrock) are demonstrated in Table 3 and Fig. 4. The graphs in Fig. 4 illustrate the average convergence behavior of each algorithm. For unimodal functions, intensification mechanism of the algorithms has a dominant role. Since PSO and RePSORVNS have inherited and imported intensification mechanisms, respectively; their results are better than the RePSO for the Spherical function. Especially, with its superior exploitation mechanism RePSORVNS achieves faster improvement (check the CPU times) and enhances average results of standard PSO almost 90 (3) order of magnitude for Spherical (Rosenbrock). It takes less time to terminate since a local search is substantially faster in new solution generation than the PSO algorithm. The standard deviation of the fitness over all reputations is quite low for RePSORVNS compared to others so, it is more robust.

Table 3. Performance of PSO, RePSO and RePSO+RVNS for unimodal problems

	PSO			RePSO			RePSORVNS		
	Fitness	Eval.	CPU	Fitness	Eval.	CPU	Fitness	Eval.	CPU
Mean	7.93E-102	199,993.1	100.68	1.68E-41	199,954.7	100.30	5.44E-194	199,725.8	72.94
Best	1.24E-107	199,960.0	98.52	5.92E-44	199,760.0	99.63	1.87E-217	198,846.0	70.55
Median	3.44E-104	200,000.0	99.75	4.80E-42	199,960.0	99.72	1.19E-206	199,792.0	72.64
Worst	1.55E-100	200,000.0	101.75	1.65E-40	200,000.0	103.77	1.41E-192	200,237.0	81.56
Std dv.	2.91E-101	15.38	2.22	4.16E-42	199,980.0	99.73	0.00E+00	297.8	2.10

a) Spherical Function

	PSO			RePSO			RePSORVNS		
	Fitness	Eval.	CPU	Fitness	Eval.	CPU	Fitness	Eval.	CPU
Mean	2.75E+01	164,038.7	80.31	1.12E+01	200,000.0	102.59	7.90E-01	200,000.0	26.41
Best	3.52E-03	99,000.0	48.36	1.19E-02	200,000.0	99.60	7.54E-05	200,000.0	23.28
Median	2.13E+01	165,100.0	80.66	1.08E+01	200,000.0	101.04	1.86E-01	200,000.0	26.42
Worst	7.69E+01	200,000.0	98.35	5.74E+01	200,000.0	103.88	4.12E+00	200,000.0	33.31
Std dv.	2.41E+01	31,955.5	15.62	1.06E+01	0.0	3.97	1.49E+00	0.0	2.04

b) Rosenbrock Function

Results for the multimodal functions are illustrated in Table 4 and Fig. 5. The results support the hypothesis that standard PSO has a premature convergence problem for multimodal functions. For most of the repetitions, the standard PSO converges and provides almost no improvement after the first 70,000-80,000 evaluations. We attack

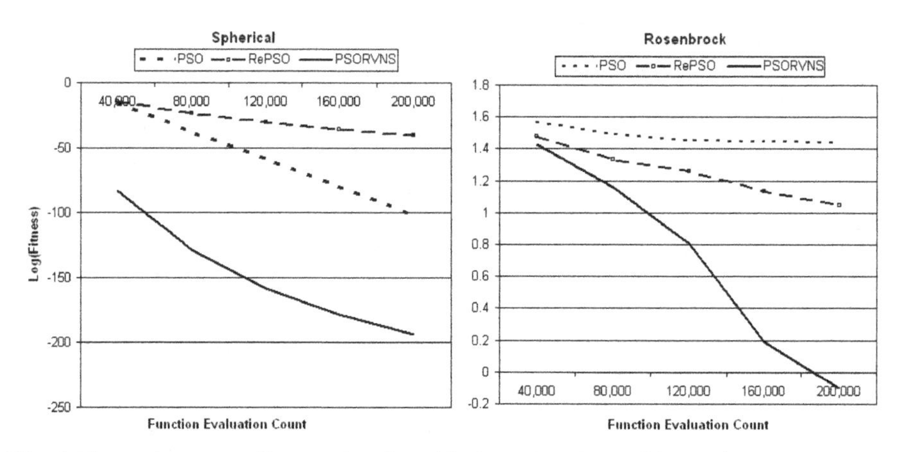

Fig. 3. Change in average fitness value (logarithmic scale) of the PSO, RePSO and PSORVNS for unimodal problems over function evaluation count

Table 4. Performance of PSO, RePSO and RePSORVNS for multimodal problems

	PSO			RePSO			RePSORVNS		
	Fitness	Eval.	CPU	Fitness	Eval.	CPU	Fitness	Eval.	CPU
Mean	5.63E-02	64,754.7	32.69	1.24E-02	88,153.3	44.65	1.15E-02	82,993.4	15.54
Best	0.00E+00	38,880.0	19.27	0.00E+00	59,240.0	30.02	0.00E+00	27,421.0	4.94
Median	2.22E-02	48,840.0	24.39	7.40E-03	72,200.0	37.33	8.63E-03	65,877.0	8.86
Worst	6.52E-01	177,760.0	88.19	7.11E-02	184,680.0	93.08	5.86E-02	189,979.0	62.59
Std dv.	1.18E-01	37,873.7	18.94	1.55E-02	34,628.7	17.46	1.29E-02	48,604.1	13.87

a) Griewank Function

	PSO			RePSO			RePSORVNS		
	Fitness	Eval.	CPU	Fitness	Eval.	CPU	Fitness	Eval.	CPU
Mean	1.28E+00	77,018.7	40.79	1.88E-14	180,545.3	96.10	7.55E-15	99,263.7	31.34
Best	7.55E-15	37,200.0	19.86	1.47E-14	150,880.0	78.02	7.55E-15	62,917.0	19.94
Median	1.50E+00	68,120.0	36.87	2.00E-14	180,560.0	96.58	7.55E-15	92,020.5	28.46
Worst	2.74E+00	200,000.0	106.76	2.89E-14	199,600.0	105.46	7.55E-15	152,282.0	47.56
Std dv.	8.71E-01	39,601.9	20.98	4.08E-15	12,417.3	6.56	3.21E-30	25,199.0	7.14

b) Ackley Function

	PSO			RePSO			RePSORVNS		
	Fitness	Eval.	CPU	Fitness	Eval.	CPU	Fitness	Eval.	CPU
Mean	1.65E+03	89,937.1	47.52	5.23E+02	199,458.7	106.41	3.40E+02	187,467.6	59.14
Best	8.29E+02	47,320.0	24.83	1.18E+02	192,200.0	102.16	1.18E+02	151,980.0	49.50
Median	1.72E+03	81,060.0	42.71	4.76E+02	199,820.0	106.15	3.55E+02	190,289.5	58.69
Worst	2.65E+03	175,680.0	93.14	1.07E+03	200,000.0	115.62	5.92E+02	200,000.0	75.54
Std dv.	4.09E+02	35,195.9	18.53	2.82E+02	1,404.0	2.97	1.42E+02	13,718.8	6.02

c) Schwefel Function

	PSO			RePSO			RePSORVNS		
	Fitness	Eval.	CPU	Fitness	Eval.	CPU	Fitness	Eval.	CPU
Mean	4.92E+01	82,636.0	48.94	2.48E+00	199,762.7	101.38	2.35E+00	199,863.5	99.10
Best	2.79E+01	34,920.0	20.47	1.26E-10	196,760.0	99.50	1.26E-10	198,960.0	96.61
Median	4.88E+01	71,060.0	42.13	1.50E+00	199,960.0	100.86	1.00E+00	199,960.0	98.92
Worst	8.36E+01	170,040.0	100.73	1.49E+01	200,000.0	103.47	1.49E+01	200,000.0	101.17
Std dv.	1.27E+01	36,111.3	21.40	2.98E+00	611.7	1.43	2.95E+00	229.7	0.80

d) Rastrigin Function

to this problem by our modifications. Both RePSO and RePSORVNS do not have such a problem, they continue to produce better solutions until termination at 200,000 function evaluations. Moreover, they provide better solutions than standard PSO for almost all evaluation criteria and better solutions compared to GCPSO in [10] based on best and average results.

The optimal solution is found only for Griewank problem. While approximately 5 repetitions produce the optimal solution using standard PSO, this number is more than 10 for RePSO and RePSORVNS. However, the corresponding graph for Griewank problem depicts that after a very short time, the algorithms can not further improve the best solution found so far. Neither use of re-initialization nor local search rectifies this problem completely.

For the remaining three multimodal problems (Ackley, Schwefel and Rastrigin) behavior of the algorithms are quite similar. RePSO starts to give better results than standard PSO after approximately 10,000 function evaluations. At the early iterations of the search, with the effect of re-initiation which hinders intensification mechanism in PSO, RePSO is not successful. However, the premature convergence problem of PSO becomes significant later and RePSO becomes better than PSO. On the other hand, RePSORVNS is almost always more successful than others in finding more promising solutions in a shorter time.

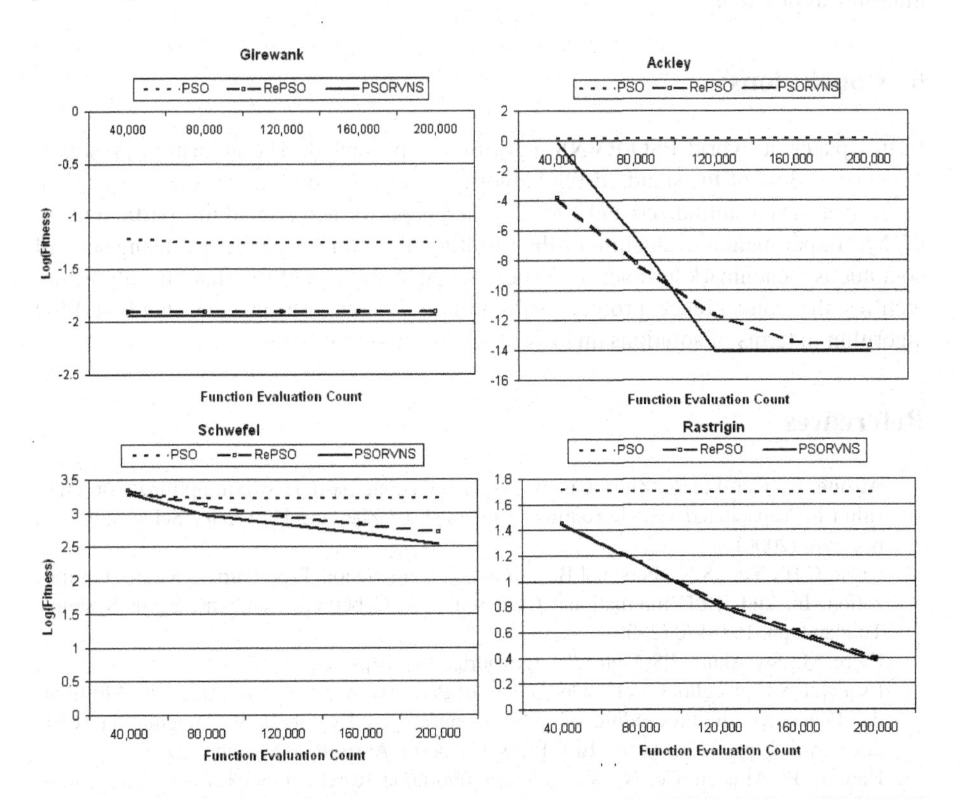

Fig. 4. Change in average fitness value (logarithmic scale) of the PSO, RePSO and PSORVNS for Girewank, Ackley, Schwefel and Rastrigin problems over the function evaluation count

Table 5. Performance of RePSO and RePSORVNS against the GCPSO,MPSO and CLPSO

	Spherical (1)	*Rosenbrock* (2)	*Griewank* (3)	*Ackley* (4)	*Rastrigin* (5)	*Schwefel* (6)	(1)	(2)	(3)	(4)	(5)	(6)
GCPSO-g [10]	3.00E-161	1.29E-02	1.62E-02	2.02E+00	7.19E+01	4.54E+03	+	-	+	+	+	+
GCPSO-l [10]	3.00E-93	6.54E-01	3.90E-03	2.79E-01	6.12E+01	4.76E+03	+	-	-	+	+	+
GCPSO-v [10]	6.00E-119	1.80E-01	1.04E-02	6.82E-01	5.58E+01	4.50E+03	+	-	-	+	+	+
MPSO-g [4]	-	-	2.53E-02	0.00E+00	4.69E+01	1.50E+03			+	-	+	+
MPSO-l [4]	-	-	1.40E-03	0.00E+00	3.00E+00	1.83E+03			-	-	+	+
CLPSO [13]	4.46E-14	2.10E+01	3.14E-10	0.00E+00	4.85E-10	1.27E-12	+	+	-	-	-	-
RePSO	1.68E-41	1.12E+01	1.24E-02	1.88E-14	2.48E+00	5.23E+02	4	1	2	3	5	5
RePSORVNS	5.44E-194	7.90E-01	1.15E-02	7.55E-15	2.35E+00	3.40E+02						

We compare our results with several other algorithms in Table 5. Termination criteria of all algorithms are the same (200,000 evaluations) and results for the other algorithms are taken from the references that are given in Table 5. Although our algorithms do not always produce the best solutions, the results are comparable and close to the best solutions for all functions either unimodal or multimodal. On the other hand, while GCPSO is more suitable for unimodal problems CLPSO is better for multimodal problems.

6 Conclusions

In this paper, a hybrid PSO/RVNS algorithm is presented. The algorithm preserves the general flow of the standard PSO algorithm but at each iteration, the best particle in the swarm is re-initialized and a local search is performed around this particle using RVNS. Experimental evaluation of the resulting algorithm is performed using several continuous benchmark problems. Experimental results confirm that the algorithm rectifies the convergence problem substantially. It outperforms the standard PSO algorithm in terms of solutions quality, efficiency and robustness.

References

1. Ai-ling, C., Gen-ke, Y., Zhi-ming, W.: Hybrid discrete particle swarm optimization algorithm for capacitated vehicle routing problem. J. of Zhejiang University Science A 7(4), 607–614 (2006)
2. Chen, C.H., Yeh, S.N.: Personal Best Oriented Constriction Type Particle Swarm Optimization. In: 2nd IEEE International Conference on Cybernetics and Intelligent Systems, Thailand, pp. 167–170 (2006)
3. Clerc, M.: Swissknife PSO, http://clerc.maurice.free.fr/pso/e
4. Esquivel, S.C., Coello, C.A.C.: On the Use of Particle Swarm Optimization with Multimodal Functions. In: Proceeding of IEEE Congress on Evolutionary Computation (CEC 2003), vol. 2, pp. 1130–1136. IEEE Press, Canberra, Australia (December 2003)
5. Hansen, P., Mladenovic, N.: Variable neighborhood search: principles and applications. Eur. J. Oper. Res. 130, 449–467 (2001)

6. Liu, H., Abraham, A., Choi, O., Moon, S.H.: Variable Neighborhood Particle Swarm Optimization for Multi-objective Flexible Job-Shop Scheduling Problems. In: Wang, T.-D., Li, X.-D., Chen, S.-H., Wang, X., Abbass, H.A., Iba, H., Chen, G.-L., Yao, X. (eds.) SEAL 2006. LNCS, vol. 4247, pp. 197–204. Springer, Heidelberg (2006)
7. Kennedy, J., Eberhart, R.C.: Particle swarm optimization. In: International Conference on Neural Networks (Perth, Australia), Piscataway, NJ, pp. 1942–1948. IEEE Service Center, Los Alamitos (1995)
8. Kennedy, J., Mendes, R.: Population structure and particle swarm performance. In: Proceeding of IEEE Congress on Evolutionary Computation CEC 2002, Honolulu, Hawaii, Piscataway, NJ, vol. 2, pp. 1671–1676. IEEE Service Center, Los Alamitos (2002)
9. Kennedy, J.: Small worlds and mega-minds: Effect of neighborhood topology on particle swarm performance. In: Proceeding of IEEE Congress on Evolutionary Computation, pp. 1931–1938 (July 1999)
10. Peer, E.S., Van den Bergh, F., Engelbrecht, A.P.: Using neighbourhoods with the guaranteed convergence PSO. In: Swarm Intelligence Symposium (Indianapolis, Indiana), Piscataway, NJ, pp. 235–242. IEEE Service Center, Los Alamitos (2003)
11. Van den Berg, F., Engelbrecht, A.: A new locally convergent particle swarm optimizer. In: Proceeding of IEEE Conference on System, Man and Cybernetics (October 2002)
12. Sevkli, Z., Sevilgen, F.E., Keles, O.: Particle Swarm Optimization for the Orienteering Problem. In: Proceedings of International Symposium on Innovation in Intelligent Systems and Applications (INISTA 2007), Istanbul, Turkey, pp. 185–190 (June 2007)
13. Liang, J.J., Qin, A.K., Suganthan, P.N., Baskar, S.: Comprehensive learning particle swarm optimizer for global optimization of multimodal functions. IEEE Transaction on Evolutionary Computation 10(3), 281–295 (2006)

Memory Based on Abstraction for Dynamic Fitness Functions

Hendrik Richter[1] and Shengxiang Yang[2]

[1] HTWK Leipzig, Fachbereich Elektrotechnik und Informationstechnik
Institut Mess–, Steuerungs– und Regelungstechnik
Postfach 30 11 66, D–04125 Leipzig, Germany
richter@fbeit.htwk-leipzig.de
[2] Department of Computer Science, University of Leicester
University Road, Leicester LE1 7RH, United Kingdom
s.yang@mcs.le.ac.uk

Abstract. This paper proposes a memory scheme based on abstraction for evolutionary algorithms to address dynamic optimization problems. In this memory scheme, the memory does not store good solutions as themselves but as their abstraction, i.e., their approximate location in the search space. When the environment changes, the stored abstraction information is extracted to generate new individuals into the population. Experiments are carried out to validate the abstraction based memory scheme. The results show the efficiency of the abstraction based memory scheme for evolutionary algorithms in dynamic environments.

1 Introduction

As a class of stochastic algorithms, evolutionary algorithms (EAs) work by maintaining and evolving a population of candidate solutions through selection and variation. New populations are generated by first selecting relatively fitter individuals from the current population and then performing variations (e.g., recombination and mutation) on them to create new off–spring. EAs have been applied to solve many stationary optimization problems with promising results. Usually, with the iteration of EAs, individuals in the population will eventually converge to the optimal solution(s) due to the pressure of selection. Convergence at a proper pace is expected for EAs to locate optimal solution(s) for stationary problems. However, many real world problems are actually dynamic optimization problems (DOPs), where changes may occur over time. For DOPs, convergence becomes a big problem for EAs because it deprives the population of genetic diversity. Consequently, when the environment changes, it is hard for EAs to escape from the optimal solution of the old environment.

For DOPs, the aim of an EA is no longer to locate a stationary optimum but to track the moving optima with time. This requires additional approaches to adapt EAs to the changing environment. In recent years, with the growing interest in studying EAs for DOPs, several approaches have been developed into EAs to address DOPs [6,9], such as diversity schemes [7,16], memory schemes [2,4,19],

M. Giacobini et al. (Eds.): EvoWorkshops 2008, LNCS 4974, pp. 596–605, 2008.

and multi–population approaches [5]. Among these approaches developed for EAs in dynamic environments, memory works by retaining and reusing relevant information, which might be as straightforward as storing good solutions.

This paper proposes a memory scheme based on abstraction. Abstraction means to select, evaluate and code information before storing. A good solution is evaluated with respect to physically meaningful criteria and in the result of this evaluation, storage is undertaken but no longer as the solution itself but coded with respect to the criteria. Thus, abstraction means a threshold for and compression of information, see e.g. [8] which proposes similar ideas for reinforcement learning. When an environment change is detected, the stored abstraction information is extracted to generate new individuals into the population.

The rest of this paper is organized as follows. In the next section, we briefly review the usage of memory for EAs for DOPs. Sec. 3 describes the dynamic fitness landscape used as the test bed in this paper. Sec. 4 presents the abstraction memory scheme. The experimental results and their analysis are given in Sec. 5. Finally, Sec. 6 concludes this paper with discussions on future work.

2 Memory Schemes for Dynamic Environments

The principle of memory schemes for EAs in dynamic environments is to, implicitly or explicitly, store useful information from old environments and reuse it later in new environments. Implicit memory schemes use redundant encodings, e.g., diploid genotype [10,12,18], to store information for EAs to exploit during the run. In contrast, explicit memory uses precise representations but splits an extra storage space to explicitly store information from a current generation and reuse it later [4,11,17].

For explicit memory schemes, usually only good solutions are stored in the memory as themselves and are reused directly. When a change occurs or every several generations, the solutions in the memory can be merged with the current population [4,11]. This is called *direct memory scheme*. It is also interesting to store environmental information as well as good solutions in the memory and reuse the environmental information when a change occurs. This is called *associative memory scheme*. For example, Ramsey and Greffenstette [13] studied a genetic algorithm (GA) for a robot control problem, where good candidate solutions are stored in a memory together with information about the current environment the robot is in. When the robot incurs a new environment that is similar to a stored environment instance, the associated controller solution in the memory is re–activated. In [19], an associative memory scheme was proposed into population based incremental learning (PBIL) algorithms for DOPs, where the working probability vector is also stored and associated with the best sample it creates in the memory. When an environmental change is detected, the stored probability vector associated with the best re–evaluated memory sample is extracted to compete with the current working probability vector to become the future working probability vector for creating new samples. Similarly, an associative memory scheme was developed into GAs for DOPs in [20], where the allele distribution statistics information of the population is taken as the representation of the current environment.

For explicit memory schemes, since the memory space is limited, we need to update information stored in the memory. A general strategy is to select one memory point to be replaced by the best individual from the population. As to which memory point should be updated, there are several strategies [4]. For example, the most–similar strategy replaces the memory point closest to the best individual from the population. Though memory schemes have been shown to be beneficial for EAs in dynamic environments, so far they are only a promising option for regular and predictable dynamics, such as cyclic or translatory, but not for irregular dynamics, such as chaotic or random dynamics. The abstraction memory scheme proposed in Sec. 4 is an attempt to overcome this shortcoming.

3 Description of the Dynamic Test Problem

As dynamic fitness landscape, we use a moving n–dimensional "field of cones on a zero plane", where N cones with coordinates c_i, $i = 1, 2, ..., N$ are initially distributed across the landscape and have randomly chosen heights h_i and slopes s_i. We introduce discrete time $k \in \mathbb{N}_0$ and consider the coordinates $c_i(k)$ to be changing with it. So, the dynamic fitness function is:

$$f(x, k) = \max\left\{0, \max_{1 \leq i \leq N} [h_i - s_i \|x - c_i(k)\|]\right\}. \tag{1}$$

Hence, the DOP is $\max_{x \in M} f(x, k) = \max\left\{0, \max_{1 \leq i \leq N} [h_i - s_i \|x - c_i(k)\|]\right\}, k \geq 0$ whose solution $x_s(k) = \arg\max\left\{0, \max_{1 \leq i \leq N} [h_i - s_i \|x - c_i(k)\|]\right\}$ forms a solution trajectory in the search space $M \subset \mathbb{R}^n$. This means that for every $k \geq 0$ the problem might have another solution, which we intend to find using an EA. In the experiments we report the dynamics of $c(k)$ that starts from randomly chosen $c(0)$ and is either regular (cyclic) or chaotic or random (normally and uniformly distributed).

The EA we use has a real number representation and λ individuals $x_j \in \mathbb{R}^n$, $j = 1, 2, \ldots, \lambda$, which build the population $P \in \mathbb{R}^{n \times \lambda}$. Its dynamics is described by the generation transition function $\psi : \mathbb{R}^{n \times \lambda} \to \mathbb{R}^{n \times \lambda}$, see e.g. [1], p.64–65. It can be interpreted as a nonlinear probabilistic dynamical system that maps $P(t)$ onto $P(t + 1)$. It hence transforms a population at generation $t \in \mathbb{N}_0$ into a population at generation $t + 1$, $P(t + 1) = \psi(P(t)), t \geq 0$. Starting from an initial population $P(0)$, the population sequence $P(0)$, $P(1)$, $P(2)$, ... describes the temporal movement of the population in the search space. Both the time scales t and k are related by the change frequency γ as

$$t = \gamma k. \tag{2}$$

For $\gamma = 1$, apparently, the dynamic fitness function is changing every generation. For $\gamma > 1$, the fitness function changes every γ generations. The change frequency is an important quantity in dynamic optimization and will be the subject of the experimental studies reported in Sec. 5.

4 The Abstraction Memory Scheme

The main idea of the abstraction based memory scheme is that it does not store good solutions as themselves but as their abstraction. We define an abstraction of a good solution to be its approximate location in the search space. Hence, we need to partition the search space. This can be obtained by partitioning the relevant (bounded) search space into rectangular (hyper–) cells. Every cell can be addressed by an element of a matrix. So, we obtain for an n–dimensional search space M an n–dimensional matrix whose elements represent search space sub–spaces. This matrix acts as our abstract memory and will be called memory matrix $\mathcal{M}$. It is meant to represent the spatial distribution of good solutions. Such ideas have some similarity to anticipating the dynamics of the DOP [3].

The abstract storage process consists of two steps, a selecting process and a memorizing process. The selecting process picks good individuals from the population $P(t)$ while the EA runs. In general, selecting has to be done in terms of (i.) the amount and choice of considered individuals, ideally sorted according to their fitness, from the population and (ii.) points in the run–time of the EA, ideally sorted according to changes in the environment detected by, for instance, a falling sliding mean of the best individual. For the individuals either only the best or a few best from the population could be used. In terms of the run–time between changes only the best over run–time or the best over a few generations before a change occurs could be taken. We define the number of the individuals selected for memorizing as well as the number of generations where memorizing is done.

In the memorizing process, the selected individuals are sorted according to their partition in the search space which they represent. In order to obtain this partition, we assume that the search space M is bounded and in every direction there are lower and upper bounds, $x_{i\,min}$ and $x_{i\,max}$, $i = 1, 2, \ldots, n$. With the grid size ϵ, which is a quantity we will examine in the numerical experiments given in Sec. 5, we obtain for every generation t the memory matrix $\mathcal{M}(t) \in \mathbb{R}^{h_1 \times h_2 \times \ldots \times h_n}$, where $h_i = \lceil \frac{x_{i\,max} - x_{i\,min}}{\epsilon} \rceil$. In the memory $\mathcal{M}(t)$ the entry of each element $m_{\ell_1 \ell_2 \ldots \ell_n}(t)$ is a counter $count_{\ell_1 \ell_2 \ldots \ell_n}(t)$, $\ell_i = 1, 2, \ldots, h_i$, which is empty for initialization. That is, $count_{\ell_1 \ell_2 \ldots \ell_n}(0) = 0$ for all ℓ_i. For each individual $x_j(t) \in P(t)$ selected to take part in the memorizing, the counter of the element representing the partition that the individual belongs to is increased by one. That is, we calculate the index $\ell_i = \lceil \frac{x_{i\,j} - x_{i\,min}}{\epsilon} \rceil$ for all $x_j = (x_{1j}, x_{2j}, \ldots, x_{nj})^T$ and all $1 \leq i \leq n$ and increment the corresponding $count_{\ell_1 \ell_2 \ldots \ell_n}(t)$. Note that this process might be carried out several times in a generation if more than one individual selected belongs to the same partition. The abstraction storage process retains the abstraction of good solutions by accumulating locations where good solutions occur. In this way, we encode and compress the information about good solutions. As the matrix $\mathcal{M}$ is filled over run–time, the memorizing process can be seen as a learning process in both its figurative and literal meaning.

After a change has been detected (for instance if the sliding mean of the best individual is falling), the abstract retrieval process is carried out. It consists of two steps. First, a matrix $\mathcal{M}_\mu(t)$ is calculated by dividing the matrix $\mathcal{M}(t)$ by the sum of all matrix elements, that is $\mathcal{M}_\mu(t) = \frac{1}{\sum_{h_i} \mathcal{M}(t)} \mathcal{M}(t)$. Hence, the sum of all elements $\mu_{\ell_1 \ell_2 \ldots \ell_n}(t)$ in $\mathcal{M}_\mu(t)$ adds up to one. Each element in $\mathcal{M}_\mu(t)$ contains an approximation of the natural measure $\mu \in [0, 1]$ belonging to the corresponding partition cell $M_{\ell_1 \ell_2 \ldots \ell_n}$ of the search space M. This natural measure can be viewed as the probability of the occurrence of a good solution within the partition over time of the dynamic environment. Secondly, we fix a number of individuals to be created by τ, $1 \leq \tau \leq \lambda$ and create these individuals randomly such that their statistical distribution regarding the partition matches that stored in the memory $\mathcal{M}_\mu(t)$. Therefore, we first determine their number for each cell by sorting the $\mu_{\ell_1 \ell_2 \ldots \ell_n}(t)$ according to their magnitude and producing the number $\lceil \mu_{\ell_1 \ell_2 \ldots \ell_n}(t) \cdot \tau \rceil$ of new individuals for high values of μ and the number $\lfloor \mu_{\ell_1 \ell_2 \ldots \ell_n}(t) \cdot \tau \rfloor$ for low values, respectively. The rounding needs to ensure that $\sum \lceil \mu_{\ell_1 \ell_2 \ldots \ell_n}(t) \cdot \tau \rceil + \sum \lfloor \mu_{\ell_1 \ell_2 \ldots \ell_n}(t) \cdot \tau \rfloor = \tau$. Then, we fix the positions of the new individuals by taking realizations of a random variable uniformly distributed within each partition cell $M_{\ell_1 \ell_2 \ldots \ell_n}$. That means the τ individuals are distributed such that the number within each cell approximates the expected value for the occurrence of good solutions, while the exact position within partition cells is random. These individuals are inserted in the population $P(t)$ after mutation has been carried out. This abstract retrieval process can create an arbitrary number of individuals from the abstract memory. In the implementation considered here we upper bound this creation by the number of individuals in the population. As the abstract storage can be regarded as encoding and compressing of information about good solutions in the search space, abstract retrieval becomes decoding and expansion.

An advantage of such a memory scheme is that it leads to a reduction of the information content, which is typical for abstraction. Naturally, this depends on the coarseness of the partitioning, which will be an important quantity to study. This also means that the number of individuals that take part in the memorizing and the number of individuals that come out of the memory and are inserted in the population are completely independent of each other. So, in contrast to explicit schemes, memory space and memory updating are not topics that need to be addressed. Another attribute of the proposed memory scheme is that in the memory matrix not the good solutions are stored but the event of occurrence of the solution at a specific location in the search space.

5 Experimental Results

We now report numerical experiments with an EA that uses tournament selection of tournament size 2, fitness–related intermediate sexual recombination, a mutation operator with base mutation rate 0.1 and the proposed abstraction memory scheme. The performance of the algorithms is measured by the Mean Fitness Error (MFE), defined as:

$$MFE = \frac{1}{R} \sum_{r=1}^{R} \left[\frac{1}{T} \sum_{t=1}^{T} \left(f\left(x_s, \lfloor \gamma^{-1} t \rfloor\right) - \max_{x_j(t) \in P(t)} f\left(x_j(t), \lfloor \gamma^{-1} t \rfloor\right) \right) \right], \quad (3)$$

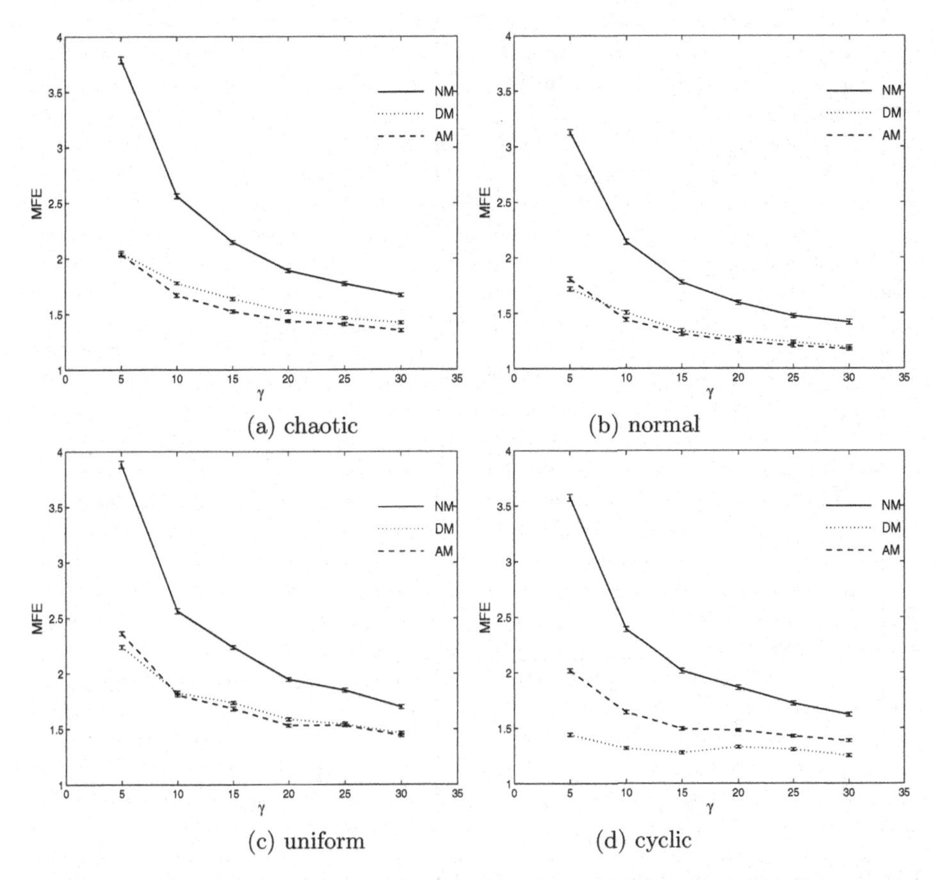

Fig. 1. Performance of the EA measured by the MFE over change frequency γ for different types of dynamics and no memory but hypermutation (NM), direct memory scheme (DM) and abstraction memory scheme (AM)

where $\max_{x_j(t)\in P(t)} f\left(x_j(t), \lfloor\gamma^{-1}t\rfloor\right)$ is the fitness value of the best–in–generation individual $x_j(t) \in P(t)$ at generation t, $f\left(x_s, \lfloor\gamma^{-1}t\rfloor\right)$ is the maximum fitness value at generation t, T is the number of generations used in the run, and R is the number of consecutive runs. Note that $f\left(x_s, \lfloor\gamma^{-1}t\rfloor\right)$ and $\max_{x_j(t)\in P(t)} f\left(x_i(t), \lfloor\gamma^{-1}\rfloor t\right)$ change every γ generations according to Eq. (2). The parameters we use in all experiments are $R = 50$ and $T = 2000$. We consider the dynamic fitness function (1) with dimension $n = 2$ and the number of cones $N = 7$. We study four types of dynamics of the coordinates $c(k)$ of the cones; (i.) chaotic dynamics generated by the Hénon map, see [14,15] for details of the generation process, (ii.) random dynamics with $c(k)$ being realizations of a normally distributed random variable, (iii.) random dynamics with $c(k)$ being realizations of a uniformly distributed random variable, and (iv.) cyclic dynamics where $c(k)$ are consequently forming a circle. As dynamic severity is an important factor in dynamic optimization, for

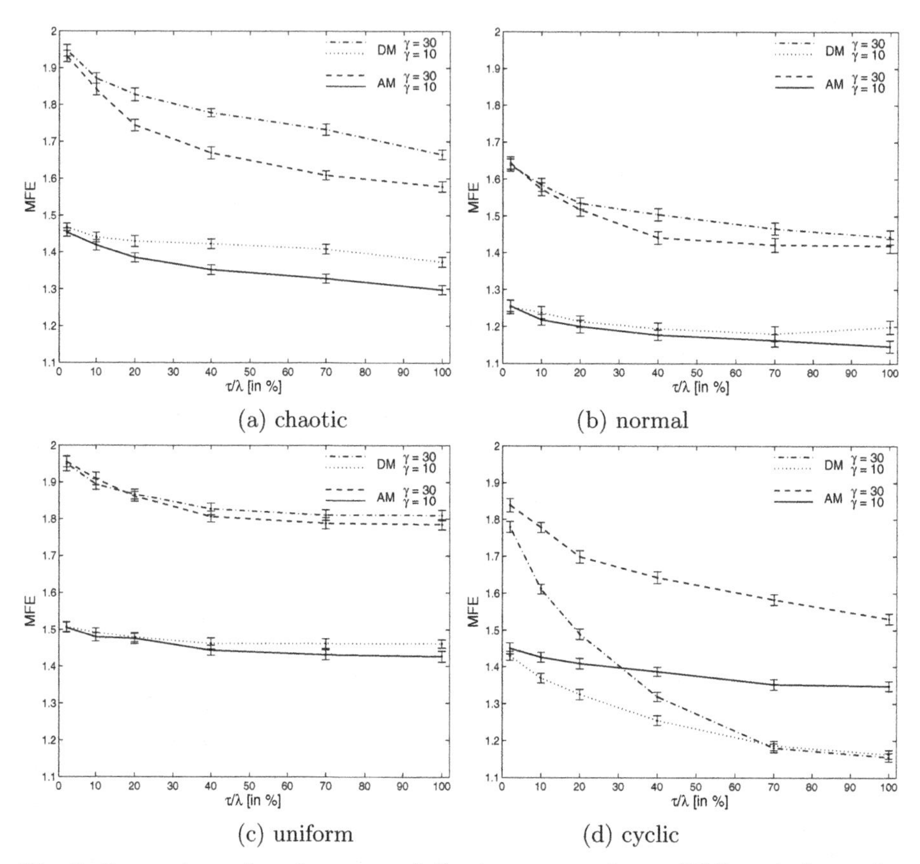

Fig. 2. Comparison of performance of direct memory scheme (DM) and abstraction memory scheme (AM) measured by the MFE over the percentage of individual inserted from the memory for different types of dynamics and $\gamma = 10$ and $\gamma = 30$

all considered dynamics, severity is normalized and hence has no differentiating influence. In a first set of experiments, the abstraction memory scheme (AM) is tested and compared with a direct memory scheme (DM), which stores the good solutions as themselves and inserts them again in a retrieval process, and with an EA with no memory (NM), that uses hypermutation with the hypermutation rate set to 30, see Fig. 1. Here, as well as in the other experiments, we fixed the upper and lower bounds of the search space at $x_{1\,min} = x_{2\,min} = -3$ and $x_{1\,max} = x_{2\,max} = 3$. The best three individuals of the population take part in the memorizing process for all three generations before a change in the environment occurs. Further, we set the grid size to $\epsilon = 0.1$. We used a fixed population size of $\lambda = 50$ and inserted $\tau = 20$ individuals in the retrieval process for one generation after the change. In Fig. 1 we give the MFE over the change frequency γ for all four types of dynamics considered. Also, the 95% confidence intervals are given. We observe that the memory schemes outperform the no memory scheme for all dynamics. This is particularly noticeable for small change frequencies γ

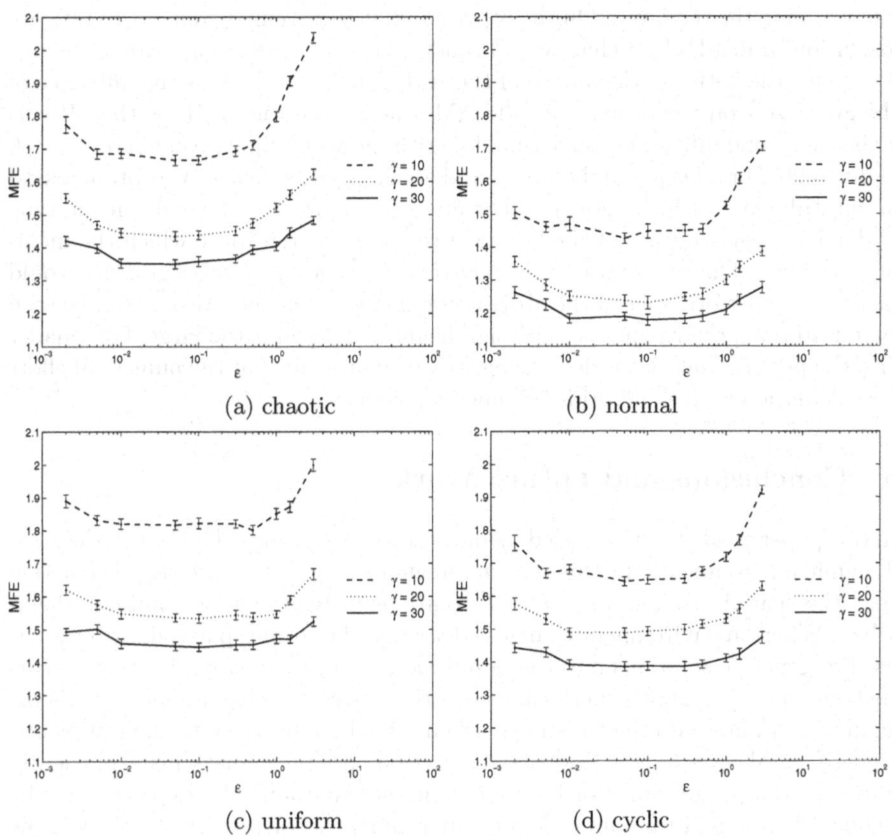

Fig. 3. Comparison of performance of abstraction memory scheme (AM) measured by the MFE for different grid size ϵ and different types of dynamics and $\gamma = 10$, $\gamma = 20$ and $\gamma = 30$

and means that by memory the limit of γ for which the algorithm still performs reasonably can be considerably lowered. Also, it can be seen that the AM gives better results than the DM for irregular dynamics, that is, chaotic and random. For chaotic dynamics, this is even significant within the given bounds. For regular, cyclic dynamics we find the contrary, with DM being better than AM. In a second set of experiments, we examine the effect of different amount τ of individuals inserted from the abstract memory, see Fig. 2, where the MFE is given for the four types of dynamics and the change frequencies $\gamma = 10$ and $\gamma = 30$, over the percentage of individuals τ retrieved and inserted in the population of size $\lambda = 50$, that is $\frac{\tau}{\lambda}$ in %. It can be seen that an increase in the number of inserted individuals leads to a better performance, but also that a certain saturation sets in, particularly for random dynamics with normal and uniform distribution. Further, it can be observed that the AM is a better option than the DM for chaotic and random, but not for cyclic dynamics. Here, we find that a large number of individuals inserted results in an even better performance, which can be attributed to the fact that for a large insertion the best solution is most likely among the retrieved. On the other hand, it is extremely unlikely to retrieve a solution from a DM for chaotic and random dynamics which appears to be one reason for the better performance of the AM. Finally, we look at the influence of the grid size ϵ on performance of the AM scheme, see Fig. 3. Here, the MFE is given over ϵ and different γ on a semi–logarithmic scale while we again set $\lambda = 50$ and $\tau = 20$. For all types of dynamics and all change frequencies we obtain a kind of bathtub curve, which indicates that an optimal grid size depends on the type of dynamics and the size of the bounded region in search space which the memory considers. This gives raise to the question whether an adaptive grid size would increase the performance of the abstraction memory scheme. Also, it can be seen that a drop in performance is more significant if the grid is too large. For smaller grid the performance is not decreasing very dramatically, but the numerical effort for calculation with small grids becomes considerable.

6 Conclusions and Future Work

In this paper an abstraction based memory scheme is proposed for EAs to address dynamic environments. In this scheme, memory is used to store the abstraction (i.e., the spatial distribution) of good solutions instead of good solutions themselves. When an environment change is detected, the stored spatial distribution is used to generate solutions into the population. A series of experiments were carried out to investigate the performance of the abstraction based memory scheme against a traditional direct memory scheme for EAs in dynamic environments. Experimental results show that the abstraction based memory scheme efficiently improves the performance of EAs for dynamic environments, especially for irregular chaotic and random dynamic environments. For regular cyclic dynamic environments, a traditional direct memory scheme seems a better choice than the abstraction based memory scheme. As there is a strong link between the proposed scheme and learning, a high–diversity scheme should be used besides the memory, in particular in the beginning of learning, that is, in the filling of the memory matrix. In the experiments, the learning curves should be studied. Also, it could be tested if after the learning, that is, the matrix filling, has reached a certain degree of maturity, memory should replace high–diversity to obtain the best results. These topics fall in our future research.

Acknowledgement

The work by Shengxiang Yang was supported by the Engineering and Physical Sciences Research Council (EPSRC) of UK under Grant No. EP/E060722/1.

References

1. Bäck, T.: Evolutionary Algorithms in Theory and Practice: Evolution Strategies, Evolutionary Programming, Genetic Algorithms. Oxford University Press, NY (1996)
2. Bendtsen, C.N., Krink, T.: Dynamic memory model for non–stationary optimization. In: Proc. of the 2002 IEEE Congress on Evol. Comput., pp. 145–150 (2002)

3. Bosman, P.A.N.: Learning and anticipation in online dynamic optimization. In: Yang, S., Ong, Y.-S., Jin, Y. (eds.) Evolutionary Computation in Dynamic and Uncertain Environments, pp. 129–152 (2007)
4. Branke, J.: Memory enhanced evolutionary algorithms for changing optimization problems. In: Proc. of the 1999 Congr. on Evol. Comput., pp. 1875–1882 (1999)
5. Branke, J., Kauß, T., Schmidt, C., Schmeck, H.: A multi–population approach to dynamic optimization problems. In: Proc. of the 4th Int. Conf. on Adaptive Computing in Design and Manufacturing, pp. 299–308 (2000)
6. Branke, J.: Evolutionary Optimization in Dynamic Environments. Kluwer, Dordrecht (2002)
7. Cobb, H.G., Grefenstette, J.J.: Genetic algorithms for tracking changing environments. In: Proc. of the 5th Int. Conf. on Genetic Algorithms, pp. 523–530 (1993)
8. Fitch, R., Hengst, B., Suc, D., Calbert, G., Scholz, J.: Structural abstraction experiments in reinforcement learning. In: Zhang, S., Jarvis, R. (eds.) AI 2005. LNCS (LNAI), vol. 3809, pp. 164–175. Springer, Heidelberg (2005)
9. Jin, Y., Branke, J.: Evolutionary optimization in uncertain environments – a survey. IEEE Trans. on Evol. Comput. 9, 303–317 (2005)
10. Lewis, E.H.J., Ritchie, G.: A comparison of dominance mechanisms and simple mutation on non–stationary problems. In: Parallel Problem Solving from Nature V, pp. 139–148 (1998)
11. Mori, N., Kita, H., Nishikawa, Y.: Adaptation to changing environments by means of the memory based thermodynamical genetic algorithm. In: Proc. of the 7th Int. Conf. on Genetic Algorithms, pp. 299–306 (1997)
12. Ng, K.P., Wong, K.C.: A new diploid scheme and dominance change mechanism for non–stationary function optimisation. In: Proc. of the 6th Int. Conf. on Genetic Algorithms, pp. 159–166 (1995)
13. Ramsey, C.L., Greffenstette, J.J.: Case–based initialization of genetic algorithms. In: Proc. of the 5th Int. Conf. on Genetic Algorithms, pp. 84–91 (1993)
14. Richter, H.: Behavior of evolutionary algorithms in chaotically changing fitness landscapes. In: Parallel Problem Solving from Nature VIII, pp. 111–120 (2004)
15. Richter, H.: A study of dynamic severity in chaotic fitness landscapes. In: Proc. of the 2005 IEEE Congress on Evolut. Comput., vol. 3, pp. 2824–2831 (2005)
16. Tinos, R., Yang, S.: A self–organizing random immigrants genetic algorithm for dynamic optimization problems. Genetic Programming and Evolvable Machines 286, 255–286 (2007)
17. Trojanowski, T., Michalewicz, Z.: Searching for optima in non–stationary environments. In: Proc. of the 1999 Congress on Evol. Comput., pp. 1843–1850 (1999)
18. Uyar, A.Ş., Harmanci, A.E.: A new population based adaptive dominance change mechanism for diploid genetic algorithms in dynamic environments. Soft Computing 9, 803–815 (2005)
19. Yang, S.: Population–based incremental learning with memory scheme for changing environments. In: Proc. of the 2005 Genetic and Evol. Comput. Conference, vol. 1, pp. 711–718 (2005)
20. Yang, S.: Associative memory scheme for genetic algorithms in dynamic environments. In: Applications of Evolutionary Computing: EvoWorkshops 2006, pp. 788–799 (2006)

A Memory Enhanced Evolutionary Algorithm for Dynamic Scheduling Problems

Gregory J. Barlow and Stephen F. Smith

Robotics Institute, Carnegie Mellon University, Pittsburgh, PA, USA
gjb@cmu.edu, sfs@cs.cmu.edu

Abstract. This paper describes a memory enhanced evolutionary algorithm (EA) approach to the dynamic job shop scheduling problem. Memory enhanced EAs have been widely investigated for other dynamic optimization problems with changing fitness landscapes, but only when associated with a fixed search space. In dynamic scheduling, the search space shifts as jobs are completed and new jobs arrive, so memory entries that describe specific points in the search space will become infeasible over time. The relative importances of jobs in the schedule also change over time, so previously good points become increasingly irrelevant. We describe a classifier-based memory for abstracting and storing information about schedules that can be used to build similar schedules at future times. We compared the memory enhanced EA with a standard EA and several common EA diversity techniques both with and without memory. The memory enhanced EA improved performance over the standard EA, while diversity techniques decreased performance.

1 Introduction

Dynamic optimization with evolutionary algorithms (EAs) lends itself to problems existing within a narrow range of problem dynamics, requiring a balance between solution fitness and search speed. If a problem changes too quickly, search may be too slow to keep up with the changing problem, and reactive techniques will outperform optimization approaches. If a problem changes very slowly, a balance between optimization and diversity is no longer necessary: one may search from scratch, treating each change as a completely new problem. Many problems do lie in this region where optimization must respond quickly to changes while still finding solutions of high fitness.

Prior work has shown many techniques for improving the performance of evolutionary algorithms on dynamic problems of this sort [1,2]. Approaches to these types of problems must consider two competing objectives: improving the fitness of solutions, and decreasing the search time necessary to reach good solutions. There are three broad categories of approaches for improving evolutionary algorithms on dynamic problems: keeping the population diverse to avoid population convergence [3], using a memory or multiple populations to maintain good solutions for future use [1,4], and anticipating changes to produce solutions that will

M. Giacobini et al. (Eds.): EvoWorkshops 2008, LNCS 4974, pp. 606–615, 2008.

be robust to change [5,6]. The first two categories address the issue of population convergence, which can hinder the ability of the EA to find better solutions after a change. Approaches in the third category produce solutions that not only have high fitness in the current environment, but are robust to environmental changes.

A variety of dynamic benchmark problems have been considered, including the moving peaks problem [4,7], the dynamic knapsack problem, dynamic bit-matching, dynamic scheduling, and others [1,2]. The commonality between most benchmark problems is that while the fitness landscape changes, the search space does not. For example, in the moving peaks problem, any point in the landscape—represented by a vector of real numbers—is always a feasible solution. One exception to this among common benchmark problems is dynamic scheduling, where the pending jobs change over time as jobs are completed and new jobs arrive. Given a feasible schedule at a particular time, the same schedule will not be feasible at some future time when the pending jobs are completely different. Previous work on evolutionary algorithms for dynamic scheduling problems have focused primarily on extending schedulers designed for static problems to dynamic problems [8,9], problems with machine breakdowns and redundant resources [10], improved genetic operators [11], heuristic reduction of the search space [12], and anticipation to create robust schedules [5,6]. While Louis and McDonnell [13] have shown that case-based memory is useful given similar static scheduling problems, there has been no work on memory for dynamic scheduling. Since the addition of memory has been successful in improving the performance of EAs on other dynamic problems, there is a strong case for using memory for dynamic scheduling problems as well.

In most dynamic optimization problems, the use of an explicit memory is relatively straightforward. Stored points in the landscape remain viable as solutions even as the landscape is changing, so a memory may store individuals directly from the population [4]. In dynamic scheduling problems, the jobs available for scheduling change over time, as do the attributes of any given job relative to the other pending jobs. If an individual in the population represents a prioritized list of pending jobs to be fed to a schedule builder, any memory that stores an individual directly will quickly become irrelevant. Some or all jobs in the memory may be complete, the jobs that remain may be more or less important than in the past, and the ordering of jobs that have arrived since the memory was created will not be addressed by the memory at all. For these types of problems that have both a dynamic fitness landscape and time-dependent constraints that shift the feasible region of the search space, a memory should provide some indirect representation of jobs in terms of their properties to allow mapping to similar solutions in future scheduling states.

In this paper, we present one such memory for dynamic scheduling, which we call classifier-based memory. Instead of storing a list of specific jobs, a memory entry stores a list of classifications which can be mapped to the pending jobs at any time. In the remainder of this paper, we will describe classifier-based memory for dynamic scheduling problems and compare it to both a standard EA and to other approaches from the literature.

2 Dynamic Job Shop Scheduling

The dynamic job shop scheduling problem used for our experiments is an extension of the standard job shop problem. In this problem, n jobs must be scheduled on m machines of mt machine types with $m > mt$. Processing a job on a particular machine is referred to as an operation. There are a limited number of distinct operations ot which we will refer to as operation types. Operation types are defined by processing times p_j and setup times s_{ij}. If operation j follows operation i on a given machine, a setup time s_{ij} is incurred. Setup times are sequence dependent—so s_{ij} is not necessarily equal to s_{ik} or s_{kj} $(i \neq j \neq k)$—and are not symmetric—so s_{ij} is not necessarily equal to s_{ji}. Each job is composed of k ordered operations; a job's total processing time is simply the sum of all setup times and processing times of a job's operations. Jobs have prescribed due-dates d_j, weights w_j, and release times r_j. The release of jobs is a non-stationary Poisson process, so the job inter-arrival times are exponentially distributed with mean λ. The mean inter-arrival time λ is determined by dividing the mean job processing time $\bar{P}$ by the number of machines m and a desired utilization rate U, i.e. $\lambda = \bar{P}/(mU)$. The mean job processing time is $\bar{P} = (\varsigma + \bar{p})\,\bar{k}$ where ς is an expected setup time, $\bar{p}$ is the mean operation processing time, and $\bar{k}$ is the mean number of operations per job. There are ρ jobs with release times of 0, and new jobs arrive non-deterministically over time. The scheduler is completely unaware of a job prior to the job's release time. Job routing is random and operations are uniformly distributed over machine types; if an operation requires a specific machine type, the operation can be processed on any machine of that type in the shop. The completion time of the last operation in the job is the job completion time c_j. We consider a single objective, weighted tardiness. The tardiness is the positive difference between the completion time and the due-date of a job, $T_j = \max(c_j - d_j, 0)$. The weighted tardiness is $WT_j = w_j T_j$. As an additional dynamic event, we model machine failure and repair. A machine fails at a specific time—the *breakdown time*—and remains unavailable for some length of time—the *repair time*. The frequency of machine failures is determined by the percentage downtime of a machine—the breakdown rate γ. Repair times are determined using the mean repair time ε. Breakdown times and repair times are not known a priori by the scheduler.

3 Evolutionary Algorithms for Dynamic Scheduling

At a given point in time, the scheduler is aware of the set of jobs that have been released but not yet completed. We will call the uncompleted operations of these jobs the set of pending operations $P = \{o_{j,k}\,|r_j \leq t, \neg complete(o_{j,k})\}$ where $o_{j,k}$ is operation k of job j. Operations have precedence constraints, and operation $o_{j,k}$ cannot start until operation $o_{j,k-1}$ is complete (operation $o_{j,-1}$ is complete

$\forall j$, since operation $o_{j,0}$ has no predecessors). When the immediate predecessor of an operation is complete, we say that the operation is schedulable. We define the set of schedulable operations as $S = \{o_{j,k} \,| o_{j,k} \in P, complete(o_{j,k-1})\}$.

Like most EA approaches to scheduling problems, we encode solutions as prioritized lists of operations. In static problems, these are permutations of all jobs; since this is a dynamic problem where jobs arrive over time, a solution is a prioritized list of only the pending operations at a particular time. Since the pending operations change over time, each individual in the population is updated at every time step of the simulator. When operations are completed, they are removed from every individual in the population, and when new jobs arrive, the operations in the job are randomly inserted into each individual in the population.

We use the well known Giffler and Thompson algorithm [14] to build active schedules from a prioritized list. First, from the set of pending operations P we create the set of schedulable operations S. From S, we find the operation o' with the earliest completion time t_c. We select the first operation from the prioritized list which is schedulable, can run on the same machine as o', and can start before t_c. We then update S and continue until all jobs are scheduled.

1. Build the set of schedulable operations S
2. (a) Find o' on machine M' with the earliest completion time t_c
   (b) Select the operation $o^*_{i,k}$ from S which occurs earliest in the prioritized list, can run on M', and can start before t_c
3. Add $o^*_{i,k}$ to the schedule and calculate its starting time
4. Remove $o^*_{i,k}$ from S and if $o^*_{i,k+1} \in E$, add $o^*_{i,k+1}$it to S
5. While S is not empty, go to step 2

The EA is generational with a population of 100 individuals. We use the PPX crossover operator [15] with probability 0.6, a swap mutation operator with probability 0.2, elitism of size 1, and linear rank-based selection. Rescheduling is event driven; whenever a new job arrives, a machine fails, or a machine is repaired, the EA runs until the best individual in the population remains the same for 10 generations.

4 Classifier-Based Memory for Scheduling Problems

The use of a population-based search algorithm allows us to carry over good solutions from the immediate past, but how can we use information from good solutions developed in the more distant past? Some or all of the jobs that were available in the past may be complete, there may be many new jobs, or a job that was a low priority may now be urgent. Unlike many dynamic optimization problems, this shifting search space means we cannot store individuals directly for later recall. Instead, a memory should allow us to map the qualities of good solutions in the past to solutions in the new environment. We present one such memory for dynamic scheduling, which we call classifier-based memory. Instead of storing prioritized lists of operations, we use an indirect representation, storing a prioritized list of classifications of operations. To access a memory entry at a future time, the pending jobs are classified and matched to the classifications in the memory entry, producing a prioritized list of operations.

A memory entry is created directly from a prioritized list of pending operations. First, operations are ranked according to several attributes and then quantiles are determined for each ranking in order to classify each operation with respect to each attribute. The number of attributes a and the number of subsets q determine the total number of possible classifications q^a. Rather than storing the prioritized list of operations, we store a prioritized list of classifications as a memory entry. To retrieve an individual from a memory entry, we map the pending operations to a prioritized list of classifications. We rank each operation according to the same attributes, then determine quantiles and classify each operation. Then, for each of these new classifications, we find the best match among the classifications in the memory entry. We assign each pending operation a sort key based on the position of its classification's best match within the prioritized list in memory. The sort key for classification x in memory entry Y is j such that $\min_{j=0,j<|Y|} \sum_{i=0}^{a} |x_i - Y(j)_i|$ where $Y(j)$ is classification j in list Y. If there is more than one best match, we use the average of the positions as the sort key. Then, we sort the pending operations by these sort keys to create a prioritized list of operations which can be used as an individual for the EA.

The basic mechanisms for interacting with the memory are the same as those for other explicit memories used for dynamic optimization with evolutionary algorithms. At every generation, we create an individual from each memory entry and insert the individuals into the population. Every φ generations and at the end of every rescheduling cycle, a replacement strategy chooses whether to insert the best individual in the population into memory. If the memory is full, the classification list of this best individual replaces a current memory entry using the *mindist2* replacement strategy [1]. To maintain diversity in the memory, we determine the two classification lists i and j that are closest together among the classification of the best individual in the population and all of the memory entries. We then choose the less fit list j as a candidate for replacement. The distance between two classification lists S and T is the sum of the differences between a classification's position in one list and the position of its best match in the other list. As before, if there is more than one best match, we use the mean of the positions. Since this is not symmetric, it is done for both lists. If S has length s and T has length t, then $d = \sum_{i=0}^{s} |i - \mathit{bestmatch}\,(S(i), T)| + \sum_{i=0}^{t} |i - \mathit{bestmatch}\,(T(i), S)|$. As long as the classification of the best individual in the population is not the candidate for replacement, we replace classification list j with the new classification list when $f_j \frac{d_{ij}}{d_{max}} \leq f_{best}$ where f_x is the fitness of the prioritized list produced by the classification list x, d_{ij} is the distance between classification lists i and j, and d_{max} is the maximum possible distance.

Figure 1 shows a simplified example. Suppose we have a memory with $q = 2$ and $a = 3$ and the following attributes: job due-date (dd), operation processing time (pt), and job weight (w). At time 400, we have a prioritized list of four operations that we'd like to store in the memory. With $q = 2$ and four operations,

At $t = 400$, store $[C, B, A, D]$	At $t = 10000$, get $[011, 000, 101, 110]$
$A = \{dd : 800, pt : 100, w : 7\} \rightarrow 101$	$W = \{dd : 10400, pt : 80, w : 1\} \rightarrow 110 \rightarrow (3)$
$B = \{dd : 450, pt : 110, w : 5\} \rightarrow 000$	$X = \{dd : 10100, pt : 70, w : 5\} \rightarrow 011 \rightarrow (0)$
$C = \{dd : 500, pt : 130, w : 9\} \rightarrow 011$	$Y = \{dd : 10500, pt : 50, w : 6\} \rightarrow 101 \rightarrow (2)$
$D = \{dd : 900, pt : 150, w : 3\} \rightarrow 110$	$Z = \{dd : 10070, pt : 60, w : 2\} \rightarrow 000 \rightarrow (1)$
$[011, 000, 101, 110] \rightarrow$ memory	$[X, Z, Y, W] \rightarrow$ population

Fig. 1. Classifier-based memory example

the lower two values for each attribute receive a classification of 0 and the higher two values a classification of 1. So job A has a due-date classification of 1, a process time classification of 0, and a weight classification of 1, for an overall classification of $class(A) \rightarrow 101$. At time 10000, we would like to use the memory entry to create a prioritized list from the four pending operations. These new operations are classified and given a score based on their best match within the memory entry, creating a new individual to be inserted into the population.

This classifier-based memory also allows new jobs to be ordered alongside older jobs that may have been available when the memory entry was created; the classifier-based memory does not store specific information about operations, only how a particular operation compares to other pending operations at a specific point in time. A memory entry may place a particular operation at different positions in the prioritized list as its due-date becomes more imminent or as the mix of pending operations changes the operation's relative importance.

In this paper, we use four attributes ($a = 4$): job due-date, job weight, operation processing time, and operation order within the job. We divide rankings into quartiles ($q = 4$) for a total of 256 possible classifications. Many other attributes exist that could easily be included, as this approach does not depend on a particular set of attributes.

5 Experiments

To examine the effects of classifier-based memory on schedule fitness and search time, we compare several common approaches. We use a standard evolutionary algorithm (SEA) as a baseline, since we don't know the optimal schedules for any of the problem instances, and we also consider the standard EA with classifier-based memory (SEAm). Prior results on benchmarks like the moving peaks problem suggest that memory-based approaches work better when combined with a diversity strategy [1]. Hence, we also consider a standard EA with 25 random immigrants [3] per generation (RI) and the same approach with classifier-based memory (RIm). Finally, we consider the memory/search approach of [1], also using the classifier-based memory. In memory/search, the population is divided into a memory subpopulation and a search subpopulation. The memory population can both store individuals to the memory and retrieve memory entries. The search population can only store items to the memory, and the population is re-initialized randomly every time the problem changes.

When creating problem instances, we select the utilization rate so that jobs arrive at approximately replacement rate, so the number of jobs available at time 0, ρ, is also the expected schedule size. We would like to be able to vary the due-date tightness to change the difficulty of the problem, so we use a due-date tightness parameter τ, the percentage of jobs we expect to meet their due-dates. The expected waiting time before job completion is the expected number of jobs in the schedule times the mean job completion time $\rho\bar{P}$. Due-dates are generated by $d_j = r_j + \bar{P} + \left[0, 2\rho\bar{P}\tau\right]$. The setup time severity is given by $\eta = \bar{s}/\bar{p}$ where $\bar{s}$ is the mean setup time and $\bar{p}$ is the mean operation processing time. The number of breakdowns per machine is uniformly distributed with the mean number of breakdowns per machine equal to $\frac{n\bar{P}}{m}\frac{\gamma}{\varepsilon}$. Breakdown times for each machine are uniformly distributed over $\left[0, \frac{n\bar{P}}{m}\right]$. Repair times are uniformly distributed over $\left[\frac{1}{2}\varepsilon, \frac{3}{2}\varepsilon\right]$.

For the experiments in this paper, we used the following settings to create problem instances. The job shop contains 2 machines each of $mt = 3$ machine types, for a total of $m = 6$ machines. There are 50 operation types, with mean operation processing time $\bar{p} = 100$ and processing times uniformly distributed over $[50, 150]$. The setup time severity is $\eta = 0.5$, so the mean setup time is $\bar{s} = 50$. The setup times are uniformly distributed over $[0, 2\bar{s}]$. The estimated setup time is $\varsigma = 35$. A problem instance consists of 500 jobs, each with $k = 3$ operations. There are $\rho = 25$ jobs with release times of 0. Job weights are uniformly distributed over $[1, 10]$. The utilization rate is $U = 0.7$, and the breakdown rate is $\gamma = 0.1$, for a total utilization of 0.8. The mean repair time is $\varepsilon = 10\bar{p} = 1000$. To control the problem difficulty, we varied the due-date tightness of the jobs. As the due-date tightness changes, the types of situations the scheduler faces also change. We tested with due-date tightnesses $\tau \in \{0.5, 0.8, 1.1\}$, from very tight due-dates where many jobs will be late, to loose due-dates where we expect most jobs to be on time. For each value of τ, we created 10 problem instances, for a total of 30 problem instances.

Rather than rebuild the memory from scratch on every problem instance during our experiments, we pre-built several seed memories using SEAm over a larger number of jobs, varying the due-date tightnesses of the jobs. Though we test over a limited number of jobs, if actually implemented in a scheduling system, the EA would work over a long time horizon, and so we are more interested in the steady state performance of the EA. By pre-building the memory, we better simulate this state of the algorithm. The memory may still change with the same replacement strategy, but after seeing a large number of jobs, the stability of the memory is much higher than if the memory was built from scratch for every problem instance. For each run of an EA with memory, one of the pre-built memories was chosen at random as a seed memory. Updating of the memory occurred as normal: memory replacement took place every $\varphi = 10$ generations and at the end of each rescheduling cycle.

We performed simulation runs for each EA variant on each of the 30 problem instances. Since this is a dynamic problem, we are interested not just in fitness improvements but in improvements in the speed of search. As in [5,6,9], we attempt to measure only the steady state performance by discarding the first 100 and last 100 jobs. We use the summed weighted tardiness of the middle 300 jobs as the fitness. We measure search in a similar way, by only including optional search generations that occur while the middle 300 jobs are among the pending jobs in the system. At the end of every rescheduling event, the scheduler is required to search for 10 generations where the best individual does not improve. Any generations per rescheduling event aside from these 10 constitute the optional search. Also, the number of rescheduling events is made up both of new job arrivals and machine breakdowns. Since the scheduler performance determines how long this period lasts, the number of machine breakdowns during this period is not fixed, so neither is the total number of rescheduling events. We can compare search more fairly by comparing the number of optional generations per event.

6 Results

Table 1 shows the percentage of improvement in average fitness over the standard EA. SEAm performs slightly better than SEA with tight and medium due-dates. When the due-dates are loose, SEAm performs much better than SEA. When diversity measures are introduced, performance actually drops. With just random immigrants, fitness worsens for all τ, but especially for medium due-dates. With RIm, performance on loose due-dates actually improves over SEA, though not over SEAm. With memory/search, performance gains are very slight for tight and loose due-dates, but performance worsens for medium tightness. The improvement (or lack thereof) for each of the approaches is worst with $\tau = 0.8$, except for SEAm where there the improvement for medium due-dates is slightly better than that for tight due-dates.

Table 1. Fitness improvement over the standard EA

	$\tau = 0.5$	$\tau = 0.8$	$\tau = 1.1$
Standard EA with memory	0.9%	1.5%	15.7%
Random immigrants	-5.7%	-49.6%	-30.3%
Random immigrants with memory	-8.3%	-51.2%	13.1%
Memory/Search	0.2%	-18.0%	2.1%

Table 2 shows the percentage improvement in average optional generations per event over the standard EA. SEAm shows good search reduction for tight and loose due-dates, with very slight improvement for medium due-dates. RIm actually improves search speed over SEAm for medium due-dates, but if we consider how much worse fitness was in this case, this improvement is not really meaningful. Of all the approaches, memory/search is the only one that fails to improve search speed for any due-date tightness. Again, medium due-dates show the worst performance in three of the four approaches, with RI as the only exception.

Table 2. Search improvement over the standard EA

	$\tau = 0.5$	$\tau = 0.8$	$\tau = 1.1$
Standard EA with memory	10.0%	2.9%	22.8%
Random immigrants	9.4%	-5.3%	-13.5%
Random immigrants with memory	9.5%	8.5%	23.4%
Memory/Search	-18.5%	-35.6%	-16.8%

For both fitness and search, the addition of classifier-based memory improved performance over the standard EA. While large improvement in fitness was only evident for loose due-dates, search improved for most problem instances. We saw improvement using SEAm for all three values of τ, but we saw the least improvement for $\tau = 0.8$. Our belief is that of the three, medium due-dates present search landscapes that are larger and more difficult to search than those for the other due-date tightnesses.

While the combination of memory and diversity techniques has yielded good results for most dynamic benchmark problems, for this dynamic scheduling problem none of the diversity approaches performed well. Perhaps due to the shape of the search landscape, diversity techniques are simply disruptive, rather than helpful in finding areas of high fitness. Memory/search, which devotes half of its population to searching for new individual to include in the memory, is at a disadvantage in the steady state environment we are interested in, though this approach might still be useful for pre-building memories, where search time is not an issue.

7 Conclusion

This paper describes a memory enhanced evolutionary algorithm approach to the dynamic job shop scheduling problem. Memory enhanced evolutionary algorithms have been widely investigated for other dynamic optimization problems, but not for problems like dynamic scheduling where changes in the fitness landscape are accompanied by shifts in the search space. We describe a classifier-based memory that enables the mapping of information about jobs at one point in time to the creation of valid schedules at another point in time. We compared several EA variants, with and without memory, on problem instances of varied difficulty. Our results show that classifier-based memory can improve both schedule fitness and the speed of search over a standard evolutionary algorithm. Our results also show that diversity techniques, which have had success on other dynamic benchmark problems, show decreased fitness and search speed for the dynamic scheduling problem we investigated.

We did not consider anticipation of robust schedules, heuristic reduction of the search space, or other approaches from previous work for improving performance on dynamic scheduling problems, because these approaches are complementary to the use of memory. We have also made no attempt to finely tune the EA used by each approach, following the example of [1]. Given the lack of prior work on memory enhanced EAs for dynamic scheduling, these experiments were an attempt to determine the potential of classifier-based memory.

As this is a preliminary investigation of the use of memory for dynamic scheduling, there are many avenues for future work in dynamic scheduling with evolutionary algorithms. Comparing the performance of classifier-based memory using different attributes, a variety of quantile sizes, larger memories, or other changes in the memory structure would shed more light on the potential of classifier-based memories for dynamic scheduling. Also, other memory types could be constructed to include ways to retain information about setup times, periodic changes in the mix of operation types over time, or other types of information that this memory cannot easily capture. Applying other approaches from the literature, like self-organizing scouts [1], to dynamic scheduling problems might also be a good area for future work.

References

1. Branke, J.: Evolutionary Optimization in Dynamic Environments. Kluwer Academic Publishers, Dordrecht (2002)
2. Jin, Y., Branke, J.: Evolutionary optimization in uncertain environments—a survey. IEEE Transactions on Evolutionary Computation 9(3), 303–317 (2005)
3. Grefenstette, J.J.: Genetic algorithms for changing environments. In: Parallel Problem Solving from Nature, pp. 137–144 (1992)
4. Branke, J.: Memory enhanced evolutionary algorithms for changing optimization problems. In: Congress on Evolutionary Computation, pp. 1875–1882 (1999)
5. Branke, J., Mattfeld, D.C.: Anticipatory scheduling for dynamic job shop problems. In: AIPS Workshop on On-line Planning and Scheduling, pp. 3–10 (2002)
6. Branke, J., Mattfeld, D.C.: Anticipation and flexibility in dynamic scheduling. International Journal of Production Research 43(15), 3103–3129 (2005)
7. Morrison, R., DeJong, K.: A test problem generator for non-stationary environments. In: Congress on Evolutionary Computation, pp. 2047–2053 (1999)
8. Lin, S.C., Goodman, E.D., William, F., Punch, I.: A genetic algorithm approach to dynamic job shop scheduling problems. In: International Conference on Genetic Algorithms, pp. 481–488 (1997)
9. Bierwirth, C., Mattfeld, D.C.: Production scheduling and rescheduling with genetic algorithms. Evolutionary Computation 7(1), 1–17 (1999)
10. Chryssolouris, G., Subramaniam, V.: Dynamic scheduling of manufacturing job shops using genetic algorithms. Journal of Intelligent Manufacturing 12, 281–293 (2001)
11. Vazquez, M., Whitley, L.D.: A comparison of genetic algorithms for the dynamic job shop scheduling problem. In: Genetic and Evolutionary Computation Conference, pp. 1011–1018 (2000)
12. Mattfeld, D.C., Bierwirth, C.: An efficient genetic algorithm for job shop scheduling with tardiness objectives. European Journal of Operations Research 155, 616–630 (2004)
13. Louis, S.J., McDonnell, J.: Learning with case-injected genetic algorithms. IEEE Transactions on Evolutionary Computation 8(4), 316–328 (2004)
14. Giffler, B., Thompson, G.L.: Algorithms for solving production scheduling problems. Operations Research 8(4), 487–503 (1960)
15. Bierwirth, C., Mattfeld, D.C., Kopfer, H.: On permutation representations for scheduling problems. In: Parallel Problem Solving from Nature, pp. 310–318 (1996)

Compound Particle Swarm Optimization in Dynamic Environments

Lili Liu[1], Dingwei Wang[1], and Shengxiang Yang[2]

[1] School of Information Science and Engineering, Northeastern University
Shenyang 110004, P.R. China
liulili1202@gmail.com, dwwang@mail.neu.edu.cn
[2] Department of Computer Science, University of Leicester
University Road, Leicester LE1 7RH, United Kingdom
s.yang@mcs.le.ac.uk

Abstract. Adaptation to dynamic optimization problems is currently receiving a growing interest as one of the most important applications of evolutionary algorithms. In this paper, a compound particle swarm optimization (CPSO) is proposed as a new variant of particle swarm optimization to enhance its performance in dynamic environments. Within CPSO, compound particles are constructed as a novel type of particles in the search space and their motions are integrated into the swarm. A special reflection scheme is introduced in order to explore the search space more comprehensively. Furthermore, some information preserving and anti-convergence strategies are also developed to improve the performance of CPSO in a new environment. An experimental study shows the efficiency of CPSO in dynamic environments.

1 Introduction

In recent years, there has been an increasing concern on investigating evolutionary algorithms (EAs) for dynamic optimization problems (DOPs) due to the relevance to real world applications, where many problems may involve stochastic changes over time. For DOPs, the goal of EAs is no longer to find a satisfactory solution, but to trace the moving optimum in the search space. This poses a great challenge to traditional EAs. To address this challenge, several approaches have been developed into EAs to improve their performance in dynamic environments, see [5,12,19,20] for examples.

Recently, particle swarm optimization (PSO), as a class of EAs, has been applied to address DOPs with promising results [10,15,17]. In this paper, one behavior of particle swarms from the domain of physics is integrated into PSO and a compound particle swarm optimization (CPSO) is proposed to address dynamic environments. Within CPSO, a number of "compound particles" are constructed as a new type of particles that have a geometric structure similar to that described in [7]. But, instead of using geometric principles, a specialized reflection scheme is introduced in CPSO in order to explore the search space more comprehensively in a new environment. Furthermore, in order to improve

M. Giacobini et al. (Eds.): EvoWorkshops 2008, LNCS 4974, pp. 616–625, 2008.

the performance of CPSO in a new environment, some information preserving and anti-convergence strategies are also developed to exploit various valuable information and avoid collision of particles respectively. An experimental study is carried out to validate the efficiency of CPSO in dynamic environments.

The rest of this paper is organized as follows. In the next section, we briefly review the usage of PSO for DOPs. Sec. 3 describes the CPSO proposed in this paper in details. The experimental results and analysis are given in Sec. 4. Finally, Sec. 5 concludes this paper with discussions on future work.

2 Particle Swarm Optimization in Dynamic Environments

Particle swarm optimization is a population based optimization technique with the inspiration from the social behavior of a swarm of birds (particles) that "fly" through a solution space [1,13]. Each particle accomplishes its own updating based on its current velocity and position, the best position seen so far by itself and by the swarm. The behavior of a particle i, can be described as follows:

$$\boldsymbol{v}_{ij}(t) = \omega \boldsymbol{v}_{ij}(t-1) + c_1\xi(\boldsymbol{p}_{ij}(t) - \boldsymbol{x}_{ij}(t)) + c_2\eta(\boldsymbol{p}_{gj}(t) - \boldsymbol{x}_{ij}(t)) \quad (1)$$

$$\boldsymbol{x}_{ij}(t+1) = \boldsymbol{x}_{ij}(t) + \boldsymbol{v}_{ij}(t), \quad (2)$$

where $\boldsymbol{v}_{ij}(t)$ and $\boldsymbol{x}_{ij}(t)$ represent the current velocity and position of particle i in the j-th dimension at time t, and $\boldsymbol{p}_{ij}(t)$ and $\boldsymbol{p}_{gj}(t)$ represent the position of the best solution discovered so far by particle i and by all particles in the j-th dimension. The inertia weight ω controls the degree that a particle's previous velocity will be kept. Parameters c_1 and c_2 are individual and social learning factors, and ξ and η are random numbers in the range of $[0.0, 1.0]$.

PSO has been widely used for stationary problems [13,14,18]. In recent years, PSO has obtained an increasing concern to solve DOPs [3]. For DOPs, an efficient PSO must show continuous adaptation to track the variation of the optimal solution. For this aim, the basic PSO needs to be modified to improve the performance due to the following reasons. First, with the increasing of iterations, particles will congregate to a local or global optimum in the search space. When the optimum changes, the slackened velocities and convergent particles will result in a low exploration ability in the changing environment. Second, each particle takes into account the information from the best particles in the present swarm, while neglecting some valuable information contained in the inferior particles. This mono-directional mechanism restricts the ability of PSO to efficiently search for a new optimum.

Therefore, there are some key considerations to improve the adaptation of PSO in dynamic environments. They are shown as follows.

- Some weak particles should fly toward a better direction (i.e., the direction that intends to have increasing fitness) as quickly as possible, in order to adapt themselves to the changed environment and explore the search space comprehensively.

- Particles should also exploit the useful information from some other particles besides the best particle, in order to accelerate the optimization process in a new environment.

In recent years, several variations of the traditional PSO have been developed in the literature for promising performance in dynamic environments. Carlisle and Dozier [6] carried out a thorough investigation of PSOs on a large number of dynamic test problems and improved the performance of PSOs in both static and dynamic environments. An adaptive PSO that tracks various changes autonomously in a non-stationary environment was proposed in [8,10]. Blackwell and Bentley [2] introduced a charged PSO for DOPs, which was then extended [4] by constructing interacting multi-swarms and using the charged sub-swarm to maintain population diversity for tracking multiple peaks in a dynamic environment. Parrott and Li [17] investigated a PSO model using a speciation scheme and employed this method to track multiple peaks simultaneously, the experiments manifested that the technique was able to track the changing trajectory.

3 Compound Particle Swarm Optimization

The concept of "compound particle" is derived from a branch of physics [9,16]. It refers to a particular sort of particles that are composed by at least two particles through the chemical bond. Such particle possesses not only the qualities of each "member particle", but also some composite characters [16]. The characteristics of CPSO lie mainly in the following three aspects: 1) having the basic framework of canonical PSO; 2) incorporating the construction and update of compound particles; 3) employing a specialized reflection strategy and an integral-moving scheme for compound particles. In the following sections, the construction and operation for compound particles are described in details.

3.1 Initialization

Initially, a number of compound particles are created. Each compound particle is created as a simple geometrical structure that consists of three particles: one particle is selected from the initial swarm randomly and the other two are randomly generated to form a triangle with the length of the interconnecting edges being L. The three particles in a compound particle are denoted as "member particles". The particles in the swarm that do not belong to any compound particle are denoted as "independent particles".

3.2 Self-adjustment

Each compound particle will adjust its internal structure in order to track the trace of the changing optimum. The essential steps involve constructing a new compound particle to explore good solutions in a new environment, and identifying a "representative particle" for each compound particle to participate in the canonical PSO. The procedures are exhibited as follows.

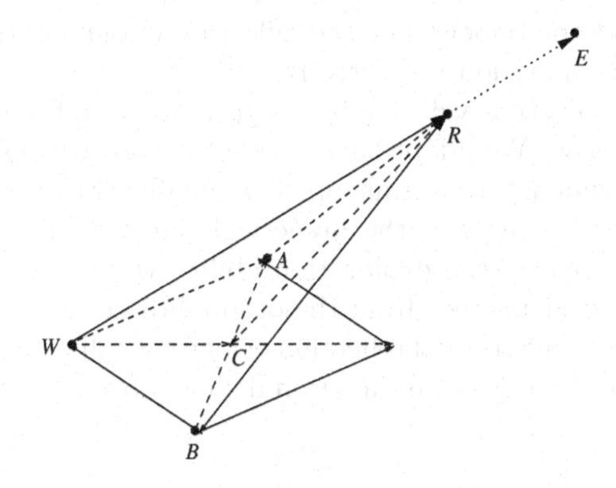

Fig. 1. Constructing a new compound particle

Velocity-anisotropic reflection scheme. The construction of a new compound particle is illustrated in Fig. 1. The position of the worst particle in a compound particle is denoted as W and the position of the central point between the other two member particles is denoted as C. Then, the compound particle is reflected in accordance with the point W to a point R. The new compound particle consists of points A, R and B. If the solution at point R is better than that at point W, an expansion is further made in that direction to point E and the compound particle is updated to consist of points A, E, and B. The reflection point R and the extension point E are calculated as follows:

$$\boldsymbol{WR} = \boldsymbol{WC} + \boldsymbol{\gamma} \times \boldsymbol{WC} \tag{3}$$

$$\boldsymbol{WE} = \eta \times \boldsymbol{WR}, \quad \text{if } f(R) > f(W), \tag{4}$$

where $\boldsymbol{\gamma}$ is the inequality-velocity reflection vector and η is the extension factor.

Since the structure of a compound particle is a triangle in 2-dimensional space, in order to ensure that particles can explore the search space in N-dimension, a velocity-anisotropic reflection (abbreviated as *VAR*) scheme is introduced in the relevant vector.

Definition: An N-dimension vector $\boldsymbol{\gamma} = (\gamma_1, \gamma_2, \cdots, \gamma_N)$ is a VAR vector, if it complies with:

$$0 < |\gamma_i - \gamma_j| \leq d, \quad i, j \in (1, 2, ..., N), \tag{5}$$

where d is the maximum difference between reflection velocities for each dimension, which determines the degree of departure from the initial direction $\boldsymbol{WC}$. The larger the value of d, the larger space of compound particles could explore.

It is clear that with the VAR vector shown in Eq. (5), $\boldsymbol{WR}$ can not be linearly represented by $\boldsymbol{WA}$ and $\boldsymbol{WB}$ in any case [11], that is, the VAR vector can drive compound particles to explore in the N-dimension search space. In this paper, each constituent in the VAR vector $\boldsymbol{\gamma}$ is generated as follows:

$$\gamma_{ij} = rand(0, e^{-|v_{ij}/v_{max}|}), \quad j \in (1, 2, \cdots, N), \tag{6}$$

where v_{ij} and γ_{ij} are the velocity and the reflection velocity of the i-th compound particle in the j-th dimension respectively.

In Eq. (6), the reflection velocity is designed to be relevant to the velocity of the member particle. We adopt such a rule for two reasons. First, when the velocities have a tendency to shrink up to a small value, especially when the population becomes convergent, the numerical range of the reflection velocity tends to be larger. Hence, the exploration ability will be enhanced adaptively because the degree of departure from the original direction is enlarged. second, the difference of each dimensional reflection velocity d will be restricted to a moderate degree in case the reflection direction deviates from the "better" direction significantly.

Identifying the representative particle. In order to maintain diversity as well as guarantee the searching precision in compound particles, two factors are integrated in identifying the representative particle: one is the fitness and the other is the total distance from one member particle to the other two particles. For each compound particle, its representative particle is identified according to the following probability:

$$P_{ci} = (1 - \beta)P_{ci}^{f} + \beta P_{ci}^{d}, \tag{7}$$

where P_{ci}^{f} and P_{ci}^{d} is the proportion of the fitness and the proportion of the total distance of the i-th member particle in the c-th compound particle respectively, β is the identification factor, and P_{ci} is the probability that the i-th member particle of the c-th compound particle becomes the representative particle.

3.3 Integral Movement

After the representative particles have updated their positions, an information preserving scheme is employed. In Blackwell and Branke's model [4], the updating of member particles all rely on their sub-swarm attractors and particle attractors. In this work, the velocity of a representative particle is conveyed to the other two member particles in the compound particle. That is, we first calculate the distance that a representative particle has moved and then move the other two member particles in the corresponding compound particle by the same distance. The reason for introducing this scheme lies in that the tendency of all member particles moving towards the local optimum is reduced and hence the convergence of the population is avoided and that, in the meantime, valuable information is preserved for the next iteration.

Fig. 2 is the pseudo-code of the CPSO we proposed.

4 Experimental Study

To test the validity of the proposed algorithm, Branke's moving peaks function [5] was used as a benchmark dynamic problem. The fitness at a point of the

```
begin
  Parameterize
  t := 0
  Initialize population (swarm of particles) P(0) randomly
  Initialize compound particles C(0) based on the value of L
  repeat
    for each compound particle do
      Perform self-adjustment according to Eqs. (3) and (4)
      Identify the representative particle according to Eq. (7)
    end for
    Create population P(t) containing independent and representative particles
    for each particle i in P(t) do
      Update v_i and x_i by Eqs. (1) and (2)
      if f(x_i) < f(p_i) then p_i := x_i
    end for
    p_g := arg min_{p_i} {f(p_i) | i = 1, ..., swarm_size}
    for each compound particle do
      Calculate the distance the representative particle has moved after update
      Move the other two member particles by the same distance
    end for
    t := t + 1
  until the termination condition is met
end
```

Fig. 2. Pseudo-code of the Compound Particle Swarm Optimization (CPSO)

fitness landscape is assigned the maximum height of all optima at that point as below.

$$F(\boldsymbol{x}, t) = \max_{i=1,\dots,M} \frac{H_i(t)}{1 + W_i(t) \sum_{j=1}^{N} (x_j(t) - X_{ij}(t))^2}, \tag{8}$$

In the experiments, we set $N = 5$, $M = 10$, and $\boldsymbol{x} \in [X_{max}, X_{min}]^5 = [0, 100]^5$. The height and width of each peak were randomly generated with a uniform distribution in $[30, 70]$ and $[1, 12]$ respectively. The locations of peaks are changed as follows:

$$\boldsymbol{v}_i(t) = \frac{s}{|\boldsymbol{r} + \boldsymbol{v}_i(t-1)|}((1-\lambda)\boldsymbol{r} + \lambda \boldsymbol{v}_i(t-1)) \tag{9}$$

$$\boldsymbol{X_i}(t) = \boldsymbol{X_i}(t-1) + \boldsymbol{v}_i(t), \tag{10}$$

where the vector $\boldsymbol{v}_i(t)$ is a linear combination of a random vector $\boldsymbol{r} \in [0.0, 1.0]^N$ and the previous vector $\boldsymbol{v}_i(t-1)$ and is normalized by the length factor s (s controls the severity of changes), and λ is the correlation parameter. In our experiment, λ is set 0, which indicates that the movement of peaks is uncorrelated.

The performance of CPSO is compared with the simple PSO model (PSO) and the speciation based PSO (SPSO) proposed by Parrott and Li [17]. In order

Table 1. Dynamic environments

τ	Scenario		
10	1	2	3
50	4	5	6
100	7	8	9
$s \rightarrow$	0.05	0.5	1.0

Table 2. The t-test results of comparing algorithms on dynamic problems

t-test results	Scenario 1	2	3	4	5	6	7	8	9
SPSO – PSO	+	+	~	+	+	+	+	+	+
R-CPSO – PSO	+	+	~	+	+	+	+	+	+
R-CPSO – SPSO	~	~	~	–	+	+	~	–	–
CPSO – PSO	+	+	+	+	+	+	+	+	+
CPSO – SPSO	+	+	+	+	+	+	~	~	+
CPSO – R-CPSO	+	+	+	+	+	+	~	+	+

to test the effect of the VAR scheme, a corresponding algorithm called R-CPSO with $\gamma_{ij} = rand(0,1)$ is involved in the experiments, which implies a random exploration within a constrained space. For all PSO models, the learning factors $c_1 = c_2 = 2.0$, the inertia weight $\omega = \omega_{max} - (\omega_{max} - \omega_{min}) * iter/iter_{max}$ ($\omega_{max} = 0.7$, $\omega_{min} = 0.5$, $iter_{max}$ and $iter$ are the max number of iterations and the current iteration respectively). Parameters in SPSO are set as [17]: $P_{max} = 20$ and $r = 20$. Parameters in CPSO are set as follows: the length of edges $L = 0.01 \times (X_{max} - X_{min}) = 1$, the extension factor $\eta = 1.25$, the identification factor $\beta = 0.5$, implying that the ingredients of fitness and distance in Eq. (7) have an equal strength. The total number of particles is set to 50 for PSO and SPSO and is set to 20 for CPSO and R-CPSO, where half of particles in the initial swarm are selected to construct compound particles, to ensure the fairness of comparisons between algorithms.

For environmental dynamics parameters, we set $s \in \{0.05, 0.5, 1.0\}$ and $\tau \in \{10, 50, 100\}$, where τ determines the speed of change (i.e., the environment changes every τ generations). This gives 9 different scenarios, i.e., 9 pairs of (s, τ). For each scenario, 20 random instances were created and the results were averaged over the 20 runs. For each run the error concerned with the best-of-period fitness was record every iteration [15]. The mean error of an algorithm is calculated as follows:

$$E_{mean} = \frac{1}{G}\sum_{i=1}^{G}(\frac{1}{N}\sum_{j=1}^{N} e_{ij}), \tag{11}$$

where $G = 10$ is the total number of changes for a run, this way, the number of iterations $iter_{max} = 10\tau$, $N = 20$ is the total number of runs, e_{ij} is the error of the last iteration at the i-th change in the j-th run.

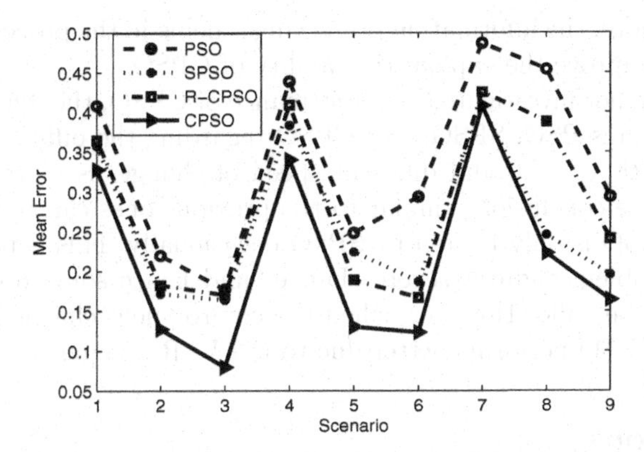

Fig. 3. The mean error of PSOs on different dynamic problems

The experimental results are plotted in Fig. 3 for different dynamic problems, which are indexed by the combination of the pairs (s, τ), as shown in Table 1. The statistical test results of comparing algorithms by one-tailed t-test with 38 degrees of freedom at a 0.05 level of significance are given in Table 2, and the t-test result regarding *Alg.* 1 $-$ *Alg.* 2 is shown as "+" or "$\sim$" when *Alg.* 1 is significantly better than or statistically equivalent to *Alg.* 2 respectively.

From Fig. 3 and Table 2, it can be seen that both SPSO and CPSO significantly outperform PSO on the dynamic problems with different environmental dynamics, CPSO outperforms SPSO on most dynamic problems, and CPSO performs significantly better than R-CPSO. This result validates our expectation of CPSO. The better performance of CPSO is because compound particles in CPSO integrate valuable information effectively, and comparing with the random exploration scheme in a guideless way within R-PSO, the VAR scheme has an intensive exploration ability and helps the compound particles to search for more optimal solutions continuously rather than converging into a solution

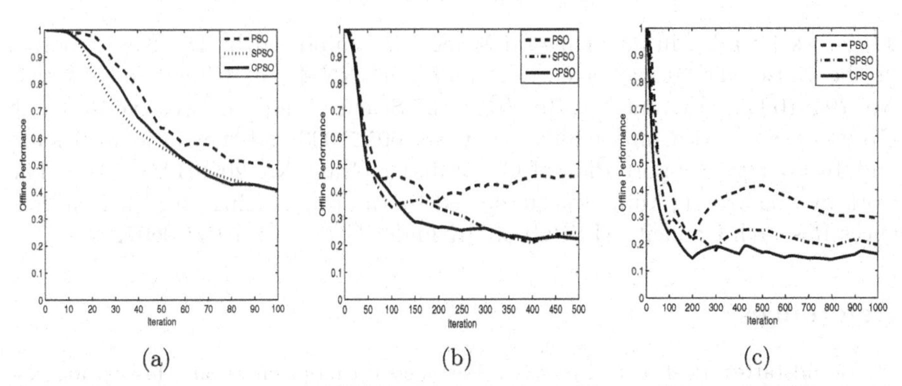

Fig. 4. Dynamic performance of PSOs on dynamic problems with $s = 1.0$: (a) $\tau = 10$, (b) $\tau = 50$, and (c) $\tau = 100$

ahead. Furthermore, the information preserving scheme in the process of Integral Movement can improve the exploitation ability of CPSO.

In order to further investigate the performance of CPSO, the dynamic performance of algorithms PSO, SPSO, and CPSO regarding the offline performance defined in [5] with $s = 1.0$ and different speed of changes is plotted in Fig. 4. From Fig. 4, some conclusions similar to the previous ones can be drawn, and, furthermore, CPSO is only beaten by SPSO occasionally. This is because when $s = 1.0$, the problem endures severe changes and hence some previous information may not be valid. However, when the environment changes slowly, e.g., when $\tau = 100$, CPSO performs better due to the VAR scheme.

5 Conclusions

This paper introduces a new kind of particles, called "compound particles", together with some specialized operating mechanisms into PSO for DOPs. The compound particles introduced can aggregate more valuable information with a simple configuration, which is an improvement over traditional PSO. A velocity-anisotropic reflection method is proposed to construct new compound particles, which can drive particles to search for a better solution especially in a new environment. Furthermore, the information preserving and anti-convergence strategies are also applied to the motion of compound particles. Experimental study over a benchmark dynamic problem shows that the proposed schemes efficiently improve the performance of PSO in dynamic environments.

For future work, it would be valuable to investigate the effect of different modifications to CPSO for DOPs. The VAR mechanism in CPSO are quite general and hence can be integrated to other optimization methods to improve their capability in dynamic environments. This is another interesting research for the future work.

Acknowledgments

The work by Lili Liu and Dingwei Wang was supported by the Key Program of the National Natural Science Foundation (NNSF) of China under Grant No. 70431003 and Grant No. 70671020, the Science Fund for Creative Research Group of NNSF of China under Grant No. 60521003 and the National Science and Technology Support Plan of China under Grant No. 2006BAH02A09. The work by Shengxiang Yang was supported by the Engineering and Physical Sciences Research Council (EPSRC) of UK under Grant No. EP/E060722/1.

References

1. Brandstätter, B., Baumgartner, U.: Particle swarm optimization: Mass spring system analogon. IEEE Trans. on Magnetics 38(2), 997–1000 (2002)
2. Blackwell, T.M., Bentley, P.J.: Dynamic search with charged swarms. In: Proc. of the 2002 Genetic and Evol. Comput. Conf., pp. 19–26 (2002)

3. Blackwell, T.M.: Swarms in dynamic environments. In: Cantú-Paz, E., Foster, J.A., Deb, K., Davis, L., Roy, R., O'Reilly, U.-M., Beyer, H.-G., Kendall, G., Wilson, S.W., Harman, M., Wegener, J., Dasgupta, D., Potter, M.A., Schultz, A., Dowsland, K.A., Jonoska, N., Miller, J., Standish, R.K. (eds.) GECCO 2003. LNCS, vol. 2723, pp. 1–12. Springer, Heidelberg (2003)
4. Blackwell, T.M., Branke, J.: Multi-swarm optimization in dynamic environments. In: Raidl, G.R., Cagnoni, S., Branke, J., Corne, D.W., Drechsler, R., Jin, Y., Johnson, C.G., Machado, P., Marchiori, E., Rothlauf, F., Smith, G.D., Squillero, G. (eds.) EvoWorkshops 2004. LNCS, vol. 3005, pp. 489–500. Springer, Heidelberg (2004)
5. Branke, J.: Evolutionary Optimization in Dynamic Environments. Kluwer Academic Publishers, Dordrecht (2002)
6. Carlisle, A., Dozier, G.: Adapting particle swarm optimization to dynamic environments. In: Proc. of the 2000 Int. Conf. on Artif. Intell, pp. 429–434 (2000)
7. Chio, C.D., Moraglio, A., Poli, R.: Geometric particle swarm optimisation on binary and real spaces: from theory to practice. In: Proc. of the 2007 GECCO conference companion on Genetic and evolutionary computation, pp. 2659–2666 (2007)
8. Eberhart, R.C., Shi, Y.: Tracking and optimizing dynamic systems with particle swarms. In: Proc. of the 2001 IEEE Congress on Evol. Comput, pp. 94–100 (2001)
9. Galasso, F.S.: Structure and Properties of Inorganic Solids. Pergamon Press, Oxford (1970)
10. Hu, X., Eberhart, R.C.: Adaptive particle swarm optimization: Detection and response to dynamic systems. In: Proc. of the 2002 IEEE Congress on Evol. Comput, pp. 1666–1670 (2002)
11. Jain, S.K.: Linear Algebra. Thomson, 4th edn., pp. 12–20 (1994)
12. Jin, Y., Branke, J.: Evolutionary optimization in uncertain environments: a survey. IEEE Trans. on Evol. Comput. 9(3), 303–317 (2005)
13. Kennedy, J., Eberhart, R.: Particle Swarm Optimization. In: Proc. of the 1995 IEEE Int. Conf. on Neural Networks, vol. 4, pp. 1942–1948 (1995)
14. Kennedy, J.: Stereotyping: Improving particle swarm performance with cluster analysis. In: Proc. of the 2000 IEEE Congress on Evol. Comput, pp. 1507–1512 (2000)
15. Li, X., Dam, K.H.: Comparing particle swarms for tracking extrema in dynamic environments. In: Proc. of the 2003 IEEE Congress on Evol. Comput, pp. 1772–1779 (2003)
16. Nakahata, S., Sogabe, K., Matsuura, T., Yamakawa, A.: One role of titanium compound particles in aluminium nitride sintered body. Journal of Materials Science 32(7), 1873–1876 (1997)
17. Parrott, D., Li, X.: A particle swarm model for tacking multiple peaks in a dynamic environment using speciation. In: Proc. of the 2004 Genetic and Evol. Comput. Conf., pp. 98–103 (2004)
18. Parsopoulos, K.E., Vrahatis, M.N.: On the computation of all global minimizers through particle swarm optimization. IEEE Trans. on Evol.Comput. 8(3), 211–224 (2004)
19. Yang, S.: Population-based incremental learning with memory scheme for changing environments. In: Proc. of the 2005 Genetic and Evol. Comput. Conference, vol. 1, pp. 711–718 (2005)
20. Yang, S., Yao, X.: Experimental study on population-based incremental learning algorithms for dynamic optimization problems. Soft Computing 9(11), 815–834 (2005)

An Evolutionary Algorithm for Adaptive Online Services in Dynamic Environment

Alfredo Milani[1,2], Clement Ho Cheung Leung[1], Marco Baioletti[2], and Silvia Suriani[2]

[1] Hong Kong Baptist University, Department of Computer Science , Hong Kong
milani@unipg.it
[2] Universitá degli Studi di Perugia, Dipartimento di Matematica e Informatica, Via Vanvitelli 1, I-06100 Perugia, Italy

Abstract. An evolutionary adaptive algorithm for solving a class of online service provider problems in a dynamical web environment is introduced. In the online service provider scenario, a system continuously generates digital products and service instances by assembling components (e.g. headlines of online newspapers, search engine query results, advertising lists) to fulfill the requirements of a market of anonymous customers. The evaluation of a service instance can only be known by the feedback obtained after delivering it to the customer over the internet or through telephone networks. In dynamic domains available components and customer/agents preferences are changing over the time. The proposed algorithm employs typical genetic operators in order to optimize the service delivered and to adapt it to the environment feedback and evolution. Differently from classical genetic algorithms the goal of such systems is to maximize the average fitness instead of determining the single best optimal service/product. Experimental results for different classes of services, online newspapers and search engines, confirm the adaptive behavior of the proposed technique.

1 Introduction

Mass services based on new information technologies represent a challenging domain for providing adaptive services to a population of anonymous customers[1]. Purely digital service/products are generated and delivered instantaneously on demand to thousands of users. For instance, the headlines of an online newspaper, web banners advertising special offers, voice menu systems proposing valuable offers to mobile phone customers, list of documents returned after a search engine query, are all example of products which are submitted to a mass of individual users, which in a short time decide to browse, to buy, to accept or to neglect the proposed product. It is worth noticing the dynamical nature of these purely digital products: the items to compose (e.g. news, available offers,indexed documents) and the customer preferences rapidly change over time. It is then required a quick adaptation of the products to the changing conditions of the market environment.

M. Giacobini et al. (Eds.): EvoWorkshops 2008, LNCS 4974, pp. 626–632, 2008.

User modeling approaches [2] are often hard to apply to this scenario for a number of reasons: short duration and anonymity of internet connection, privacy reasons, and customers changing preferences do not allow to build a significant user model. Moreover a model of the many services or products cannot be built since they are short lived and disappear very quickly. Some works have introduced "real world" issues into the genetic algorithm (GA) loop. For instance [8] proposes an interactive approach where the user provides the fitness function, [3,4] also interacts with the external world in order to optimize a behavior. Most approaches to adaptation based on GA adopt a classical machine learning "offline" approach [5,6,7,9,10]): in a first "training phase", they evolve the solutions in a "virtual training environment" eventually returning the best one. The offline method fails when the domain of services is dynamically evolving over time and the customer responses can vary as well.

The evolutionary schema we propose is both interactive and adaptive [11], it extends the GA framework with fitness from the external environment, exploiting the optimization [4] and adaptive features of GA operators.

2 The Online Service Provider Problem

Let C a domain of components, S the set of services which can be assembled with elements from C,R a compositional structure $R : C^m \rightarrow S$ where $s = R(c_1, ..., c_m)$ is a service assembled by instantiating the compositional structure R with m components $c_i \in C$, let $f : N \times S \rightarrow [0, 1]$ a *dynamic fitness* function, defined over nonnegative integer N and services, where $f(t, s)$ represents the customers satisfaction for service s at time t, let n the number of service requests per time instant and S_t the set of services delivered at time t where $|S_t| = m$. The Online Service Provider(OSP) problem consists in devising a production strategy which maximizes the overall fitness over a given interval of time $[0, t_{max}]$

$$f_{tot} = \sum_{i=0}^{t_{max}} \sum_{s \in S_i} f(i, s)$$

where the function f is unknown to the systems, i.e. the evaluation of service s at time t is made by interacting with the external agents to which the services/products are delivered. We assume that all the external agents give the same product the same evaluation at a given time. Different hypotheses about availability of components, and constraints on the compositional structure apply to different classes of online services. In the following we will assume that the compositional structure is fixed and the components are available in unlimited number of copies, as usually pure digital products are. Moreover the product evaluation is assumed to be component *additive*. The agents evaluate a service s at time t with a dynamic fitness equal to the sum of satisfaction degrees of the components c_i, $i = 1, \ldots, m$ of s.

3 An Algorithm for Online Service Providers

The algorithm we have designed for OSP problem is based on a standard evolutionary scheme driven by selection, crossover and mutation operators. The service provider assembles components and grows up a population of n services, moreover it alternates an evolution phase, in which the service provider generates the next population of n service instances by means of genetic operators, with an interactive fitness evaluation phase. Note that the system can evaluates a service only by the satisfaction degrees expressed by the external agents. In general the system cannot compute the satisfaction degree of a product which is not delivered, or the satisfaction degree of the single components.

3.1 Classes of Online Service Providers

The two classes of OSPs which have been analyzed are *search engine* and *online newspaper* . Despite of the apparent differences their structures are quite similar: they both verify the hypothesis of fixed compositional structure with a fixed size (newspaper have usual a fixed structure and search engines return a fixed number of links per page, they can be easily represented by vectors). In both cases the fitness model is mainly additive: the more relevant the news, the more relevant is the newspaper front page; the more relevant the objects are to query q, the more relevant is the search engine answer list.

Search Engine. Components domain: in the case of the search engine it consists of the set of available documents/objects which can be potentially returned as an answer to a query term q. Compositional structure R is the list of m objects returned as answer. Constraints on R: all the objects links in an answer list must be distinct. Population: is the set of n answer lists returned to the users which have submitted query q during the time instant.Representation: each individual answer list is easily represented by an ordered vector of m entries, where each entry represent a link to an object relevant to the query. The feedback reflects the relevance of the answer list with respect to the user query. The feedback acquisition phase can be realized to monitor the numbers of clicks received by the answer list, the user session duration, user activity patterns, the amount of data exchanged.

Online Newspaper. Components domain: it consists of the set of available news, and for each single news the set of possible candidate titles and/or pictures. The compositional structure R is a set of m news, distributed in the front page, the constraints on R require that: all news in a front page must be distinct; news are grouped in categories (i.e. sports, politics, internal affair etc.); news can appear only in a category to which they belong, news categories can overlap (e.g.the same news can potentially appear under different categories). Population: is a set of n front pages most recently distributed to the customers. Representation: each individual newspaper is represented by an ordered vector of m entries, where each entry contains news is, title/headline and picture. A technique of feedback acquisition similar to the previous one can be used. The feedback reflects the interest of the reader on the delivered front page.

3.2 Representation and Algorithm Structure

An instance of the dynamical OSP problem is characterized by a tuple (C, R, n) where C is the set of components, R is the compositional structure which defines the constraint of the consistent services, n is the amount of service requests per time instant. The algorithm is also parametrized by the classical pair (p_c, p_m),i.e. probabilities of crossover and mutation, note that n also represents the population size.

The representation of components and services is straightforward. The components (i.e. the domain set for genes) are represented by unique identifiers in $\{1, \ldots, N\}$. The services (i.e. query answer lists, or newspaper front pages) are represented by sequences of m component identifiers, with no replications, services correspond to chromosomes of the population.

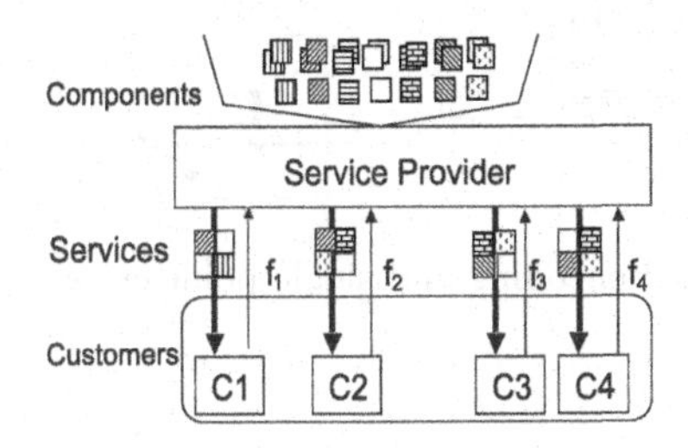

Fig. 1. The Online Service Provider Scenario

The evaluation phase is realized by a procedure, external to the system, which returns for each service s_i in the population at time t, the values of $f(t, s_i)$. These values are used as the fitness of the chromosome.

The evolution phase uses a *selection* procedure to select the chromosomes for the crossover phase, a standard roulette–wheel method is applied, where the fittest individuals are more likely to be selected.

3.3 Crossover and Mutation

The *crossover operator* used in our algorithm is a one–point crossover. The gene sequence of the two chromosome parents are cut in two subsequences of r and $m - r$ elements each, where r is a random number between 1 and $m - 1$. Each of the two chromosome outbreeding is created by taking the union of two subsets coming from different parents. A repairing phase can be necessary after crossover in order to guarantee distinct components in the outbreeding services: if a child has one or more duplicated genes, the replacing components are randonly selected from the parents, or by mutation operators when not possible.

Three different *mutation operators* are used. Explorative mutation operator replaces a component with a randomly chosen component which does not appear in any other chromosome. Full randomized mutation operator replaces a component with a randomly chosen component among all the available components.

Exploitative mutation operator try to use the knowledge acquired in the evaluation phase in order to prefer the components which are likely to give a higher contribution to the chromosome fitness. Since this information is not available to the system, a rough estimate of a quantity proportional to $f_c(t, c_i)$, for each component c_i, is computed by averaging the degree of all the chromosomes in which c_i appears. In the case of online newspaper the mutation operator also takes into account that components are triple and they are organized into topics or categories, i.e. the domain of mutation is the category, and triple elements, like pictures and titles, can also mutate independently.

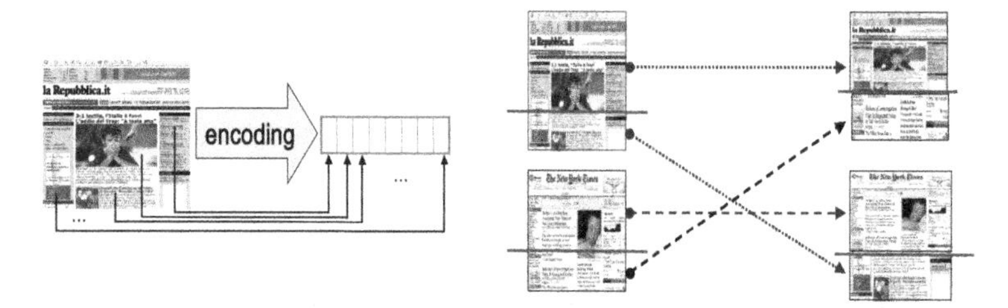

Fig. 2. Component based encoding and single point crossover for online newspaper

4 Experiments

Experiments have focused on the *search engine* and *online newspaper* OSP problems. A special module has been designed to simulate the evaluation given by the external agents and the dynamical changes of the domain by a random uniform distribution of hidden values. The rate of domain evolution over the time is directed by two parameters: probability of change,p_d, and percentage of change, p_a, which have been then varied systematically.

For both classes of experiments have been used the same general parameters: a component domain size of $|C| = 200$ components, each service made of $m = 10$ components (chromosome length), the service provider receives/delivers $n = 20$ service requests per time instant (population size), and optimizes on a time interval $t_{max} = 2000$ (number of generations).

After an initial phase of parameters tuning the probability of crossover and probability of mutation parameters has been fixed to $p_c = 0.15$ and $p_m = 0.09$. The results obtained from the simulations are encouraging and do not significantly differ for both OSP classes. The graph in Figure 3 shows the performances of the system when $p_a = 0.05$. The values of the parameter p_d are shown in correspondence of the line depicting the mean fitness values we obtained. The line for $p_d = 0$ represents the static case, while the line for $p_d = 1$ represents the opposite case when the 5% of fitness is changed at each generation. It is worth to noticing that also in this case the algorithm is still performant. The mean value for the fitness is 0.67 with respect the initial mean value equal to 0.6, with an improvement of more than 10%. An interesting result obtained in the experiments is that the overall fitness obtained by the algorithm appears to be

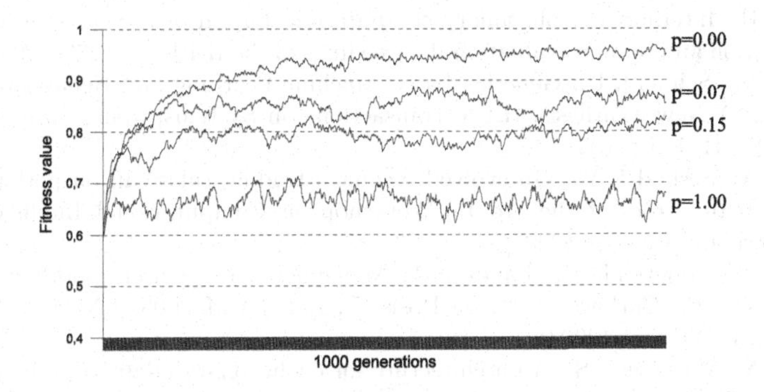

Fig. 3. Results for the online newspaper dynamic OSP problem

invariant with respect to the product $p_d \cdot p_a$, i.e. when the probability of a change in a single component is constant. A possible explanation of these phenomenon is that since p_d is directly related with the time rate of dynamical change, a low value of p_d gives the algorithm more time to adapt to the changes.

5 Conclusions

A general interactive and adaptive evolutionary scheme has been introduced, for two classes of dynamic Online Service Provider problems characterized by component additive fitness: *search engine* and *online newspaper*. The goal of OSP is to maximize the average user fitness while producing and adapting services in a dynamical environment. Experimental results show that the technique converges and adapts to changes for both OSP classes. An interesting properties of the adaptive behavior of the scheme is the ability to recover a given amount of changes despite of their distribution over the time, while preserving the average fitness. Future work will regard different classes of interactive OSP problems where the external fitness is not additive with respect to the components or it is context dependent.

References

1. Binder, J., Koller, D., Russell, S., Kanazawa, K.: Adaptive probabilistic networks with hidden variables. Machine Learning 29(2–3), 213–244 (1997)
2. Kobsa, A., Wahlster, W.: User models in dialog systems. Springer, London (1989)
3. Holland, J.: Adaptation in Natural and Artificial Systems. University of Michigan Press (1975)
4. Whitley, D.: An overview of evolutionary algorithms: practical issues and common pitfalls. Information and Software Technology 43(14), 817–831 (2001)
5. Masui, T.: Graphic object layout with interactive genetic algorithms. In: Proc. of IEEE Workshop on Visual Languages, pp. 74–87 (1992)
6. Peñalver, J., Guervós, J.: Optimizing web page layout using an annealed genetic algorithm. In: Proc. Parallel Problem Solving from Nature V, pp. 1018–1027 (1998)

7. Takagi, H.: Interactive evolutionary computation: Fusion of the capabilities of ec optimization and human evaluation. Proceedings of the IEEE 89, 1275–1296 (2001)
8. Dorigo, M., Schnepf, U.: Genetics–based Machine Learning and Behaviour Based Robotics: A New Synthesis. IEEE Transactions on Systems, Man and Cybernetics 23(1), 141–154 (1993)
9. Becker, A., Seshadri, M.: Gp-evolved technical trading rules can outperform buy and hold. In: Proc. of the 3rd Int.Workshop on Computational Intelligence in Economics and Finance (2003)
10. Kay, J., Kummerfeld, R., Lauder, P.: Managing private user models and. In: Brusilovsky, P., Corbett, A.T., de Rosis, F. (eds.) UM 2003. LNCS, vol. 2702, Springer, Heidelberg (2003)
11. Milani, A., Marcugini, S.: An architecture for evolutionary adaptive web systems. In: Deng, X., Ye, Y. (eds.) WINE 2005. LNCS, vol. 3828, pp. 444–454. Springer, Heidelberg (2005)

A Study of Some Implications of the No Free Lunch Theorem

Andrea Valsecchi and Leonardo Vanneschi

Department of Informatics, Systems and Communication (D.I.S.Co.)
University of Milano-Bicocca, Milan, Italy
a.valsecchi8@campus.unimib.it
vanneschi@disco.unimib.it

Abstract. We introduce the concept of "minimal" search algorithm for a set of functions to optimize. We investigate the structure of closed under permutation (c.u.p.) sets and we calculate the performance of an algorithm applied to them. We prove that each set of functions based on the distance to a given optimal solution, among which trap functions, onemax or the recently introduced onemix functions, and the NK-landscapes are not c.u.p. and thus the thesis of the sharpened No Free Lunch Theorem does not hold for them. Thus, it makes sense to look for a specific algorithm for those sets. Finally, we propose a method to build a "good" (although not necessarily minimal) search algorithm for a specific given set of problems. The algorithms produced with this technique show better average performance than a genetic algorithm executed on the same set of problems, which was expected given that those algorithms are problem-specific. Nevertheless, in general they cannot be applied for real-life problems, given their high computational complexity that we have been able to estimate.

1 Introduction

The No Free Lunch (NFL) theorem states that all non-repeating search algorithms, if tested on all possible cost functions, have the same average performance [1]. As a consequence, one may informally say that looking for a search algorithm that outperforms all the others on all the possible optimization problems is hopeless. But there are sets of functions for which the thesis of the NFL theorem does not hold, and thus talking about a "good" (or even the "best") algorithm on those sets of functions makes sense. In this paper, we investigate some of those sets of functions, we define for the first time the concept of minimal algorithm and we investigate some of its properties.

The sharpened-NFL [2] states that the thesis of the NFL theorem holds for a set of functions F if and only if F is *closed under permutation* (c.u.p.). For these particular sets of functions calculating the performance of an algorithm is relatively simple. For instance, for c.u.p. sets of functions with a constant number of globally optimal solutions, a method to estimate the average performance of an algorithm has been presented in [3]. In this paper, we try to generalize this result and to give for the first time an equation to calculate the average performance of algorithms for any c.u.p. set of functions.

To prove that a set of cost functions is *not* c.u.p., it is possible to use some properties of c.u.p. sets. For instance, over the functions that belong to c.u.p. sets, it is not

M. Giacobini et al. (Eds.): EvoWorkshops 2008, LNCS 4974, pp. 633–642, 2008.

possible to define a non-trivial neighborhood structure of a specific type, as explained in [4]. Furthermore, a set of functions with description length sufficiently bounded is not c.u.p. [5]. In this paper, we prove that some particular sets of problems, which are typically used as benchmarks for experimental or theoretical optimization studies, are *not* c.u.p. In particular, we focus on problems for which the fitness (or cost) of the solutions is a function of the distance to a given optimum. These problems include, for instance, trap functions [6], onemax and the recently defined onemix [7] functions. We also consider the NK-landscapes. As a consequence of the fact that these sets of functions are not c.u.p., and thus the NFL does not hold for them, we could informally say that it makes sense to look for a "good" (or even the "best") optimization algorithm for them. In this paper, we present a method to build a "good" (although not necessarily minimal) search algorithm for a given set of problems and we apply it to some sets of trap functions and NK-landscapes. Those algorithms are experimentally compared with a standard Genetic Algorithm (GA) on the same sets of functions.

This paper is structured as follows: in Section 2 we briefly recall the NFL theorem. In Section 3 we define the concept of minimal search algorithm and we prove some properties of its performance; furthermore, we give an equation to estimate the performance of an algorithm applied to any set of c.u.p. functions. In section 4 we prove that each set of functions based on the distance to an optimal solution is not c.u.p.; successively, we prove the same property for NK-landscapes. In section 5 we present a method to automatically generate an optimization algorithm specialized for a given set of functions. In section 6, the performance of some of the algorithms generated by our method are compared with the ones of a GA. Finally, in section 7 we offer our conclusions and discuss possible future research activities.

2 No Free Lunch Theorem

Let $\mathcal{X},\mathcal{Y}$ be two finite sets and let $f:\mathcal{X}\to\mathcal{Y}$. We call *trace* of length m over $\mathcal{X}$ and $\mathcal{Y}$ a sequence of couples: $t=\langle(x_1,y_1),\cdots,(x_m,y_m)\rangle$ such that $x_i\in\mathcal{X}$ and $y_i=f(x_i),\forall i=1,2,\cdots,m$. If we interpret $\mathcal{X}$ as the space of all possible solutions of an optimization problem (search space), $f(.)$ as the fitness (or cost) function and $\mathcal{Y}$ as the set of all possible fitness values, a trace t can be interpreted as a sequence of points visited by a search algorithm along with their fitness values.

We call *simple* a trace t such that: $t[i]=t[j]\Rightarrow i=j$, i.e. a trace in which each solution appears only once. Let T be the set of all possible traces over the sets $\mathcal{X}$ and $\mathcal{Y}$. We call *search operator* a function $g:T\to\mathcal{X}$. A search operator can be interpreted as a function that given a trace representing all the solutions visited by a search algorithm until the current instant (along with their fitness values) returns the solution that will be visited at the next step. We say that g is *non-repeating*[1] if $\forall t\in T, g(t)\notin t^{\mathcal{X}}$. We can observe that, if $t\in T$ is simple and g is non-repeating, then also $t'=t\parallel(g(t),f\circ g(t))$ where $\parallel$ is the concatenation operator is a simple trace.

A deterministic search algorithm A_g is an application $A_g:((\mathcal{X}\to\mathcal{Y})\times T)\to T$ with $\forall t\in T\quad A_g(f,t)=t\parallel(g(t),f\circ g(t))$ where g is a search operator. We say that A_g is

[1] Given a trace $t=\langle t_1,t_2,...,t_m\rangle$, with $\forall i=1,2,...,n: t_i=(x_i,y_i)$, we use the notation $t^{\mathcal{X}}$ to indicate the set $\{x_1,x_2,...,x_m\}$. and the notation $t^{\mathcal{Y}}$ to indicate the set $\{y_1,y_2,...,y_m\}$.

non-repeating if g is non-repeating. From now on, we use the notation A to indicate a search algorithm (i.e. omitting the search operator g used by A). Furthermore, except where differently specified, with the term "algorithm" we indicate a deterministic non-repeating search algorithm. We indicate $A^m(f,t)$ the application of m iterations of algorithm A to trace t, defined as: $A^0(f,t) = t, \quad A^{m+1}(f,t) = A(f, A^m(f,t))$. Finally, we define $A^m(f) = A^m(f, \langle\rangle)$, where $\langle\rangle$ is the empty trace.

A set $F \subseteq \mathcal{Y}^{\mathcal{X}}$ is called *closed under permutation (c.u.p.)* if $\forall f \in F$ and for each permutation σ of $\mathcal{X}$, $(f \circ \sigma) \in F$. Then, we can enunciate:

Proposition 1 (Sharpened NFL [2]). *Let A, B be two deterministic non-repeating search algorithms and let $F \subseteq \mathcal{Y}^{\mathcal{X}}$ be c.u.p. Then $\forall m \in \{1, \cdots, |\mathcal{X}|\}$: $\{(A^m(f))^{\mathcal{Y}} : f \in F\} = \{(B^m(f))^{\mathcal{Y}} : f \in F\}$*

3 The Minimal Algorithm and Its Performances

Given an algorithm A and a function $f : \mathcal{X} \to \mathcal{Y}$, we define *minimum fitness evaluations for optimum* as: $\phi_A(f) = \min\{m \in \mathbb{N} \mid A^m(f)$ contains an optimal solution$\}$. Given $F \subseteq \mathcal{Y}^{\mathcal{X}}$ the average of such value on F is: $\bar{\phi}_A(F) = \frac{1}{|F|} \sum_{f \in F} \phi_A(f)$.

If $\mathcal{X}, \mathcal{Y}$ are finite, then also the number of search algorithms is finite (each algorithm produces a simple trace over $\mathcal{X}$ and $\mathcal{Y}$ and the number of possible simple traces over $\mathcal{X}$ and $\mathcal{Y}$ is finite if $\mathcal{X}$ and $\mathcal{Y}$ are finite). Let $\{A_1, A_2, ..., A_n\}$ be the finite set of all possible algorithms. Then the set $\{\bar{\phi}_{A_1}(F), \bar{\phi}_{A_2}(F), ..., \bar{\phi}_{A_n}(F)\}$ is also finite and thus has a minimum. Let $\bar{\phi}_{A_{min}}(F)$ be that minimum, i.e. $\bar{\phi}_{A_{min}}(F) = min\{\bar{\phi}_{A_1}(F), \bar{\phi}_{A_2}(F), ..., \bar{\phi}_{A_n}(F)\}$. We call an algorithm like A_{min} a *minimal algorithm* for F [2]. From now on, we use the notation $\Delta(F)$ to indicate $\bar{\phi}_{A_{min}}(F)$ for simplicity.

Proposition 2. *If F is c.u.p. and all the functions in F have k optimal solutions, then: $\Delta(F) = (|\mathcal{X}| + 1)/(k + 1)$.*

The proof of this proposition can be found in [3].

Let F be c.u.p. and let $\sim$ be a relationship defined as follows: $\forall f, g \in F, f \sim g \iff \exists \sigma \in S_{\mathcal{X}} : f = g \circ \sigma$, where $S_{\mathcal{X}}$ is the set of all permutation of $\mathcal{X}$. We remark that, given a c.u.p. set F and any two functions f and g in F, $f \sim g$ does not necessarily hold. $\sim$ is an equivalence relationship and $[f]_\sim$ can be written as $\{f \circ \sigma \mid \sigma \in S_{\mathcal{X}}\}$. Thus $[f]_\sim$ is c.u.p. Furthermore, all functions in $[f]_\sim$ have the same number of optimal solutions. According to the sharpened NFL, for each algorithm A, $\Delta(F) = \bar{\phi}_A(F)$ (all the algorithms have the same average performance over a c.u.p. set of functions) and thus also $\Delta([f]) = \bar{\phi}_A([f])$ holds. Let R be a set of class representatives of $F/\sim$, then the following property holds:

$$\Delta(F) = \bar{\phi}(F) = \frac{1}{|F|} \sum_{f \in F} \phi(f) = \frac{1}{|F|} \sum_{r \in R} \sum_{f \in [r]} \phi(f) =$$

$$= \frac{1}{|F|} \sum_{r \in R} |[r]| \bar{\phi}([r]) = \frac{1}{|F|} \sum_{r \in R} |[r]| \Delta([r]) = \frac{1}{|F|} \sum_{r \in R} |[r]| \frac{|\mathcal{X}| + 1}{\mathrm{op}(r) + 1}$$

$$= \frac{|\mathcal{X}| + 1}{|F|} \sum_{r \in R} \frac{|[r]|}{\mathrm{op}(r) + 1} \tag{1}$$

where $\text{op}(r)$ is the number of optimal solutions of r. Considering the image of r as a multiset, we have $\text{Im}(r) = \{a_1, \ldots, a_{|\mathcal{Y}|}\}$ (i.e. $a_i = |\{x \in X : r(x) = i\}|$). It follows that:

$$||[r]|| = \binom{|X|}{a_1, \ldots, a_{|\mathcal{Y}|}} = \frac{|X|!}{(a_1)! \cdots (a_{|\mathcal{Y}|})!}$$

Equation (1) allows us to estimate $\Delta(F)$ for any set of c.u.p. functions.

4 Some Sets on Non-c.u.p. Functions

The objective of this section is to prove that some sets of functions that are often used as benchmarks in experimental or theoretical optimization studies are *not* c.u.p. As a consequence, for these sets of functions it makes sense to look for a "good" (or even for the minimal) algorithm. In particular, in this work we focus on the set of all the problems where fitness can be calculated as a function of the distance to a given optimal solution (like for instance trap functions [6], onemax and the recently defined onemix [7] functions) and the NK-landscapes [8].

4.1 Functions of the Distance to the Optimum

Here we consider the set of functions of the distance to a unique known global optimum solution. If we normalize all fitness values into the range $[0,1]$ and we consider 1 as the best possible fitness value and zero as the worst one, we can characterize these functions as follows: Let $G = \{g : \{0, \ldots, n\} \to [0,1] \mid g(k) = 1 \iff k = 0\}$. We say that $f_{o,g} : \{0,1\}^n \to [0,1]$ is a *function of the distance* if $\exists o \in \{0,1\}^n$, $\exists g \in G$ such that $f_{o,g}(z) = g(d(z,o))$, where d is a given distance, z is a (binary string) solution and o is the unique global optimum. In this work we focus on Hamming distance. We call *trivial* a function $f_{o,g}$ such that $g \in G$ is constant over $\{1, \ldots n\}$. In other words, we call trivial a "needle in a haystack" function. We want to prove that each set of functions of the Hamming distance containing at least one non-trivial function is *not* c.u.p.

Proposition 3. *Let $n \geq 1$ and let $F = \{f_{g,o} \mid f(z) = g(d(z,o)),\ z \in \{0,1\}^n,\ o \in \{0,1\}^n,\ g \in G\}$, where d is the Hamming distance. For each $F' \subseteq F$, if a non-trivial function $f_{g,o} \in F'$ exists, then F' is not c.u.p.*

Proof. If $n = 1$, the thesis is trivially true. The proof of the case $n = 2$ is not reported here to save space (the case of binary strings of length 2 is probably not very interesting). Let us consider $n > 2$ and let $z = z_1 \cdots z_n \in \{0,1\}^n$. We call $\text{flip}(z,i) := \bar{z}_1, \ldots, \bar{z}_i, z_{i+1}, \ldots, z_n$ and $\text{backflip}(z,i) := z_1, \ldots, z_{n-i}, \bar{z}_{n-i+1}, \ldots, \bar{z}_n$. Then: $d(\text{flip}(z,i), z) = i$ and $d(\text{backflip}(z,i), z) = i$.

Let *ab absurdo* F' be a c.u.p. set. An $f_{g,o} \in F'$ non-trivial exists by hypothesis. Thus $\exists i, j$ with $0 < i < j \leq n$ such that $g(i) \neq g(j)$. Let σ a permutation of $\{0,1\}^n$ such that $\sigma(o) = o$, $\sigma(\text{flip}(1,o)) = \text{flip}(i,o)$ and $\sigma(\text{backflip}(1,o)) = \text{flip}(j,o)$. Since F' is c.u.p., we can say that $g' \in G$ and $o' \in \{0,1\}^n$ exist such that $(f_{g,o} \circ \sigma) = f_{g',o'}$. Then the following property holds: $g'(d(o,o')) = f_{g',o'}(o) = f_{g,o}(\sigma(o)) = f_{g,o}(o) = g(d(o,o)) = g(0) = 1$. By definition, it follows that $d(o,o') = 0$ i.e. $o' = o$. Then the following properties hold: $g'(1) = g'(d(\text{flip}(1,o),o) = f_{g',o'}(\text{flip}(1,o)) = f_{g,o}(\sigma(\text{flip}(1,o))) =$

$f_{g,o}(\text{flip}(i,o)) = g(d(\text{flip}(i,o),o)) = g(i)$ and $g'(1) = g'(d(\text{backflip}(1,o),o) = f_{g',o'}(\text{backflip}(1,o)) = f_{g,o}(\sigma(\text{backflip}(1,o))) = f_{g,o}(\text{flip}(j,o)) = g(d(\text{flip}(j,o),o)) = g(j)$. From these properties, we can deduce that $g(i) = g'(1) = g(j)$, which is against the hypothesis. □

4.2 NK-Landscapes

In this section, we prove that the set of all the NK-landscapes [8] with adjacent neighbourhood is *not* c.u.p.

Proposition 4. *Let $n \geq 1$ and let $0 \leq k < n$. Then the set of functions $NK = \{F_\phi \mid \phi : \{0,1\}^{k+1} \to \{0,1\}\}$ where $F_\phi(x) = \frac{1}{n}\sum_{i=1}^{n} \phi(x_i,\ldots,x_{i+k})$ is not c.u.p.*

Proof. In order to prove that this set of functions is not c.u.p. we consider a function that belongs to this set and we show that one of its permutations does not belong to this set. Let $k < n-1$, $\phi = XOR$ and σ a permutation such that $\sigma(0^n) = 10^{n-1}$. We show that a g such that $F_\phi \circ \sigma = F_g$ does not exist. The following properties hold: $(F_\phi \circ \sigma)(0^n) = F_\phi(10^{n-1}) = (k+1)/n$ and $F_g(0^n) = \frac{1}{n}\sum_{i=1}^{n} g(0^{k+1}) = g(0^{k+1}) \in \{0,1\}$. It follows that: $(k+1)/n \in \{0,1\}$ which is absurd, since $k+1 < n$.

Let now $k = n-1$, $\phi = XOR$ and σ such that $\sigma(10^{n-1}) = 0^n$, $\sigma(0^n) = 10^{n-1}$ and $\sigma(x) = x$ for the remaining elements. We have: $(F_\phi \circ \sigma)(0^{n-1}1) = F_\phi(0^{n-1}1) = \frac{1}{n}[\phi(0^{n-1}1) + \ldots + \phi(10^{n-1})] = n/n = 1$; thus: $F_g(0^{n-1}1) = \frac{1}{n}[g(0^{n-1}1) + \ldots + g(10^{n-1})] = 1$; and thus: $(F_\phi \circ \sigma)(10^{n-1}) = F_\phi(0^n) = 0$; but: $F_g(10^{n-1}) = \frac{1}{n}[g(10^{n-1}) + \ldots + g(0^{n-1}1)] = F_g(0^{n-1}1) = 1$, which is an absurd. □

5 Automatic Generation of an Optimization Algorithm

We now introduce a method to automatically generate a specific algorithm for a given set of functions F. Let us consider the following "game" with two players A and B: player A chooses a cost function f in F (without revealing its choice to player B) and player B must find an optimal solution of this function. At each step of this game, player B can choose a solution i in the search space X and player A is forced to reveal the fitness of that solution, i.e. the value of $f(i)$. The objective of player B is to find an optimal solution in the smallest possible number of steps. We assume that player B knows all the values of all the possible solutions for all the functions in F (for instance, we could imagine that they have been stored in a huge database). Here we propose a strategy that player B may use to identify the function f chosen by player A. Once player B has identified this function, since he knows all the cost values of all the possible solutions, he can immediately exhibit the optimal solution at the subsequent step. According to this strategy, in order to identify the function f chosen by player A, at each step of the game player B should perform the action that allows him to "eliminate" the largest possible number of functions among the ones that can be candidate to be function f. We call F_c such a set of candidate functions. The strategy initializes $F_c := F$. Then, for each point k of X, we create a "rule" $s_1 \mid \ldots \mid s_m$ where s_i is the number of functions in F_c such that $f(k) = i$. The most suitable rule is the one that minimizes the *average* of $|F_c|$ after its application, i.e. the one that minimizes $(\sum_1^m (s_i)^2)/|F_c|$, or more simply $\sum_1^m (s_i)^2$, given that $|F_c|$ is the same for all the rules before their application. Thus, at each step we choose point b with the most suitable rule, we ask to player A the value of $f(b)$ and we eliminate from the set F_c all the functions f' with $f(b) \neq f'(b)$.

Example 1. Let $\mathcal{X} = \mathcal{Y} = \{0,1,2,3,4\}$ and let us consider the set of functions $F = \{f_1,...,f_8\}$ with: $f_1 = (1,2,3,4,0)$, $f_2 = (3,1,2,0,0)$, $f_3 = (4,1,1,2,3)$, $f_4 = (0,0,0,1,2)$, $f_5 = (0,0,0,1,1)$, $f_6 = (0,0,0,0,1)$, $f_7 = (4,1,0,2,3)$, $f_8 = (0,1,0,0,0)$; where $\forall i = 1,2,...,8, f_i$ has been defined with the notation: $f_i = (y_0,y_1,y_2,y_3,y_4)$ where $\forall j = 0,...,4\ f_i(j) = y_j$. Let $f_1, f_2,..., f_8$ be functions to be maximized. Let us suppose that the target function (the one chosen by player A) is f_6. Our method begins initializing the set of candidate functions: $F_c := F$. Then a loop over all the solutions begins. Let us consider first solution $x = 0$. We have to generate a rule $s_1 \mid \ldots \mid s_m$ where s_i is the number of functions in F_c such that $f(0) = i$. The number of functions f_i in F for which $f_i(0) = 0$ is equal to 4, thus $s_0 = 4$; the number of functions for which $f_i(0) = 1$ is equal to 1, thus $s_1 = 1$. Iterating this process, we obtain a rule for solution $x = 0$ equal to $4|1|1|2$. Now we have to calculate a measure of the "cost" of this rule. We define it as $\sum_1^m (s_i)^2$, and thus it is equal to 22. We now repeat the same process for all the solutions in the search space and we obtain the data reported in table 1. The two rules

Table 1. The rules produced for all the possible solutions at the first iteration in the example introduced in the text

solution	rule	cost
0	4 | 1 | 1 | 2	22
1	3 | 4 | 1	26
2	5 | 1 | 1 | 1	28
3	3 | 2 | 2 | 1	18
4	3 | 2 | 1 | 2	18

that minimize the cost are the ones associated with solutions 3 and 4. Thus player B asks to player A the value of the chosen function for the first of those solutions, i.e. solution 3. The answer of player A is $f(3) = 0$. Now, player B can modify the set of candidate solutions F_c, by eliminating those functions f_i such that $f_i(3) \neq 0$. Thus: $F_c := \{f_2, f_6, f_8\}$.

If we iterate this process, we can see at the second iteration one of the solutions associated with the rule with the minimum cost is 0 and at the third iteration it is 1. After that, $F_c = \{f_6\}$ and the algorithm terminates, since player B has identified the function chosen by player A. All she has to do is to return the solution that maximizes function f_6, i.e. $x = 4$.

To automatically produce a search algorithm inspired by this strategy, it is sufficient to store the choices that have been done in a decision tree. Such a tree has the nodes labelled with solutions $x \in \mathcal{X}$ and arcs departing from a node labelled with x are labelled with all possible values of $f_i(x)$ for each $f_i \in F_c$. The son of a node labelled with x and linked to x by an arc labelled with $f_i(x)$ is the solution related to rule with minimum cost after that all functions f_j with $f_j(x) \neq f_i(x)$ have been eliminated from F_c. Figure 1

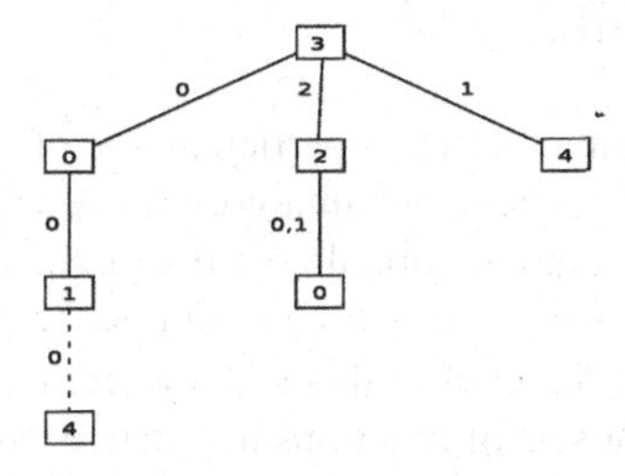

Fig. 1. The decision tree generated by our algorithm for the example introduced in the text. The dashed line refers to the last step of the algorithm, where the function has been identified and thus the optimal solution can be returned.

represents the decision tree generated by our method for the previous example. The first solution that we have considered (the one that minimizes the cost in Table 1) was 3. Successively, we have considered solutions 0 and 1 and we have been able to return solution 4. We remark that the leaves of this tree are labelled with points that are optimal solutions for at least one function in F.

We call "RULE" this method to automatically generate algorithms[2] for a given set of functions F. The following property holds:

Proposition 5. *RULE has a computational complexity of* $\Theta(|\mathcal{X}||F|^4)$.

Proof. The algorithm is executed on each solution in $\mathcal{X}$ and for each function in F. Thus, the generation of the sequence of rules has a complexity of $\Theta(|\mathcal{X}||F|)$. At each application of the rules, at least one function is eliminated from the set of candidate ones, thus one iteration has a complexity of: $|\mathcal{X}||F|\sum_{i=0}^{|F|}|F|-i = |\mathcal{X}||F|\frac{(|F|+1)|F|}{2}$. Thus the total complexity is $\Theta(|\mathcal{X}||F|^4)$. □

Proposition 6. *The decision tree produced by RULE has a number of nodes of* $O(|\mathcal{X}||F|)$.

Proof. For each $f \in F$ the search for an optimal solution consists in examinating a sequence of points of $\mathcal{X}$ with no repetitions. □

It is possible to show that RULE does *not* necessarily generate the minimal algorithm for a given set of functions. For instance, one may imagine a set of functions that all share the same set of optimal solutions $\text{Opt} = \{o_1, o_2, ..., o_q\}$; for this set of functions, a minimal algorithm is clearly one that outputs a solution $o_i \in Opt$ at the first step, and not the algorithm generated by RULE. For this reason, we informally say that RULE generates "good", although not necessarily minimal, algorithms.

Furthermore, from proposition 5, we can deduce that RULE is clearly too complex to be used for real-life problems. Nevertheless, in the next section we show how RULE (and the algorithms generated by RULE) can be used for interesting (although rather "small") instances of trap functions and NK-landscapes.

[2] Strictly speaking, RULE does not generate "search algorithms", since they are specialized for a particular set of functions and domain; nevertheless, we continue calling the output of rule "algorithm" for simplicity.

6 Experimental Results

Propositions 5 and 6 show that given a particular set of functions F defined over a domain $\mathcal{X}$, executing RULE to generate an algorithm A and then running A on all the functions in F has a larger computational cost than exhaustively examinating all the possible solutions for each function in F (whose cost is clearly $O(|\mathcal{X}||F|)$). For this reason, it must be clear that the goal of this study is *not* trying to produce a technique to efficiently solve particular sets of functions in practice. Nevertheless, we think that it might be interesting to quantify the theoretical performance improvement of a problem-specific algorithm, compared to a "general" one, like for instance a GA. For this reason, in this section the algorithms produced by RULE are compared with a standard GA. The performance measure used for this comparison, consistantly with our definition of minimal algorithm for a set of functions F (see Section 3), will be $\bar{\phi}(F)$. Since GAs are *repeating*, we count its fitness evaluations without repetitions. For GAs, we have used the following set of parameters: population size of 100 potential solutions, standard single-point crossover [9,10] with rate equal to 0.9, standard point mutation [9,10] with rate equal to 0.01, tournament selection with tournament size equal to 2, elitism (i.e., the best individual is copied unchanged into the next population), maximum number of generations equal to 200.

The sets of functions that we use in our experiments are partitioned into two groups, each one composed by three sets of functions. The first group contains three sets of trap functions. Trap functions [6] are a particular set of functions of the distance (as defined in Section 4.1) that depend on the values of two costants: B (the width of the attractive basin for each optimum) and R (their relative importance). The three sets of trap functions used for our experiments are respectively composed by 100, 250 and 500 "randomly chosen" trap functions. For "randomly chosen" trap function we mean a trap function where the B and R constants and the (unique) optimal solution have been chosen randomly with uniformly distributed probability over their domains (the range $[0,1]$ for B and R, the search space for the optimal solution). The second group of functions that we have used contains three sets of NK-landscapes functions. NK-landscape functions [8] are completely defined by the value of two constants (N and K) and one "kernel" function $\phi : [0,1]^{K+1} \rightarrow [0,1]$ The sets of functions we have used are respectively composed by 100, 250 and 500 "randomly generated" NK-landscapes, i.e. NK-landscapes where K and ϕ have been generated uniformly at random. For all these functions, the search space $\mathcal{X}$ that we have chosen is composed by binary strings of 8 bits (thus $N = 8$ for NK-landscapes).

Table 2 shows the results obtained by the GA. The first column represents the set of functions F on which the experiments have been done (for instance "Trap p" means a set of p "randomly chosen" trap functions). The second column reports the average number of evaluations with no repetitions that have been spent by the GA for finding an optimal solution with their standard deviations; more in particular, for each $f \in F$ we have executed 100 independent GA runs and only for those runs where the optimal solution has been found (before generation 200) we have calculated the number of evaluations without repetitions that have been performed before finding the optimum. Then, we have averaged all those numbers over the 100 independent runs. The result that we report is the average of all those averages over all functions in F. The third

Table 2. Results returned by the GA. Each line reports the results for a different set of functions.

F	$\bar{\phi}_{GA}(F)$	Avg Total FE	SR
Trap 500	145.18 ($\sigma = 13.6$)	3633.41 ($\sigma = 7232$)	0.82
Trap 250	145.52 ($\sigma = 13.7$)	3607.35 ($\sigma = 7200.7$)	0.83
Trap 100	145.64 ($\sigma = 13.4$)	4128.53 ($\sigma = 7641.6$)	0.85
NK 500	141.61 ($\sigma = 12.5$)	804.15 ($\sigma = 3024.8$)	0.98
NK 250	142.05 ($\sigma = 12.9$)	886.54 ($\sigma = 3267.2$)	0.97
NK 100	141.86 ($\sigma = 12.5$)	754.18 ($\sigma = 2867.6$)	0.98

column reports the average number (calculated as above) of evaluations (also counting repetitions) that have been spent by the GA for finding an optimal solution with their standard deviations. Finally, the fourth column reports the success rate, i.e. the number of runs where an optimal solution has been found divided by the total number of runs that we have performed (100 in our experiments) averaged over all functions in F.

Table 3 reports the results of the algorithms generated by RULE on the same sets of problems. The first column identifies the set of functions F on which the experiments have been done; the second column reports the average (calculated over all functions in F) number of evaluations spent to find an optimal solution with their standard deviations. An optimal solution has always been found for each one of these executions (thus we do not report success rates).

Table 3. Results returned by the algorithms generated by RULE. Each line reports the results for a different set of functions.

F	$\bar{\phi}_{rule}(F)$
Trap 500	2.99 ($\sigma = 0.22$)
Trap 250	2.95 ($\sigma = 0.21$)
Trap 100	2.75 ($\sigma = 0.43$)
NK 500	4.57 ($\sigma = 0.62$)
NK 250	4.22 ($\sigma = 0.52$)
NK 100	3.81 ($\sigma = 0.49$)

Comparing results in Tables 2 and 3 we can clearly see that the algorithms generated by RULE have a remarkably better performance than the GA. This was expected since these algorithms are problem-specific, i.e. they have been generated to solve those particular problems.

7 Conclusions and Future Work

We have defined the concept of minimal search algorithm for a given set of problems. We have also introduced an equation to calculate the average performance of an algorithm over a closed under permutation (c.u.p.) set of functions. Furthermore, we have proven that some particular sets of functions are *not* c.u.p. In particular, we focused on any set of functions of the distance to a given optimal solution (this set contains some well known benchmarks, like trap functions, onemax and onemix) and on NK-landscapes. Not being c.u.p., for those sets the No Free Lunch theorem does not hold and thus it makes sense to look for a minimal algorithm. Inspired by this, we have presented a method to build a specific (not necessarily minimal) search algorithm for a given set of functions to optimize. We have experimentally shown that the algorithms generated by such a method remarkably outperform a standard Genetic Algorithm on some "small" instances of trap functions and NK-landscapes. This was expected given that the generated algorithms are problem-specific. Our method cannot be applied to real-life applications, given its complexity, which we have estimated as a function of the size of the search space and of the cardinality of the considered set of functions.

In the future, we plan to prove other interesting properties of the minimal algorithm, to prove whether other interesting sets of functions are c.u.p. or not and to improve the RULE algorithm, eventually employing some concepts of Rough-Sets.

References

1. Wolpert, D.H., Macready, W.G.: No free lunch theorems for optimization. IEEE Transactions on Evolutionary Computation 1(1), 67–82 (1997)
2. Schumacher, C., Vose, M.D., Whitley, L.D.: The no free lunch and problem description length. In: Spector, L., Goodman, E.D., Wu, A., Langdon, W.B., Voigt, H.-M., Gen, M., Sen, S., Dorigo, M., Pezeshk, S., Garzon, M.H., Burke, E. (eds.) Proceedings of the Genetic and Evolutionary Computation Conference (GECCO-2001), 7-11 2001, pp. 565–570. Morgan Kaufmann, San Francisco (2001)
3. Igel, C., Toussaint, M.: Recent results on no-free-lunch theorems for optimization. CoRR: Neural and Evolutionary Computing cs.NE/0303032 (2003)
4. Igel, C., Toussaint, M.: On classes of functions for which no free lunch results hold. Inf. Process. Lett. 86(6), 317–321 (2003)
5. Streeter, M.J.: Two broad classes of functions for which a no free lunch result does not hold. In: Cantú-Paz, E., Foster, J.A., Deb, K., Davis, L., Roy, R., O'Reilly, U.-M., Beyer, H.-G., Kendall, G., Wilson, S.W., Harman, M., Wegener, J., Dasgupta, D., Potter, M.A., Schultz, A., Dowsland, K.A., Jonoska, N., Miller, J., Standish, R.K. (eds.) GECCO 2003. LNCS, vol. 2724, pp. 1418–1430. Springer, Heidelberg (2003)
6. Deb, K., Goldberg, D.E.: Analyzing deception in trap functions. In: Whitley, D. (ed.) Foundations of Genetic Algorithms, vol. 2, pp. 93–108. Morgan Kaufmann, San Francisco (1993)
7. Poli, R., Vanneschi, L.: Fitness-proportional negative slope coefficient as a hardness measure for genetic algorithms. In: Thierens, D., et al. (eds.) Genetic and Evolutionary Computation Conference, GECCO 2007, pp. 1335–1342. ACM Press, New York (2007)
8. Altenberg, L.: Nk fitness landscapes. In: Back, T., et al. (eds.) Handbook of Evolutionary Computation, Section B2.7.2, p. 2. B2.7:5 – B2.7:10 IOP Publishing Ltd and Oxford University Press (1997)
9. Goldberg, D.E.: Genetic Algorithms in Search, Optimization and Machine Learning. Addison-Wesley, Reading (1989)
10. Holland, J.H.: Adaptation in Natural and Artificial Systems. The University of Michigan Press, Ann Arbor, Michigan (1975)

Negative Slope Coefficient and the Difficulty of Random 3-SAT Instances

Marco Tomassini[1] and Leonardo Vanneschi[2]

[1] Information Systems Department, University of Lausanne, Lausanne, Switzerland
[2] Dipartimento di Informatica, Sistemistica e Comunicazione, University of Milano-Bicocca, Milan, Italy
marco.tomassini@unil.ch, vanneschi@disco.unimib.it

Abstract. In this paper we present an empirical study of the Negative Slope Coefficient (NSC) hardness statistic to characterize the difficulty of 3-SAT fitness landscapes for randomly generated problem instances. NSC correctly classifies problem instances with a low ratio of clauses to variables as easy, while instances with a ratio close to the critical point are classified as hard, as expected. Together with previous results on many different problems and fitness landscapes, the present results confirm that NSC is a useful and reliable indicator of problem difficulty.

1 Introduction

Several algorithm-independent statistical measures have been introduced in the last two decades with the aim to characterize the intrinsic difficulty of searching a given problem space, usually called a *fitness landscape*, for optimal or nearly optimal solutions. Among these, we may mention Fitness-Distance Correlation (FDC) analysis [2], Landscape Correlation Functions [11], Negative Slope Coefficient (NSC) [7,8], and several others, see for instance [3]. The goal of all these methods is to characterize the hardness of a given problem space with a single number in such a way that landscapes that are classified as "difficult" will in general be difficult to search by any algorithm or heuristic, and those that are "easy" will be solvable with limited computational resources. By their very nature, all the above measures are of a statistical character, i.e. except for the simplest and smallest landscapes, one needs to sample the landscape in some way to calculate them. Thus the results are always an approximation to the true situation and these approaches can all be deceived by particular features of the landscape, natural or artificially introduced [3]. These problems notwithstanding, some of these measures have proved useful as they give a global view of the hardness of a given problem or problem class, and they can help suggest the use of particular neighborhoods and search techniques as being more appropriate than others for a given landscape. In our own work, we have defined and used the NSC difficulty measure [8,9,6]. Since NSC, as well as the other measures, is not dependent on any algorithmic or problem representation issue, it can in principle be used for studying any fitness landscape of interest. In the present work we have applied it to important class of landscapes: those corresponding to randomly generated instances of the 3-SAT problem.

M. Giacobini et al. (Eds.): EvoWorkshops 2008, LNCS 4974, pp. 643–648, 2008.

The propositional satisfiability problem (SAT) is an important constraint satisfaction problem. Here we use a particular class of SAT problem instances called 3-SAT; we briefly explain the notation in the next section. Our main goal in the present work is to find out whether NSC gives coherent results for hard and easy instances of 3-SAT before and near the solubility phase transition region. If this is the case, together with the previous results mentioned above, we shall have grounds to believe that the NSC indicator is generally useful to gauge the intrinsic difficulty of fitness landscapes having very different structures.

2 Negative Slope Coefficient

Evolvability quantifies the ability of genetic operators to improve fitness quality. One possible way to study it is to plot the fitness values of individuals against the fitness values of their neighbours, where a neighbour is obtained by applying one step of a genetic operator to the individual. Such a plot has been presented in [10] and it is called a *fitness cloud*.

Since high-fitness points tend to be much more important than low-fitness ones in determining the behaviour of evolutionary algorithms, an alternative algorithm to generate fitness clouds was proposed in [8]. The main steps of this algorithm can be informally summarised as follows: (1) Generate a set of individuals $\Gamma = \{\gamma_1, ..., \gamma_n\}$ by sampling the search space and let $f_i = f(\gamma_i)$, where $f(.)$ is the fitness function. (2) For each $\gamma_j \in \Gamma$ generate k neighbours, $v_1^j, \ldots, v_k^j$, by applying a genetic operator to γ_j and let $f_j' = \max_j f(v_j)$. (3) Finally, take $C = \{(f_1, f_1'), \ldots, (f_n, f_n')\}$ as the fitness cloud. In this paper we use the Metropolis-Hastings algorithm [5] to sample the search space and tournament selection (with tournament size equal to 2) to sample neighborhoods.

The fitness cloud can be of help in determining some characteristics of the fitness landscape related to evolvability and problem difficulty. But the mere observation of the scatterplot is not sufficient to quantify these features. The Negative Slope Coefficient (NSC) has been defined to capture with a single number some interesting characteristics of fitness clouds. It can be calculated as follows: let us partition C into a certain number of separate ordered "bins" $C_1, \ldots, C_m$ such that $(f_a, f_a') \in C_j$ and $(f_b, f_b') \in C_k$ with $j < k$ implies $f_a < f_b$ [9]. Next we consider the average fitnesses $\bar{f}_i = \frac{1}{|C_i|} \sum_{(f,f') \in C_i} f$ and $\bar{f}_i' = \frac{1}{|C_i|} \sum_{(f,f') \in C_i} f'$. The points $(\bar{f}_i, \bar{f}_i')$ can be seen as the vertices of a polyline, which effectively represents the "skeleton" of the fitness cloud. For each of the segments of this we can define a *slope*, $S_i = (\bar{f}_{i+1}' - \bar{f}_i')/(\bar{f}_{i+1} - \bar{f}_i)$. Finally, the negative slope coefficient is defined as $NSC = \sum_{i=1}^{m-1} \min\,(0, S_i)$.

The hypothesis proposed in [8] is that NSC should classify problems in the following way: if NSC$=0$, the problem is easy; if NSC<0 the problem is difficult and the value of NSC quantifies this difficulty: the smaller its value, the more difficult the problem. The justification put forward for this hypothesis was that the presence of a segment with negative slope would indicate a bad evolvability for individuals having fitness values contained in that segment as neighbours would be, on average, worse than their parents in that segment [7]. A more formal justification for NSC has recently been given in [6].

The results reported in [8,9,6] confirmed that the NSC is a suitable hardness indicator for many well known GP and GA benchmarks and synthetic problems, including various versions of the symbolic regression problem, the even parity problem of many different orders, the artificial ant problem on the Santa Fe trail, the multiplexer problem, the intertwined spirals, the GP Trap Functions, Royal Trees, the Max problem, Onemax, Trap, and Onemix, which naturally capture some typical features of easy and difficult fitness landscapes.

3 The 3-SAT Problem

The satisfiability problem in propositional logic (SAT) is a decision problem in which, given a formula $F(x_1,x_2,\ldots,x_n)$ of n Boolean variables $\{x_i\}_{i=1}^n$, $x_i \in \{0,1\}, \forall i \in [1..n]$, it is to be determined if there exists an assignment of the variables such that F evaluates to 1 (true). If this is the case then the formula is satisfiable, otherwise it is unsatisfiable. Formulas are usually given in conjunctive normal form, i.e. as a conjunction of m clauses each one of which is a disjunction of a finite number n of variables: $F = \bigwedge_{k=1}^m C_k$, where each C_k is of the type $(x_i \vee x_j \vee \ldots \vee x_l)$. In the k-SAT version of the problem, each clause contains exactly k variables. In this paper, we consider $k = 3$.

The problem is normally studied under its optimization interpretation rather than as a decision problem. In this case, if a formula F is satisfiable, the optimal solution, not necessarily unique, is an assignment of the n variables such that the number of unsatisfied clauses is equal to 0. The problem is thus a minimization one and the evaluation function for a given assignment is the number of unsatisfied clauses: the smaller this number the better the solution. The solution space is finite and comprises all the possible 2^n different assignments of the n Boolean variables. The common one-bit flip operator is used to generate the neighborhood of a solution.

3-SAT is NP-complete and its optimization version is NP-hard. However, it has been shown that not all 3-SAT instances are equally hard. Indeed, for 3-SAT formulae generated uniformly at random, there is a *solubility phase transition* as a function of a critical parameter $\alpha = m/n$, where m is the number of clauses in F and n is the number of variables [4]. The really hard problems are only found near the critical point $\alpha_c \simeq 4.26$, while problem instances with α less or greater are typically easier to solve. At the same time, as α passes through its critical value and grows larger, it quickly becomes more and more difficult to generate 3-SAT instances that are satisfiable. The phase transition, i.e. the divergence in solution cost and in the probability of generating satisfiable formulas, becomes crisper as $n \to \infty$. A large amount of randomly generated 3-SAT instances with various values of m and n can be downloaded from SATLIB, an online repository of benchmark problems and solvers for SAT [1]. In the next section, we present the results of our systematic study of the fitness landscapes of hard and easy 3-SAT instances from SATLIB using the NSC hardness indicator.

4 Results and Discussion

The experiments reported in this section have been performed over a set of satisfiable SAT instances taken from SATLIB. In particular, we have taken 100 instances with

[1] `http://www.satlib.org`

Table 1. The rate of negative and equal to zero values of the NSC for various satisfiable random instances of SAT with a given value of $\alpha = m/n$. Values are averages over 100 instances each.

# variables (n)	**# clauses** (m)	$\alpha = m/n$	$rate(NSC < 0)$	$rate(NSC = 0)$
200	600	3.0	0.0	1.0
200	700	3.5	0.0	1.0
200	800	4.0	0.09	0.91
200	860	4.3	0.90	0.10
225	700	3.11	0.0	1.0
225	800	3.55	0.0	1.0
225	900	4.0	0.06	0.94
225	960	4.27	0.92	0.08
250	800	3.2	0.0	1.0
250	900	3.6	0.0	1.0
250	1000	4.0	0.01	0.99
250	1065	4.26	0.98	0.02

$n = 250$ and $m = 1065$, 100 instances with $n = 225$ and $m = 960$ and 100 instances with $n = 200$ and $m = 860$. The ratio $\alpha = m/n$ indicates that all these instances should be difficult to solve. To verify if NSC confirms this hypothesis, we have calculated its value for all these instances, using samples composed by 40000 solutions. Furthermore, we also wanted to test the NSC on a set of easy instances, in order to check if its behavior is different. Thus, we have taken the same instances as before, but we have considered only a subset of the clauses, thus making the ratio m/n smaller. In particular, we have taken the same 100 instances with 250 variables considered before, but we have selected only 1000 of the 1065 existing clauses. These 1000 clauses have been chosen randomly with uniform distribution over all the clauses. In this way, we have obtained 100 satisfaisable instances of SAT with $n = 250$ and $m = 1000$. Analogously, we have been able to obtain 100 instances with $n = 250$ and $m = 900$, $n = 250$ and $m = 800$, $n = 225$ and $m = 900$, $n = 225$ and $m = 800$, $n = 225$ and $m = 700$, $n = 200$ and $m = 800$, $n = 200$ and $m = 700$, $n = 200$ and $m = 600$. All these new instances should be easy to optimize. Results are reported in Table 1.

When the ratio m/n is smaller than $\alpha_c \simeq 4.26$, NSC is almost always equal to zero, as expected. In contrast, when the ratio m/n is approximately equal to α_c, NSC is almost always negative (on 98 instances over 100 for $n = 250$ and $m = 1065$, on 92 instances over 100 for $n = 225$ and $m = 960$ and on 90 instances over 100 for $n = 200$ and $m = 860$).

In table 2 we report the average and standard deviations of the NSC values over all the instances where m/n is approximately equal to α_c.

As expected, the NSC is smaller for the instances with a higher number of variables and clauses and larger for the instances with a lower number of variables and clauses. Furthermore, average differences seem to be statistically significant.

Figure 1 shows the fitness clouds and segments for two randomly chosen instances among those studied; one difficult (a), and one easy (b). For the sake of consistency with the graphical representation used in [7,8,9], we have transformed the fitness function turning the problem into a maximization one to draw these figures; in other words, fitness represented in these figures is equal to the number of satisfied clauses in the

Table 2. Averages and standard deviations of the NSC values for all the instances for which NSC< 0 in table 1

# variables	# clauses	# instances	NSC average	NSC std. dev.
250	1065	98	-1.0347	0.0487
225	960	92	-0.8732	0.0248
200	860	90	-0.6232	0.0031

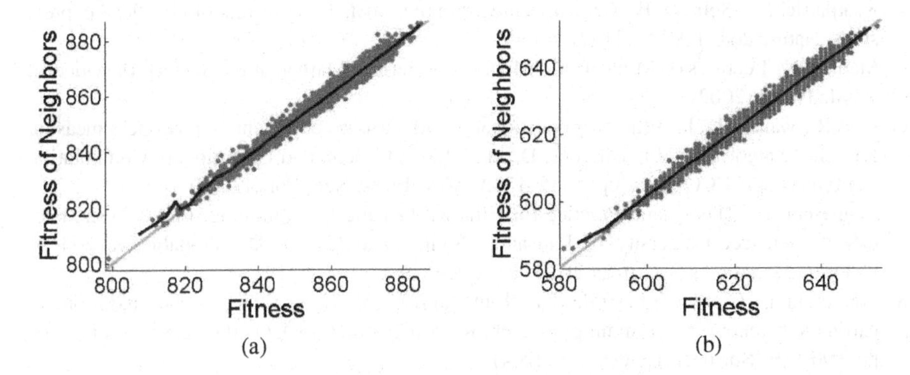

Fig. 1. Fitness clouds and segments for a randomly chosen satisfiable instance of SAT. (a): $m = 225$ and $n = 960$; (b): $m = 225$ and $n = 700$.

given solution. We can see that in all cases, clouds are quite similar to each other, i.e. they appear distributed along the axis bisector. This means that, for the 3-SAT problems, the fitness of a given solution is often similar to that of its neighbors and a large amount of neutrality characterizes 3-SAT fitness landscapes. This finding agrees with previous observations [1]. Nevertheless, we can see that for the instance where the ratio n/m is approximatedly equal to α_c (a), segments of negative slopes appear.

5 Conclusions

We have used the NSC hardness statistic to characterize the difficulty of the 3-SAT fitness landscape for randomly generated problem instances. Results agree with what is known about this problem class. In particular, NSC correctly classifies problem instances with a low ratio of clauses to variables as easy, while instances with a ratio close to the critical point are classified as hard. Furthermore, although the exact numerical values are not significant in themselves, the classification is crisp i.e., NSC goes from one, or nearly one, to almost zero when going from easy to difficult instances, and the statistic is significant as judged by the small standard deviations. Together with previous results on many completely different problems and fitness landscapes, the present results confirm that NSC is a useful indicator of problem difficulty, although it might fail in particular cases or for contrived problems.

References

1. Hoos, H., Stützle, T.: Stochastic Local Search, Foundations and Applications. Morgan Kaufmann, San Francisco (2005)
2. Jones, T., Forrest, S.: Fitness distance correlation as a measure of problem difficulty for genetic algorithms. In: Eshelman, L.J. (ed.) Proceedings of the Sixth International Conference on Genetic Algorithms, pp. 184–192. Morgan Kaufmann, San Francisco (1995)
3. Kallel, L., Naudts, B., Rogers, A. (eds.): Theoretical Aspects of Evolutionary Computing. Springer, Heidelberg (2001)
4. Kirkpatrick, S., Selman, B.: Critical behavior in the satisfiability of random Boolean expressions. Nature 264, 1297–1301 (1994)
5. Madras, N.: Lectures on Monte Carlo Methods. American Mathematical Society, Providence, Rhode Island (2002)
6. Poli, R., Vanneschi, L.: Fitness-proportional negative slope coefficient as a hardness measure for genetic algorithms. In: Thierens, D., et al. (eds.) Genetic and Evolutionary Computation Conference, GECCO 2007, pp. 1335–1342. ACM Press, New York (2007)
7. Vanneschi, L.: Theory and Practice for Efficient Genetic Programming. Ph.D. thesis, Faculty of Science, University of Lausanne, Switzerland (2004), Downlodable version at, http://www.disco.unimib.it/vanneschi
8. Vanneschi, L., Clergue, M., Collard, P., Tomassini, M., Vérel, S.: Fitness clouds and problem hardness in genetic programming. In: Deb, K., et al. (eds.) GECCO 2004. LNCS, vol. 3103, pp. 690–701. Springer, Heidelberg (2004)
9. Vanneschi, L., Tomassini, M., Collard, P., Vérel, S.: Negative slope coefficient. a measure to characterize genetic programming fitness landscapes. In: Collet., P., et al. (eds.) EuroGP 2006. LNCS, vol. 3905, pp. 178–189. Springer, Heidelberg (2006)
10. Vérel, S., Collard, P., Clergue, M.: Where are bottleneck in NK fitness landscapes? In: CEC 2003: IEEE International Congress on Evolutionary Computation, Canberra, Australia, pp. 273–280. IEEE Computer Society Press, Piscataway (2003)
11. Weinberger, E.D.: Correlated and uncorrelated fitness landscapes and how to tell the difference. Biol. Cybern. 63, 325–336 (1990)

A Memetic Algorithm for the Team Orienteering Problem

Hermann Bouly[1,2], Duc-Cuong Dang[1], and Aziz Moukrim[1]

[1] Université de Technologie de Compiègne
Heudiasyc, CNRS UMR 6599, BP 20529, 60205 Compiègne, France
[2] VEOLIA Environnement, Direction de la Recherche
17/19, rue La Pérouse, 75016 Paris, France
{hermann.bouly,duc-cuong.dang,aziz.moukrim}@hds.utc.fr

Abstract. The Team Orienteering Problem (TOP) is a generalization of the Orienteering Problem (OP). A limited number of vehicles is available to visit customers from a potential set. Each vehicle has a predefined running-time limit, and each customer has a fixed associated profit. The aim of the TOP is to maximize the total collected profit. In this paper we propose a simple hybrid Genetic Algorithm (GA) using new algorithms dedicated to the specific scope of the TOP: an Optimal Split procedure for chromosome evaluation and Local Search techniques for mutation. We have called this hybrid method a Memetic Algorithm (MA) for the TOP. Computational experiments conducted on standard benchmark instances clearly show our method to be highly competitive with existing ones, yielding new improved solutions in at least 11 instances.

1 Introduction

The Team Orienteering Problem first appeared in Butt and Cavalier [4] under the name of the Multiple Tour Maximum Collection Problem. The term TOP, first introduced in Chao, Golden and Wasil [5], comes from a sporting activity: team orienteering. A team consists of several members who all begin at the same starting point. Each member tries to collect as many reward points as possible within a certain time before reaching the finishing point. Available points can be awarded only once. Chao, Golden and Wasil [5] also created a set of instances, used nowadays as standard benchmark instances for the TOP.

The TOP is an extension to multiple-vehicle of the Orienteering Problem (OP), also known as the Selective Traveling Salesman Problem (STSP). The TOP is also a generalization of Vehicle Routing Problems (VRPs) where only a subset of customers can be serviced. As an extension of these problems, the TOP clearly appears to be NP-hard.

The assumption shared by problems of the TSP and VRPs family is that all customers should be serviced. In many real applications this assumption is not valid. In practical conditions, it is not always possible to satisfy all customer orders within a single time period. Shipping of these orders needs to be spread

M. Giacobini et al. (Eds.): EvoWorkshops 2008, LNCS 4974, pp. 649–658, 2008.

over different periods and, in some cases, as a result of uncertainty or dynamic components, customers may remain unserviced, meaning that the problem has a selective component which companies need to address.

Recently Feillet, Dejax and Gendreau [6] have reviewed the TOP as an extension of TSPs with profits. They focus both on travel costs and selection of customers, given a fixed fleet size. They discuss and show that minimizing travel costs and maximizing profits are opposite criteria. As far as we know, the most recent paper dealing with solution methods for the TOP is [7]. Most of the metaheuristics shown to be effective for the TOP are Tabu Search (TS) and Variable Neighborhood Search (VNS) [1,12]. The Memetic Algorithm (MA), first introduced by Moscato [8], is a recent technique that has been shown to be competitive for VRPs [9]. An MA consists in a combination of an Evolutionary Algorithm with Local Search (LS) methods. In this paper we propose an MA that makes use of an Optimal Split procedure developed for the specific case of the TOP. An Optimal Split is performed using a modified version of the Program Evaluation and Review Technique/Critical Path Method (PERT/CPM). We have also developed a strong heuristic for population initialization that we have termed Iterative Destruction/Construction Heuristic (IDCH). It is based on Destruction/Construction principles described in Ruiz and Stützle [10], combined with a priority rule and LS. Computational results are compared with those of Chao, Golden and Wasil [5], Tang and Miller-Hooks [12] and Archetti, Hertz and Speranza [1].

The article is organized as follows. Section 1 gives a formal description of the TOP. Section 2 describes our algorithm with employed heuristics, an adaptation of the PERT/CPM method yielding an Optimal Split procedure and the MA design. Numerical results on standard instances are presented in Section 3. At the end we put forward some conclusions.

2 Problem Formulation

The TOP can be modeled with a graph $G = (V, E)$, where $V = \{1, 2, ..., n\}$ is the vertex set representing customers, and $E = \{(i, j) \mid i, j \in V\}$ is the edge set. Each vertex i is associated with a profit P_i. There is also a *departure* and an *arrival* vertex, denoted respectively d and a. A tour r is represented as an ordered list of $|r|$ customers from V: $r = (r_1, \ldots, r_{|r|})$. Each *tour* begins at the departure vertex and ends at the arrival vertex. We denote the total profit collected from a tour r as $P(r) = \sum_{i \in r} P_i$, and the total travel cost or duration $C(r) = C_{d,r_1} + \sum_{i=1}^{i=|r|-1} C_{r_i,r_{i+1}} + C_{r_{|r|},a}$, where $C_{i,j}$ denotes the travel cost between i and j. Travel costs are assumed to satisfy the triangle inequality. A *solution* is a set of m (or fewer) feasible tours in which each customer is visited only once. A tour r is feasible if its length does not exceed a pre-defined limit L. So a solution is feasible if $C(r) \leq L$ for any tour r. The goal is to find a collection of m (or fewer) tours that maximizes the total profit while satisfying the pre-specified tour length limit L on each tour.

3 Resolution Methods

Genetic Algorithms (GA) are classified as Evolutionary Algorithms: a *population* of solutions evolves through the repetitive combination of its *individuals*. A GA *encodes* each solution into a similar structure called a *chromosome*. An encoding is said to be *indirect* if a decoding procedure is necessary to extract solutions from chromosomes. In this paper we use a simple indirect encoding that we denote as a *giant tour*, and an Optimal Split procedure as the decoding process. Optimal Split was first introduced by Beasley [2] and Ulusoy [13], respectively for the node routing and arc routing problems. The splitting procedure we propose here is specific to the TOP.

To insert a chromosome in the population and to identify improvements, it is necessary to know the performance of each individual in the population through an *evaluation* procedure. In our algorithm, this evaluation involves the splitting procedure corresponding to chromosome decoding. The combining of two chromosomes to produce a new one is called *crossover*. A diversification process is also used to avoid homogeneity in the population. This diversification is obtained through a *mutation* operation and through conditions on the insertion of new chromosomes in the population.

A Memetic Algorithm (MA) is a combination of an Evolutionary Algorithm and Local Search techniques. This combination has been shown to be effective for the VRP in Prins [9]. Our MA is a combination of GA and some LS techniques.

3.1 Chromosome Encoding and Evaluation

As mentioned above, we do not directly encode a solution, but an ordered list of all the customers in V, which we term a *giant tour*. To evaluate the individual performance of a chromosome it is necessary to split the giant tour to identify the multiple-vehicle solution and unrouted customers.

The giant tour is encoded as a *sequence*, i.e. a permutation of V that we denote as π. We extract m tours from the giant tour while respecting the order of the customers in the sequence and the constraint on the length of each tour (referred to from now on as the L-constraint). We consider only tours whose customers are adjacent in the sequence, so that a tour can be identified by its starting point i in the sequence and the number of customers following i in π, denoted $l_i \geq 0$, to be included in the tour. A tour corresponds to the subsequence $(\pi[i], \ldots, \pi[i + l_i])$ and is denoted as $\langle i, l_i \rangle_\pi$.

The maximum possible value of l_i for a feasible tour, given a sequence π, depends on L. A tour of maximum length is called a *saturated* tour, meaning that all customers following i in π are included in the tour as long as the L-constraint is satisfied, or until the end of the sequence is reached. Customers remaining unrouted after splitting can only be located between tours in π. We denote as $l_i^{max,\pi}$ the number of customers following i in the sequence starting with $\pi[i]$ such that $\langle i, l_i^{max,\pi} \rangle_\pi$ is saturated, i.e. the tour represented by $\langle i, l_i^{max,\pi} + 1 \rangle_\pi$ is infeasible, or the end of the sequence has been reached.

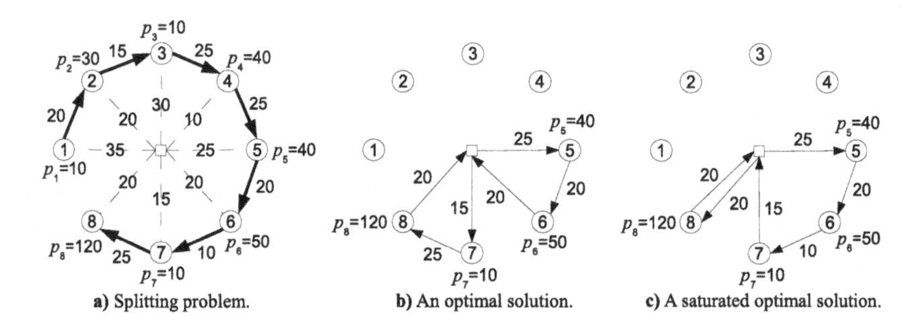

Fig. 1. A giant tour with 8 customers and two optimal solutions for $L = 70$

A π-solution $S^\pi = (\langle i_1, l_{i_1}\rangle_\pi, \ldots, \langle i_k, l_{i_k}\rangle_\pi)$ is such that $k \leq m$, $\langle i_p, l_{i_p}\rangle_\pi$ respects the L-constraint for each p, and $i_{q+1} > i_q + l_q$ for each q in $1, \ldots, k-1$. A π-solution is optimal if the sum of the profits from customers in the subsequences, denoted $P(S^\pi)$, is such that there exists no π-solution yielding a greater profit. A π-solution $(\langle i_1, l_{i_1}\rangle_\pi, \ldots, \langle i_k, l_{i_k}\rangle_\pi)$ is said to be saturated if each tour $\langle i_p, l_{i_p}\rangle_\pi$ in S^π is saturated for $p < k$. Figure 1 describes an instance with 8 customers. Profits from these customers are respectively $10, 30, 10, 40, 40, 50, 10, 120$. We consider $\pi = (1, 2, 3, 4, 5, 6, 7, 8)$. Two optimal solutions, one of which is saturated, are shown in Figure 1.

The splitting problem consists in identifying a π-solution that maximizes the collected profit. Given these notations, we make the proof (that cannot be placed here because of space restriction) that an optimal splitting of the giant tour is obtained through consideration of only the saturated tours.

Proposition 1. *Let π be an arbitrary sequence of the vertices of an instance of the TOP. Then there exists a saturated optimal π-solution.*

Therefore, the splitting can be done considering only saturated tours. Consequently, we are only interested in finding saturated π-solutions.

Optimal Evaluation. The splitting problem can be formulated as finding a path on an acyclic graph $H = (X, F)$, where $X = \{d, 1, 1', 2, 2', ...n, n', a\}$. An arc linking nodes x and x' represents a saturated tour starting with customer $\pi[x]$. The weight $w_{x,x'}$ of this arc is set to the value of the collected profit of the corresponding tour. An arc linking nodes x' and y with $y > x + l_x$ shows that the tour starting with $\pi[y]$ can commence after the tour starting with $\pi[x]$. These arcs are weighted by $w_{x',y} = 0$. The graph construction is such that any path in the graph from departure to arrival nodes is $2q + 1$ arcs long and contains q compatible tours. Figure 2 shows the graph corresponding to the splitting problem in Figure 1. Values of $l_i^{max,\pi}$ for each starting customer i are $0, 2, 1, 1, 2, 1, 1, 0$.

The splitting problem is finding the longest path in the new graph H that does not use more than $2m + 1$ arcs (m is the maximum number of vehicles). This can be done by modifying the well-known PERT/CPM method as follows.

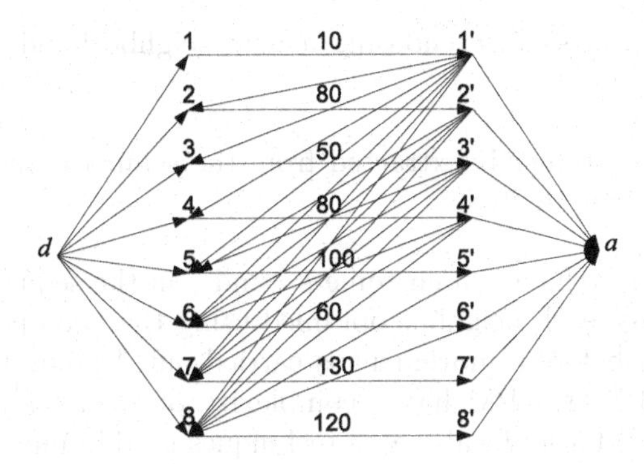

Fig. 2. Graph representation for the splitting problem

For each node k in H (except the departure node d) we create two arrays μ_k and γ_k of fixed size $2m+1$. Component $\mu_k[i]$ memorizes the maximum profit collected within a path of i arcs long from d to k and $\gamma_k[i]$ memorizes the predecessor of k matching the corresponding maximum profit. As H does not have any cycle, the idea is to visit nodes from d to a and to fill the two arrays μ_k and γ_k for each node k. At the end of the procedure, the largest component of μ_a represents the maximum profit that can be reached by splitting the sequence. Next, a backtrack is performed on γ_k in order to determine the corresponding tours.

Nodes are visited in the order $1, 1', 2, 2', ..., n, n', a$. We denote as $\Gamma^-(i)$ the set of the predecessors of i in H. We compute $\mu_k[i]$, the component i of the vector μ at node k, as follows: $\mu_k[i] = \max_{j \in \Gamma^-(k)} \{\mu_j[i-1] + w_{j,i}\}$. At the end of the procedure, the greatest value of μ_a indicates the final node of the longest path. We can use array γ to rebuild that path, and finally translate non-zero weighted links into their corresponding tours. The complexity of this modified PERT/CPM method for Optimal Splitting of the giant tour is $O(m \cdot n^2)$.

Fast Evaluation. We may also use a faster evaluation technique, which we call Quick Split, as a splitting procedure. This simple split uses the assumption that if the tour p ends with the customer $\pi[x]$, the next tour begins with customer $\pi[x+1]$. This assumption that there are no unrouted customers between two different tours in the sequence and considering only saturated tours enable us to obtain an approximate evaluation where the first tour begins with $i_1 = 1$. The complexity of this method is $O(n)$.

3.2 Local Search as Mutation Operator

To complete the memetic algorithm, local search techniques are used as mutation operators with probability pm. We use different neighborhoods during mutation. The procedure selects these neighborhoods in random order. Once an improvement is found, the search within the current neighborhood is stopped, and we restart the procedure choosing a new neighborhood, until no further improvements are found.

Shift operator. A customer is extracted from the sequence and its insertion in all other positions is evaluated.

Swap operator. An exchange of customers i and j in the sequence is evaluated.

In TSP problems, each neighbor obtained using the two operators described above is evaluated before a movement is carried out, leading to $O(n^2)$. As the evaluation using PERT/CPM has a complexity of $O(m \cdot n^2)$, and to keep a complexity of $O(n^3)$ for all local search techniques used in mutation, we decided to use Quick Split in association with these two operators. A *compressed* version of the current chromosome is produced at the beginning of these local search techniques to left-shift tours within the sequence so that solutions produced by the Shift operator and the Swap operator can be evaluated by the Quick Split quickly and efficiently. At the end of these local search techniques, when a better neighbor has been selected, unrouted customers are redistributed along the chromosome in the same order as in the initial chromosome.

Destruct and Repair operator. The idea is to remove a small part of the solution with a view to rebuilding an improved solution (see Destruction/Construction, [10]). This local search is applied to the solution given by PERT/CPM method on the current chromosome. A certain number (selected randomly between 1 and n/m) of customers are removed from tours and redeclared as unrouted customers. The solution is reconstructed using a parallel version of the Best Insertion algorithm [11]. This constructive method evaluates the insertion cost $(C_{i,z} + C_{z,j} - C_{i,j})/P_z$ of any unrouted customer z between any couple of customers i and j in a tour r so that j directly follows i in r. The feasible insertion that minimizes the cost is then processed, and the method loops back to the evaluation of the remaining unrouted customers. If more than one possible insertion minimizes the insertion cost, one of them is chosen at random. This process is iterated until no further insertions are feasible, either because no tour can accept additional customers, or because all customers are routed (the solution is optimal in this case). The complexity is $O(n^3)$, since all customer insertions in all positions have to be evaluated and the process is iterated at most n times to insert all customers.

3.3 Algorithm Initialization

To create some good solutions for an initial population, we developed an Iterative Destruction/Construction Heuristic (IDCH) based on the *Destruct and Repair operator* and some diversification components. The key idea of this heuristic is that the more difficult it is to insert an unrouted customer into a solution, the more this customer will be considered for insertion. Starting with an empty solution, we use the parallel version of the Best Insertion [11] to build a first solution. On following iterations a small part of the current solution is destroyed by removing a limited random number of customers from tours, and a 2-opt procedure is used to reduce the travel cost of tours. A reconstruction phase is then processed using a parallel prioritized version of the Best Insertion. The destruction and construction phases are iterated, and each time a customer remains unrouted after the construction phase its priority is increased by the value of its associated profit. At each construction phase the subset of unrouted customers with the highest priority is considered for insertion. When no more of these customers can be inserted, unrouted customers with lower priorities are considered, and so on. The procedure stops after n^2 Destruction/Construction iterations without improvement. After each n iterations without improvement we apply the diversification components. This involves destroying a large part of the solution, applying 2-opt to each tour to optimize the travel cost, and finally performing the reconstruction phase.

3.4 Memetic Algorithm

The algorithm starts with an initialization in which a small part of the population is created with an IDCH heuristic and the remainder is generated randomly. At each iteration a couple of parents is chosen among the population using the Binary Tournament, which showed more efficient than random selection and the Roulette-Wheel procedures. The LOX crossover [9] operator is used to produce a child chromosome. New chromosomes are evaluated using the Optimal Splitting procedure described in the previous section. They are then inserted into the current population using a simple and fast insertion technique to maintain a population of constant size, avoiding redundancy between chromosomes. The population is a list of chromosomes sorted lexicographically with respect to two criteria: the profit associated with the chromosome and the total travel cost. If a chromosome with the same profit and the same travel cost exists in the population, it is replaced with the new one. Otherwise, the chromosome is inserted and the worst chromosome of the new population is deleted. A child chromosome has a probability pm of being mutated, using a set of Local Search techniques repeatedly while improving. The stop condition of the Memetic Algorithm is a bound on the number of iterations without improvement of the population, that is to say the number of iterations where the child chromosome simply replaces an existing chromosome in the population, or where its evaluation is worse than the worst chromosome in the current population. At the end of the search the chromosome at the head of the population is reported as the best solution.

4 Numerical Results

We tested our MA on standard instances for the TOP from Chao, Golden and Wasil [5]. Instances comprise 7 sets containing different numbers of customers. Inside each set customer positions are constant, but the number of vehicles m varies between 2 and 4, and the maximum tour duration L also varies so that the number of customers that can really be serviced is different for each instance.

We set parameter values for our algorithm from a large number of experiments on these benchmark instances. The population size is fixed to 40 individuals. When the population is initialized, five individuals are generated by IDCH. Other individuals are generated randomly. The mutation rate pm of the MA is calculated as: $pm = 1 - \frac{iter_{ineffective}}{iter_{max}}$. The algorithm stops when $iter_{ineffective}$, the number of elapsed consecutive iterations without improvement of the population, reaches $iter_{max} = k \cdot n/m$ with $k = 5$.

For each instance, results are compared to results reported by Chao, Golden and Wasil [5], Tang and Miller-Hooks [12] and by Archetti, Hertz and Speranza [1]. These results, as well as benchmark instances, are available at the following url: `http://www-c.eco.unibs.it/~archetti/TOP.zip`.

Archetti, Hertz and Speranza [1] proposed different methods: two TS and two VNS. For each method, they reported, for each instance, the worst profit z_{min} and the best profit z_{max} obtained from three executions. The difference $\Delta z = z_{max} - z_{min}$ is presented as an indicator of the stability of each method. Other results (Chao, Golden and Wasil [5] and Tang and Miller-Hooks[12]) are given for a single execution. In order that our method may be measured against the algorithms presented by Archetti, Hertz and Speranza [1], all of which have been shown to be very efficient, we report results of the MA the same way: we consider z_{min} and z_{max} for three executions of the MA.

Because of space limitations we only report the sum of the differences between the best known value of the profit and z_{max} (resp. z_{min}) for each instance: $\Delta_{Best}^{z_{max}} = Best - z_{max}$ (resp. $\Delta_{Best}^{z_{min}}$). The $Best$ value we consider is the best known profit of an instance, including our results. More detailed results are available at: `http://www.hds.utc.fr/~boulyher/TOP/top.html`.

Table 1 reports $\Delta Z_{max} = \sum \Delta_{Best}^{z_{max}}$ and $\Delta Z_{min} = \sum \Delta_{Best}^{z_{min}}$ for each method. The difference $\Delta Z = \Delta Z_{max} - \Delta Z_{min}$ between these two values is also given as an indicator of the stability of each method.

Column headers are as follows: $TS_{Penalty}$, $TS_{Feasible}$, VNS_{Fast} and VNS_{Slow} are the four methods proposed by Archetti, Hertz and Speranza [1], TMH denotes the Tabu Search of Tang and Miller-Hooks [12] and column CGW denotes the heuristic of Chao, Golden and Wasil [5]. The column UB corresponds to the upper bound of the profit obtained with an exact algorithm, if known. As far as we know, the only existing upper bound for the TOP is that described by Boussier, Feillet and Gendreau [3].

Our Memetic Algorithm produced solutions that improve on the best known solutions from the literature for 11 instances of the benchmark set. Profits 965, 1242, 1267, 1292, 1304, 1252, 1124, 1216, 1645, 646 and 1120 have respectively been reached for instances $p4.2.j$, $p4.2.p$, $p4.2.q$, $p4.2.r$, $p4.2.s$, $p4.3.q$, $p4.4.p$, $p4.4.r$, $p5.2.y$, $p7.2.j$ and $p7.3.t$.

A comparison of profits with the upper bound of Boussier, Feillet and Gendreau [3] shows that profits reached by $TS_{Penalty}$, VNS_{Slow} and CGW exceed the upper bound on a subset of seven instances. It seems abnormal, and these instances are consequently not included in the results of Table 1, and details about

Table 1. Overall performance of each algorithm

	$TS_{Penalty}$	$TS_{Feasible}$	VNS_{Fast}	VNS_{Slow}	TMH	CGW	MA
ΔZ_{min}	2370	1178	1430	421	2398	4334	428
ΔZ_{max}	975	393	346	78	N/A	N/A	**74**
ΔZ	1395	785	1084	343	N/A	N/A	354

Table 2. Results for instances for which some profits exceed the upper bound

	$TS_{Penalty}$		$TS_{Feasible}$		VNS_{Fast}		VNS_{Slow}		TMH	CGW	MA		$Best$	UB
$file$	z_{min}	z_{max}	z_{min}	z_{max}	z_{min}	z_{max}	z_{min}	z_{max}			z_{min}	z_{max}		
$p1.3.h$	70	70	70	70	70	70	70	70	70	**75**	70	70	70	70
$p1.3.o$	205	205	205	205	205	205	205	205	205	**215**	205	205	205	205
$p2.3.h$	165	**170**	165	165	165	165	165	165	165	165	165	165	165	165
$p3.4.k$	350	350	350	350	350	350	350	**370**	350	350	350	350	350	350
$p5.3.e$	95	95	95	95	95	95	95	95	95	**110**	95	95	95	95
$p6.4.j$	366	366	366	366	366	366	366	**390**	366	366	366	366	366	366
$p6.4.k$	528	528	528	528	528	528	528	528	522	**546**	528	528	528	528

Table 3. Average and maximal CPU times for the different instance sets

	TMH	CGW	$TS_{Penalty}$		$TS_{Feasible}$		VNS_{Fast}		VNS_{Slow}		MA	
	cpu	cpu	avg	max	avg	max	avg	max	avg	max	avg	max
1	N/A	15.41	4.67	10.00	1.63	5.00	0.13	1.00	7.78	22.00	1.31	4.11
2	N/A	0.85	0.00	0.00	0.00	0.00	0.00	0.00	0.03	1.00	0.13	0.53
3	N/A	15.37	6.03	10.00	1.59	9.00	0.15	1.00	10.19	19.00	1.56	3.96
4	796.7	934.8	105.29	612.00	282.92	324.00	22.52	121.00	457.89	1118.00	125.26	357.05
5	71.3	193.7	69.45	147.00	26.55	105.00	34.17	30.00	158.93	394.00	23.96	80.19
6	45.7	150.1	66.29	96.00	20.19	48.00	8.74	20.00	147.88	310.00	15.53	64.29
7	432.6	841.4	158.97	582.00	256.76	514.00	10.34	90.00	309.87	911.00	90.30	268.01

these instances are given in Table 2. Bold values identify profits that exceed the upper bound of Boussier, Feillet and Gendreau [3].

Table 3 finally reports CPU time for each method and for each instance set from 1 to 7. We denote *cpu* the CPU time if a single execution was performed and *avg* and *max* respectively the average and the maximal CPU time if three executions were performed. Computers used for experiments are as follows:

- CGW: run on a SUN 4/730 Workstation,
- TMH: run on a DEC Alpha XP1000 computer,
- $TS_{Penalty}$,$TS_{Feasible}$,VNS_{Fast} and VNS_{Slow}: run on an Intel Pentium 4 personal computer with 2.8 GHz and 1048 MB RAM,
- MA: run on a Intel Core 2 Duo E6750 - 2.67 GHz (no parallelization of the program) with 2 GB RAM.

These results clearly show our Memetic Algorithm compares very well with state of the art methods. MA outperforms the VNS Slow algorithm of Archetti,

Hertz and Speranza in term of efficiency and is quite equivalent in term of stability. A comparison of computational times using similar computers shows, however, that MA outperforms VNS Slow on this point.

5 Conclusion

We propose a new resolution method for the TOP using the recent Memetic Algorithm approach. It is the first time that an Evolutionary Algorithm has been used for this problem. We also propose an Optimal Split procedure as a key feature of this method especially intended for the TOP. Our method proved very efficient and fast compared with the best existing methods, and even produced improved solutions for some instances of the standard benchmark for the TOP.

These results show, first, that population-based algorithms can efficiently be applied to the Team Orienteering Problem. Secondly, the use of the Optimal Splitting procedure shows that further research into specialized methods is a promising direction in addressing the Team Orienteering Problem.

References

1. Archetti, C., Hertz, A., Speranza, M.G.: Metaheuristics for the team orienteering problem. Journal of Heuristics 13(1), 49–76 (2006)
2. Beasley, J.E.: Route-first cluster-second methods for vehicle routing. Omega 11, 403–408 (1983)
3. Boussier, S., Feillet, D., Gendreau, M.: An exact algorithm for the team orienteering problems. 4OR 5, 211–230 (2007)
4. Butt, S., Cavalier, T.: A heuristic for the multiple tour maximum collection problem. Computers & Operations Research 21, 101–111 (1994)
5. Chao, I.-M., Golden, B., Wasil, E.A.: The team orienteering problem. European Journal of Operational Research 88, 464–474 (1996)
6. Feillet, D., Dejax, P., Gendreau, M.: Traveling salesman problems with profits. Transportation Science 39(2), 188–205 (2005)
7. Khemakhem, M., Chabchoub, H., Semet, F.: Heuristique basée sur la mémoire adaptative pour le problème de tournées de véhicules sélectives. In: Logistique & Transport, Sousse, Tunisie, pp. 31–37 (November 2007)
8. Moscato, P.: New Ideas in Optimization, chapter Memetic Algorithms: a short introduction, pp. 219–234 (1999)
9. Prins, C.: A simple and effective evolutionary algorithm for the vehicle routing problem. Computer & Operations Research 31(12), 1985–2002 (2004)
10. Ruiz, R., Stützle, T.: A simple and effective iterated greedy algorithm for the permutation flowshop scheduling problem. EJOR 177, 2033–2049 (2007)
11. Solomon, M.: Algorithms for the vehicle routing and scheduling problems with time window constraints. Operations Research 35, 254–265 (1987)
12. Tang, H., Miller-Hooks, E.: A tabu search heuristic for the team orienteering problem. Computer & Operations Research 32, 1379–1407 (2005)
13. Ulusoy, G.: The fleet size and mixed problem for capacitated arc routing. European Journal of Operational Research 22, 329–337 (1985)

Decentralized Evolutionary Optimization Approach to the p-Median Problem

Stephan Otto[1] and Gabriella Kókai[2]

[1] method park Software AG, Am Wetterkreuz 19b, Erlangen, Germany
stephan.otto@methodpark.de
http://www.methodpark.de

[2] University Erlangen-Nuremberg, Martensstr. 3, Erlangen, Germany
kokai@informatik.uni-erlangen.de
http://www2.informatik.uni-erlangen.de

Abstract. The facility location problem also known as p-median problem concerns the positioning of facilities such as bus-stops, broadcasting stations or supply stations in general. The objective is to minimize the weighted distance between demand points (or customers) and facilities. In general there is a trend towards networked and distributed organizations and their systems, complicating the design, construction and maintenance of distributed facilities as information is scattered among participants while no global view exists. There is a need to investigate distributed approaches to the p-median problem. This paper contributes to research on location problems by proposing an agent oriented decentralized evolutionary computation (EC) approach that exploits the flow of money or energy in order to realize distributed optimization. Our approach uses local operators for reproduction like mutation, recombination and selection finally regulated by market mechanisms. This paper presents two general outcomes of our model: how adaptation occurs in the number and strategies of agents leading to an improvement at the system level. The novelty of this approach lies in the biology-inspired bottom-up adaptation method for inherent distributed problems. It is applied to the uncapacitated p-median problem but is also intended to be general for a wide variety of problems and domains, e.g. wireless sensor networks.

1 Introduction

Today's IT systems like the internet, global supply chains, sensor networks or grid applications are large distributed systems with a huge number of elements working in collaboration in order to fulfill requirements from customers, service providers, organizations and other systems. These systems cannot be fixed in their structure, design and behavior in order to cope with a highly dynamic and unpredictable environment. For this reason effective positioning of facilities is crucial. Current approaches are not easily adapted to for distributed systems which have no central coordinator. This is based on two reasons. At first elements

M. Giacobini et al. (Eds.): EvoWorkshops 2008, LNCS 4974, pp. 659–668, 2008.

are limited by how much they can communicate and process [1]. The second reason is information hiding, which means, that not all information can be given to a central control due to intellectual property or security reasons [2,3,4]. Throughout this paper the term decentralized system for systems without central control is used.

From a general point of view, we understand a decentralized system as "distributed solution" to a distributed problem. The overall solution (system strategy) is scattered amongst participating system elements, each with part of the solution. In order to adapt distributed systems to distributed problems, both an on-the-fly system size and strategy adjustment is needed. In [5,6] an approach is presented to show decentralized coordination in distributed economic agent systems. The investigation in [7] proposes evolutionary algorithms with on-the-fly population size adjustment as a rewarding way of parameter regulating. As in natural environments population sizes of species change and tend to stabilize around appropriate values according to some factors such as resources provided by the carrying capacity of the environment. Technically seen, population size is used as the most flexible parameter in natural systems. As a combination of these two approaches we present in this paper an artificial life like optimization strategy using a decentralized economic-agent based evolutionary algorithm that offers both properties needed. First results of a prototypic implementation using the examples of two p-median problems are shown. This paper seeks an answer to the question whether it is possible and rewarding to solve distributed problems using a decentralized evolutionary algorithm.

The remainder of this paper is organized as follows. In section 2, we review previous work related to p-median problem and evolutionary computation and evolutionary multi-agent systems. In section 3, we present our model of a market-based agent system. Section 4, illustrates our decentralized optimization method by combining evolutionary computation and multi-agent systems. Using this model two emergent effects are specified that occur within such a distributed system formed by agents: adaptation of the number of agents and adaptation of agent's strategies on the p-median problem. We show how the number of agents is regulated by the flow of money and how resource efficient strategies dominate. In section 5, we describe our current implementation and present experimental results. Finally, the conclusion and future plans are mentioned in section 6.

2 Related Work

Consider a set of locations $L = \{1 \ldots l\}$ for customers and a set of potential locations for facilities $P = \{1 \ldots p\} \subseteq L$ and $l \times l$ matrix D_{ij} of transportations costs for satisfying the demands w_i of the customers from the facilities. The weighted distance matrix is $W_{ij} = w_i D_{ij}$. The p-median problem is to locate the p facilities at locations of L in order to minimize the total transportation cost for satisfying the demand of the customers. Each customer is supplied from the cheapest open facility according to W. The uncapacitated p-median problem can be expressed as follows:

$$minimize \sum_{j=1}^{l} \sum_{i=1}^{p} w_{ij} \quad (1)$$

The p-median problem has attracted a great deal of interest over the last decades in evolutionary research. Based on our search, Hosage and Goodchild [8] published the first research, followed by [9,10] and many others. However, they use the classical evolutionary approach with a central optimization loop and central operators. Classical generation-based evolutionary algorithms with a nonoverlapping population model where the entire population is replaced at each generation is not generally desirable in adaptive applications, where a high level of on-line performance is needed [11]. Also the central control of traditional evolutionary algorithms is inappropriate for distributed problems. A decentralized EC-enabled multi-agent-system (MAS) framework is presented in [12] where only local selection occurs. Based on [12] Smith and Eymann [6,5] investigate negotiation strategies in a supply chain for the production of cabinets but concentrate merely on self-organizing coordination effects. Explicit optimization is not investigated which is not suitable to the given problem.

To the best of our knowledge, there is no approach that applies a distributed evolutionary agent model to the p-median problem. The flow of money is used to direct evolutionary search. This paper describes a scalable multi-agent approach without central control providing distributed optimization.

3 Model

This section describes the model and generic outcomes used throughout the paper. In order to model and study generic decentralized systems, the term multi-agent system is used and defined as follows.

3.1 Multi-Agent System

Definition 1 (Multi-Agent System (MAS)). *A multi-agent system* $MAS = (A, E)$ *consists of a finite set of agents* $A = \{a_1, ..., a_z\}$ *embedded in an environment* E.

An agent is defined as follows:

Definition 2 (Agent). *An agent* $a = (I, O, m, s, c)$ *consists of a finite set of sensory inputs* $I = \{i_1, \ldots, i_r\}$*, a finite set of effector outputs* $O = \{o_1, \ldots, o_q\}$*, a function* $m : I \rightarrow O$ *which maps sensory inputs to effector outputs, the strategy vector* $s = (s_1, s_2, \ldots, s_r)$ *determining or parameterizing* m*, and the agents current funds* $c \in \mathbb{R}$.

As a basic element of a decentralized system, an agent a is an autonomous sensor effector entity with function m_a. The strategy vector s_a represents a set of parameters that are assigned to a. We consider the agent's function m_a to be determined by s_a and thus, the strategy s_a of agent a determines the behavior. Access to a particular element in the strategy vector is given by the following convenience notation: $s_a(parameter)$ denotes *parameter*. Note that the dimension of s may vary between different agents.

3.2 Distributed Optimization Problem

The task of optimizing can be formulated as follows:

Definition 3 (Distributed Optimization Problem (DOP)). *A distributed optimization problem is given by:*

$$\begin{aligned} \textit{minimize} \quad & f(s_A), \\ & \textit{subject to } \gamma(s_A) \end{aligned} \tag{2}$$

where $s_A = (s_1^{a_1}, s_2^{a_1}, \ldots, s_t^{a_1}, \ldots, s_1^{a_z}, s_2^{a_z}, \ldots, s_u^{a_z}) \in S_A^*$ *is the strategy vector of MAS and* S_A^* *is the search space. The objective function is* f *and there are* v *constraints* $\gamma_i(s_A), i = 1, \ldots, v$ *imposed on* s_A.

The strategy vector s_A is distributed among A and the objective function f is unknown to A and therefore cannot be calculated by any single agent. The set $\gamma(s_A)$ contains constraints of the DOP distributed over A. The combination of local strategy vectors of all agents s_A forms the strategy vector and is also the distributed solution of the DOP. If constraints vary over time t, the DOP becomes even harder. The dimensions of s_A are not fixed, rather they may vary over time, as agents enter or leave the system.

3.3 Integrating an Economic Perspective

In our model we assume a discrete timeline, where all actions take place at consecutive steps. Further, agents have to pay a tax $\mathcal{T}_a(t)$ to the environment E at every time t for their actions and according to their strategy parameters s_a. Given the tax, we can calculate the profit π_a an agent a receives at time t by

$$\pi_a(t) = \mathcal{R}_a(t) - \mathcal{P}_a(t) - \mathcal{T}_a(t) \tag{3}$$

where $\mathcal{P}_a(t)$ denotes the payment a has to pay to other agents or the environment in order to follow its strategy s_a, $\mathcal{T}_a(t)$ denotes tax turned over to the environment and finally a may have receipts $\mathcal{R}_a(t)$ from previously executed actions. Based on the profit $\pi_a(t)$, an agent accumulates funds c_a over time expressed by:

$$c_a(t+1) = c_a(t) + \pi_a(t) \tag{4}$$

where $c_a(t+1)$ denotes the funds of agent a at time $(t+1)$, $c_a(t)$ denotes a's funds at time t and $\pi_a(t)$ is the profit of a at time t. Money cannot be 'created' by the agent, rather it is provided by the environment E representing the demand $\mathcal{D}$ of the market since it provides a specific amount of funds in return to services offered by the set of agents A.

4 Evolutionary Computation as Decentralized Adaptation Method

The aim of this paper is to show an decentralized optimization approach for distributed problems lined out in section 3.2. Our assumptions on a system that applies this method are 1) Agents have the ability to communicate and to sense their environment at least locally 2) There is no central manager, instead the DOP and its solution is distributed among A 3) There is a limited amount of money ($\mathcal{D}$) that will be provided by the environment E.

The previous section has shown a market-based multi-agent system that consists of economic agents who have to solve a distributed problem DOP collaboratively and fulfill previous prerequisites. In this section we re-use evolutionary algorithm theories of Holland [11,13] concurrently with a decentralized economic agent perspective given in [5] as well as our economic market perspective as we believe that a decentralized market mechanism can be a profound approach to replace the central fitness calculation in evolutionary algorithms. There are a variety of evolutionary computational models that have been proposed and studied that we will refer to as evolutionary algorithms [11,13]. In our agent based approach we turn the typical evolutionary algorithm software design on its head by breaking up any central instance and move the genetic representation as well as the operators to the agents respective individuals itself as proposed in [12].

Local reproduction by an agent includes all steps and operations necessary to produce new offspring, such as local search, recombination and mutation. We introduce an agent specific variable θ that serves as a threshold and enables agents to reproduce. Whenever the funds of an agent exceed this threshold ($c_a \geq \theta$) it will reproduce. Advantages of local versus global reproduction are discussed in [14] in detail. As we want to focus on emerging global effects of local reproduction, based on [15] an agents algorithm motivated by modeling ecologies of organisms adapting in environments is formulated as follows:

Algorithm 1. Main loop, every agent executes asynchronously

Require: s and c from parent
1: initialize m
2: **loop**
3: get sensory inputs I
4: execute m
5: set effector outputs O
6: calculate c using equation (4)
7: **if** $c \geq \theta$ **then**
8: reproduce and split c with child
9: **else if** $c < 0$ **then**
10: die
11: **end if**
12: **end loop**

An agent a is set up with its strategy s_a and funds c from its parent. The initial population is created by assigning a random strategy and an initial reservoir of funds to every agent. Once started, an agent will continuously follow his strategy by executing m. Whenever c reaches θ the agent will reproduce (line 7-8) and create a new agent. In the opposite case, if an agent dies, his funds drop below zero ($c < 0$, line 9), construed to signify the bankruptcy of an inefficient agent. Therefore the notion of 'generation number' does not exist and instead a perpetual reproduction and replacement of agents takes place. This mechanism makes survivor selection unnecessary and population size an observable. This can be regarded as a combination of biological and economical ideas [5], because no rational economic agent would create its own competition. However, in biology, it is the case.

Given the model in chapter 3 and algorithm 1 it is clear that the distribution of money within the system differs among the agents. Any agent selfishly tries to exploit the limited money based on its own strategy; however we assume an agent has no particular intelligence about how this strategy relates to its own success. In the following sections a short description of the two basic properties of our model is given: adaptation in the number of agents and their strategy. A more extensive investigation is given in [16].

4.1 Adaptation of the Number of Agents

Following the statement in [7] in natural environments the population size tend to stabilize around the carrying capacity of the ecosystem we define the carrying capacity as demand $\mathcal{D}$. As outlined in [11] the system can only be stable for bounded input when $\mathcal{D}$ does not grow in magnitude of its own accord.

For further analysis we approximate the creation of agents from discrete by a continuous adaptation in population growth. For a particular agent a the ratio $\frac{\pi_a(t)}{\theta}$ is an expression of the estimated creation of new agents at time t. In fact, the number of agents increase/decrease stepwise and not continuously, depending on the distribution of $\Pi(t) = \sum_{a \in A} \pi_a(t)$ and the funds in A.

Dividing the overall profit $\Pi(t)$ by θ we get the average number of new/dying agents at time $t+1$ (again, averaged for sufficient long runs) in equation 5

$$|A(t+1)| = |A(t)| + \frac{\Pi(t)}{\theta} \tag{5}$$

Thus $|A(t+1)|$ vary by factor $\frac{\Pi}{\theta}$ compared to $|A(t)|$, where $|A(t)|$ is the number of agents in A at time t. If $\frac{\Pi}{\theta}$ is positive (negative), the number of agents will increase (decrease) depending on Π. This emergent behavior can be observed in real scenarios [2] where actors enter and leave the market. The rate at which agents enter or leave the market is directly correlated with the overall profit (see equation 5) and the market supply $\mathcal{S}$ and demand $\mathcal{D}$ in such a scenario. Even without any central control this effect can be observed.

4.2 Distributed Problem Optimization by Spread of Successful Strategies

Since in our model there is no central instance that can rank and compare the agents funds, no central selection based on fitness can be calculated and performed. Therefore the system must evolve fitter strategies in an emergent self-organizing way.

For the following discussion we assume that strategies were reproduced in a pure way without disturbance of mutation or recombination. The proportion of strategy s_a in $A(t)$ is $p_a(t) = \frac{1}{|A(t)|}$, where $|A(t)|$ denotes the number of agents. Based on evolutionary pressure the agent population profit Π reaches zero after a sufficient amount of time and for long running simulation we can set $\lim_{t\to\infty} \Pi(t) = 0$. According to equation 5 the number of agents in the next time step can be rewritten by $|A(t+1)| = |A(t)|$ and the expected proportion of s_a at time step $t+1$ is as follows:

$$p_a(t+1) = \frac{1 + \frac{\pi_a(t)}{\theta}}{|A(t)|} \tag{6}$$

where $1 + \frac{\pi_a(t)}{\theta}$ is the original strategy plus the estimated additional quantity of s_a that together forms the number of samples of strategy S_a in $A(t+1)$. It follows, that based on the proportion $p_a(t+1) > p_a(t)$ for a positive profit the part of s_a in A grows. In words, the proportion of a particular strategy s_a grows as the ratio of the average profit π_a of agent a. Strategies causing a positive/negative profit will receive an increasing/decreasing number of replications in the next time step.

It follows that strategies inducing a positive profit on the one hand may have to pay less tax compared to below zero profit strategies due to the usage of less resources or more efficient resource utilization. Both are needed to streamline logistic networks while simultaneously improving service to the customer. Based on equation 3 one can see that tax T is that part of an agents profit determining variable which is not explicitly related to the flow of goods. With tax the models basic conditions can be set respective controlled and the agent population will adapt to it, if tax is not too high. Therefore, an intrinsic property of systems using our model is the constant search for better resource utilization.

5 Case Study: The p-Median Problem

We include a case study that illustrates the successful application of decentralized EC-enabled economic agents. It consists in positioning facilities on a discrete map by using two p-median sample problem sets presented in [9,10]. The set of locations L is given by the problem set and facilities are represented by agents. The agents strategy vector consists two values: the location $s_a(location) \in L$ and the profit factor $s_a(profit) \in \mathbb{R}$. An agent a offers service to customer at location $j \in L$ according to $cost_{aj} = w_{s_a(location)j} * s_a(profit)$. Customers choose always the agent with lowest $cost$ and the selection is expressed as

$$\sigma_{aj}(t) = \begin{cases} 1 & \text{customer } j \text{ is served by agent } a \text{ at time } t \\ 0 & \text{otherwise} \end{cases}$$

Multiple agents can offer their service at the same location l and the 'cheapest' agent with lowest *cost* at location l is defined as $\dot{a}^l$. The set of cheapest agents is defined as $\dot{A} \subseteq A$ with $s_{\dot{a}_i}(location) \neq s_{\dot{a}_j}(location), \dot{a}_i, \dot{a}_j \in \dot{A}$ and considered as the set of facilities P that form the solution to the p-median problem given in equation 1. For simplicity reasons the set of $\dot{A}$ is denoted with A, if not stated otherwise. The p-median problem specific DOP is given as

$$\text{minimize } f(s_A) = \sum_{j=1}^{l} \sum_{a=1}^{A} w_{s_a(location)j} \tag{7}$$

$$\text{subject to} \quad \pi_a(t) = \mathcal{R}_a(t) - \mathcal{P}_a(t) - \mathcal{T}_a(t)$$

$$\mathcal{R}_a(t) = \sum_{j=1}^{l} \sum_{a=1}^{A} cost_{aj}(t) * \sigma_{aj}(t) \tag{8}$$

$$\mathcal{P}_a(t) = 0 \tag{9}$$

$$\mathcal{T}_a(t) = 10 + \sum_{j=1}^{l} \sum_{a=1}^{A} w_{aj} * \sigma_{aj}(t) \tag{10}$$

According to equations 8 and 10 the profit $\pi_a(t)$ is basically dependent on $s_a(profit)$. The fixed tax of 10 currency units is necessary to slowly remove agents with no income, e.g. for $a \notin \dot{A}$, otherwise they would remain in the system for ever and consuming resources. As the fixed tax is negligible, a profit factor $s_a(profit) > 1$ is important and induces an evolutionary pressure on agents to evolve a strategy with $s_a(profit)$ slightly over 1. Otherwise an agent will be removed based on a negative profit as well as turning inactive $(a \ni \dot{A})$. There is no direct ressource transfer among the agents (eq. 9) in this particular application but money is transferred between customers and agents. Splitting threshold θ is dynamic calculated as the average income over the last two time steps since a fixed θ would need to be adjusted for every problem.

The initial setting is a population size of 20 agents, a mutation rate of 0.03, 10000 time steps and a profit factor of 2. It is important to set the initial profit factor high enough compared to expected payments in order to get a running system. During the simulation an asymptotic convergence of the profit factor to 1 is expected. The agents strategy is represented as an integer/float vector that can be evolved using the common variable-length binary valued crossover and mutation operators [11]. We use uniform crossover and an implementation of the Breeder Genetic Algorithm [17] for mutation throughout all runs. Selection is disabled and results are obtained by averaging 50 simulation runs.

In figure 1 the results of two different problem sets are shown and costs for different p values are compared to optimal values and random search. At the end of each independent run all pareto-optimal values explored during the run are averaged (The averaged values are not necessary pareto optimal). During simulation the agent population explores different values of p as the population

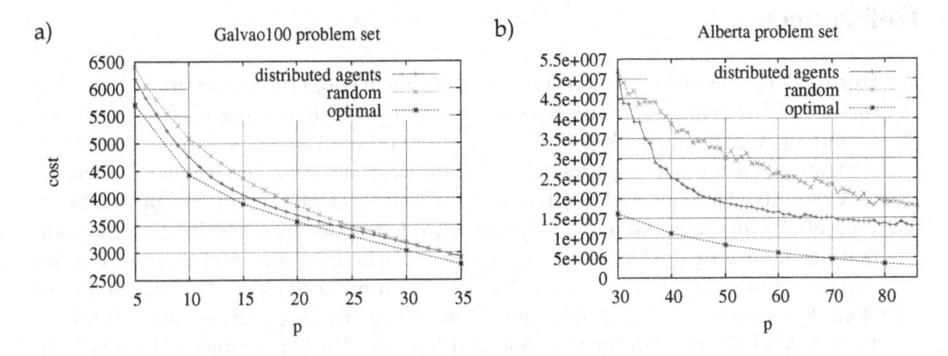

Fig. 1. Distributed agents approach compared with random search and optimal values for multiple p on two problem sets (Galvao: 100 nodes, Alberta: 316 nodes)

size is changing. The graphs show clear differences between distributed agents approach and random search. There are, however, significant differences in search space exploitation for different p values in the distributed agents approach that needs further investigations in different tax and starting conditions.

6 Conclusion and Further Directions

This paper has addressed the potentials of distributed evolutionary systems whereas a market-based multi-agent approach is used. All necessary steps of the evolutionary algorithm have been moved to the agents itself which allows a fully decentralized approach. We have shown with our approach, that even in a decentralized system without central control an adaptation occur, caused by two basic properties. First the number of agents $|A|$ adapt according to the demand provided from environment E and second, successful strategies spread in the population of agents which form together a distributed and optimized solution. The optimization is based on funds provided by E that spread within the system and allows local selection and reproduction by the agents. The two effects indicate that an agent population and their resource utilization adapt to a pre-defined demand and will constantly continue to adapt thereby satisfying the demand with lower resource utilization. We have applied our method on the p-median problem. Then, it was compared to random search. The results show, that the distributed approach is clearly outperforming random search and is also a very promising perspective for distributed problems.

Further research in this line will pursue the development of a generalized model of complex adaptive systems using an economics-enabled market-based evolutionary approach. One aspect that we have not addressed in this paper is the comparison of the results against classical and multi objective evolutionary algorithms. Next, we will look for relations to bi-level or multi-level optimization/adaptation. We also hope to address other useful application fields for this method, e.g. in the context of grid computing and agent systems, artificial immune systems, life-cycle management and other logistics problems.

References

1. Durfee, E.H., Lesser, V.R., Corkill, D.D.: Trends in cooperative distributed problem solving. IEEE Transactions on Knowledge and Data Engineering 1(1), 63–83 (1989)
2. Graudina, V., Grundspenkis, J.: Technologies and multi-agent system architectures for transportation and logistics support: An overview. In: International Conference on Computer Systems and Technologies - CompSysTech, Varna, Bulgaria (2005)
3. Cai, W., Turner, S.J., Gan, B.P.: Hierarchical federations: an architecture for information hiding. In: PADS 2001: Proceedings of the fifteenth workshop on Parallel and distributed simulation, pp. 67–74. IEEE Computer Society, Washington (2001)
4. Hohl, F.: A framework to protect mobile agents by using reference states. Technical Report 2000/03, Institut für Parallele und Verteilte Höchstleistungsrechner (IPVR) (March 2000)
5. Smith, R.E., Bonacina, C., Kearney, P., Eymann, T.: Integrating economics and genetics models in information ecosystems. In: Proceedings of the 2000 Congress on Evolutionary Computation CEC 2000, La Jolla Marriott Hotel La Jolla, 6-9 2000, pp. 959–966. IEEE Press, California (2000)
6. Eymann, T.: AVALANCE - Ein agentenbasierter dezentraler Koordinationsmechanismus für elektronische Märkte. PhD thesis, Universität Freiburg (2000)
7. Eiben, A., Schut, M.: New Ways to Calibrate Evolutionary Algorithms. In: Advanced in Metaheuristics and Optimization, Springer, Heidelberg (2007)
8. Hosage, C.M., Goodchild, M.F.: Discrete space location-allocation solutions from genetic algorithms. Annals of Operations Research 6(2), 35–46 (1986)
9. Alp, O., Drezner, Z., Erkut, E.: An efficient genetic algorithm for the p-median problem. Annals of Operations Research 122(1-4), 21–42 (2003)
10. Galvão, R.D., ReVelle, C.: A lagrangean heuristic for the maximal covering location problem. European Journal of Operations Research 88, 114–123 (1996)
11. Goldberg, D.E.: Genetic Algorithms in search, optimization, and machine learning. Addison Wesley Longmann, Inc., Reading (1989)
12. Smith, R.E., Taylor, N.: A framework for evolutionary computation in agent-based systems. In: Looney, C., Castaing, J. (eds.) Proceedings of the 1998 International Conference on Intelligent Systems, pp. 221–224. ISCA Press (1998)
13. Holland, J.H.: Hidden Order: How Adaptation Builds Complexity, vol. 9. Addison Wesley Publishing Company, Reading (1996)
14. Menczer, F., Degeratu, M., Street, W.: Efficient and scalable pareto optimization by evolutionary local selection algorithms. Evolutionary Comp 8(3), 223–247 (2000)
15. Menczer, F., Belew, R.K.: Adaptive retrival agents: Internalizing local context and scaling up to the web. Mach. Learn. 39(2-3), 203–242 (2000)
16. Otto, S., Kirn, S.: Evolutionary adaptation in complex systems using the example of a logistics problem. International Transactions on Systems Science and Applications 2(2), 157–166 (2006)
17. Mühlenbein, H.: The breeder genetic algorithm - a provable optimal search algorithm and its application. In: Colloquium on Applications of Genetic Algorithms, IEEE, London, vol. 67, p. 5/1– 5/3, IEEE, London (1994)

Genetic Computation of Road Network Design and Pricing Stackelberg Games with Multi-class Users

Loukas Dimitriou[1], Theodore Tsekeris[2], and Antony Stathopoulos[1]

[1] Department of Transportation Planning and Engineering, School of Civil Engineering, National Technical University of Athens, Iroon Polytechniou 5, 15773 Athens, Greece
lucdimit@central.ntua.gr, a.stath@transport.ntua.gr
[2] Centre for Planning and Economic Research, Amerikis 11, 10672 Athens, Greece
tsek@kepe.gr

Abstract. This paper deals with the problems of optimal capacity and pricing decisions in private road networks. These problems are described as a class of design and pricing Stackelberg games and formulated as nonconvex, bilevel nonlinear programs. Such games capture interactions among the decisions of system designer/operator, government regulations and reactions of multi-class users on optimal toll-capacity combinations. The present class of games applies to a realistic urban highway with untolled alternative arterial links. In contrast with the mostly used continuous representations, the highway capacity is more intuitively expressed as a discrete variable, which further complicates the solution procedure. Hence, an evolutionary computing approach is employed to provide a stochastic global search of the optimal toll and capacity choices. The results offer valuable insights into how investment and pricing strategies can be deployed in regulated private road networks.

Keywords: Road Pricing, Equilibrium Network Design, Regulated Private Highways, Stackelberg Games, Genetic Algorithms.

1 Introduction

The growing demand for mobility and subsequent needs for new or expanded transportation infrastructure provision, have led many governments worldwide to adopt privatization mechanisms to ensure the financing of this infrastructure. In particular, the Build-Operate-Transfer (BOT) scheme allows private investors to build new highways and operate them by collecting toll charges for a given number of years (concession period), sufficient to attain an agreed level of investment benefit, and then transfer them to the government. For the successful implementation of such schemes, a profit-maximizing firm should consider the tradeoff between the cost of road capacity provision and the revenues generated by the toll charges imposed on users. Namely, the private-sector operator should jointly adapt the level of capacity and prices in order to determine the expected volume of highway users and the toll revenues. Also, this adaptation should commonly encounter government regulation on the level of tolls and, possibly, the service to be provided to the users.

M. Giacobini et al. (Eds.): EvoWorkshops 2008, LNCS 4974, pp. 669–678, 2008.

Despite the obvious need for jointly addressing the optimal road capacity (part of network design) and toll pricing decisions, these issues have mostly been treated separately in the current literature (e.g. see [1], [2]). Moreover, the few studies considering these choices simultaneously are largely restricted to simple networks, having either parallel or serial links [3], [4]. The joint capacity and pricing optimization problem has been also examined for general inter-urban road networks in [5] and [6] under alternative project objectives and market conditions. These models have considered a standard, deterministic user equilibrium (DUE) traffic assignment procedure with elastic demand to represent the responses of users on optimal tolls and capacity investments. Despite that these studies have showed the need for jointly considering toll and capacity decisions, several of their modeling assumptions are unduly simplified, departing from the realistic conditions underlying the design and pricing of BOT road networks, and the behavior of their users.

This paper offers a number of extensions to the class of network design and pricing problems, to better capture interactions between operator, regulator and users. The proposed framework aims at designing optimal (here, revenue-maximizing) strategies able to handle unilateral or joint pricing and capacity allocation tactics, taking into account the responses of multi-class users, budget and toll constraints, and service requirements. In contrast with the other studies, which treat traffic as homogeneous, this study recognizes the existence of heterogeneity through multiple classes of users, according to their income and trip purposes. Such a treatment circumvents potential problems of underestimating the benefit of road pricing for the users. Another extension refers to the modeling of variations in the perceived generalized travel cost and route choice uncertainty, which provide more realistic assumptions on the assignment of demand into the network. These features give rise to a multi-class stochastic user equilibrium (SUE) assignment procedure with elastic travel demand.

Furthermore, this framework is able to handle variable toll pricing strategies, which allow the spatial (here, entry-based) differentiation of toll charges, in comparison to the common, flat toll policy that ignores spatial variations in origin-destination (O-D) patterns. Another important contribution of the study relates to the relaxation of the assumption commonly made in the literature that provided road capacity is adjustable in continuous increments. This relaxation gives rise to a simultaneous toll-capacity optimization problem, referred to here as the joint discrete network design and pricing problem (JDNDPP). Provided that, in reality, the number of lanes or links is discrete, the expression of capacity as a discrete variable provides a solution that is infrastructure related, which may involve greater physical intuition and bearing in the road network design process, in comparison to the solution of the corresponding continuous version of the problem.

Nonetheless, the adoption of discrete units of capacity leads to more complex solution procedures, since it yields a mixed-integer nonlinear problem. Hence, an evolutionary computation procedure is adopted to provide an efficient solution for the complex JDNDPP. The set of road network design and pricing problems is generally expressed as a class of Stackelberg games with multiple players (the system designer/operator, the government/regulator and the users), where different situations can be modeled through altering the components (objectives and constraints) of the problem setup. Section 2 presents a game-theoretic formulation of the capacity and pricing optimization framework in road networks with multi-class users. Section 3 describes the solution procedure, Section 4 reports the computational experiments and results obtained from a real-world urban network application of the model, and Section 5 concludes.

2 Description of the Design and Pricing Stackelberg Games

The processes of road network design and pricing can be considered as a class of two-stage Stackelberg games with perfect information among alternative players, namely, the system operator(s), the regulator and the users. These games can be expressed as bilevel programs, where at the upper level the system operator (the 'leader'), taking into account various constraints, integrates within the design and pricing tactics the non-cooperative responses (typically, those of departure time and route choice) of users of different classes (the 'followers'), which are performed at the lower level. These responses are captured here through a multi-class SUE assignment with elastic demand, which is expressed as an unconstrained minimization problem [7]. The private highway market is regarded as monopolistic and subject to governmental control on the level of tolls and service. In this way, a third player is added, i.e. the state (or regulator), whose decisions affect the performance of the other two players. Also, it is assumed that the set (number and location) of the highway links has already been identified before the problem is formulated and solved. Thus, in this class of games, the system operator imposes alternative (uniform or entry-based) toll charges and/or modifications on the number of link lanes, attempting to attract the largest possible number of users to the toll highway, which competes with free alternative roads.

Consider a network $G(N, A)$ composed of a set of N nodes and A links, which connect the origin zone r with destination zone s, and q_m^{rs} be the demand of the users of class m (that is, a user group with similar socio-economic and travel characteristics) for moving between the O-D pair r - s. Also, consider the link travel time function $t_a(x_a)$ as being positive and monotonically increasing with traffic flow x_a at each link $a \in A$. Then, the complete form of the bilevel optimization framework, which refers to the JDNDPP, for the design and pricing of a private highway that constitutes part of the network can be expressed as follows:

Upper-Level Problem:

$$\max_{p,y} R(p_a, y_a) = \sum_{a \in \hat{A}} E\{p_a f_a(p_a, y) - \lambda V_a(w_a)\} \tag{1}$$

$$\text{Subject to} \qquad p_{\min} \leq p_a \leq p_{\max}, \qquad \forall a \in \hat{A} \tag{2}$$

$$w_a \in \{0, 1, \ldots, \ell\}, \qquad \forall a \in \hat{A} \tag{3}$$

$$\sum_{a\in\hat{A}}(V_a(w_a)) \leq B, \qquad \forall a \in \hat{A} \tag{4}$$

$$x_a(p_a, y_a)/y_a \leq L \quad \forall a \in \hat{A} \tag{5}$$

Lower-Level Problem:

$$\begin{aligned}\min_{x,q} Z(x,q) = &\sum_{rsakm} \mathrm{VOTT}_m\, \delta^{rs}_{a,km}\, t_a(x_a)\, x_a - \sum_{rsakm} \mathrm{VOTT}_m\, \delta^{rs}_{a,km} \int_0^{x_a} t_a(w)\mathrm{d}w + \\ &\sum_{rsm} \mathrm{VOTT}_m\, D^{-1}_{rsm}(q^{rs}_m)\, D_{rsm}\left(S^{rs}_m(x)\right) - \sum_{rsm} S^{rs}_m(x)\, D_{rsm}\left(S^{rs}_m(x)\right) + \\ &\sum_{rsm} \int_0^{q^{rs}_m} \mathrm{VOTT}_m\, D^{-1}_{rsm}(q)\mathrm{d}q - \sum_{rsm} q^{rs}_m\, \mathrm{VOTT}_m\, D^{-1}_{rsm}(q^{rs}_m)\end{aligned} \tag{6}$$

In the upper-level problem, R denotes the expected profit function of the private-sector highway designer/operator, which is the objective function of the complete form of the JDNDPP. In this function, the first component refers to the toll revenues raised from the toll charges, p_a, imposed on the volume f_a of users entering the highway through the entry node of the access link $a \in \hat{A} \subset A$ with capacity y_a, where $\hat{A}$ is the set of highway links with total capacity y. The second component corresponds to the monetary expenditures $V_a(w_a)$ for capacity provision at link $a \in \hat{A}$, where λ is a parameter that transfers the capital cost of the project into unit period cost, depending on the number of years of operating the project by the private sector, and w_a is an integer decision variable which determines the number of lanes in link $a \in \hat{A}$, up to a physical threshold ℓ, as shown in relationship (3). The scalars $p_{\min}$ and $p_{\max}$ in relationship (2) are the minimum and maximum allowable toll charges, regulated by the government. The budgetary restrictions are denoted in inequality (4), where B is the total available highway construction budget. Relationship (5) introduces the regulatory control of the government on the minimum required level of service, where L is the maximum allowable flow-to-capacity (x_a/y_a) ratio for each highway link $a \in \hat{A}$.

At the lower-level problem, Z expresses the objective function of network users of different classes m, in terms of their value of travel time VOTT_m, who seek to minimize their perceived generalized travel cost. The traffic assignment procedure assumes that users have elastic demand and their route choice behavior is consistent with the SUE link flow conditions, in the sense that no traveler can improve his/her perceived travel time by unilaterally changing routes [8]. The binary variable $\delta^{rs}_{a,km}$ takes the value 1, if link a is part of the path $k \in K^{rs}$ of the feasible path set followed by users of group m, or 0 otherwise. Assuming that the demand function

D_{rsm} is nonnegative and strictly decreasing with respect to the cost of paths between r - s, then $q_m^{rs} = D_{rsm}(S_m^{rs})$ and $S_m^{rs} = D_{rsm}^{-1}(q_m^{rs})$, where D_{rsm}^{-1} is the inverse demand function and S_m^{rs} is the perceived travel cost function. The latter function is expressed in relation to the expected value E of the total path travel cost C_{km}^{rs} as follows:

$$S_m^{rs}(x) = E\left[\min_{k\in K^{rs}} C_{km}^{rs} \,\middle|\, C^{rs}(x)\right], \text{ with } \frac{\partial S_m^{rs}(C^{rs})}{\partial C_{km}^{rs}} = P_{km}^{rs} \tag{7}$$

where P_{km}^{rs} denotes the probability that users of class m select path k between r - s pair. Then, the measure of probability P_{km}^{rs} depends on the following utility function:

$$U_{km}^{rs} = -\theta C_{km}^{rs} + \varepsilon_{km}^{rs} \tag{8}$$

where U_{km}^{rs} expresses the utility of users of class m selecting path k between r - s pair, θ is the path cost perception parameter and ε_{km}^{rs} is a random error term, independent and identically distributed (iid) for all routes, which is here assumed to follow a Gumbel distribution, hence, yielding a logit model. The path travel cost C_{km}^{rs} is expressed in monetary terms, as a composite function of the value of travel time (VOTT) and toll charge:

$$C_{km}^{rs} = \sum_{a\in A} \text{VOTT}_m\, \delta_{a,km}^{rs}\, t(x_a) + \sum_{a\in \hat{A}} \delta_{a,km}^{rs}\, p_a \tag{9}$$

The estimation of the demand responses of users to changes in path travel cost is based here on the following relationship [9]:

$$D_{rsm}^{(n)} = D_{rsm}^{0} \exp(uC_{rs}), \ \forall\, r,s,m \tag{10}$$

where D_{rsm}^{0} refers to the initial demand level of users of class m for the r - s pair and u is a scaling parameter (here, a relatively large elasticity value $u = -0.3$ is used, since the path cost is expressed in monetary terms).

3 The GA-Based Solution Procedure

The present formulation of the network design and pricing games results in a *NP*-hard mixed-integer programming problem of increased computational complexity. This is because of the nonlinearity of the functions involved in the upper and lower-level problems, and the combinatorial selection of the highway link lanes. The nonlinearity and nonconvexity portend the existence of local solutions, which imply that it might be difficult to solve for a global optimum (or adequately near-optimum) solution. In view of the difficulty in applying the standard algorithmic approaches for search of the global optimum, this study first adopts a well-established evolutionary computation approach for such a class of games, based on a genetic algorithm (GA).

The GAs employ population-based stochastic, global search mechanisms suitable for the solution of bilevel mixed-integer programs with multiple constraints, requiring information only about the performance of a 'fitness' function for various candidate states [10]. A GA was employed in [6] for solving the continuous design and pricing problem with single-class deterministic assignment of users into the network. The pseudo-code below shows the steps of the current iterative solution procedure:

```
Step 1. (Initialization)
Produce an initial random population of candidate
feasible solutions (link lanes and toll charges) and
select the properties of the genetic operators
DO UNTIL CONVERGENCE:
Step 2. (Lower-level problem)
Perform path enumeration and produce the SUE link flow
pattern for each candidate solution
Step 3. (Performance evaluation)
Check for the consistency of constraints and estimate
the fitness function for each candidate solution
Step 4. (Upper-level problem)
Produce a genetically improved population of candidate
solutions through the stochastic selection of the
'fittest' solution set, crossover operation among the
selected individuals, elitist movement of individuals
from previous generations and mutation of individuals
```

The GA population is composed here of 50 individuals and each of them corresponds to alternative binary codings of number of lanes and toll charge combinations. In computational terms, the CPU time required for each individual is 25 sec, which yields about 20 min for each generation. After the random selection of an initial population, in accordance with the problem constraints, the expected profit function (1) is mapped into a fitness function. Subsequently, the solution of the multi-class SUE assignment model with elastic demand provides the equilibrium state of the system for each candidate solution. Then, the fitness function is evaluated for each candidate solution, and the genetic operations are employed to produce an improved population. The reproduction operator employs a tournament selection between 3 candidate individuals. The crossover operation uses a relatively high crossover rate equal to 80%, augmented with an elitism strategy (10% of the best individuals from the 10 preceding generations feeds the current generation), which enhance the probability of the individuals with good performance to exchange genetic information. A mutation rate equal to 5% is used to diminish the probability of finding a false peak. The convergence of the GA is considered to be achieved when the average of the population performance will not be significantly improved for 10 successive generations, or a maximum number of 50 generations has been performed.

4 Computational Experiments and Results

The proposed model applies to a suitably selected part of the urban road network of Athens, Greece, that is composed of primary and secondary roads, which are linked with a closed urban highway, called Attiki Odos. The network (see Fig. 1) covers the most densely populated region along the highway, where the heaviest daily traffic volumes are observed. It is composed of 54 links servicing the demand represented by a 10×10 O-D matrix. Attiki Odos operates under a BOT concession scheme by a private-sector operator, which imposes uniform toll charges (2.70 €/private car) at the highway access points. The study identifies two VOTT user classes, based on their trip purpose and income level. The first class has an hourly $VOTT_A$=4.0 €, representing commuters and travelers of higher income, while the second class has an hourly $VOTT_B$ = 1.5 €, representing more elastic trips and lower income travelers. The minimum and maximum toll levels are set equal to $p_{\min}$ =0 € and $p_{\max}$ =7 €. The optimal toll and capacity choices are examined for the private highway links (coded in green) based on the demand pattern in a representative (design hour) travel period. In order to calculate the travel time t_a at link a, the well-known Bureau of Public Roads (BPR) function is used, as follows:

$$t_a(x_a) = t_a^0\left(1+\mu\left(\frac{x_a}{G_a}\right)^{\beta}\right), \ \forall a \in A \qquad (11)$$

where t_a^0 is the link travel time at free-flow conditions, μ and β are parameters referring to local operating conditions (in this study, $\mu = 0.15$ and $\beta = 4$) and G_a is the maximum traffic capacity at link a.

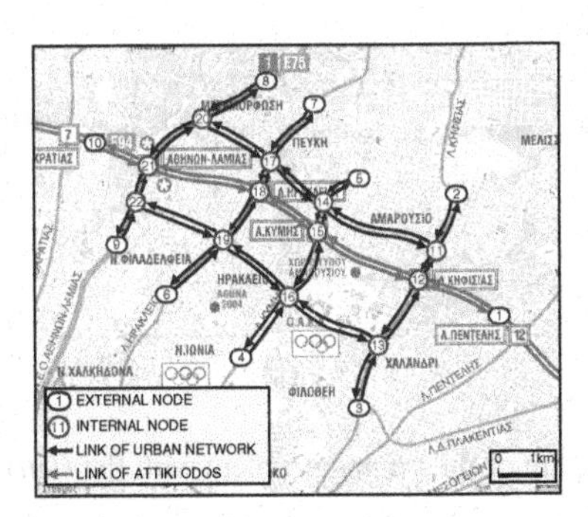

Fig. 1. Configuration and coding of the urban network and tolled highway of the study

Initially, the highway configuration is considered as fixed with 3 lanes per direction, and only pricing decisions are made, ignoring service requirements. Two alternative pricing strategies are considered, i.e. uniform and variable (entry-based) toll charges. The latter strategy allows to the 'leader' of the game (the operator) a greater flexibility to identify those critical access points (and links) for which the users have increased willingness-to-pay, and it facilitates the control of the prevailing network traffic conditions, thus enhancing the demand for highway travel.

Table. 1. Results of the alternative network design and pricing games

Problem	Expected profit (€)	Total travel cost (€)
Flat toll pricing	20 719	37 508
Variable toll pricing	23 447	35 788
Variable toll and network design	22 781	36 001
Variable toll and network design with service requirements	22 504	35 905

The results (Table 1) clearly indicate the increased profits obtained from the variable toll pricing, in comparison to the uniform toll pricing. This is because flat tolls charge the same amount for both short and long-distance trips, resulting in the diversion of a significant portion of users (especially those with an assumed lower VOTT) towards free arterial routes and, in turn, dropping toll revenues and increasing the network total travel cost. On the contrary, the variable toll strategy is more profitable, since it recognizes the spatial significance of each highway access point to both the network operating conditions and the revenues, giving rise to a more fair treatment of network users. Particularly, the efficiency gain of the variable toll strategy reflects the 13% increase of the expected profits and the 5% reduction of total travel costs (see Table 1). From the computational point of view, the convergence of the flat toll pricing is evidently much faster than that of the variable toll pricing, since it is a univariate optimization problem, constrained to a narrow 0-7 € decision set, whose near-optimal solution can be approximated with an increased probability by a population of 50 individuals within the 3-4 first generations, while the rest generations are used to fine tune it. In contrast, the variable toll pricing is a multivariate problem requiring an increased computational effort with an average of 20 generations to converge (Fig. 2).

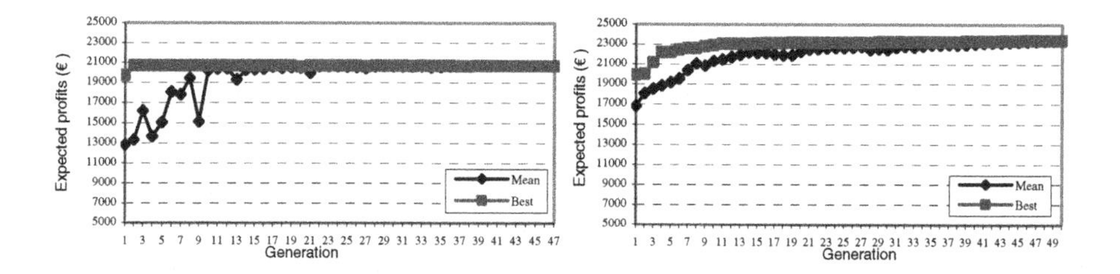

Fig. 2. Convergence diagram of the pricing problem with flat (*left*) and variable (*right*) tolls

Then, the problem extends to the joint optimization of the (more efficient) variable toll charges and network design, in terms of the number of lanes, to examine the strategic pricing and investment behavior of the system operator in the long run. This game yields a mixed-integer bilevel program, as that formulated in Section 2. The first JDNDPP setup ignores constraint (5) on service requirements. The estimation of the construction and maintenance cost normalized for the peak hour and per lane and kilometer is made with the following equation:

$$E^k = \frac{K^k}{L_m}\left(\frac{r_o(1+r_o)^n}{(1+r_o)^n - 1}\right) \tag{12}$$

where E^k is the normalized peak hour construction and maintenance cost per kilometer, K^k is the total construction and maintenance cost for the whole period, $L_m = 365 \times n \times 24 \times p^h$ are the daily hours of operation normalized in peak hours by the factor p^h and r_0 is the interest rate for the payback period n (here equal to 10 years). The construction (plus land acquisition) and maintenance costs have been assumed to amount to 50 M€/Km. Hence, the monetary expenditures for each highway link are calculated as $V_a = Lngh_a \times \ell \times E^k$, with $Lngh_a$ be the length of link *a*.

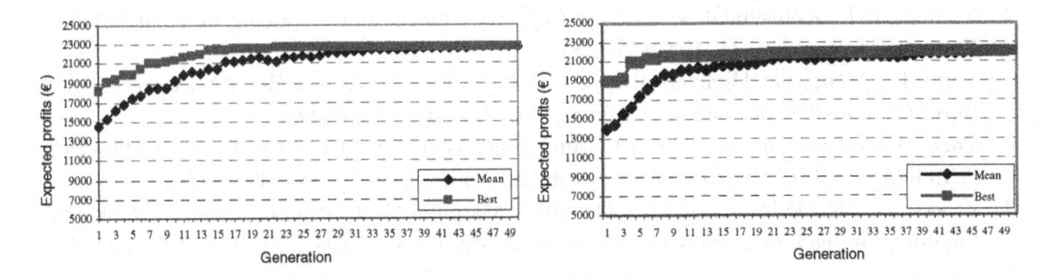

Fig. 3. Convergence diagram of the network design and variable toll pricing problem without (*left*) and with (*right*) service requirements

Table 1 shows that the expected profit is dropped, due to the inclusion of investment costs in the objective function (1), while the total travel cost is increased since the optimal number of lanes found is less than those *a priori* considered in the variable toll pricing problem. Finally, the problem considers the government regulation on the level of service, by requiring a flow-to-capacity ratio lower than 90% ($L = 0.90$) for each highway link. Such a regulation entails additional capacity provision and, as it was expected, is found to decrease both the expected profit and total travel cost (see Table 1). The joint pricing and capacity problem requires about 20 generations for convergence (see Fig. 3), similar to the problem of variable toll pricing.

5 Conclusions

This paper described a class of design and pricing Stackelberg games and their GA-based solution for the case of mixed-ownership (public and regulated private) general road networks with multi-class users. These games are formulated as a continuous bilevel program, when ignoring network design, or a mixed-integer bilevel program, when considering both decisions on pricing and network design. The results show the significant benefits of imposing variable (entry-based) toll charges on highway users for the profitability of the system designer/operator, in contrast with the case of flat tolls. The joint consideration of variable toll charges and capacity (number of lanes) was found to provide useful insight into the tradeoff between the strategic pricing and investment decisions of the private-sector operator, and the government regulations, which influence the public acceptability of BOT schemes.

References

1. Dimitriou, L., Tsekeris, T., Stathopoulos, A.: Evolutionary Combinatorial Programming for Discrete Road Network Design with Reliability Requirements. In: Giacobini, M. (ed.) EvoWorkshops 2007. LNCS, vol. 4448, pp. 678–687. Springer, Heidelberg (2007)
2. Shepherd, S.P., Sumalee, A.: A Genetic Algorithm Based Approach to Optimal Toll Level and Location Problem. Networks Spat. Econ. 4, 161–179 (2004)
3. Keeler, T.E., Small, K.A.: Optimal Peak-Load Pricing, Investment and Service Levels on Urban Expressways. J. Polit. Econ. 85, 1–25 (1977)
4. Verhoef, E.T., Rouwendal, J.: Pricing, Capacity Choice, and Financing in Transportation Networks. J. Reg. Sci. 44, 405–435 (2004)
5. Yang, H., Meng, Q.: Highway Pricing and Capacity Choice in a Road Network under a Build-Operate-Transfer Scheme. Transp. Res. 34A, 207–222 (2000)
6. Chen, A., Subprasom, K., Ji, Z.: A Simulation-based Multi-Objective Genetic Algorithm (SMOGA) for BOT Network Design Problem. Optim. Eng. 7, 225–247 (2006)
7. Zhao, Y., Kockelman, K.M.: On-Line Marginal-Cost Pricing across Networks: Incorporating Heterogeneous Users and Stochastic Equilibria. Transp. Res. 40B, 424–435 (2006)
8. Daganzo, C.F., Sheffi, Y.: On Stochastic Models of Traffic Assignment. Transp. Sci. 11, 253–274 (1977)
9. Yang, H., Bell, M.G.H.: Traffic Restraint, Road Pricing and Network Equilibrium. Transp. Res. 31B, 303–314 (1997)
10. Goldberg, D.E.: Genetic Algorithms in Search, Optimization and Machine Learning. Addison-Wesley, Reading (1989)

Constrained Local Search Method for Bus Fleet Scheduling Problem with Multi-depot with Line Change

Kriangsak Vanitchakornpong[1], Nakorn Indra-Payoong[1], Agachai Sumalee[2], and Pairoj Raothanachonkun[1]

[1] College of Transport and Logistics, Burapha University, Chonburi, Thailand
[2] Department of Civil and Structural Engineering, The Hong Kong Polytechnic University, Hong Kong
kriangsv@buu.ac.th, nakorn.ii@gmail.com, ceasumal@polyu.edu.hk, pairoj.iang@gmail.com

Abstract. This paper proposes a bus fleet scheduling model with multi-depot and line change operations with the aim to reduce the operating costs. The problem is constrained by various practical operational constraints, e.g. headway, travel time, and route time restrictions. A constrained local search method is developed to find better bus schedules. The method is tested with the case study of the Bangkok bus system with nine bus service lines covering around 688 trips per day. The test result shows that around 10% of the total operating costs could be saved by the optimized schedule.

Keywords: Bus scheduling, Multi-depot and line change, Constrained local search.

1 Introduction

Bus scheduling is a complex decision problem for transit operators. Generally, there are four planning steps involved: i) bus routes and headway determination, ii) bus timetabling, iii) vehicle scheduling, and iv) crew scheduling. These steps are highly interdependent. (i) and (ii) are rather long-term plans whereas the other two are short-term decisions. In this paper, we focus on the vehicle scheduling problem of bus services with multi-depot and line change operations. We also focus on developing a flexible algorithm that can accommodate additional side constraints commonly found in real cases. Since the multi-depot vehicle scheduling problem is NP-hard, we propose a local search method that uses a simple local move to obtain a good quality solution within a viable time. The proposed model and algorithm are evaluated with the real data from the bus system in Bangkok.

The paper is organized as follows: the next section reviews the literature on bus scheduling; the following two sections describe the model formulation and the test results from the Bangkok case study respectively. The conclusions and future research are then discussed in the last section.

2 Literature Review

Relevant literature on the bus fleet scheduling problem can be categorized as: i) problem formulation, ii) solution algorithm development; and iii) applications to real

M. Giacobini et al. (Eds.): EvoWorkshops 2008, LNCS 4974, pp. 679–688, 2008.

life problems. Since the multi depot vehicle scheduling problem (MDVS) is NP-hard, most of the algorithms proposed thus far in the literature are heuristic. Bertossi *et al.* [1] proposed a multi-commodity matching model for MDVS and developed two heuristics to solve the problem. Ribeiro and Soumis [2] formulated MDVS as a set partitioning (SP) problem. They also proposed a method based on the column generation (CG) and branch and bound methods. Lobel [3, 4] tackled MDVS with the problem size of around 7,000 trips. The problem is formulated as a multi-commodity flow problem, and the Lagrangian pricing method was adopted. The proposed algorithm has been employed by three bus companies in Berlin and Hamburg cities. Hadjar *et al.* [5] proposed a branch and cut algorithm for MDVS. They successfully solved the problem with around 800 trips. Pepin *et al.* [6] and Kliewer *et al.* [7] formulated the MDVS as a time-space model to reduce the number of decision variables. With the proposed model, any off-shelf optimization software can be adopted to solve the problem. The proposed method is tested with the bus data from the Munich city.

In this paper, we propose a different algorithm for MDVS. The method is based on a special type of local search algorithm which operates simple local moves to improve the solution. The algorithm proposed is arguably more flexible, and can easily accommodate additional side constraints. The users can easily introduce and/or modify the constraints without any changes to the algorithm structure. For instance, a strict time-window constraint can be easily converted to a time-window constraint with some tolerances.

3 Problem Formulation

The vehicle scheduling process will be analyzed after the bus timetable has been determined. The vehicle scheduling problem for bus systems involves assigning different buses to different scheduled trips so as to minimize the total operating costs whilst satisfying several practical constraints including (i) only one bus is assigned to a scheduled trip, and (ii) other constraints (e.g. time-window or route restriction).

In this paper, the MDVS is formulated as a set partitioning problem as shown in Fig. 1. Each row represents a scheduled trip, and each column represents a vehicle.

In Fig. 1, there are five scheduled trips with two different service lines: L1 and L2. Vehicle i1 and i2 belong to L1, and i3 and i4 belongs to L2. A feasible solution is shown in Fig. 1 in which for L1 bus i2 departs at 8:00 and 9:00, and i1 departs at 8:30; for L2, bus i3 and i4 depart at 8:15 and 8:35 respectively.

Timetabled trips		L1		L2	
		i1	i2	i3	i4
L1	8:00	0	1	0	0
	8:30	1	0	0	0
	9:00	0	1	0	0
L2	8:15	0	0	1	0
	8:35	0	0	0	1

Fig. 1. Set partitioning formulation for vehicle scheduling problem

To facilitate the discussion, the following notations are used: $Z = \{1,...,z\}$ denotes a set of depots, $L = \{1,...,l\}$ is the set of bus lines and $L'_z = \{1,...,l_z\}$ is the set of bus lines for depot z, $W = \{1,...,w\}$ is the set of scheduled trips, $W'_l = \{1,...,w_l\}$ is the set of trips of line l, $V = \{1,...,v\}$ is the set of buses, $V'_l = \{1,...,k_l\}$ is the set of buses of line l, $D_i = \{1,...,d\}$ is the ordered set of the trips served by bus i, Q_i is the allowable number of trips for bus i, c_{ij} is the fixed cost of bus i serving trip j, $x_{ij} = 1$ if bus i serving trip j; $x_{ij} = 0$ otherwise, e_{ij} is the starting time of trip j for bus i, t_{ij} is the trip time for bus i servicing trip j, s_l is the number of buses serving line l, w_l is the total trips of line l, k_l is the number of dedicated buses of line l, $f_i = 1$ if bus i services at least 1 trip; $f_i = 0$ otherwise, $a_{ij} = 1$ if bus i is used for the other lines when serving trip j; $a_{ij} = 0$ otherwise, $b_{ij} = 1$ if bus i is shared between depots in serving trip j; $b_{ij} = 0$ otherwise.

The bus vehicle scheduling problem can be considered as a constraint satisfaction problem in which the violation of hard constraints (operational constraints) is prohibited, and the objective function is converted into a soft constraint. In general, the constraint violation (v) can be written as:

$$Ax \le b \Rightarrow v = \max(0, Ax - b) \quad . \tag{1}$$

where A is coefficient value, x is decision variable, and b is a bound. A solution for the problem is achieved when i) the hard violation (H) equals to zero, and ii) the soft violation (S) is minimized.

3.1 Hard Constraints

In this paper, the hard constraints are further categorized into the basic and side constraints. The basic constraints are mainly concerned with the consistency of the solution, e.g. one trip should be assigned to one vehicle. On the other hand, the side constraints are related to additional operational constraints considered, e.g. time-window constraint on arrival time of the vehicle back to the depot.

Basic constraints: Constraint (2) below states that each scheduled trip j must be assigned to only one bus. h_p is the constraint violation level for this constraint.

$$h_p = \max\left(0, \sum_{j \in W} \left| \sum_{i \in V} x_{ij} - 1 \right| \right) \quad . \tag{2}$$

Constraint (3) ensures that the bus can only start the trip after finishing the previous assigned trip.

$$h_e = \max\left(0, \sum_{i \in V} \sum_{j \in D_i} \left[\left(e_{i,j-1} + t_{i,j-1} \right) - e_{ij} \right] \right) \quad . \tag{3}$$

Side constraints: These constraints are mainly for route-time constraint, line change and vehicle transfer operations, and relaxed hard time windows.

In general, the route-time constraint may be associated with vehicle range (due to fuel limit), maintenance period, and maximum driver's working hours which can be written as:

$$h_r = \max\left(0, \sum_{i \in V} \left(e_{i,m^*+1} - \left(e_{im^*} + t_{im^*} + bk\right)\right)\right) \quad . \tag{4}$$

Where $m^* = j \bmod m \;\; \forall j \in D_i$, m is the longest continuous trip served by vehicle i, bk is the break time (unit: min), and h_r is the violation for the constraints.

Constraint (5) ensures that the bus cannot run longer than the daily allowable maximum working hours (its violation level is h_c):

$$h_c = \max\left(0, \sum_{i \in V} \left| \sum_{j \in W} x_{ij} - Q_i \right| \right) \quad . \tag{5}$$

Line change and vehicle transfer operations are proposed to reduce the operating costs by efficiently utilizing the current resources. Typically, the vehicles are assigned to specific lines (routes) and depots. With this fixed plan, some vehicles may be underutilized and some lines may be lack of vehicles at some periods due to the fluctuation of travel times and demands during a day. The line change operation is the assignment of a vehicle belonging to one line to serve a trip of another line. The multi-depot operation (vehicle transfer) allows the vehicles to arrive or depart from different depots from their original depots. Thus, the vehicle scheduling with multi-depot and line change operations can potentially increase the efficiency of the bus utilization. Nevertheless, some buses may not be available for this operation due to some contractual/practical issues (e.g. advertisement media contract on a particular line). The violation for the constraints can be formulated as:

$$h_m = \max\left(0, \sum_{l \in L} \left[\sum_{i \in V} \left(\sum_{j \in W_l'} x_{ij} \right) m_{il} \right] \right) \quad . \tag{6}$$

$$h_t = \max\left(0, \sum_{z \in Z} \sum_{i \in V} \left[\sum_{l \in L_z'} \left(\sum_{j \in W_l'} x_{ij} \right) n_{iz} \right] \right) \tag{7}$$

where m_{il}=1 if bus i on line l is prohibited to serve other lines, n_{iz}=1 if bus i on depot z cannot be transferred between depots. h_m and h_t are the constraint violation levels for the constraints on line change and vehicle transfer respectively.

An additional relaxed time-window constraint is introduced to allow for journey time variability. The late time-windows of ω in (3) are relaxed; where ω is the set of scheduled trips. The time-window constraint is converted to a soft constraint as:

$$h_{e^*} = \max\left(0, \sum_{i\in V}\sum_{j\in D_i}\left[\left(e_{i,j-1}+t_{i,j-1}\right)-e_{ij}\right]\left(1-y_j\right)\right) \quad . \tag{8}$$

$$s_e = \max\left(0, \sum_{i\in V}\sum_{j\in D_i}\left[\left(e_{i,j-1}+t_{i,j-1}\right)-e_{ij}\right]y_j\right) \tag{9}$$

Note that (3) is substituted by (8). (9) is a soft constraint y_j=1 if $j \in \omega$; otherwise y_j=0.

3.2 Soft Constraint

The objective function of the problem is to minimize the total operating cost which can be formulated as a soft violation, s, as:

$$S = \max\left(0, \sum_{i\in N}\left[\sum_{j\in N}\left(c_{ij}x_{ij} + LC_i a_{ij} + VT_j b_{ij}\right) + FV_i f_i\right] + s_e\right) \tag{10}$$

FV_i is the fixed cost of running bus i. LC_j is the fixed cost of line change operation, and VT_j is the fixed transfer cost of the vehicles between depots. From (2) – (8), the hard violation can be defined as $H = h_p + h_r + h_c + h_m + h_t + h_{e^*}$, and the total constraint violation is $V = H + S$. Note that for the Bangkok bus system, the depot is used for the first and end bus stops; thus the deadhead trip is not considered.

4 Solution Methods

Typically, the bus travel time is defined as a fixed parameter. However, in a very congested traffic network (like Bangkok) there exists a large variation of bus travel time. Fig. 2 illustrates the travel time in a typical day for a bus service in Bangkok. The travel time variability can seriously affect the vehicle schedule. In this study, the travel time forecast model is based on the historical travel time in hourly timeslot of the day and the day of the week with some expert rules. The detail of this model will not be discussed here to the limited space.

The pre-processing step must be carried out by sequencing scheduled trips in an ascending order, i.e. $e'_{j-1} < e'_j \quad \forall j \in W$; where e'_j is the departure time for scheduled trip j. This process helps decreasing the complexity and size of the problem. Then, the constrained local search (CLS) [8] is used to solve the vehicle scheduling problem. The hard and soft constraints are represented in a constraint matrix. CLS employs the random variable selection strategy and simple variable flip as a local move. The move quality is assessed by its total constraint violation, V. The violation scheme and redundant constraints are also used to guide the search into more promising regions of the search space. In addition, the constraint propagation is

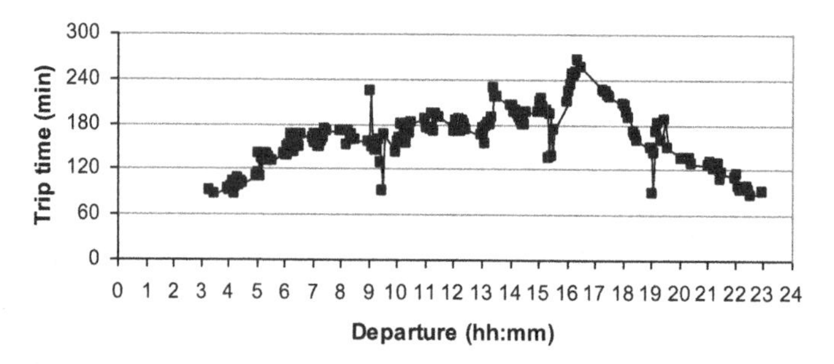

Fig. 2. Travel time variation

	i1	i2	i3	i4
j1	1			
j2			1	
j3				1
j4		1		
j5	0	←	1	
j6	1			
j7		1		
j8			1	

	i1	i2	i3	i4
j1	1			
j2			1	
j3				1
j4		1		
j5	1		0	
j6	1	→	→	0
j7		1		
j8			1	

	i1	i2	i3	i4
j1	1			
j2			1	
j3				1
j4		1		
j5	1			
j6	0			1
j7		1		
j8	0	←	1	

	i1	i2	i3	i4
j1	1			
j2			1	
j3				1
j4		1		
j5	1			
j6				1
j7		1		
j8	1		0	

Fig. 3. Variable flip operations

maintained, i.e. once a bus is assigned to a particular scheduled trip the flip operations associated with infeasible trips (e.g. those overlapping with the current assigned trip) will be banned. This reduces the size of the search space and speed up the computation time. Fig. 3 illustrates the variable flip operations.

Rows in the matrix represent the scheduled trips and columns represent the vehicles. The flip operation is designed to ensure that a new solution will satisfy the hard constraints (e.g. one trip is assigned to one vehicle). Thus, the consistency of these constraints is always maintained. The overall pseudo-code of the CLS is depicted in Fig. 4.

Referred to Fig 4, the procedure of CLS can be summarized into three steps:

- **Step 1:** Constraint selection. After the initial assignment, A, the number of columns (nc in Line 5) are randomly selected in accordance with the violated constraints (columns) are given the priority. (Line 1-5)
- **Step 2:** Variable selection. CLS selects the row (variable assigned to 1) to perform a trial flip with other variables in the same row. (Line 8-12)
- **Step 3:** Move acceptance. CLS chooses the best V' amongst the flipped variables in every single iteration and assigns the best V' to the current solution, $A_l \leftarrow A'$. (Line 15-20)

Adaptive upper bound. Since the set partitioning formulation has a large number of columns and rows in the constraint matrix, the algorithm may spend too much computational time to obtain a feasible solution. It is possible to improve the CLS by introducing memory-based search, guided information obtained from linear

```
// Algorithm: CLS
// INPUT: soft and hard constraints
// OUTPUT: a best feasible solution found
1    A:=initial solution assignment
2    A_b:=best solution
3    V_b:=the best violation
4    WHILE (iteration < stopping criterion) DO
5      nc:=Random(numcol)
6      V_g:=MAX_INTEGER
7      FOR i<-1 TO nc
8        C_fst:=select-columns(A)
9        P:=select-variables(C_fst)
10       V_l:=MAX_INTEGER
11       FOR j<-1 TO Random(numcol)
12         A':=flip(C_fst, C_j, P)
13         H':=total hard violation(A')
14         V':= total violation(A')
15         IF H'=0 AND V'<V_b
16           A_b <- A'
17           V_b <- V'
18         END IF
19         IF V'<V_l THEN
20           A_l <- A'
21         END IF
22       END FOR
23       IF V_l<V_g THEN
24         A_g <- A_l
25         V_g <- V_l
26       END IF
27     END FOR
28     A <- A_g
29   END WHILE
```

Fig. 4. The procedure of CLS

programming relaxation, or other search intensification techniques. In this study, the number of busses (columns) is gradually decreased and fixed iteratively when CLS find a better feasible solution; thus intensifying the search in more promising regions.

5 Computational Experiments

5.1 The BMTA Case Study

We test the proposed model and algorithm with the data from the Bangkok mass transit authority (BMTA), Thailand. There are currently 3,535 buses operating 108 routes (lines) clustered into 8 zones in the Bangkok city. In general, there are 3 - 4 depots for each zone, and each depot operates 5-10 bus lines. The bus schedules are currently determined by the BMTA planner manually once every few months without any computer-based tool. The schedule plan is rather ad hoc and often completely different from the actual services. For example, most of the buses are stuck in the traffic during rush hours creating the bus bunching problem.

5.2 Potential Cost Savings with Multi-depot and Line Changes

The proposed solution applies to the BMTA data with nine bus lines and 125 buses located in three depots. The model assumes that most buses are independently operated, i.e. can be utilized for the other lines or depots with the penalty cost (soft violation). The parameters in the model are set as follows: the fixed cost of each bus is 5,000 Baht/vehicle/day (1 Euro = 46 Baht). The line change cost is 100 Baht. The vehicle transfer cost is 500 Baht, which is excluded from running cost. The vehicle operating cost per Km is 30 Baht. The computational results are shown in Table 1.

The first column shows the tested lines and their depot and number of vehicles. The second column shows the number of scheduled trips for each line. The column Pure-line presents the data regarding the current operation including the number of buses required and the operating cost (in Baht). The column Line change (LC) reports the results based on the optimized vehicle schedule which allows for the line change operation. The next column then reports additional results when both line change and multi depot operations are considered. The last column shows the maximum reduction in terms of the required fleet size and total operating cost for each line. Time (sec) represents the computational time for particular test cases.

From Table 1, the schedules with line change can decrease the vehicle size from 85 (in pure line case) to 27 vehicles in total. In addition, when allowing the multi-depot operations the fleet size can be further reduced to 41 vehicles (almost 60% reduction). In terms of the operating costs, the multi-depot with line change operations can save around 91,000 Baht/day or 2.73 million Baht/month. From the computational point of

Table 1. Multiple-depot with line change operations

Line/	Trips		Pure-line		Line changes (LC)		Multi-depot with LC		Savings %	
#Bus		#Bus	Cost	Time(sec)	#Bus	Cost	#Bus	Cost	#Bus	Cost
Depot 1										
4 [18]	75	8	80687.5	2.8	7	82287.5	4	75687.5	50.00	6.20
72 [5]	78	8	80950	2.6	7	83450	7	86550	12.50	-6.92
205 [12]	76	11	94900	2.5	4	64700	4	75600	63.64	20.34
552 [15]	62	10	134630	2.5	5	115230	3	119630	70.00	11.14
					Time(sec) : 23.4					
Depot 2										
62 [14]	79	9	94770	2.8	10	104370	5	92870	44.44	2.00
77 [18]	101	12	155445	3.3	8	140445	4	131145	66.67	15.63
					Time(sec) : 10.8					
Depot 3										
12 [10]	66	7	69650	2.2	3	52450	3	64850	57.14	6.89
137 [18]	98	12	114880	3.1	8	101880	7	95780	41.67	16.63
551 [15]	53	8	91940	2.1	6	87240	4	84740	50.00	7.83
					Time(sec) : 16.3		Time(sec) : 128.4			
Total [125]	688	85	917852.5		58	832052.5	41	826852.5	51.76	9.91

Table 2. The schedule details

Scheduling Plan	KPI
Buses	41
Lines	9
Trips	688
Total Distance (Km)	15421.5
- Transfer KM	1340
- Trip KM	14081.5
Total Cost (Baht)	826852.5
- Vehicle	205000
- Line Change	43600
- Vehicle Transfer	38500
- Service KM	539752.5

view, the multi-depot with line change considering 688 trips in three depots; CLS provided a good quality solution in 128.4 seconds, which is practically acceptable.

Note that this is only the result with nine service lines. If a similar rate of improvement is applied to all 108 lines, the total cost reduction can be around 33 million Baht/month. The scheduling plan can also be assessed by the KPI (key performance indices) as shown in Table 2. The tests show that the proposed solution can potentially improve the bus fleet scheduling operation in this particular case.

6 Conclusions

The paper considered the bus fleet scheduling problem with multi-depot and line change operations, and proposed the constrained local search algorithm to solve the problem. The computational experiments were performed to evaluate the performance of the proposed method. The data sets from the Bangkok mass transit authority were used for the case study. By scheduling nine bus lines with 688 scheduled trips, the results indicated the potential operating cost savings by around 9.91 percent compared to the current schedule (manually prepare). The future research will attempt to apply the real-time information obtained from the automatic vehicle location system to generate more reliable and robust feet schedules. In addition, the integrated bus and crew schedule with improved algorithm will also be considered.

Acknowledgments. This research has been supported in part by a grant from the National Electronics and Computer Technology Center (NECTEC), Grant No. 09/2550 NT-B-22-IT-26-50-09). The authors are grateful to the BMTA for help and support.

References

1. Bertossi, A.A., Carraresi, P., Gallo, G.: On some matching problems arising in vehicle scheduling models. Networks 17, 271–281 (1987)
2. Ribeiro, C.C., Soumis, F.: A column generation approach to the multiple-depot vehicle scheduling problem. Operations Research 42, 41–52 (1994)

3. Lobel, A.: Optimal vehicle scheduling in public transit. PhD thesis, Technische University, Berlin (1997)
4. Lobel, A.: Vehicle scheduling in public transit and Lagrangian pricing. Management Science 4, 1637–1649 (1998)
5. Hadjar, A., Marcotte, O., Soumis, F.: A branch and cut algorithm for the multiple depot vehicle scheduling problem. Operations Research 54, 130–149 (2006)
6. Pepin, A.S., Desaulniers, G., Hertz, A., Huisman, D.: Comparison of heuristic approaches for the multiple depot vehicle scheduling problem. Report EI2006-34, Econometric Institute, Erasmus University Rotterdam (2006)
7. Kliewer, N., Mellouli, T., Suhl, L.: A time-space network based exact optimization model for multi-depot bus scheduling. European Journal of Operational Research 175(3), 1616–1627 (2006)
8. Indra-Payoong, N., Kwan, R.S.K., Proll, L.: Rail Container Service Planning: A Constraint-based Approach. In: Kendall, G., Petrovic, S., Burke, E., Gendreau, M. (eds.) Multidisciplinary Scheduling: Theory and Applications, pp. 343–365. Springer, Heidelberg (2005)

Evolutionary System with Precedence Constraints for Ore Harbor Schedule Optimization

André V. Abs da Cruz, Marley M.B.R. Vellasco, and Marco Aurélio C. Pacheco

Applied Computational Intelligence Lab
Electrical Engineering Department
Pontifical Catholic University of Rio de Janeiro
Rua Marquês de São Vicente, 225, Gávea
Rio de Janeiro – Brazil
{andrev,marley,marco}@ele.puc-rio.br

Abstract. This work proposes and evaluates an evolutionary system using genetic algorithms and directed graphs for the optimization of a scheduling problem for ore loading of ships in a harbor. In this kind of problem, some tasks are constrained in such a way that they must be planned or executed before others. For this reason, the use of conventional evolutionary models, such as genetic algorithms with an order-based representation, might generate invalid solutions which can not be penalized, needing to be discarded or corrected, leading to a loss in performance. To overcome this problem, we use a hybrid system, based on directed graphs, to allow better handling of the precedence constraints. Results obtained show performances almost 3 times better than a non-trivial search algorithm.

1 Introduction

Planning and scheduling are tasks that bring important economical results but pose difficult problems for computers. Scheduling problems can be found in several application areas, like production processes in industries, computational processes in operating systems, cargo transportation, refinery plants operation and ore loading in harbors, which is the chosen case study for this work.

In the latest years, several efforts have been made to investigate and explore the application of evolutionary computation to several scheduling problems [1,2,3]. One of the main difficulties is to determine an appropriate chromosome representation which leads to a straightforward generation of schedules. Additionally, some specific problems present extra difficulties, like the need to plan or schedule a task before another. This kind of difficulty occurs, for instance, on problems like the "Dial-a-Ride" one, where a bus must get passengers along several cities and take them to other ones, without going through the same city more than once and attending objectives like minimizing fuel consumption, distance and passengers waiting time [1,4]. These precedence constraints make it

M. Giacobini et al. (Eds.): EvoWorkshops 2008, LNCS 4974, pp. 689–698, 2008.

harder to find an appropriate representation for the chromosome which avoids the creation of schedules that violate precedences.

The problem of optimizing the scheduling of ore loading in ships usually involves this kind of precedence constraining: trains transport ore from the mine and the unloading process must be planned before stocking the ore on the harbor areas available for that, before the arrival of the ship responsible for transportation; on the other hand, the ship mooring must be executed before the loading itself. All these tasks require the use of several different equipments and must be done in such a way that several objectives can be optimized, for instance, minimizing the fine (called *demurrage*) that is payed by the harbor when the ship has to wait more time than the one specified on contract.

Several different approaches have been used to solve precedence constraints issues. In [5] an heuristic solution was used to solve the Travelling Salesman Problem (TSP) using precedence constraints. In [6] dynamic programming strategies were used for the TSP together with time slots and precedence constraints. In [7] a model to optimize navigation scheduling was done as a *shortest path* graph problem. In [2] a genetic algorithm associated to a directed graph was used to control the precedence constraints.

This work is based on [2] and proposes the use of a hybrid genetic algorithm with directed graphs to solve the problem of optimizing the ore loading of ships in a harbor.

This article is organized as follows: Section 2 presents a description of the proposed method. Section 3 presents the case study, describing in details the problem to be optimized and presenting results. Finally, Section 4, presents conclusions, remarks and proposals for future works.

2 The Hybrid Evolutionary System with Precedence Constraints

2.1 Chromosome Representation

As previously mentioned, an appropriate representation for optimizing scheduling problems with precedence constraints must generate, during the evolutionary process, only valid solutions. This principle is vital for the system's performance since it avoids the need for penalization, discard or repair of chromosomes created by the genetic algorithm.

In this sense, the model with order-based chromosomes [8] does not trivially satisfies this requisite, due to the fact that precedence constraints creates a whole set of invalid solutions. For instance, in a problem with 3 different tasks A,B and C and given that task B must be scheduled before task C, we can define a set of valid solutions $V = \{(A, B, C), (B, C, A), (B, A, C)\}$ and a set of invalid solutions $I = \{(A, C, B), (C, A, B), (C, B, A)\}$. For the latter, when the evaluation function is presented with one of the invalid solutions, it has either to correct or discard it.

To overcome this problem, this work proposes a hybrid model, based on genetic algorithms and directed graphs, which can solve this particular situation.

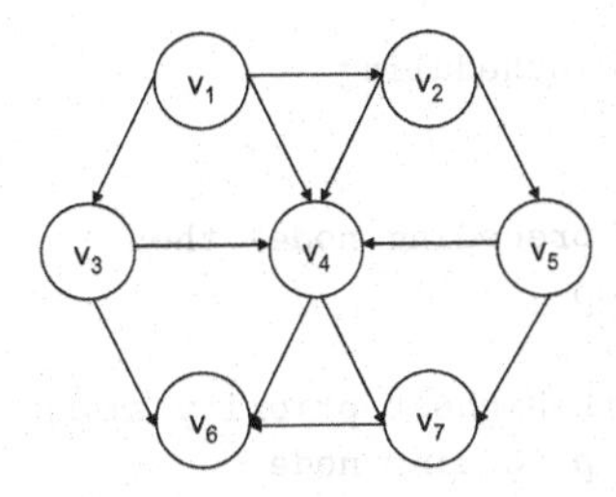

Fig. 1. Example of a graph defining precedences

The hybrid methodology is the same used for the Travelling Salesman Problem and is detailed below.

The approach used is based on a topological sort (TS) which allows the GA to generate only valid solutions during the evolutionary process. The topological sort is a node ordering in a directed graph such as that if there is a path from a node v_i and a node v_j, then v_j appears after v_i in the ordering. An example of a directed graph is shown in Figure 1.

The procedure to sort nodes is trivial and consists on selecting and storing any node which has not an arrow pointing to it (in the example on Figure 1 only the node v_1 satisfies this). After this is done, the node is removed from the graph, as well as the arrows leaving from it. Regarding scheduling, one can think of nodes as the tasks to be scheduled and of arrows as the precedence constraints. Thus, the path $< v_i, v_j >$ in the directed graph shows that task v_i must be executed or scheduled before task v_j. Also, it is clear that topological sort is only feasible if there is no cycle over the graph.

Another issue to consider is how to select a node (task) when two or more of them have no preceding tasks. To solve this issue we must create a list which shows the priorities for scheduling the tasks. To define the best order for the tasks that are free of precedence constraints in this list, we can use a genetic algorithm. This is accomplished by applying an order-based genetic algorithm in which the chromosome represents the priority list. Thus, the chromosome has n genes, each one holding a different value between 1 and n, representing one of the n tasks or nodes from the graph [2]. Table 1 shows an example of a chromosome that can be used to optimize the priorities for the tasks shown in the graph from Figure 1.

Using the chromosome and the precedence graph, we can use the algorithm shown in Figure 2 to generate a valid schedule.

From the graph shown in Figure 1 and using the chromosome from Table 1, the first task that can be scheduled is v_1, as this is the only task without preceding constraints. After scheduling v_1, the task is removed from the graph together

Table 1. Chromosome with priorities for a graph with 7 nodes

Task (Vertex)	v_1	v_2	v_3	v_4	v_5	v_6	v_7
Priority	5	1	7	2	4	6	3

```
procedure: generate valid scheduling
  input: directed graph
  while exists a node
    if (all nodes have a preceding node) then
      invalid graph - stop
    else
      select a node v with highest priority among
        the ones without preceding node
    end if
    schedule v
    remove v from the graph and all arrows leaving from it
  end while
end procedure
```

Fig. 2. Algorithm for generating valid schedules from a graph and a chromosome

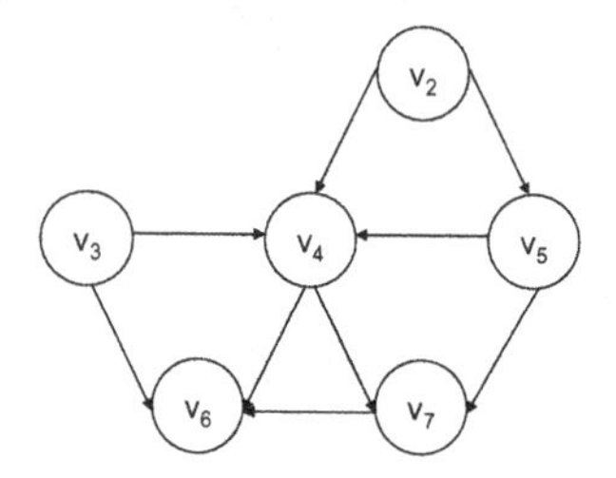

Fig. 3. Graph after scheduling the first task v_1

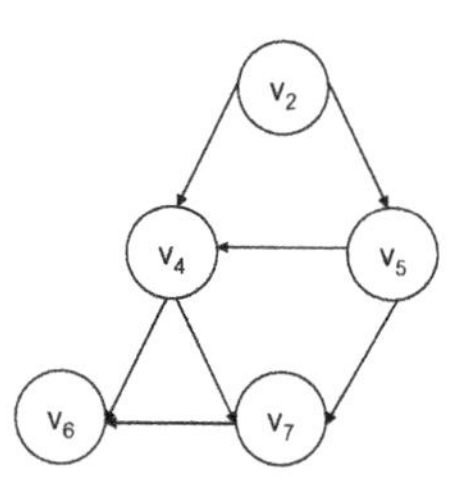

Fig. 4. Graph after scheduling task v_3

with the connections $< v_1, v_2 >$, $< v_1, v_3 >$ and $< v_1, v_4 >$. The resulting graph becomes like the one shown in Figure 3.

Next, both nodes v_2 and v_3 are free from precedence constraints. However, according to Table 1, node v_3 has a highest priority than node v_2; thus, it must be scheduled first. After doing that, we can remove task v_3 from the graph which takes the form shown in Figure 4.

Following this procedure until there is no more nodes in the graph, the following sequence will be obtained: $< v_1, v_3, v_2, v_5, v_4, v_7, v_6 >$. This sequence respects all precedence constraints specified by the graph from Figure 1.

2.2 Evaluation Function

The proposed evaluation function decodes the chromosome and precedence graph, creating a complete and valid schedule according to the goals that must be achieved. When optimizing a schedule, not only the task that must be executed is optimized, but also, which resources must be used to accomplish that task. The strategy used in this work to allocate resources is a very simple heuristic, which searches a fix list of resources that can be used to carry on a task, taking the first available resource to execute the given task at a given time. If there is no available resources to carry on the given task at the specified time, the algorithm will pick up the first resource to become available, and the task will be postponed to start on that time. This algorithm is detailed in Figure 5.

```
procedure: select resource
  r_list: resource list that can be used for the task to be scheduled
  i := 0
  while (true)
    repeat
      if (r_list[i] is available from the beginning time
          to the end time of the task) then
        use r_list[i] as a resource
        terminate
      else
        i := i + 1
    until (i = size(r_list))
    shift task starting time to match time of first available resource
    shift task ending time by the same amount
  end while
end procedure
```

Fig. 5. Algorithm for resource selection

3 Case Study

The main objective of this work is to optimize the schedule for ore loading of ships in a harbor. There are several tasks that must be accomplished in order to correctly schedule this:

- Unload trains carrying the ore from the mines;
- Stockpiling the ore on the specific areas on the harbor to wait for the ship which will transport it;
- Organize the mooring of the ships along the available piers;
- Moving the ore from the storage area to the ships moored on the piers.

The objective to optimize is the *demurrage cost.* The demurrage cost is the fine payed by the harbor to its clients when there is a delay in attending the ship in relation to what is established in contract. If the harbor can attend

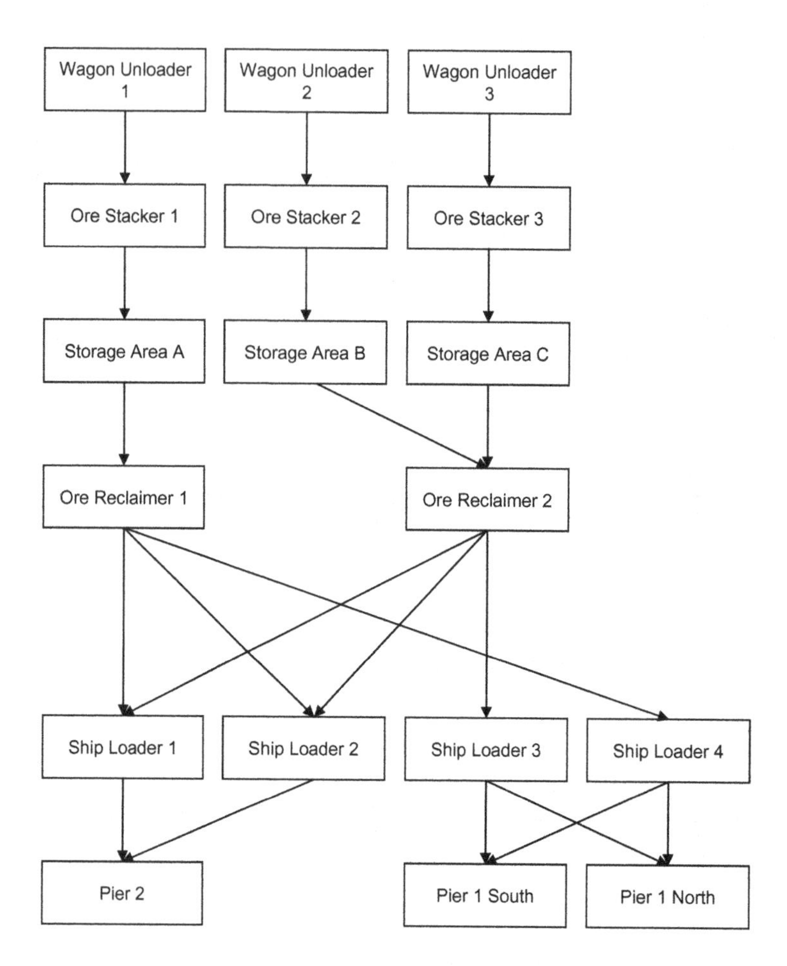

Fig. 6. Schematic diagram depicting the equipments available in the harbor as well as the connections between them, which are done by conveyor belts

the ship faster than a certain amount of time, also specified by contract, the clients pay the harbor for the agility on attending them. Thus, the optimization goal is to minimize the fine or maximize the prize.

The harbor has 3 wagon unloaders; 3 ore stackers, used to stockpile the ore arriving by train in the areas reserved for storage; 3 storage areas; 2 ore reclaimers, used to retrieve ore from the stockpiles and place them over conveyor belts; 4 ship loaders; 3 piers and several conveyor belts that are used to move the ore from one place to another over the harbor. A schematic diagram depicting the harbor is shown in Figure 6.

The precedence constraints are needed due to several different issues:

- Some storage areas might not have a connection with certain piers through conveyor belts. Thus, it is important to firstly schedule where the ship is going to be moored and then, where the ore for that ship is going to be stored;

- Some wagon unloaders might not be connected through conveyor belts to some storage areas. Thus, we first need to schedule where the ore is going to be stored and then to schedule which wagon unloader is going to be used for that cargo;
- We can only schedule the departure time of the ship after knowing when the ore loading is finished;
- The order in which the ships arrive at the harbor, usually determines the order they must be attended (except if the ship is too big, in which case we must use a heuristic to choose the pier to be used – this issued is discussed further ahead). Thus, in order to moor a ship, we must know when a previous ship has departed.

Besides precedence constraints, the simulation must consider several rules that also constrains the schedule. These rules are listed below:

- All calculations concerning ore volume are made based on ore density. Among others, the quantity of iron ore that fits inside a wagon and the quantity of ore between two storage delimiters are examples of values that depend on ore densities. The density values used here have no real physical meaning, they are just used as a reference value (so, a density value equal to 1 does not mean the ore has the same density as water);
- Each train can transport a batch with 20000 tons of ore. Thus, it is usually necessary to have several batches to load a ship;
- The train takes 12 hours to travel from the ore mines to the harbor;
- Each storage area on the harbor is divided in 50 stockpiles. Each stockpile can hold 20000 tons of ore (a batch) with density equal to 1;
- If a ship is moored on a pier which cannot receive ore from certain storage areas due to the lack of conveyor belt routes from the storage area to the pier, then this area has to be ignored when scheduling the ore storage for that ship;
- Heaviest ships (more than 250000 tons) can only be moored on pier 2. Medium size ships (between 150000 and 250000 tons) can be moored on pier 2 and pier 1 north). Small ships (less than 150000 tons) can be moored on any pier;
- All equipment on the harbor have an operation rate specified in *tons/hour*. As the ore is transported through several different equipments coupled to each other, these equipments must operate using the rate of the slowest equipment;
- Ore mines have a limited rate of production. Thus, for each type of ore, we specify a daily production rate. This rate must be respected and thus the algorithm responsible for simulating all harbor restrictions must take it into consideration when scheduling the request of ore to the mine given the ship's estimated date of arrival.

The precedence graph can be generated using the following steps:

1. Check the type of the ship and on which piers it can be moored;
2. Insert a mooring and unmooring task for that ship, in each pier where it can be moored;

3. Insert a task that reserves enough stacks on the storage area for the ore the ship must be loaded with. This task must be preceded by the mooring task from step 2;
4. Insert tasks for unloading ore batches from the wagons. The numbers of tasks is equal to the number of batches needed to complete the ore cargo. All those tasks must be preceded by the task from step 3;
5. Insert a task for loading the ship with the ore stored in the harbor. This task must be preceded by all tasks from step 4;
6. Create the graph connections for mooring and unmooring. The following rules should be used for that:
   (a) If two different ships have the same size, all the unmooring tasks for the first ship to arrive must precede the mooring tasks for the other one;
   (b) Ships that can only be moored to certain piers, have precedence over ships that can be moored to smaller piers;

For the tests, different configurations were used: one with the estimated time of arrivals for 35 ships and the other with the estimated time of arrivals for 20 ships. The crossover operators used were the PMX and CX ones [8] and the mutation operators were the PI and Swap ones [8]. We conducted 5 experiments with 40 generations in each one. The populations have 100 individuals and the number of individuals replaced on each generation was equal to 20% of the total size of the population. All those parameters were chosen by experimentation.

To establish a comparison, we used a non-trivial search algorithm (a mutation-based hill climber) which evaluated the same number of individuals as the genetic algorithm. Figure 7 shows the values of demurrage for the genetic algorithm and the non-trivial search along the evolution process for the 35 ships experiment.

Figure 8 shows a comparative chart for the genetic algorithm and the non-trivial search algorithm for the 20 ships experiment.

These results show that, after 40 generations, the genetic algorithm was able to obtain a significantly better result than the non-trivial search. Also, all

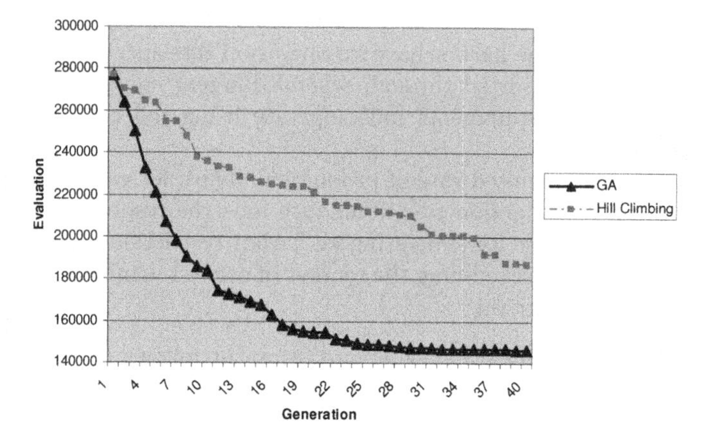

Fig. 7. Comparative chart of the evolution process for the 35 ships experiment

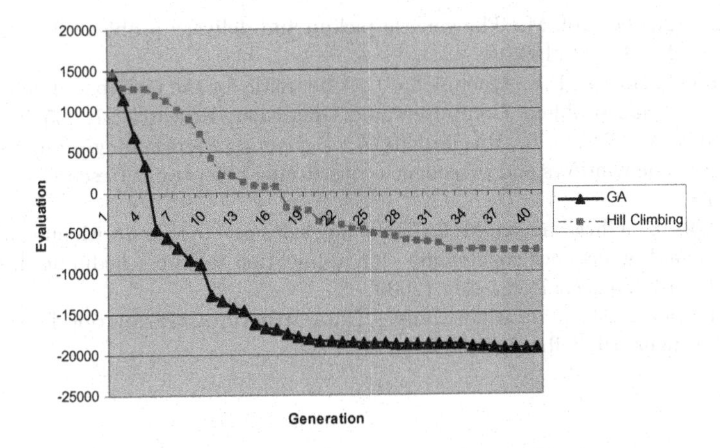

Fig. 8. Comparative chart of the evolution process for the 20 ships experiment

solutions created by the genetic algorithm were valid ones. Therefore, the algorithm's performance was improved, and no heuristics for correcting or penalizing the solutions nor discarding them were necessary. The CPU times for both experiments are 50 minutes for the experiment with 35 ships and 10 minutes with 20 ships.

4 Conclusions and Future Works

This work presented an hybrid evolutionary algorithm for optimizing the schedule of task on an ore harbor taking precedence constraints into consideration. This model uses an order-based representation, together with a directed graph to handle the precedence constraints, avoiding the need to discard invalid solutions or to use correction algorithms.

Results demonstrate that the use of the genetic algorithm brings a significant improvement regarding the optimization goals.

As future work, we intend to use a co-evolutionary model to optimize the resource allocation for the tasks to be scheduled, instead of using a fixed list of resources.

References

1. Bergvinsdottir, K.B., Larsen, J., Jorgensen, R.M.: Solving the dial-a-ride problem using genetic algorithms. Technical report, Danmarks Tekniske Universitet (2004)
2. Moon, C., Kim, J., Choi, G., Seo, Y.: An efficient genetic algorithm for the traveling salesman problem with precedence constraints. European Journal of Operational Research 140, 606–617 (2002)
3. da Cruz, A.V.A.: Schedule optimization with precedence constraints using genetic algorithms and cooperative co-evolution. Master's thesis, Pontifical Catholic University of Rio de Janeiro (March 2003)

4. Savelsbergh, M., Sol, M.: The general pickup and delivery problem. Transportation Science (29), 17–29 (1995)
5. Renaud, J., Boctor, F.F., Quenniche, J.: A heuristic for the pickup and delivery travelling salesman problem. Computers and Operations Research 27, 905–916 (2000)
6. Mingozzi, A., Bianco, L., Ricciardelli, A.: Dynamic programming strategies for the tsp with time windows and precedence constraints. Operations Research 45, 365–377 (1997)
7. Fagerholt, K., Christiansen, M.A.: Travelling salesman problem with allocation, time window and precedence constraints – an application to ship scheduling. Intl. Trans. Operations Research 7, 231–244 (2000)
8. Michalewicz, Z.: Genetic Algorithms + Data Structures = Evolution Programs, 3rd edn. Springer, Heidelberg (1996)

Author Index

Lecture Notes in Computer Science

Sublibrary 1: Theoretical Computer Science and General Issues

For information about Vols. 1–4641
please contact your bookseller or Springer

Vol. 4974: M. Giacobini et al. (Eds.), Applications of Evolutionary Computing. XXV, 701 pages. 2008.

Vol. 4973: E. Marchiori, J.H. Moore (Eds.), Evolutionary Computation, Machine Learning and Data Mining in Bioinformatics. X, 213 pages. 2008.

Vol. 4972: J. van Hemert, C. Cotta (Eds.), Evolutionary Computation in Combinatorial Optimization. XII, 289 pages. 2008.

Vol. 4971: M. O'Neill, L. Vanneschi, S. Gustafson, A.I. Esparcia Alcazar, I. De Falco, A. Della Cioppa, E. Tarantino (Eds.), Genetic Programming. XI, 375 pages. 2008.

Vol. 4963: C.R. Ramakrishnan, J. Rehof (Eds.), Tools and Algorithms for the Construction and Analysis of Systems. XVI, 518 pages. 2008.

Vol. 4962: R. Amadio (Ed.), Foundations of Software Science and Computational Structures. XV, 505 pages. 2008.

Vol. 4961: J.L. Fiadeiro, P. Inverardi (Eds.), Fundamental Approaches to Software Engineering. XIII, 430 pages. 2008.

Vol. 4960: S. Drossopoulou (Ed.), Programming Languages and Systems. XIII, 399 pages. 2008.

Vol. 4959: L. Hendren (Ed.), Compiler Construction. XII, 307 pages. 2008.

Vol. 4957: E.S. Laber, C. Bornstein, L.T. Nogueira, L. Faria (Eds.), LATIN 2008: Theoretical Informatics. XVII, 794 pages. 2008.

Vol. 4934: U. Brinkschulte, T. Ungerer, C. Hochberger, R.G. Spallek (Eds.), Architecture of Computing Systems – ARCS 2008. XI, 287 pages. 2008.

Vol. 4927: C. Kaklamanis, M. Skutella (Eds.), Approximation and Online Algorithms. X, 289 pages. 2008.

Vol. 4921: S.-i. Nakano, M.. S. Rahman (Eds.), WALCOM: Algorithms and Computation. XII, 241 pages. 2008.

Vol. 4919: A. Gelbukh (Ed.), Computational Linguistics and Intelligent Text Processing. XVIII, 666 pages. 2008.

Vol. 4917: P. Stenström, M. Dubois, M. Katevenis, R. Gupta, T. Ungerer (Eds.), High Performance Embedded Architectures and Compilers. XIII, 400 pages. 2008.

Vol. 4915: A. King (Ed.), Logic-Based Program Synthesis and Transformation. X, 219 pages. 2008.

Vol. 4912: G. Barthe, C. Fournet (Eds.), Trustworthy Global Computing. XI, 401 pages. 2008.

Vol. 4910: V. Geffert, J. Karhumäki, A. Bertoni, B. Preneel, P. Návrat, M. Bieliková (Eds.), SOFSEM 2008: Theory and Practice of Computer Science. XV, 792 pages. 2008.

Vol. 4905: F. Logozzo, D.A. Peled, L.D. Zuck (Eds.), Verification, Model Checking, and Abstract Interpretation. X, 325 pages. 2008.

Vol. 4904: S. Rao, M. Chatterjee, P. Jayanti, C.S.R. Murthy, S.K. Saha (Eds.), Distributed Computing and Networking. XVIII, 588 pages. 2008.

Vol. 4878: E. Tovar, P. Tsigas, H. Fouchal (Eds.), Principles of Distributed Systems. XIII, 457 pages. 2007.

Vol. 4875: S.-H. Hong, T. Nishizeki, W. Quan (Eds.), Graph Drawing. XIII, 402 pages. 2008.

Vol. 4873: S. Aluru, M. Parashar, R. Badrinath, V.K. Prasanna (Eds.), High Performance Computing – HiPC 2007. XXIV, 663 pages. 2007.

Vol. 4863: A. Bonato, F.R.K. Chung (Eds.), Algorithms and Models for the Web-Graph. X, 217 pages. 2007.

Vol. 4860: G. Eleftherakis, P. Kefalas, G. Păun, G. Rozenberg, A. Salomaa (Eds.), Membrane Computing. IX, 453 pages. 2007.

Vol. 4855: V. Arvind, S. Prasad (Eds.), FSTTCS 2007: Foundations of Software Technology and Theoretical Computer Science. XIV, 558 pages. 2007.

Vol. 4854: L. Bougé, M. Forsell, J.L. Träff, A. Streit, W. Ziegler, M. Alexander, S. Childs (Eds.), Euro-Par 2007 Workshops: Parallel Processing. XVII, 236 pages. 2008.

Vol. 4851: S. Boztaş, H.-F.(F.) Lu (Eds.), Applied Algebra, Algebraic Algorithms and Error-Correcting Codes. XII, 368 pages. 2007.

Vol. 4848: M.H. Garzon, H. Yan (Eds.), DNA Computing. XI, 292 pages. 2008.

Vol. 4847: M. Xu, Y. Zhan, J. Cao, Y. Liu (Eds.), Advanced Parallel Processing Technologies. XIX, 767 pages. 2007.

Vol. 4846: I. Cervesato (Ed.), Advances in Computer Science – ASIAN 2007. XI, 313 pages. 2007.

Vol. 4838: T. Masuzawa, S. Tixeuil (Eds.), Stabilization, Safety, and Security of Distributed Systems. XIII, 409 pages. 2007.

Vol. 4835: T. Tokuyama (Ed.), Algorithms and Computation. XVII, 929 pages. 2007.

Vol. 4800: A. Avron, N. Dershowitz, A. Rabinovich (Eds.), Pillars of Computer Science. XXI, 683 pages. 2008.

Vol. 4783: J. Holub, J. Žďárek (Eds.), Implementation and Application of Automata. XIII, 324 pages. 2007.

Vol. 4782: R. Perrott, B.M. Chapman, J. Subhlok, R.F. de Mello, L.T. Yang (Eds.), High Performance Computing and Communications. XIX, 823 pages. 2007.

Vol. 4771: T. Bartz-Beielstein, M.J. Blesa Aguilera, C. Blum, B. Naujoks, A. Roli, G. Rudolph, M. Sampels (Eds.), Hybrid Metaheuristics. X, 202 pages. 2007.

Vol. 4770: V.G. Ganzha, E.W. Mayr, E.V. Vorozhtsov (Eds.), Computer Algebra in Scientific Computing. XIII, 460 pages. 2007.

Vol. 4769: A. Brandstädt, D. Kratsch, H. Müller (Eds.), Graph-Theoretic Concepts in Computer Science. XIII, 341 pages. 2007.

Vol. 4763: J.-F. Raskin, P.S. Thiagarajan (Eds.), Formal Modeling and Analysis of Timed Systems. X, 369 pages. 2007.

Vol. 4759: J. Labarta, K. Joe, T. Sato (Eds.), High-Performance Computing. XV, 524 pages. 2008.

Vol. 4746: A. Bondavalli, F. Brasileiro, S. Rajsbaum (Eds.), Dependable Computing. XV, 239 pages. 2007.

Vol. 4743: P. Thulasiraman, X. He, T.L. Xu, M.K. Denko, R.K. Thulasiram, L.T. Yang (Eds.), Frontiers of High Performance Computing and Networking ISPA 2007 Workshops. XXIX, 536 pages. 2007.

Vol. 4742: I. Stojmenovic, R.K. Thulasiram, L.T. Yang, W. Jia, M. Guo, R.F. de Mello (Eds.), Parallel and Distributed Processing and Applications. XX, 995 pages. 2007.

Vol. 4739: R. Moreno Díaz, F. Pichler, A. Quesada Arencibia (Eds.), Computer Aided Systems Theory – EUROCAST 2007. XIX, 1233 pages. 2007.

Vol. 4736: S. Winter, M. Duckham, L. Kulik, B. Kuipers (Eds.), Spatial Information Theory. XV, 455 pages. 2007.

Vol. 4732: K. Schneider, J. Brandt (Eds.), Theorem Proving in Higher Order Logics. IX, 401 pages. 2007.

Vol. 4731: A. Pelc (Ed.), Distributed Computing. XVI, 510 pages. 2007.

Vol. 4728: S. Bozapalidis, G. Rahonis (Eds.), Algebraic Informatics. VIII, 291 pages. 2007.

Vol. 4726: N. Ziviani, R. Baeza-Yates (Eds.), String Processing and Information Retrieval. XII, 311 pages. 2007.

Vol. 4719: R. Backhouse, J. Gibbons, R. Hinze, J. Jeuring (Eds.), Datatype-Generic Programming. XI, 369 pages. 2007.

Vol. 4711: C.B. Jones, Z. Liu, J. Woodcock (Eds.), Theoretical Aspects of Computing – ICTAC 2007. XI, 483 pages. 2007.

Vol. 4710: C.W. George, Z. Liu, J. Woodcock (Eds.), Domain Modeling and the Duration Calculus. XI, 237 pages. 2007.

Vol. 4708: L. Kučera, A. Kučera (Eds.), Mathematical Foundations of Computer Science 2007. XVIII, 764 pages. 2007.

Vol. 4707: O. Gervasi, M.L. Gavrilova (Eds.), Computational Science and Its Applications – ICCSA 2007, Part III. XXIV, 1205 pages. 2007.

Vol. 4706: O. Gervasi, M.L. Gavrilova (Eds.), Computational Science and Its Applications – ICCSA 2007, Part II. XXIII, 1129 pages. 2007.

Vol. 4705: O. Gervasi, M.L. Gavrilova (Eds.), Computational Science and Its Applications – ICCSA 2007, Part I. XLIV, 1169 pages. 2007.

Vol. 4703: L. Caires, V.T. Vasconcelos (Eds.), CONCUR 2007 – Concurrency Theory. XIII, 507 pages. 2007.

Vol. 4700: C.B. Jones, Z. Liu, J. Woodcock (Eds.), Formal Methods and Hybrid Real-Time Systems. XVI, 539 pages. 2007.

Vol. 4699: B. Kågström, E. Elmroth, J. Dongarra, J. Waśniewski (Eds.), Applied Parallel Computing. XXIX, 1192 pages. 2007.

Vol. 4698: L. Arge, M. Hoffmann, E. Welzl (Eds.), Algorithms – ESA 2007. XV, 769 pages. 2007.

Vol. 4697: L. Choi, Y. Paek, S. Cho (Eds.), Advances in Computer Systems Architecture. XIII, 400 pages. 2007.

Vol. 4688: K. Li, M. Fei, G.W. Irwin, S. Ma (Eds.), Bio-Inspired Computational Intelligence and Applications. XIX, 805 pages. 2007.

Vol. 4684: L. Kang, Y. Liu, S. Zeng (Eds.), Evolvable Systems: From Biology to Hardware. XIV, 446 pages. 2007.

Vol. 4683: L. Kang, Y. Liu, S. Zeng (Eds.), Advances in Computation and Intelligence. XVII, 663 pages. 2007.

Vol. 4681: D.-S. Huang, L. Heutte, M. Loog (Eds.), Advanced Intelligent Computing Theories and Applications. XXVI, 1379 pages. 2007.

Vol. 4672: K. Li, C. Jesshope, H. Jin, J.-L. Gaudiot (Eds.), Network and Parallel Computing. XVIII, 558 pages. 2007.

Vol. 4671: V.E. Malyshkin (Ed.), Parallel Computing Technologies. XIV, 635 pages. 2007.

Vol. 4669: J.M. de Sá, L.A. Alexandre, W. Duch, D.P. Mandic (Eds.), Artificial Neural Networks – ICANN 2007, Part II. XXXI, 990 pages. 2007.

Vol. 4668: J.M. de Sá, L.A. Alexandre, W. Duch, D.P. Mandic (Eds.), Artificial Neural Networks – ICANN 2007, Part I. XXXI, 978 pages. 2007.

Vol. 4666: M.E. Davies, C.J. James, S.A. Abdallah, M.D. Plumbley (Eds.), Independent Component Analysis and Signal Separation. XIX, 847 pages. 2007.

Vol. 4665: J. Hromkovič, R. Královič, M. Nunkesser, P. Widmayer (Eds.), Stochastic Algorithms: Foundations and Applications. X, 167 pages. 2007.

Vol. 4664: J. Durand-Lose, M. Margenstern (Eds.), Machines, Computations, and Universality. X, 325 pages. 2007.

Vol. 4661: U. Montanari, D. Sannella, R. Bruni (Eds.), Trustworthy Global Computing. X, 339 pages. 2007.

Vol. 4649: V. Diekert, M.V. Volkov, A. Voronkov (Eds.), Computer Science – Theory and Applications. XIII, 420 pages. 2007.

Vol. 4647: R. Martin, M.A. Sabin, J.R. Winkler (Eds.), Mathematics of Surfaces XII. IX, 509 pages. 2007.

Vol. 4646: J. Duparc, T.A. Henzinger (Eds.), Computer Science Logic. XIV, 600 pages. 2007.

Vol. 4644: N. Azémard, L. Svensson (Eds.), Integrated Circuit and System Design. XIV, 583 pages. 2007.